THE EVOLUTION OF
CHILDHOOD

THE EVOLUTION
OF CHILDHOOD

Relationships, Emotion, Mind

Melvin Konner

THE BELKNAP PRESS OF
HARVARD UNIVERSITY PRESS

Cambridge, Massachusetts
London, England

2010

Library of Congress Cataloging-in-Publication Data

Konner, Melvin.
The evolution of childhood : relationships, emotion, mind / Melvin Konner.
p.   cm.
Includes bibliographical references and index.
ISBN 978-0-674-04566-8 (alk. paper)
1. Childhood.   2. Child development.   3. Children—Anthropometry.
4. Emotions in children.   5. Human evolution.   I. Title.
GN482.K66 2010
305.231—dc22   2009050775

Ann

Contents

Part I. Evolution: *The Phylogenetic Origins of Childhood*
wherein we learn how the laws of evolution produced the
shape of human social and emotional development

Part II. Maturation: *Anatomical Bases of Psychosocial Growth*
wherein we see how neural and endocrine systems guide
the paths of development called for by natural selection

Part III. Socialization: *The Evolving Social Context of Ontogeny*
wherein we discern the contributions of social life
to developing relationships and emotions

Part IV. Enculturation: *The Transmission and Evolution of Culture*

wherein we come to understand what culture changes

Part V. Conclusion
wherein we see, as through a glass darkly, how human relationships and emotions may actually emerge

Prologue

On a partly sunny afternoon many years ago, when I began this book, I sat near a cottage in the Adirondack Forest in New York, looking through an opening in a stand of conifers, maples, and birches, over a meadow full of daisies, yarrow, vetch, and black-eyed Susans, to a still, small, lily-bordered lake; birds brayed noisily, and occasionally a frog croaked or a mayfly buzzed my ear. Sun and shadow alternated pleasantly. In such a setting one can only be impressed by the sometimes thrilling, sometimes calming interdependence of life.

Yet nothing outside the human circle of the cottage held a candle, for me, to what was in it. Nearby, a six-month-old slept peacefully, and his four-year-old sister, picking flowers, tried to resist the impulse to disturb him. Nothing in nature is more marvelous than children's emotions and behavior—the dark as well as the pleasant—toward each other and others around them. Not even the seasons are more beautiful or interesting than the growth of those feelings. And even the adult figure in the scene—moodily musing, scribbling, chewing his pencil—was in his way as worthy of wonder as any butterfly.

Despite the range of species and the spectrum of behavior and biology covered in this book, its only goal is to elucidate some principles and facts of biology relevant to the development of human behavior and mind. But the most powerful principles in nature rest on the broadest empirical base. Our understanding of attachment depends, in part, on observations of the following behavior of goslings; our understanding of brain plasticity on changes in the pyramidal neurons of jewelfish; our understanding of kinship on the social lives of honeybees; and our understanding of the long, slow human childhood on the rise and fall of populations of flowers.

Academic psychology in the United States has had two major phases, reflecting two concerns: learning and, more recently, cognition. These concerns stemmed in part from the relationship of developmental psychology to education and in part from transcendentalism—that superbly optimistic American striving toward complete equality, a continuum from everyman to Emerson. There has always been theorizing about social and emotional development, mainly by psychodynamic clinicians or by thinkers in the fields of nursing and child care advice. They have been viewed by the academic mainstream with sufferance or derision. Indeed, students of emotional development, however careful their work, were disdained—"good mother types" was one phrase applied to them—until recently.

At midcentury, social learning theory substituted for the study of emotion. The clear results of reward and punishment in the child's social world made it seem picky to ask what the reinforcers were, why they reinforced, how they changed with growth, and how maturation limited learning or left previous behaviors behind. Cognitive psychology wanted to change all that. The black box of the mind was to be filled with apparatus that shaped perception and action and that transformed over time according to laws of its own. Elegant natural processes—like constructing the visual world through feature extraction or retrieving an object obscured by a cloth—yielded models for the mind. Computer programs also stood in for mental function, a triumph of abstraction over common sense. Finally, in an inversion of the genesis of the brain's structure and function, the most rational mental functions were held to be the bases for relationships, and the hybrid "social cognition" was born. This project, which seemed to ignore emotion and motivation, modeled social behavior as an epiphenomenon not just of rational thought but of rational thought about other minds.

To a biologist or anthropologist, it seemed that the most vital features of complex behavior, animal or human—parent-offspring relations, courtship and mating, aggression and dominance—had been permanently relegated to second-class status. Yet these determine most of the variance in the reproductive success of higher organisms, which is tantamount to saying that these features govern their evolution. The great evolutionary geneticist Theodosius Dobzhansky famously said that nothing in biology makes sense except in the light of evolution; we can say now that nothing in childhood does either. And yet, until very recently, most of psychology was carried out as if evolution had not happened.

Still, evolution was implicit in the background. For methodological reasons, learning psychologists usually studied rats and other animals under conditions of hunger, thirst, pain, and fear, simulating the motivated conditions in which most of life goes forward. And cognitive psychologists, weary of the "black box," proposed "cognitive structures," or "competences," that might explain behavior or performance. Most glossed over the neurobiology of these abstractions, and some even viewed it as irrelevant, granting cognitive structures a life of their own. Although many cognitive psychologists accept that some competences reflect innate or a priori knowledge, this provides cold comfort for biologically oriented behavioral theorists. For us, despite promising signs (Ellis and Bjorklund 2005; Simpson and Belsky 2008), a rapprochement between developmental psychology and evolutionary biology remains awkward and elusive; thus the perceived need for this book. I hope to use the relevant biological facts and principles to set the stage for a true reconciliation in the realm of psychosocial growth.

One response would be to ask, "Don't you mean psychosocial *development?*" No. Even Piaget was reluctant to take a stand on the innate, while his successors have backed away from it even further, obfuscating genetic and maturational factors or systematically discounting them. Yet cognitive development is, in evolutionary terms, largely an epiphenomenon of psychosocial growth. Students of developing social cognition and, to some extent, social neuroscience posit mental functions that must reach a given level of complexity before correspondingly complex social functions are affected. Even if we accepted this stance, "mental functions" would have to include not just the cool, dry, hard ones studied by cognitive scientists, but also the soft and messy sorts of functions studied by psychodynamic theorists, ethologists, and evolutionary psychologists, who are truer descendants of Darwin.

But in evolution the soft takes precedence. Human mental functions were formed in social contexts: in mate selection and reproduction, in sometimes violent competition for resources, in cooperation and coalition, in increasingly intricate communications, and in perhaps the supreme challenge—parenting. These, not the inanimate objects, puzzles, and abstractions of cognitive science, shaped the evolution of the mammalian brain, the higher primate brain, and finally the human brain.

This is increasingly the view of researchers in the evolution of intelligence (Whiten and Byrne 1997; Cosmides and Tooby 2000) and neuroscientists who find that reason itself requires an emotionally functioning brain (Damasio 1994, 1998) and that little understanding can be gained about the developing mind and brain without placing relationships at the center (Nelson, Leibenluft, et al. 2005; Cohen Kadosh and Johnson 2007; Meltzoff 2007; Steinberg 2008). Even "theory of mind" research in cognitive development points to an increasing grasp of the centrality of social functions to cognitive ones (Bruene and Bruene-Cohrs 2006; Blakemore 2008). But relationships are not just central to the development of intelligence, they are crucial to its evolution. Social cognition drives the phylogeny of mind; social life is where the biobehavioral rubber meets the evolutionary road.

The Structure of This Book

This book presents, in four parts, findings relevant to the question of how the laws and facts of biology underlie normally developing social behavior. I do not say "whether" this is so, since I would not be writing this particular book unless I had already answered that question; nor do I ask "how much" this is true, since, to the extent that I consider that question meaningful and answerable, it is merely preliminary. I say "how" because the possibilities for answering this question have expanded enormously since it was singled out by a prescient psychologist as the most useful one (Anastasi 1958). Fortunately, after decades of ideologically motivated neglect, this question has become the focus of some leading child development scientists (Plomin and Rutter 1998; Kagan 2003; Rutter 2007; Belsky, Bakermans-Kranenburg, et al. 2007). It is the question of the future.

I embarked on these biological explorations as a philosophically committed environmental determinist, and in a key sense I remain one. I consider improving the human condition to be the aim of scientific activity, and many critical aspects of social and emotional development are due to known and unknown environmental causes.

Why would I write if I didn't believe in learning? Another baby awakes, sits up, and comes alive to the world. Among the objects at her disposal are a yellow rattle shaped like a telephone handpiece, a green rubber hedgehog, a plastic block with a rod through it, her own thumb, her fa-

ther's pad and pencil, his free hand, and his face. As I watch her appreciate these things in the stimulus world—assimilate them, know them—
I would not care to ask whether maturation or learning is in ascendance;
I would find that question foolish. Yet in every case—the deliberate
shake of the rattle, the look and grasp at the rubber toy, the ominous
stare at the pencil and pad, the suck at the thumb and then at a toe,
the look, and smile, and laugh at a cherished, familiar face—I would
be able, if I were not busy giggling back at her, to delineate some nontrivial, purely biological foundations of the action and, I believe, am
thereby freed (should I so choose) to influence that action the more securely.

The four parts that follow describe the foundations of psychosocial
growth as revealed by four research domains—evolution, maturation,
socialization, and culture. But the overarching goal is to make possible
a firmer grasp of the basis for modification. Accordingly, the argument
flows toward a treatment of the facts and laws of modification as viewed
from a biological stance, within the context of culture. Each part of the
book could stand on its own. However, I have tried to integrate them
and present the basis for a very general outlook: a behavioral biology of
psychosocial development—a developmental sociobiology. I conceived
the book as a seamless intellectual web, connecting the lines of research
that converge on this subject with arguments that dissolve disciplinary
boundaries. In the event, I hope it will at least reduce the temptation for
practitioners in these disparate fields to ignore one another's findings.

The introduction describes past and recent biological trends in child
development research. Part I, about evolution, begins by defending
studies in the evolution of behavior and its development and disposes
of some misconceptions that help explain past opposition. I introduce
some lines of theory and research in evolution that bear directly on behavioral development, and I mention some key results. All life is the
evolution of ontogeny, and without evolution, ontogeny is incompre-
hensible; life history theory is crucial to explain it. Subsequent chapters
describe our current understanding of the evolution of the brain and
behavior and their ontogeny. Transition 1 between Parts I and II considers, against the background of phylogeny, some brain models relevant
to psychosocial functioning.

Part II, which has an anatomical and physiological focus, begins with
the general basis of developing social competence in the neural and en-

docrine systems. Subsequent chapters treat selected phases in psychosocial development: the emergence of sociality in the first months; the growth of attachment and the social fears after the first half-year; the growth of language competence and the dawn of self-consciousness; the emergence of gender identity and differences in aggression; the transition to middle childhood, or the "five-to-seven shift"; and the growth of reproductive behavior and other transformations of adolescence. Transition 2 between Parts II and III considers the nature and evolution of plasticity.

Part III, about socialization, uses comparative approaches to reconstructing phylogeny and history, emphasizing the difference between them, with special attention to higher primates and human hunter-gatherers. An early chapter establishes the power of social experience, especially early in life, to change neural and neuroendocrine systems. The central chapters reconstruct the evolution of that experience—the social context of development—focusing on the phylogeny and adaptive functions of the mother-infant bond; the diffusion of nurturance and the functions of families; the role of fathers and other adult males; relations among the young; the phylogeny of play, social learning, and teaching; the causes and consequences of changing pubertal dynamics; and the central roles of stress and resilience in children's lives. The last chapter establishes the baseline pattern of children's experience in hunting-and-gathering cultures—childhood, as it were, after humanity but before history. Transition 3 between Parts III and IV discusses and answers the question: Is there culture in nonhuman animals?

Part IV, about enculturation and gene-culture coevolution, considers the nature of human cultural variation and how it is transferred to and transforms every child. It introduces the main paths of inquiry relevant to enculturation and discusses how anthropologists and psychologists have tried to understand how mind is embedded in culture. It deals with the claims of the "culture and personality" school and proposes a model to explain the more plausible ones. The next four chapters offer brief accounts of how culture modifies the developmental biogram in our species, from fetal life to adolescence and beyond. The next-to-last chapter in Part IV, on evolutionary culture theory, discusses the mechanisms of culture change that include, but are not limited to, transmission during childhood. And the final chapter in this section considers the nature of universals of human behavior and culture and goes on to outline a culture acquisition device, using known processes of behavior modification, emotional engagement, symbol systems, and intersubjective learning.

Part V, the concluding section, briefly considers the *Urpflanz,* Goethe's model of growth and form that would be at once natural and ideal and serve as a paradigm for all life on earth. It goes on to suggest that this model has been found, that it has implications like those expected by Goethe, and that it is central to all future research on behavioral development. Part V then proposes, at least half seriously, a law of psychogenetic inertia—a developmental analogy to Newton's first law of motion. It goes on to propose more seriously a very general theory, based on chance, chaos, emergence, challenge, and selection, that may apply to various levels of evolution and development. The book ends with a reprise, followed by an epilogue.

Of all the research touched upon here, none is more urgent or more neglected than the cross-cultural. Modernization on the positive side and conflict and oppression on the negative side are rapidly altering unique and irretrievable opportunities for study. Although anthropologists are making use of some of them, they must be joined by others who bring different ideas and methods to these relatively exotic settings. We still devote almost all our resources to subjects, phenomena, and settings that will be available for study in similar form many years from now, while ignoring cross-cultural research opportunities approximated by an exponentially decaying function.

Whatever the reason—disciplinary blinders? habit? attachment to the comforts of home?—the loss to human knowledge will be permanent. Some excellent cross-cultural rescarch has been done by psychologists and psychiatrists on infancy (Leiderman, Tulkin, et al. 1977; Field, Sostek, et al. 1981; Super 1987) and childhood (Super 1991; Cole 1997; Rogoff 1997; Greenfield, Keller, et al. 2003; Callaghan, Rochat, et al. 2005), and anthropologists have proceeded with such investigations (Whiting and Edwards 1988; Schlegel and Barry 1991; Shweder 1993; Edwards 1994; LeVine, Dixon, et al. 1994; Hewlett and Lamb 2005; Lancy 2008). There has been considerable activity in the area of child development and its social context among hunter-gatherers. When I first wrote on hunter-gatherer infancy (Konner 1972), I challenged colleagues in all disciplines to seize the remaining few chances for research on hunter-gatherer childhood.

Fortunately, some stalwart investigators needed no such urging. Patricia Draper and Heide Sbrzesny had already done excellent studies of San children, and there were soon to be many more, studying a wide range of hunter-gatherer childhoods: Edward Tronick, Gilda Morelli, Paula Ivey, Barry Hewlett, Bonnie Hewlett, Hillary Fouts, Michael Lamb, Hilliard Kaplan, Kim Hill, Magdalena Hurtado, Nicholas Blurton Jones, Kristen Hawkes, Frank Marlowe, Agnes Estioko-Griffin, Bion Griffin, Douglas Bird, Rebecca Bliege Bird, Akira Takada, Bram Tucker, Alyson Young, and Michael Gurven, among others (Hewlett and Lamb 2005). Still, there is much more to be done. Many opportunities for hunter-gatherer research have disappeared; others, however, still exist, and I now repeat the challenge.

Six Paradigms

There are no longer many environmentalist hard-liners; it has become too implausible a position. But views still range widely, from positions that ignore or belittle biology to those that insist on it very strongly.

At one extreme are the *modern Vygotskians,* who declare, with admirable enthusiasm, that mind is pervaded by culture. Inspired by Lev Vygotsky, the brilliant Russian psychologist of the 1920s, they seem unaware that these ideas have pervaded anthropology since Vygotsky was a boy and for generations before his work became widely known. Franz Boas (1938, 1940), Lucien Lévy-Bruhl (1925), George Herbert Mead (1934), Edward Sapir (1994), Ruth Benedict (1934), Margaret Mead (1932), and Claude Lévi-Strauss (1963, 1968), among many others (Stocking 1968, 1974; Spindler 1979; D'Andrade and Strauss 1992; Shore 1996), showed that human mental life is inescapably cultural. But neither they nor Vygotsky nor his recent disciples have shown that culture is an acid bath that dissolves structures provided by the genes. The structures, by evolutionary design, invite culture in, and culture bathes them roughly or gently, but not dismissively. Mind is bathed in culture because biology makes it so, and biology does that with clear guidelines.

Next are what might be called the *cautious interactionists,* such as Gilbert Gottlieb (1992, 1995), who studied behavioral development in birds; Esther Thelen (Thelen and Smith 1994), who analyzed human infant motor development; and Robert Siegler (1995, 1996), who pioneered the microanalysis of children's cognitive change. Nodding toward biological

determinants, they and many others chose to emphasize the complicated and gradual nature of change and the relationship of everything
in development to everything else. First in the language of open systems
theory and later in that of complexity and emergence, they have resisted
causality.

One can say, in physics, that the behavior of a gas in a confined space
has to do with pressure, temperature, volume, and the number of molecules. We can make wonderfully precise observations of how it is now
temperature, now pressure, that seems to affect volume, and that volume, in turn, affects temperature, which affects pressure, and so on. But
it was not until Boyle, Avogadro, and others decided to study cause and
effect that we began to have laws (relating temperature to volume at
constant pressure, for example). Nor could the universal gas law, which
precisely and predictively relates each variable to the others, have been
discovered without isolating individual causes. Physicists do not say,
"Pressure, temperature, volume, and molecular number are inextricably intertwined, and it makes no sense to try to tease them apart." They
are interested in the whether, how much, and how, and they answer
these questions quantitatively for each causal element.

Third are the *connectionists* and *neuroconstructivists,* who are also interactionists (Johnson and Karmiloff-Smith 1992; Bates and Elman 1993;
Karmiloff-Smith 1995; Elman, Bates, et al. 1996; Karmiloff-Smith 2009).
Building on neuron network models, they created a theory for development based on a new kind of artificial intelligence (AI). Unlike classical
AI, this theory does not model directly the outcomes of learning and
problem solving, but rather the process; that is, it builds neural units
capable of stimulating and inhibiting one another in small networks.
These (simulated) networks are then exposed to (simulated) experience.
They learn, they solve problems using what they have learned, and in
simple ways they act like parts of brains. But in real brains neurons
come in scores of shapes and chemical varieties, with myriad functions
to match, and are bathed in hormones and other compounds that alter
their action. All this is cobbled together by evolution in a pastiche of different organs interacting in puzzling ways. So the main concern about
connectionism resembles that about the old AI: you can make a machine
produce the results achieved by brains, but that does not mean you have
found out how brains achieve them.

Fourth are the *new behavior geneticists,* who have brought to their work greater statistical sophistication, a true longitudinal and developmental dimension, and the power of molecular genetics (Eaves, Eysenck, et al. 1989; Kagan 2003; Plomin 2005; Rutter 2007). Some of their discoveries are discomfiting, but they believe that the truth is the only acceptable starting point. Some work in developmental behavior genetics elucidates environmental influences more clearly than does any other paradigm (Scarr and McCartney 1983; Scarr 1993; Kagan and Neville 1996; Plomin and Rutter 1998). Indeed, it could be said that anyone seriously interested in establishing the power of the environment in psychological development should be planning intervention studies with identical-twin pairs. Although few such studies are done, behavior genetics research, in drawing attention to some processes, such as the within-family power of environments to make siblings different or the genetically guided self-selection of environments, is focusing research on environmental influences in new ways.

Fifth, the *developmental social neuroscientists* have created a new paradigm based on new brain-imaging technologies and electroencephalographic techniques that could be as revolutionary as genomic technology has been in behavior genetics. Whether studying the neurobiology of vision in early infancy (Dubois, Dehaene-Lambertz, et al. 2008b), the development of face perception (Cohen Kadosh and Johnson 2007), the neuropsychology of attachment (Fox and Hane 2008), the myelination processes underlying the emergence of language (Dehaene-Lambertz, Hertz-Pannier, et al. 2006), the brain circuitry of imitation and empathy in ten-year-olds (Pfeifer, Iacoboni, et al. 2008), or the transformations of adolescent brain and behavior (Nelson, Leibenluft, et al. 2005; Lenroot and Giedd 2006; Steinberg 2008; Blakemore 2009), they are adding a vast new domain of data to the established facts of developing behavior and mind. They understand that they are not explaining away, or even always explaining, the psychological facts, but they are providing a dimension without which those facts cannot be understood. Some aspects of brain development are under genetic control, and others are substrates for the impact of experience; either way, we need that new dimension.

Sixth, the *evolutionary psychologists* have a take-no-prisoners approach to the application of their paradigm (Belsky, Steinberg, et al. 1991; Chisholm 1993; Bjorklund and Pellegrini 2002; Ellis 2004; Simpson and Belsky 2008). They are not as interested in the role of culture as they are in the role of adaptation through natural selection (Betzig, Mulder, et al. 1988; Betzig 1997; Cosmides and Tooby 2000). They have a very strong theory, and they apply it vigorously, not always with full attention to other levels of causal explanation or to the non-adaptive factors that may contribute to the outcomes they study. This is a good strategy for them, since they are under constant criticism from others who are adequately committed to accounting for causality at more proximate levels, and those critics highlight the limits of neo-Darwinian theory. Beyond doubt, the theory explains important facts of human psychology and behavior, including some facts about parents and children. Its light can cut through the turbulent cloud of other causes, and it cannot be brushed aside just because they also matter. However, evolutionary psychology is in the stage of paradigm formation in which it usually seems enough to show that an effect exists; the size of these effects has yet to be determined.

The last four of these paradigms go beyond description to make child development a science complete with classical causality—which, contrary to some claims, is not negated by the pervasive effects of interaction, chaos, and complexity. The paradigms are imperfect and challenging to reconcile. None holds more than a piece of the truth. But with these and other approaches, we can dispel part of the darkness and reveal decisive and specific roles for biology in the development of relationships, emotion, and mind. Then we will at last be able to see culture for what it is: a set of influences that have, within definable limits, great variety and power.

1
Introduction

Evolution and development have in common an uncertain interaction of three clear principles: chance, order, and selection. The ultimate source of chance or randomness is entropy, which, according to the second law of thermodynamics, must steadily increase. Yet, as if in defiance of entropy, order surrounds us, inspiring scientists to find the laws that govern it—the laws of physics and, in the living realm, those of genetic and metabolic molecular interaction. While at one time it seemed that these laws would enable the derivation of laws at all higher levels of order, that is no longer plausible (Anderson 1972). Laws at all levels of organization must be consistent with lower-level laws, but that does not mean that they can be derived from them. On the contrary, it is increasingly apparent that each higher level of order requires at least some new laws of its own, and the search for such laws has become a new field of science, the branch of nonlinear dynamics often called, simply, complexity (Kauffman 1993).

But whether orderly organizations stabilize because of fundamental physical laws or because of nonlinear dynamical interactions, they must do more than stabilize to be living; they must be part of a reproducing system. In a world of finite resources, the competition not just for reproductive options but also for the sun's energy in its various forms is another source of instability. Thus, some systems will stabilize longer, be thermodynamically more efficient, sequester energy more effectively, and reproduce more frequently. These, in each generation, leave a legacy of systems like themselves. This process, *natural selection,* takes orderly systems beyond thermodynamic, nonlinear dynamical, emergent, or other physical stability into *organic* stability and ultimately into the *phylogenetic* stability that reflects reproductive success.

The most significant manifestation of the second law of thermody-namics—disorder continually licks at the edges of orderly organic sys-tems—is seen in the form of reproductive errors, or replicative stum-bling. Most of these blunders are physically unstable, and whether we view their dissolution as a kind of natural selection or not is unimpor-tant; most of the remaining stable formations will not survive to repro-duce despite their adequate physical and developmental stability. These more stable blunders, which we commonly associate with mutation, are more interesting than those that simply disintegrate, and their lessened ability to reproduce is known as *stabilizing selection.* It favors the "nor-mal," tends to keep things the same, and has been the commonest pro-cess in the history of life.

But some replication blunders survive to reproduce, and these are the ultimate source of change, providing the first of three components of evolution: *variation* in form and function. The other two are some degree of *heritability* in that variation and some degree of *differential re-productive success* tied to that heritable variation.

So the essence of evolution is different reproduction, and natural se-lection and its sister concept, fitness, constitute neither more nor less than a ratio of the reproductive success of a given variant to that of the population on average (Lewontin 1974). Over time, populations contain more of the high-ratio variants. Since a species is just a collection of populations, this process can eventually produce a new species. Specia-tion happens infrequently but is the essence of phylogenetic change (Mayr 1963, 2001; Venditti, Meade, et al. 2010).

No one denies that there is an element of tautology in this line of rea-soning. Falsification in evolutionary biology is not like falsification in embryology or experimental psychology, where theory is modest and local, control of phenomena is possible, and experiment, at least in prin-ciple, can decisively put aside hypothesis and theory. It is more like fal-sification in astronomy or geophysics, where theory is global, control is impossible, and observation dominates experiment. The criteria for acceptance of theory in these fields have to do with the orderliness brought to data, the perceived fit between theory and phenomena, and the prediction of observations not yet made. Einstein's general relativ-ity predicted that the light of a distant star would bend in the sun's gravitational field; under other theories, gravity should have no effect on light. During the 1919 total eclipse, Einstein's theory could be tested, since it predicted a slightly changed position for the star, which appeared near the darkened sun, relative to other stars in the sky. After the prediction was made, the observed change in position was measured, and the prediction was confirmed to seven decimal places, the theory of general relativity gained very wide acceptance. We could colloquially call this a natural experiment, but it was just a focused observation. Such observations are the bread and butter of evolutionary biology. Only rarely are these observations as decisive as those in astronomy or geophysics, but countless new ones are made yearly that strengthen the truth-claims of evolutionary theory, both in general and in many particulars.

Some Premises

Among these observations and experiments are those that confirm or disconfirm hypotheses arising from one branch or another of evolutionary theory related to social and emotional development. This book tries to integrate the findings most relevant to biological bases of psychosocial growth. It accepts as proven, and does not attempt to prove, certain premises:

1. *All life on earth is the product of organic evolution from inorganic molecules by a process including, but not limited to, mutation, recombination, selection, genetic drift, and migration.* This process is enabled and constrained by physicochemical law, emergence and complexity being special cases of physicochemical law.

2. *On a long-term scale, the resulting history of life includes both gradual and rapid evolution, or punctuated equilibrium.* Punctuations are periods of rapid change (in relative terms only) that require no new assumptions about the process. They result from population founder effects and mass extinctions, some of which may have been periodic. All such punctuations are compatible with standard microevolutionary processes.

3. *Natural selection, while not the only evolutionary process, is the main explanation for the adaptedness of life* (Darwin 1859/1958; Fisher 1958; Williams 1966, 1992; Mayr 2001). Natural selection includes but is not limited to differential survival, differential reproduction, sexual

selection, kin selection, inclusive fitness, cooperation, and, at the population level, evolutionarily stable strategies. Group selection plays a modest role. Limits imposed on adaptedness by physical and chemical law, complexity, organic constraint, the coherence of the genome, and the phylogenetic history of the organism are important, but do not negate the pervasive role of selection (Maynard Smith, Burian, et al. 1985; Maynard Smith 1993).

4. *Natural selection works with the materials of ontogeny, which translate genome into phenotype, acting at every point from conception to senescence* (Rose 1991; Finch and Rose 1995). A coherent life history is a complex compromise between survival at any time point, survival at later and earlier time points, reproductive success, and inclusive fitness, which is the ultimate summary measure of adaptedness and the essence of organic evolution (Charnov 1993; Sibly and Brown 2007, 2009).

5. *Human ontogeny cannot be understood except in relation to ancestral ontogenies and those that competed with them for the limited chances afforded living forms* (Gould 1977, 1988; McKinney and McNamara 1991; McNamara and McKinney 2005). There is no coherence to human development without reference to this background (Shea 1983a, 1988; Bogin 1994, 1999; Godfrey and Sutherland 1996; Minugh-Purvis and McNamara 2002).

6. *Human behavior and its development, including all of mental life, are in their entirety biological manifestations.* This does not mean that they are not plastic or culturally influenced, but that all such effects must be understood as plasticity of and influence on biological materials (West-Eberhard 1989, 2003, 2005). Plasticity is limited by boundary conditions, the resistance of biological materials to change, and processes known to evolutionists as reaction norms and facultative adaptations, which channel change along particular, limiting pathways.

7. *As with ontogeny, and because of it, the brain has evolved by accretion.* Rather than replacing older structures with newer ones, brain evolution has been a patchwork of additions (Martin 1983, 1990; Deacon 1990, 1997b; Allman 1999; Striedter 2005). These additions are always intimately—if not always smoothly—integrated with older

elements, but do not replace them and do not necessarily control them.

8. *Constraints on plasticity are greater than they were assumed to be in late-twentieth-century psychology and social science.* Although only a few relevant genes are known (Enard, Przeworski, et al. 2002; Evans, Anderson, et al. 2004), genetic influence is clear for universal features of the human behavior developmental plan; for some sex differences in behavior; and for substantial proportions of the variation in behavioral traits, including sociability, aggressiveness, anxiety, depression, verbal ability, mathematical ability, nurturance, sexual drive, and sexual orientation.

9. *Neither these nor any other human traits can be understood except in terms of brain function.* The history of the technical concept of "mind" is largely over, and cognitive psychology can progress only by becoming neuropsychology. This change has no necessary implications for plasticity, since the need to use brain function as an explanation for psychological process and its development is orthogonal to the dimension of plasticity (Kaas 1995; Merzenich 1998). But to the extent that human ontogeny has behavioral invariants, brain development, including neuroendocrine development, is the only plausible substrate for these invariants.

10. *Despite limits to plasticity, cultural influence on human development is strong because its power is biologically assured* (Lumsden and Wilson 1981; Durham 1991; Shore 1996). The same state-of-the-art personal computer on the desk of an architect, a Bible scholar, a mathematician, and a football-obsessed twelve-year-old has unvarying hardware but varied output. In the brain even the hardware is plastic, the software all the more so. To solve the apparent paradox of the powers of biology and culture is a central task of developmental research.

11. *Socialization is not the same as enculturation.* Socialization is widely shared among birds and mammals, especially higher primates. This process, not culture, explains such states as love, lust, fear, rage, joy, and grief, not to mention social inhibition, social "scripts," and some intersubjectivity (Konner 2002). These are not specifically human

and do not require culture, although they are certainly changed by enculturation (Kruger and Tomasello 1996; Shore 1996).

To say that these are premises is not to say that they are assumptions, but rather that, although they are provable, proof is not offered here; passing citations must suffice. Those readers who more or less accept these premises, or at least do not react with hardened opposition, will benefit most from this book's demonstration of how they illuminate human development. In doing so, it will form the basis for a new developmental psychology—one that is, finally, cognizant of biology.

Some History

In recent decades the evolution of behavior has recaptured the attention of biologists and social scientists, and this has struck some observers as novel. But when Darwin, Wallace, and Huxley unveiled evolution, the biology of human behavior was high on the agenda. Both Darwin's pioneering 1872 work on the subject, *The Expression of the Emotions in Man and Animals,* and the ill-conceived distortions of social Darwinists (Eiseley 1958; Chorover 1979) stemmed from the same realization: that the evolutionary theory could help explain human behavior.

Darwin's procedure (1872), still followed in efforts to build such a theory (Ekman and Rosenberg 1997), was the same as that used in modeling the anatomical evolution of animals with poor fossil records. Features of groups are compared to determine their relatedness, fossil evidence is considered, and evolutionary history is reconstructed in the light of what we know about how evolution works. In some cases behavioral phylogenies were so compelling as to force reconsideration of evolutionary histories (Tinbergen 1964; Lorenz 1941/1971).

Despite its auspicious, literally Darwinian beginning, the evolution of behavior suffered neglect and even hostility from both biological and behavioral scientists. One argument was that most behaviors are so variable and subject to such extensive developmental modification as to make behavioral traits inaccessible to systematic zoology. But we learned that many behavioral traits are highly species-specific and sometimes more invariant within a species than morphological traits (Mayr 1958; Hinde and Tinbergen 1958; Eibl-Eibesfeldt 1975; Lorenz

1981). We know too that although virtually all behavioral traits can be environmentally modified, this modification can be so narrowed by the genetic programming of development that behavior so "learned" is indistinguishable, for practical purposes, from "instinct" (Lorenz 1965).

The near-absence of a fossil record for behavior was another complaint. Paleontology is the cornerstone of evolution, but suppose that (by the act of some vengeful deity?) all fossils disappeared. Evolutionary biology would suffer greatly, but it would proceed, through studies in comparative morphology and embryology, ecological adaptation, molecular genetics, ongoing natural selection, and so forth. The study of behavioral evolution proceeds similarly—although of course its models must be consistent with fossil evidence. Meanwhile, studies in comparative neurology, neurophysiology, endocrinology, and especially genetics make the physiological substrates of behavioral evolution continually more discernible.

As Anne Roe and G. G. Simpson put it in 1958 (the same year Anne Anastasi stressed the question "How?"): "It should by now be . . . obvious that there is, indeed, a general theory of behavior and that theory is . . . evolution, to just the same extent and in almost exactly the same ways in which evolution is the general theory of morphology" (Roe and Simpson 1958:2). The great evolutionary geneticist Theodosius Dobzhansky said that nothing in biology makes sense except in the light of evolution, and the same can now be said about the branches of biology known as behavioral and social science.

Conversely, to see the value of behavioral studies for evolutionary biology we have only to consider that behavior often mediates between an organism and its environment and that a behavioral change often first enables a population to enter a new ecological niche, or even a major adaptive zone (Romer 1958; Colbert 1958; Mayr 1958, 1988), introducing the organism to selective pressures that force morphological changes at the population level (Tinbergen 1964). Adaptive behavior often leads morphological and biochemical evolution; in the words of the leading twentieth-century evolutionist, Ernst Mayr (1988:408), "Behavior is the pacemaker of evolution."

Ethologists, sociobiologists, and behavioral ecologists, usually zoologists by training and inclination, had as their ultimate goal understanding the evolution of behavior. Although they often dealt with the evolution of particular behaviors in specific groups of animals (Tinbergen

1964; van Hoof 1972), their energies were first directed to two prior problems. The first one, "How does the animal behave?" seemed an obvious question, but it could not be answered through laboratory studies or anecdotal reports from zoos, a conclusion ultimately echoed by American students of animal behavior (Beach 1950; Lockhard 1971). The field study as a research tool gave us our first real grasp of the adaptive significance of behaviors, the manner and sequence in which they appear in the life cycle, and our first view of an animal's total behavioral repertoire, or ethogram. Where field study is not feasible or does not tell us enough, artificial environments approximate natural conditions as closely as possible (Gouzoules, Gouzoules, et al. 1998).

Evolution and Modification of Behavior

But the second stumbling block before behavioral evolution is more central to this book: just what in behavior can be called "innate" or "instinctual," and what do these terms mean? Before we can discover modes of phylogenetic change, we must know something about how the genome programs any (relatively) fixed patterns and their emergence in the life cycle. Deprivation experiments addressed this problem by withholding a specific aspect of environmental input or, in the case of a movement pattern, the chance to perform it (Eibl-Eibesfeldt 1960; Lorenz 1965).

The pitfalls of the deprivation experiment contributed to errors of both the champions and critics of ethology. The former sometimes failed to realize that, while withholding some information, it is possible inadvertently to provide other experiences from which the needed input can be derived. Early experiments on "innate" maternal behavior in primiparous rats (Beach 1950) were a good example, and the contributions of prehatching experience to the behavior of newly hatched birds were another (Gottlieb 1991a, 1991b, 1992, 1996).

The classic American experimentalists, who believed that behavior is overwhelmingly learned (Skinner 1938), were encouraged by two flaws in deprivation experiments (Lorenz 1965). First, isolation may inhibit normal growth, and consequently the behavioral failure could have resulted from general physiological inadequacy rather than from the specific environmental deprivation. Second, the test situation may be so foreign to the species that even a normally reared wild animal would fail to show the behavior in question. Experiments on nesting behavior in isolated rats are a good example of this error. But both errors were more avoidable as fieldwork unveiled the natural environment and life cycle of the organism.

American comparative psychologists especially studied one kind of learning that involved the shaping (through reinforcement) in animals of motor responses to stimuli, either internal (drives) or external (cues). They did not ask why reinforcers reinforce, whether some stimulus-response connections are easier to establish than others, or whether the behaviors they studied have any significance for the animal in its natural surroundings. Their subjects' performances were often more a tribute to the experimenter's ingenuity and patience than an elucidation of animal behavior. They also focused on the albino Norway rat to the exclusion of most other organisms, yet few took an interest in the Norway rat in nature, despite published accounts (Barnett 1975).

Still, their emphasis on learning proved a good foil for ethologists' emphasis on inherited patterns. Konrad Lorenz (1965) explained his own emphasis: in any system whose function consists of the interaction of many parts, the least changeable are the best point of departure, because they appear most often as causes. But he also suggested that phylogenetic change in the ontogeny of behavior has been a major process during the whole history of life. For their part, ethologists had little to say about the evolution of learning, to which psychologists made major contributions (Harlow 1958a; Bitterman 1965, 1975; Rumbaugh and Pate 1984; Rumbaugh, Washburn, et al. 1996; Rumbaugh 1997; Rumbaugh and Washburn 2003).

So it is useful to catalog and compare relatively fixed patterns of behavior and to determine what is programmed in the genome. But it is equally essential to know how behavior is modified and to construct phylogenies of modification processes. Comparative neurology and experimental genetics become more relevant to the evolution of behavior as we learn how the genetic programming of behavior is decoded in ontogeny. Evolutionary biology, neurobiology, anthropology, ethology, and experimental psychology all contribute. But before we take them up systematically, consider four widely held misconceptions.

Misconception 1: Ontogeny recapitulates phylogeny. Overall, this claim, known in the nineteenth century as Haeckel's biogenetic law (Mayr 1982), is false, but properly recast, it holds a piece of the truth: evolution is opportunistic, building on what it has. Therefore many embryologic and other developmental events are conserved, while evolution builds on later phases of the organism. In other words, ontogeny recapitulates not the *adult* phases of ancestral forms but, to some extent, the early ontogeny of those forms—an idea already known before Darwin as von Baer's law (Gould 1977): the early developmental phases of related forms resemble one another more than the adult phases of these forms. Von Baer's law is as widely true a generalization as can be made in the evolution of ontogeny, and it formed part of the evidentiary basis for evolution that Darwin reviewed in *The Origin of Species* (1859/1958). In vertebrates, it turns out, the earliest embryonic phases are also variable, but there soon comes a point, the *phylotypic stage,* when all embryos are similar; this might be called von Baer's bottleneck.

We cannot expect characteristics in human children that are much like those of *adults* of other animals, yet this expectation strongly influenced early-twentieth-century psychology. It resulted in the contention that Western children's mental functioning resembles that of adults in "primitive" populations (Werner 1948), a contention not only disproved by early cross-cultural experiments (Mead 1928) but also inconsistent with all we know about the biology and evolution of human populations.

Misconception 2: The higher the animal, the slower its development, the less developed it is at birth, and the more plastic is its behavioral repertoire. All these relationships are commonly thought to help explain human plasticity, yet none is strictly valid. First, there is no accepted way of arraying animals in a hierarchy (Lockhard 1971; Gould 1989); what is interesting about animals is how and why they differ, not how they rank. Second, the degree of plasticity is extremely variable, even in closely related animals, and is not always linked to slow development. Of two subspecies of North American deer mice (*Peromyscus maniculatus*), the one that developed faster was more affected by early experience (Rosen and Hart 1963). Studies of sensory-motor development in two prosimians, one Old World monkey, and two New World monkey species revealed no trends in the rate of emergence of the behavior and no relation to the wide differences in learning ability among these animals in adulthood (Ehrlich 1974; Elias 1977). Variation within a species in time of emergence of the behaviors proves in some instances greater than variation between them.

Very broadly speaking, behavioral plasticity increases as we come closer to humans—whether within vertebrates, within mammals, or within primates—using reasonable definitions of plasticity and several different phylogenetic scales (Lorenz 1965; Rumbaugh, Washburn, et al. 1996; Rumbaugh 1997). But plasticity does not correlate well with rate of development and may vary with the stage of the life cycle in unexpected ways.

Also, there is no simple way to predict the state of development at birth, which is determined not just by broad phylogenetic trends but by pressures specific to individual species, such as the level of predation, the possibility of nest building, the time and energy available for gestation, and the overall length of life. Lifestyles as well as life histories adapt. State of development at birth need not be correlated with overall rate of development or ultimate behavioral plasticity, and it may not be consistent among behavioral and organ systems. Newborn kangaroos and opossums, otherwise the most altricial (underdeveloped) of all mammals, have highly developed forelimbs and an orienting system that carry them from the uterus, across a great expanse of belly, into the pouch and to the teat, with no assistance (Eisenberg 1988). Closely related species of ungulates differ in whether the newborn is able to follow the mother immediately; following occurs in highly migratory species such as wildebeest, while in a number of others the calf is cached in a shifting hiding place and the mother returns periodically until it can follow well (Janis and Jarman 1984). Hares and rabbits differ as dramatically in their state of development at birth as virtually any other pair of mammals (Bourliere 1955). Humans and other higher primates are precocial (more mature at birth) in most sensory systems and altricial (more helpless) in most motor systems.

Thus it seems that state of development at birth, which as we will see influences many other aspects of infant and parent behavior, is in turn the result of specific adaptation—in effect a selection funnel in ontogeny—and follows no grand phylogenetic logic or relation to plasticity (Ewer 1968; Eisenberg 1981). This misconception too has had unfortunate effects. African motor precocity—dubious as an empirical claim —has been viewed as consistent with an alleged African intellectual inferiority in adulthood because of some erroneous notions about phylogeny. There is no good empirical evidence for any correlation, positive or negative, between infant motor development and later intellectual capacity, but in addition, there is no basis in evolutionary theory for expecting such a relationship.

Misconception 3: If a behavior is phylogenetically widespread, it must be a "fixed action pattern" or "instinct" and thus genetically based; in this event, there is no sense trying to change it. Each step of this syllogism is wrong.

First, analogy is not homology. Widely disparate animals may face similar puzzles posed to them by natural selection, and their solutions may look similar and serve similar functions, without having similar mechanisms. The wings of flies come from thorax, the wings of birds from forelimbs, the wings of bats from fingers, and the wings of people from aircraft manufacturers. Attachment formation in human infants and imprinting in precocial birds serve similar adaptive functions, but we need know only that the latter is completed in several days and the former requires months to know that their underlying mechanisms must differ.

Second, there are many paths to behavioral stability. Fixed action patterns, the building blocks of what used to be called instincts (Lorenz 1981), are arrived at by as many routes as there are species to exhibit them. To accept that a fixed action pattern or a response to a given stimulus (a releasing mechanism) is not learned, we look to deprivation experiments, in which numerous features of the behavior of any organism prove innate—or at least to say that they are is "less inexact than the statement that a steam locomotive or the Eiffel Tower are built entirely of metal" (Lorenz 1965:27).

In human children, crying, smiling, bipedal walking, language competence, and sexual feelings after puberty are examples of behavioral or mental characteristics not meaningfully dependent on learning for their *emergence.* However, some quite stereotypic action patterns and releasing mechanisms, in a wide variety of species, may depend in important ways on experience. Among these are food choice in moths, habitat choice in damselflies, homing in salmon, sexual choice in some pair-bonding birds, and sexual performance in rhesus monkeys. The same evidently applies to mouse killing by rats (Denenberg, Paschke, et al. 1968) and maternal behavior in rhesus monkeys (Arling and Harlow 1967).

It has even been argued that the pecking behavior of chicks immediately after hatching is acquired during gestation through the training of head movements by the beat of the fetal heart (Lehrman 1953). This case may not be convincing (Lorenz 1965), but there is no doubt of the role of embryonic experience in the behavior of avian hatchlings (Gottlieb 1982, 1991a, 1991b), and we know that activity shapes the embryonic brain (Changeux and Danchin 1976; Changeux 1985). Overall, the concept is important because it makes a further distinction: behaviors (and releasers) that are innate need not be genetically determined. The specific determination of each pattern must be investigated individually (Moltz and Robbins 1965).

Third, stability, even genetically bounded stability, does not imply unchangeability. Proper discussion of this point must be deferred to Parts III and IV, but because the specter of biology seems to cast a shadow over public policy, I mention it in passing here. The belief is held by some psychologists that if a behavior, ability, or behavior disorder is genetically determined, it is not promising to try to change it (Jensen 1969; Herrnstein and Murray 1994). Wearers of eyeglasses and takers of insulin can testify to the folly of this view. Although these are palliative measures and not cures, medical institutes gamely forge ahead from day to day searching for ways to alter conditions in people who have them because of gene action. Behavioral scientists can do the same (Rutter and Plomin 1997; Plomin and Rutter 1998).

Misconception 4: If animals are so variable, we may as well focus on our closest relatives, since they will teach us the most. This view guides the funding of research on monkeys and apes and the comparative thinking of many social scientists. It is partly true, but closeness of phylogenetic relationship is only one of at least four bases for comparing two species. Three others are similarity of reproductive strategy, similarity of ecological adaptation, and similarity of major sensory channels used in communication. The chimpanzee, our closest relative and a species surely worthy of study, differs from humans living under natural conditions in having a much smaller territorial or home range, doing much less hunting, dramatically displaying ovulatory status, and exhibiting little or no pair bonding (Goodall 1968, 1986; Wrangham, McGrew, et al. 1994). Mammals that hunt, such as lions, exhibit much more human-like patterns of sharing behavior (Schaller 1972)—and some elements of teaching behavior (Schenkel 1966; Caro and Hauser 1992)—than do any higher primates. Ring-necked doves (Lehrman 1955, 1965; Lehrman, Brody, et al. 1961; Erickson 1986; Cheng 1992) and prairie voles (Insel 1997; Carter 1998) have proved to be superb models of the physiology of pair bonding, which is nonexistent in any great ape. Foxes are pair-bonding, hunting mammals and a potentially good model for certain aspects of human parent-offspring relations, but have been studied only a little, mainly in field settings (Fox 1975; Baker, Robertson, et al. 1998).

Rhesus monkeys (*Macaca mulatta*), a model on which we have relied heavily for laboratory manipulation of the growth of social relations, may be inappropriate in some important respects. Under natural conditions, rhesus monkeys live in very large troops, are highly promiscuous, and carefully keep their infants away from other individuals (Southwick, Beg, et al. 1965; Breuggeman 1973; Maestripieri 1995; Berman, Rasmussen, et al. 1997). These important behaviors distinguish them from human foragers. Indeed, in two closely related species—bonnet macaques (*Macaca radiata*) and pigtail macaques (*M. nemestrina*)—the response of infants to the removal of their mothers when the pair are living in a social context relates to just such differences in the species' normal guarding of infants (Rosenblum and Kaufman 1968; Rosenblum 1971, 1978). The bonnet infant is quickly adopted by another female and is soon behaving more normally; the pigtail infant enters a profound depression from which it never fully recovers. In these and other respects, the pigtail resembles the rhesus, our most common subject for study; the bonnet, studied in only one or two laboratories, resembles humans. Ingenious and informative experiments on bonnet mother-infant pairs have varied spatiotemporal patterns of food availability (though not nutrition itself) in an attempt to simulate foraging stress effects on maternal care (Coplan, Smith, et al. 2006).

A final example illustrates the principle of comparison based on similarity in the sensory processes used for communication. The rhesus monkey relies largely on olfactory signaling for communication in courtship (Michael and Keverne 1968, 1969). Similar, although much weaker, effects in humans have been demonstrated experimentally and should receive more attention (McClintock 1998; Stern and McClintock 1998). Yet it may be of greater interest to study the ring dove, which relies overwhelmingly (as people do) on visual and vocal-auditory signaling in courtship (Lehrman 1971; Cheng 1992; Cheng, Peng, et al. 1998). Thus a close relative, the rhesus, with its much better sense of smell than ours, may be of less value as a model than a remote relative, the ring dove, which, like us, has a poor sense of smell and excellent vision and, in addition, is pair-bonding. Similar considerations may apply to the development of attachment in infancy. However, human infants at the age of six days do show preferential head-turning toward their own mother's breast pads, as opposed to another mother's, using olfactory discrimination (MacFarlane 1974), and a role for infant odors in influencing maternal responses has also been demonstrated (Fleming, Corter, et al. 1993).

Evolution of Ontogeny in the Human Animal

Beginning in the 1960s, Konrad Lorenz (1965), John Bowlby (1969), Jerome Bruner (1972), and Niko Tinbergen (Tinbergen and Tinbergen 1972) all made major claims about the importance of evolution for the study of human behavioral development. Lorenz and Tinbergen, leaders of ethology, saw in human development a logical practical application for their expertise on the apparent confrontation of nature and nurture. Skirting this ancient battlefield, they called special attention to the role of species-specific behavior and to the functions of play in the growth of competence. Bowlby and Bruner, leaders of psychoanalysis and cognitive psychology, respectively, saw in evolution the wisdom of natural function, which gave rise to the phenomena they dealt with; they saw in it too a potential way out of the maze of learning theory, in which both attachment and cognition had been lost. They thus harked back to their turn-of-the-century forebears, who were deeply interested in evolution.

Those forebears, as we have seen, also pursued comparative developmental studies. Though evolutionary forces act upon individuals and modify populations, what connects all living things is the genetic material; studying adult organisms is thus far removed from the substance of continuity. The comparative study of development filled this gap (Gould 1988; McKinney and McNamara 1991; Raff 1996). Eric Lenneberg (1967:247) discussed how development is interposed between adaptations and the genes ultimately under selection. In evolutionary biology today we are increasingly able to study genomic transformations (Nee, Mooers, et al. 1992; Carroll 2003; Evans, Anderson, et al. 2004), but the mapping of adult forms and functions onto the molecular level remains obscure for most traits. The logical conclusion is that an appropriate area for study is development, the process of mapping.

A second reason for such study is that while ontogeny does not recapitulate phylogeny in content, they may share some features in process (Lorenz 1965; Skinner 1966). For example, a partial functional analogy exists between trial-and-error learning and the way in which part of the progeny is risked in the "experiment" of mutation. Probably one of the chief functions of sexual reproduction is to disseminate new information, as stored in DNA, quickly through a population. Further, learning apparently does often increase the survival value of the behavior mechanisms it modifies.

These parallels have significance beyond analogy. They show that a developing organism must be viable in its social and ecological milieu *at every point in its development,* in addition to passing through a series of events calculated to produce a viable adult. These two considerations reflect the two processes through which natural selection determines the development of organisms.

Levels of Causation in the Explanation of Behavior

Having considered some important caveats about attempts to relate evolution and ontogeny in the explanation of behavior, we can proceed to the simple model that is at the heart of this book.

Most explanations of behavior occur at one level only. But as was pointed out by Tinbergen (1963), the question "Why did the animal do that?" can be answered at different levels, four of which were identified in his classic paper: phylogenetic, ecological, developmental, and eliciting. These can be exemplified by categories of answers to a question about a short flight of a bird—say, a jay rising from a holly bush up to a longleaf pine. It flies because it is a bird; because flight gave it an advantage (over predators, prey, or competitors) in its environment of evolutionary adaptedness; because its ontogeny gave it light bones, wings, feathers, and a motor neuron circuit oscillator for flight, through a genetically determined maturational pattern shaped by nutrition, exercise, and practice; and because a fox is chasing it.

I break these explanations down further into nine levels in three categories of causes, which are briefly described in the remainder of this introductory chapter. Levels 1 and 2 are the same as Tinbergen's; his category 3 would include levels 3 through 8 in this model, which breaks out and expands the ontogenetic causes; and his category 4 is the same as level 9 here. These nine levels can be aggregated into three overarching kinds of causes.

Levels 1 to 3: Remote or Evolutionary Causation

1. *Phylogenetic constraints.* Because an organism is of a certain broad taxonomic type, it is constrained to some extent in the way it can solve the problems posed by its environment, even under fairly aggressive selection; its deep evolutionary history is relevant.

2. *Ecological/demographic causes.* Because the organism faced a certain set of adaptive problems in a particular environment, its fitness was in effect maximized for that environment; studying it in that environment should be illuminating.

3. *Genome.* The result of the first two causes, the individual's genome falls within a certain spectrum of variation for its species, population, or sex. It is the result of the phylogenetic and ecological/demographic causes, and in turn the cause of all further *possibilities* in the life cycle, although not all further outcomes.

Levels 4 to 6: Intermediate or Developmental Causation

4. *Embryonic/maturational processes.* Given the normal expectable environment or ontogenetic niche (West and King 1987) of the species, the genome does not merely start the events of ontogeny, but guides them; birth (hatching, pupation) may be an important event, but ontogenetic mechanisms operate throughout life.

5. *Formative early-environment effects.* These "critical" or "sensitive" period effects, which constitute important developmental directions set by the environment, are either facultative adaptations (developmental options shaped by natural selection) or maladaptive consequences of deprivations.

6. *Ongoing environmental effects.* These are factors such as nutrition, stress, and reinforcement contingencies that operate similarly at different life stages, in a time frame of days to years; in principle

they are more reversible than formative early effects, although major trauma at any stage of life can be irreversible.

Levels 7 to 9: Proximate or Functional Causation

7. *Longer-term physiology.* Though mainly hormonal, longer-term physiology also accounts for other metabolic effects (energy flow, muscle fatigue, toxic substances), in a time frame of minutes to days, as outcomes of gene expression in response to environmental contingencies.

8. *Short-term physiology.* Behavioral output is mediated by short-term physiology, mainly through neural circuits and their sensorimotor "peripherals," on a time course of milliseconds to minutes, which are the immediate internal causes of behavior.

9. *Elicitors or releasers.* The immediate external causes of behavior, elicitors are the events in the stimulus envelope that precipitate the behavior; ethologists call this the releasing mechanism, and to the learning psychologist it is the conditioned or unconditioned stimulus.

Clear thinking about the causes of relationships, emotion, mind, and other aspects of behavior requires keeping multiple levels of explanation in play and, preferably, separate for the purposes of analysis. Innumerable unhelpful controversies arise when experts on different kinds and levels of explanation argue with each other. There is often no justification for these arguments, which commonly result from a failure to appreciate the levels of causality, each of which is valuable, indeed essential, in explanation. Complete explanation requires integration among the levels.

I

Evolution: *The Phylogenetic Origins of Childhood*

**wherein we learn how the laws of evolution produced the
shape of human social and emotional development**

It should, by now, be . . . obvious that there is, indeed, a general
theory of behavior and that that theory is . . . evolution, to just
the same extent and in almost exactly the same ways in which
evolution is the general theory of morphology.
—Anne Roe and G. G. Simpson, *Behavior and Evolution* (1958)

2

Paradigms in the Evolution of Development

We begin with eight paradigms that emerged separately in the history of biology—eight paths to an evolutionary developmental psychology. Although they remain partly independent modes of inquiry, their separation is artificial, and integration is needed. They are (1) neo-Darwinian theory; (2) life history theory; (3) evolutionary allometries; (4) heterochrony in the phylogeny of development; (5) evolution of developmental genes ("evo-devo"); (6) phyletic reorganization in brain evolution; (7) developmental ethology; and (8) evolutionary developmental psychology.

1. Neo-Darwinian Theory

Adaptationist models of social evolution were originally associated with the work of Charles Darwin (1872), then with that of Ronald Fisher and J. B. S. Haldane (Fisher 1958), later with that of William D. Hamilton (1964), John Maynard Smith (1982), and Robert Trivers (1972, 1974, 1985), and more recently with that of Tim Clutton-Brock (2002, 2009), Martin Nowak (2006a, 2006b), and many others. The basic insight is that natural selection has shaped behavior to maximize reproductive success. The basic data include the social actions of organisms, the differential reproductive success associated with different behaviors, and the degree of relatedness among the individuals, ideally determined from DNA.

Typical observations: Nurturance and prosocial risk taking can be predicted from the degree of relatedness, which is the likelihood that any genes affecting these behaviors will be identical by descent; males and females behave as if they have different reproductive interests; and conflict may reach such extremes as brood parasitism, siblicide, and competitive infanticide. These models grade naturally into those of the second paradigm.

2. Life History Theory

This paradigm was associated with the work of Robert MacArthur, Edward O. Wilson, and Eric Pianka (MacArthur and Wilson 1967; Pianka 1970), later with that of Stephen Stearns (1992) and Eric Charnov (1993), and most recently with Richard Sibly and James Brown (2007, 2009), Derek Roff (2007), and many others. Its basic insight is that life cycles, not organisms, evolve, and that an organism *is* a life cycle, adapting to local ecological conditions by changing its trajectory and trading off one adaptive advantage against another, now or later. The paradigm's basic data are the demographic (especially reproductive and mortality) statistics of populations and species and the threats and demands of their environments.

   Typical observation: Under certain ecological conditions, a species of bird in an English wood will in time adjust its age at first reproduction, brood size, offspring size, and longevity in a way that maximizes individuals' reproductive success. Locally in time and space these are mostly reaction norms or facultative adaptations, both of which are aspects of plasticity, but they grade into genetically based differences with population variation and so eventually become the stuff of phylogenetic allometry in paradigm 3.

3. Evolutionary Allometries

Evolutionary allometrics was long associated with J. B. S. Haldane (1953/1990) and later with the work of Paul Harvey and his colleagues (Harvey and Pagel 1991) and other investigators (West, Brown, et al. 1997; Brown and Sibly 2006). Its basic insight is that there are mathematical (usually induced rather than deduced) relationships among the main features of the life cycles of species or higher-level taxa, often in relation to their size. The basic data are comparative statistics on longevity, age at sexual maturity, gestation length, brain size, and metabolic rate for species and higher taxonomic categories.

   Typical observation: After correcting statistically for body size, a pair of other features of the life cycle (say, brain size at birth and gestation length) are correlated, suggesting a fundamental organismal constraint.

Thus species, genera, or families of organisms that depart from the (usually logarithmic) regression line can be interpreted in relation to ecological adaptation. Humans are sometimes outliers on these plots, but not always.

4. Heterochrony in the Phylogeny of Development

Interest in development as a key to evolution precedes Darwin. It was associated originally with the work of Karl Ernst von Baer and Ernst Haeckel, and much later with that of Stephen Jay Gould (1977, 1988, 2000, 2002), Michael McKinney and Kenneth McNamara (1991; McNamara and McKinney 2005), and Rudolph Raff (Raff and Raff 1987; Raff 1996). Its basic insight is that evolutionary change must work through the *Bauplan,* or building-plan, for the organism, changing the genetically regulated course of development. The basic data are paleontological and comparative (anatomical or molecular).

Typical observation: Some adult organisms resemble, in part, a juvenile phase of an ancestral or descendant species. Classically called heterochrony, or "different time," this kind of process has long been thought central to human evolution. With advancing molecular biology of developmental genes, heterochrony will gradually merge with comparative developmental genetics in the evolution of development ("evo-devo").

5. Evolution of Developmental Genes (Evo-Devo)

This new field emerged from fruit fly genetics, which by systematically exploring mutant "monsters" began to elucidate the genetic control of body plans, and from embryology, where a goal has been constructing fate maps to identify cells in the early embryo that have specific anatomical destinies in the adult. The fact that embryology has a number of different model organisms—the roundworm, the fly, the zebrafish, the frog, the domestic chick, and the mouse, among others—set the stage for a broad phylogenetic comparison of developmental genes when they were discovered. Evo-devo is associated with the work of Walter Gehring (1999, 2005), Christiane Nüsslein-Volhard (2008), Sean Carroll (2003, 2005), and many others (for example, Angelini and Kaufman 2005). Its basic insight is that there are deeply conserved homologies in the genetic control of embryonic development vital to our understanding of both evolution and ontogeny. Its data are the comparative genetics of early embryonic development, supplemented and informed in significant ways by the fossil record (Raff 2007).

Typical observation: The regulatory gene *PAX6* is involved in eye development in widely differing types of animals (humans and flies, for example) despite profound differences in the anatomical structure and development of the eye (Gehring 2005).

6. Phyletic Reorganization in Brain Evolution

This mode was associated with the work of C. U. Ariëns Kappers and James Papez (Ariëns Kappers, Huber, et al. 1936; Papez 1937), later with the work of Paul MacLean (1952, 1985, 1990), and more recently with that of Ann Butler and William Hodos (2005), Barbara Finlay and Richard Darlington (1995), John Allman (1999), Todd Preuss (2001), George Striedter (2005), Jon Kaas (2006), and James Rilling (2008a, 2008b). Its basic insight is that new structures must be understood as transformations of older ones. Its data are those of comparative neuroanatomy, neuroimaging, and gene expression, supported where possible by endocranial casts of fossil skulls, which partly simulate the surface anatomy of brains. These data are combined with the inferences of phylogenetic (cladistic) taxonomy, which uses molecular and morphological divergences to generate phylogenetic trees, and with comparative behavioral observation and experiment, which reflect the evolving functional capacity of brains.

Typical observation: The brain structures shared by all tetrapods became specialized through accretion in reptiles and diverged to form the different brain morphologies of birds and mammals. Ontogenetic (Striedter 1997), neurogenetic (Smith Fernandez, Pieau, et al. 1998), and gene expression comparisons (Preuss, Caceres, et al. 2004) now play a central role in this line of research, which will merge with paradigm 4 and with developmental neurogenetics.

7. Developmental Ethology

The study of animal behavior as adaptation originated with Darwin (1872) and was extended by Konrad Lorenz (1935/1970, 1941/1971, 1981) and Niko Tinbergen (1975), and later by Patrick Bateson (1983, 1991), Robert Hinde (Hinde and Tinbergen 1958; Hinde and Spencer-Booth 1971), Irenaus Eibl-Eibesfeldt (1960, 1973a, 1973b, 1988), Nicholas Blurton Jones (1972b), Jeanne Altmann (1979), Robert Fagen (1981, 1993), and many others. Its basic insight is that evolution has produced ethograms (species-specific behavioral repertoires), which develop predictably under and for the conditions the species evolved in. Its data are: behavior observed in fieldwork under natural conditions; field or laboratory experiments to discover the stimuli that elicit certain behaviors or to test hypotheses about their adaptedness; and deprivation experiments to establish the minimal environmental conditions that allow a behavior to develop.

Typical observation: The attachment behaviors of nonhuman higher-primate young (Harlow 1963; Bowlby 1969, 1973, 1980), which include clinging, distress calls, maintaining eye contact, and fleeing to the mother under threat, are widespread among Old World monkeys and apes, both in field studies and in laboratory conditions of deprivation, and can be largely restored after prolonged isolation rearing (Suomi and Harlow 1972; Mason and Kenney 1974).

8. Evolutionary Developmental Psychology

This new paradigm arose from a convergence of neo-Darwinian and life history theory with developmental ethology. Pioneered by Daniel Freedman (1974) and others, it has been associated with the work of Nicholas Blurton Jones (1978, 1993), Patricia Draper (Draper and Harpending 1982, 1987), Jay Belsky (Belsky, Steinberg, et al. 1991), James Chisholm (1993), Sarah Blaffer Hrdy (1976, 1999, 2009), and more recently with that of Glenn Weisfeld (1999; Weisfeld and Woodward 2004), David Bjorklund and Anthony Pellegrini (2002), Bruce Ellis (2004; Ellis and Bjorklund 2005), and Mark Flinn (2006). A parallel branch of this tradition has to do with empirical studies of hunter-gatherer childhood and has been associated with the work of Barry Hewlett (1991a, 1991b, 2005), Gilda Morelli (Morelli and Tronick 1992), Edward Tronick (Tronick, Morelli, et al. 1987), Ronald Barr (Barr, Konner, et al. 1991), several of the above-mentioned authors (Draper 1976; Blurton Jones, Hawkes, et al. 1994), and myself (Konner 1972, 1981, 2005), among many others. Hunter-gatherer childhood studies began as attempts to approximate children's experience in the environments of evolutionary adaptedness (EEAs), but were increasingly influenced by neo-Darwinian theoretical considerations.

Typical observation: Hunter-gatherer childhood, and by inference that of our ancestors, is characterized by high physical contact, frequent nursing, late weaning, a socially supportive environment for parenting, a transition to a multi-aged, mixed-sex child group, varying degrees of responsibility for food gathering and child care, late puberty, and low restriction on premarital sex (Konner 2005).

Each of these eight paradigms will now be considered in greater detail. We revisit paradigms 1, 2, 7, and 8 at the beginning of Part III.

Neo-Darwinian Theory—The Adaptationist Paradigm

Since the 1970s, influential extensions of evolutionary theory have emerged, known variously as neo-Darwinian, adaptationist, and optimization theory, or, as applied to behavior, sociobiology and evolutionary psychology (Williams 1966, 1992; Wilson 1975). Despite some resistance, this approach was adopted by most behavioral ecologists (Krebs and Davies 1991) and many ethologists (Wickler and Seibt 1977), and it has also influenced anthropologists and psychologists (Daly and Wilson 1983; Barkow, Cosmides, et al. 1992; Buss and Schmitt 1993; Betzig 1997; Cosmides and Tooby 2000), including developmental psychologists (Belsky, Steinberg, et al. 1991; Bjorklund and Pellegrini 2002; Bjorklund and Smith 2003; for overviews and history, see Alcock 2001 and Sagerstråle 2000). The principles are:

1. *An animal is a gene's way of making another gene.* More strictly, it is a way for thousands of genes to copy themselves; maintaining itself (strictly speaking, its information) in an ongoing succession is the first function of any gene (Williams 1966, 1992; Dawkins 1987). To the extent that a gene influences development or behavior, it can only persist if that influence maintains or increases its copies in the next generation's gene pool. Contrary to common error, the cohesiveness of the genome—multigenic traits, epigenetic effects, pleiotropy, local optima in the adaptive landscape, incomplete penetrance or expressivity, and so on—has only quantitative and temporary, not qualitative or permanent, bearing on this principle.

2. *Genes increase in number by enhancing reproductive success (RS).* The relevance of this principle for behavior has been demonstrated in many species (Clutton-Brock 1988), including ours (Betzig, Mulder, et al. 1988;

Chagnon 1988; Betzig 1997; Irons 1997). Improving survival is relevant where survival enhances reproduction, but where they conflict, RS prevails. Formally, fitness has no meaning except as relative gene frequency; it is a tautological dimension of RS and is not the same as medical, athletic, social, or ethical fitness, any or all of which can decrease while technically defined fitness increases. Theory notwithstanding, the human species has for centuries lowered its fertility as conditions improved (Vining 1986; Lutz and Qiang 2002). Neo-Darwinians have addressed this puzzle, but without much success.

However, recent demographic analyses suggest that in some (but not all) of the richest countries, the trend to smaller family size may be reversing (Myrskyla, Kohler, et al. 2009). Some neo-Darwinians predicted that this would happen, reasoning that reduction in family size by the rich was a temporary measure and that sufficient accumulation of wealth would begin to reverse the trend. This would be consistent with within-country data showing a small increase in fertility in the very wealthiest families compared to the middle class.

3. *Fitness is inclusive.* Genes influence their frequency not just through the survival and reproduction of each individual but also through the survival and reproduction of related individuals carrying the same gene by common descent. This is the concept introduced almost whimsically by J. B. S. Haldane and developed by W. D. Hamilton (1964), who offered the first explanation of altruism in animals using the mathematics of evolutionary genetics. Altruism theoretically should have been culled by natural selection, but inclusive fitness and kin selection dissolve the paradox. In Haldane's account, if I die to save my identical twin, the frequency of any gene that helped predispose me to that action will (all else being equal) be unaffected; if my risk of death is lower than his, the gene will be favored over time.

In general, genes predisposing us to sacrifice will be favored when $b/c > 1/r$, where b is the benefit to the recipient, c the cost to the altruist, and r the degree of genetic relatedness—the likelihood that a gene found in the altruist is identical to the same gene in the recipient *by common descent.* In practice, the calculation of relatedness is more subtle than in Haldane's example; just as fitness is defined against the background of population fitness, so relatedness is most meaningful as a ratio of the genetic closeness of one's kin to that of random members of the population (Grafen 1991). Inclusive fitness and kin selection explain the self-sacrifice of soldier ants for the colony (Seger 1991), similar cooperation in naked mole rats (Sherman, Jarvis, et al. 1991), alarm calls in birds and mammals (Sherman 1977; Clutton-Brock 2002), mutual aid (Chapais, Gathier, et al. 1997; Moore 1992; Clutton-Brock 2009) and most adoptions of infants and juveniles (Thierry and Anderson 1986) in monkeys and apes, and many adoptions (Silk 1980, 1987a, 1987b), not to mention nepotism, in people (Struhsaker and Leland 1987; Borgerhoff Mulder 1988, 1991; Fairgrieve 1995). However, inclusive fitness cannot explain all seemingly altruistic acts (Clutton-Brock 2002, 2009). Other explanations include reciprocal altruism, locally generalized altruism, and Prisoner's Dilemma–type models of cooperation, which do not require relatedness (Nowak 2006a). These are clearly important in societies where most social networks involve nonrelatives, but they neither negate nor replace kin selection, and they function even in kin-based societies.

4. *In sexual species, the sex that invests more in offspring will be a scarce resource for which the other sex competes* (Bateman 1948; Trivers 1972). In most mammals and many birds, females invest more, but direct male parental investment is very high in some species. These pair-bonding species, including eight thousand species of birds and a small minority of mammals, contrast with tournament species—so-called because some have annual seasonal breeding tournaments, or leks, where males compete intensely for females. These species tend to have high sexual dimorphism for fighting or display (canine teeth, antlers, peacock feathers); low tendency for pair formation; low direct male investment in offspring; and high variability in male reproductive success. In the elephant seal *Mirounga angustirostris,* 4 percent of males account for 85 percent of copulations during the breeding season, a skewing that can accelerate evolution and has produced extreme sexual dimorphism (Le Boeuf 1974). Males invest no care in offspring, and pups may be killed during male fights.

This extreme is uncommon, but there is evidence that a milder form of this kind of sexual selection among males has operated to cause rapid evolution of male genes in primates, including humans (Wyckoff, Wang, et al. 2000). In many primates, however, reproductive success includes male-female cooperation, female-female competition, and biparental nurturance (Drea 2005); in marmosets (small South American monkeys), males resemble females, fight little, and carry infants most of the time. Females are not passive in male-male competition, and often (as

in peacocks) female preference matters much more than male prowess; as Darwin understood, female choice guides the evolution of males
(Lailvaux and Irshick 2006).

The continuum also extends beyond pair bonding to a mirror image
of male-male competition: multiple mating by females, including polyandry, which can also have advantages (Simmons 2005; Clutton-Brock
2007). In jacanas and phalaropes, males brood the eggs, while females
compete for males and are larger than males. Such size reversals occur
in mammals as well (Ralls 1976). Even in species without such reversals,
including many mammals, primates among them, female competition
for reproductive opportunities is intense and sometimes violent (Hrdy
1981; Clutton-Brock, Hodge, et al. 2006; Pizzari, Birkhead, et al. 2006),
and the optimal number of "fathers" is not always one (Drea 2005; Hrdy
2000). Sexual selection theory does not preclude this; it only suggests
that in species where males invest less in offspring, they will compete
with each other more and diverge from females anatomically and behaviorally.

Humans as a species are near but not at the pair-bonding point of the
continua in sexual dimorphism, degree of direct male parental care, and
marriage forms (Daly and Wilson 1985; Murdock 1967). Polygyny (one
man, two or more women) is allowed in most cultures in the anthropological record (708 of 849, or 83 percent), while the converse, polyandry,
is rare (4 of 849) and usually involves brothers. In polygynous societies,
both men and women tend to mate at least in part according to neo-
Darwinian predictions (Borgerhoff Mulder 1988, 1991). A double standard of sexual restriction is common across cultures; still, most human
marriages have been mainly monogamous, owing either to environmental constraint or cultural principle. Modern cultures are monogamous in principle, but both adultery and serial monogamy are common.
In at least thirty-seven countries, men express preference for women
several years younger than themselves and place more emphasis on appearance, while women prefer men several years older and emphasize
status and wealth (Buss 1989, 1998; Buss and Schmitt 1993). But in humans, as in other primates, females hedge their bets, and female choice
is pervasive (Borgerhoff Mulder 1990; Irons 1997).

5. *Parent-offspring conflict is natural and unavoidable, not incidental
or pathological.* Adaptationist models of parent-offspring conflict have
deep implications for the nature of the family (Trivers 1974; Mock 2004).

Weaning conflict is very common in mammals, and birds have an equivalent, including tantrums. If the evolutionary goals of mother and offspring were identical, they should "agree"—should have been selected to act as if they agree, implying no conscious intent—on the level and duration of investment, after which the mother should turn to her next potential offspring. However, even if the current offspring and its unborn sibling have the same father, the current one's RS will be twice as great if it acts selfishly. Eventually the offspring's diminishing need for further investment is outweighed by the inclusive fitness advantage it gains through a subsequent sibling, but this occurs later for the offspring than for the mother, who is equally related to both. A naive model of the family assumes that it was designed to function as a harmonious unit; it was not. Like the breeding pair, it consists of individuals with overlapping but distinguishable reproductive purposes. Natural conflict is not the result of friction in what should be a smoothly functioning system, but is intrinsic.

6. *Competition can be extreme.* Virtually all animal species for which there is sufficient evidence show extremes of violent conflict in the wild, including the killing of conspecifics (Wilson 1975; Demaret 1999). The belief that human beings are rare among animal species in that we kill our own kind is false. Competitive infanticide, as in wild chimpanzees (Goodall 1977), fulfills a prediction of neo-Darwinian theory. The paradigmatic case is that of the Hanuman langur monkey of India, *Presbytis entellus* (Hrdy 1977; Boggess 1979, 1984; Mohnot 1980; Hausfater 1984). In langur troops, a core of related females and their offspring associate for up to a few years with unrelated immigrant males. Occasionally new males arrive and challenge the resident males. If the newcomers win, they drive the other males away and systematically kill infants less than six months old. The mothers come into estrus again much sooner than if the infants had survived and continued to nurse, and they are impregnated by the new males.

Controversy over whether this is normal behavior or a response to social stress (Boggess 1979, 1984; Vogel and Loch 1984) misses the point: in either case the behavior enhances the new males' RS. Life is stress; evolution is stress. It does not change the implications of infanticide to attribute it to stress. Similar systematic killing of infants by new males has been shown in other monkeys (Struhsaker and Leland 1987; Soltis,

Thomsen, et al. 2000) and lions (Packer, Herbst, et al. 1988), and killing of infants by unrelated females has been shown in chimpanzees (Goodall 1977; Arcadi and Wrangham 1999), wild dogs, and other mammals under natural conditions. Similar patterns occur in many species (Hrdy 1979; Hausfater and Hrdy 1984; Struhsaker and Leland 1987; Fairgrieve 1995; van Scheik and Janson 2000).

7. *Beyond relatedness, cooperation will evolve when the payoff matrix is suitable.* In the Prisoner's Dilemma model of real-time cooperation, gains are set to be greater than losses if a player cooperates while his partner cooperates, but losses are greatest for the player who cooperates while the partner defects. The greatest immediate gains come to the player who selfishly defects while the partner cooperates, but soon the partner mimics the player and both defect and lose consistently. Multiple iterations yield stable behavior and can, under some conditions, produce consistent cooperation based on selfish motives (Axelrod and Hamilton 1981; Axelrod 1984; Nowak 2006a, 2006b).

Delayed reciprocity, or reciprocal altruism, among unrelated individuals is unusual (Clutton-Brock 2009), but can evolve by individual selection if the species has a long enough life span and a memory of others' help (Trivers 1971). Since humans are such a species, delayed reciprocity evolved as a fundamental economic mode in most preindustrial societies (Mauss and Halls 1990; Cashdan 1989; Plattner 1989) and remains a vital process in modern ones (Gouldner 1960; Pollock and Dugatkin 1992; Cronk 1994; Reeve 1998). The key role played by reciprocity enables the evolution of cheating and its detection (Trivers 1991; Cosmides and Tooby 2000). As noted earlier, in addition to kin-based altruism, cooperation, and reciprocal altruism, indirect reciprocity may operate through the enhanced reputation of altruists and may extend to network reciprocity and, under certain conditions, group selection (Nowak 2006a, 2006b; Bowles 2009).

8. *Evolutionarily stable strategies (ESSs) help explain population adaptations over time.* In a game theory model of the evolution of competition, individuals (in any species) called Doves and Hawks are set to respond by conceding or fighting, and a payoff matrix sets gains and losses in encounters, as in the Prisoner's Dilemma (Maynard Smith 1982; Daly and Wilson 1988b). If a population of Doves is invaded by Hawk mutants, Doves become less numerous because Hawks consistently defeat

them. But as hawkishness spreads, Hawks fight mainly other Hawks, sustaining heavy losses, and Doves do better. The resulting equilibrium is an ESS. If ecological conditions change the payoff matrix, the equilibrium point will change. There could also be a developmental switch rather than a committed genetic difference, and a plausible developmental pathway for producing Hawks and Doves has been proposed for mammals, depending on maternal care and the resulting stress-reactivity of offspring (Ellis, Jackson, et al. 2006).

Similar considerations apply to the balance of cooperators and defectors in a population engaged in the Prisoner's Dilemma, or to the balance of reciprocators and cheaters in a population practicing reciprocal altruism. Cultural as well as genetic evolution may produce an ESS, and much of what we consider distinctive about a particular human culture *may be* its ESS, expressed through the complexities of social structure and cultural rules. In most cultures, not all individuals benefit equally from the equilibrium solution, and some individuals or groups may impose heavy fitness costs on others (Durham 1991; Borgerhoff Mulder, Bowles, et al. 2009).

9. *Facultative adaptations or norms of reaction are compromises between genetically fixed and open solutions.* Even simple organisms have evolved plasticity to respond to the different conditions encountered in the environment of evolutionary adaptedness (West-Eberhard 2003). The *lac* operon, a genetic switch in the intestinal bacterium *Escherichia coli*, responds to predictable environments. Genes produce three enzymes for lactose digestion. Absent that sugar, these genes are switched off by a repressor protein that binds DNA, preventing enzyme synthesis. The repressor is always present, but with milk in the intestinal habitat, a lactose molecule binds to each repressor protein and releases its grip on the gene promoter. Thus the genes of this bacterium direct a change in "behavior" geared to environmental conditions.

Facultative adaptations are common in nature, including social behavior. In the redwing blackbird (*Agelaius phoeneceus*), males singing on richer territories mate with several females instead of one (Harding and Follett 1979). The mechanism of this facultative adaptation may be quite different from that of the human parallel (Borgerhoff Mulder 1990; Irons 1997). Flight is adaptive in humans, but it is a facultative adaptation, dependent on cultural evolution. The same is true of some adaptations in social behavior. Other human social behaviors have a narrower range of reaction, and some are essentially fixed. The classic example of incest avoidance, a human facultative adaptation, is highly instructive and is discussed in Interlude 6.

Life History Theory

While it is a useful approximation to think of individuals as the units, selection works on life-cycle trajectories (Bonner 1965, 1993), each with its own reproductive success. Life history theory attempts to explain and predict life cycles as trade-offs or balances in the optimization of energy expenditure and risk (Lessells 1991; Stearns 1992; Stearns, Pereira, et al. 2003; Charnov 1993; Brown and Sibly 2006). These trade-offs in turn determine the timing of birth or hatching, litter or clutch size, the pace of growth, the amount of parenting, and aging. Plasticity in development enables all these possible trade-offs and many more (West-Eberhard 2003).

The variety of animal life cycles is impressive and puzzling (Stearns 1992). Salmon grow slowly, postpone breeding until late in life, and after enormous effort spawn once, producing millions of offspring, most of which soon die. Elephants grow slowly and produce just one offspring at a time, but can repeat the process a number of times in a life course. The little English perching birds called great tits grow quickly and produce a clutch of offspring annually. Madagascar chameleons hatch in November, mature in January, fight over mates, breed, and die by February—a species consisting of just eggs for nine months a year, with the shortest recorded life span of a tetrapod (Karsten, Andriamandimbiarisoa, et al. 2008). Each of these life cycles is a successful answer to such adaptive questions as how big to be, how much to invest in each offspring, and how to allocate energy between growth and breeding.

Life history theory also relates trade-offs to ecologies. Evolution can only work with the materials at hand, and the phylogenetic legacy imposes constraints on modification (Maynard Smith, Burian, et al. 1985), yet life cycles should adjust—through selection on their controlling genes (Roff 2007; Lande 1982) and also nongenetically—as facultative adaptations. But in practice life history trade-offs have been difficult to demonstrate.

The r versus K Paradigm

Initially it was useful to classify reproductive strategies into two major ecological types (MacArthur and Wilson 1967; Pianka 1970), named *r-selected* and *K-selected* (or *r* and *K*) after terms in the equations describing population growth. The simplest is the rate of change equation for unconstrained growth, $dN/dt = rN$, where N is the population size at any time *t*, and *r* is equal to the birth rate minus the death rate. When a group of N individuals of a species first enters a rich and suitable environment—in the ideal case, an island free of predators—it should be limited only by its intrinsic rate of natural increase, *r*, "the Malthusian parameter." The growth is exponential, as experienced by our species during the past century. When the same population is close to the maximum that the island (or planet) can accommodate (the carrying capacity, *K*), the rate of increase depends on *r* but also on the difference between the current population size and that carrying capacity: $dN/dt = rN(K − N/K)$.

That is, the rate of growth slows as the environment fills up. In theory, populations in severely fluctuating environments crash frequently and then expand again (a sawtooth pattern), spending most of their time in rapid growth (*r*-selected), and they differ from those stabilized near carrying capacity (*K*-selected). The cluster of life history features thought to result from *r*-selection includes: many offspring per litter or clutch; short periods of parental investment, short time to reproductive maturity, and brief overall life span; and a small body size both at birth and at maturity (Pianka 1970, 1988). *K*-selected species should be at the other end of the continuum in all these features. A population need not be on an island to be *K*-selected; limited resources in any environment can serve. This type of selection may have operated in higher-primate, especially hominin, evolution. As for *r*-selection, it can occur when resources are unlimited or, more realistically, when environments fluctuate widely.

Enthusiasm for this model has weakened, partly because it is an oversimplification, but also because *within* species the proposed causal chain (for *r*-selection, fluctuating/unpredictable environments → periodic catastrophic mortality → the cluster of life history variables) has been often disconfirmed, although it holds up better comparing different kinds of organisms. In the 1980s life history theory was largely replaced by demographic models of life history that were relatively agnostic about population density, mortality, and other environmental conditions that might cause a cluster of life history adaptations (Stearns 1992). However, in many specific situations, density dependence, resource availability, and environmental instability could have causal roles reminiscent of *r* versus *K* (Reznick, Bryant, et al. 2002).

In addition, to whatever extent mortality is deemed by life history theorists to be a governing factor in life-cycle evolution (Charnov 1993), one initial intuition of *r* versus *K* theory has persisted: adult mortality may be a sound theoretical starting point for predicting age at sexual maturity and other features of the life cycle.

Adult Mortality as a Shaper of Life Histories

The Charnov model derives the flow of life history consequences from the adult mortality rate:

> *Adult mortality* → *Age at maturity* → *Adult weight* →
> *Fecundity* → *Juvenile mortality*

The model says that high adult mortality causes postponement of maturity, which causes higher adult weight, a defense against mortality. Large body size must in turn keep fecundity low, which means juvenile mortality cannot be high. The causal arrows may be speculative, but the covariation is real, and although the ecological factors that cause this covariation are unclear, the cluster has proved robust in broad phylogenetic analyses. For example, over a range of organisms spanning seven orders of magnitude in size, size correlates highly with generation time (which varies over six orders of magnitude). The intrinsic rate of natural increase (*r*) offers more opportunity for selection (faster reproducers win), and it too tracks body size over this range.

Thus large size does predict long life span, slow development, iteroparity (repeated reproduction), small litter or clutch size, and lengthened parental care. But much of this covariation is allometric; that is, large adult size tends physiologically to require slow development, which in turn requires more parental care, smaller litter size, and so on. These mechanistic relationships raise doubts about the importance of selection per se in the evolution of any one of the dimensions. Consequently, it is useful to hold body size constant before describing covariation among other life history variables (Harvey and Pagel 1991; Stearns 1992; Sibly and Brown 2007). Many relationships do persist with body weight held constant, so pure allometry in the sense of physiological constraint cannot explain all of the covariation.

It is also important, particularly for our purposes, to recognize that adult mortality is not the only possible starting point in determining life histories. Other investigators have found juvenile mortality to be an independent source of variance (Sibly and Brown 2007, 2009) in, for example, either neonatal or adult brain size (Gittleman, Luh, et al. 2000), and "brain size seems to be a pacemaker in mammalian life histories" (Parker 2000:291). In this view, juvenile mortality can act as a selection funnel, so that juvenile survival requires special adaptations that may have consequences for morphology and behavior that are not necessarily optimal for the resulting adult. Adult mortality could be part of a trade-off rather than the start of the causal chain.

The Logic of Trade-offs

How do populations, species, or higher taxa allocate energy among life history strategies? Frequently studied trade-offs include: current reproduction versus current survival; current versus future reproduction; reproduction versus growth; reproduction versus condition; and number versus quality (say, size) of offspring. Some species succeed by "burning the candle at both ends," others by "slow and steady wins the race." One organism figuratively or even literally casts its seed on the wind, while another carefully nurtures a few dependent offspring.

Such trade-offs have been studied (Stearns 1992). Beeches (*Fagus sylvatica*) are among many tree species that alternate years of heavy seed production with years of little seeding, inversely correlated with the width of the growth ring for the given year. In field grasshoppers (*Chorthippus bigattulus*) clutch size is inversely related to egg size, and their covariance is predictable from environmental conditions. Male neotropical frogs of various species must croak to attract mates, but in doing so they measurably risk predation by bats. Older lactating female red deer (*Cervas elephas*) have smaller fat reserves and higher winter mortality than age-mates not nursing young.

In *facultative* trade-offs, an individual adjusts its energy allocation or risk in response to environmental conditions within a genetically specified range or norm of reaction. The recent historical lengthening of the human female reproductive span at both ends in response to improved nutrition and decreased infectious disease is probably a facultative adaptation or physiological trade-off (Worthman 1999; Eaton, Pike, et al. 1994). But some trade-offs are *genetically* controlled (Roff 2007), so that competing individuals differ in their (fixed) relative distribution of energy to growth versus reproduction, or offspring size versus offspring number, or parental versus courtship investment. Such microevolutionary trade-offs can lead to phylogenetic changes such as the lengthening of growth during human evolution. It is possible that facultative postponement of puberty paved the way for genetic change in the same direction, with the facultative option preserved after genetic prolongation. In some cases an ESS for a population may include two or more quite different life history pathways (Charnov 1993). There also may be macroevolutionary trade-offs, such as the phylogenetic commitment of mammals to viviparity and lactation; despite the costs and risks, it would be difficult for a species to diverge from these adaptations, which belong to the foundation-plan of mammalian life history—a constraint on evolution (Maynard Smith, Burian, et al. 1985).

Life history trade-offs imply invariants in some underlying parameter that the life cycle is "trying" to stabilize or is constrained by, and it has proved both elegant and heuristically valuable to attempt to derive or infer these (Charnov 1993). For example, a regression of the average annual number of offspring (clutch or litter size) on age at maturity in 112 species of mammals has a slope of −1.0 and a correlation coefficient of about 0.8, with primates clustered in the upper two-thirds of the distribution in age at maturity and in the lower two-thirds of that in litter size. Also, for mammals (42 species), the average female adult life span (reflecting mortality after maturity, not maximum life span) is proportional to age at maturity, with a slope of 1.42, while the corresponding slope for birds is 2.47. This means that for a given age at maturity, birds have much longer adult life spans (much lower adult mortality) than mammals do; they develop fast without a high level of adult mortality, probably owing to the protection afforded by flight.

For reptiles the slope is 0.8, and for fish 0.5; thus there is a counterintuitive phylogenetic series in these slopes, the order of increase being fish, reptiles, mammals, birds. Birds have the highest postmaturity life span and the lowest adult mortality for a given age at maturity, or the slowest maturation for a given adult mortality, and humans are closer to the bird than the mammal line. Another way to think about it is to say that fish must invest more heavily in the prereproductive growth of offspring for smaller gains in average adult life span, while bird species gain a great deal more in adult survival—even more than mammals do—by lengthening offspring development. So birds and humans live especially long lives for their age at maturity—the former protected by flight, the latter by brains.

If the slope were 1, the ratio of adult survival to age at maturity would be constant and would qualify as a life history *invariant,* or *symmetry* (Charnov 1993). Empirical departure from this invariance in both directions by different classes of vertebrates constitutes *asymmetry* and probably reflects some underlying principle intrinsic to the class. Perhaps postponing maturity is riskier or more costly for birds than for mammals, demanding greater gains in adult survival for a given postponement. Fish may have the lowest slope because their growth continues throughout life. Such relationships invite investigations of the physiological mechanisms underlying systematically different trade-offs in different kinds of animals.

Genetic and Physiological Mechanisms

We can move these trade-offs from the theoretical to the empirical realm by specifying the physiological and genetic bases for switching mechanisms (Ketterson and Nolan 1992; Finch and Rose 1995; Zera and Harshman 2001) within the context of plasticity theory more generally (West-Eberhard 2003). Individual genes, gene families, and gene clusters may control different aspects of metabolism, growth, and reproduction at different stages of life. A gene can transcribe different messenger RNAs if promoted by different transcription factors with different binding locations in the same DNA chain, or in some cases with opposite transcription directions. Variations in RNA splicing, assembly of receptors and binding proteins, receptor up- or down-regulation, and other kinds of plasticity afford many devices by which the same gene, gene cluster, or noncoding RNA could have different phenotypic consequences for different simultaneous or sequential aspects of growth or reproduction.

Such genes allow natural selection to resolve life history trade-offs, switching allocations of energy in response to environmental signals and sometimes altering trade-offs through rapid evolution. For example, variation in the promoter sequence of the genes for brain vasopressin receptors determines whether male voles (*Microtus* spp.) invest more in promiscuous mating or in direct care of their young (Hammock and Young 2004; Lim and Young 2004), and variation in the transporter gene for the neurotransmitter serotonin determines the degree of vulnerability to early life stress in rhesus monkeys (Barr, Newman, et al. 2004) and influences how much adult monkeys invest in social conflict (Suomi 2003). These genes can be viewed as life history trade-off switches, the first resulting in fixed outcomes in different species and the second in a facultative allocation within a species, depending on environmental information.

The genetic mechanisms for life history trade-offs may be a form of pleiotropy (multiple effects of one gene). In higher primates, including humans, testosterone (T) exhibits such multiple-mechanism temporal pleiotropy (Goy 1970, 1996; Goy, Bercovitch, et al. 1988; Collaer and Hines 1995). In early fetal life it causes regression of incipient female reproductive organs. Later it produces male external genitalia, but this action depends on an enzyme, 5-α-reductase, which converts it to dihydrotestosterone (DHT) in certain cells. Also at this time, it influences the differentiation in some hypothalamic cells, and this action depends on two characteristics of T: its ability to cross the blood-brain barrier and its conversion to estradiol by an aromatase enzyme found in some hypothalamic and other brain cells.

At puberty, T causes male secondary sex characteristics by direct action, but only when levels are high enough; this is fortunate, since females also have a large increase in T (from adrenal sources) at puberty; they are protected from masculinization partly by end-organ unresponsiveness but also partly by the threshold, which is between female and male T levels. T stimulates long-bone growth but also closes growth plates, framing the pubertal growth spurt and constituting a switch for the trade-off between age at sexual maturity and ultimate body size. Carol Worthman (1999) applied the Charnov model to cross-cultural variation in the timing of human puberty; the hormonal mechanisms of this life history transition respond to nutritional and activity factors (Frisch and Revelle 1971; Frisch 1978; Frisch, Wyshak, et al. 1980), and complex alterations of pubertal patterning have been seen cross-culturally (Worthman 1986, 1993). Bruce Ellis and his colleagues applied the West-Eberhard model of plasticity and facultative adaptation to age at puberty, using stress hormone activation as an intervening mechanism during development, in the context of *r*- versus *K*-selection theory (Boyce and Ellis 2005; Ellis, Jackson, et al. 2006; Ellis and Boyce 2008).

Both the male and the female sex drive may be enhanced or at least enabled by the *activational* effects of T, and its decline in older men and in menopausal women plays a role in behavioral aspects of reproductive aging. Some other aspects or risks of aging, such as male pattern baldness and prostate diseases, also appear to be T-dependent. No direct role in life history trade-offs is proven, but T could be part of a switching mechanism for the allocation of resources to reproductive activity or to later aging (Finch and Rose 1995). To the extent that the same genes regulate the balance of investment of energy in divergent life-cycle choices, they are said to exhibit antagonistic pleiotropy and are suitable substrates for life history trade-offs (Lessells 1991).

Thus, because of genetically coded differences, one hormone can mediate several life history trade-offs and even be part of the switch determining sex. Other developmental processes are more complex but utilize mechanisms, such as the stress response system, that may serve "fight or flight" responses in adulthood and mediate life history trade-offs during development (Ellis, Essex, et al. 2005; Ellis, Jackson, et al. 2006). This kind of efficient evolutionary conservatism is apparent from insects to primates; Part II of this book is largely concerned with how human development has solved the problems posed by natural selection.

Body Size and Lifestyle—A Two-Dimensional Model

In concluding their review of the history of the *r* versus *K* model, including an attempt to discern its ongoing value in recent work, David Reznick, Michael Bryant, and their colleagues (2002:1518) write of "the disparity between theoretical concepts and empirical realities that continues" despite the existence of many elegant models; they conclude: "The challenge now . . . is to overcome this disparity so that our understanding of life histories continues to progress." Citing Robert Ricklefs's claim of a "muddle in life-history thinking," they make the point that an organizing and motivating simplification can do much good even if it turns out to be wrong.

Such a helpful simplification may now be at hand. Whereas Pianka's *r* versus *K* model and Charnov's adult mortality cascade are both one-dimensional causal pathways, Richard Sibly and James Brown have a two-dimensional model that is conceptually elegant and empirically satisfying (Dobson 2007; Sibly and Brown 2007, 2009). As F. Stephen Dobson comments, we have a classic continuum from the mouse, which breeds fast and dies young, to the elephant, which takes its time with both. Not surprisingly, this fast-slow continuum is correlated with body size; it takes longer to build an elephant than a mouse, and the elephant (in evolution, big is later) has to find its own way. But when we are done explaining all we can with body size, there is still a lot of unexplained variation. What to do?

What Sibly and Brown do is find another slow-fast continuum independent of body size. Slow-versus-fast (also a key dimension of the *r* versus *K* paradigm) remains powerful, even after body size is accounted for. Bats, small and short-lived among mammals, are very long-lived for their size: a mouse-sized bat lives much longer and reproduces more slowly and carefully than a mouse does. These slow-for-their-size animals—we primates are another—cover the whole range of mammal sizes, as do their fast-for-their-size counterparts. What explains these size-corrected differences? Lifestyle.

Other than body size (the first dimension), there are two key ideas here: if you have an easy-to-get, reliable resource base, you are in one situation; if you are liable to be preyed on a lot, you are in another. Reproduction in this model is represented not by *number* of offspring but by total neonatal *mass* produced, per year, per unit of adult female body weight. This measure is predicted across species in relation to body weight in one simple equation: $R = R_1 M^{-b}$, where M is body mass, b is an empirically inferred exponent (between ¼ and ⅓ for mammals), and R_1 is a normalization constant. For 637 mammal species, the fitted regression line is $0.98\ M^{-0.28}$, and it accounts for 59 percent of the variation. This equation accounts for the body size dimension in the model and is a straightforward allometric relationship.

However, some orders of mammals deviate upward or downward from the line, and the second, more interesting dimension—lifestyle—

explains this residual variability. Insectivores, taken as the mammalian baseline, are representative of the relationship for mammals generally. But whales and dolphins, seals and sea lions, rabbits, and ungulates are above the line (higher normalization constants, R_1), while bats and primates are below it (lower R_1). Sibly and Brown account for these differences by arguing that grazing and browsing leaf-eaters (folivores) and marine mammals are *more* productive of offspring for their body size because they have rich and reliable resource bases, whereas *lower*-than-average production rates occur where species have found ways to lower predation: bats because they are airborne, primates because they are arboreal, and moles and gophers because they burrow underground.

Comparisons within these groups are also instructive. Extremely large browsers such as elephants and rhinos are less productive on the size dimension but deviate downward even from that expectation because their size also protects them from predation. Within the rodent order, folivores are more productive than others, while sea-dwelling carnivores are much more productive than land carnivores. Individual cases confirm the rule: sea otters and crab-eating foxes are more productive than their land-based counterparts. Other explanations are invoked in some cases: the naked mole rat and the African wild dog rely heavily on cooperative breeding, which may account for their added productivity. We return to this possibility in Chapter 17. Sibly and Brown do not analyze the human case specifically, but we consider it in relation to their model in Chapter 4.

Evolutionary Allometries

The allometric growth concept was developed to describe the relationship between the size and weight of organ systems whose growth rates are not related linearly (*iso*metrically) but are still related predictably (Huxley 1932/1993; Martin, Genoud, et al. 2005). Allometry provides (usually) simple exponential equations of the form $Y = bX^k$, where Y and X are the respective organs or body parts and b and k are constants (Sinclair 1973; McMahon and Bonner 1983), or in logarithmic form, $\log Y = \log b + k \log X$, visualized as a linear plot with slope k and intercept b. Such equations describe a wide variety of the shape changes of organisms with evolution or development, as well as variation within or among species. (The Sibly and Brown life history model is an example, but with neonatal body weight rather than an organ on the Y axis.) At least three types of allometry are definable (Fleagle 1985; Martin 1990:184; Harvey and Pagel 1991):

1. *Ontogenetic or growth allometries:* Regularities in the differential growth of morphological fields or organs during individual development

2. *Intraspecific or static allometries:* Predictable changes in organ or field size as overall body size varies among adults of a species

3. *Phylogenetic or interspecific allometries:* Body shape change regularities across a range of species

The significance of allometries for phylogeny is that major shape changes during evolution may be attributable to overall body size changes and may need no separate adaptive explanation. In horse evolution, the face lengthened faster than the body size increased because face length is determined by the area of tooth surface needed for chewing, which, since it tracks the amount eaten, depends on body mass. Body mass increases roughly as the cube of body length, and tooth surface as the square of face length, so face length must increase faster than body length, a prediction confirmed by the fossil record and interspecies comparisons. Similar reasoning applies to brain, behavioral, and life-cycle evolution. In a sense, life history theory is about the evolution of ontogenetic allometries.

While brain size scales to body size in many groups of higher animals, the scaling is not one-to-one (Martin 1981, 1990; Pagel and Harvey 1989). In vertebrates, brain weight increases more slowly than body size, and in many studies this relationship takes a linear form on a log-log plot, as in the basic allometric equation, with a slope less than 1. We can think of brain evolution at the species level as deviation from allometric lines considered basic to the taxonomic group (Jerison 1973; Allman 1999; Striedter 2005). Unfortunately for simplicity, the slope of the brain weight–body weight relationship is correlated with taxonomic level (Pagel and Harvey 1988, 1989); typically, although not always (Harvey and Pagel 1991), the broader the collection of mammals examined, the higher the slope (Martin and Harvey 1985).

Still, it is commonly agreed that the slopes for broader ranges of taxa are functionally meaningful and worth explaining. Initially it seemed that the slope of log brain weight as a function of log body weight approximated 0.67, suggesting brain size scales on the square of the linear dimension of body size and body weight scales on the cube of the same dimension (Jerison 1973). This would be plausible if brain weight had to increase with body surface area (not volume) in order to provide an adequate neural representation of the body's periphery, and would yield a logarithmic exponent of $2/3$.

However, better methods applied to a broader mammalian database have shown that the 0.67 exponent is too low (Martin 1981, 1983; Martin and Harvey 1985). Most recent estimates center on 0.75, with 95 percent confidence intervals *excluding* 0.67. The slope of the logarithmically drawn line for primates, 0.75, is indistinguishable from that for nonprimates, 0.74, so that the lines are parallel. However, the difference in intercept on the log scale (2.06 versus 1.7) shows primate brains to be 2.3 times larger for body size than nonprimate brains, a *grade shift*. A similar difference in intercept distinguishes primate, porpoise and dolphin, and other mammalian *fetal* brains (Allman 1999:162, after Martin 1990:422). This scaling difference reflects the early primate departure in brain size from the mammal baseline, apparent by 40 million years ago (LeGros Clark 1959; Radinsky 1974, 1975; Martin 1990) and implemented in fetal brain ontogenies. Porpoises and dolphins made an intermediate grade shift.

But the constancy of the slopes reflects some deep underlying constraint, such as would be imposed on brain size evolution by basal metabolic rate (BMR), which scales on the $3/4$ power of body mass (Kleiber's law): $BMR = kW^{0.75}$, with weight measured in kilograms and BMR in kilocalories per day. One persuasive derivation of the 0.75 exponent is based on the branching geometry of the vascular tree needed to nourish the tissue (West, Brown, et al. 1997), but there are other theories. As an empirical relationship, Kleiber's law is well established and does not depend on any particular functional theory.

Max Kleiber's own data on 26 mammal species yielded the empirical relationship $BMR = 67.6W^{0.756}$, with the correlation coefficient $r = 0.998$. Robert Martin's (1990) extension and analysis of a larger sample of species ($N = 149$) confirms the fit of the distribution and explains some of the residual variation as a distinction between marsupial and placental mammals. For 23 marsupial species, $BMR = 47.4 \ W^{0.75}$ kcal/day, while for 126 placental mammals, $BMR = 75.7 \ W^{0.76}$ kcal/day. The similarity of the slopes shows the validity of Kleiber's law, while the marked difference in intercept, illustrating the concept of allometric grade, shows that marsupials use about 35 percent less energy at rest. Thus the grade shift of BMR from marsupial to placental mammals, like that from reptiles to mammals, supported more active behavior and larger brains. The relationship for primates does not differ from the overall relationship for placental mammals, but nocturnal primates tend to be smaller and relatively more energetic than diurnal ones simply because of their position on the regression line (Martin 1990). Recent analyses comparing many orders of mammals, but using statistical refinements that take phylogenetic history into account, suggest a greater variety of slopes, mainly clustering between 0.67 and 0.75 (White, Blackburn, et al. 2009). This suggests that the truth for BMR and body weight could be somewhere between a geometric estimate and the purely metabolic prediction.

Still, the evidence that *brain* size in mammals scales on the ¾ power of body mass suggests a metabolic constraint, consistent with the finding, across a range of mammals, that adding basal metabolic rate to body size in the equation predicting brain size accounts for most of the residual variance unexplained by body size alone (Armstrong 1983, 1985). This constraint may be ultimately developmental, derived from the allometry of life-span components, especially gestation and infancy; in the "maternal energy" hypothesis, maternal metabolism most determines brain size in a species (Martin 1981, 1983, 2007), consistent with the fact that gestation length across mammal species is much better predicted by brain weight than body weight (Sacher and Staffeldt 1974). Such allometries yield a satisfying convergence between life history theory and the empirically observed phylogenetic scaling patterns considered here (Charnov and Berrigan 1991; Harvey, Pagel, et al. 1991; Berrigan, Charnov, et al. 1993).

Although body size and life span are strongly related, especially across a broad range of organisms, brain size is more highly correlated with life span than is body size in mammals, suggesting that the life cycle and its components evolved primarily in mutual constraint with brain size (Allman, McLaughlin, et al. 1993). For 49 species of haplorhine primates (tarsiers, monkeys, apes, and humans), life-span residuals (departures from the regression line of life span on body weight) are correlated with brain weight residuals (departures from the regression of brain weight on body weight), at $r = 0.657$, $P < 0.001$, slope $= 1.39$. Plotting only the points for gorillas, orangutans, chimpanzees, and humans, the four points lie virtually on this regression line ($r = 0.989$, $P = 0.017$, slope = 0.61, although much of this effect is due to humans).

The relationship suggests that the genetic mechanisms controlling life span are tied to those controlling brain growth (Allman, McLaughlin, et al. 1993). Either senescence is set later in large-brained animals because they more effectively avert premature adult mortality (conferring a selective advantage on long life), or adult brain size depends on the length of time for brain growth either during gestation (Sacher and Staffeldt 1974) or postnatal development. In normal human fetuses resulting from elective abortions, brain weight is the best predictor of gestational age, and little is gained by adding other variables (Burn, Birkbeck, et al. 1975). In mammals, the length of time available for hyperplastic brain growth (increased cell number) is the main predictor of brain size (Finlay and Darlington 1995). If growth time is key, life span may be an epiphenomenon of the time required during development to generate a brain of a given size.

It has also been shown experimentally that brain size–body size scaling can be driven by artificial selection for body size (Riska and Atchley 1985). In either rats or mice, such selection yields an allometric increase in brain size with a log-log slope between 0.2 and 0.4, the same range found in allometries among closely related species or within species. This unexplained but much lower slope suggests a different mechanism for these taxonomically "local" (microevolutionary) allometries as compared to macroevolutionary ones. Ontogeny may explain the differences: since both brain and body growth are hyperplastic—dependent on cell division and proliferation—in embryonic or early postnatal life, genes pleiotropically affecting both brain and body hyperplasia may mediate steeper slopes for macroevolutionary allometry, with metabolic rate–to–body size scaling setting an upper limit on the slopes. The central phylogenetic role of developmental genes is considered later, but we first review heterochrony, which has been prominent in thinking about macroevolution and the evolution of childhood.

Heterochrony in the Phylogeny of Development

Heterochrony ("different time") refers to changes in developmental plans that figure in evolutionary changes; it is strongly grounded in anatomy, offering descriptive generalizations about fossil evidence and inferences about process from the comparative ontogeny of living forms (Gould 1977; Shea 1988; Raff 1996; McNamara and McKinney 2005). Increasingly, the direct microscopic and X-ray study of fossil embryos (Donoghue, Bengston, et al. 2006) and juveniles from dinosaurs to hominins (Carpenter, Hirsch, et al. 1994; Alemseged, Spoor, et al. 2006) make direct hypothesis-testing possible.

For example, study of the limbs of juvenile *Ichthyostega*, an amphibian of 365 million years ago (mya) among the first to walk on land, shows how developmental changes made this possible (Callier, Clack, et al. 2009). It remained aquatic in its early development, but ontogenetic evolution allowed terrestrial activity. Going 20 million years further back, juveniles of the lobe-finned fish *Eusthenopteron* are found to have the most tetrapod-like developmental patterns in their forelimbs, consistent with the era of the first footprints on land in Poland and Australia; their form suggests that the heterochronic process of peramorphosis (progressing beyond the ancestral adult form before development stops) was involved in the transition to walking (Long 1990).

Also, studies show both the experimental power of selection to produce microevolutionary heterochronies (Riska and Atchley 1985) and the analytical power of life history theory to help explain macroevolutionary ones (McKinney and McNamara 1991; McNamara and McKinney 2005). These advances have eroded the false distinction between external or extrinsic factors in evolution, especially Darwinian natural selection, and internal or intrinsic factors associated with older theories of heterochrony. There are no intrinsic *driving* factors in evolution, but there are intrinsic constraints and canalized paths along which either evolution or development may more easily proceed (Maynard Smith, Burian, et al. 1985; Gould 2002). Both random and competitive factors within the genome (Burt and Trivers 2006), as well as its internal coherence, influence developmental processes and constrain the impact of natural selection on life cycles. Finally, as heterochronies are increasingly understood through a molecular taxonomy of homeotic and other developmental genes, they have become the targets of new and potentially generative genetic theory (Kenyon 1994; Raff 1996; Slack and Ruvkun 1997; Amores, Force, et al. 1998; Kim, Kerr, et al. 2000).

Stephen Jay Gould (2002) proposed the concept of *positive* constraints, suggesting an internal tendency of an organism to evolve in a particular direction based on deep homologies, but this seems semantic, since he explicitly denied that they are intrinsic driving factors. We might distinguish *stabilization constraints,* in which the components of the organism, the life cycle, and/or the genome cohere, resisting selection for particular adaptations, from *canalization constraints,* which make evolutionary change under selection relatively easier along some pathways than others. This is suggested by new evidence of deep genetic homologies.

But the foundations of the approach are morphological. After long neglect, Gould (1977) restored the subject of heterochrony to scientific standing, synthesizing previous ideas in a new model and testing it against modern data in ontogeny and phylogeny, especially of marine and other invertebrates. Unfortunately, one legacy of the past—a heavy burden of unwieldy terminology—has continued to inhibit progress. Many coinages have provoked both substantive and semantic arguments (McKinney and McNamara 1991; Raff 1996; Ridley 1997), not least of all with regard to human evolution (Shea 1992; Godfrey and Sutherland 1996; Minugh-Purvis and McNamara 2002). Leaders in this field regularly decry its ambiguous use of terms and semantic arguments (Alberch and Blanco 1996; Gould 2000).

But one general principle is unambiguous: *dissociability.* The speed of growth in specific organ systems or parts of organs, the age at termination of growth in those systems, the age at sexual maturity, the length of gestation, and the length of life all may evolve independently of one another. The principle of dissociability, shown in part by exceptions to phylogenetic allometry, is central to the role of ontogeny in phylogeny, and so to evolution itself. As we have seen, developmental trajectories, not organisms at a moment in time, are the units of evolution; traits do not evolve, development evolves. New traits may appear at one phase of development, then shift to another or extend over more than one; old traits may fade in similar phase-specific fashion. Sexual maturation or senescence may be advanced or delayed against the background of body growth or changes in body shape. Individual organs may evolve growth rate changes or may advance or delay their onset or offset of growth, also dramatically affecting adult morphology.

One heterochronic concept, *recapitulation,* and its leading proponent, Ernst Haeckel, dominated evolutionary thought during the late nineteenth and early twentieth centuries (Haeckel 1905; Mayr 1982). In this model the evolution of ontogeny was simply cumulative, with new structures tacked on to the end of the ontogeny of an earlier form, an iterative routine called *terminal addition.* This process of accretion, often followed by acceleration and compression, became known as Haeckel's biogenetic law (Breidbach and Ghiselin 2007). The ontogeny of later-evolving forms was viewed as a shorthand record of evolutionary changes, with the overarching conclusion that ontogeny recapitulates phylogeny. Such observations as the transient presence of gill arches and tails in human embryos seemed to support this principle, and it is in some ways valid. But the literalness with which it was taken by Sigmund Freud, Heinz Werner, and others had a distorting influence on ideas about psychological development. It has been more intelligently applied recently in an attempt to use Piagetian developmental stages to explain the cognitive status of human ancestors (Donald 1991), but this effort remains highly speculative.

A more conservative formulation, articulated by Karl von Baer in the early nineteenth century, has survived the rise and fall of Haeckel's law (Breidbach and Ghiselin 2007). Von Baer's law holds that organisms begin from a similar starting point in development and diverge from there, adding specific developmental transformations (not just accretions) as they evolve. This *phylotypic stage,* the first embryonic phase that looks like an animal, is easy to find in major categories of organisms (vertebrates, for example), but it is not the earliest stage; processes of fertilization, implantation, and the first few cell divisions vary more than does the phylotypic stage, which is indeed remarkably constant.

Strangely, pervasive recapitulation theory persisted alongside evidence for *neoteny,* which means holding on to newness or youth and which some recent theorists believe may have operated in human evolution (Gould 1977, 1988; Godfrey and Sutherland 1996). In this process the rate of general body growth becomes slowed relative to the timing of sexual maturation, thought of (not quite accurately) as the termination of growth, so the adult of the descendant resembles the juvenile of the ancestor. In fact, human adults resemble ape infants and juveniles in various physical characteristics more than we resemble adult apes; a leading proponent of human neoteny, Louis Bolk, called our species "a sexually mature fetal ape" (Gould 1977:361).

Before we can evaluate this view, we need to distinguish the evolutionary *result* from the three different processes that could have generated it. The result, called *paedomorphosis,* is the anatomical (and possibly also physiological and behavioral) resemblance of the adult to the young form of the ancestor. This condition can be attained by: *progenesis,* or early termination of body growth, which can be achieved by accelerating gonadal maturation; *neoteny,* or slowing of body growth relative to sexual maturation; and *postdisplacement,* starting the relevant growth process later. The opposite condition, *peramorphosis,* a sort of hyper-adult morphology, can likewise be achieved in three ways: *hypermorphosis,* or later growth offset; *acceleration,* or an increased body development rate relative to sexual maturation; and *predisplacement,* or starting sooner.

Vigorous argument persists regarding the degree of relevance of these processes to human evolution, or to evolution generally, even in an era when developmental genes will soon allow direct observation of developmental processes and the construction of molecular phylogenies of these processes. These debates will not be easily resolved. Meanwhile, we might simplify the terminology. That experienced investigators in a specialty disagree is no surprise, but the lack of a common language suggests a need for clearer terms. Accordingly, a newer, English-language terminology is proposed here, juxtaposed to the corresponding terms in a chart drawn by Kenneth McNamara (1997:41). With apologies to the nineteenth-century biologists who spent their youth on Greek texts, we need a less forbidding vocabulary. The basic phenomena (with the corresponding older terms) are as follows:

1. *Juvenility (paedomorphosis).* This is not a process, but a state ("juvenilized") in which the adult of a descendant species resembles the young of an ancestor. It is generally agreed to characterize some species, such as the axolotl or Mexican salamander, and may be partly applicable to us. It may result from one of at least three different, but not mutually exclusive, processes:

   a. *Truncation (progenesis).* Size growth and shape development progress on the same timetable as in the ancestor but are cut short because of earlier offset or cessation of growth and development, often due to earlier sexual maturity.

   b. *Slowing (neoteny).* Shape development slows and size growth either slows or remains unchanged, while sexual maturity is attained on the old chronological timetable. Cessation, chronologically unchanged, occurs before slowed shape development has attained the old adult form.

c. *Postponement (postdisplacement).* The time of onset of development of a particular organ or system is delayed compared to the ancestor. (This would seem to require a deceleration preceding it.)

2. *Overdevelopment (peramorphosis; recapitulation).* Overdevelopment is also a state, not a process, and was once thought to be the most pervasive evolutionary phenomenon, affecting all complex organisms. It may result from one of at least three different processes:

a. *Delayed offset (hypermorphosis).* Conceptualized as allowing size growth and/or shape change to go on longer by delaying the end of growth, resulting in the terminal addition classically associated with recapitulation.

b. *Acceleration.* The same result may be achieved by an increasing developmental rate without a change in the time of offset of development.

c. *Early onset (predisplacement).* Growth rate and timing of growth offset are unchanged, but the growth starts sooner. (As in postponement, predisplacement seems to require a rate change—in this case, acceleration—preceding the growth phase in question.)

In addition to its relevance for human ontogeny, heterochrony may affect our models of evolutionary process. If a minor genetic change can alter a given growth rate, and if differential growth rates in turn account for most of the morphological differences between humans and apes, then heterochrony fits the model of phylogeny called *punctuated equilibrium* (Eldredge and Gould 1972). In his initial interpretation of the model, Gould attempted to revive the notion of *saltation*—sudden, major morphological change ("the hopeful monster" of Richard Goldschmidt) instead of gradualism—to account for the emergence of new species. Such a drastic change is plausible with the asexually reproducing marine invertebrates Gould studied; they could mutate and go about their solitary reproductive business. But for sexually reproducing species, the microevolutionary scenario is as puzzling as it was when Goldschmidt suggested it: how would the hopeful monster find a mate?

Fortunately, heterochrony, including its hominid versions, is compatible with polygenic gradualism. The fossil record for many organisms includes long periods of stasis and much shorter periods of change, and it was this observation that suggested punctuated equilibrium (Eldredge and Gould 1972). But a "punctuation" may include thousands of generations—a period easily long enough for conventional, cumulative, Darwinian microevolutionary change (Jones 1981). Large variations in evolutionary rates, including prolonged stasis due to stabilizing selection, were part of the "modern synthesis" in evolutionary theory (Simpson 1944); gradual evolution over very long time scales may be more difficult to detect (Breton 1996; Sheldon 1997). Both stasis and rapid change require explanation, but that explanation is unlikely to be outside the bounds of the modern synthesis (Lande 1980; Mayr 1982, 2001; Gingerich 1983).

In his last book, Gould (2002:778–79) wrote, "The theory of punctuated equilibrium . . . asserts no novel claim about modes or mechanisms of speciation; punctuated equilibrium merely takes a standard microevolutionary model and elucidates its expected expression when properly scaled into geological time," and "conventional mechanisms of speciation scale into geological time as the observed punctuations and stasis of most species." So in macroevolutionary theory, Eldredge and Gould's contribution was to increase the emphasis on *relatively* short periods of Darwinian evolution between much longer periods of stasis.

Also, they appropriately emphasized the central role of speciation—the splitting of one species into two—in evolutionary change (Mayr 1963; Venditti, Meade, et al. 2010). Probably the divergence of the ancestor of the great apes from that of the hominins 6 or 7 million years ago was such an event, coinciding with the evolution of erect posture in the early hominins. The subsequent divergence of robust australopithecines from the hominin line was probably similar. But the word *event* suggests suddenness, whereas if the split took 100,000 years—shorter than the level of resolution of the early hominin fossil record—some 4,000 hominin generations would have passed, easily allowing change to be Darwinian and gradual.

Speciation is especially clear in the fossil record after mass extinction episodes, when many adaptive niches are quickly vacated and the opportunity for evolution in the remaining species greatly increases. The resulting adaptive radiations are among the most creative episodes of evolution by natural selection. The early radiation of the primate order, based largely on advances in vision and brain expansion (in part an instance of overdevelopment), exemplifies this process. How heterochrony figures in speciation and adaptive radiation remains to be seen.

The Evolution of Developmental Genes (Evo-Devo)

The intriguing history of heterochrony notwithstanding, controversies over the evolution of ontogeny will ultimately be resolved by direct analysis of the functions of developmental genes in comparative perspective. The twenty-first century has seen a revolution in this field, and the discoveries point to one broad conclusion: there are stunning homologies in the structure and function of developmental genes across a very wide range of organisms, and they make up a genetic tool-kit without which neither development nor evolution can be understood (Shubin, Tabin, et al. 2009).

Homeobox Genes

A new, molecular embryology has grown up around the discovery that some genes that control early development have wide and deep family resemblances across the animal kingdom. These homeobox (*Hox*) genes share a conserved sequence of DNA, the homeodomain, or homeobox, first described in the fruit fly *Drosophila melanogaster*. DNA sequences *of* and *within* these genes are conserved from the half-millimeter-long translucent roundworm *Caenorhabditis elegans* through the fruit fly to the mouse, and they regulate the development on the longitudinal axis of these three very different organisms, although in somewhat different ways (Kmita and Duboule 2003).

This is a deep homology, a common genetic mechanism for building overall animal body plans and then, often in duplicated form elsewhere in the genome, for building limbs or central nervous systems (Reichert and Simeone 1999; Reichert 2002). The mechanism arose early in the evolution of multicellular animals. The genes are found in a characteristic linear order on the chromosome, corresponding to the anatomical order of their influence in the embryo ("the *Hox* clock") (Ruddle, Bartels, et al. 1994; Kmita and Duboule 2003). It is clear that these genes help determine the regional organization of the central nervous system, with additional regulatory genes, common to fruit flies and humans, encoding the forebrain (Lichtneckert and Reichert 2005; Reichert 2009).

The *Hox* genes specify proteins that act as transcription factors (promoters and silencers) for genes in embryonic cells, causing differential growth along the longitudinal axis. They are only one group of players in the vastly complex interaction of genes and their transcription factors, including noncoding RNAs; they themselves must be turned on and off, and some of the genes they influence promote or silence *other* genes, and so on in a cascade of molecular events that generates and transforms an embryo. Mathematical modeling of these intricate cascades of interacting switches will play an important role in decoding development, a problem orders of magnitude more difficult than decoding the genome.

Several themes figure in current thinking about the evolution of *Hox* genes and their influence on ontogeny. First, unexpectedly, the longest strings of *Hox* genes were not the last to evolve. Specialization in various lines of organisms was achieved by *deletion* of some of them, restricting established gradients and in some cases apparently derepressing the development of structures. Second, the evolution of the vertebrates entailed two *duplications* of the original array of homeobox genes, preserving the overall sequence in all four resulting sets. The new sets are applied to laying down the linear anatomy of limbs. However, following these two duplications much of the rest of the evolution of vertebrate body plans can be accounted for by deletions within the four arrays, just as the evolution of invertebrate body plans can largely be accounted for by deletions within the single original array.

Among the most interesting discoveries in this vein relates to control of central nervous system (CNS) development. At first it was thought that the homologues of the *Hox* genes for the *Drosophila* CNS specified only the postcranial CNS (the brain stem and spinal cord) of vertebrates. But even the genes that determine initial events in mammalian *brain* development, different from the *Hox* genes, show deep homologies. These "cephalic gap" genes specify the forebrain of mammals and the anterior end of the nervous system of fruit flies (Reichert and Simeone 1999). In general, "comparative studies of brain development in vertebrate and invertebrate model systems demonstrate remarkable similarities in expression and action of developmental control genes during embryonic patterning, neural proliferation and circuit formation in the brain" (Reichert 2009:112). The implications for the relevance to ourselves of the behavior and development of organisms only distantly related to us are substantial, since it means that the basic plan of the axis of the CNS (including the brain) in complex animals goes back some 600 million years.

Hox genes do their specifying, in part, through differential sensitivity to retinoic acid, the active form of vitamin A, which is present throughout the embryonic neural axis. The greater sensitivity of anterior *Hox* genes over posterior ones sets up a head-to-tail timing pattern with profound anatomical consequences. Interestingly, as we will see, the head-to-tail sequence is repeated postnatally in myelination and motor control. In addition to the cephalic gap genes, other genes known as *BF-1 (FOXP1)* and *BF-2* are needed to specify the forebrain structures of mammals, but these are part of a family of very ancient "terminal" genes involved in shaping the ends of the embryo, including both the nervous system and the gut. We return to this key gene family in later chapters.

Genetic Control of Eye Development

A further deep homology involves the gene *PAX6*, a member of the *Pax* gene family, which has evolved by duplication to serve various functions in neural development. *PAX6* signals the embryo to begin eye formation in a very wide range of animals, including fruit flies, lampreys, frogs, and mammals (Pichaud and Desplan 2002; Gehring 2005). Upstream regulation of *PAX6* by the gene signal Notch is also widely shared, and the signals are experimentally interchangeable between flies and frogs (Onuma, Takahashi, et al. 2002). In humans and some other species, *PAX6* also functions in the olfactory system and some brain regions (van Heyningen and Williamson 2002). The binding properties of the transcription factor do distinguish lineages such as mice and primates (Zhou, Zheng, et al. 2002), suggesting that modification of binding sites may explain the evolution of different functions, as the same developmental genes (or duplications of them) are recruited for somewhat different structural roles.

These findings have surprised observers who believed that such widely separated lineages would have to have evolved eyes separately and independently, especially since the respective mature eyes have very different structures and operate on substantially different principles. It has also led to a rethinking of the definitions of parallel and convergent evolution (West-Eberhard 2003). However, it would be unlikely for the common ancestor of mammals and flies to have been blind, since it stood (or rather, swam) at the end of 3 billion years of evolution in ambient light. If it had an organ responsive to light, however minimal, we could expect, based on conservative evolution and von Baer's law, that its embryonic signal system for forming that organ would be part of its legacy to both insects and mammals. The subsequent evolution of far more complex and very different organs serving similar functions would still be an instance of impressive parallel or convergent evolution under natural selection.

We know too that the fact that two similar phenotypic adaptations are arrived at through similar genetic and metabolic pathways does not in itself rule out convergent evolution. African and Asian leaf-eating monkeys independently evolved the same amino acid substitutions in the genes for an enzyme that allows them to digest leaves (Zhang 2006). Similar genetic-metabolic convergences have occurred in the cases of color vision in primates, ultraviolet vision in different bird lineages, and other adaptations (Prud'homme, Gompel, et al. 2007). Natural selection takes the path of least resistance, and when it leads through the same biochemistry, that is where it goes.

Other lines of investigation suggest that the wiring patterns of the nervous system are also influenced by homeotic genes. For example, it has been suggested that the interaction of *Hox* and *Pax* proteins in the embryonic neural axis leads to the differential expression of certain cell adhesion molecule (CAM) genes. These genes produce protein products that appear on the surfaces of migrating neurons and the growing tips of axons, allowing them to recognize cells and extracellular surfaces and navigate among them, ultimately determining where they go and what connections they make. Experiments with transgenic mice suggest that *Hox* and *Pax* genes can regulate CAM genes (Edelman and Jones 1998).

Sequencing genomes is simple compared to decoding development. Gene products interact in a vast web of mutual influence and hierarchical cascades of specification. Beyond them, the environment has similarly large powers and enters into the causal chains both early and late in development, not least by turning genes on and off, as regulatory genes do. For the first time we are starting to discern the deep genetic transformations in evolution that underlie developmental change. For example, Celine Gomez, Ertuğrul Özbudak, and their colleagues (2008) have shown that the difference between a snake, with far more vertebrae and associated nerves, and a mouse or other shorter-bodied vertebrate is determined by a heterochronic shift in the clock governing segment formation in the embryo, probably under the control of *Hox* genes. Research on the limbs of the first land animals has combined fossil evidence with comparative developmental genetics among living forms to show that changes in the timing of expression of just one of the *Hox* genes might suffice to change one type of fin into a walking limb (Shubin, Tabin, et al. 2009). A similar synthesis of fossil and genetic evidence suggests that heterochrony was involved in the evolution of the definitive mammalian middle ear from the reptilian jaw (Ji, Luo, et al. 2009). Studies like these test heterochronic hypotheses and give substance to the truism that to change organisms, evolution must change development.

Phyletic Reorganization in Brain Evolution

Evolution being conservative, traces of earlier structures are often found in later ones, and it is not meaningless to rank organisms by a well-defined criterion—body size, limb length ratios, relative brain size, metabolic rate, number of neuronal types, or behavioral repertoire. Although sometimes marred by a naive progressivism, twentieth-century efforts to study comparative neuroanatomy with models involving phylogenetic accrual were not wasted (Jerison 1973, 2001; MacLean 1985, 1990).

However, traditional models were based on descriptive accounts combined with attempts to differentiate brain regions by cellular architecture. In the 1960s histochemistry further differentiated brain circuits by their neurotransmitters. In the 1970s fiber-tracing methods made it possible to redefine the anatomical areas of the nervous system by their connections—a key advance—and in the 1980s it became possible to demonstrate the distributions of the messenger RNAs associated with particular circuitry or neuronal types. In the 1990s the increasing resolution of magnetic resonance imaging (MRI) and positron-emission tomography (PET) and the introduction of functional MRI and diffusion tensor imaging (DTI) led not only to the imaging of living brains but also to the visualization of their ongoing functions, in comparative context (Rilling and Insel 1999; Semendeferi and Damasio 2000). And in the twenty-first century the measurement of gene expression and noncoding RNAs in the brain in humans and other animals is revolutionizing our knowledge of brain function, development, and evolution (Gray, Fu, et al. 2004; Preuss and Caceres 2004; Sun, Patoine, et al. 2005; Pollard, Salama, et al. 2006).

At the same time the use of tracing methods involving the labeled products of immediate early genes such as *fos* and *jun* have allowed the delineation of the neural pathways activated by particular functions (Keller, Meurisse, et al. 2004; Levine, Youngs, et al. 2007). The refinement of methods for measuring messenger RNA has facilitated the comparative study of gene expression in the brain, and temporary suppression of gene expression (knock*down*) using short interfering RNAs (siRNA) has dramatically expanded the possibilities of gene knock*out* by making it reversible (Salahpour, Medvedev, et al. 2007). Finer resolution in brain imaging is shrinking the gap between imaging and histology, and functional imaging is driving a revolution in all the behavioral sciences.

In Chapters 3 through 6 we consider in greater detail the processes of brain evolution and how they reorganized the vertebrate and ultimately the primate and human brains.

Developmental Ethology

Darwin's great work *The Expression of the Emotions in Man and Animals* (1872) laid the foundations for the evolutionary study of behavior. His procedure, still followed in efforts to build such a field (Ekman and Rosenberg 1997), was the same as that used in generating hypotheses about the morphological evolution of animal groups for which fossil evidence is scanty. We compare characteristics of living organisms to estimate their degree of relatedness, we consider fossil evidence if available, and we try to see evolutionary history in the light of what we know about how evolution works.

There were three main components to the ethological enterprise that Darwin founded (Tinbergen 1963; Lorenz 1981; Eibl-Eibesfeldt 1975). First was the field study of animal behavior under natural conditions, a systematic natural history adding quantitative methods and unobtrusive experiments. Evolution and adaptation were always the intellectual foundations of ethological fieldwork, and a basic presumption was that observing the animal in its environment of evolutionary adaptedness would reveal the functions of behavior. Another goal was to establish the behavioral repertoire of a species, the ethogram, including what are called in English fixed action patterns (FAPs)—a mistranslation of *Erbkoordination,* German for "legacy coordination"—and innate releasing mechanisms (IRMs). It turned out that many FAPs were flexible and that some IRMs required experience, but both belonged to the ethogram in the normal expected (natural) environment.

Second, grasping the evolution of behavior would lead to generalizations across species that would enhance understanding of both adaptation and phylogeny. Konrad Lorenz published a monograph in 1935, *Der Kumpan in der Umwelt des Vogels* (roughly, *The Companion in the Bird's World*), that introduced the relationships seen among birds—parents, infants, mates, social partners, and siblings (Lorenz 1935/1970). Using examples of each from several species, he inferred the expectations apparently wired into their nervous systems: the partner's appearance, what it will probably do, what to do in return. Such relationships are reliable (up to a point), often lasting, crucial to survival or reproduction, and highly patterned (stereotyped); experience guides their emergence, but major features are independent of learning.

Comparative analysis of behavior also aided the construction of phylogenies based on exact descriptions of the FAPs in ethograms, and in some cases behavioral phylogenies were so compelling as to force rethinking of evolutionary histories (Tinbergen 1964)—for example, Lorenz's systematic comparison of displays in the Anatidae (ducks and geese), which made it possible to reconstruct the evolutionary history of the behaviors just as one would do in comparing bones, brains, or DNA sequences (Lorenz 1941/1971). Later this approach was applied to the facial expressions of higher primates (Preuschoft 1995; van Hooff 1972), the displays of stickleback fishes, and other phylogenetic analyses.

Third, and most relevant for our purposes, ethologists studied behavior in *developing* animals, often through laboratory experiment. Investigation of innate behavior began with Darwin's observational studies of one of his own infants (Darwin 1877/1956), and fieldworkers routinely documented infant, juvenile, and parental behavior as part of the ethogram. Field experiments augmented these observations. For example, Tinbergen (1960) established the innate releasers of pecking on the parents' bills in herring gulls, which in turn triggers parental feeding. Wolfgang Schleidt's (1961) experiments on the fearful crouching of chicks in response to a hawk's shadow passing overhead showed that the response was not simply innate but occurred because the hawk's shadow was rarely seen; chicks also crouched in response to a goose-shaped shadow when it was made rare. The study of filial imprinting in birds (Bateson 2003) became a central theme (discussed in Chapter 9). The deprivation experiment, removing experiences that might shape "innate" behavior during development, showed that some are not innate but others are—or as Lorenz (1965:27) put it, to say that they are innate is about as much of an exaggeration as to say that the Eiffel Tower is made of metal.

The growing field of neuroethology (Pflueger and Menzel 1999; Bullock 1999; Bradshaw and Schore 2007) is incorporating neurobiology and neurogenetics, giving credence to the claim that many behaviors in ethograms are "wired-in"—although they may be "soft-wired" rather than "hardwired," requiring some experience (Marler 1991; Bateson 2003). Finally, more than a century after Darwin's infant diary, ethological approaches were once again applied to infants and children (Blurton Jones 1972b; Konner 1972; McGrew 1972).

Evolutionary Developmental Psychology

It was apparent to Darwin that his own children came into the world with behavioral adaptations that were products of natural selection and phylogenetic legacies, as shown in his infant diary and other writings (Darwin 1872, 1877/1956). A century later, a few students of child development began to extend this idea, addressing the adaptive value of innate infant behavior (Freedman 1974, 1997), child behavior (Blurton Jones 1972b), cognitive development (Bruner 1972), children's learning (Fishbein 1976), and adolescent behavior and development (Weisfeld 1999). As we have seen, parent-offspring conflict theory predicts a dissymmetry between infant or child and parental goals, as well as sibling and other intrafamily conflicts, within the context of the care, nepotism, and cooperation predicted by inclusive fitness theory.

In addition, a new body of theory has grown up with respect to human childhood, grounded in late-twentieth-century life history models. Starting from life history theory, this approach assumes that different life plans are adaptive in different environments and that patterns of early attachment are *designed* to produce different kinds of adults. There have been significant contributions by Patricia Draper and Henry Harpending (1982, 1987), James Chisholm (1993, 1999b), Jay Belsky and his colleagues (Belsky, Steinberg, et al. 1991; Simpson and Belsky 2008), Sarah Blaffer Hrdy (1999, 2009), and Bruce Ellis (2004), among others. Overall, this approach seems to reassess as normative some variations in child care that have previously seemed pathological. But the clinical and evolutionary models are not in contradiction; in fact, they must be reconciled. We cannot derive "ought" from "is," but we have to know what is.

All species have life histories designed to maximize reproductive success, but many have more than one (Pianka 1970; Stearns 1992). They have flexible life histories that change direction depending on environmental signals about consistency and quality. They may reproduce earlier or later, try for a few offspring of high quality or many more dispensable young, even change sex with temperature, food availability, or the stress of dominance challenges. Darwinian theory predicts that life histories will adjust to maximize reproduction in different environments. The result may be within-species trade-offs between growth and sexual maturity, between number and quality of offspring, between sex and parenting, between a short but intense life and a long but stable one (Sibly and Brown 2007). A species may evolve a range or norm of reaction, a predictable relationship between environment and the life course. Facultative adaptation does not mean that "anything goes," but only that *within a given species* different life cycles are adaptive in and activated by specific environments.

Through the lens of this theory, different patterns of child care and attachment are not necessarily better or worse for infants; they are strategies for maximizing reproductive success in different environments. If a poor or unstable environment makes for a relatively rejecting mother and the developmental result is a sexually promiscuous or delinquent adolescent, that may be a way of maximizing reproductive success in that environment (Belsky, Steinberg, et al. 1991; Chisholm 1993; Ellis and Essex 2007). Abuse, neglect, favoring males over females, and such milder variants as multiple caretaking and wet nursing can be seen as adaptations that parents make to demanding environments. This paradigm highlights the needs of parents that are traditionally ignored, especially those of mothers, in favor of children's needs (Hrdy 1999, 2009). But in addition, the infant's brain may be designed to detect environmental conditions that demand different reproductive strategies. The amygdala could help serve this function (Chisholm 1999b), a hypothesis consistent with some findings on the lasting biological effects of early stress. One of testosterone's functions as a life history regulator may depend on its ability to decouple the amygdala from the orbitofrontal cortex, as demonstrated experimentally in healthy women (van Wingen, Mattern, et al. 2010). This probably reduces regulation of the amygdala by the cortex, as is also seen in normal men. Testosterone levels could be involved in signaling environmental quality as detected early in life.

Interlude 1

Thinking about Birdsong

Songbirds, we now know, are not rhapsodizing about the beauty of the spring, and if they are singing about love, it is a very insistent, demanding, boastful, possessive sort of love, mixed with no small dose of lust. It is the gift of natural selection to (mostly) the males of many species, and what they are singing—or more exactly, crowing—about is something like "Mate with me, babe, I've got the best genes and the sweetest swatch of greenery this side of the Rockies," and also, to any other males in earshot, "Don't look for trouble, pal, take your lame erotic dreams somewhere else." The crow's caw, the rooster's cock-a-doodle-doo, and the finch's melodious trill all amount to pretty much the same self-publicizing story (Marler and Slabbe-koorn 2004).

Males do the crooning in most species, and in seasonal climates they do it mainly in the spring. The lengthening sunlit days signal the brain to signal the testes to produce and release testosterone, and in some species the brain's song center waxes and wanes with millions of new neurons made to order for the yearly thaw. It sounds pretty mechanical, and in some species the call or song is basically wired-in. But in others, including some with the most elaborate and musical riffs, the path from genes to songs is quite circuitous.

In the white-crowned sparrow, songs are almost cultural (Marler 1991; Marler and Slabbekoorn 2004). Experiments show that males must, during a critical period in infancy, listen to the song sung by their fathers in order to produce the species-specific sequence and melody. In the absence of this in-put, they will sing only a grotesque and feeble shadow of the proper song. This is the template, and in some species it is very close to the adult song or call and can be thought of as instinctual; in the white-crowned sparrow, how-ever, only a sketch of the song is innate.

More remarkably, the young male must practice in adolescence, when he begins to sing imperfectly under the impulse of his first testosterone surges.

And he must hear himself sing; this allows him to progress so that his own song gets closer and closer to the one he heard his father sing when he was a hatchling—yet without hearing it again in adolescence. He has somehow stored his father's song, and as he practices he moves his own from the crude innate sketch toward the shape of the family heirloom.

Is this a tradition? Well, there are dialects within the species. If you travel up the California coast near San Francisco Bay, you find the song changing significantly—those all-important receptive females can tell the difference—depending not on genes but on the behavioral variations passed from father to son over many generations. Is this culture? I would say not. The white-crown is a one-trick bird in the cultural sphere, and there is no evidence that these traditions are symbolic, creative, or cumulative.

Nevertheless, it is quite a trick, and it gives us great insight into how natural selection strikes a balance between the innate and the flexible in behavior. Even in a behavior that must be stable over generations, even in a small perching bird, plasticity is allowable as long as there is a normal expectable environment. Genes, in the context of such an environment, can do the job they need to do for adaptation with much less than full specification in the brain. If this is true for the white-crowned sparrow, how much more so for us?

We have learned much that is new. Birdcalls, as opposed to birdsong, were thought to be not just simpler but simply innate; now we know that learning is involved in those too (Marler and Slabbekoorn 2004). In fact, although vocal learning is unusual in mammals, around half of all bird species have it. Aside from the fact that birdsong can involve massive genesis of new neurons in adulthood, we know quite a lot now about the brain circuitry involved. The high vocal center (HVC) is a critical part of the circuitry, and the ratio of female to male HVC volumes predicts the extent to which females also sing (Ball and MacDougall-Shackleton 2001). But a different, more specialized brain area is required for juvenile babbling, which persists even when HVC is inactivated (Aronov, Andalman, et al. 2008). So the brain separates practice from performance.

Females of course must move to the song, and female choice seems to drive song evolution, as it does with so many other things puffed-up males take pride in. Males with more variety and faster repetition of their songs are more likely to be chosen in several species (Podos, Huber, et al. 2004). Male swamp sparrows use some of the same HVC neurons to *listen* to other males sing the species-typical song as they do to sing it themselves, suggesting that those neurons function as do mirror neurons in primates (Prather, Peters, et al. 2008), an important subject to which we will return.

Central Park, my old stomping ground, is dense with obsessive bird-watchers, and not far from it a young scientist named Olga Feher, at my alma mater, the City University of New York, raised some zebra finch males with no chance to hear their fathers sing (Feher, Wang, et al. 2009). Predictably, some very strange noises ring out when those boy-chicks in the 'hood grow up and blow their pipes like kings of the hill. But here's the news: take one of those males, get a female to hold her ears and do the deed with him, and his sons will sing more normally than he does. His grandsons sing better still, and his great-grandsons approach the normal wild song—which none of these boys has ever heard. Some collective cultural evolution puts the song right again. It's hard to avoid a comparison with the transition from pidgins to creoles: a first generation of conquered people speak a crude, broken version of the colonizers' language, but their children create a unique, more fully developed version (Bickerton 1990, 1998). It's a strange kind of evolution, recruiting genes for cultural change.

But my favorite birdsong story is about the dull English robin, set down in detail by one of the twentieth century's greatest naturalists. Studying robins in an English wood, David Lack (1943) mapped out the territories of five pairs in a curved row, the males in each pair singing away. One ill-fated fellow was taken by a cat, yet after a while the silence on his territory was broken. It was not another male, however, but his widow who took up his serenade, and she went on singing, staking claim to the patch of wood, until another male came along and chose *her*.

It's one of those delicious observations that wake us up to the rude fact that nothing in biology is 100 percent. If it's a biological law, it's fuzzy around the edges; all of the rules are probabilities. And it's why some of us got bored with math and physics and turned to scientific subjects that are marvelously messier. That would be, well, life.

3

Brains Evolving

Although ontogeny does not really recapitulate phylogeny, the ontogeny of related organisms passes through a developmental bottleneck (the phylotypic stage) of great similarity, so we need to understand developmental homologies. We could start at the origin of life 3.5 billion years ago, or at 620 million years ago, when the common ancestor of vertebrates and insects had already established the profound homologies of the homeobox gene sequence, but we are interested in the plan of human development, so we will start with a small swimming creature that, at 530 million years ago (mya), is close to the root of all vertebrates. But first, some general principles of brain evolution.

Expansion and Organization in Brain Evolution

As we saw in Chapter 2, allometric analyses produce regression equations that predict brain size from body size, basal metabolism, gestation length, and other measures. Prolonged human brain growth yields a product roughly triple the size of those of the apes or of our australopithecine ancestors, and (with apologies to dolphins) ape brains would be the pinnacle of nervous system evolution if not for ours. But mere measures of size, even for outliers like us, do not reveal the anatomical changes during brain evolution, and beyond the brain, evolution has increased structural size and functional efficiency in peripheral as well as central nervous systems (Butler and Hodos 2005). These changes defy any simple overarching explanation, and most of them are beyond the scope of this book. Nevertheless, however humbling the challenge, we attempt to trace some large evolutionary trends and the ontogenetic changes involved in them.

Considering Complexity

Many have tried to array animal nervous systems in order of complexity, size, or other features. At first the goal was a *scala naturae* of brains, each better than the last; later—after Darwin—the goal was reconstructing bushier phylogenies. But the latter effort never completely freed itself from the former, and lingering progressivism has distorted many models. Although humans have the best brains by several reasonable criteria—relative size, regional differentiation, neuronal differentiation, and others—the history of life was not a prologue to that achievement. Our brain is a particular offshoot of a particular lineage, not a culmination—and certainly not a necessary culmination—of what went before (Gould 1989, 2002).

Still, increasing nervous system complexity—measured, for instance, by the number of different neuronal groups (Wicht and Northcutt 1992)—like increasing overall size and body tissue complexity, is a general feature of evolutionary process (Yarczower and Hazlett 1977; Young 1981; Mayr 2001). The last few pages of J. Z. Young's classic, *The Life of Vertebrates* (1981), deal helpfully with these issues. He cites an "increasing turn-over of energy and more and more complicated mechanisms for insuring homeostasis," which "in turn depend on an increasing store of instructions in the genotype" (Young 1981:584). This tendency "is carried to its extreme in the higher vertebrates, giving them the power to maintain life under varied and unpropitious conditions" (584). Young thought it reasonable to call this process progressive. Ernst Mayr (1997:197–98), emphasizing that he meant "a purely mechanistic path toward progress," said, "Most Darwinians have discerned a progressive element in the history of life on earth."

Certain trends are seen again and again in different evolutionary lines and eras and reassert themselves after mass extinctions. Since some are engineering problems, the laws of physics ensure that some of the solutions will be essentially inevitable. If life on other planets is someday studied, these trends—larger size, greater complexity, more information storage—are likely to appear there too. Microscopic size, simplicity, and brainlessness are also adaptive and persist throughout the history of life—microbes may vanquish us yet—but it is in the nature of evolutionary competition to yield repeated increases in these measures in some lines of organisms. Simplicity may outsurvive complexity after some future catastrophe, but that does not negate the role of increasing complexity in the history of life.

Even two comparative neuroanatomists who reject progress in evolution concede that "there can be no doubt that evolution has resulted in some dramatic improvements in the design of brains and bodies," and they conclude:

> The adaptive advantages conferred by an organized nervous system as opposed to a nerve net, by a variety of specific sensory input systems, by the gain of some new connections and the loss of some established connections, by the formation of multiple new nuclei through elaboration or duplication, and by regionally specific cell proliferation in many different parts of the brain have determined the course of brain evolution among vertebrates. (Butler and Hodos 2005:473)

This underscores Mayr's point: purely Darwinian selection can make things work better. Still, the history of life is full of nonlinearities, including *decreases* in complexity, and many organisms evolved much greater complexity than we have in some systems (olfaction, echolocation, the perception of electrical charges). Evolution in those systems is progressive, but it points toward bloodhounds, bats, and electric fish, not toward us.

Also, convergent evolution—the emergence of functionally and even structurally similar adaptations in lines separated before their appearance—is very common in nervous systems. The eye exists in some forty different animal lines (Salvini-Plawen and Mayr 1977), at every level of complexity from a cluster of light-responsive cells to the octopus, fly, or mammal eye, and although it makes use of some of the same developmental (*PAX6*) genes and cellular elements, it has evolved independently in several major forms (Nilsson 2004). Because there are limits to developmental possibilities, a common ontogeny and even a common genetic determination can no longer be viewed as clear evidence that the trait existed in the common ancestor (West-Eberhard 2003; Mundy, Badcock, et al. 2004). This kind of convergence also applies to other sensory and central nervous system (CNS) features (Allman 1999; Nishikawa 2002; Striedter 2005), and small changes in neural structure can have functional consequences, so that gradual microevolution could explain big functional advances (Nilsson and Pelger 1994; Nishikawa 2002).

Despite convergent evolution, the human brain has features, some unique, that emerged only recently—over the last 3 million years and especially the last half-million—built on an established higher-primate foundation (Martin 1983; Deacon 1997a, 1997b; Striedter 2005). Likewise, the higher-primate brain, perhaps 25 million years (my) old, evolved from brains at the root of the primate order before the extinction of the dinosaurs at 65 mya (Tavare, Marshall, et al. 2002; Martin, Soligo, et al. 2007). These in turn built on brains that had been mammalian for over 150 my and reptilian for about 100 my before that (Carroll 1997; Striedter 2005).

Mechanisms of Brain Evolution

We will shortly turn to the specializations introduced at various stages, but consider first the processes of overall vertebrate brain evolution. A comprehensive overview distilled at least eleven mechanisms by which natural selection has altered brain structure (Butler and Hodos 2005). The main mechanisms can be defined as follows:

1. *Induction.* Differentiated and juxtaposed embryonic tissues use chemical messages to produce further changes in either or both tissues. For example, in the early embryo, part of the outer or ectodermal layer is induced to become neural tissue by contact with underlying mesoderm. Later, a more specific piece of ectoderm is induced to become the lens of the eye by contact with an underlying bulge of the neural tube that becomes the retina. The brain may evolve by generating new opportunities for induction, as through mutations in induction factors.

2. *Homeobox gene patterning.* Homeobox gene arrangements and their phylogeny have already been discussed, including their critical role in specifying the linear anatomy of the neural tube. Changes in these genes have changed the central nervous system, and other gene families have emerged to build brains and eyes.

3. *Invasion.* Evolution can modify neural connectivity. This may involve invasion of cell territories by projections from adjacent or distant cell groups and subsequent competitive replacement of invading projections from one source with those from another. Mutations

in genes for cell and surface adhesion molecules (CAMs and SAMs) could produce evolving patterns.

4. *Parcellation.* This makes diffusely organized collections of neurons and pathways more subdivided by function (Ebbesson 1984). In primitive mammals, for example, the regions of the cerebral cortex for somatosensory and motor processing widely overlap, but in advanced mammals these functions are non-overlapping. Other parcellation processes have been identified at lower brain levels and in earlier stages of evolution.

5. *Loss of connections.* Specificity can also be achieved by a *loss* of connections that narrows the focus of a brain nucleus. In vertebrate brain evolution, connections from the thalamus to the opposite side of the cerebral cortex were gradually lost, increasing lateralization. Similarly, in the evolution of the primate sense of smell, relaxation of selection apparently allowed a loss of connections from the olfactory bulbs to some central brain areas (LeGros Clark 1959). These are necessary for intermale aggression in mice (Edwards, Nahai, et al. 1993), but the sense of smell is not critical to aggression in higher primates.

6. *Duplication of neural regions.* This may be a main mechanism of neocortical evolution, based on functionally coherent vertical columns that act as units of cortical architecture, as well as a mechanism of its development (Rakic 1988). As the cortical sheet expanded, columns proliferated while remaining largely constant in size, although there are significant species differences in their internal organization (Buxhoeveden and Casanova 2002). Duplication of master regulatory genes for a column may have multiplied columns to expand cortical mass without changes in thickness, which would compromise function. Column duplication may have led to parcellation or regional differentiation (Kaas 1995), but in any case it is neuron number, not neuron size, that increased (Herculano-Houzel, Collins, et al. 2007). Since columns show plasticity of function (Kaas 1995), they may participate in a Baldwin-like effect in which behavioral niches are pioneered through plasticity and secured over generations through column duplication, a neural manifestation of the behavior-as-pacemaker concept (Mayr 1988).

7. *Developmental differentiation of multiple nuclei.* This may be a special case of parcellation or duplication followed by elaboration and expansion, but with dramatic functional consequences. Unique, large, novel cell groups for the control of specialized electromotor organs in electric fish and for the production of song in finches, sparrows, and other songbirds may have originated through such a process. Since the relevant nuclei in some songbirds expand in size by proliferating neurons seasonally in response to hormones, this may be another way for plasticity within one life cycle to introduce populations into new niches where natural selection will then produce convergent genetic changes.

8. *Regionally specific changes in neuron proliferation.* This has been a crucial mechanism of brain phylogeny, weighting some structures much more heavily than others. Proliferation can involve changes in neuronal distribution patterns, such as lateral or radial migration of populations of neurons, or even reversal of layering patterns, as has occurred in the evolution of the mammalian cortex. It can also involve differentiation of neuronal phenotypes (as opposed to the differentiation of nuclei), distinctive in shape, size, and/or neurotransmitter or neuromodulator specificity.

Vertebrate Body Plans and Behavioral Advances

Over the course of about 600 million years, these and other mechanisms have produced the human brain and spinal cord from a simple tubelike structure. This structure itself was much older and resulted when a linear fold on the external surface of one of our ancestors became cylindrical, bringing the responsive cells facing the outside world into the center of the nervous system—a process recapitulated in human development as the embryonic ectoderm folds and forms a tube.

A genetic dissection of the central nervous system of the marine worm *Platynereis,* related to the earthworm and resembling our common ancestor, shows remarkable continuities with our nervous system (Denes, Jekely, et al. 2007). Thus the genetics and chemistry governing major CNS domains was in place, as were those generating the major neuronal types, by more than half a billion years ago. This was one of many basic animal body and nervous system plans that diversified in "the Cambrian explosion," an adaptive radiation of varied animal life, possibly following a mass extinction (Conway Morris 1989, 1998).

The modern *Amphioxus,* a favorite of biology labs, is thought to resemble the ancestral chordate, between the ancestral worm and the first vertebrates. *Amphioxus* homeobox genes yield insights into the origins of vertebrate ontogeny (Murakami, Uchida, et al. 2005). The draft sequence of the *Amphioxus* genome (Putnam, Butts, et al. 2008:1070) "provides conclusive evidence for two rounds of complete genome duplication on the jawed vertebrate stem"—the line leading to us—and suggests that this "was a formative event in the early history of vertebrates that provided genomic flexibility . . . for the emergence of developmental, morphological, and physiological novelties such as . . . neural crest cells . . . hindbrain patterning, finely graded nervous system organization, and the elaborated endocrine system of vertebrates."

The tubelike central nervous system of the early chordates had homologues of our midbrain, hindbrain, and spinal cord, but the forebrain, the paired sense organs, and the whole forward part of the head were left for the early vertebrates to evolve, using different regulatory genes (cephalic gap genes, *BF-1,* and so on). In a hypothetical primitive vertebrate brain, *Hox* genes govern the postcranial CNS, while newer, vertebrate-specific genes govern the formation of the brain. An endoskeleton protected an increasingly complex central nervous system. The absence of an exoskeleton meant far fewer limitations on size and flexibility, so that the evolution of large, supple, motile, strong animals with concentrated and protected central nervous systems remained possible. Vertebrate phylogeny was not goal-directed or preparatory, but key early phylogenetic events in the vertebrate lineage left open the possibility of some interesting later ones.

First Vertebrates

A fossil from the early Cambrian of China, around 530 mya, represents a transition to the earliest vertebrates. It is known from many specimens and can be analyzed in detail. Named *Haikouichthys* (*"Haikou"*), it is more like the lamprey—a jawless fish but a true (if primitive) vertebrate—than the simpler *Amphioxus* (Shu, Morris, et al. 2003). Or to be exact, it is more like the lamprey larva, suggesting that modern lampreys evolved from *Haikou* in part through terminal addition or prolongation, in which delayed offset of growth allowed newly "adultified" forms to develop. Still, even the "less developed" *Haikou* had a head and therefore a brain; eyes, nasal sacs, and ear capsules; and a notochord with separate vertebral segments. It was one of the earliest animals with a brain.

Advances came through changes in ontogeny. From comparative studies now strongly confirmed by *Haikou*, vertebrates, but not simpler chordates, evolved (1) a neural crest—cells that break off from the neural tube as it closes, destined to form much of the peripheral nervous system—and (2) sensory placodes, or specialized areas of ectoderm that can be induced to form parts of sense organs (the first of the eight processes cited earlier), with some of this induction caused by neural crest cells. Major parts of the thalamus, as well as the hypothalamus, the epithalamus, and some new specializations of the roof of the midbrain, appeared with the vertebrates, mainly to interpret information detected by new sense organs, especially olfaction, light-detection by primitive retinas, and a light-sensitive pineal gland just under the roof of the skull. At the dawn of the vertebrates, information in the environment was driving brain evolution.

But why a great advance just then? Compared with primitive chordates, the simplest vertebrates are more mobile and efficient predators, an important recurrent theme in brain evolution. Thus the most parsimonious account: selection for better predation (and evading predators) made vertebrates out of simple chordates by altering *induction*. It also caused *changes in* Hox *genes* that made the front end of the neural tube bulge, producing a forebrain, while the neural crest offered many new developmental possibilities. *Invasion* (the third process) was also involved, since projections from the new sense organs would have penetrated the forward end of the former chordate neural tube. These projections were probably mainly olfactory.

Thus a fundamentally new brain can arise from simple changes in neural development. Indeed, most of the eight processes can be seen as heterochronic, but in a much more specific sense than the global heterochronies discussed earlier. If, as is likely, ancestral chordate nervous systems were plastic, then here too individual plasticity could have pioneered in ecological niches that then produced parallel evolutionary changes in gene-regulated maturation, through mutation and selection (the Baldwin effect).

Rudiments of not only the thalamus, hypothalamus, and epithalamus but also the basal ganglia, amygdala, and other core structures of the mammalian forebrain were present in early vertebrates (Butler 1995), so the evolution of the brain in the first jawed fish, the first tetrapods (amphibians), the first amniotes (reptiles), the first mammals, and even advanced mammals must have entailed invasion, expansion, duplication, parcellation, and other processes that transformed existing structures into larger and more differentiated ones with novel interconnections.

In the jawed fishes, the next major advance in vertebrate history, motor control was enhanced through the elaboration of the cerebellum and a new level of differentiation of the midbrain roof as a topographic map of the external world. In the evolution of the first amphibians, some 400 mya in the Devonian seas, a crossopterygian fish used its strong, lobular fins to scuttle along the floors of shallow pools and, occasionally, to crawl through mud from pool to pool (Carroll 1988, 1997). And in the evolution of reptiles, a massive new wealth of sensory information reached the dorsal part of the forebrain, enhancing survival in the terrestrial environment (Butler and Hodos 1996).

The Emergence of Mammalian Brain and Behavior

By 500 mya, some fishes had vertebrate sensory organs and brains as well as a key embryonic component, neural crest cells. There may have been some forms of parental care of eggs and/or young, since such behavior is common in modern fishes and frogs (Gittleman 1981; Crump 1996). Amphibious (partly terrestrial) vertebrates arose by around 400 mya (Radinsky 1987; Niedźwiedzki, Szrek, et al. 2010). But the reptiles emerging at 310 mya were the first amniotes—vertebrates with embryonic coverings. Amniote mothers must store enough food in their protected eggs to carry their young through the larval stage so that they can be more developed at birth—a pattern called direct development. Amniotes can be completely terrestrial because they re-create in each life cycle an underwater environment for their embryos.

Mammal-like reptiles diverged soon after reptiles appeared, and the earliest mammals emerged by 200 mya (Luo 2007). In addition to new skeletal features, the first mammals had relatively larger brains and (by inference from their body size and proportions) the higher metabolic rates required for warm-bloodedness (Wang, Wang, et al. 2003). Some hallmarks of mammalian life emerged. Reptiles, like fish and amphibians, replace teeth continuously and grow throughout life; mammals restrict replacement to two sets—deciduous ("milk") and permanent teeth—and stop growing around sexual maturity. We can think of the early mammals as partly juvenilized versions of cynodonts, the mammal-like reptiles they evolved from (Allman 1999). Cynodonts were far larger and had many successive sets of teeth, while their early mammalian descendants were small and had only two. (One mammal dated at 195 mya had an estimated body weight of 2 grams [Luo, Crompton, et al. 2001].) Thus a process of partial neoteny (truncation) could have juvenilized them.

The platypus genome reflects the common ancestor. Like it, the first mammals were monotremes (egg-laying mammals), but the platypus has a typically mammalian suite of milk protein genes (Warren, Hillier, et al. 2008); thus feeding of the young was a feature of the earliest mammals. The opossum genome, reflecting the transition from egg-laying to marsupial mammals tens of millions of years later, has confirmed the strong role played by conserved noncoding RNA in mammalian evolution (Mikkelsen, Wakefield, et al. 2007). A growing consensus among evolutionary geneticists holds that most significant steps in evolution involved changes in noncoding regulatory elements, not changes in protein-coding genes (Carroll 2005).

Mammal Brains

Even if the first mammals were juvenilized cynodonts, the mammalian brain cannot be a juvenile feature. Compared to reptiles, early mammals show a five- to tenfold increase in relative brain size, and two specific changes deserve attention.

First, mammals increased the complexity of the reptilian sensory cortex: a granule cell layer inserted *within* the pyramidal cell layer yielded the two pyramidal cell layers we now have. Mathematical models suggest that this change allowed a better integration of "what" with "where" information in the representation of the periphery (Montagnini and Treves 2003). In general, the monotreme neocortex has much in common with that of other mammals, suggesting that the main features of cortical organization were established very early in mammalian history, before live birth (Krubitzer 1998).

Second, relative brain size increased. A log-log plot of brain size against body weight for mammals has the same slope but a substantially higher intercept than that for reptiles—a grade shift whose main function may have been to package "extra" neurons for the sense-specific analysis of stimuli (Jerison 1973, 1976, 2001). Since the rise of early mammals paralleled that of the dinosaurs, our much smaller ancestors probably had to occupy nocturnal or burrowing niches (Damiani, Modesto, et al. 2003), which would have favored senses other than vision—an inference supported by the mammalian features of ear bones, vibrissae, and small eye sockets. We know from fossil endocasts and comparative neuroanatomy that both the olfactory apparatus and the auditory cortex expanded, accounting in part for increased relative brain size compared to mammal-like reptiles. This second aspect of mammalian brain expansion may have been basically a solution to a packaging problem (Jerison 1973, 1976). In the visual system, many primary neural cells are outside the brain in the retina. But in the auditory and olfactory systems, comparable processing is done at higher levels—in the case of hearing, perhaps because temporal, not just spatial, integration is needed. The visible evolution of mammalian middle ear bones from precursors in the reptilian jaw angle parallels brain expansion.

Finally, from comparison of reptile brains with those of the most primitive, smallest living mammals, the transition elaborated the brain structures present in reptiles in a more rudimentary form (MacLean 1990, 1993). These structures have been called the "paleomammalian" brain, but we now know that the limbic system structures referred to are much older than mammals, probably dating to the common ancestor of tetrapods (Bruce and Neary 1995; Lanuza, Belekhova, et al. 1998). Still, Paul MacLean's main point is consistent with current research: these structures are ancient legacies that remain embedded in the human brain and mediate emotion and memory (Salloway, Malloy, et al. 1997; Reep, Finlay, et al. 2007).

Still, there is good evidence that von Baer's law, not Haeckel's law, best explains brain evolution. Using measures of regulatory gene expression in successive developmental stages of frog, turtle, chick, and mouse embryos, it is possible to identify a phylotypic stage for the forebrain (Smith Fernandez, Pieau, et al. 1998) with distinct patterns of gene expression early in forebrain development. Embryonic gene expression reflects deep homologies among these four groups of vertebrates, and that justifies the extrapolation backward to what these authors call the "archetypic embryonic stage."

Social and Emotional Advances

What did early mammals add to the well-established reptilian limbic system? One contribution was a more differentiated hippocampus, which is vital to learning and memory; another was an elaborated cingulate cortex, along with its thalamic connections (Kunzle and Radtke-Schuller 2001; Rodriguez, Lopez, et al. 2002). MacLean (1985, 1990) emphasized that the early mammals evolved the capacity for maternal behavior by advancing cortical connections of the limbic system, notably the thalamocingulate projection. Perinatal removal of the great majority of the neocortex spares maternal and other reproductive behaviors in Syrian golden hamsters, suggesting that the limbic system as elaborated by the early mammals is sufficient to maintain these behaviors (Murphy, MacLean, et al. 1981). Further experiments in adult rats showed that neurochemical damage to the thalamocingulate connection specifically abolishes maternal behavior, direct support for MacLean's model of early mammalian brain evolution (Peredery, Persinger, et al. 1992). In an important convergence, this circuit is also implicated in mammalian infants' isolation calls (Newman 2003).

Evolution in parental care and offspring dependence is consistent with these changes. The mother-offspring relationship probably evolved earlier (Oftedal 2002), but was almost certainly in place by the mammals' first appearance around 225 mya. "Mammal" (from *mammae,* or mammary glands) implies feeding the young from modified sweat glands, which in egg-laying mammals like the platypus are spread over the mother's ventral skin surface instead of in pairs of teats, a later feature; monotremes show, however, that physiological feeding long preceded live birth.

The same is probably true of maternal retrieval in response to infant separation calls (MacLean 1985), another key to the mother-infant relationship. Based on comparative studies of living mammals, these calls coevolved with the middle ear bones, which fossil evidence shows were derived from the reptilian jaw angle by 195 mya (Luo, Crompton, et al. 2001). Also, the mammalian cochlea has specialized outer hair cells that fine-tune the inner ones (Brownell, Bader, et al. 1985), enabling mammals to hear frequencies an order of magnitude higher than those that can be heard by reptiles, which are deaf to mammalian infants' distress calls. This would have added a separate adaptive advantage (aside from nocturnal habitat) favoring a larger auditory brain.

Thus, by the middle of the Mesozoic, 125 mya, mammals had evolved the reciprocal attachments of offspring and mother (Ji, Luo, et al. 2002), along with at least rudimentary versions of the associated emotions. By the Cretaceous, 100 mya, when flowers first appeared and spread, the major forms of modern mammals emerged in the shadow of the dinosaurs (Luo 2007)—many of them in the trees (Gibbons 1998a; Kumar and Hedges 1998). Subsequent mammalian advances, especially in certain lineages, involved elaboration of the neocortex, which greatly expanded learning capacity and increased the complexity of behavioral repertoires (Striedter 2005). There is evidence that a marked rise in oxygen levels in earth's atmosphere facilitated the evolution of the earliest mammals, the transition to live birth, and the adaptive radiation of mammals after the dinosaur extinctions, which included further advances in brain size and complexity (Falkowski, Katz, et al. 2005).

Gestating Mammalian Brains

The neocortex accounts for almost all the variation in relative brain size in mammals. Two compatible models explain this. The first links expansion to the gestation time available to generate cortical neurons (Finlay and Darlington 1995). The time needed to double the neuron number varies modestly, so lengthening gestation yields an exponential increase. In fact, even though each division of neural progenitor cells takes *longer* in monkey embryos, they have time for twenty-five doublings, while mice have only eleven (Kornack and Rakic 1998)—a prolongation rather than an acceleration of brain growth. In mouse knockout and overexpression experiments the effect of the *BF-1* gene in generating these doublings, specific to the forebrain, is dramatic (Xuan, Baptista, et al. 1995; Dou, Li, et al. 1999; Ahlgren, Vogt, et al. 2003). In all mammals the last doubling seems designed to allow selection within the cortex, since half the final number of neurons generated die, in part depending on connectivity, activity, and stimulation. The scope here for environmental influence is clear, and in humans the process continues for years postnatally.

The second model identifies cortical columns as coherent units not only of function and embryogenesis but also of evolutionary expansion (Kaas 1995; Rakic and Kornack 2001; Finlay and Brodsky 2007). In this model the cortical sheet expands by phylogenetically adding columns as units, which are then recruited for new functions. This expansion could involve duplications of regulatory genes, each at the top of a developmental cascade generating a column. So a neocortex repeatedly augmented by column duplication may be another evolutionary achievement of mammals.

Yet it remains true that removal at birth of the hamster neocortex is compatible with survival and development; the resulting adults mate and care for their young, retrieving pups and nursing until weaning (Murphy, MacLean, et al. 1981). This finding is less surprising when it is considered that more primitive mammals also limited in neocortex, such as the platypus, perform such functions very well. This supports an association of these behaviors both with particular structures (the extended limbic system) and with a particular phylogenetic advance (the emergence of mammals). Subsequent neocortical expansion broadened behavioral repertoires and enhanced learning, but these higher cortical functions did not replace limbic ones.

Conserving Brain Development

The sequence of embryonic brain maturation and neurogenesis is highly conserved across mammals, with later-developing structures growing larger than earlier ones (Finlay, Hersman, et al. 1998), and it is mainly the pace and length of development that distinguish different mammal brains. The heterochronic process of terminal addition through prolongation seems to apply: prolong neurogenesis to create more complex structures.

But conserved brain development patterns ensure (with important exceptions) that brain structures coevolve as coherent and constrained subgroups, resisting selection on individual circuits (Finlay, Darlington, et al. 2001). This pattern would not prevent selection, but it does require, say, expansion of one circuit in the group to "drag along" the others if it is strongly favored. The exceptions (olfactory reduction, for one) result from heterochronic change in the rates, onset, duration, and offset of the growth of brain organs or systems. In primates, later offset of neo-cortical growth makes the subcortical limbic system a smaller proportion of brain size than in other mammals (Finlay, Hersman, et al. 1998; Reep, Finlay, et al. 2007), although no less central.

We saw that brain weight scales to roughly the ¾ power of body weight in mammals, probably owing to energy constraints. However, it may be the *mother's* metabolic rate that, by regulating both fetal and infant brain growth, accounts for the relationship (Martin 1981, 2007; Isler, Christopher, et al. 2008). The fact that mammalian brain growth slows at birth and is largely completed by weaning supports the maternal energy flow hypothesis (Striedter 2005). Between closely related animals, there is also a striking relationship between brain size and maturity at birth. In relative brain size, precocial mammals exceed altricial ones by a factor of 4.5, but in altricial mammals brain size increases 7.5-fold from birth to adulthood, while in precocial ones the increase is only one-third as great. Primates, although dependent, fall in the latter category—except for humans, whose neonatal brain weight roughly quadruples, putting us between the general primate and the altricial mammal value.

Developmental Keys to Psychosocial Evolution

Primates are one among at least twenty-one orders of living mammals, some others being marsupials, rodents, bats, carnivores, and elephants (Nowak 1991). The primates vary greatly and have classically been defined by trends, not categorical absolutes (LeGros Clark 1959), since not all primates meet every criterion. (Cognitively, primates are a fuzzy set, although phylogenetically—cladistically—they are a coherent group.) The trends are:

1. A generalized limb structure, including a five-digit hand

2. Enhanced digital mobility

3. Flattened nails and sensitive digital pads

4. Abbreviation of the snout and the apparatus of smell

5. Elaboration of vision, including binocular perception

6. Loss of some primitive mammalian teeth, and preservation of a simple molar cusp pattern

7. Expansion of the brain, especially the cerebral cortex and related structures

8. Prolonged gestation serving one or at most two young

The last two trends carry two mammalian evolutionary achievements to a higher order of complexity, but the first preserves limb bones going back to the lobe-finned fishes, and the fourth extends a common mammalian trend toward olfactory reduction.

Exceptions can be found to most of the trends, but overall they describe both the extant primates and their main evolutionary history (Napier and Napier 1985; Martin 1990; Fleagle 1999). It is often said that the primates succeeded by remaining generalized mammals. This is certainly true of the skeleton, especially the forelimb. The human hand—a superb evolutionary achievement—has not only the same number of digits but the same number of bones, and the same pattern of muscle insertions, as the primitive primate *Notharctus* did 45 mya (Napier and Napier 1985). Consider horses, whales, and bats to see how generalized our forelimb is. But primates crucially depart from generalized mammals in brain size and function, sensory adaptations, learning ability, litter size, and parental care. Indeed, "many aspects of primate life histories are of general theoretical interest due to their uniqueness or extreme nature among the mammals" and "particular combinations of traits setting them far enough apart from other mammalian orders to require specific evolutionary explanations" (Stearns, Pereira, et al. 2003).

The primates originated in the Cretaceous, before the mass extinction that included the dinosaurs (65 mya); molecular phylogenies put the last common ancestor of living primates at more than 80 mya (Martin, Soligo, et al. 2007). The early primates had two successive adaptive radiations, a smaller one after the dinosaur extinction and a major one 10 my later. After this (Eocene) radiation, primates already had relative brain sizes far above those in most contemporary mammals. Endocasts show that this owed much to an expanded occipital cortex that was largely devoted to vision (Radinsky 1977). Skeletal features supporting improved vision include the forward rotation of the orbits of the eyes, allowing depth perception through greater overlap of the visual fields, and reduction of the snout and olfactory apparatus. *Teilhardina asiatica,* a prosimian of 55 mya described as "a small, diurnal, visually oriented predator," demonstrates these points (Ni, Wang, et al. 2004).

Primates can be divided into two main groups, prosimians (lemurs, lorises, and tarsiers), and anthropoids (monkeys, apes, and humans).* The earliest anthropoid we have is either *Darwinius,* a 47-mya cat-sized creature from Germany (Franzen, Gingerich, et al. 2009), or the more widely accepted *Eosimias* ("dawn monkey"), at 45 mya in China. Like many pioneers in brain evolution, *Eosimias* was a small (human hand–sized) predator that fed on insects and fruit (Gebo, Dagosto, et al. 2000). From roughly this point of origin, the higher primates arose.

Primate Ontogeny

Compared with other mammals, primates, especially anthropoids, have a prolonged juvenile period for their body size, requiring more parental care. This may be due to a trade-off between the increased food intake necessary with faster growth and the greater risk of predation during development; in this model, reduced food needs trump predation risk (Janson and Van Schaik 1993). In the Sibly and Brown (2007) lifestyle model (see Chapter 3), primates are less productive for their body size because arboreal lifestyles protect them from predation. But in this context longer growth can yield a larger relative brain size, since we know that dental development (Godfrey, Samonds, et al. 2001), age at reproductive maturity (Leigh 2004), and life span (Allman, McLaughlin, et al. 1993) predict adult brain size and that neonatal and adult brain

---

* Recent classification often draws the main line on the other side of the tarsiers, which may have a closer evolutionary affinity with the higher primates. The main division under this arrangement is between strepsirhines (lemurs and lorises) and haplorhines (higher primates including tarsiers). Tarsiers share with the anthropoids a type of placentation, the hemochorial placenta, and a relatively long gestation—180 days—among other features. However, tarsiers remain quite distinct from the other haplorhines, still formally referred to as anthropoids. In this book, the phrase "higher primates" refers to the anthropoids: monkeys, apes, and humans.

size predict gestation length and life span, respectively, more than does body size at either time point (Sacher and Staffeldt 1974; Pagel and Harvey 1988, 1990).

Still, much of primate brain growth occurs in infancy or before birth, and brain size is not perfectly correlated with time to maturity (Pereira and Leigh 2003). Brain development is not just about size; slow growth allows more time for synaptogenesis, synaptic remodeling, myelination, and other maturational change, and also for experience effects; play peaks around the time that synapse number does in several primate species, a strong opportunity for play to influence synapse survival (Fairbanks 2000).

Lemurs and squirrel monkeys stop their growth for a time after weaning, perhaps so that they can learn to forage successfully, and dental development is accelerated in species that experience especially high postweaning nutritional stress, enabling them to forage better immediately (Godfrey, Samonds, et al. 2001). Primate folivores (leaf-eaters), facing relatively low foraging demand, grow faster than same-size frugivores (fruit-eaters), which need more time (and better brains) to learn to find food (Leigh 1994). If we compare yellow baboons (*Papio cynocephalus*) and vervet monkeys (*Cercopithecus aethiops*) living in essentially the same environment (Bolter and Zihlman 2007), baboons take two years longer to develop, during which time they learn to exploit many of the 250 species of foods they will use over a lifetime (Altmann 1998); vervets race to maturity, exploit far fewer foods, and suffer population crashes during scarcity (Lee and Hauser 1998), when baboons are much more resilient. Thus life history theory is supported by key aspects of primate development (Stearns, Pereira, et al. 2003).

In theory there is a trade-off between the energy consumed by growth between weaning and reproductive maturity and the amount invested in offspring after it (Charnov and Berrigan 1993); this may explain why higher primates have long life spans and few offspring, consistent with K-selection and lifestyle theory. In fact, it may be "that the long primate lifespan follows from the cost, rather than some cognitive benefit, of having a big brain" (Charnov and Berrigan 1993:193), which would mean that the brain must somehow be worth the cost. Puberty entails a growth spurt in many higher primates (Pereira and Leigh 2003), suggesting that the last phase of growth is a vulnerable time that should be passed through quickly. In any case, it helps generate sexual dimorphism in size, which varies greatly among species. Sexual dimorphism may ultimately result from female choice, but if it is realized in part through relative slowing of males' growth, it may protect them from competition longer. Young males have high mortality in many primates, especially where they must change groups to reproduce, so both prolonged prereproductive growth and rapid pubertal growth could be protective, while leading to larger adult size.

The Primate Brain

Primates have large relative brain sizes—a grade shift upward on the log-log plot, beyond the shift from reptiles to mammals—most of all in the higher primates (Martin 1990, 2007). As with mammalian brain evolution generally, expansion of the higher-primate neocortex occurred through increase in neuron number (Kaas 2006; Herculano-Houzel, Collins, et al. 2007), perhaps involving duplication of developmentally and functionally coherent microcolumns (Rakic and Kornack 2001; Buxhoeveden and Casanova 2002; Finlay and Brodsky 2007). As we have seen, neural progenitor cells in the monkey neocortex divide twenty-five times compared with eleven times in rats, despite longer cell cycle duration (Kornack and Rakic 1998), giving the neocortex a greatly disproportionate role in evolutionary brain expansion (Kaskan and Finlay 2001).

Like the heroes of several prior evolutionary advances, early primates were small, active predators that chased insects (and ate plant foods) in an arboreal niche. Their forward-facing eyes, shown by the bony orbits of fossil skulls (Radinsky 1970, 1975), enabled depth perception for predation and rapid movement on smaller tree branches. The olfactory system shrank and the visual cortex greatly expanded, as seen in endocasts and living prosimians, which also suggest other visual system adaptations emerging in the major primate radiations: increased density of photoreceptors in the central retina, an enlarged optic nerve and tract, a two-hemisphere destination of optic fibers for depth perception, a specialized temporal lobe area for rapidly processing moving visual stimuli, and a specialized cortical motor control area for visual guidance of movement (Allman 1999; Kaas 2005, 2006).

Indeed, it is the visual system that best predicts relative brain size in primates. The volumes of the lateral geniculate body (LGN), the visual way station in the thalamus, and of the primary visual cortex are well correlated with overall relative brain size. A specialized pathway within this system, through the parvocellular (small-cell) layer, is strongly implicated; neuron number in this but not other LGN layers correlates with brain size and also with diet and group size (Barton 1998). The parvocellular pathway, processing color and other specific visual cues, has more neurons in the relatively large brains of frugivorous species, enabling them to find and select berries and other fruits (Stevens 2001).

But group size separately predicts visual brain evolution, probably because social information in primates is so visual. Overall relative brain size varies with group size in primates (Dunbar 1998), mainly owing to neocortical size. Larger groups mean more faces to remember and more subtleties of facial expression to flag their moods and likely behavior. Higher primates have a much wider range of expression and more complex facial muscles than other mammals, and while there is no general increase in the size of the facial nucleus with group size, cortical specializations for orofacial control distinguish apes and humans from monkeys (Sherwood, Holloway, et al. 2003, 2004; Sherwood 2005). Facial expression complexity coevolved with visual processing and visual pattern learning. One of the main trends in primate brain evolution, the extension of the temporal pole and increasing definition of the lateral sulcus due to temporal lobe expansion, was once thought related to auditory processing. However, more of the higher-primate temporal lobe is dedicated to higher-order visual processing and visual pattern learning, including face recognition.

In addition, quantitative cross-species analysis shows that observations of "social learning, innovation, and tool use frequencies are positively correlated with species' relative and absolute 'executive' brain volumes" (Reader and Laland 2002:4436). Primates have an unusual scaling ratio between the size of the frontal lobe and that of the rest of the neocortex: it increases more steeply for primates than even for the (also large-brained) carnivores (Bush and Allman 2004). Although evolution is never predestined or relentlessly directional, higher primates laid the foundation for what would become the human brain.

The primate brain is also unique among mammals in its development (Martin 1983, 2007). Brain to body weight ratios for primate fetuses are tightly distributed along a (logarithmic) linear regression line with an allometric exponent of 1, and this distribution does not overlap with the one for nonprimate mammals (whales and dolphins being intermediate). Brain weight is about 12 percent of body weight for all primate fetuses up to one kilogram of body weight, but about 6 percent for nonprimates (Sacher 1982); primates have proportionally twice as much nervous system tissue from the outset. They also have very low rates of fetal *body* growth, so the twofold-greater brain to body weight ratio compared to nonprimates is due mainly to evolutionary slowing of fetal body growth while the general mammal rate of brain growth held steady. Since both prosimians and anthropoids show the pattern, this heterochronic change would have happened early. Early primates were small, highly mobile predators; slowing fetal body growth while maintaining brain growth may have been a key adaptation.

But there is a further complication: *absolute* brain size may matter after all. After a century in which we compared animal brains only after correcting for body size to focus on the "extra" neurons, it turns out that absolute brain size also strongly predicts intelligence (Striedter 2005). In retrospect, it stands to reason that this would be so. Since an elephant's neurons are not much larger than a mouse's, then with the same relative brain size the elephant would have many more neural elements. Furthermore, brains reorganize as they get bigger, regardless of body size; for example, the cortical sheet becomes more convoluted. A large brain is good to have regardless of body size. There may even be evolutionary lines where a larger brain was needed and the simplest way to get one was to grow a larger body; in such cases, natural selection may have dragged the body along behind the brain.

Anthropoid Behavior and Relationships

Higher primates are social animals with great learning capacity, and the mother-offspring bond is almost always the center of social life (Strier 2006; Campbell, Fuentes, et al. 2007). This bond is prolonged, as is individual development—each life cycle phase as well as the whole life span (Leigh and Blomquist 2007). Laboratory and field studies reveal complex social cognition and social learning (de Waal and Tyack 2003; Panger 2007), including nongenetic transmission of communication, rank, knowledge of food sources, and in some species tool-using and even (in capuchin monkeys, *Cebus apella*, and chimpanzees) tool-making techniques (Wrangham, McGrew, et al. 1994; Whiten, Goodall, et al. 1999; Moura and Lee 2004; Ottoni, de Resende, et al. 2005).

As we saw, higher-primate brain size, especially in visual areas, is correlated with group size, partly because of the need to recognize faces and their expressions; although largely innate, facial expressions can be highly complex in usage and context, especially the context of a particular face. Growing evidence also shows that anthropoid acoustic communication is far more complex than anticipated, and that since it may be deceptive, listeners must up to a point be skeptical (Gouzoules and Gouzoules 2002). Although calls are generally quite species-specific (Gouzoules and Gouzoules 2000), they are meaningful.

Vervet monkeys (*Chlorocebus pygerythrus*) have three different alarm calls proven to be warnings of snakes, leopards, and predatory birds, respectively, and they elicit three different appropriate responses in listeners (Cheney and Seyfarth 1986, 1990). Pigtail macaques have four acoustically distinct recruitment screams (calls for help) that are reliably associated with different types of aggressive encounters, depending on the rank of the opponent and the severity of the attack (Gouzoules and Gouzoules 1995), and clearly the considerations we invoked in relation to face recognition, facial expression, and group size also apply to these complex interpersonal calls. Moreover, there is a developmental course: as juveniles grow to adulthood, their agonistic screams become more situation-specific (Gouzoules and Gouzoules 1995). The remarkable fact is that monkey calls have fine shades of meaning and a component learned during socialization, despite being in some ways innate.

Social play characterizes all anthropoid species, particularly during development, and is an important opportunity for learning and for shaping brain development (Fagen 1993; Pellegrini and Smith 2004). Play is found in most mammals and birds, but the higher-primate emphasis on both the mother-infant bond and play intensifies the pattern established by early mammals. We take these matters up again in Parts III and IV.

4

Ape Foundations, Human Revolution

We have seen the major advances made by vertebrates, mammals, and primates; though we do not know whether we humans are the founders of a new major lineage of species, our evolution from the apes is revolutionary, even though it took millions of years. As in the last chapter, we focus here on brain, behavior, and development.

Ape Evolution and Behavior

If *Eosimias* (the "dawn monkey," 45 mya) was close to the ancestral line of all higher primates, a similar role for the catarrhines—Old World monkeys, apes, and humans—is played by *Aegyptopithecus zeuxis,* which lived about 30 mya in the Fayum basin of Egypt (Simons 1995), then wooded and laced with rivers. It was a macaque-sized tree-climber with a monkeylike skull and limbs but apelike teeth, and it had forward-facing eyes in nearly complete bony orbits. The arboreal predatory niche of the early primates favored depth perception through binocular vision (Fleagle 1999), and the widened optic foramen suggests an enlarged optic nerve for detailed perception (Kay and Kirk 2000). A remarkably preserved female skull of this species (Simons, Seiffert, et al. 2007) shows that it had very marked sexual dimorphism; this suggests serious male-male competition and probably a polygynous mating system with a potentially large role for female choice. However, it had a prosimian-like brain in an anthropoid body—a cortex that was not very large except in the visual areas.

So there was a shift with significance for the evolution of mind between 60 and 30 mya—from chemical signals, which are identifiable but have limited patterning, to visual ones, whose patterns are orderly and complex. At some point, reliance on vision created certain neural features special to higher primates—stereoscopic and color vision, advanced focusing and movement detection, eye-hand coordination, and visual learning.

During the Miocene (about 25 to 5 mya) a new radiation produced all forms of apes, including, late in that period, the upright ones ancestral to us (Fleagle 1999; Klein 1999; Wood 2005). Gibbons and siamangs diverged first, then orangutans, then gorillas; the line leading to chimpanzees, bonobos, and hominins* remained primitive, smaller, and anatomically unspecialized. On molecular evidence these three lines are more closely related than any is to the gorilla, and orang is still more distant (Sarich 1970, 1992). We share at least 96 percent of our DNA sequence with chimpanzees and bonobos, and since we have only 22,000 to 25,000 protein-coding genes, not many of those distinguish us. Several of them are known, and we consider them in the next chapter, along with equally or more important differences in nonprotein-coding functional RNA.

Among the extremely varied Miocene apes (Andrews and Kelley 2007), our most likely ancestor is *Proconsul africanus* (about 18 mya); it had a small face, with neither long canines for tearing nor massive molars for grinding (Klein 2009). *Proconsul*-like fossils have been found in Europe, Asia, and Africa, suggesting great success (Walker 1997; Fleagle 1999). It was probably adapted to tropical forest or open savanna and may have been in transition, at home in the trees or on the ground. If open savanna, its small size would have made it vulnerable to predation, a critical change, so savanna-living chimpanzees have been proposed as a model for behavior and social order in these ancestral apes (Moore 1996). If this is right, they may have moved in the open in large, somewhat hierarchical groups (Melnick and Pearl 1987), although these occur in some forest settings as well (Mitani 2006). In baboons, coalitions help

---

* Most recent literature uses "hominin" to mean all upright, walking, apelike species since the common ancestor of apes and humans. This usage was meant as a demotion from family ("hominid") to subfamily status, implying a closer relationship between apes and humans. Based on molecular evidence, this definition made sense until recently; new appreciation of the role of rapidly evolving, noncoding RNAs may confirm the intuition of prior generations that gave us and our ancestors family status based on other criteria. This book, however, follows the current convention.

establish dominance, males must often befriend females to be accepted into a new troop, and individual variation in behavior can be dramatic (DeVore 1965a; Smuts 1985; Sapolsky 2001). Chimpanzees are even more social, or at least more subtly social (de Waal 1989), and Machiavellian intelligence—outsmarting competitors by anticipating their moves, deceiving them, shifting alliances, and so on—helps explain ape brain expansion (Whiten 1988; Whiten and Byrne 1997). Dominant chimpanzee males allow mating opportunities to subordinate males who support them politically (Duffy, Wrangham, et al. 2007).

After *Proconsul,* other apes close to our ancestral line evolved in approximate succession: the more terrestrial *Equatorius* at 15 mya (Ward, Brown, et al. 1999), *Kenyapithecus* at 14 mya, and an array of other species that spread throughout Africa, Europe, and Asia (Andrews and Kelley 2007). After this, the geographic range of apes shrank to a much smaller Afro-Asian distribution, in tropical or subtropical forests where they depended on fruit. Since we know that fruit-eating (as opposed to leaf-eating or grazing) is associated with greater brain size and cognitive ability, this ecological change may have selected for the apes' advances in mental function (Potts 2004). From a species in this successful group, upright-walking hominins emerged some 6 or 7 mya; they (we) would end with by far the longest childhood and the largest brain.

Ape Brains Evolving

The hominoids—apes and humans—had advanced brain function even before the splitting off of the human lineage, but although apes are more intelligent than monkeys by various measures (Rumbaugh, Washburn, et al. 1996; Swartz, Sarauw, et al. 1999; Rumbaugh and Washburn 2003), it has not been easy to find unique features of their brains. As a group, they do not differ from monkeys in relative brain size, although chimpanzees lie on or above, and gorillas below, the anthropoid regression line.

Primates generally show a strong relationship between group size and the ratio of neocortex to the rest of the brain (Dunbar 1995, 1998, 2003), but a marked difference between the regression lines for monkeys and hominoids reflects a grade shift in the relationship (Dunbar 2003). The ratio is higher for all hominoids than for monkeys of similar group size but increases less as group size increases, probably because social behavior is more subtle and complex (McGrew, Marchant, et al. 1996). This does not rule out life history factors—group size may be one of the mortality-reducing lifestyles in the Sibly and Brown (2007, 2009) model, since the young are more protected (Janson and van Schaik 1993)—but social complexity may independently select for neocortex. In fact, group size explains 45 percent of the variation in primate neo-cortical ratios, while life span explains 47 percent, with only 3 percent of group size variation explained by life span (Allman 1999); thus almost 90 percent of the variation in primate neocortex ratio is explained by these two mutually independent variables. Furthermore, the "excess" of nonsensory, nonmotor neocortex shows an inflection point as the brains of apes and then humans are attained (Dunbar 2003).

In other ways too the primate brain did not always evolve propor-tionally. The cerebellum is 45 percent larger relative to brain size in hominoids than in monkeys (Rilling and Insel 1998), owing mainly to a marked increase in the relative size of the lateral or neocerebellum—another grade shift from the regression line for monkeys (MacLeod, Zilles, et al. 2003)—by the ancestor of apes and humans, perhaps 30 mya. The role of the neocerebellum in visual-spatial skills, planning complex movements, procedural learning, attention switching, and sen-sory discrimination in manipulation would have allowed ancestral apes to find and eat fruit, even while suspended from tree branches. These neocerebellar functions could easily be recruited in insect hunting, so-cial interaction, tool use, and other cognitive tasks.

Also, the ratio of cerebral white to gray matter increases with over-all brain size in anthropoids (Rilling and Insel 1999). This is predictable, since neuron number largely determines brain size, and connections should increase exponentially with neuron number. In fact, the white matter (axons) cannot keep up, so large brains have to be less internally connected (more modular and lateralized) than smaller ones. Ape brains are more lateralized than monkey brains (LeMay and Geschwind 1975; LeMay 1976; Gannon, Holloway, et al. 1998; Hopkins and Marino 2000); the planum temporale, strongly lateralized in humans and implicated in language, is also larger on the left in apes, but the arcuate fasciculus, a major fiber bundle connecting the temporal and frontal language ar-eas, projects to the temporal lobe only in humans (Rilling, Glasser, et al. 2008).

Remarkably, great apes and humans share a particular spindle-shaped neuron type in layer 4b of the anterior cingulate cortex (Nim-chinsky, Gilissen, et al. 1999), a part of the brain that has been implicated in consciousness and decision making. No other primate has been found to have these cells, and their volume correlates with encephalization; dolphins, whales, and elephants also have them. Although the precise function of spindle cells is unclear, that function seems to be shared by great apes, humans, and some other large-brained species, but not by other primates. Among primates, some features of the brain are shared by chimpanzees and humans, while others are unique to humans; this makes it possible to extrapolate back to the last common ancestor (LCA) of chimpanzees, bonobos, and humans. A comparison of the ape to the human and monkey brains is shown in Table 4.1 (Sherwood, Subiaul, et al. 2008 and other sources).

Behavior does not leave many fossil traces, but studies of living apes hint at the behavioral range of ancestral ones. Wild chimpanzees (*Pan troglodytes*) in the Gombe Stream Reserve in Tanzania, the Tai Forest of Ivory Coast, and elsewhere in Africa use and make tools in complex learned patterns (McGrew 1992; Wrangham, McGrew, et al. 1994). Approaching a termite mound, they strip the small branches from a twig and use it to fish out termites, a highly desired food. Juveniles' close observation of adults appears to facilitate learning. Also, adults crumple leaves by moderate chewing, then use the leafy mass as a sponge to extract water from the hollow of a tree or to get the last bits from the in-

Table 4.1. Brain Differences among Monkeys, Apes, and Humans

Feature	Monkey	Ape	Human	Adaptive Value
Visual motion cortex	✓	✓	✓	Track moving prey, etc.
Mirror-neuron system	✓	✓	✓	Imitation, empathy
Encephalization (brain/body)	✓	✓	✓✓✓	Cognition, learning
Cerebellum relative size		✓	✓	Movement integration
Facial motor nucleus		✓	✓	Facial expression
Orbitofrontal asymmetry		✓	✓	Emotion regulation
Frontal lobe relative size		✓	✓(✓)	Inhibition, memory
Planum temporale asymmetry		✓	✓✓	Language, cognition
Anterior cingulate spindle neurons		✓	✓✓	Decision making?
Temporal lobe relative size			✓	Language, memory
Frontal lobe gyrification			✓	Inhibition, memory
Nucleus ambiguus cortical projection			✓	Autonomic control

Note: Checkmarks indicate an increase over the group of origin (monkeys > prosimians, apes > monkeys, humans > apes). Adaptive values are simplified and indicate innovations or improvements.
Source: Based on Sherwood, Subiaul et al. (2008), Allmann (1999), and other sources.

side of a baboon skull after a kill. Related behaviors in the Tai Forest include using a pair of stones to crack nuts. These traditions in tool use and toolmaking have been called cultural (Whiten, Goodall, et al. 1999, 2001).

In an early experiment that demonstrated both tool use and cooperation, a life-sized model of a leopard roaring and turning its head was introduced to wild chimpanzees (Kortlandt 1967). They mounted a group assault on the model using large sticks as weapons, attacked either simultaneously or alternately, and threw their arms around each other in excitement after landing a blow. The emergence of hominin weapons may have been similar.

Whether or not these and other transgenerationally transmitted behaviors qualify as culture, they suggest some first steps our ancestors took toward it. *Ardipithecus ramidus, Australopithecus afarensis* (the "Lucy" species), and other early bipedal hominins may have had similar toolmaking and tool-use capacities, since they had chimplike cranial capacities, around 350 to 450 cubic centimeters (cc). It was long thought that upright posture and increased brain size coevolved (Darwin's view), but bipedalism preceded the start of hominin brain expansion by 3 to 4 million years. We do not know how to construe some size differences seen in similar hominin fossils. In some areas australopithecines could be male and female of one species with marked sexual dimorphism, as in gorillas, but most evidence points toward a low level of dimorphism in our ancestors, consistent with low male competition (Plavcan and van Schaik 1997; Reno, Meindl, et al. 2003). This is particularly true of the earlier *Ardipithecus ramidus,* in which males and females resemble each other in size and many other traits, including canine teeth, which are small in males as well as females (Lovejoy 2009; White, Asfaw, et al. 2009). If so, they may have been pair-bonding (Trivers 1972; Reichard and Boesch 2003), since such species typically have little dimorphism and bimaturism (the sex differences in growth that generate dimorphism), both moderate in our own species.

In chimpanzees, contrary to the predominant mode in Old World monkeys, adolescent females change groups to seek reproductive opportunities (Goodall 1968, 1986). Thus the female kin group, unlike in most primate groups, is not the core of chimpanzee society—although in one group three female generations were able to stay together, resisting female transfer. Because of monthly estrus with its dramatic genital display, the female plays a cyclical, active role in soliciting sex. She may go off with an individual male during estrus, by choice or coercion, and these "consortships"—like "sexual friendships" in baboons (Smuts 1985)—have implications for the evolution of human pair-bonding. After several years of adolescent infertility, females give birth at roughly five-year intervals, each birth beginning an intense, prolonged maternal relationship.

Cooperating groups of chimpanzee males hunt, kill, share, and eat mammals such as gazelles, small monkeys, and infant baboons; so, probably, did our ape ancestors (Stanford 1996, 1999; Stanford and Wrangham 1998). Tai Forest chimps and their main prey species, red colobus monkeys, have strongly influenced each other's evolution; for the apes, hunting plays a critical role (Stanford 2002). Kibale chimps, living in large groups, have especially active and varied hunting patterns (Mitani and Watts 1999; Pobiner, DeSilva, et al. 2007). Even a baboon group hunted for several years (Harding and Strum 1998). Hunting does not, however, support a "killer ape" model of human evolution (Ardrey 1961). Killing within species has little to do with predation; herbivores like red deer and bison often fight, and carnivore predation can resemble play.

Chimpanzees show violence toward conspecifics at both the individual and group levels; this, not hunting, is relevant to the origins of human violence (Wrangham and Peterson 1996; Watts, Muller, et al. 2006). But an equally close ape relative of ours, the bonobo, *Pan paniscus*, behaves differently (Stumpf 2007). Their population in the bend of the Zaire River has been studied for decades (Kano 1992, 1996), and so have captive groups (de Waal 1989). Anatomically bonobos may be closer to us than chimpanzees (Zihlman, Cronin, et al. 1978; Zihlman 1996), although the relatedness must ultimately be resolved genomically (Szamalek, Goidts, et al. 2006; Varki and Nelson 2007). Bonobos resemble chimps in some behaviors: females change groups and have long, close mother-offspring relationships, but unlike chimps, bonobo females form strong, eroticized relationships, including genital-genital rubbing (Paoli, Palagi, et al. 2006). Male-female ties are more persistent and benign in bonobos, and copulations more common and promiscuous (Takahata, Ihobe, et al. 1996); consortships do not occur (Stumpf 2007). Violence occurs at the individual and group levels but is far less frequent and extreme than in chimpanzees, and male dominance hierarchies, although present, are less significant, probably owing to the strength of female coalitions

(Kano 1992; White and Wood 2007). Hunting and toolmaking are not seen in the wild, although bonobos readily learn toolmaking skills from human teachers. (For a systematic comparison of chimp and bonobo behavior, see Stumpf 2007.)

Hominin Evolution and Behavior

After 7 mya, we have thousands of hominin fossils from at least nine different species. Molecular clocks suggest that our ancestor diverged from that of chimpanzees and bonobos between 6 and 8 mya (Rannala and Yang 2007), soon producing several upright species. At least three from this period—*Sahelanthropus, Orrorin,* and *Ardipithecus kabbaba*—are not human ancestors and are probably closely related despite their disparate names (Semaw, Simpson, et al. 2005). Just over 4 mya, there are at least two East African species. *Ardipithecus ramidus,* on or near the line leading to *Homo,* was apelike in some features but upright in posture and was probably a forest-dwelling, bipedal walker that could still easily climb trees with the aid of its grasping feet (White, Suwa, et al. 1994; Lovejoy 2009; White, Asfaw, et al. 2009). *Australopithecus anamensis* may also be in our lineage; it was apelike in both the upper and lower jaw, but the lower leg bones show it was bipedal (Leakey, Feibel, et al. 1998). The wrist bones suggest that it was recently descended from a knuckle-walking quadrupedal ape, suggesting that our lineage may have been terrestrial before it was bipedal (Richmond and Strait 2000). However, *Ardipithecus ramidus,* known as "Ardi," is represented by a much larger number of specimens (one quite complete), studied in great detail (Lovejoy 2009; White, Asfaw, et al. 2009). Ardi's skeleton gives no support to the idea of a knuckle-walking stage and in fact suggests that the living apes are much modified evolutionary offshoots from the last common ancestor between us and them.

Australopithecines Evolving

At Laetoli in Tanzania at about 3.6 mya, in what may be evidence of an australopithecine family, footprints in the hardened mud resulting from a deposit of volcanic ash indicate two or three upright-walking hominins (Leakey and Hay 1979; White 1980), an adult and one or perhaps two juveniles, one of which may have left footprints inside the adult's. The footprints as well as many fossil bones in the general area belong to *Australopithecus afarensis,* also known from Ethiopia and, like *A. anamensis,* almost fully bipedal (Johanson and White 1979).

The adaptive functions of upright posture are uncertain; hypotheses include freeing the hands to carry food or infants, seeing farther in the savanna, exposing less body surface to direct midday sunlight, and sexual display of breasts or genitals (Klein 2009). Human bipedal walking uses less energy than the chimpanzee's four-limbed gait (Rodman and McHenry 1980). Causes aside, the consequences were formative. The long, slender ape pelvis shortened and thickened for weight bearing, resulting in difficult births due to cephalopelvic disproportion, a nexus for natural selection through maternal and infant mortality (Trevathan 1987; Rosenberg 1992; Rosenberg and Trevathan 2001, 2002). This change played a major role in the evolution of hominin females and of our species (Zihlman 1981; Hrdy 1999; Jolly 1999).

Upright posture does enable carrying, which is largely restricted to humans among living primates. Apes occasionally walk on two legs for short distances holding food or infants, but for early hominins this would have been routine. Ape mothers can walk tripedally, supporting very young infants with one hand; this posture suggests an advantage for hominins, especially if body hair was sparse and the infant's clinging ability weak. In fact, ape newborns do not cling as well as their monkey counterparts, and the transient tripedalism of ape mothers may have been a transition to upright walking.

We have no clear evidence that *Australopithecus afarensis* used tools, but the natural tool use and toolmaking of chimpanzees, along with bonobos' ability to learn how to make stone tools, suggests that early hominins had some toolmaking capacity. Chimps and bonobos present very different possible models of the social behavior, hunting, and tool use of the apes ancestral to hominins; probably both species and other apes as well are relevant. Savanna-dwelling chimps may be an appropriate model (Moore 1996); this context would have presented growing challenges in protecting the young from predation. Larger groups of chimpanzees show distinctive patterns of social behavior, aggression, and hunting, and these are probably relevant for understanding ancestral hominins (Mitani, Watts, et al. 2002; Mitani 2006). Toolmaking often involves observational learning by juveniles; local variation can emerge, become stable over generations, and form traditions (McGrew 2002; Whiten 2007).

Homo *Rising*

From 2.6 mya, we have definite evidence of stone toolmaking in chopper or pebble-tool assemblages, along with deliberately made cut marks on mammal bones (Semaw, Rogers, et al. 2003), probably preceding but fostering the earliest emergence of the genus *Homo.* The prevailing view for some time was that our early ancestors were mainly scavengers and that hunting was not an important influence in subsequent hominin evolution (Shipman 1986). But most carnivores, including lions, hyenas, and human hunter-gatherers, do both, and chimpanzees hunt avidly in the wild (Teleki 1973; Stanford and Wrangham 1998; Pobiner, DeSilva, et al. 2007), with groups of males coordinating ambushes of small monkeys and other prey. Sharing of meat occurs mainly among males immediately after a kill and may just be tolerated theft, but it could have led to human economic exchange (Stanford 2002). The occasional use of meat by males to obtain sexual favors may suggest the beginnings of the human division of labor by sex, and perhaps of pair bonding.

The numerous hominin fossils from 3 to 1 mya range from species with large, winglike cheekbones, huge molar teeth—*Paranthropus,* successful for over a million years—to a relatively gracile, or slight-boned, form with a bigger brain and delicate face that belonged to our genus, *Homo* (Johanson, Edgar, et al. 2006; Klein 2009). Its earliest members shaped the pebble tools—oval rocks with one rough edge made by two or three blows of another rock. Modern experiments show that these simple tools make butchering a large animal carcass far easier (Isaac and McCown 1976; Pfeiffer 1978) and reveal the substantial cognitive and neural capacities required to make them (Toth, Schick, et al. 1993; Stout 2002; Stout and Chaminade 2007; Stout, Toth, et al. 2008). They may have first been used in butchering, which was crucial to either scavenging or hunting; several million years of chimpanzeelike group hunting could have led gradually, through selection for toolmaking ability, to the pebble-tool tradition (Wynn 2002).

At least five different hominin species lived between 2 and 3 mya (Johanson, Edgar, et al. 2006; Klein 2009). Two, *Australopithecus afri-*

canus and *Paranthropus robustus,* represent the gracile and robust australopithecines—upright walkers with ape-sized brains. Both had small brains (under 500 cc) and low brain to body weight ratios, but *Paranthropus* had much larger jaws and teeth for grinding and crushing leaves and seeds, with a crest on the top of the skull for the attachment of large temporalis (chewing) muscles. Dietary variation within each of these species (Scott, Ungar, et al. 2005) supports the hypothesis that hominins were adapted to variability itself (Potts 1998). Climatic variation was continual in this period in Africa; thus hominins were not designed for one climate but for transformations between wet and dry, between hotter and cooler climates. These fluctuations profoundly affected the spectrum of available plants and animals. But it was another species, *Homo habilis,* or "able person" (Leakey, Tobias, et al. 1964), with a small jaw and relatively flat face but a larger cranial capacity of 650 cc (Holloway 1974), that was the likely stone toolmaker and direct human ancestor.

There were still three species in East Africa just after 2 mya, but while the small-brained forms had changed little, *Homo* had evolved further. Cranial capacity had increased another 200 cc, and the face was more modern; this species, *Homo ergaster* ("working person"), gave rise to *Homo erectus,* the last major species before *Homo sapiens.* It ultimately spread throughout Africa, Europe, and Asia, had a cranial capacity close to 1,000 cc, and was highly skilled in stone toolmaking—not crude pebble tools but the much more sophisticated hand axes. Invented by *Homo ergaster* in Africa, they stayed almost uniform in design throughout the world for a million years. By this time in evolution, infant brains had enlarged enough to greatly exacerbate the already established cephalopelvic disproportion of early hominins (Rosenberg and Trevathan 2002). Some models place more species on the direct human line, but they need not concern us. Fossils punctuate the record, but it is continuous. As the record improves, we may discover how stagelike the process was; a transition invisible in the known fossil record, however, could still have taken thousands of generations, so the naming of "missing links" is arbitrary.

The rest of the sequence is simpler (Wood and Richmond 2000; Wood 2005; Klein 2009); by 1 mya, *Homo erectus* had a range throughout the Old World. With one possible exception—an apparent dwarf variant found on the island of Flores in Indonesia (Brown, Sutikna, et al. 2004)—this species was relatively uniform in brain size, facial form, body size, and stone tool culture. Around 500,000 years ago (500 kya) the earliest humans emerged, sometimes called *Homo heidelbergensis* (Rightmire 2004). Found in Africa, but also at Swanscombe, England; Steinheim, Germany; and elsewhere, these hominins had shorter faces, smaller brows, and a thinner, higher skull than *Homo erectus.*

The European forms led, 100,000 years later, to the quite different, skeletally robust Classic Neanderthals (Stringer and Gamble 1993; Trinkaus and Shipman 1993; Tattersall 1998). Long-bone diameters and muscle insertion points indicate a body made to bear great stresses. The legs were adapted for running, climbing, and weight bearing, and the arms for throwing or striking blows. Sturdiness is evident in Neanderthal skeletons by age five, so these features were not the result of a strenuous life, but a genetic adaptation. The only thinner bones are in the pelvis, probably a response to selective mortality from cephalopelvic disproportion, which had already worsened for 3 to 4 million years and was exacerbated in Neanderthals by infant robustness.

Meanwhile, in Africa, *H. ergaster* gradually evolved into anatomically modern *H. sapiens* (McBrearty and Brooks 2000; McBrearty 2007), which was established by 200 kya and left Africa less than 100 kya, replacing *H. erectus.* This, our species, had spread across Asia and into Australia by 45 kya, and by 30 kya the Neanderthals were replaced in Europe by fully modern humans from Asia, within only 5,000 or 10,000 years. By at least 15 kya, modern humans in Siberia had extended their population into North America, and by 10 kya throughout the Americas.

So modern humans had evolved in Africa by 200 kya or earlier and migrated throughout the world between 100 and 10 kya. This out-of-Africa model, based on mitochondrial and Y-chromosome DNA and confirmed with whole-genome studies (Cavalli-Sforza and Feldman 2003; Kidd, Pakstis, et al. 2004), underscores the genetic unity of the species. Since mitochondria are transmitted only through females, it has been argued that all humans carry genes from an African woman, known informally as Mitochondrial Eve. This version of the claim is dubious (Ayala 1995; Satta and Takahata 2002), but clearly our species passed through a bottleneck of no more than a few hundred thousand individuals (Ayala and Escalante 1996), and possibly many fewer. Other genetic data support the unity model. Genetic research has consistently shown that more than 90 percent of all human variation is contained within any major racial group (Lewontin 1972; Brown and Armelagos 2001). These findings have important implications for the search for universals of human childhood behavior and culture, as we will see in Parts II and IV.

Hominin Brain Evolution

Great apes have brains around 450 to 500 cc, and the australopithecines are in this absolute size range. It is not clear whether the increase in brain size in early *Homo erectus* (Leonard, Robertson, et al. 2003; Rightmire 2004) over the last 600,000 years was gradual (Cunnane and Crawford 2003; Lee and Wolpoff 2003), punctuated—little change followed by a steep increase—or both, but brain size did expand to the current human average of 1,350 cc (Klein 2009). Although variations in normal brain size in the modern human population only weakly predict intelligence (Tramo, Loftus, et al. 1998; Pennington, Filipek, et al. 2000), such predictions are valid when comparing different species, especially after correcting for body weight. But since in our present view absolute size also matters (Striedter 2005), the increase in brain size *with* body size over the first million years of the genus *Homo* may have made more psychological difference than we would guess from relative brain size.

The brain reorganized as it expanded (Holloway 1966; Holloway, Broadfield, et al. 2003b; Kaas 2006), as it had many times before in brain evolution (Striedter 2005). It is clear that there are significant differences in neocortical organization (Preuss 2001) and cerebral gene expression (Preuss, Caceres, et al. 2004) between extant apes and humans, and we will return to them. The convolutions and blood vessels of hominin brains leave impressions inside skulls, and some features absent from ape brains are found in *Australopithecus,* suggesting that *A. africanus* could have been ancestral to *Homo* (Falk, Redmond, et al. 2000). Sophisticated imaging is gleaning much more information about hominin brains from fossils (Falk 2004a).

Comparative imaging of living nonhuman primate and human brains yields interesting results and controversy. On MRI, the human neo-cortex is large for our brain size (Rilling and Insel 1999); it has been ar-gued that the frontal lobe is not (Semendeferi and Damasio 2000), but rather follows general primate scaling: a larger-than-expected frontal to other neocortex ratio than is the case for other mammals (Bush and Allman 2004), even the relatively large-brained carnivores. However, there is growing evidence that the human brain is not just an allometri-cally scaled-up version of the anthropoid or even the ape brain (Rilling 2006).

The human frontal lobe is more convoluted than expected for brain size (Rilling and Insel 1999) and may exhibit specific cytoarchitectonic departures (Sherwood, Stimpson, et al. 2006; Sherwood, Subiaul, et al. 2008). Although apes have smaller temporal lobes for their brain size than monkeys, humans exceed apes in relative overall temporal lobe size and especially in the ratio of temporal white to gray matter, prob-ably owing to greater demand for visual pattern learning, social learn-ing (including face recognition and facial expression reading), and lan-guage (Rilling and Seligman 2002). Compared with monkeys, humans and other apes share an elaboration of the cerebellum and the frontal lobes and probably of the circuits connecting them (Rilling 2006). Hu-mans add to that a larger proportion of neocortex, with disproportion-ate further enlargement of prefrontal and temporal association areas; increased connectivity of the cognitive association areas generally with the cerebellum; and probably increased intracortical connectivity in the prefrontal areas.

Fossil endocasts can also shed some light on brain function. For ex-ample, the primary visual cortex of higher primates is demarcated by a groove called the lunate sulcus; in australopithecines it is not as far for-ward as in average apes, but humanlike in its position, implying more association neurons (Holloway, Broadfield, et al. 2003a). Examination of two chimpanzee brains that differed from their species norm in the human direction show that the internal organization of their primary visual cortex also differs in a human direction, so the reorganization could have begun by the australopithecine stage. We can describe ana-tomical advances made by the modern human brain over the extrapo-lated LCA that we share with bonobos and chimpanzees.

Molecular analysis using oligonucleotide microarrays shows that human brains have substantial up-regulation of genes compared with ape brains, suggesting that humans have considerably more neuronal activity (Preuss, Caceres, et al. 2004). There are 169 genes with distinguishable expression levels, 91 of which are unique to humans (Caceres, Lachuer, et al. 2003). Other differences appear in other tissues, but these are brain-specific. A review of five studies using this method (Preuss, Caceres, et al. 2004) concludes that: (1) the rate of change in brain gene expression has accelerated in human evolution; (2) 2 to 4 percent of genes expressed in the brain are affected; and (3) the changes mainly involve increased expression.

Several up-regulated genes are involved in synaptic transmission (two in excitatory glutamate synapses), one in the role of astrocytes supplying lactate to neurons for energy, several in lipid metabolism, several in protecting neurons (potentially more important in a long-lived species), one in learning and memory, and two in mutations linked to mental retardation (Caceres, Lachuer, et al. 2003). As we will see, other genes distinguishing humans from apes include two that influence brain size and another involved in vocalizations and possibly speech. But most distinctively expressed human genes have unknown functions; many are probably involved in brain development, with specific functions in generating neurons and assembling circuits. However, some up-regulated genes appear to have the function of raising the brain's energy level above that expected for so large a brain. As Todd Preuss (2009) has put it, the human brain, compared with the ape brain, now appears to be not just much larger, but "rewired and running hot."

It seems clear that the evolutionary expansion of the hominin brain required an increase in energy-processing efficiency that depended on diet. The first step toward this improvement would have been the restriction of the range of apes to environments where fruit-eating was essential (Potts 2004), and the second would have been an increasing reliance on animal foods (Leonard, Robertson, et al. 2003). This second step may have greatly increased in importance with the invention of cooking, which probably had occurred by 790 kya, when we have the first evidence of fire under hominin control (Goren-Inbar, Alperson, et al. 2004) and the use of advanced butchering techniques (Rabinovich, Gaudzinski-Windheuser, et al. 2008), but it could have been earlier and may have had as much to do with preparing plant foods as flesh (Wrangham and Conklin-Brittain 2003; Wrangham 2009). Long-chain polyunsaturated fatty acids, much higher in lake and ocean fauna and in the birds and mammals feeding on them than in inland species, may have specifically been needed for the development of a large hominin brain, especially at fetal and infant growth rates (Broadhurst, Wang, et al. 2002); hominins often lived on lakeshores and spread out of Africa along coastlines.

We know that the human gut is almost as reduced for a primate of our size as the human brain is increased (Aiello and Wheeler 1995). Extrapolating from primate regression lines, our gut should weigh about 1,800 grams (g), but in fact weighs about 1,100; our brain should weigh less than 500 g but weighs over 1,300. The "expensive tissue" hypothesis of Leslie Aiello and Peter Wheeler posits a necessary trade-off between these two tissues, both in size and function. Gut size could be reduced only as foods increased in quality, energy, and digestibility—first fruit, then flesh, then cooked flesh, and finally aquatic animals—and such foods can be obtained only by animals with better brains. (We also have larger-than-expected hearts, which may compensate for the difference between gut loss and brain gain, especially given the brain's disproportionate energy demands during infancy.)

Recall that the Sibly and Brown life history model calls for higher productivity of offspring mass per year with a richer and more stable food base; this was part of the basic hominin adaptation. However, *Homo erectus* and *Homo sapiens* extended this, in part through the invention of cooking (Wrangham 2009). *Homo sapiens,* which even when hunting and gathering was more productive of offspring than great apes, went on to exploit coastal food resources (shellfish) at least by 164 kya (Marean, Bar-Matthews, et al. 2007) at the southern tip of Africa; they must have been thoroughly adept at using these resources as they spread along the Asian coast 100,000 years later. Sibly and Brown (2007) note that mammalian marine carnivores are much more productive of offspring than their land counterparts, and they attribute this difference to the food base; they also point out that the crab-eating fox is highly productive compared to other canines. Shellfish provide densely concentrated energy from protein along with valuable omega-3 fatty acids, so both cooking and marine food exploitation may have been crucial to our species' exceptional reproductive and geographic success. As we will see in Chapter 26, children probably played an important role in this process.

Evolving Human Life Histories

Compatible with the Sibly and Brown model (despite having preceded it) is an important conceptualization of the evolution and uniqueness of the human life cycle, that of Hilliard Kaplan, Kim Hill, Jane Lancaster, and Magdalena Hurtado (2000). We know that among anthropoid primates generally, nonmaternal care is correlated with earlier weaning, shorter birth spacing, and higher reproductive success (Mitani and Watts 1997; Ross and MacLarnon 2000), so it seems reasonable to expect this to be true of apes and humans. Kaplan and his colleagues synthesize previous work and thought with new data on hunter-gatherers to extend an earlier model, which held that parental provisioning after weaning is the key to our species' success (Lancaster and Lancaster 1983; Lancaster and Lancaster 1987). Combined with the cooperative breeding hypothesis (Hrdy 2009; see Chapter 17), it explains key differences between humans and chimpanzees. Postweaning survival in several hunter-gatherer populations is much higher. That gap in postweaning mortality combined with our earlier age of weaning and shorter birth spacing accounts for the far greater success of our species. They argue

> that the shift to calorie-dense, large-package, skill-intensive food resources is responsible for the unique evolutionary trajectory of the genus *Homo*. The key element in our theory is that this shift produced co-evolutionary selection pressures, which, in turn, operated to produce the extreme intelligence, long developmental period, three-generational system of resource flows, and exceptionally long adult life characteristic of our species. (Kaplan, Hill, et al. 2000:161)

Their overview of the dietary change is straightforward. As we saw in the last chapter, fruit-eating primates have larger relative brain sizes than leaf-eaters; not only are chimpanzees excellent fruit foragers, but they add termites and some meat. Human hunter-gatherers go much further in their dependence on extracted foods and game. To the dimensions of nutrient density and skill requirement we add increasing brain size: "Although, as compared to humans, chimpanzees engage in relatively little extractive foraging and hunting, they do much more than monkeys. In this sense, their superior intelligence and greater encephalization . . . illustrates the same evolutionary forces that separate humans from apes" (176).

Because of their heavy reliance on data from neotropical foragers, Kaplan and his colleagues underestimate the importance of plant foods (Eaton and Konner 1985; Eaton, Eaton, et al. 1997), the contribution of grandmothers to subsistence and child care (see Chapter 17), and the contribution of children and adolescents to their own subsistence (Chapter 26); they correspondingly overestimate the importance of men and meat. However, these do matter. It seems implausible that data on extant hunter-gatherers will ever be able to resolve these controversies decisively, but a cooperative breeding model that combines the varied contributions of mothers, fathers, grandmothers, and children within a framework emphasizing the importance of provisioning and the acquisition of skill during a long childhood seems to explain the most important features of human evolution.

Hominin Behavior, Social Organization, and Culture

It is difficult to make inferences about the social and cultural lives of the australopithecines, or even about the tools they used or made. None of the extant chimpanzees' tools—termite sticks, leaf sponges, nut-cracking stones, or logs as clubs—would be preserved as fossils, and some of the most important tools used by modern hunter-gatherers— the digging stick, for instance, a sturdy, smoothed, pointed stick for getting at buried roots, the sack for carrying vegetables, and the infant sling—are perishable. Such tools were at least as pivotal in the emergence of culture as stone tools and weapons, and the low profile of women in human origins research due to neglect by male scholars has been abetted by the fragility of their artifacts (Dahlberg 1981; Lancaster 1991).

The notion that tool use and toolmaking are a cause more than a consequence of human brain evolution was hinted at by Darwin (1871/1981) and embraced by many anthropologists, who argued that the use, manufacture, and design of tools advantaged the more intelligent (Washburn 1960; Lancaster 1968). Some have mastered stone toolmaking in order to understand hominin tool assemblages and make inferences about their mental abilities, which at some stage exceeded those of apes

(Schick and Toth 1993; Toth, Schick, et al. 1993). This point is underscored by work with an enculturated, language-using bonobo named Kanzi (Savage-Rumbaugh and Lewin 1994). Bonobos have not been seen to make tools in their natural habitat, but archaeologist Nick Toth taught Kanzi to make stone flake tools when he needed them and to use them instrumentally, such as in cutting a rope to get to a banana. However, Kanzi has not modified the process he was taught.

There is little doubt that at some point in the evolution of toolmaking, teaching and language became critical to cultural transmission. Dietrich Stout's research on the acquisition of stone toolmaking skills and use of positron-emission tomography (PET) imaging on their neural representation have shown important differences in brain activation between naive and skilled stone knappers and between making the simplest chopper tools and making a hand ax. Naive knappers can make chopper tools without activating planning circuits in the frontal cortex (Stout and Chaminade 2007), but more advanced toolmaking by expert knappers involves parietal-frontal circuits bilaterally as well as the right hemisphere homologue of Broca's language area: "The observed patterns . . . suggest that toolmaking and language share a basis in more general human capacities for complex, goal-directed action . . . consistent with coevolutionary hypotheses linking the emergence of language, toolmaking, population-level functional lateralization and association cortex expansion in human evolution" (Stout, Toth, et al. 2008:1939).

Together with ethnographic evidence from present-day traditional stone adze makers in Irian Jaya (Stout 2002), who teach the skill with a constant stream of talk—direction, correction, instruction, explanation—and with the dangers inherent in stone knapping (flying chips, bleeding hands, broken fingers), the brain-imaging evidence makes it extremely likely that not only speech but teaching had to have evolved before or as the first complex stone tools were made. This is consistent with evidence that teaching occurs in all human cultures (see Chapter 27) and with the notion that our species could be called *Homo docens* because we have something like an instinct to teach (Barnett 1968), which could have emerged with the genus *Homo* some 1.6 mya.

Since there are bones bearing cut marks of crude chopper tools dating from 2.6 mya, and since chimpanzees hunt, it seems clear that meat eating preceded the genus *Homo* and the making of complex tools. However, australopithecines may have gotten their meat more by scavenging than hunting. If our ancestor *A. afarensis* had low levels of sexual dimorphism, this species may have been mainly pair-bonding 3 mya (Plavcan and van Schaik 1997; Reno, Meindl, et al. 2003). Since the considerably earlier *Ardipithecus ramidus,* at 4.4 mya, clearly has low sexual dimorphism, the possibility of pair-bonding should apply equally or more to them (Lovejoy 2009; White, Asfaw, et al. 2009). If so, sharing scavenged or hunted meat and provisioning the young may already have been in place.

Hunting was prominent by the time of *H. erectus,* whose pursuit of large game with an effective but limited tool-kit could only have been accomplished through cooperative hunting. They probably had a low population density with flexibly organized nomadic groups and slowly developing young that needed intensive care. Almost certainly mothers were responsible for that care, as well as for gathering wild fruits and vegetables, but males would probably have contributed meat. At some point, fire came under control and was used in cooking (Wrangham 2009). This must have softened and detoxified some plant foods and increased the appeal of meat, making certain parts of the carcass more easily chewed and digestible and relaxing natural selection on the hunter's teeth, jaws, and face. Controlled fire also dramatically increased protection from predators, especially at night, and may have been crucial to our ancestors' success. As language emerged, fire would have enabled nightly discussion of a day's events, plans for the next day, important events in the personal and cultural past, myths and tales, and future possibilities for the band, as in recent hunter-gatherers (Marshall 1976, 1999; Shostak 1981; Biesele 1993). All of this went on in the presence of children—alert or drowsy, frightened and comforted, watching and learning. At some point, control of fire also meant an advance in human culture: rituals centered on a steady, managed blaze (Lee 1982; Konner 1985). Extending cultural life beyond the daylight hours must have favored intelligence, speech, and thought. Fire also helped create habitable spaces; in what is now Israel, hearths were in use 790 kya (Goren-Inbar, Alperson, et al. 2004), 500 kya before the dawn of our own species.

There is evidence of collective, large-scale big-game hunting in Spain at around that time (Howell 1968; Klein 2009). The bones of many woodland elephants, accumulated quickly, reflect human killing and butchering. Together with evidence of grass fires, this suggests collective elephant driving. It also suggests higher population densities than among most hunter-gatherers (consistent with Sibly and Brown's productivity model) and therefore the possibility of social hierarchies, as among Native American foragers of the Northwest coast, who also had an exceptionally rich resource base of marine animals (Codere 1950).

Meanwhile, early *H. sapiens* in Africa was creating an ever-more-complex tool technology in a gradual process beginning around 300 to 200 kya—the real human revolution. Earlier generations of anthropologists believed that a revolution occurred in Europe between 45 and 35 kya, but in fact it took not 10,000 years but more than 100,000. In contrast to the old Eurocentric view, our African cultural revolution was a slow burn that gradually produced more finely worked stone and bone tools, exploitation of aquatic resources, and the increasing use of pigment for decoration and art (McBrearty and Brooks 2000; McBrearty 2007). An early human group collected shellfish 164 kya in southern Africa; they also had a variety of pigments, suggesting the possibility of symbolic or ritual use, and a fairly sophisticated tool-kit (Marean, Bar-Matthews, et al. 2007). In addition, they had begun to use fire (subdued by hominins long before) to glaze or otherwise prepare stone for more effective flaking (Brown, Marean, et al. 2009).

According to some models, cultural advances should be lost below a certain population density but preserved above it, as a critical number of skilled and knowledgeable individuals is reached (Henrich 2004; Powell, Shennan, et al. 2009; Mace 2009). This level was attained in Europe about 45 kya, but also once before, in Africa, at 90 to 100 kya; remarkably, this was a period of cultural advance with many of the technological features found in Europe much later. But populations of African hunter-gatherers waned thereafter owing to declining environmental quality, and so the cultural florescence there may have been transient, as it was much more recently among Tasmanian hunter-gatherers (Henrich 2004). But it was a logical extension of the long prior process of African cultural evolution to produce a major advance when the population threshold was crossed.

Neanderthal stone tools were also an advance over *Homo erectus* hand axes (Bordes 1968; Stringer and Gamble 1993). Statistical analysis reveals the complexity of tool assemblies, and scanning electron microscopy shows how they were used; the skills were as advanced as those of many modern humans, but their bodies were exceptionally robust. Their activities would have included hunting big game, carrying heavy loads, walking steep terrain, and probably engaging in combat, any of which could also account for a second distinctive feature of Neanderthals: the frequency of injuries. This is especially true at the Shanidar site in Iraq (Trinkhaus 1978; Trinkhaus and Howells 1979). Although the source of most of the injuries is unknown, modern rodeo cowboys have similar patterns of injury (Berger and Trinkhaus 1995), suggesting that close encounters with large, powerful animals could have done the damage.

But fighting is another possibility. At least one Shanidar male skeleton shows evidence of violent death (Trinkhaus 1995): a partly healed scar on the left ninth rib caused by a sharp object thrust into the chest, either an accident or a deliberately inflicted spear wound. It is the clearest such evidence for Neanderthals, and along with their high rate of injury it raises the possibility that violence was common. Cannibalism occurred among Neanderthals at 100 kya (Defleur, White, et al. 1999); it may be much older than that (Fernandez-Jalvo, Diez, et al. 1999), and it has persisted in recent humans (Sanday 1986; White 1992; DeGusta 1999; Billman, Lambert, et al. 2000). Some models of human evolution posit that warfare was common among Pleistocene hunter-gatherers (Bowles 2009). Cannibalism need not be associated with conflict, but it often is.

Modern hunter-gatherer cultures have elaborate hunting beliefs and rituals (Marshall 1999). Among the Mbuti of the Ituri Forest, young men learn to kill elephants; a successful hunt promotes solidarity and occasions a day of ritual and narrative (Turnbull 1962). The Siriono of eastern Bolivia rub their bodies with the feathers of a harpy eagle they have killed, to absorb its power (Holmberg 1969). An insightful book on !Kung myths was called *Women Like Meat* (Biesele 1993), and there is evidence from two other hunter-gatherer groups, the Hadza and the Ache, that the appeal of meat leads women to choose successful big-game hunters, even when the meat is widely shared (Hawkes 1991a, 1991b). Even many non-hunter-gatherer cultures made hunting central to rituals of manhood (Gilmore 1990), and the hunt is celebrated in the art and literature of great civilizations (Cartmill 1993).

Neanderthals may also have had systematic rituals (Maringer 1960; Bordes 1968), including the first deliberate burials of the dead (Trinkhaus 1978; Trinkhaus and Howells 1979; Klein 2009). At La Ferrassie, a rock shelter dated to 60 kya was apparently a family cemetery, with remains including those of a man, a woman, two children about five years old, and two infants. The man and woman were laid head to head, the children set at the man's feet, and one of the infants, a newborn, interred with three well-crafted and no doubt valued flint tools. At one Near Eastern site a boy's grave was ringed by pairs of goat horns, and at another a man with a crushed skull may have been buried with flowers (Solecki 1971).

Analysis of five complete Neanderthal mitochondrial genomes (Briggs, Good, et al. 2009) indicates that at about 40 kya, when they encountered modern humans, they were a fairly homogeneous population of about 70,000 spread across Europe. Biological and cultural evolution had continued in Africa and, after 100 kya, beyond it, when some modern humans left Africa and began to populate the world. They would have displaced almost all the remaining populations of *H. erectus* and, after 40 kya, the Neanderthals, with or without violence. Their spread across Asia, probably along coastlines, was relatively rapid; their coastal subsistence ecology—some fishing, shellfish collecting, and exploitation of the land fauna feeding on the fish—did not need to change dramatically during the 20,000 years or less it took to reach Australia.

Modern humans entering Europe (probably after a time in Asia) had an unprecedented array of tools and weapons of stone, bone, antler, and ivory: spear points, lances, knives, chisels, needles, tools for scraping, perforating, sawing, whittling, pounding, even iron pyrite for making fire. They had sophisticated ivory and bone flutes capable of making good music (Conard, Malina, et al. 2009), and most impressively, they created cave art. Neanderthals had no major plastic art other than ornaments, but the people who replaced them did.

In our passion for wildlife we have nothing over our ancestors. Altamira in northern Spain has a herd of colorful bison on the ceiling, along with a horse, two deer, and some wild boar (Breuil 1952); at Font de Gaume in France a reindeer male sniffs the head of a kneeling female; at Trois Frères a chaotic group of big game is engraved, along with a pair of snow owls guarding their chick and an odd figure with antlers, staglike ears and body, human hands and feet, a prominent human phallus, a horse's tail, and striking eyes in a mouthless, bearded face. Chauvet in southern France, the oldest site at 30 to 33 kya (Chauvet, Deschamps, et al. 1996; Gombrich 1996), shows realistic lionesses, bison, and rhinoceroses, and the theme of large, dangerous animals links it to another cave at Arcy (Balter 1999), but there are also stereotyped sketches of wooly mammoths and parts of humans (two pubic triangles with vulvas) as well as stripes and circles of dots of unknown meaning. Lascaux has a yellow and black horse beset by flying arrows, a herd of ornately antlered little stags, a horse with a fluffed mane, two bison tail-to-tail, a large red dappled cow, and a five-meter-long, big-eyed black bull.

The meaning of these works is unknown, but speculation has focused on the small size of the handprints that occur in many places; perhaps the paintings were made with the help of children and adolescents, or at least some rituals involving them may have occurred in the painted caves. Gaining access to these caves took great effort—some could be reached only by crawling through cramped spaces—so it has been proposed that the caves may have been part of a kind of ritual theater. Perhaps an adolescent initiation rite took place there, with a torch-lit, obstacle-filled approach as the prelude (Pfeiffer 1982). But inaccessibility may have had more to do with secrecy—hunting magic, totemism, shamanism?—or just decoration (Sieveking 1979). This novel art required no new advance in brain function, but it was predicated on the prior existence of a fully modern human brain (Klein and Edgar 2002). Still, it entailed a cultural level that took even *Homo sapiens* many thousands of years to arrive at and required higher population densities, collaborative learning, and cumulative traditions. This was also, perhaps, a level at which teaching children important life lessons had become central to their development and worthy of great art, among the most precious of resources.

5

The Evolution of Human Brain Growth

Having reviewed the relevant scientific paradigms and the history of human brain and behavior, we proceed to the heart of the evolutionary story for our purposes, the evolution of brain growth and development. As the pioneering cognitive psychologist Jerome Bruner (1972:1) once said, "The nature and uses of immaturity are themselves subject to evolution, and their variations are subject to natural selection."

The classic description of the nature of human growth, by James Tanner (1962), was an idealized picture of four curves with very different shapes corresponding to four categories of organ systems. General body growth and reproductive organ growth have different S-shaped curves, the brain and head have a power curve, and the lymphatic system looks like a partial inverted V. It is likely that these are heterochronically independent, adjusted in evolution to change the adult outcome as well as adaptedness at each developmental stage. The lymphatic system waxes into middle childhood as it meets antigenic challenges; after they are met, it partly involutes. The result is declining vulnerability to infection from birth to adulthood. Reproductive growth lags body growth, enabling us to end larger and leaving more time for learning. The steep brain growth curve results from a relatively shortened gestation superimposed on a declining capacity for neuronal division as connectivity stabilizes.

Neonatal Status and Early Brain Growth

Brain growth in mammals takes place mainly during fetal and early postnatal life, but it is variable. As we saw in Chapter 3, the main determinant of the size of a mammal's brain is the number of doublings it goes through in fetal life (Finlay and Darlington 1995). This observation helped explain a finding made two decades earlier. In an important refinement of the allometric associations of gestation length, the cube root of brain weight was shown to be a substantially better predictor of gestation length than the cube root of body weight (Sacher and Staffeldt 1974). Regression of the cube root of neonatal brain weight on gestation time yields a markedly tighter distribution for ninety-one species of mammals. Fetal body growth rates vary over a tenfold range in this sample, but fetal brain growth rates vary only one-fourth as much, or two and a half times. Multivariate analysis including neonatal body weight, neonatal brain weight, litter size, and the advancement factors for neonatal body and brain weight—that is, the respective neonatal to adult weight ratios—confirms the special role of neonatal brain size in explaining gestation length.

In addition to absolute neonatal brain weight, the proportion of adult brain weight attained by birth—the *advancement factor,* correlated with precociality—explains some of the residual variation, while placental anatomy does not. Although the lower limit of gestation time for a given neonatal brain weight is quite constrained in evolution, ecological conditions can lengthen gestation to produce relatively precocial newborns with high brain advancement factors (Sacher and Staffeldt 1974). Given the altriciality of monotremes, marsupials, and insectivores, altriciality must have been the ancient mammalian condition, with derived precociality evolving in some lines, including primates, and secondary altriciality occasionally evolving against that precocial background, as in humans. However, as we saw in Chapter 1, the altricial-precocial dimension can vary even in closely related species and is not tightly linked to taxonomic group or adult brain size.

So we can view the timing of birth as a separate trait, partly independent of rates and durations of development. Birth may occur in a mammal before (rats and rabbits), during (pigs and humans), or after (guinea pigs, sheep, and monkeys) the main brain growth spurt. Kathleen Gibson (1991) showed that the brain advancement factors of newborn mammals range from less than 10 percent for newborn tigers, bears, dogs, tree shrews, and mice to more than 50 percent for guinea pigs, zebras, rhesus monkeys, and gorillas; humans are in the middle of the range.

This large difference in postnatal brain growth multiples could have important implications for nutrition and other kinds of vulnerability and plasticity in early life (Dobbing and Sands 1979).

For example, squirrel-like rodents (including rats, mice, chipmunks, and marmots) are almost always altricial, while beavers and porcupine-like rodents (including cavies and capybaras) are precocial (Nowak 1991), yet the precocial cotton rat (*Sigmodon hispidis*) is an exception to even this generalization (Sacher and Staffeldt 1974). Aquatic carnivores like seals, sea lions, and walruses are very precocial, while the land carnivores are altricial. The highly altricial rabbit newborn, which expands its brain sevenfold after birth, and the highly precocial hare, which is at birth almost halfway to adult brain size, are sister genuses (Nowak 1991). Humans, with our intermediate multiple, are precocial in sensory but altricial in motor systems.

Humanizing Anthropoid Brain Growth

Looked at more closely, how does the human brain growth curve compare with other primates and mammals generally? Building on Portmann (1945), Count (1947), Holt, Cheek, et al. (1975), and others, Gould (1977) persuasively argued that a key difference between humans and other higher primates is that the steep prenatal slope of log brain weight plotted against log body weight continues postnatally. In the other higher primates the slope drops around birth; in humans it continues as steeply for around another year. To paraphrase Portmann, humans should be born about twelve months later than we are, in terms of brain growth rate, after twenty-one months of gestation—almost as long as that of the elephant, which also gestates a very large newborn brain in absolute size. This would put human birth at a stage when walking is about to begin, making humans no more altricial than apes. Looked at another way, chimpanzee brains reach half their adult size at birth, human brains at thirty-six weeks after birth; if human birth occurred then, it would be at a stage when the infant is capable of forming attachments.

Such reasoning suggests that infant behavior development in humans is at least as much a kind of postnatal neuroembryology as a learning process (Konner 1991; see also Part II). It also raises questions about our animal models of human infant behavior and deprivation—although with appropriate caution and phylogenetic context, these models are useful. It may be that prolonged rapid brain growth increases the human capacity to recover from deprivation, or that it makes the human infant's brain more vulnerable to certain impressions of experience. Objectively both effects are seen in different features of brain and behavioral development (see Transition 2).

Ontogenetic Allometry and Social Behavior

Ontogenetic allometry has a dual significance for psychosocial development: (1) predictable changes in brain to body size ratios and (2) predictable changes in parental behavior in response to shape changes. Humans, like other mammals and birds, have ontogenetic allometries that give the young a characteristic shape: limbs short in relation to the torso, a very large head, a flat face, and large eyes in relation to the face. In addition to small size and behavioral awkwardness, these features are releasing mechanisms for parental behavior—the "cute" response (Lorenz 1965)—and their universality in mammals and birds may explain cross-fostering of species, including human affection for the young of other species. Normal growth allometries dissolve the "cute" features, reducing the child's power to release parental nurturance. There are further consequences: among other problems faced by children learning to walk, their center of gravity is higher than it will be later, and among other problems faced by adolescents, their feet and hands grow faster than the rest of their bodies.

Hominoid Ontogeny

The lengthening of the time required for human sexual maturation, accompanied by an even greater lengthening of body maturation to produce our paedomorphic or juvenilized adult appearance, probably occurred through a series of small genetic changes. From the dentition of juvenile fossil apes (for example, a roughly three-year-old *Afropithecus turkanensis* with preserved molar eruption patterns) we know that, by 17.5 mya, the ape lineage already had the developmental plan of chimpanzees and bonobos (Kelley and Smith 2003), longer than that of monkeys. This persisted in the early hominins (Anemone, Mooney, et al. 1996; Dean 2000). Among living great apes, heterochronic changes in both the rate and duration of growth help account for large differences in adult body size as well as sexual dimorphism, although species differ in whether the rate or duration of growth is more important (Shea 1983a; Leigh and Shea 1996). Higher ecological risk predicts slower or longer growth (Janson and van Schaik 1993; Sibly and Brown 2009), suggesting a trade-off between juvenile energy needs and the onset of reproduction; growing more slowly reduces starvation risk during development. Pubertal processes vary substantially, although only female chimps appear to lack a pubertal growth spurt, perhaps owing to very high ecological risk (Leigh and Shea 1996).

Hominin Ontogeny

Gestation and Birth

A crucial early advance in the shaping of human ontogeny was upright posture, which at least by 4 mya created a pelvis that looks not at all apelike but very human, although detailed analysis shows that three different heterochronic changes in pelvic growth were still to come (Berge 1998, 2002). Birth is short and easy in apes but long and very difficult in humans, owing to cephalopelvic disproportion (CPD) (Trevathan 1987; Rosenberg 1992; Rosenberg and Trevathan 2001, 2002). Bipedal locomotion alone did not increase CPD to modern levels, since brains remained ape-sized, but upright posture must have caused significant problems; the australopithecines began a hominin trend toward greater lumbar lordosis in females, resulting from the challenges of carrying a pregnancy to term with vertical posture and an awkward center of gravity (Whitcome, Shapiro, et al. 2007).

Since upright posture presented serious adaptive challenges, it must have conferred advantages. One, the ability to carry infants unable to cling, set the stage for the first increases in hominin brain size, achieved by prolonging fetal brain growth rates. To the extent that fetal brains were enlarging, maintaining roughly constant head size at birth meant producing a less mature (more altricial) neonate, in turn putting pressure on mothers to provide more support and care. This caregiving may have been one of the key selection pressures for the evolution of higher intelligence. In fact, modern human gestation is only about 12 to 20 percent longer than ape gestation, even though the prolongation of overall

growth and life span is more like 50 percent; this difference clearly indicates the difficulty of prolonging gestation beyond nine months. CPD
is bad enough as it is, and stabilizing selection has reached equilibrium
between the risks of immaturity and the risks of childbirth.

There is also a lower limit to the length of human gestation given
brain growth allometries. One of the great risks of immaturity is having
to sustain outside the uterus a brain that uses too much of its owner's
total energy (Epstein 1973)—in the case of the modern human newborn, 74 percent (Cunnane and Crawford 2003). Humans are not only
by far the largest-brained of the hominoids at birth, but also the fattest,
with brain and fat each amounting to about 14 percent of total weight.
Polyunsaturated fatty acids (PUFAs), stored in body fat, are crucial for
rapidly growing brains, since they are needed for brain lipid synthesis.
The brain and its supporting fat stores would have to consume an even
larger proportion of total metabolic resources if the fetus were born
earlier, and especially given its unstable homeothermy, survival would
be jeopardized.

As we have seen, based on brain growth rates as well as on size attained as growth slows, human gestation "should" be as long as twenty-
one months, an impossibility without a drastic redesign of the human
female. Alternatively, based on the fact that we have stretched the ape
life span and its main components by 50 percent, our gestation should
be twelve months, but even this posed intolerable intrapartum risks. (It
has been proposed that Neanderthals, based on their pelvic anatomy,
may have had a twelve- to fourteen-month gestation period, but this
hypothesis has been strongly criticized [Anderson 1989].) So the upper
limit of gestation is set by CPD, and much about human behavioral development in the first postnatal year, and especially in the first few
months, can be better understood if we see it as postnatal gestation.

In addition to the limit set by CPD, there may have been a process
of parent-offspring conflict involved in ending gestation thirty-eight
weeks after conception. Recall that mother and infant have non-
identical reproductive goals, which can produce weaning conflict (Trivers 1974); there may be a similar conflict over the timing of birth. Data
on large numbers of human births, even under modern conditions,
show that the average weight at birth is lower than the weight at which
neonatal mortality is lowest (Blurton Jones 1978), suggesting that expulsion occurs at a time that is not ideal for the neonate but perhaps pro-

vides some later reproductive advantage for the mother. Our ancestors had to trade off gestation length and maturity at birth to avert maternal and infant mortality.

Also, as we will see in Part II, by two or three months of age a suite of behavioral changes makes the infant more attractive: early crying has peaked, while social smiling and mutual gaze increase dramatically. Mothers often say that their feelings for the infant become much stronger at that time—the infant seems like a person, more rewarding to interact with and care for. Why did selection not give the infant this critical survival equipment at birth? Perhaps because infant mortality is so high for the first month or two after birth that it is in the mother's interest not to become too attached during this period; since the mortality is mainly from infection, it would not be much different if the infant were born a month or two later. This explanation, along with CPD, resolves the paradox.

The next life history milestone is weaning. A key human divergence from the apes is shorter lactation, about three rather than five years; this shorter duration is possible because only humans among primates provision their young after weaning (Lancaster and Lancaster 1983; 1987; Kennedy 2005; Konner 2005). Hunted or scavenged meat in substantial quantities greatly facilitated this change, and bone marrow might have been an ideal weaning food. Although prechewing of food by mothers, observed during weaning in hunter-gatherers and others, would have served an important function, the impact of cooking, by making food easier to eat and digest, could have been particularly important for weanlings and other juveniles (Wrangham and Conklin-Brittain 2003; Wrangham 2009). Earlier weaning liberates maternal gonadal function from prolactin suppression, shortening interbirth intervals. This alone could account for the fact that *Homo,* eventually one species, was the only survivor from a large field of hominins, and that we enormously outbred the (other) apes.

Australopithecus *Juveniles*

Eruption schedules of permanent teeth have been an inroad into hominin development, and debates have been vigorous (Mann, Lampl, et al. 1990; Smith, Crumett, et al. 1994). However, australopithecines now appear to have had a pattern of dental maturation similar to that of the great apes, showing little of the delayed maturation that emerged in human evolution (Conroy and Kuykendall 1995; Dean 2000). Since australopithecine brain size is in the great ape range and adult brain size correlates with the duration of life history phases, it would not be surprising if the lengthening of development came later.

At this stage of hominin evolution we have direct evidence of the developing brain—notably the well-preserved skeleton of a child about three years old of the species *A. afarensis,* probably in our direct ancestry, found in Dikika, Ethiopia, and dated to 3.3 mya (Alemseged, Spoor, et al. 2006). Like Lucy, the defining adult of her species, "Lucy's child" was probably female. Her age was estimated using an ape model of dental development, based on the fact that all her deciduous teeth had erupted and her unerupted permanent teeth included fully formed first molars and partly formed second molars in addition to other unerupted permanent teeth. Her cranial capacity was between 275 and 330 cc, comparable to chimpanzees of similar dental age, and roughly three-fourths of the 400 cc of an adult female of her species. Like the adults of her species, she had a lower body characteristic of bipedal walkers, but somewhat apelike hands, arms, and shoulders, suggesting some continued tree-climbing. Like the skull of the more recent Taung child, her skull has an exposed natural endocast that will reward further study. Her brain growth trajectory may have made a small step toward the human plan, but it is not possible to conclude that from this single specimen.

A fossil that has been much better studied is the first *Australopithecus* found, and dated to around 2.5 mya—the so-called Taung child, who was an almost-four-year-old with a naturally formed and exposed endocast (Dart 1925). It now seems likely that the child, weighing an estimated twenty-six pounds, was killed and eaten by a large predatory bird, highlighting the importance of protecting the young from predators during our evolution (Berger 2006), at a time when adult mortality may have been reduced by size and, perhaps, the use of stones as weapons. Scanning electron microscopy of the teeth suggests a period of nutritional stress at two and a half years of age, which could have been due to weaning (Lacruz, Ramirez Rozzi, et al. 2005); if so, this would represent a marked reduction in weaning age compared to the living great apes and, probably, the last common ancestor. For reasons we return to, this may reflect an increase in aid to the mother, perhaps the initial step in hominin cooperative breeding.

Ralph Holloway (1984) observed in the Taung endocast that the lunate sulcus, separating the visual cortex from the association cortex anterior to it, is farther back than it is in ape brains, suggesting an expanded association area in the parietal lobe. This claim has been criticized by Dean Falk (1983, 1989) and has been defended (Holloway 1991), but the controversy lingers. However, when Falk reconstructed the endocast using (very accurate) three-dimensional computed tomography (CT), she concluded that projecting the remaining growth of the brain would yield an adult with a larger-than-ape-sized temporal lobe (Falk and Clarke 2007), the culmination of a trend in higher-primate evolution and a defining feature of humans among extant primates (Rilling and Seligman 2002). If either the parietal or temporal lobes of the cortex — serving association, thought, language, and memory in humans — or both are larger in the Taung child's brain than in apes, then the brain was small but starting to be reorganized. Also, the venous drainage of the Taung child's brain was within the range of the genus *Homo* but significantly different from earlier australopithecines and from their more robust contemporaries (Tobias and Falk 1988).

Nevertheless, as with the Dikika child, and despite the possibly early weaning age, there is no clear evidence from the brain, skull, or teeth to suggest that the Taung child reflects a lengthening of childhood such as emerged later in evolution (Conroy and Vannier 1987). The timing of this lengthening has also been controversial (Conroy and Vannier 1988; Mann 1988; Wolpoff, Monge, et al. 1988), but the current consensus is that substantial slowing of human growth did not precede the genus *Homo*.

We know that large brain size in mammals is correlated with lengthy growth; both increased during human evolution. Hominins were at some point selected for longer childhoods in part so that they could play, observe, learn more, grow larger brains, and turn those brains to cultural purposes (Bogin 1997, 1999). It now seems unlikely that the australopithecines had humanlike growth, but their descendants were under selection for larger brains with more complex functions: puzzle-solving ability, needed to track game without a good sense of smell; information storage for memory-guided gathering of extracted foods; language and planning for group hunting; intersubjectivity for teaching skills like toolmaking; foresight for the protection and training of offspring during a long childhood; Machiavellian intelligence for predicting the intentions of rivals; and abstract thought for reckoning the complexities of kinship, marriage, and economic exchange. Brain size and cultural complexity increased in parallel over the last two million years, possibly with two accelerated periods of increase—one in the early evolution of the genus *Homo,* the second in the emergence of *Homo sapiens.*

Childhood in Early Homo

From dental maturation in fossils, using a method based on the rate of enamel formation, it appears that specimens of early *Homo,* probably *H. habilis,* show the partial lengthening of development missing from australopithecines, but even the much later *H. erectus* did not show the fully human pattern (Dean, Leakey, et al. 2001). This suggests the possibility of a two-stage evolutionary process. The fully modern pattern seems to be present in fossils representing a transition to early *H. sapiens* (Bermudez de Castro, Rosas, et al. 1999), dated to between 325 and 205 kya (Pares, Perez-Gonzalez, et al. 2000).

One specimen of *H. ergaster* dating to 1.6 mya at Turkana was a nine- to eleven-year-old boy projected to grow over six feet tall, suggesting that his species may have been taller than modern humans. Enamelization of his teeth suggested a shorter process of dental maturation than ours (Dean, Leakey, et al. 2001), confirming that fully human development came late in our evolution. Judging from the size of the spinal canal in the thoracic region, he may not have been able to speak. Modern humans have a bulge in the thoracic spinal cord believed to reflect denser innervation and voluntary control of the diaphragm to facilitate speech, which requires control of the stream of exhaled air. The Turkana boy, like most nonhuman primates, lacked that bulge, suggesting that speech evolved later (MacLarnon 1993; MacLarnon and Hewitt 1999). However, for a variety of reasons, especially the anatomy of hominin endocranial casts, it is likely that *H. ergaster* had substantial symbolic capacity even if speech came later.

Another child from the same era was found at Modjokerto, Indonesia (Coqueugniot, Hublin, et al. 2004; Balzeau, Grimaud-Hervé, et al. 2005), and classified as *H. erectus,* although it probably resembled *H. ergaster.* The significance of this child's cranial capacity hinges on its age at death, which is uncertain (Leigh 2006). If it was a year or so old, its brain size would mean that at this stage we had already evolved the postnatal extension of rapid brain growth rates that characterizes our own species;

if older, the brain would reflect an apelike pattern of growth. Since the adult cranial capacity at this stage of evolution was only about two-thirds of ours, the younger age estimate would imply that between then and now there has been an acceleration or prolongation of brain growth later in childhood.

But based on adult cranial capacity, reduction in tooth size, and tool-making, *H. ergaster* was a major step toward modern humans. It has been suggested that 850 cc of adult cranial capacity was a kind of cerebral rubicon above which it was no longer possible to increase adult brain size merely by increasing fetal brain growth rate (Martin 1983); rather, it required for the first time an extension of fetal growth rates into postnatal life, a possibility supported by measurements of the pelvic canal in this species (Walker and Ruff 1993). This transition has also been called "a turning point in the evolution of hominid life history and behavioral development" (Parker 2000:314), based on three changes from the early *Homo* background: (1) secondary neonatal altriciality based on increased neonatal brain size, which exacerbated existing CPD; (2) a longer childhood, demanding more care by more relatives (cooperative breeding); and (3) a larger suite of more complex subsistence skills to learn. The last may have demanded a level of cognitive function called concrete operations (Piaget 1970), which under this model would have been reached by *H. ergaster* only in the early teenage years (Parker 2000).

But despite the evidence of a narrow spinal canal, the Turkana boy may have had some language capability. As we have seen, brain imaging and other evidence strongly suggests that complex stone toolmaking coevolved with language and teaching (Stout 2002; Stout, Toth, et al. 2008). Although he may have been too young to knap stone himself, we can imagine this boy watching a master teach an older boy the skill, listening to the corrections, and, with a stray, not-too-valuable hunk of flint, perhaps out of sight and hearing of his elders, giving it a try.

The Modern Human Developmental Plan

The final stage of human evolution yielded another increase in brain size of roughly 300 to 500 cc, with or without an increase in relative frontal lobe size but with an increase in temporal lobe size. The overall size alone would have produced a neonatal brain requiring yet another, perhaps tertiary, increase in altriciality. This may account for the exceptional immaturity of the human newborn and the dramatic changes at two to three postnatal months (Watson 1972; Watson and Ramey 1972; Rochat 1999). This transition from *H. ergaster* to *H. sapiens* also entailed a final lengthening of development. It has been argued that a prolonged juvenile period, defined as the time between weaning and sexual maturity, is a hallmark of our species and an adaptation to the great complexity of cultural information and skill that must be transmitted to the young (Bogin 1997). But because of the imprecision of age at weaning, it would be useful to find an actual characteristic of growth itself rather than parental care to mark the beginning of this period.

For several reasons, what psychologists call the five-to-seven shift is a logical candidate (Sameroff and Haith 1996). This transition subsumes the behavioral and psychological changes noted by several branches of child psychology; every major theory posits important new capacities emerging at this age. Equally significant are the physical growth markers: uniquely in humans, the permanent incisor, canine, and first molar teeth emerge more or less in concert at this age, while the first two of these appear relatively much later in apes (Conroy and Kuykendall 1995). Also, the midgrowth spurt occurs at this age, an increase or at least a pause in the decrease in growth velocity between ages six and eight (Meredith 1978; Tanner 1981). As we will see in Chapter 11, it is possible to model this suite of changes comprehensively as a single developmental change unique to our species (Hochberg 2008).

What makes this interesting is that this is the (absolute) age at which chimpanzees begin puberty, not long after their late weaning. It has been hypothesized that the midgrowth spurt is an evolutionary vestige of the beginning of puberty in our ancestors. A difficulty with this hypothesis is that the midgrowth spurt is associated with adrenarche, a surge in androgens from the adrenal cortex at that age, which is completely independent of puberty. Nevertheless, adrenarche is associated with the onset of gonadal puberty in apes and is actually later than puberty in monkeys, suggesting a gradual evolutionary shift of gonadal maturation upward in age and away from both adrenarche and the midgrowth spurt.

This would mean that *our* developmental phase of middle childhood, between the five-to-seven shift and puberty, could have been an evolutionary interpolation, a period of quiescence designed to allow ample time for enculturation. This *could* have happened before *H. sapiens,* but the increase in cultural complexity with early *H. sapiens* was great. Not only did stone tools become markedly more various and sophisticated, but the stability of tool technology over the previous million or more years was supplanted by a changing and improving tool culture. This "ratchet effect"—cumulative cultural change (Tomasello 1999a, 1999b)—probably required a more prolonged period of developmental quiescence for enculturation. In any case, the human adolescent growth spurt is very delayed in absolute time compared with other primates (Leigh 1996), and it may be that an absolute lengthening is all that is needed for learning culture.

Heterochrony in Hominin Evolution

Heterochrony has both ancient and recent relevance to the human lineage. In some respects the earliest mammals were juvenilized (paedomorphic) versions of their ancestors, the mammal-like reptiles, and in at least some features humans seem clearly juvenilized. Adult humans resemble young apes in their lack of lower face protrusion (prognathism) and in the ratios of upper to lower face size, head to body size, and upper to lower limb length. As apes develop, they lose these humanlike proportions, and when *their* sexual maturation arrives (according to the thesis), they are at a point that is farther along a path of face and body development and that humans no longer reach.

There are problems with this analysis, but consider it on its own terms. Neoteny does not posit a similarity between adult humans and any particular *living* infant ape, nor does it require exact correspondence. Rather, it proposes a parallel to the early ontogenetic stage of our ape ancestor, modified by developmental slowing and such particular adaptive departures from overall juvenilization as natural selection may have called for. A useful way to view it is to compare the *transformations* between infancy and adulthood in humans, on the one hand, and in an ape, on the other.

D'Arcy Thompson (1942/1992) approximated the multidimensional growth of form before multivariate statistics by drawing skulls or bodies on grids and then stretching the grid. He could thus visualize growth (or evolution) as a differential stretching process by preserving the placement of anatomical landmarks while redrawing the organ in its transformed shape. This deformation of the previously perpendicular planar coordinates was an attempt to show what happens in evolution. In our case, growth takes the infant skull (similar in ape and human) to a much less transformed adult end point than in chimpanzees.

Thompson's assumption was that simple heterochronic changes due to one or a few genes could stretch or bend the grid, parsimoniously producing the needed anatomical changes. Formally, heterochrony separates (at a minimum) the growth of size, the development of shape, and the age at sexual maturity. If we imagine the human adult's grid as having been slowed in its rate of deformation, with no change in its rate of size increase or age at sexual maturity, the result would be a juvenilized ape. Although no grid-stretching genes have been found, multivariate statistics confirm the graphic model for human evolution. Studies of ape, human, and fossil skulls at various stages of development show that: (1) all the *juveniles,* ape or human, fossil or modern, cluster toward one end of the spectrum of variation (a postnatal instance of von Baer's law); and (2) human *adult* skulls cluster toward that end, with fossil protohuman adults occupying intermediate positions.

So human adults are objectively juvenilized on multivariate analysis as well as on inspection. However, the implicit assumption is that a simple genetic change underlay grid stretching or compression in one dimension or another—a satisfying biological simplicity matching the mathematical simplicity of the graphic. Since developmental genetics gives a very prominent role to the genes that establish biochemical gradients, the notion of grid-stretching mutations seems plausible. However, it is also possible, as some have argued, that *many* heterochronic changes are necessary to explain any substantial evolutionary change (McKinney and McNamara 1991; Raff 1996, 2007).

One difficulty with this argument is that as the number of heterochronies multiplies, the result is tantamount to specifying particular and independent selective action on the developmental trajectories of individual traits, and the simplifying power of the models is lost. No one doubts that traits must have developmental histories, but if each trait, with its special ontogeny, must be independently teased out of the overall developmental process, then what is gained from heterochronic models—other than the mundane truth that to change a trait selection must change its ontogeny?

But even if we accept multiple *specific* juvenilization processes, there are other problems with the neoteny model. In view of the absolute slow rate of human gonadal maturation compared with apes, neoteny cannot be the only mechanism by which our (partial and apparent) juvenilized condition (paedomorphosis) was attained. Indeed, the facts of the comparative development of humans and other higher primates (Schultz 1963; Tanner 1962; Shea 1983a, 1988, 1992; Bogin 1999) require an absolute slowing or postponement of gonadal maturation *exceeded by* an even greater absolute deceleration in shape development, with no change in the rate of size growth.

An alternative is to view some aspects of human development, such as brain size and leg length, as overdeveloped, or hypermorphic (McKinney and McNamara 1991). Thus Stephen Jay Gould, on the one hand, and Michael McKinney and Kenneth McNamara, on the other, look at the same human feature—prolonged brain growth—and interpret it in terms of diametrically opposed heterochronic processes and results. Since to understand human uniqueness we must understand the evolution of brain ontogeny, Laurie Godfrey and Michael Sutherland's (1996) attempt to reconcile the two seemingly opposed views is valuable.

Models must be as complex as the underlying reality, and the human brain is a case where, as in the elephant's trunk, the porcupine's spines, the garden spider's web, or indeed the human pelvis, no parsimonious general heterochronic explanation may do the job. Seeing fetal brain growth rates persisting later in development, just as large eyes and small jaws do, Gould, following Portmann, called the human brain a paedomorphic or juvenilized feature achieved through neoteny, or the slowing of its growth rate. McKinney and McNamara (1991), following Shea (1988, 1992), see large brain size as a result of postponing the offset of brain growth (prolongation, or hypermorphosis), which takes the brain to a size that an ape ancestor could attain only by evolving a longer duration for brain growth; therefore they call the human brain a hypermorphic or overdeveloped feature. Shea has pointed out that both Gould and his critics are partly right: the human infant is overdeveloped in the sense that fetal brain growth rates have been extended postnatally, but the human adolescent is neurologically juvenilized in the sense that the decline in the ratio of brain size to body weight with growth after infancy is not allowed to go as far as it does in apes.

Critics of human neoteny point out that humans and chimps grow in size at the same rate, and from this they conclude that human development cannot have been slowed. But Gould and other neoteny theorists were not so naive as to miss this fact; indeed, Gould was always careful to distinguish size growth from shape change and made the difference very explicit in his clock models (Gould 1977).

Gould conceptualized heterochrony as differential change in one or more of three different developmental processes: size growth, shape change (somatic development), and gonadal maturation. Human neoteny called for an absolute delay of gonadal maturation, a much greater absolute slowing of shape development, and little or no change in an absolute size-growth rate that would nevertheless lead to a larger size at the new, later time of sexual maturity. Looked at slightly differently, compared to our smaller *Proconsul*-like ape ancestor, we take longer to mature and grow to a larger size, but we have so slowed somatic development (shape change) relative to the other two measures that we are large and sexually mature while still resembling juvenile apes.

Could this have happened? Yes. Can it explain all major features of human anatomy? No. A pastiche of different processes is still required —*heterochronic mosaicism,* we might call it. McNamara (1997) offers two examples of this kind of process in the human case. He suggests basically that the paedomorphic appearance of the human head is a coincidence, resulting from an overdevelopment of brain and cranial size combined with a juvenilization of jaw size. Thus, against the ape background, we have hyper-adult brains and cranial vaults but juvenilized jaws. The overdeveloped brain required (among other things) a delay in the closure of the cranial sutures, an instance of overdevelopment by hypermorphosis, or late finishing. Similarly, the human foot (as compared with the more handlike ape foot) can be viewed as a mosaic of underdeveloped (shorter) toes with overdeveloped (longer) foot bones.

In addition, as we saw in the last chapter, hominids, at least beginning with the genus *Homo,* needed a smaller (underdeveloped) gut to offset the greatly increased energy requirements of the larger brain (Aiello and Wheeler 1995); thus the human gut may be juvenilized as well. This change required a shift to a partly carnivorous diet that supplied energy far more efficiently and without the large processing surfaces needed to absorb a high-fiber all-carbohydrate diet (McNamara 1997). Finally, a variety of heterochronic models have been applied to the evolution of the human brain. Perhaps the simplest suggests that proportional lengthening of all growth processes during hominin evolution, while maintaining all growth rates in the different organ systems and alterations in growth rates in different developmental stages, would suffice to produce the human brain (Vrba 1998), but many other models have been offered (Bogin 1997; Shea 2000).

Much of this debate is about terminology, and much of the terminology dates back to the early nineteenth century. Godfrey and Sutherland (1996) find serious fault with the way these terms are used by McKinney and McNamara (1991) and even by Gould, who was also, not very plausibly, viewed by McKinney and McNamara as having no grasp of the basic concepts of heterochrony; Gould (2000) retaliated with his own semantic analysis.

What do these antagonists agree on when the semantic clouds are lifted? Human development as compared to ape development is characterized by:

1. Approximately the same rate of size growth and attained adult size as the chimpanzee or bonobo

2. Lengthening of all life-cycle phases, not necessarily proportionately

3. Sexual dimorphism and bimaturism less than the chimpanzee, much less than the gorilla, and approximating that of the bonobo

4. Brain growth at high rates for a much longer period than in any ape, resulting in a tripled adult brain size

5. Leg growth to a much greater attained adult length than in the apes, with many specific adaptations for walking

As genetic analysis takes the mystery out of the control of these growth processes, it will become apparent how much mosaicism there is in the evolution of human development and to what extent it makes sense to invoke any general change in growth rates. In time classical heterochrony will be replaced by a comprehensive genetics of comparative ontogeny. In the meantime this five-part, commonsense model tells us the core of what we need to know about what happened in human evolution.

The Developmental Genetics of the Evolving Hominin Brain

Constant advances in our understanding of brain evolution are coming from research on brain genes. As discussed earlier, human brains have considerable up-regulation of genes compared with ape brains, suggesting considerably more neuronal activity and perhaps more energy-hungry brains (Caceres, Lachuer, et al. 2003; Preuss, Caceres, et al. 2004). However, most of the genes that are differentially expressed have unknown functions, and many of these will turn out to have functions important in or even limited to embryonic brain development. In fact, some of the differently expressed genes of known functional categories are regulatory genes that produce transcription factors or genes that code cell adhesion molecules on cell surfaces for cell-cell recognition; both play critical roles in determining structure and circuitry in the embryonic brain.

When the gene sequences of humans and other species are directly compared, a ratio can be calculated to give the proportion of genes in the nervous system with nucleotide changes that do (nonsynonymous, K_a) or do not (synonymous, K_s) result in amino acid changes in their protein product (Dorus, Vallender, et al. 2004). In protein-coding DNA, only nonsynonymous substitutions have functional consequences that can be selected for, while synonymous ones are indicative of the baseline, random, functionally neutral mutation rate. The resulting ratio, K_a/K_s, reflects the intensity of selection in particular evolutionary lines in a phylogenetic analysis. Based on 214 genes with known functions in the nervous system, primates have a ratio more than 30 percent higher than that in rodents, indicating a more rapid evolutionary process, and this difference rises to 50 percent when only genes that affect neural development are considered. These differences are due to many genes, not just a few, and most of these are involved in either brain development or behavior. Non-nervous-system genes show much less disparity. Furthermore, the disparity in nervous system K_a/K_s ratios is even more pronounced in the lineage leading to humans.

Of particular interest for our purposes are five developmentally significant genes that, if mutated in humans, produce microcephaly, a condition with hundreds of genetic causes, defined as a head circumference two standard deviations or more below the mean (Gilbert, Dobyns, et al. 2005). Microcephalic children who survive to adulthood have brains in the size range of australopithecines. High-functioning and low-functioning subgroups of microcephalics have been clinically defined, the former having no obvious neuroanatomical abnormalities other than very small brain size. The two groups have distinct sets of mutated genes and, remarkably, the three genes mutated in the high-functioning subgroup are the same three that are associated with adaptive evolution in the human lineage by the K_a/K_s ratio criterion.

Two of the genes involved in high-functioning microcephaly, *ASPM* and *MCPH1,* have been directly studied in a phylogenetic comparison (Gilbert, Dobyns, et al. 2005); both show accelerated evolution in the primate lineage and even more acceleration in the human lineage. However, they have accelerated at different stages of hominoid brain evolution, with *MCPH1* evolving more rapidly between the monkey and ape stages and *ASPM* much more rapidly in the human lineage after the split from our common ancestor with the chimpanzee. Both genes have physiological roles that would enable them to influence markedly the process of cell division in embryonic neural progenitor cells; this would be the most straightforward way to influence brain size without causing marked structural changes—in effect, a heterochronic gene change.

Finally, direct genetic investigations of specific language impairment (SLI)—an inherited disorder of neural processing that mainly affects speech and language—have implicated a family of genes known as *FOXP,* especially *FOXP2* (Fisher and Scharff 2009). A particular mutation in *FOXP2* was found in all members of a kindred that express SLI and in none who do not, and this mutation was also found in an unrelated SLI individual (Lai, Fisher, et al. 2001). It was expressed in embryonic development in areas involved in the motor control of speech (Lai, Gerelli, et al. 2003), corresponding to abnormalities found in affected adults. In some studies, individuals expressing the mutation have less than normal activity in Broca's speech area (Liegeois, Baldeweg, et al. 2003), although other motor system abnormalities have been found in both mature and developing brains (Vargha-Khadem, Gadian, et al. 2005).

Comparative genetic analysis shows that *FOXP2* evolved very slowly since our common ancestor with the mouse, and insertion of human *FOXP2* into mice alters the calls of pups (Enard, Gehre, et al. 2009). Some evidence suggested that rapid change in the gene took place recently in the human lineage, between 120 and 200 kya (Enard, Przeworski, et al. 2002), which could mean a very late evolution of the full human capacity for speech. However, the subsequent discovery that Neanderthal DNA includes these same *FOXP2* variants would put the mutations back to at least 400 kya, the estimated time of divergence between us and Neanderthals (Krause, Lalueza-Fox, et al. 2007), in addition to suggesting that Neanderthals could speak. Advanced symbolic capacity and lesser degrees of speech capacity could have evolved much earlier. But even if FOXP2 is the basis of a fairly specific language impairment, it is far from being a language-specific gene. On the contrary, it is a regulatory gene that has wide functions in many genomic and metabolic pathways (Spiteri, Konopka, et al. 2007), so it cannot be used as a sole indicator of events or rates of language evolution.

A Heterochronic Mosaic

In outline, between our ape ancestor and ourselves, the following major changes occurred in general and brain ontogeny:

1. Overall lengthening of life span (roughly 50 percent), with roughly proportional lengthening of time to sexual maturity

2. Only slight lengthening of gestation (10 to 15 percent), with the consequence of secondary altriciality against the higher-primate background

3. Marked prolongation of rapid brain growth, with relative suppression of body growth

4. Specific relative leg lengthening and pelvic shape changes

5. Juvenilization of some other anatomical features by stretching global body development while postponing sexual maturation somewhat less

Other, more speculative ontogenetic changes during hominid evolution might include:

6. Retention of juvenile playfulness and behavioral plasticity in later development and adulthood

7. Specific lengthening of one period of development, middle childhood, by insertion of a period of quiescent growth as a years-long interruption in puberty

Finally, the future of "evo-devo" belongs to genetics and genomics, and we already have findings that are tantalizing in their implications:

8. Many genes are expressed in humans that are not expressed in apes, and this difference is most marked in the brain; some of these are involved in assembling neural circuits.

9. Genes that affect nervous system function (but not other organs) have been under stronger selection in primates than in other mammals, and more so in humans than in other primates; this is especially true of genes that affect brain development.

10. Of two genes that can produce microcephaly if mutated, one evolved rapidly between monkeys and apes, the other between apes and humans; either can affect neuronal proliferation during brain development.

11. A gene that if mutated causes specific language impairment accelerated its evolution in the human line.

12. Noncoding regulatory elements, mainly micro-RNAs, play a much larger role in evolution and development than we thought until very recently. Ongoing discoveries in this domain undermine old convictions and theories about "junk" DNA and weaken the case for genomic similarity between chimpanzees and humans, thus strengthening the case for human uniqueness.

TRANSITION 1

Neurological Models of Psychosocial Function

Neurological models are a pivotal reference point relating evolutionary, developmental, and cultural accounts of psychosocial functioning. Promising attempts, not independent or mutually exclusive, that are likely to be components of any future integrative model include: (1) the limbic system model; (2) the orbitofrontal cortex and the somatic marker hypothesis; (3) the polyvagal model; (4) the mirror-neuron system; and (5) lateralized emotional functions. Each has been applied to research on infants and children, and we return to them explicitly and implicitly throughout Part II. Their development is the end point (so far) of the evolution of human psychosocial growth and the substrate of socialization and enculturation.

The Limbic System Model

The "limbic lobe" (*grande lobe limbique*) was defined by Paul Broca (1878) as the rhinencephalon, or smell-brain, and comprised certain structures he saw as the fringe (*limbe*) of the cerebral hemispheres (Schiller 1979), but it was not until the 1930s that James Papez proposed an emotional function for these structures. He laid what he called the "stream of feeling" on a core of circuitry, including especially the hippocampus, fornix, mammillary body, mammillothalamic tract, anterior nucleus of the thalamus, cingulum bundle, and cingulate cortex (Papez 1937). It later became clear that the main outgoing pathways of the amygdala—the ventral amygdalofugal path and the stria terminalis, various parts of the hypothalamus, and the septal area—all belong in the proposed emotional circuitry (MacLean 1952, 1990; Nauta and Domesic 1980). It also became clear that the hypothalamus was both a kind of hub for limbic circuits and a major mediator between brain and body.

In addition, Walle Nauta formulated three major extensions of the system beyond these primarily diencephalic and older cortical nuclei and circuits: (1) a two-way communication with the frontal lobes as "the neocortex of the limbic system" (Nauta 1971); (2) direct connections with the striatal and pallidal circuitry (the basal ganglia) via the ansa lenticularis and fields of Forel (Nauta and Mehler 1966), which help mediate species-specific displays of the emotions (MacLean and Ploog 1978; Ploog 1988; Baxter 2003); and (3) connections with the "limbic midbrain" over the mammillary peduncle, the dorsal longitudinal fasciculus, the habenulo-interpeduncular tract, and other pathways (Nauta 1958; Nauta and Domesic 1978; Morgane and Mokler 2006). Important parts of the non-endocrine effector output of the system—especially, but not exclusively, in the visceral realm—the connections of the limbic system and its relation with other brain circuitry are quite well known (Nauta and Domesic 1980; Nauta and Feirtag 1986; Kupferman 1991; Mega, Cummings, et al. 1997; Heimer and Van Hoesen 2006), and recent formulations of the "extended" limbic system include the medial prefrontal cortex, the insula, the limbic midbrain, and parts of the brain stem and spinal cord (Morgane and Mokler 2006).

Data from stimulation studies, lesion studies, and, to a lesser extent, psychosurgical practice (for reviews, see Salloway, Malloy, et al. 1997; Panksepp 1998; Mashour, Walker, et al. 2005; Morgane and Mokler 2006) indicate the involvement of portions of the amygdala, the cingulum, the hypothalamus, and the limbic midbrain in the mediation of fear and anxiety. In classic experiments on the consequences of temporal lobe lesions, Heinrich Klüver and Paul Bucy (1939) found that removal of large portions of the temporal lobe in monkeys resulted in a syndrome that included fearlessness, tameness, tendency to approach and mouth objects indiscriminately ("psychic blindness"), and hypersexuality. Since then, many human cases have been reported from a wide variety of brain illnesses causing temporal lobe damage (Jha and Patel 2004; Janszky, Fogarasi, et al. 2005). The syndrome has been parsed into components attributable to the hippocampus (Isaacson 2002), the amygdala (Williams, McGlone, et al. 2005), and the inferotemporal cortex (Gross 2005), but the general relationship of limbic system structures to social behavior and emotion has been strongly confirmed (Heimer and Van Hoesen 2006).

Most resistance to the limbic system concept dissolved after the discovery of limbic system–associated membrane protein (LAMP), which selectively marks structures that are remarkably consonant with classical descriptions (Horton and Levitt 1988). LAMP proved to be a cell adhesion molecule (CAM) involved in building the circuitry in the embryonic brain (Pimenta, Reinoso, et al. 1996) and to be coded by a gene conserved between mice and humans (Pimenta and Levitt 2004). It has led to some interesting extensions of the limbic system circuitry (Shu, Bao, et al. 2003) and has even been identified in pigeon brains, helping to define what may be equivalent circuits in the structurally very different avian brain (Pimenta and Levitt 2004). Overall, the LAMP findings strongly vindicate the limbic system concept as put forward by Paul MacLean, Paul Yakovlev, Walle Nauta, and others, and they do so anatomically, developmentally, phylogenetically, functionally, and now genetically.

The neocortex of the rostral limbic system can be roughly divided into two main functional units, each of which links part of the frontal and part of the cingulate cortex with part of the temporal lobe. Crudely, one functional unit has more to do with memory and involves the hippocampus, the posterior cingulate gyrus, and the dorsolateral frontal cortex. Developmentally, it forms the basis of emergence of object permanence and delayed response performance during the last third of the first postnatal year (Diamond 1991; Diamond, Werker, et al. 1994). The second functional unit has more to do with emotion and involves the amygdala, the anterior cingulate gyrus, and the ventromedial (partly synonymous with orbitofrontal) cortex. Developmentally, it may underlie the emergence of fear of separation and fear of strangers, of secondary intersubjectivity, and of attachment itself, during a similar time period. Hormonal interactions with the limbic system also help explain the sex differentiation of physical aggression and the behavioral problems associated with puberty.

The Orbitofrontal Cortex and the Somatic Marker Hypothesis

The orbitofrontal cortex, which overlaps with the ventromedial prefrontal cortex (vmPFC), is the part of the frontal lobes between and overlying the bony orbits of the eyes. It figures centrally in the somatic marker hypothesis, a neurological model of emotion and thought proposed by Albert Damasio and his colleagues, but its role in emotion does not depend on this hypothesis. The orbitofrontal cortex has been studied in many clinics and laboratories, and its role in emotional function, in the experience of pleasure, and in both reward and punishment has been demonstrated (Kringelbach 2005). But the Damasio hypothesis and its associated brain model attempt to integrate current knowledge about the brain's mediation of emotion with the psychology of decision making, on the one hand, and the streaming perception of the body, on the other (Nahm, Tranel, et al. 1993; Damasio 1994, 1995).

The somatic marker hypothesis proposes circuitry that serves and extends the James-Lange theory of the emotions, according to which emotion requires perception of one's bodily state. It locates emotional consciousness not in a Cartesian theater, still less in a single locus or circuit, but in multimodal association areas in three parts of the cortex: (1) the ventral frontal/anterior cingulate complex, relying in part on the insula and the somatosensory cortex for ongoing signals from the body; (2) the left parietal language cortex; and (3) the inferior temporal lobe cortex. Without the first, emotional value cannot be assigned to thoughts, plans, and choices, and an otherwise cognitively "normal" brain cannot produce a socially normal life; inappropriateness is its hallmark, as in the famous case of Phineas Gage (Damasio, Grabowski, et al. 1994) and several modern counterparts (Saver and Damasio 1991; Damasio 1994). The dorsolateral portion of the frontal cortex is involved in delayed response performance, object permanence, working memory, and non-emotional planning (executive functions), while the ventral or orbital portion is inherently emotional and embodied in its plans, yet without a normally functioning orbitofrontal cortex, sound judgment completely fails, despite intact analytical skills.

Second, the parietal association, including the language cortex, usually located in the left hemisphere, serves as a convergence zone for linguistic labels for and interpretations of memories, thoughts, and feelings that are first of all properties of other parts of the brain. A word, in this model, is a node of convergence of all cognitive functions. Language is not the only convergence process, but it is an important one, and one that is centrally involved in metacognition—for example, through inner speech. To the extent that metacognition—thinking about thinking —*is* higher consciousness, the left parietal cortex subserves that consciousness, in concert with the other two convergence zones.

Third, the inferior temporal and entorhinal cortex (adjacent to and overlying the hippocampus) draws input from neurons throughout the neocortex and delivers its own output into the hippocampus, a key part of the "save" mechanism for explicit memory. The hippocampus bilaterally is essential for storing and to some extent retrieving declarative memories, as opposed to reflex, procedural, and narrative memories mediated wholly or partly by other structures. In one interpretation, after hippocampal processing, the fornix, a massive fiber tract connecting the hippocampus to the septal area and the mammillary bodies, carries the processed sensory and perceptual information that converges on the entorhinal cortex and hippocampus from all over the cortex to more primitive regions of the limbic system, where it takes on emotional and possibly physical valence and thus enhances memory.

However, it is increasingly understood that emotional memory and even emotion perception—especially but not exclusively in the domain of fear—depend on a separate, parallel circuit in which the amygdala plays the role of the hippocampus and the stria terminalis that of the fornix. An intricate, ancient, increasingly delineated circuitry within the amygdala receives input from both the subcortical (for example, midbrain tectal and thalamic) and cortical avenues of perception, processes the impulses, and sends output to the hypothalamus (specifically the bed nucleus of the stria terminalis, or BNST) to enhance emotional memory (Phelps and LeDoux 2005). In both rats and humans the interaction between the amygdala and the orbitofrontal cortex, which are directly and intimately connected, helps explain the maintenance of fearful emotion after learning (Phelps, Delgado, et al. 2004).

The amygdala's role in processing emotion is increasingly understood at the level of molecular events and detailed synaptic connections in local circuits (Johnson, Farb, et al. 2005; Schafe, Doyere, et al. 2005). For example, fear conditioning in rats drives the protein profilin into dendritic spines in the lateral amygdala, where it enters postsynaptic densities and probably plays a role in forming fearful memories (Lamprecht, Farb, et al. 2006). However, at the macroscopic level, what we know about the amygdala's interaction with the orbitofrontal cortex strengthens both the (completely compatible) amygdalofocal and orbitofrontal models of emotion.

The Polyvagal Model

The functions of the vagus nerve (the Xth cranial nerve) have been central to the understanding of heart rate variation in relation to psychological variables (Kagan, Reznick, et al. 1988; Kagan, Snidman, et al. 2007). However, only recently has a phylogenetic model suggested a resolution of some central paradoxes of vagal function (Porges 2003, 2007). The vagus nerve is an evolutionary mixture of fibers with different sources and physiological roles. Its division into an essentially reptilian branch and an essentially mammalian branch is a simplification, but a useful one. Reptiles have low metabolic rates and small brains with relatively limited response options; they react to threats mainly by behavioral freezing and prompt, marked slowing of heart rate (Porges, Riniolo, et al. 2003). With high metabolic rates to support their large brains, mammals have many response options; they react to threat with transient attentive freezing, followed by fight or flight, both of which require increased heart rate. Mobilization for fight or flight is mediated by the sympathetic component of the autonomic nervous system and by the hypothalamic-pituitary-adrenal axis. But the advanced component of the vagus nerve, in the counterbalancing parasympathetic component, modulates fight or flight to generate social behavior.

The reptilian branch, mostly unmyelinated fibers, originates in the dorsal motor nucleus of the vagus, which lies lengthwise along the roof of the medulla in the brain stem, just under the floor of the fourth ventricle. It is responsible for the marked, prolonged slowing of heart rate that accompanies reptilian behavioral freezing and for the transient slowing involved in mammalian orienting—probably an evolutionary vestige of the reptilian response.

Mammals have sympathetic nervous system activation for their own responses that may follow transient fleeing: fight or flight. But mammals also have complex social behavior that reptiles lack, especially social affiliation; they cannot be fleeing or fighting in response to every social challenge. Enter the mammalian branch of the vagus, mostly myelinated fibers, originating in the nucleus ambiguus, located below the dorsal nucleus. It is responsible for the normal rhythmic change in heart rate that accompanies breathing (respiratory sinus rhythm) and for a normal tonic braking of heart rate during the routine ambient state. It is the removal of this brake (due to a decrease in the firing rate of this newer, mammalian branch of the nerve) that allows mammals greatly to increase their heart rate for fight or flight.

The nucleus ambiguus also contains cells that control muscles in the face, neck, throat, and larynx. These allow orientation and a coordination of aggressive or fearful displays with the increase in heart rate if necessary. Since the vagus nerve also carries sensory impulses, it has profound implications for our understanding of emotion, and particularly for the James-Lange theory and the related somatic marker hypothesis. One serious challenge to the James-Lange theory has always seemed to be the emotional life of quadriplegics. People with transected cervical spinal cords are cut off from most sensations in their bodies, which would seem to limit the role for somatic sensation in the generation of emotion (Chwalisz, Diener, et al. 1988). One way around this difficulty has been to give somatic sensation a developmental role, suggesting that memories of it sustain emotion in quadriplegics (Damasio 1994). But the vagus carries visceral sensation past the spinal injury, and other cranial nerves such as the trigeminal carry somatic sensation from the head and neck—also parts of the body, above the injury.

The polyvagal model, vagal reactivity, and autonomic nervous system function generally are increasingly used in studies of infants' responses to patterns of care (Feldman and Eidelman 2003), gene interactions with maternal sensitivity (Propper, Moore, et al. 2008), and the stability of patterns of reactivity as the child grows (Doussard-Roosevelt, Montgomery, et al. 2003). A major line of longitudinal research on shy, timid, or inhibited children has established a role for autonomic reactivity within the context of regulation by the limbic system, especially the amygdala, in maintaining the stability of that trait throughout the whole span of human development (Kagan, Snidman, et al. 2007).

The Mirror-Neuron System

One of the more exciting recent discoveries in higher brain function has been the mirror-neuron system (MNS), a region of cells in the lateral frontal cortex that are active in monkeys or humans watching others perform actions (Rizzolatti and Craighero 2004; Gallese, Rochat, et al. 2009). MNS neurons that fire when an action is observed also fire when that action is performed. The system appears to be unique to higher primates (Nudo and Masterton 1990). Broca's speech area in humans may be a partial analogue of this area and may serve (in the auditory channel, mirrored in the premotor apparatus of speech) in language acquisition. More fundamentally, mirror neurons play a key role in the perception of action itself; it may be necessary to begin to mirror action (at the premotor level) to understand it. Finally, in both monkeys and humans the MNS is active when the observed individual is *about* to perform an action, suggesting that it could also serve as a detector of others' intentions (Iacoboni 2005).

The description of this system has led to a proposed circuitry of imitation, which consists of the ventral premotor area, the adjacent region of the inferior parietal lobule (also part of the MNS), and an area in the superior temporal sulcus that processes visual input and transmits the visual perception of another's movement to the mirror neurons (Iacoboni 2005). Transcranial magnetic stimulation of the inner portion of the inferior frontal gyrus (but not of the occipital cortex mediating vision) disrupted imitation of an observed action of a finger tapping keys (but not that of a red dot tapping the same keys). Human brain-imaging studies further show that when the subject is observing and trying to imitate an action not previously learned (imitative learning), the core imitation system described here is expanded to include the dorsolateral prefrontal cortex (involved in working memory) as well as motor preparation areas closer to the primary motor cortex. But when the mirroring is emotional—as when the facial expression of emotion is being observed and mimicked—the limbic system and the insula (the neocortex of the viscera) are recruited along with the core imitation circuitry.

But there is a paradox: If the mirror-neuron circuitry is present in monkeys, why don't they show true imitation? What are they using the MNS for? In view of growing evidence that, under appropriate motivating circumstances, nonhuman primates act as if they understand the perspective and perhaps even the intentions of other individuals (Hare, Call, et al. 2000; Tomasello, Call, et al. 2003; Flombaum and Santos 2005), it seems likely that mirror neurons play a role in making this possible (Lyons, Santos, et al. 2006). Indeed, it is likely that both the phylogeny and ontogeny of intersubjectivity begin with the ability to interpret the actions of others through our own incipient action or preparedness for action, a less purely cognitive but more realistic view of intersubjectivity (Gallese, Rochat, et al. 2009). Extending this insight subcortically into the limbic system will deepen our understanding.

The MNS must be part of any understanding of the development of imitation and is certainly involved in true imitation in childhood. Some studies find impairment of MNS function in children with autism (Martineau, Cochin, et al. 2008), but others do not (Hamilton, Brindley, et al. 2007). Typically developing ten-year-olds were scanned with functional magnetic resonance imaging (fMRI) while observing and imitating emotional expressions (Pfeifer, Iacoboni, et al. 2008). They showed activation in the frontal lobe component of the MNS, but also in the anterior insula and amygdala, and in all regions the activation was correlated with their empathic behavior and interpersonal skills. A three-year-old being studied for epilepsy with subdural electrodes on her left frontoparietal cortex showed overlapping activations while drawing or watching someone drawing (Fecteau, Carmant, et al. 2004), confirming earlier findings with normal two- to eight-year-olds using surface electrodes.

Andrew Meltzoff (2007) has put the MNS at or near the core of a theory of developing social cognition in which perceptions of social similarity and imitation are central from the earliest days of life (a theory to which we return). But there are qualitative as well as quantitative changes in development, and the surface similarity of infant to later behavior does not always reflect psychological or neuropsychological continuity (Kagan 2008; Campos, Witherington, et al. 2008). Whether what is called neonatal imitation utilizes an unmyelinated MNS or relies on subcortical circuitry remains unclear. Since the amygdala and anterior insula are involved, along with the MNS, in observing the facial expressions of emotion, limbic system involvement in the emotional analogues of observing hand movements is likely.

Lateralized Higher Functions

There is growing evidence for lateralization in the nervous systems of many vertebrate species, and it appears that the left brain is more devoted to routine perception and the right to novelty or danger (Vallortigara and Rogers 2005; MacNeilage, Rogers, et al. 2009). The best-known lateralization in the human brain is the high likelihood of language functions being located in the left hemisphere, and there are anatomical asymmetries in known language regions (Geschwind, LeMay, et al. 1978; Geschwind 1992); some lateralization in these areas is found in ape brains (Gannon, Holloway, et al. 1998; Hopkins and Marino 2000). But there are peculiarities of lateralization in the human brain that are barely shared by even our closest relatives. For example, on diffusion tensor imaging (DTI) the arcuate fasciculus, which is involved in some aphasias, connects the frontal to the temporal cortex prominently in humans but not in apes or monkeys (Rilling, Glasser, et al. 2008).

We have seen that as brains expand phylogenetically they become more regionally specialized, the last stage of which for hominins and possibly great apes appears to have been the lateralization of cerebral function. To some extent this is a packaging problem. The number of possible connections among brain cells increases geometrically as cell number increases arithmetically, so the increase of neuron number in the last few million years would have required, with no change in functional organization, a proliferation of connections (white matter) too large for the skull. The solution, as has happened in smaller ways throughout brain evolution, was to specialize brain areas so that they function in part independently and report to each other after their respective tasks are done.

Perhaps the first evidence for laterality was provided by Paul Broca (1861; Geschwind 1972). In patients unable to speak, there was frequently a brain lesion in the area just in front of the primary motor control area for the face and throat. A few years later he observed that in the vast majority of these cases of aphasia the damage in question was on the left side (Geschwind 1972). These "Broca's aphasics" can produce an effortful, telegraphic version of speech: "New York . . . bus" may stand for a larger idea about a trip that would normally be expressed in a complex sentence. Most difficult for these patients are small transitional words essential in grammar. However, they often show understanding of speech by carrying out requests.

In 1874 another neurologist, Carl Wernicke, described patients who were capable of fluent and rapid but largely meaningless speech, with some of the content words mistaken or missing, and with loss of comprehension despite normal hearing. These patients had damage in a different region, now called Wernicke's area, adjacent to the primary auditory cortex on the left. Wernicke (1874) hypothesized that this area analyzed sound patterns for language, while Broca's translated thought into speech. Research has vindicated that concept of Broca's area. Working with patients who had electrodes implanted in the region preparatory to epilepsy surgery, Neil Sahin, Steven Pinker, and their colleagues (2009) decoded precisely how Broca's area sets up an utterance. Patients had to read a noun or verb silently, pluralize it or change the tense, and utter it silently; these three parts of the task occurred in sequence consistently at 200, 320, and 450 milliseconds across several millimeters of brain. This is the beginning of our understanding of how the brain understands and generates language at a fine-grained level in time and space.

By the late twentieth century the concept of two specialized hemispheres was well established (Gazzaniga, Bogen, et al. 1992; Gazzaniga 2008). Handedness, language, music, space perception, and certain emotional functions are among the traits that are processed preferentially in one hemisphere or the other. As in all vertebrates, each side of the brain controls the opposite side of the body, and in many species one side or the other is dominant in strength and speed. But in humans the great majority of individuals have right-sided dominance, and in them the left brain is dominant and is also where language is localized, while musical ability, spatial perception, and some emotional functions are processed more in the right brain. In left-handed people the situation is more ambiguous, but two-thirds still have left-brain language dominance (Obler and Gjerlow 1999).

In normal adult human brains the left planum temporale near Wernicke's area is about one-third larger than the right (Geschwind 1972; Galaburda, LeMay, et al. 1978), and the left lateral sulcus a centimeter longer. The difference is visible to the naked eye and confirmed by microscopy of cell structure (Scheibel 1991). It exists in newborns with no experience of language, even in thirty-one-week-old fetuses (Witelson and Pallie 1973), and as early as twelve weeks of gestational age human fetuses express twenty-seven genes differently on the left side of the brain than on the right (Sun, Patoine, et al. 2005). A bias toward lateralization is clearly encoded at conception, although unilateral damage during fetal or child life may be compensated by the other hemisphere. Experiments on subtle aspects of lateralization show functional differences in the youngest infants. Such marked specialization is unique to humans and figures prominently in the development of language.

Artistic, spatial, and some aspects of musical ability are preferentially right-hemisphere functions, as are prosodic and some gestural aspects of speech (Ross and Mesulam 1979; Weintraub, Mesulam, et al. 1981; Ross 1993) and key components of emotional processing (Davidson 2004; Demaree, Everhart, et al. 2005). The right and left frontal lobes preferentially process negative and positive emotions, respectively, one basis of developing emotion in infants and children (Fox and Davidson 1991; Da-

vidson, Lewis, et al. 2002; Buss, Schumacher, et al. 2003). Some cases
of dyslexia are tied to anatomically demonstrated neuronal migration
defects in the left-hemisphere cortical regions connecting visual input
with linguistic analysis (Galaburda, Sherman, et al. 1985; Galaburda
1993), and clinical evidence suggests that some cases of impaired social
interaction may be tied to subtle developmental anomalies in the right
hemisphere (Weintraub and Mesulam 1983). Human emotion may owe
as much to the right hemisphere as language owes to the left, although
we can also say that negative emotions are predominantly attributable
to the right frontal cortex, and subjective well-being to the left.

Imperfect Models

The models described here are simplifications and are not an exhaustive account of how the brain generates psychosocial functions. Parts
of the occipital and superior temporal cortex form a circuitry for face
recognition, gaze following, and (with the amygdala) interpretation of
facial expressions. This circuit, which includes the fusiform gyrus and
the cortex of the superior temporal sulcus, may be innately wired for
face perception, become expert at it through long experience during development, or, most likely, both (Cohen Kadosh and Johnson 2007; Johnson, Grossman, et al. 2009). It changes throughout development (Nelson, Leibenluft, et al. 2005; Blakemore 2008).

So do the circuits involved in "mentalizing" or "theory of mind"—
efforts to understand the thoughts and intentions of others—which
overlap substantially with the circuits described here, including the orbitofrontal and medial prefrontal cortex. Some neuroscientists studying the development of social cognition are giving long overdue attention to emotions and their substrates in the limbic system (Nelson,
Leibenluft, et al. 2005). So these models are imperfect, and we will not
adhere to them rigidly, but they are clearly articulated and valid, and
they help form the basis of the developmental explorations on which we
now embark.

II

Maturation: *Anatomical Bases of Psychosocial Growth*

---

wherein we see how neural and endocrine systems guide the paths of development called for by natural selection

---

We must recollect that all our provisional ideas in psychology will presumably one day be based on an organic substructure.
—Sigmund Freud, "On Narcissism" (1914)

6

Paradigms in the Study of Psychosocial Growth

Having set out in Part I the laws and processes of evolution that have shaped human childhood, we now examine the developmental mechanisms that implement the results of phylogeny. In complex nervous systems, genes set development in motion, but their role continues throughout life. Of the estimated 25,000 genes in the human genome, many or most are never activated in any given cell, and perhaps 20 percent are active at any moment. The genetic uniformity of the zygote drops off after the first few cell divisions. Through processes involving position, distribution of cell contents, cell surface signal molecules, and other local features, particular genes are turned on in some cells and off in others, directing them to become nerve, bone, or liver, or particular nerve cell types. Sequencing the genome is only a small step toward understanding how those genes help shape what we are; developmental research aimed at this question will take centuries.

But we have some basic principles. Genes guide the generation and specification of neurons and the major connections of the circuitry; they bias the number of neurons performing a function, their positioning, their interconnections, and overall neural anatomy. Even in the early embryo, however, there is underdetermination of development by genes, and some developmental patterns are unpredictable because of sensitivity to initial conditions—*chaos*, in the formal sense. These may resolve later in embryonic life through self-stabilization or self-organization, leading to new predictability at a higher level of order and generality—*complexity*, in the formal sense. Thus in identical-twin embryos only weeks old, there is significant individuality, based not just on environmental shaping but on the sensitivity of patterns of self-stabilization in cellular networks to barely measurable differences in initial developmental steps.

But in addition, the fetal milieu does shape the growth of neural circuitry. In many mammals androgens influence the organization of circuitry in the male hypothalamus, but there are also subtler effects of the hormones supplied by the mother, the placenta, the fetus itself, or adjacent fetuses. With the effects of nutrition, drugs, toxins, umbilical blood flow, oxygen supply, infection, length of gestation, and the physical forces of labor and birth, hormone effects are an additional influence on nervous system development that is both genetic and environmental. Quantitative genetics should consider prenatal sources of environmental variance (Asbury, Dunn, et al. 2006). Since behavior genetic studies often estimate the environmental contribution by subtracting genetic from total variance, the resulting estimate can have any environmental source (or be variance due to gene-environment interaction), and some of it may be formally chaotic and unknowable. The inference that it is mainly in the realm of learning is not always sound.

And despite formal chaos and the power of the environment, genetic influence continues, canalizing development both pre- and postnatally and limiting divergence. Neurons are born through cell division, migrate until diverted or stopped by physical or chemical signals—cell adhesion and surface adhesion molecules (CAMs and SAMs)—and commit to a particular shape. Many more neurons are born than will survive, so there is developmental selection, which depends in part on functional differences.

All this repeats for connections, as neurons extend processes toward possible target cells. Guiding and stopping the tips of the extensions depends on signaling molecules like those that governed migration. Here too overproduction means that most connections are pruned, and influence from the cells contacted affects survival. Then, with electrical activity, functioning links escape pruning, while inactive circuits fail and die, as do inactive cell bodies. Still, programmed cell death (PCD) figures in neuron fates, and although activity provides a route for experience effects, it too is under genetic guidance. Thus the impact on circuitry is in part self-organization rather than passive experience. Nonetheless, there are proven effects of intrauterine learning.

Experience shapes circuitry postnatally as well. Some neurons are generated after birth (and throughout life) in higher primates, but many more die, shaping the brain by deletion. Trillions of connections are made, and many are pruned; in areas of the human cerebral cortex, synaptic density peaks months to years after birth. As in the fetus, but more so, experience regulates activity and selects among connections. Yet early postnatal brain growth extends the mechanisms of late fetal life. Genetic regulation, canalization, self-stabilization, and other dynamics of an unfolding program continue to operate side-by-side with learning.

Indeed, the way experience works is biased by genes. Although there are general laws of learning with clear neuronal substrates, there are exceptions to these laws. Some associations are much easier to make, and there are species-specific patterns, which we might call ease-of-learning landscapes. For many species the landscape changes throughout life, and special learning may occur preferentially in early life. But while some environmental influences or injuries are more permanent if they occur early, other early insults are easier to recover from. The power of culture in mind depends in part on the special character of early experience effects (Part IV).

A classic visual metaphor is Conrad H. Waddington's in *The Strategy of the Genes* (1957). In the familiar drawing, which he called "the epigenetic landscape," he depicted development as a ball rolling down a corrugated plane slightly inclined toward the viewer. If the plane were flat, the ball would have an infinite number of possible developmental pathways, and the actual forces would be completely environmental, like breezes, winds, and hurricanes blowing across the plane. In reality the pathways are limited, as represented by the corrugations down which the ball tends preferentially to roll. This is essentially what we have called facultative adaptation, a common result of the evolution of plasticity.

Waddington called the tendency for growth and development to stay on or return to certain paths *homeorhesis*, an analogy to homeostasis, and the resulting developmental process *canalization*; following an environmental perturbation, development tends to return to the path it has been in. A classic example is catchup growth: a child grows faster after an illness or deprivation and returns to the line on the growth curve (say, the sixtieth percentile) she was at before the deprivation (Golden 1998). If the canals are shallower at the outset, it should be easier to divert the ball earlier on—and more difficult, although not impossible, to get it out of the deeper canals further along. But catchup growth can become impossible if deprivation lasts long enough.

For the second part of Waddington's visual metaphor, we go around, behind, and under the corrugated plane. The perspective is like being under the bleachers at a high school football game; we get to see the apparatus of canalization. We find ourselves among crisscrossed wires leading up to the plane from stakes in the ground that are the genes; these guide wires (gradients of gene expression) give the canals their shape. Many stakes have multiple lines coming out of them, representing pleiotropy (one gene, many effects), and multiple lines converge on points that represent multigenic traits (one effect, many genes)—by far the most common. *FOXP2*, the regulatory gene, helps in canalizing speech, but also in hundreds of other developmental pathways; speech, for its part, relies on many genes besides *FOXP2*. The genes tug on the wires throughout life; they also respond to the environment, but that does not negate their power.

Genes do not stop operating at conception, birth, or any time in life. They do not set the organism going like a windup toy and then let experience take over guidance. Their continual activation is of the essence of development and of life. Some, like the one that causes Huntington's disease, a progressively degenerative brain syndrome that may first appear as emotional symptoms, delay their effects until about age thirty-five; the effects of others, such as Alzheimer's disease, may not appear until age seventy or later. Some genes for *normal* phenotypes increase their effects with age, as shown by the lifelong convergence of identical twins in health and mental traits. The twentieth-century postulate that increasing experience cumulatively buffers us against the effects of genes is now in question.

As with evolutionary inquiry, paradigms in behavioral development have been separate. They include: (1) neurogenetics, (2) neuroembryology, (3) developmental neuroendocrinology, (4) postnatal brain development, (5) developmental behavior genetics, and (6) neurological individuality.

The Neurogenetics of Animal Models and Human Disease

Psychologists throughout the twentieth century, with contempt or respect, used terms like "wired-in" or "hardwired" to describe behavioral functions built into the brain by the genes. The last few decades took these notions out of the abstract through studies of simple nervous systems, like those of the roundworm *Caenorhabditis elegans* and the sea slug *Aplysia californica,* more complex ones like the grasshopper and fruit fly, and very complex ones like the mouse. We are learning how genes regulate the developmental plan, guide neurons and extensions to their places, build ion channels and neurotransmitter receptors, and code for the enzymes that catalyze the synthesis and removal of neurotransmitters and hormones, or hormones themselves.

We also glimpse a hierarchy of gene regulation in which genetically coded DNA-binding transcription factors produce cascading effects. They change cell populations from extreme multipotentiality (pluripotent stem cells) to increasingly committed states, with some genes switched on or off by transcription factors that bind to promoters upstream in the cascade. We now know that they are not all proteins; many, perhaps most, are noncoding RNAs, including micro-RNAs (Mattick 2005; Amaral, Dinger, et al. 2008; Grosshans and Filipowicz 2008). Two suites of genes and transcription factors—the homeobox sequence genes that control the linear plan of the body and limbs and the *PAX6* genes that control eye development—have been mentioned. With external signals, including neural cell and surface adhesion molecules, the genetically governed synthesis of regulatory molecules plays a central role in determining cell fate over a very wide phylogenetic range, with a common ancestor over 600 mya.

Some single genes that alter the nervous system are partially understood, and the gap between disease and animal models will narrow. It can seem that human neurological diseases are gross abnormalities not relevant to normal function, but science begins with the most obvious things and proceeds to subtler ones. Many laboratories are seeking genes for human disease traits, and the more subtle genetics of behavior are also coming to light (Pfaff, Berrettini, et al. 2000; Dorus, Vallender, et al. 2004; Caspi and Moffitt 2006). Evolutionary considerations make even simple animal models relevant. The following examples of genetic effects illustrate the available approaches and demystify behavior genetics by tracing the causal chain from gene through neural structure and function to behavior.

Caenorhabditis elegans

This half-millimeter-long translucent roundworm has 18,000 genes (over two-thirds the human number) but only 959 cells (including 302 neurons), each with a known identity, function, embryonic origin, and genetic control, often shared with mammalian nervous systems (Hobert 2003). Over 70 of its genes affect ion channels in nerve cell membranes, over 50 code for over 200 neuropeptides, and at least 17 are genes that affect learning in other species and the worm's own behavioral plasticity. It has programs for movement, feeding, defecation, egg-laying, and sex, all strongly genetic, yet modified by experience. Male sexual neurons are individually identified and their development followed from transcription through neurogenesis and circuit wiring to behavior (Emmons and Lipton 2003). Feeding varies from social to solitary, depending on single-amino-acid variants of the gene for the neuropeptide Y receptor (de Bono 2003) among others. This simple system with few neurons but many genes shared by mammals illuminates development, behavior, and plasticity (Rankin 2002).

Aplysia californica

A large snail without a shell, *Aplysia* has about 20,000 central neurons —far more than *C. elegans* but far fewer than an insect. Egg-laying, a stereotyped behavior, involves a pattern of head undulations, then weaving and tamping with inhibited locomotion as eggs are deposited. A family of genes codes neuropeptides that control the behavior (Scheller, Kaldany, et al. 1984). First, egg-laying hormone (ELH) oozes from specialized bag cell neurons in the abdominal ganglion, a main nerve center (Bernheim and Mayeri 1995). A prohormone made by genes in the bag cells is spliced to make ELH and other neuropeptides (Li, Garden, et al. 1999). The cells up-regulate ELH production under electrical stimulation, by increasing translation from RNA, while other protein synthesis declines (Wayne, Lee, et al. 2004). During egg-laying, the slug releases waterborne pheromones named attractin, enticin, temptin, and seductin because—in combinations—they attract mates (Cummins, Nichols, et al. 2005).

Yet *Aplysia* is better known for learning; much of what we know about learning at the cellular level began with it (Kandel 2001). The defensive reflex of gill withdrawal can be habituated (diminished with repeated harmless stimuli), sensitized (heightened with harmful stimuli), and classically conditioned (to respond to a signal repeatedly paired with tail shock). We know a lot about the molecular mechanisms that explain all three forms of plasticity. Massed exposure to noxious stimuli over a short time produces short-term sensitization lasting hours, but distributed stimuli produce changes lasting weeks. These depend on protein synthesis through known gene activation pathways. So this animal with about one-millionth of our neuron number and some stereotyped, gene-controlled behaviors learns at the cellular level much as we do.

Drosophila melanogaster

Some *Drosophila* neurobehavioral mutants have genes that alter ion channels (Wicher, Walther, et al. 2001). Some channel components are conserved across many species (Zwingman, Neumann, et al. 2001), whether squid, snail, crayfish, frog, or mammal (Hille 1992). Potassium and sodium channels, both vital to nerve conduction, can have mutations that make behavior abnormal with single-amino-acid substitutions in channel proteins. *Shaker* mutants have leg tremors because their nerve cell membranes after firing do not repolarize; they are missing one component of the outward potassium flow (Butler, Wei, et al. 1989; Doyle, Morais Cabral, et al. 1998).

Some fly mutants relate directly to human behavior. A dominant human gene on the short arm of chromosome 20, similar to the *shaker* potassium-channel gene, is altered in an inherited form of epilepsy in which young infants have unprovoked seizures but usually outgrow them within a few months of birth. RNA from the implicated human gene was injected into frog eggs, creating channels with impaired potassium flow (Biervert, Schroeder, et al. 1998). Yet potassium channels are also altered nongenetically; stress hormones affect how component RNAs are spliced (Xie and McCobb 1998), despite normal genes.

Other fly mutants with memory impairments have single-gene defects in enzymes of the energy cascade involved in synaptic plasticity (Zhong and Wu 1991). And many fruit fly behavioral syndromes, such as "hyperkinetic" and "sluggish," are caused by single genes. Their influence can be traced through fate maps from cells of the early embryo to brain, nerve, and muscle abnormalities that explain the behaviors (Hotta and Benzer 1972; Benzer 1973; Greenspan and Ferveur 2000). Some are severe. A recessive gene on the X chromosome riddles the brain with holes. Subtler mutants have alterations in phototaxis, climbing against gravity, or the twenty-four-hour behavioral cycle. *Easily shocked* has a seizure in response to a mechanical jolt. *Comatose* succumbs to heat and recovers slowly, owing to depressed nerve function.

Other genes affect courtship and sex (Greenspan and Ferveur 2000;

Manoli, Foss, et al. 2005): one male mutant courts lackadaisically, and another pursues males *and* females; *stuck* cannot disengage after intercourse, and *coitus interruptus* cannot complete it. Male courtship tunes, played by wing vibrations, change tempo and other features in mutants with altered mating success (Hall, Kulkarni, et al. 1990; O'Dell and Kaiser 1997). We know the chemical paths from gene to behavior, and some have known evolutionary relevance (Kyriacou and Hall 1986). Genes also explain different female receptivity to variations in male song and changes in receptivity with experience (Hall, Kulkarni, et al. 1990). Not surprisingly, learning also matters: normal male flies can, with punishment, be conditioned to court less (Kamyshev, Iliadi, et al. 1999).

Finally, we know the genetics of a normal mutation that is adaptive in the wild (Osborne, Robichon, et al. 1997; Sokolowski 2001). Some fly larvae roam widely in their search for food, while others lag, tendencies stable from larva through metamorphosis to adulthood; in nature seven of ten are *rovers,* and the rest are *sitters.* Many generations of selective breeding yield different strains; a gene on chromosome 2 codes for an enzyme that mobilizes energy in cells, affecting behavior. Inserting *rover* genes into *sitter* eggs makes them *rovers* by the larval stage.

An adaptive explanation has been tested. The *sitter-rover* difference is brought out after feeding, so sitting should be adaptive where a fly can be sure of its next meal, but under scarcity, eating should not stop the food quest. Wild populations were captured, kept in mixed jars of either fifty or a thousand flies, and the numbers kept steady for seventy-three generations. The result was more *rovers* in the crowded jars. Thus a clearly adaptive, polymorphic, single gene affecting normal behavior was located, sequenced, cloned, used to change behavior through gene transfer, shown to work through a known enzyme, and made to evolve through seminatural selection (Osborne, Robichon, et al. 1997; Sokolowski 2001).

Mouse Neurological Mutants

The domestic mouse (*Mus musculus*) has long been a favored neurogenetic model. Although now superseded by transgenic animals, the production of 99-percent-homozygous inbred lines provided a uniform genetic background for experiment and led to many studies of strains with characteristic mutations, including single-gene defects that produce neurodevelopmental syndromes (Caviness and Rakic 1978; Wimer 1990; Rakic and Caviness 1995); these, although grossly abnormal, were a good starting point for mammalian developmental neurogenetics (Caviness and Rakic 1978; Wimer 1990). Radioactively labeled thymidine incorporated into DNA revealed the sequence of nerve cell birthdays, in turn set by regulatory genes (Sidman 1970).

Cell birth sequence determines structure. In the cerebral cortex, earlier-born cells stop in the first, inner layer, and later-born cells move past them to form the outer layers; elsewhere in the central nervous system (CNS) the reverse is true. Violating these patterns destroys function. The *reeler* mutant has abnormal gait and severe memory problems. The heterozygote has a small cerebellum, but the homozygote has inverted layering in the cerebral and cerebellar cortex (Caviness and Sidman 1973) and distorted hippocampal architecture (Sekiguchi, Nowakowski, et al. 1995). Cell birthdays are normal, but the earliest-born cells migrate to the outer, not the inner, layers. *Reelin,* the defective gene product, is a membrane protein on migrating neurons; normally it combines with receptors on certain other cells, triggering an enzymatic cascade that allows neurons to complete migration (Jossin 2004).

Studies with chimeras (combinations of a normal and a mutant embryo at the eight-cell stage that integrate to form one embryo) illuminate some puzzles (Wimer 1990). In the normal mouse, glial guide fibers lead the granule cells to their destined positions after migrating through the cortex (Rakic 1972). Chimeras show that the *reeler* phenotype results from a failure of cerebellar purkinje cells to detach from the guides when migration should end. *Staggerer,* by contrast, has a defect intrinsic to those pivotal cerebellar neurons. Granule cells degenerate as a secondary result, and if they are surrounded in a chimera by normal instead of *staggerer* purkinje cells, they are rescued. But in the *weaver,* the granule cells seem to bear the degenerative message.

Some human disorders are not in principle more complicated than these (Rakic and Caviness 1995; Lacbawan and Muenke 2002). Mouse mutants with defective calcium channels have spike-wave EEG patterns linked to behavioral arrest, pointing to a possible molecular-genetic mechanism in human epilepsy (Fletcher and Frankel 1999). Also, the brains of some dyslexics show abnormal cell layering, apparently due to an abnormal sequence of cell births, in the cortical region where language and visual circuits meet (Galaburda, Sherman, et al. 1985). Some schizophrenics have similar defects of cell migration (Bullmore, Woodruff, et al. 1998). Cell layering defects in mice and rats, based on both genetic and environmental causes, shed light on human brain disorders (Denenberg, Sherman, et al. 1996; Rosen, Sherman, et al. 1996).

In the hippocampus, input to the pyramidal cells varies across inbred strains, and differences in behavior and learning have been neurophysiologically explained. A knockout mouse with a missing homeobox gene has specific hippocampal abnormalities that cause learning and memory impairments, but the mouse's behavior is otherwise normal (Paylor, Zhao, et al. 2001). Functionally different cortical areas are also genetically specified (Mallamaci and Stoykova 2006). Two brain genes, *Emx2* and *PAX6*, are expressed in longitudinal gradients, with the highest concentrations of *Emx2* in the occipital area where visual and auditory sensory cortex develop and the highest *PAX6* in the frontal, motor cortex. If *Emx2* is knocked out, the sensory areas are abnormally small and the motor cortex large, while in *PAX6* knockouts the reverse is true. A mutation in *Emx2* in early primates could perhaps explain their enlarged visual cortex.

Mouse Metabolic and Hormonal Mutants

In early studies, genetically different mice raised in identical environments showed differences in behavior linked to metabolic differences (Bruell 1970; Wade 1998). Thyroid hormones were related to activity level, serotonin and other neurotransmitters to emotionality and reaction to shock (Ray and Barrett 1975), liver enzymes to alcohol preference (Rogers, McClearn, et al. 1963), adenosine triphosphate (ATP) and glutamate to noise-induced seizures (Schlesinger and Griek 1970), and androgen sensitivity to sexual intercourse (McGill and Tucker 1964). In the 1980s gene technology superseded this approach. Hypogonadal mice had a defective gene for gonadotropin releasing hormone (GnRH), but correction of the germ line in the parental generation led to correct neuronal expression of the hormone in the offspring and normal reproductive behavior and biology (Mason, Pitts, et al. 1986; Mason, Hayflick, et al. 1987).

Two decades after this landmark experiment, genes are routinely manipulated to change behavior. Consider the classic studies. Thyroid hormones predicted activity; now we know the genetics of several thyroxine receptors, variants due to RNA splicing differences, distributed appropriately in the brain and expressed at key points in development so that they can influence neuronal branching patterns, with consequences for activity and learning (McEwen 2000). Serotonin levels predicted emotionality; now we know that knockout mice without the gene for a serotonin transporter ($5HT1_B$) are more likely to attack intruders (Maxson 2000).

We also know more about the genetics of alcohol susceptibility (Crabbe 2002). Alcohol dehydrogenase converts ethanol to aldehyde, which is then converted to acetate by aldehyde dehydrogenase. In humans this enzyme has two genes, each with several variants. In any of them, one amino acid substitution changes alcohol metabolism and predicts different dependency risks (Enoch and Goldman 2000). A *Drosophila* gene, *hangover,* influences both tolerance to ethanol and resilience under stress (Scholz, Franz, et al. 2005). For noise-induced seizures, glutamate is implicated by knockout mice. *Grik2* knockouts have one altered glutamate receptor, increasing seizure resistance. *Gria* knockouts, with altered neuronal calcium influx, have more seizures (Ferraro and Buono 2000). Also, because many androgen effects on the brain require conversion to estradiol, its receptor plays a key role in male sex. Disrupting the receptor gene affects male mounting and intromission and prevents ejaculation (Ogawa, Krebs, et al. 2000).

These molecular-genetic functional dissections of the brain leave no doubt about the role of genes, but they also illuminate learning. Altering genes for the n-methyl-d-aspartate (NMDA) glutamate receptor in the mouse forebrain improves age-related learning and cognition. Young mice express a particular NMDA subtype, NR2B, and learn better than the old. Inserting multiple copies of *NR2B* into normal mice improves learning by six different measures (Tang, Shimizu, et al. 1999). You can teach an old mouse new tricks if you give it young brain genes.

Single-Gene Human Disorders

Thousands of defined genetic conditions affect the human nervous system, and in many the path from gene to behavior is now largely or fully known. A few illustrate key principles:

> *Tay-Sachs disease.* In this homozygous recessive form of mental retardation that strikes mainly the infants of Ashkenazic Jews, the newborn is normal, but the dendritic branches of brain neurons lose

their spines over the first year (Purpura and Susuki 1976; Purpura 1979). The gene causes malfunction in the enzyme ß-hexosaminidase A, which allows a usually harmless ganglioside of neuronal membranes to build to toxic levels, distending the dendrites and denuding them of spines. There is no treatment, and death comes usually before age two.

Phenylketonuria. PKU is a severe but treatable form of mental retardation due to a homozygous recessive gene that may mutate in several ways (Scriver and Clow 1980), producing different degrees of impairment (Enns, Martinez, et al. 1999). It codes for phenylalanine hydroxylase, which converts phenylalanine to tyrosine. Both are common in normal diets, but in PKU phenylalanine builds up behind the block, owing to the defective enzyme. The disease is treated by withholding phenylalanine (Weglage, Pietsch, et al. 1999), and universal neonatal screening for high phenylalanine levels allows this treatment from birth (Wappner, Cho, et al. 1999). Here a strong and direct genetic determination is negated by environmental change.

Huntington's disease. This neurodegenerative process, primarily affecting the basal ganglia and marked by involuntary movements, and usually beginning in the thirties or forties (often with psychiatric manifestations), is caused by a dominant gene on the short arm of chromosome 4 (Gusella, Tanzi, et al. 1984). Unlike the point mutations—one substituted base pair—in many defective genes, the Huntington's gene has triplet repeats, each a sequence of three bases (CAG). Above thirty repeats the risk for Huntington's crosses a threshold, and the number thereafter is inversely correlated with age of onset. Each CAG triplet codes the amino acid glutamine, and if the protein (*huntingtin*) has too many glutamines, it binds other identified protein molecules, creating molecular garbage that poisons neurons (Martin 1999; Bates 2003). Huntington's shows that a genetic trait can first appear late in life.

Alzheimer's disease (AD). This most common dementia is linked in various kindreds to genes on (at least) chromosomes 1, 14, 19, and 21; three increase ß-amyloid, found in the characteristic plaques containing dead neurons, while the fourth prevents its removal (Selkoe 2004). If ß-amyloid is strongly implicated, so is the tau protein, found in neurofibrillary tangles, the second AD signature. Neuron loss is greatest in the hippocampus, amygdala, and association cortex. Emotional and behavioral changes can accompany or precede the memory loss. Like Huntington's, only more so, AD depends on genes that lie silent for at least five decades.

5-α-reductase deficiency. This single-gene form of pseudo-hermaphroditism, found most often in four inbred villages in the Dominican Republic, results from defects in 5-α-reductase, which converts testosterone (T) to dihydrotestosterone (DHT). The Dominican mutation substitutes thymidine for cytosine and tryptophan for arginine in the enzyme, although other mutations in the same gene can cause the syndrome (Cai, Zhu, et al. 1996). Because the normal growth of male external genitals requires DHT, affected children have ambiguous or feminine bodies until puberty, when, owing to surging T levels, they usually (seventeen of eighteen cases in the Dominican studies) become physically and psychologically masculine (Imperato-McGinley, Peterson, et al. 1979, 1980; Imperato-McGinley and Zhu 2002). Their ability to function as adult men after being raised as girls may be due to T effects on the fetal brain.

These findings leave no doubt that genes can alter complex behavior through biochemical pathways that affect the nervous system, and they illustrate some genetic mechanisms as well. As the next section shows, fetal activity and stimulation shape brain circuits, and several classic models explain how (Purves and Lichtman 1980; Changeux 1985; Edelman 1987; Purves 1988). But this does not mean that genes are unimportant. Consider:

1. A great deal of neuron loss is due to apoptosis, or programmed cell death. PCD can depend on molecular signals from other cells, but it can also precede contact (de la Rosa and de Pablo 2000); genetic signals intrinsic to these neurons help precipitate their deaths. Several PCD genes are known (Bredesen 1995); eliminating one creates a mouse with too many neurons in an abnormally large brain (Kuida, Hayda, et al. 1998). Trans-synaptic stimulation sustains neurons experimentally and in computer models, and it probably matters in fetal life.

2. Mammals lose around half the neurons they create, but this equals only the last cycle of cell division after many doublings. Since the time for and number of doublings determines mammalian brain

size (Finlay and Darlington 1995), overproduction may be insurance against underdevelopment, but it is also an adaptation for activity-molded circuitry. The signal to stop neurogenesis, whether intrinsic (say, telomere reduction) or extrinsic (a density-dependent external signal), may be set to stop one division past the minimum number needed.

3. The quantitative underdetermination of the brain—in humans fewer than 10^5 genes generate 10^{11} neurons (versus 302 in the roundworm) and 10^{15} synapses—could make it impossible for genes to direct wiring, but not if thousands of axons follow the same path. Major pathways do not number 10^{11}, but are fewer than brain genes. If an axon pioneers the path and a bundle follows, then the role of experience may be mainly fine-tuning; this is important, but it does not build the circuitry. To reconcile these interacting causes we turn to neuroembryology.

Neuroembryology

In contrast to *C. elegans,* many axons connect two points in mammals. Genetic specification may collectively govern the bundle, one or a few pioneer axons, or glial guide fibers, which could then direct other cell extensions and migrations. Still, there are too many degrees of freedom in the developing mammalian brain for comprehensive genetic determination. The fact of underdetermination has led to several important models.

Chaos, Emergence, and Complexity

Models formally motivated by nonlinear dynamical systems (complexity) theory have been used in embryology (Kauffman 1993; Goodwin 1994a, 1994b). Embryonic development is conceived as "a robust natural process" that needs relatively little input from genes to keep it on track (Goodwin 1993). This model starts with observations of chemical systems, which produce predictable patterns after a lot of less predictable—formally chaotic—activity. Computer models of some developmental processes simulate forms like those in nature, with a few reasonable assumptions about the physics and chemistry of the system (Kauffman 1987, 1993).

The key claim is that some developmental outcomes are far more likely than others, whatever the genes are doing. There may be an ensemble effect by which a suite of mutually regulatory genes emerges regularly from unknown (but not simply reductionist) laws of self-organization. Each ensemble is an attractor in initially chaotic gene interactions (Kauffman 1993). Natural selection stabilizes the ensemble in its attractor basin by eliminating less viable departures. Strong selection or drift would occasionally displace the ensemble from one attractor basin to another. This resembles an older model of a species' adaptation as a multidimensional dip in an adaptive landscape—such a dip is difficult to get out of even if a superior adaptation (a deeper dip) is just over the rim of yours, because you would have to become *less* adapted first (Levins 1964; Maynard Smith, Burian, et al. 1985). These local dips are stable because they resist small changes. It is not a new idea, but is recast in nonlinear dynamic models, developed in detail. For example, differentiated cell types are viewed as attractors, or "recurrent asymptotic behaviors of the genome," of which (in simulations) only a few hundred actually occur out of the stunningly large number of possible combinations of 105 genes (Kauffman 1993).

One application in brain development is to the striate (striped) structure of the lateral geniculate, a way station between retina and visual cortex. The layering of representations from the visual fields can be simulated as an emergent process in formally chaotic cell proliferation. Such self-organization has been proposed as a key mechanism of development and the solution to the puzzle of underdetermination by the genes. If it is indeed an important process in early development, it would also produce individual differences, attributable neither to genes nor to knowable environmental causal chains, but to sensitivity to initial conditions.

Waddington's model can be reconciled with chaos and emergence if the stabilized "canals" are partly consequences of laws of self-assembly, with genes helping to maintain or switch among them. The claim that this is something that requires us to rethink natural selection (Goodwin 1993, 1994b; Kauffman 1993) is based on a misunderstanding. No one thinks that selection circumvents physicochemical law; on the contrary, it is simply the preservation of those organic forms that make the best use of it in turning solar energy into offspring. Some complexity theorists claim to have new laws that operate at higher levels of organization; if this is true, then natural selection can work only within those laws, just as it works within the laws of conservation of energy, gravity, momentum, electrochemistry, chemical bonding, and so on. But complexity theory is more effective in showing why complex systems cannot be reduced to lower-level laws than it is in discovering new laws (Ruelle 1991, 2001).

Conventionally Deterministic Processes

Chaos aside, the development, ordering, and patterns of circuitry of cell populations in the mammalian nervous system include well-described deterministic processes (Sanes, Reh, et al. 2006). Some are familiar from our earlier discussions, but here we are interested not in how transformations of them yield phylogenetic change, but in how the unfolding of them in each life yields a reliable neural and behavioral ontogeny. In Chapter 5 we saw the roles of genetic regulatory cascades, differential embryonic gene expression, and chemical gradients that determine linear and regional differentiation of the nervous system. These processes pervade all that follows, but here we focus on anatomical changes.

Table 6.1 shows them in rough order, but there is extensive overlap: (1) the birth of neuronal progenitor cells (*neurogenesis*); (2) the *migration* of these cells to their destined places; (3) the *differentiation* of neurons and neuronal types from progenitors; (4) the *extension* of neural processes to form axons and dendrites; (5) the formation of functional connections between neurons (*synaptogenesis*); (6) the *cell death* of excess neurons; (7) the *pruning* of excess connections; and (8) the wrapping of neurons in fatty sheaths that improve conductivity (*myelination*).

1. *Neurogenesis.* New neurons enable brain expansion and repair, to some extent throughout life. Some cells are committed as future neurons or glia at the outset (Luskin 1993, 1994). Among glial precursors, future astrocytes stand out, as do future pyramidal cells, among neural precursors. Before migration, cells are committed to migrate to a specific layer of cortex (Luskin 1994), suggesting substantial genetic control within the cell. Postnatally, glial cells multiply and differentiate much more than neurons do, and some play a direct role in learning (Henneberger, Papouin, et al. 2010).

2. *Migration.* Neuronal precursor cells multiply in the ventricular zone (VZ) lining the fluid-filled space formed when the neural fold be-

Table 6.1. Anatomical Processes in Brain Development

Neurogenesis	Birth of new neurons
Migration	Chemically guided progress to destinations
Differentiation	Development of progenitors into neuronal types
Extension	Putting out neurites to form axons and dendrites
Synaptogenesis	Formation of synaptic connections
Cell death	Death of about half of neurons
Pruning	Dying back of many connections
Myelination	Wrapping of axons in fatty sheaths that speed conduction

comes the neural tube. In evolutionary terms, the ventricles contain seawater from the world outside the organism, and the cells facing into them are the ones that, embryonically and phylogenetically, once faced the external world. Amoeba-like, cells leave the VZ in order, for places partly destined in advance (Sun, Takahashi, et al. 2002). CAMs and SAMs, resembling the antibodies that recognize invaders, direct the migration (Thiery 2003), and *reelin* helps regulate developing brain regions (Curran and D'Arcangelo 1998). Other genes act at different levels between the surface receptor for the signal and the intracellular machinery that directs cell movement (Couillard-Despres, Winkler, et al. 2001). Migrating neural stem cells follow radial glial cells arranged vertically throughout the cortex (Chanas-Sacre, Rogister, et al. 2000), and migration errors cause anomalies.

3. *Differentiation.* The final division of a precursor commits one or both daughter cells to become neurons; their type depends on their birthdays, the local molecular environment where they come to rest, and the neurons they synapse with. Since the progenitor's chemical gradients are symmetrical in one plane but polarized in the other, those that cleave vertically (parallel to the radial glia) have two similar daughter cells that remain nearby, while horizontal cleavage creates separated daughter cells of different neuronal types.

4. *Extension.* New neurons produce neurites—future axons and dendrites. Their tips extend by addition, stimulated and guided by molecular signals outside the cell (McFarlane 2000). These molecules overlap with those that direct migration (Uyemura, Asou, et al. 1996), but the semaphorin family helps guide axons (He, Wang, et al. 2002). In the grasshopper nervous system, gradients of semaphorins and

other molecules (Legg and O'Connor 2003) guide axons through dramatic 90-degree turns, yielding a ladder-shaped nervous system. As the neurite approaches, the target cell secretes neurotrophic factors that help keep its new neighbor alive (Sendtner, Pei, et al. 2000). In the developing mouse brain, Semaphorin-3A and Neuropilin-1 interact to sort axons into orderly bundles before they reach their targets (Imai, Yamazaki, et al. 2009).

5. *Synaptogenesis.* Next, synapses, the basis of neuronal communication, form (Munno and Syed 2003) and spines arise on the dendrites, enhancing synaptic input (Ethell and Pasquale 2005). Dendritic arbors branch out (Cline 2001). But far more synapses are formed than persist, and activity influences both their formation and maintenance (Changeux and Danchin 1976; Herrera and Zeng 2003). Synapses can form in circuits deprived of external stimulation, although with some abnormalities (Bourgeois and Rakic 1996). Known genes regulate synapse formation (Featherstone and Broadie 2000), but the environment helps determine which survive.

6. *Cell death.* As we have seen, cell death is central and much of it is programmed—a genetically predetermined molecular cascade, sometimes triggered by external stimuli. Extracellular neurotrophic factors can prevent it, including those from target cells (Sendtner, Pei, et al. 2000). It may not matter which are selected and which die if the connection shapes the partner cell's destiny, but there may be subtle differences in gene expression between a chosen and a rejected cell, which could make the process of competition adaptive.

7. *Pruning.* Given synaptogenesis, circuits may be active, which helps some survive massive pruning. The kind of synaptic function that resists pruning may resemble that in normal adults. Since synapse formation continues throughout life and peaks long after birth in many species, synaptic recruitment and survival can be greatly influenced by experience (Huttenlocher 1979, 1994; Huttenlocher and de Courten 1987; Huttenlocher and Dabholkar 1997). And since synapse number peaks at different ages in different cortical regions, there may be domain-specific sensitive-period effects (Huttenlocher 2003).

8. *Myelination.* Myelin is a fatty sheath that insulates many mammalian axons, a space-saving and energy-efficient adaptation for rapid con-

duction without the large fiber diameter of invertebrate axons. It resembles typical cell membranes, but with a higher ratio of lipid to protein and of glycolipids to phospholipids and cholesterol. The proteins, embedded in the lipid bilayer, include myelin basic protein, glycoprotein, and enzymes (Simons and Trotter 2007). The extent to which myelination is (1) a valid index of regional brain development, (2) required for behavioral functions, and (3) responsive to experience is considered later.

How much do these embryological processes change after birth? All continue postnatally and most throughout life. Birth itself entails no dramatic qualitative transition in brain development, as it does in heart and lung development (Adolph 1968), and this is especially true in humans, whose fetal brain growth rates continue for a year.

Plasticity in Neuroembryology

The quantitative underdetermination of the mammalian brain by the genes and the large role played by cell death and pruning led to models of brain development that give activity and environment a strong role. Three of these are selective stabilization (Changeux and Danchin 1976; Changeux 1985), the trophic theory of neural connections (Purves and Lichtman 1980; Purves 1988), and the neuronal group selection model (Edelman 1987). In all three, synaptic connections compete for survival, partly through activity, which in turn may depend on external stimuli. The ratio of peak to final number is greater for connections, and the peak is reached much later in development. This suggests that the developing nervous system of higher animals is plastic and environmentally responsive. Plasticity in the embryonic brain comes from at least four sources:

1. A neuron can influence the survival of others connected to it, along with the connections themselves. The exercise of circuits in spontaneous activity and response to stimulation fosters competitive elimination of neurons, connections, and perhaps neuronal groups.

2. Vertebrates from birds to humans are capable of learning before hatching or birth (Gottlieb 1982, 1991a, 1991b; DeCasper and Spence 1986; Krueger, Holditch-Davis, et al. 2004), although its importance is debated (Moon and Fifer 2000). Such learning would strengthen

existing connections, as adult learning does, but could also affect cell death and pruning.

3. Teratogens and nutrient deficits divert embryogenesis (Rice and Barone 2000). Thalidomide at a critical period causes agenesis of limbs; diethylstilbestrol alters the cervix, increasing cancer risk decades later; ethanol causes anomalies of face and brain; folate deficiency causes neural tube defects; smoking stunts body and brain growth; and heroin-using mothers have addicted fetuses that go through withdrawal after birth. Subtler consequences contribute to individual variation.

4. Hormones come from the environment, fetus, placenta, maternal circulation, and other fetuses. Androgens influence hypothalamic maturation, and since hormone levels affect neuronal migration patterns, other aspects of nervous system individuality could be affected (Geschwind and Galaburda 1985, 1987). They may help explain left-handedness, which runs in families without a Mendelian pattern. Also, androgens may cause the neuronal migration defects in some cases of dyslexia (Galaburda, Sherman, et al. 1985; Galaburda 1993) and influence autoimmune and other gender-biased diseases.

Much of this falls outside the variance explained by genes. We commonly think of nongenetic variance as dependent on psychological environments, but an unknown proportion is due to sensitivity to initial conditions, toxins, maternal hormones, other prenatal and birth effects, and developmental "noise" (which might be called developmental drift) rather than to nurturing, culture, education, modeling, reinforcement schedules, or other behavioral factors. Much *environmental* variation, after removing genetic sources, is biological.

Developmental Neuroendocrinology

Despite the proven effects of gender labeling, gender cognition, and gender enculturation (Golombok and Fivush 1994; Maccoby 1998; see also Chapters 10, 21, and 26), it is likely from evolutionary and physiological considerations that some sex differences in behavior are in part biological (Becker, Berkley, et al. 2008). It is also likely that some individual differences among males and among females (such as sexual orientation) are in part functions of the same kinds of variables that contribute to sex differentiation.

Mammals begin life with a basically female body plan; individuals without a Y chromosome are mainly female, and those with one are mainly male, regardless of how many X chromosomes they have. Although genes on the X chromosome contribute to female development, the major signal for sex comes from a gene in the sex-determining region of the Y chromosome (SRY), which makes the SRY antigen, or testis determining factor (TDF) (Goodfellow and Badge 1993), a genetic switch for the development of testes. The testes in turn produce a peptide that inhibits the development of female internal reproductive organs, while male structures emerge, and they make androgens, which account for most of the rest of maleness.

Research on sex differentiation of brain and behavior is extensive (Breedlove 1994; Collaer and Hines 1995; Hines 2005). Androgen effects are usually considered *organizational*—structural changes induced by hormones pre- or perinatally (Goy 1970; McCarthy and Arnold 2008)— or *activational,* such as hormonal enhancement or inhibition of firing in certain circuits in adulthood (Kendrick and Drewitt 1979). But since puberty also involves structural change in the hypothalamus, there may be three categories of reproductive hormone effects: organizational, activational, and pubertal.

Sex differences are seen in the hypothalamus and some other CNS regions in marsupials (Gilmore 2002), rats (Raisman and Field 1971; Pak and Handa 2008), mice (Toran-Allerand 1976), hamsters (Greenough 1977), monkeys (Ayoub, Greenough, et al. 1983), and humans (Swaab and Fliers 1985; Swaab 2004), among others (Gorski 2002). Some of these differences depend on circulating androgens that organize neural structure during fetal or early postnatal life. They may be subtle and microscopic or readily visible (Gorski, Gordon, et al. 1978). The preoptic area (POA) of the hypothalamus, involved in reproductive cycling and behavior, shows such differences, particularly in the sexually dimorphic nucleus (SDN-POA).

Paradoxically, the immediate cause of some of the alterations in males is an estrogen, estradiol-17ß (E_2), to which T is converted (Toran-Allerand 1976, 1996; Naftolin 1994); T also acts directly in some neurons (Wilhelm and Koopman 2006), and some genetic sex differences do not depend on hormones (Davies and Wilkinson 2006). E_2 does circulate in female fetuses, but protein binding keeps it mainly outside the brain; in males T enters and is converted to E_2 in neurons containing an aromatase enzyme, but other genetic background factors help prepare the hypothalamus for these effects (Carrer and Cambiasso 2002). Outside the nervous system, T causes differentiation of the male external genitalia, but only after it is converted to dihydrotestosterone (DHT) by the enzyme 5-α-reductase in appropriate target cells. The central effects are called androgenization or masculinization of the brain. There are also indirect CNS effects, as with the center controlling the rat bulbocavernosus muscle at the base of the penis. In males T enlarges the muscle, and this enlargement, in turn, maintains more neurons in the lumbar spinal cord to control it (Breedlove and Arnold 1980; Nordeen, Nordeen, et al. 1985).

Both these sexually dimorphic areas have analogues in humans, although the picture for us is less clear and the evidence that the sex differences result from organizational effects of fetal hormones is indirect (Swaab 2004). Still, interesting differences have been found in some hypothalamic nuclei (Swaab and Fliers 1985), and one shows a difference between heterosexual and homosexual men in the same direction as the difference between men and women (LeVay 1991). At present it cannot be ruled out that these differences result from life experiences. It is also unlikely that individuals with various sexual orientations merely correspond to points on a continuum of masculinization. But some differences among men and among women may result from differential exposure to naturally varying fetal hormones (see Chapter 10).

Although the concept of a basically female body plan is useful, it is insufficient. Since estrogens and progestins have their own organizing CNS effects, a two-dimensional model requiring feminizing and demasculinizing by ovarian hormones as well as masculinizing and defeminizing by testicular hormones has been proposed (Collaer and Hines 1995). The concept of a single gene and a single switch for sex differentiation is undoubtedly too simple. Added to key signal genes on the Y and X chromosomes are autosomal genes for hormone receptors, binding proteins, enzymes of synthesis and removal, and DNA binding sites for gene regulation. Divergent sex differentiation is caused by alterations in some of these, and subtler ones could influence differences in the normal range.

After infancy in humans, gonadal hormones decline to a low ebb, and reproductive maturation is delayed by hypothalamic sensitivity to feedback from those hormones. Pituitary gonadotropins stay too low to stimulate the gonads. Hypothalamic maturation causes this sensitivity to decline, so that de-repressed GnRH makes the pituitary secrete more gonadotropins for a given androgen or estrogen level, stimulating the gonads to make more. They in turn stimulate the development of internal reproductive organs, external genitalia, mammary tissue, the long bones, and secondary sex characteristics (Chapter 12). Given their impact on other systems, structural effects of sex hormones on the brain during puberty would not be surprising. The fact that androgens alter dendrite length in some rat neurons even in adulthood blurs the distinction between organizational and activational effects (Kurz, Sengelaub, et al. 1986). In some songbirds T causes large-scale neurogenesis in males each spring, but no such effect has been found in mammals.

Activational effects of gonadal hormones certainly occur after puberty, and rising levels may make them stronger. Androgens activate or enable both male and female sexual behavior and male aggressive behavior. (Female aggressive behavior is apparently not androgen-dependent.) Estradiol enhances female sexual behavior in some studies. Functional mechanisms include gene activation and increased sensitivity or higher conduction velocity of central or peripheral nerves (Kendrick and Drewitt 1979; Breedlove and Arnold 1980).

Postnatal Brain Development

In recent years, techniques have emerged for monitoring the development of brain structure and function non-invasively and safely in the growing child (Chugani, Muzik, et al. 2001; Evans and Brain Development Cooperative 2006; Almli, Rivkin, et al. 2007; Cascio, Gerig, et al. 2007; Picton and Taylor 2007; Grossman and Johnson 2007; Burnett and Blakemore 2009; for a review, see Marsh, Gerber, et al. 2008). These techniques include computed tomography (CT), magnetic resonance imaging (MRI), functional MRI (fMRI), positron-emission tomography (PET), diffusion tensor imaging (DTI), single-photon emission computed tomography (SPECT), multivariate electroencephalography (EEG), event-related potentials (ERP), near-infrared imaging (NIRI), magneto-encephalography (MEG), and other methods. For practical and ethical reasons, only the last four have typically been used to study normal children below the age of six or seven. However, retrospective studies of infants and young children whose brains were scanned under anesthesia to rule out disorders or trauma have also contributed to the imaging of normal brain development in living children. Overall, these methods are revolutionizing child development research.

Still, none of these techniques has anything like the resolution of microscopic study of postmortem brains, nor is any such technique likely to be developed for many years, so classical neuroanatomy has ongoing value. Some basic processes are almost complete by birth; neurons continue to be generated, but at far lower rates, and most have migrated to their destinations (Purves and Lichtman 1985; Rakic 1985; Gould, Tanapat, et al. 1998). Other processes proceed apace: synapse proliferation, dendrite branching, changes in dendritic spine density and connectivity, proliferation of glial support cells, myelination, and neurochemical maturation (Rakic 1985; Johnson 1993; Bell and Fox 1994; Chugani 1998). These processes are influenced by experience (Merzenich, Kaas, et al. 1983; Jenkins, Merzenich, et al. 1990; Elman, Bates, et al. 1996), but in some ways are independent of it. Human synapses are generated throughout life, peaking years after birth in some brain areas.

Gene expression studies are yielding unprecedented pictures of brain development (Wilhelm and Koopman 2006), as are measures of enzyme activity, receptor distribution, and other cellular chemistry. PET shows regional brain energy consumption tracking regional synaptic growth in children (Chugani, Phelps, et al. 1987; Chugani 1998) and measures enzyme activity and receptor distribution in living brains (Chugani and Chugani 2000; Chugani, Muzik, et al. 2001). However, a crucial guide to brain maturation remains one of the oldest, that of myelination sequences, and MRI has both confirmed and enhanced its value (Dubois, Dehaene-Lambertz, et al. 2008b).

Myelination Sequences in Development

Myelin-stained brain and spinal cord sections were a mainstay of classical neuroanatomy because they permit easy visualization of neuronal pathways (Yakovlev 1970). For example, the projection of the optic radiation from the thalamus to the occipital cortex becomes stainable (acquires myelin) during the immediate postnatal months in the human brain, making it stand out from the background of as-yet-unmyelinated pathways. Such processes can be readily seen with the light microscope (if not the naked eye) and described much more easily than counting synapses, and with much finer resolution than any form of brain imaging, although DTI has confirmed both the myelination sequence in the visual system and its visual importance during these first months (Dubois, Dehaene-Lambertz, et al. 2008b).

Microscopically, myelination begins after axons approach or reach their targets, with marked glial cell proliferation (Simons and Trotter 2007). In the peripheral nervous system Schwann cells and in the CNS oligodendrocytes, under genetic guidance, wrap a double thickness of cell membrane repeatedly around a section of axon to form a tight fatty sheath. This insulation forces ions to flow only through the nodes between myelin sections, increasing conduction speed ten- to one-hundred-fold. Although myelin develops rapidly, attaining the mature level of density may take months to years. CNS pathways myelinate at distinctly different times (Dubois, Dehaene-Lambertz, et al. 2008a; Blakemore 2008).

Attempts have long been made to correlate myelin with behavioral maturation (Flechsig 1920). Studies of the human cortex by J. Leroy Conel (1939–63; Rabinowicz 1979) showed that various measures of development exhibit considerable synaptic growth: the thickness of the cortex and its layers; the number, size, and density of neurons; intercellular components such as chromophil substance and neurofibrils; the number, size, and form of neuronal processes; and dendritic spine density. Synapse number and dendritic branching are correlated developmentally with myelination (Huttenlocher and Dabholkar 1997). Rank ordering of the relative maturity of cortical regions at a given age is similar with all these indices. To the extent possible using new imaging methods, myelination has been developmentally correlated with other measures like cortical folding and glial cell migration (Childs, Ramenghi, et al. 2001), and molecular markers like myelin-specific lipids and proteins confirm myelin staining measures (Hasegawa, Houdou, et al. 1992; Kinney, Karthigasan, et al. 1994; Haynes, Borenstein, et al. 2005).

But the most convincing demonstration of the importance of myelination sequences is coming now, with the capacity to delineate them in living infants who can also be functionally studied. For example, Jessica Dubois, Ghislaine Dehaene-Lambertz, and their colleagues (2008b), as part of an ongoing study of myelination in the brains of healthy infants, used diffusion tensor imaging to demonstrate asynchrony in myelin in different major fiber tracts. During the period from one to four months of age there are marked differences among tracts in myelin deposition, confirming in general the asynchrony found in older postmortem studies.

Animal experiments also correlate myelin with other indices of regional brain development. Changes in cell bodies during myelination include increased cell volume and changes in neuron packing density, nuclear and nucleolar diameters, and distribution of chromophil substance within the cell (Martinez and Friede 1970). In addition, myelination correlates with the growth of axon diameter, which may trigger it (Matthews 1968); a molecular basis for this has been described (Nave and Salzer 2006; Simons and Trajkovic 2006). So myelination sequences can serve provisionally as an index of regional brain development.

Some early investigators viewed myelination as a sine qua non of function (Tilney and Casamajor 1924; Windle, Fish, et al. 1934; Keene and Hewer 1931; Langworthy 1933)—an "absolute index of behavioral capability" (Angulo y Gonzalez 1929)—but in fact function precedes it. Vital circuits like the autonomic nervous system normally function without myelin, as do many invertebrate nervous systems, and demyelination does not invariably result in the expected loss of function in clinical syndromes like multiple sclerosis. But a large body of clinical data on demyelination and delayed myelination, evidence from animal models, and experiments on membrane function with and without myelin all support a significant, if imperfect, relationship between myelination and function (Krarup 2002; Caliandro, Stalberg, et al. 2007; Padua, Caliandro, et al. 2007).

Axons demyelinated pathologically or experimentally experience (1) decreased conduction velocity; (2) increased refractory period; (3) more frequent conduction failure; (4) temporal dispersion of impulses; (5) increased susceptibility to inadvertent electrical modification by neighboring axons; and (6) increased susceptibility to mechanical, thermal, and other extraneous influences (Ritchie 1984; Waxman and Bangalore 2004; Lee, Soong, et al. 2005). In addition, *re*myelination predicts the reappearance of a normal, or approximately normal, conduction latency and refractory period (Tomita, Kubo, et al. 2007).

Although these are abnormalities, the few existing studies of normal function before myelination appear to corroborate these findings. Peter Huttenlocher (1970) followed the developing cat pyramidal tract from three to five postnatal weeks. Although several functional capabilities preceded myelination, the ability of the fiber to conduct impulses under a high-frequency stimulus train did not—repetitive firing rates as low as 40/sec caused conduction block. Since rates of 50–110/sec are involved in normal pyramidal tract function during voluntary contraction of hand and forearm muscles in monkeys, premyelinated neurons would have subnormal capacity. Also, the active membrane surface of unmyelinated axons is two to three orders of magnitude greater than that of myelinated axons, making high firing rates very costly.

Experience Effects on Myelination

Although much is known about the events leading up to and associated with normal myelination (Nave and Salzer 2006; Melli and Hoke 2007), there is not yet a complete account of what causes it. Could function be a cause?

Early studies of experience and exercise effects on myelination showed effects that were either transient or, if lasting, about 10 to 20 percent (Gyllensten and Malmfors 1963; Kingsley, Collins, et al. 1970). The most severe stimulus deprivation, such as rearing in darkness, total occlusion of an eye, or sciatic nerve section, were compatible with the eventual acquisition of normal myelin in the great majority of fibers, strongly suggesting genetic control.

Later studies confirmed the experience effects. Stimulating or blocking the activity of optic nerve neurons in culture enhances or inhibits myelin formation (Demerens, Stankoff, et al. 1996). This finding led to a dramatic study that contrasted the white-matter density in certain fiber tracts in professional pianists with that in nonmusicians; the study showed that total hours of practice predicts white-matter density—most strongly if the practice occurs in childhood (Bengtsson, Nagy, et al. 2005), suggesting a sensitive-period effect. The differences are comparable in magnitude to the normal increase in white-matter density after age six—functionally significant, but much smaller than the earlier maturational changes (Schmithorst, Wilke, et al. 2002).

A considerable literature supports genetic control (Griffiths 1996; Kamholz 1996; Wegner 2000). Among the infantile diseases of the cerebral white matter, at least six have simple Mendelian inheritance patterns (Lyon, Fattal-Valevski, et al. 2006; Boespflug-Tanguy, Labauge, et al. 2008). These syndromes are characterized by global delays in motor development consistent with the delay or lack of myelination and are unresponsive to exercise (Wibom, Lasorsa, et al. 2009). They and many others, including inherited peripheral demyelinating syndromes, are now studied with molecular methods and modeled in transgenic mice (Scherer and Wrabetz 2008).

Animal models point to genetic control. In two mouse mutants— *jumpy,* an X-linked mutant with severe hypomyelination in CNS but not PNS, and *quaking,* an autosomal recessive allele on chromosome 17 causing global hypomyelination—messenger RNA transcription for myelin basic protein (MBP) is normal, but MBP-related translation products become particularly relevant at different stages of myelin synthesis (Carnow, Carson, et al. 1984). Transgenic models suggest that axon loss is important for many symptoms (Meyer Zu Hörste and Nave 2006). Myelination, then, whether pre- or postnatal, is a genetically controlled process mainly intrinsic to brain growth. Its influence is large compared with the (also significant) reciprocal influence of experience.

Human Myelination

Paul Flechsig (1920) described myelination sequences in the human cerebral cortex, inspiring early efforts to correlate myelination with behavioral development, particularly prenatally (Tilney and Casamajor 1924; Windle, Fish, et al. 1929; Keene and Hewer 1931; Langworthy 1933). Later studies extended these. Paul Yakovlev and André Roch Lecours (1967) studied myelin-stained sections of two hundred neurologically normal whole human brains, from early gestation to old age. Their method, not repeated since, was to use whole-brain serial sections to reconstruct anatomy at a resolution even now unattainable with imaging.

Postnatal changes are dramatic, especially in the first few years. Brain sections arranged by age show clearly that myelination, like other indices of brain development, is regionally specific from birth, and this is confirmed by DTI (Dubois, Dehaene-Lambertz, et al. 2008a); such asynchronies explain the development of behavior and mind, not as a uniform increase in processing power, but as particular circuits coming on-line in a genetically guided maturational sequence. The "cycles of myelination" in human development as roughly quantified by Yakovlev and Lecours were confirmed in most details by subsequent research.

For example, Lucy Balian Rorke and Helena Riggs (1969) studied the brains of 107 infants of varying birth weights, representing a range of gestational ages. Of 23 full-term newborns, 87 percent showed "essentially similar" myelination, reflecting consistency in normal neonatal brains. In their study of 323 brains ranging in estimated age from 20 to 48 postconceptional weeks, with several cases at each age, Floyd Gilles, Alan Leviton, and Elizabeth Dooling (1983) quantified myelination (0 = no myelin, 1 = microscopic only, 2 = just visible, 3 = intensity approaching mature myelin). Betty Ann Brody, Hannah Kinney, and their colleagues extended this scale to the first two postnatal years in their study of 171 more cases; they assessed 62 anatomical sites and added a fourth degree to the scale for the densest level reached (Brody, Kinney, et al. 1987; Kinney, Brody, et al. 1988). For each site they reported the median myelination by age and the ages at which 10, 50, and 90 percent of infants showed level 3. The 62 sites fell into eight subgroups by myelination pace and timing, strongly supporting circuit-specific differentiation of CNS white matter and in most patterns confirming earlier findings. In another sample, postmortem measurement of sphingomyelin, myelin basic protein, cerebrosides, and relevant brain lipids confirmed the anatomical findings for ten brain sites and again strongly upheld regional differences (Kinney, Karthigasan, et al. 1994).

During the 1990s MRI—superior to CT in distinguishing white from gray matter—came into common use, and developmental studies of normal children began (Dietrich, Bradley, et al. 1988). These studies have confirmed at a gross level in living brains the basic findings of the postmortem myelination work (Paus, Collins, et al. 2001; Parazzini, Baldoli, et al. 2002; Carmody, Dunn, et al. 2004); they have also confirmed a correlation between myelination and brain activity (Olesen, Nagy, et al. 2003) and between myelination and behavioral and cognitive function (Nagy, Westerberg, et al. 2004; Dubois, Dehaene-Lambertz, et al. 2008a). In addition, they have confirmed and extended the observation of Yakovlev and Lecours that in some brain systems significant further myelination occurs throughout later childhood, adolescence, and beyond (Paus 2005; Lenroot and Giedd 2006). These studies have been important in diagnoses of demyelinating conditions and assessments of developmental delay (Barkovich 2000) and prematurity (Counsell, Maalouf, et al. 2002).

Recently the National Institutes of Health (NIH) MRI Study of Normal Brain Development was initiated to "provide a basis for characterizing healthy brain maturation in relationship to behavior and serve as a source of control data for studies of childhood disorders." The database, to be made available to the scientific and medical communities, will consist of MRI, clinical, and behavioral data from about five hundred children age seven days to eighteen years in a mixed cross-sectional and longitudinal design (Evans and Brain Development Cooperative 2006). This is an invaluable resource for ongoing research of the kind that Paul Yakovlev envisioned, but the brain-behavior correlations, although at much lower anatomical resolution, will have different meaning in living brains (Almli, Rivkin, et al. 2007; Waber, de Moor, et al. 2007). It is a cornerstone of twenty-first-century child development research.

Most of the anatomists had little direct interest in behavioral maturation; nevertheless, their work provides a basis for reasonable inferences about it. As noted by Gilles, Leviton, and Dooling (1983), there is substantial agreement with Yakovlev and Lecours (1967), certainly in the sequence of myelination and to a lesser extent in the precise timing, across studies done at different times, by different investigators, using different methods, and with different research goals.

Myelination and the Development of Behavior

Myelination sequences have specific behavioral consequences. For example, the system serving the detection of postural orientation and vestibular stimulation is fully myelinated before birth. A basic consequence is the suite of vestibular reflexes found in all normal human newborns (Peiper 1963; Prechtl and Beintema 1964). As Annelise Korner (1972) noted, it may also explain the unique effectiveness of rocking stimulation in quieting the newborn, as well as the positive effect of upright posture on alertness (Gregg, Haffner, et al. 1976).

Major visual system tracts begin myelination just before birth and complete it rapidly in the first few months of life, along with the rapid attainment of visual function. Two detailed models have related the growth of visual perceptual ability in early infancy to visual system myelination, one using postmortem descriptions (Bronson 1974, 1982), the other using DTI and correlating myelination with visual evoked potentials in the living infants (Dubois, Dehaene-Lambertz, et al. 2008b). This contrasts markedly with auditory system myelination (Lecours 1975; Lecours, Lhermitte, et al. 1983). Components of the subcortical auditory system are myelinated before or shortly after birth, but the acoustic radiation to the cerebral cortex has a much longer myelination course, taking at least one to two years and corresponding to the slow development of the major function of the human auditory analyzer, language comprehension. This too is confirmed by MRI studies (Pujol, Soriano-Mas, et al. 2006).

Myelination of the very long neurons of the corticospinal tracts predicts the dramatic gains in neuromuscular function during the first year of life. A human newborn's spinal cord resembles that of a patient with corticospinal damage, because of the absence of myelin in those tracts (Rorke and Riggs 1969). Reflex and motor development in the normal newborn resembles what would be expected in a neurological patient recovering from such an injury (Chapter 7).

Cortical association areas continue to gain myelin up to age thirty years to an extent consistent with the known effects of experience. MRI confirms the protracted development in association areas (Benes 1998; Lenroot and Giedd 2006). The description of important changes during adolescence in particular (Sowell, Peterson, et al. 2003), in the hippocampus and in the cortex (Benes, Turtle, et al. 1994), has suggested to many that criminal responsibility should not be assigned before adolescent brain development is complete (Scott and Steinberg 2008).

In a microscopic analysis, Kathleen Gibson (1977, 1981, 1991) compared the myelination of inputs and outputs of different cortical layers in rhesus monkeys and humans in relation to cognitive development and found a reliable sequence of myelination among the six layers in any given cortical region. Just as the primary sensory and motor projection areas of the cortex develop before the association areas, the layers mediating communication with the brain stem and spinal cord (I, IV, V, and VI) myelinate in advance of those that serve communication within the cortex (II and III). Layers IV, V, and VI in the association cortex become myelinated between fifteen and twenty-four months or later in humans, when advances requiring high-order cognitive integration occur.

Other striking correspondences between myelination and behavioral development have received little or no attention. Some major limbic system tracts do not begin to myelinate until weeks or months after birth. The cingulum, linking the frontal lobe to the limbic system (Nauta 1971), myelinates between two and ten months. The fornix, a massive projection from the hippocampus, myelinates in the second half of the first year and later, as do other major limbic system tracts. Finally, the corpus striatum and globus pallidus, along with their fiber tracts, myelinate postnatally. These structures participate in the initiation and modulation of movement (Evarts 1975), but also in the cortical-subcortical circuits central to all higher brain function (Alexander 1994), and they play an important role in learning (Graybiel 2005). Moreover, stimulation within them can produce highly ritualized, species-specific fixed action patterns that serve as social displays in monkeys (MacLean and Ploog 1978); human social displays such as facial expressions could be controlled in part through similar sites.

Imaging studies, at lower resolution, also find correlations between myelination and higher cognitive functions. In DTI, for example, frontal lobe myelination corresponds to the development of working memory, while only left temporal lobe myelination correlates with reading ability (Nagy, Westerberg, et al. 2004). We consider more such studies as we move through the developmental sequence.

Developmental Behavior Genetics

There are at least eight ways to explore the genetic basis of behavior in complex animals:

1. Find behaviors that are universal for a species, on the assumption that universality often (but not always) points to a genetic basis.

2. Do breeding experiments with nonhuman species, preferably highly inbred strains, to discover the mode of inheritance of specific behaviors.

3. Create strains of animals with particular genes deleted, temporarily suppressed, or added and study the loss or gain in behavioral function.

4. Compare genetically identical and non-identical human twins on behavioral measures to calculate degrees of similarity.

5. Compare adopted children with their genetic and adoptive parents and with the latter's genetic children.

6. Study hereditary defects that run in families.

7. Use molecular genetics to screen genomes for the genes associated with behavior.

8. Trace the biochemical pathways from genes to behavior.

Breeding "experiments" have been done for at least ten millennia, often to change behavior, and were crucial in Darwin's thinking about natural selection. Dogs have been bred for sense of smell, pointing, soft-mouth bite, retrieving ability, intelligence, trainability, herding ability, aggressiveness, reliability (as in seeing-eye dogs), and gentleness (as in children's pets). Similarly, horses have been bred for sprint speed, long-range endurance, dressage, calmness, and trainability; cows, sheep, and goats for docility; and Spanish fighting bulls for bravery, speed, persistence, pain tolerance, and straightness of charge.

Classic laboratory experiments extended this history. Studies on the social behavior of dog breeds confirmed traditional efforts and showed the effects of crosses and backcrosses between breeds similar to effects for physical traits (Scott and Fuller 1965). Aggressiveness, learning ability, and tail wagging were not single-gene traits, but clearly genes influenced them. Trainability, response to punishment, and vulnerability to social isolation differed markedly across dog breeds (Fuller 1955; Freedman 1958; Fuller and Clark 1968), demonstrating genotype-environment interactions of the sort now studied in humans (Jaffee, Caspi, et al. 2005; Kendler, Kuhn, et al. 2005). Mice and rats have been bred for aggressiveness and timidity, maze-learning, food and alcohol consumption, sex drive, seizure susceptibility, and general motor activity. Early crossbreeding experiments elucidated the genetic determination, and in alcohol preference and seizure susceptibility in mice the biochemical path from gene to behavior was shown, but now molecular-genetic dissection of behavior is routine (Pfaff, Berrettini, et al. 2000). Research in eight mouse strains revealed two hundred genes expressed differently in different strains in the hippocampus alone (Fernandes, Paya-Cano, et al. 2004), and some of these differences explain strain differences in behavior (Martin, Churchill, et al. 2009).

Genotype-environment interaction is also critical. Studies of aggressiveness in fourteen mouse strains, tested by putting four previously isolated males together for an hour and counting instances of attack, chase, and fight, revealed large differences between the most aggressive and least aggressive strains in response to social isolation (Southwick and Clark 1968). Such interaction effects have been shown many times (Benus 1999). The experience of being reared in isolation or handled in infancy is at least as important in determining the degree of aggressiveness in the tested mice as the genes are, but behavior is best predicted by knowing both.

Research on the genetic influence on behavior has become highly quantitative and molecular (Pfaff, Berrettini, et al. 2000; Flint 2003). In nonhumans, genetic strains are randomized to different conditions to apportion the resulting variation to genes, environment, or their interaction; human studies substitute twin, family, and adoption paradigms. But almost all human studies are correlational and subject to statistical biases, including genotype-environment correlations (the equivalent of biased assignment of genotypes to environments in animal studies) and differential restriction of either the genetic or the environmental variation, which can bias calculations of the apportionment of variance to genetic versus environmental causes. Despite these pitfalls, we have learned a lot.

Twin Studies

If genes strongly influence a trait, then identical (monozygotic, MZ) twins should be much more similar to each other than non-identical (dizygotic, DZ) twins, MZ twins reared apart should still be very similar, and DZ twins should resemble each other about as much as ordinary siblings. There are many reasonable questions. MZ twins, because they look alike, could be treated more similarly than DZ twins. With MZ twins reared apart, the preferences of parents and the rules of adoption agencies tend to ensure placement in similar environments, sometimes even with relatives living in the same neighborhood, so we can ask to what extent being reared apart really means being reared differently. There are also questions about the mutual influence of co-twins of different twin types when they are together, about gene-gene interaction effects, and so on. Some of these concerns have been effectively dealt with; others persist.

Genes exert a significant influence on such complex human behavioral tendencies as friendliness, aggressiveness, impulsiveness, and sociability, tests of intellectual ability, and mental illnesses such as schizophrenia and depression (Plomin, Owen, et al. 1994; Plomin and McGuffin 2003; Butcher, Kennedy, et al. 2006). But because these traits are so multigenic, it is difficult to identify individual genes. Ideally, some must be found and their effects must be traced biochemically from the gene(s) through one or more brain or hormonal systems to the behavior—the task of behavioral genetics in this century.

An example is a study of more than one thousand people with autism spectrum disorders (ASDs), showing that ASDs associate with two genes coding for cadherins (Wang, Zhang, et al. 2009); these are among the cell adhesion molecules mentioned earlier, which help establish connections between neurons in the developing brain. This finding should lead to an understanding of how neurons go embryonically awry in some children, but unfortunately ASDs are genetically and environmentally heterogeneous. Similarly, schizophrenia involves thousands of genes with very small effects (International Schizophrenia Consortium 2009; Shi, Levinson, et al. 2009; Stefansson, Ophoff, et al. 2009). Unless a family can be found with simpler inheritance, it will be very difficult to gain a molecular understanding of the disorder—if indeed it is one disorder. This will also be true of variation in "normal" behavior.

Although individual differences have partly genetic bases, there is little evidence that behavioral differences between human groups do. Even in biochemical traits, such as blood types or the genes themselves, each major racial group holds over 90 percent of the variation found in our species (Lewontin 1972; Brown and Armelagos 2001). If this is true for biochemical traits with simple and fixed genetic determination, it is probably truer for such plastic traits as complex behaviors. In the past human behavioral genetics was marred by racial and political bias and scientific fraud (Kevles 1985; Matheny 1990), but this is not true of most recent work. Overall there is little support for genetically based racial and ethnic differences in behaviors (Part IV), but strong support for genetic influence on within-group individual differences.

Explicitly Developmental Studies

Much of modern behavior genetics has been explicitly developmental (Wilson 1978; Scarr and McCartney 1983; Eaves, Eysenck, et al. 1989; Hahn, Hewitt, et al. 1990; Scarr 1992, 1993; Reiss, Neiderhiser, et al. 2000; Plomin 2005), and direct measures of behavior and responsiveness in the developing child have supplemented older measures of personality and intelligence (Emde, Plomin, et al. 1992; Plomin, Emde, et al. 1993). Paradoxically, such studies often show that twin correlations, including separated twins, increase during development, whether in childhood (Wilson 1978; Knafo and Plomin 2006) or adulthood (Rose, Koskenvuo, et al. 1988). Similar late effects of genes on behavior are seen in animal models (Cairns 1993).

This contradicts the belief that we start life with genetic endowments and become increasingly differentiated by environments, but it could have been anticipated from genetic nervous system diseases like Huntington's and Alzheimer's, which show their effects only late in life. With normal variants, measures in infancy and early childhood may be too disturbed by development itself—a kind of statistical noise around the stability of traits—to reliably reflect the genotype, which may emerge stably only later (Kagan 1984). Twin studies show a significant genetic contribution to a wide range of adolescent psychopathologies (Eaves, Silberg, et al. 1997), but adolescence is also a time of rapid and varied development, and teenage delinquency is much less heritable than that of adults (Lyons, True, et al. 1995).

The quality of method and theory in behavior genetics can be criticized, but the general dismissal of it on philosophical and methodological grounds is easily answered (Baumrind 1993; Jackson 1993; Scarr 1993; Gottlieb 1995, with commentary). No matter how strong one's commitment to the epigenetic, environmental, or self-organizing mechanisms in development, it does not make sense to deny that population-based, probabilistic, statistical approaches have value; these two kinds of approaches must proceed in tandem.

It can seem paradoxical to consider genetic variation in emotional and social traits. If they are so involved in survival and reproduction, why didn't natural selection drive them to fixation, like bipedal walking? First, there could be more variability now due to a relaxation of selection outside the environment of evolutionary adaptedness. Second, there may be ongoing evolution in human personality. Third and most likely, different personalities pursue different adaptive strategies, achieving reproductive success by different behavioral means. If (also likely) there were frequency-dependent selection acting on such traits, an evolutionarily stable strategy (ESS) could emerge and maintain variability indefinitely.

The last point is suggested by cross-population consistency in patterns of variation. The five-factor model of personality, based on answers to hundreds of questions, identifies dimensions called neuroticism, extroversion, openness, agreeableness, and conscientiousness (McCrae and Costa 1999). All five have substantial and roughly comparable heritability and are valid and stable from adolescence to middle age (McCrae, Costa, et al. 2000), although the first three decline and the last two increase somewhat over this period. They emerged in seven countries representing at least five major language groups (McCrae and Costa 1997; Costa, McCrae, et al. 2000), showed consistent patterns in fifty cultures (McCrae, Terracciano, et al. 2005), and were seen in psychiatric samples in different cultures (Bagby, Costa, et al. 1999; Yang, McCrae, et al. 1999); the personality disorders diagnosed on Axis II of the *Diagnostic and Statistical Manual* may be end points of continua of personality variation.

An earlier three-factor model was also consistent across cultures. These findings are not contradictory (the number of factors is arbitrary), but rather support each other in showing that responses sort similarly into personality dimensions in different cultures. This does not mean that each culture has the same distribution of personalities—although the differences are small and nations do not conform even to their own stereotypes (Terracciano, Abdel-Khalek, et al. 2005)—but that traits vary coherently within populations and that the same factors explain variation in different cultures. (We return in Part IV to the possible role of personality in culture.)

Longitudinal Stability of Childhood Timidity

Biologically influenced personality traits, often called temperaments, are easily identified in infancy but may not be stable. By early childhood, however, many children have distinctive traits. The best evidence for genetic influence on temperament is in behavioral inhibition, also called shyness or timidity (Kagan, Snidman, et al. 2007). In children as young as fourteen months, an unusually bold and an unusually timid or inhibited subgroup can be identified. Those who at that age were behaviorally inhibited in a strange or novel situation grew up to be four-, five-, and seven-year-olds who tended to be restrained and socially avoidant with unfamiliar children and adults and to show strong fears—of television violence, for instance, or of being alone in their rooms at night. As adolescents, they were more likely "to be subdued in unfamiliar situations, to report a dour mood and anxiety over the future, to be more religious, to display sympathetic tone in the cardiovascular system, to combine a fast latency with a large magnitude of the evoked potential from the inferior colliculus, and to show shallower habituation of the event-related potential to discrepant visual events" (Kagan, Snidman, et al. 2007:1). In fact, at four months of age, ten months before timidity per se could be identified, these infants were more reactive to a novel stimulus; this predicted timidity for the next seventeen years. Using different measures in a different culture, inhibited and aggressive children in Munich at ages four to six were shown to have similar traits at age twenty-three, with the aggressive children showing high levels of delinquency and other externalizing problems (Asendorpf, Denissen, et al. 2008).

Timidity is also associated with biological measures, in particular those reflecting more sympathetic tone—a higher and less variable resting heart rate (HR); a greater increase in HR and higher blood pressure in strange situations; greater pupil dilation during questioning; more tension in the vocal cords; higher cortisol levels; and more norepinephrine activity—pointing to a global but almost certainly polygenic cause. Attempts to link shyness to specific genes, notably the polymorphic gene for the serotonin transporter promoter (5HTT), have produced conflicting results (Battaglia, Ogliari, et al. 2005), but the search continues.

Animal experiments on sympathetic tone suggest that the limbic system may have a lower arousal threshold in inhibited children, possibly involving norepinephrine enzymes. Vagal tone and respiratory sinus arrhythmia (RSA) are also related to shyness (Doussard-Roosevelt, Montgomery, et al. 2003), as predicted by polyvagal theory. Direct measurement of sympathetic and parasympathetic nervous system (measured by RSA) function shows stability in individuals from six to twelve months old (Alkon, Lippert, et al. 2006). Relating these findings to the polyvagal model (see Transition 1), inhibited children seem to have a different pattern of function in the mammalian fast component of the vagus, thought to modulate fear in the process of normal social behavior. Both timidity and boldness can be adaptive in Darwinian terms, and the balance in a population could easily be an ESS. There are no doubt other biologically related stable traits in children, most of which await comparably thorough study.

External events can cause timidity against varied genetic backgrounds; for example, very-low-birth-weight infants without measurable abnormalities in infancy are more likely to be shy as adults (Schmidt, Miskovic, et al. 2008). Since we can be sure of environmental effects only by controlling for genotype, the best insights into those effects often come from behavior genetic studies. For example, two-thirds of uninhibited children are firstborns, while two-thirds of timid ones are later-born. Since birth order does not reflect genes, there must be an environmental influence interacting with genes; perhaps older siblings are intimidating, or at least inhibiting (Kagan 1995).

Understanding the Environment by Studying Genes

A more general fact about the environment emerges from behavior genetics: the *shared* family environment contributes surprisingly little to most similarities observed between siblings after genetic similarity is factored out (Plomin and Daniels 1987; Rowe 1994). Whatever the environment is doing, its main effect for most traits is to make children in the same family different from one another. Identical twins are often more alike in personality, and on average no less alike, if raised apart than if raised together. Parents usually strive to provide equally for all their children—religion, schooling, sports, television, bedtimes, diet, household rules, toys—so it seems puzzling that between-family differences contribute little to environment effects. Yet siblings create unique environments for each other, and parents—despite their claims—do treat their children somewhat differently.

Birth order is often said to be important, and some studies find effects of it (Sulloway 1997), but meta-analysis has shown surprisingly little explanatory power for this variable in most psychological traits (Ernst and Angst 1983; Rodgers 2001; Steelman, Powell, et al. 2002). Still, somehow siblings in one family get different environments. Parents could be differentiating children's environments on some other grounds than birth order—say, similarity to or difference from the parent in looks or temperament. Every pairing of parent and child produces a unique relationship, a unique environment. Socially and in other ways, it is also likely that children's genotypes lead them to choose their microenvironments partly from innate inclinations (Scarr and McCartney 1983). And, children affect each other. MZ twins in the same family may try to assert their own identities, becoming more different than separated MZ twins, whose shared genes make them converge. Ordinary siblings often present each other with marked personality differences for genetic reasons alone; a plain, quiet girl with a flamboyant and pretty younger sister has a very different environment from the same girl with a doting older brother.

Recall that restricting environmental variation (say, by studying middle-class families) can lead to underestimates of between-family differences. Restricting *genetic* variation leads us to underestimate genes, as with species-specific universal traits. Because the genes underlying such traits have been selected to fixation, any variety is environmentally caused, and we could mistakenly infer that genes contribute little. In fact, the constancy and universality of species-wide traits is part of the case for their genetic basis, and we will be learning more and more about how these genes work. Contrary to the claim that the nature-nurture debate is fruitless (Spencer, Blumberg, et al. 2009), it is now in its most exciting phase (Spelke and Kinzler 2009).

Neurological Individuality

Children have different brains. We readily acknowledge this when we see marked departures from standard paths of cognitive and language development (Tager-Flusberg 1994). But although we are gaining knowledge of brain anomalies, causality remains obscure. Theories of conditions such as autism, schizophrenia, depression, and dyslexia are interesting, but except in individual cases brain abnormalities have not been decisively shown. To get some idea of the extent of neurological individuality—genetic or environmental—consider variation on the margins of "normal." Of children randomly sampled from American communities, around 12 percent have diagnosable problems (Costello, Egger, et al. 2005; Costello, Foley, et al. 2006); in Canada the figure is 14 percent (Waddell, Offord, et al. 2002). These are conservative estimates that exclude children with general developmental disorders like mental retardation or specific learning impairments like dyslexia, as well as children whose problems were probably transient or otherwise "normal."

For example, at least 5 percent of children have attention deficit disorder (ADD), and many are also hyperactive (ADHD). That stimulants help these children gives insight into the biological mediation of ADHD, if not its cause. Methylphenidate, amphetamine, and related newer agents in wide and growing use stimulate attention, reducing restlessness. More children have oppositional or conduct disorders, which, like ADHD, are much more frequent in boys than girls. (Depression, anxiety, and behavior problems like bedwetting are common in both.)

Functional imaging implicates brain reward systems, including the nucleus accumbens, frontal lobes, cerebellum, and amygdala, in the impulsivity of ADHD (King, Tenney, et al. 2003), but other brain regions are involved as well (Hinshaw 2003; Biederman and Faraone 2005; Seidman, Valera, et al. 2005). Dopamine transporter density in some brain areas is elevated in ADHD, and both prescription stimulants and nicotine reduce it (Krause 2008). The worldwide prevalence has been estimated at more than 5 percent (Polanczyk, de Lima, et al. 2007), but in evolutionary perspective it is important to recall that this is in part a disorder of discordance; most diagnoses occur in connection with school, and although some impairment would have resulted from it even among hunter-gatherers, the number of children practically affected would have been much smaller.

The environment also contributes to these symptom patterns. Pregnant women with anxiety in weeks 12 through 22 of gestation have children with more ADHD and externalizing problems than average at ages eight and nine years, but mothers with comparable anxiety in weeks 32 through 40 do not (Van den Bergh and Marcoen 2004); genes cannot explain this. Some dyslexics have neuronal migration defects on postmortem study (Galaburda 1993), but both the phenotypes and brain manifestations are heterogeneous (Grigorenko 2001). Difficulty learning to read (also not a problem in the deep human past) often results from adverse learning environments, and it will still (like every behavior) have functional brain correlates. And even neuronal migration defects can be environmental in origin—they may be the result, for example, of early intrauterine toxins (Rice and Barone 2000). However, behavior genetic studies show high heritability for ADHD, and molecular genetic studies are under way (Hinshaw 2003). Dyslexia is genetically heterogeneous; there are linkages to markers on six different chromosomes in various kindreds, each of which shows a fairly clear pattern of inheritance (Kaminen, Hannula-Jouppi, et al. 2003).

Even within the normal range, children have distinctive brains, and not just owing to genetic or other prenatal causes. Head injury, lead poisoning, microbes, high fevers, and other insults of childhood, not to mention learning and culture, may also contribute. Research on "multiple intelligences" challenges the idea of general cognitive powers and recognizes the distinctiveness of children's brains (Gardner 2004). The model defines at least six kinds of intelligence: linguistic, musical, logical-mathematical, spatial, bodily-kinesthetic, and personal; in Darwinian terms, "human beings have evolved to exhibit several intelligences and not to draw variously on one flexible intelligence" (Gardner 2004:xi). If we contemplate "all the roles or 'end states'—vocational or avocational—that have been prized by cultures during various eras," it is difficult to imagine that identical brains would lead to success in each of them.

Similar things can be said about emotional predispositions. Gardner's "personal intelligence"—with interpersonal and introspective subtypes —strongly implicates emotions, and others have identified emotional intelligence as a distinct category (Goleman 1994), with known brain correlates (Killgore and Yurgelun-Todd 2007). In the right-hemisphere analogue of left-brain-based developmental dyslexia mentioned earlier, (Weintraub and Mesulam 1983), children have left-sided neurological signs suggesting mild right-hemisphere damage, along with decreased social interaction and emotional expression. Like dyslexia, this syndrome could be the end of a continuum of variation in brain structure and function within the margins of "normal."

We have in general understated intrinsic individuality, in part because we have underestimated genes. This reluctance stemmed from a grotesque exaggeration in the past of genetic explanations for group differences. But we now know that the two main thrusts of genetic analysis are to help explain human universals and individual differences, while group differences are overwhelmingly cultural. Ongoing progress in behavioral genetics and in other biological explanations of behavior will benefit individuals and enhance our understanding of universals, but as we will see in Part IV, we have to look elsewhere to explain group differences.

Interlude 2

Thinking about Bipedal Walking

Bipedal walking, a hallmark of humanity and a milestone in every normal infant's life, evolved independently in (at least) dinosaurs and birds, kangaroos, and hominins, requiring skeletal and neurological modifications. The transition occurred at least 6 mya in our lineage; the adaptation allowed us to carry infants, tools, and food, see greater distances, conserve energy or promote cooling during walking, and display breasts or genitals (Harcourt-Smith and Aiello 2004; McHenry 2004). As always, the changes were achieved by altering species-specific maturational plans; early research showed experience effects, but they are minor compared to the temporal map of developmental milestones. Myrtle McGraw (1935) embraced the idea of a species-specific neuromuscular development sequence and described it in detail for a longitudinal sample; Arnold Gesell and Catherine Amatruda (1947) provided similar descriptions, augmented with photographs, for a wider range of infant behavior.

J. Leroy Conel (1939–63) began to rely on these descriptions in functional interpretations of the regional maturation of the cerebral cortex. For example, at six months of age, by all measures, Area FAγ of the precentral gyrus (the motor cortex) in the region of the hand is more developed than other areas of the motor strip, corresponding to the early maturation of hand control. In contrast, another part of FAγ, representing the lower extremities, is developmentally behind the rest of the motor cortex even at fifteen months; hence the continuing weakness of lower limb control at that age. Recent research confirms the relative timing and extends these findings; for example, specific changes in the motor cortex correspond to the emerging alternate-phase limb actions in mature walking (de Jong, Leenders, et al. 2002).

Cross-cultural research on motor development has used the Bayley scale of infant development (Bayley 1965, 1969), a similar test used in Britain (Griffiths 1954), the Gesell scales (Gesell and Amatruda 1965), and the

McGraw scale (McGraw 1943); results from these tests are correlated, but
not identical (Ramsay and Fitzhardinge 1977). There are interesting variations in the timing of motor and sensorimotor milestones within and among
populations (Werner 1972; Super 1981; Mayson, Harris, et al. 2007). Although genetic explanations of between-group differences cannot be ruled
out, the range of variation within samples considerably exceeds that among
population means worldwide for these milestones, despite greater variation
among populations in environmental influences. This consistency of population averages means suggests a species-wide timing of events; independent
sitting and visually directed reaching appear in the middle of the first year,
independent rising to stand later that year, and independent walking and
thumb-to-finger fine grasp early in the second year.

   Consider bipedal locomotion, a species-typical, centrally organized neuromotor action pattern shown by all normal adults—indeed by all normal
two-year-olds. The mean age of taking three steps without hands held is just
over a year in many samples, usually falling between eleven and fourteen
months. Large samples in five European cities had means within six weeks of
each other (Hindley, Filliozat, et al. 1966). Precocity for infants in developing
countries, especially Africa, has been claimed (Geber and Dean 1957), but
carefully designed and conducted studies failed to show any difference (Super 1976), and a critical review of a large number of studies concluded that
African infant precocity had not been demonstrated (Warren 1972). A review of studies of psychomotor development in non-Western cultures did
suggest a modest precocity for such cultures in general, but no dramatic departures from the species-typical pattern (Werner 1972).

   The means for African and other samples in developing countries for
age at independent walking typically fall within the American and European
range just mentioned, environmental differences notwithstanding. Among
the !Kung San, hunter-gatherers of Botswana, deliberate efforts are made to
accelerate the development of walking by means of seemingly appropriate
tactics. These include extensively holding the infant in a sitting or standing
posture long before independent maintenance of these postures is possible,
exercising rudimentary walking capabilities, and so on. The curve for development of independent walking in these infants falls within the range of the
corresponding curves for American infants (Konner 1973, 1977b). Other
cross-cultural variations in infant care that might be expected to alter motor
development also had modest effects. The Hopi (Dennis 1940; Dennis and
Dennis 1940) and Navajo Indians (Chisholm 1983) traditionally restricted

their infants much of the time by tightly swaddling them against cradle-boards; this did not substantially delay the age of first walking.

Finally, deliberate attempts to accelerate this maturational pattern experimentally under relatively controlled conditions have usually met with little success (McGraw 1935). Studies of the effects of sleep position and equipment such as a walker show modest effects at most (Pin, Eldridge, et al. 2007). One intervention study (Zelazo, Zelazo, et al. 1972) did succeed in producing an eight-week advancement of first independent walking by systematically exercising the neonatal automatic march reflex for the first eight weeks of life. Although this suggests that extraordinary environmental modifications may alter the rate of development, more typical variations in rearing conditions, even those that support a role for learning, have smaller effects.

In addition to regularities of developmental rate, a preponderance of evidence points to a species-typical developmental sequence. From the neonatal automatic march reflex, a centrally organized subcortical motor stereotypy responsive to visual cues (Barbu-Roth, Anderson, et al. 2009), to the mature gait of the two-year-old, with heel-toe progression directly under the hips and synchronous alternate arm swinging, the developmental history of the pattern is for the most part characteristic and universal, and the timing of its major transitions is narrowly defined (Annick Ledebt 2000; Ivanenko, Dominici, et al. 2007).

Underlying this developmental process is a plan of neural development. Classical studies linked the waning of the automatic march reflex as well as the emergence of true walking to progressive development (including myelination and other growth changes) in the corticospinal tract from late prenatal life to the end of the second postnatal year (Minkowski 1955; André-Thomas, Chesni, et al. 1960). These changes result in increasing cortical inhibition of primitive spinal reflexes, stabilization of the postural stretch reflex, and voluntary control of coordinated limb movements. Accompanying the myelination of the corticospinal tracts are associated changes in the precentral (motor) gyrus, particularly the lower limb regions, over a similar time course (Conel 1939–63). Functional evidence from clinical lesions shows that adults suffer loss of lower limb control in disease of the corticospinal tracts. Specifically, they exhibit the extensor plantar reflex of the neonate (a splaying outward and upward of the toes when the sole of the foot is stroked), known since Babinski's 1896 description to be a hallmark of corticospinal tract conduction failure (Alexander, Crutcher, et al. 1990; Ghez and Krakauer 2000). Thus, the maturation of function in this tract as a result of myelination is a likely neural basis of the waning of the Babinski reflex and the growth of walking.

Modern concepts of the cortical and subcortical control of voluntary movement have greatly expanded the range of neuroanatomical structures involved in maturing locomotor skills. According to these concepts, both the basal ganglia and the cerebellum are implicated in goal-corrected movement (Alexander 1994). Briefly, the nonmotor cortex (the great majority of the cortex) initiates voluntary movement not primarily through a direct communication to the motor cortex (although this may play a role) but through basal ganglia and cerebellar circuits. Nonmotor cortex projects to the basal ganglia and cerebellum, which project, via the ventral anterior and ventral lateral thalamus, back to the motor and premotor cortex.

This widely accepted model directs our attention to the myelination sequences of the major efferent pathways from the globus pallidus (the ansa lenticularis and field H1 of Forel); of the middle cerebellar peduncle, which carries cerebral cortical information to the cerebellum; and of the intrinsic fibers of the basal ganglia. Collectively, these structures exhibit a course of myelination highly consistent with that of the corticospinal tract and its somatosensory feedback pathway, the somesthetic radiation. This complex circuitry thus exhibits a growth pattern consistent with the maturation of the behavior.

There have been interesting attempts to apply systems theory to the development of walking in infancy (Thelen 1993, 1995), and these certainly suggest that learning, practice, self-organization, and interaction with the affordances of the environment play their roles in the emergence of this behavior (Annick Ledebt 2000; Adolph, Vereijken, et al. 2003). But neither the overall conception nor the individual studies undermine the data and argument presented here, which lead to the conclusion that this is a species-specific, evolved, genetically controlled, maturational process. It is hard to understand why there should not be other evolved features of human behavior and mind, including social behavior and emotions, that would similarly be under strong genetic and maturational control.

7

The Growth of Sociality

Pediatrics, neurology, and psychiatry proceed on the assumption that key facts about normal psychological and behavioral development are known. These assumptions form part of the basis for clinical assessment and treatment, including recommendations about primary prevention that are staples of pediatric and psychiatric advice. Some of these assumptions go beyond the evidence, and they typically lack cross-cultural and evolutionary context. For example, many clinicians believed for some time that immediate postnatal contact between mother and infant was essential for "mother-infant bonding" (Klaus and Kennell 1976), although the evidence for this assertion was weak (Svejda, Pannebecker, et al. 1982). It is often stated that infants should "outgrow" night waking after about three months of age, the suggestion being that this transition reflects neurological maturation (Spock 1968); elaborate treatment programs are then invoked to extinguish it (Spock and Parker 1998; Ferber 2006). But many cultures consider night waking normal and expectable until as late as three years (Konner and Worthman 1980; Konner and Super 1987).

Since the 1970s doubt has been cast on many common assumptions about normal and abnormal psychological development (Kagan and Klein 1973; Kagan, Klein, et al. 1979; Kagan 1984; J. Harris 1995; Plomin and Rutter 1998; Bruer 1999). Many now hold that the importance of infancy has been greatly exaggerated and that early *human* experience has not been clearly shown to have lasting important effects on the course of psychological development. Despite much evidence from animal models showing lasting effects of early experience, decisive clinical evidence of comparable effects in humans has been wanting.

The main purpose of cross-cultural and cross-population studies has

been to demonstrate the flexibility of the human psychological repertoire and, by inference, the underlying neural and endocrine functions
(Valsiner 2000), but much of psychological development is relatively
inflexible despite marked cross-cultural variation in rearing environments (Kagan 1981b; Super 1981; Eibl-Eibesfeldt 1988; Konner 1991). Still,
lacking in such studies has been attention to the underlying events of
regional neurological and neuroendocrine development or to the evolutionary background and context. Recently, however, advances in relevant fields have paralleled those in behavioral development, which has
become increasingly systematic and precise in its methods of measurement, taking advantage of large sample sizes to analyze large volumes of behavioral data with unprecedented statistical sophistication.

Changes in human brain structure and function during the first few
years of postnatal life are rapid and large, a unique feature of our species. In the first year the brain more than doubles in volume, reaching
60 percent of adult size, and the growth rate declines only gradually
during the second year (Yakovlev 1962; Blinkov and Glezer 1968). As we
saw in Chapter 5, this high rate of postnatal brain growth is unique to
our species among the higher primates and one of the most distinctive advances of human evolution. Chapter 6 described the profound
structural changes in the nervous system during this phase of growth.
Fixed maturational sequences related to these changes have long been
accepted in motor development and are increasingly established in cognitive and social development. Chapters 7 through 12 review evidence
for relatively fixed sequences of psychosocial development, drawing on
cross-cultural and cross-population studies, and relate these behavioral
data to neural and neuroendocrine development in infants and children.

The "Fourth Trimester" and the Presocial Baseline

It is not quite correct to refer to the human neonate as presocial, and
infant psychologists reject that characterization. By the time of full-
term birth, in addition to the propensity to be comforted by physical
contact, feeding, and rocking, the infant has a modest ability to orient
to the human face and voice (Aldridge, Braga, et al. 1999) and to learn
mildly but significantly preferential responses to a primary caregiver in
several sensory modalities (MacFarlane 1974; DeCasper and Fifer 1980;

Bushnell, Sai, et al. 1989). Nevertheless, the neonate's relative lack of social responsiveness and skill can disappoint parents who harbor unrealistic expectations. In an effort to highlight the special status of infants in this period, some pediatricians have called it the fourth trimester (for example, Karp 2002).

Human neonatal status is certainly altricial—sensorily, socially, and especially motorically—compared with that of our closest primate relatives, although in sensory modalities, especially vision, we and other higher primates are precocial compared to carnivores or rodents. As we have seen, our altricial status is partly due to the need to pass a large-brained infant through an inflexible pelvis at a relatively early stage of development compared with anthropoid apes, and perhaps partly due to parent-offspring conflict; selection may have protected mothers from strong attachment during the peak period of infant mortality. Average birth weight is lower than the weight at which mortality is lowest (Blurton Jones 1978). This "premature" average expulsion time is superimposed on the fundamental hominin condition of altriciality (Portmann 1945; Gould 1977; Martin 1990) and is one of five great changes in the evolution of human childhood.

Neurobehavioral Assessment of Human Neonates

There are many ways to assess neonatal behavior and responsiveness (Worobey 1990). Two used in cross-cultural studies are the Prechtl neurological evaluation (Prechtl and Beintema 1964; Beintema 1968) and the Brazelton Neonatal Behavioral Assessment Scale (BNBAS) (Brazelton 1973). The Prechtl assessment scores more than 150 items, including behavior, color, muscle tone, joint mobility, a wide variety of specific signs (hiccups, cough, and so on), and a comprehensive array of neurological reflexes, both normal and abnormal. The Brazelton Scale has 27 behavioral subscales (alertness, irritability, hand-to-mouth facility, and so on) scored from 1 to 9. With training, both tests have good inter-observer and test-retest reliability if done after three days of age with proper attention to state variables.

Such instruments characterize human neurological status at birth (Minkowski 1955; Beintema 1968) and show substantial continuity from prenatal to postnatal life (Prechtl 1984; Kurjak, Stanojevic, et al. 2004), as expected since there are no abrupt changes in brain function. The array of reflexes and movements, some essential for survival and continuous with mature functions (for example, the sucking and withdrawal reflexes), others reminiscent of the signs of some neurological disorders and destined to wane with normal growth (such as athetoid movements and the Babinski reflex), has suggested to various observers that the neonate is a "competent" or alternatively a decorticate, decerebrate, or midbrain organism. In fact there is some degree of higher brain function, but it is modest and uneven, owing to gradual and asynchronous late prenatal and early postnatal brain growth (Rorke and Riggs 1969; Gilles, Leviton, et al. 1983; Dubois, Dehaene-Lambertz, et al. 2008a).

But to describe *human* neurobehavioral status at birth we must go beyond European and American populations. Some studies claimed marked departures, notably the report by Marcelle Geber and Rex Dean (1957) suggesting that Baganda neonates in Uganda lacked some reflexes considered normal in European neonates: the automatic march reflex, the stepping reflex, the placing reflex, and the scarf sign. These and other authors held that African neonates had already developed past the point at which these reflexes can be elicited, a difference in developmental status on the order of four to six weeks. But this was not confirmed in other studies (Warren 1972); one review concluded that "there is no reliable corroboration of Geber and Dean's claim of neurological precocity in African newborns. Substantial contradictory evidence now exists, as well as doubt about the original methodology" (Super 1981:188).

This does not negate all population differences in neonatal behavior. There is evidence of temperamental differences between East Asian and European newborns, the former being less active and less irritable (Freedman and Freedman 1969; Freedman 1974). There are minor but statistically significant differences in different populations as measured by the Brazelton Scale (Sameroff 1978), as well as gender and individual differences (Lundqvist and Sabel 2000). Maternal smoking produces more irritable neonates (Stroud, Paster, et al. 2009). And differences in behavior are related to mothers' mood in pregnancy; newborns of depressed mothers had higher cortisol and norepinephrine levels and lower dopamine levels and scored more poorly on the orientation, reflex, excitability, and withdrawal clusters of the Brazelton Scale. Moreover, mothers' prenatal norepinephrine and dopamine levels significantly predicted their newborns' neurotransmitter levels and Brazelton scores (Lundy, Jones, et al. 1999).

But these results do not challenge the innate reflexive and behavioral repertoire of human newborns. Illustrative are the results of the Prechtl neonatal neurological examination for a small sample of neonates in one African population, the !Kung San, nine infants examined between eight and twelve days of age. Subtleties of threshold and amplitude of responses were noted, but most infants unambiguously exhibited all standard reflexes. The proportions were within the range expected by chance for European and American samples. Reflexes once reported to be absent in Baganda neonates after four days are present in !Kung neonates at ten days. Other investigators obtained similar results in African populations based on larger samples of neonates, as well as in several Asian populations, among Australian aborigines, and among Native Americans (Warren 1972; Super 1981). Comparisons of newborns of small-stature Efe foragers in Africa with another African group and a U.S. sample "suggest that neurobehavioral organization has a universal form" (Tronick and Winn 1992:421).

The state of development of the neonatal nervous system accounts for the universals. The behavioral data suggest a level consistent with full development of the peripheral nervous system, extensive development of the spinal cord and medulla, substantial development of the midbrain, and partial (mainly subcortical) development of the forebrain. Individual reflexes reflect the state of development of specific structures (Minkowski 1955; Peiper 1963). For example, deviation of the head and eyes in the direction of body displacement, the doll's-eye reflex, and the labyrinthine righting reflex reflect the relative maturity of the vestibular portion of the eighth cranial nerve and its projections to the brain-stem nuclei mediating the responses. The stepping and automatic march reflexes directly parallel similar responses in decorticate and decerebrate animals and reflect the relative maturity of the spinal cord and medulla in the absence of functional descending cortical pathways. The Babinski reflex (see Interlude 2) is also seen in adult patients with damaged corticospinal tracts and reflects the absence of myelin from the same tracts in neonates (Brain and Wilkinson 1959).

The Moro, named for a German pediatrician who called it the clinging reflex (*Umklammerungsreflex*), is a wired-in, complex, organized reflex, or innate coordination—the *Erbkoordination* of classical ethology, usually translated as "fixed action pattern." Newborns experiencing a loss of support under their heads throw out their arms and legs (extension) and then, in a second, flexion phase, bring them back close to their bodies. The flexion phase resembles what monkey infants do when their mothers get up and walk: they grasp the mother's fur by their hands and feet to avoid falling.

Common sense as well as scientific observation suggested that a reflex beginning with the sudden extension of the arms and legs could not be related to clinging (Parmelee 1964). But ingenious electromyographic studies by Heinz Prechtl showed that if the baby is grasping the examiner's fingers and there is traction in the arm and hand, then when the reflex is elicited the extension phase is suppressed; thus it is indeed the electromyographic equivalent of the higher-primate clinging reflex (Prechtl 1965)—or at least a behavioral vestige of it, a neurobehavioral version of the coccyx or the appendix. In addition, a newborn infant grasping an examiner's fingers can support its own weight for several seconds.

More relevant to human infant survival is the set of complex reflexes ("innate coordinations") that include rooting, sucking, stripping, and swallowing. Stroking a cheek results in turning toward the stimulus, opening the mouth, and, if the stimulus is suitable, locking on and sucking. If it is the nipple of a bottle or breast, the behavior is more challenging. In a coordination of movements that cannot be learned, the infant uses the tongue, lips, and gums to extract milk, not omitting to swallow and breathe in precisely interwoven patterns, where aspiration would be life-threatening. This requires many millions of motor and sensorimotor neurons to fire rhythmically, responding to the unique shape and character of the breast, the texture of the milk, the state of the airway, and the fullness of the mouth or stomach.

In addition, the neonate can get its fingers into its mouth, suck its thumb, stop itself from crying, protect itself by withdrawing various body parts from pains and irritations, lift its face away from obstacles to breathing, study the visual scene with great concentration, burp, spit up, urinate, defecate, coo, stop crying at the sound of a human voice, sleep twenty hours a day without developing bedsores, and probably dream—among other competencies. The human newborn is the product of ten thousand generations of human (and many more of protohu-

man) evolution, which prepared it, despite its early expulsion and secondary altriciality, to survive and thrive against daunting odds.

Visual capacity is relatively advanced (Bronson 1974, 1982; Dubois,
Dehaene-Lambertz, et al. 2008b). Classic research recording infants' eye
movements and gaze direction showed that sharp boundaries between
light and dark are especially interesting to newborns (Fantz 1958, 1963);
checkerboards and bull's eyes also compel attention, if the patterns are
not too fine. Generally, the neonate's visual preferences can be described by "a few programmed statements to get him on his visual way
. . . (1) if alert, and light is not too bright, open; (2) if eyes open, but see no
light, search; (3) if see light but no edges, keep searching; (4) if see edges,
hold and cross" (Haith 1973:322). This feature-extraction model characterized neonatal vision in a general nonsocial realm, but it understated
how much a social percept like a face could be "wired-in," bypassing
feature extraction to secure a vital social adaptation (Gross 2005).

Not surprisingly, the heart rate of newborns increases when they
hear a white noise (in proportion to loudness), but a group led by Peter
Eimas (1997, 2000) showed that one-month-old infants increase their
rate of sucking—another indicator of attention—when the sound they
are listening to changes from *pa* to *ba*. In effect this is an innate phonemic boundary, and its recognition suggested a very early propensity to
distinguish sounds basic to speech. The finding was later extended to
newborns, and a heart-rate response to changing the order of a pair of
syllables was shown even for fetuses (Lecanuet, Granier-Deferre, et al.
1987).

Although infant perception is socially biased, it remains reasonable
to think of the neonate as functionally largely subcortical (Bronson
1982). But despite limited cortical function, newborn infants modify
their behavior with experience. If a stimulus—say, a light flash—is repeated, the response—a reflex eye-blink—gets smaller with each equal
flash; this change, habituation, is very primitive, but other forms of behavior modification are well known. Newborns can be trained to suck
upon hearing a tone previously paired with a nipple to suck on—classical conditioning (Rovee-Collier and Lipsitt 1982; Lipsitt 1990).

But the social bias is clear (Als 1977). At an average age of nine minutes, newborns turn their eyes and their heads to follow a sketch of
a face more than they do with a similar sketch with upside-down or
scrambled features, and still more than they do with a face shape with

no features (Slater 2000). At a few hours old they prefer their mother's voice to the voices of strangers, probably because of intrauterine learning (DeCasper and Fifer 1980). There is no such preference for the father's voice, and female voices in general are reinforcing to newborns whereas male voices are not (DeCasper and Prescott 1984). Simple stories with striking language, such as *The Cat in the Hat* by Dr. Seuss, read to babies during the last weeks of pregnancy, influence their attentiveness to those stories after birth (DeCasper and Spence 1986). By age forty-five hours, newborns look longer and differently at their mother's face than at the faces of unknown women presented to them in the same way, reflecting association and reinforcement learning (Bushnell, Sai, et al. 1989).

Classic experiments showed that neonates are, to a limited extent, capable of mimicry (Meltzoff and Moore 1977) as well as intermodal transfer—exchange of information among different sensory modalities, a capability they were long thought to lack (Meltzoff and Borton 1979; Meltzoff 1990). For example, newborns have some capacity to respond to mouth movements as if they expect speech (Aldridge, Braga, et al. 1999). Whether this reflects a surprising level of functioning for poorly developed intracortical association areas—notably the mirror neurons of the premotor cortex—or, more likely, a role for subcortical structures in these responses is unknown. But by two or three weeks, newborns mimic *two* facial gestures—tongue protrusion and mouth opening (Meltzoff and Moore 1999)—suggesting an isometry between perception and action that is supported by neurobiology; by six weeks, they can delay their response for twenty-four hours and react to the passive face of the investigator with the facial expression they saw on it the day before (Meltzoff and Moore 1994).

These findings, along with extensive later research on the development of imitation in the second year, have become the core of a new and intriguing theory of childhood social cognition (Meltzoff 2007). In it, the mirror-neuron system (MNS) is seen as establishing relationships in early infancy based on similarity perceived by the infant ("like me") and on the inclination to build on that similarity with further interaction and further imitation. This is an interesting theory despite the fact that the MNS is in a very primitive state in early infancy. The responses cited could be the result of subcortical processing, which we know to be involved (along with the MNS) in imitation and observation of facial expressions at age ten years (Pfeifer, Iacoboni, et al. 2008), and the amygdala, the anterior insula, and the frontal part of the MNS are not only activated but correlated with individual differences in expressed empathy and social skills. There may be other subcortical structures involved, and they may be more important in infants.

In a related line of research, the fact that newborns react negatively to the cessation of social stimulation and its replacement by a still face has led one investigator to speak of innate intersubjectivity (Nagy 2008). This is consistent with a trend in infant cognition research to try to show that the youngest infants can do some version of what only older children were thought to be able to do. Jerome Kagan (2008) and other leading theorists (Campos, Witherington, et al. 2008) have called attention to the naïveté of such interpretations, conflating as they do the surface manifestations of (ostensibly) similar behavior with psychological states that must differ greatly at widely different ages. Continuities are interesting, but development is real. On the other hand, development has to start somewhere, and even the neonate can do many things.

In the light of evolution, neonates *should* be adapted for human stimuli. In hunter-gatherer communities, full of dangers and devoid of protective technologies, they need that inborn orientation. The human touch, voice, and face command attention that fosters social interaction and relationships; human, and indeed higher-primate, neonates are designed to form social bonds. Brain maturity is roughly the same in any group of human newborns regardless of climate, culture, or lifestyle. Though subtle differences exist—much more between individuals than between populations—the basic behavioral repertoire is constant, owing to universal features of the nervous system at this stage of growth. Still, competence notwithstanding, the human newborn is altricial in social behavior, and most social functions remain to be developed. The caregivers must meet the newborn far more than halfway, compensating for still missing functions. We now consider the growth of some of those functions.

The Rise and Fall of Early Crying

Crying is a vital built-in survival mechanism, one form of the infant distress call that is widespread in mammals (Buchwald 1985; Newman 1985; Hofer 2002; Soltis 2004). It was an early component of the phyletic emergence of mammals, along with lactation, maternal care, and limbic system expansion (MacLean 1985). It is clearly an adaptation for infant survival and so is highly consistent with the Janson and van Schaik (1993) and Sibly and Brown (2007) models in which the survival of the young is pivotal in understanding life histories. In the meerkat (*Suricata suricatta*), a small mongoose of the Kalahari in southern Africa, burrowing and cooperative breeding protect the young, but they also have their own resources; 40- to 60-day-old infants emit cries with sound features that strongly elicit feeding in experiments, while the same infants at 100 to 120 days (the juvenile phase) have lower-pitched cries that provoke less feeding (Madden, Kunc, et al. 2009). Under natural conditions, this change gradually prompts juveniles to forage for themselves.

Human crying, although it later functions in separation and stranger protest, is in the first few months not specifically social except in its effect; it largely reflects physical distress (Golub and Corwin 1985). It is a rough indicator of a need for some kind of care and so has great adaptive value (Zeifman 2001; Soltis 2004). Wulf Schiefenhövel's (1988) research on birth among the Eipo of New Guinea strongly suggests that crying by a deliberately abandoned newborn can bring the mother back, preventing infanticide. Observers have long noted distinguishable features of hunger and pain cries (Wolff 1969; Konner 1972), and technologically sophisticated studies support the distinction (Lenti Boero, Volpe, et al. 1998). The "hunger" cry may be the baseline form, the most common sustained crying (Wolff 1987), but it is not always due to food deprivation. Studies with computer-generated cries show that caregiver responses differ depending on pitch and duration (Dessureau, Kurowski, et al. 1998).

Despite universal features, the amount of crying at one, six, and twelve weeks is predictable from earlier fetal movement, suggesting that individual differences reflect neural biases (St. James-Roberts and Menon-Johansson 1999)—a concept confirmed by follow-up studies of children who cried excessively as infants (St. James-Roberts, Conroy, et al. 1998; Canivet, Jakobsson, et al. 2000). Newborns respond more strongly (with facial expressiveness or sucking rates) to other infants' tape-recorded cries than to their own (Dondi, Simion, et al. 1999). This resonance, probably subcortical, may help form the basis of infants' later empathy.

Intense crying may indicate the vigor and health of an infant, but excessive crying ("colic") can be maladaptive. It predicts later difficulties in the parent-infant relationship (Papousek and von Hofacker 1998) and is implicated in fatal infant abuse (Brewster, Nelson, et al. 1998). Victims fleeing or hiding from the Nazis occasionally had to smother their crying infants to save the rest of their families. In the environments of evolutionary adaptedness, excessive prolonged crying could have attracted predators or enemies, so infant crying was probably subject to stabilizing selection, culling both extremes. However, different patterns of care in those environments led to less crying (Fouts, Lamb, et al. 2004; Konner 2005; Kruger and Konner, forthcoming).

The shape of human development gives crying a special adaptive role (Lummaa, Vuorisalo, et al. 1998; Soltis 2004). Our newborns are less competent motorically, socially, and even perceptually than ape newborns. Nevertheless, in the normal expectable environment crying has social consequences, and social interventions such as physical contact and rocking (Korner and Thoman 1970), feeding (Konner 1972; Barr 1990a), and the sound of a human voice (Brazelton 1973) are often effective in halting it. Although these can be broken down into nonsocial stimuli that are also probably effective (Korner and Grobstein 1966; Korner and Thoman 1970; Korner, Kraemer, et al. 1975), and even though some nonsocial stimuli such as upright posture (Korner 1972; Gregg, Haffner, et al. 1976), swaddling (Chisholm 1983), and oral sucrose (Smith, Fillion, et al. 1990) can often stop crying, these interventions are usually versions of stimulus components of the evolutionarily relevant social responses. In a randomized controlled trial, crying from a pinprick to the heel was soothed by a sucrose solution or by holding, and the effects of the two together were additive (Gormally, Barr, et al. 2001). The sucrose effect (a supernormal stimulus compared with milk) is comparable to that of lidocaine (Abad, Diaz-Gomez, et al. 2001), and skin-to-skin contact also dramatically reduces crying during painful procedures (Gray, Watt, et al. 2000); crying is designed to be stopped by nutritive substances and physical proximity, which sustains life, reduces predation risk, and aids in temperature maintenance.

Whether an honest signal of robustness or an honest or dishonest signal of need, crying is central to the natural history of caregiver-infant relations (Lester and Boukydis 1985; Wolff 1987; Barr 1990a). Given the phylogenetic antiquity of the separation call, it is no surprise that nonhuman-primate crying plays a role in many ways similar to that in humans (Bard 2000; Maestripieri, Jovanovic, et al. 2000). Cross-cultural regularities are often described, but given the different levels of response to crying in different cultures (Whiting and Child 1953), we might not expect cross-cultural universals in its development. Since rates of crying are influenced by caregiver intervention—although by what mechanism has long inspired controversy (Ainsworth, Bell, et al. 1974; Bell and Ainsworth 1972; Gewirtz 1972)—different cultural regimes of care should produce different developmental trajectories.

In North American samples, an inverted U-shaped function describes crying in the first months of life, with a peak at two or three months (Emde, Gaensbauer, et al. 1976; Barr 1990a, 1990b)—one phylogenetically "appropriate" time for human birth. Humans, uniquely among primates, keep fetal brain growth rates for a year after birth, leading some to propose eighteen or even twenty-one months of gestation as expected, were this not anatomically impossible. But eleven or twelve months would have been a gestation length proportional to our later maturity and longer life compared with other apes. The shortening of this "logical" gestation by two or three months could explain why many important infant capacities emerge at that age (Rochat 1999), and why we can speak of a fourth trimester.

Initially it seemed that the inverted U-shaped curve could be an artifact of Western infant care, since regimes of care do matter (Korja, Maunu, et al. 2008). In a randomized controlled trial, increased parental carrying significantly reduced crying—by as much as half at the peak of the crying curve (Hunziker and Barr 1986; Barr 1990a, 1990b). This intervention also changed the age curve to a monotonic decline, suggesting that the inverted U might be specific to cultures with low levels of contact and carrying.

But analysis of crying in early infancy among the !Kung, who have extremely high levels of infant carrying and parental response to crying —92 percent of fussing or crying episodes are responded to by a caregiver within fifteen seconds—did not confirm cultural specificity (Barr, Konner, et al. 1991). !Kung infants cried less overall during the first months, but *without* a change in the age curve, suggesting that the curve is cross-culturally valid and independent of the amount of crying. Since the lessened crying is due to shorter bouts rather than lower frequency, an adaptive model was proposed combining the benefit of mother-infant proximity maintenance with the need to avoid prolonged crying that might alert predators (Barr 1990a, 1990b).

Crucially, premature infants show a rise and fall of early crying in a time frame that reflects their gestational age, not their postnatal experience: it peaks at the gestational age of forty-six weeks, just as in full-term infants, strong evidence that the age curve "is a robust maturational feature . . . and may be universal to human infancy" (Barr, Chen, et al. 1996:345). Remarkably, the most maladaptive end point of the continuum of response to crying—shaken baby syndrome (SBS)—shows a similar age curve. In 591 infants publicly reported to have been dangerously shaken, "the curves of age-specific incidence started at 2–3 weeks, reached a clear peak at about 9–12 weeks of age, and declined to lower more stable levels by about 29–32 weeks of age. . . . These curves have similar onsets and shapes and a slightly later peak compared to the normal crying curve" (Lee, Barr, et al. 2007:288). In another sample, infants *hospitalized* for SBS, the age curve has "a similar starting point and shape to the previously reported normal crying curve but the peak occurs about 4–6 weeks later. . . . There are numerous explanations for the lag in the peaks," such as multiple shakings before hospitalization (Barr, Trent, et al. 2006:7).

The Barr model places the crying curve within the adaptive, proximity-maintenance model of attachment (Bowlby 1969). In terms of developmental mechanism, the initial rise in crying could simply be due to a growing ability to mobilize energy or to lung maturity. The decline is probably due to the maturation of cortical inhibition of brain-stem circuits (Herschkowitz, Kagan, et al. 1997) and the growth of other signal systems, such as smiling and noncry vocalizations. But in adaptive terms, early crying is a developmental bridge between a (phylogenetically) "premature" birth and the first phase of attachment beginning at five months.

Parent-offspring conflict theory is relevant to this bridging function (Trivers 1974; Clutton-Brock 1991; Clutton-Brock and Godfray 1991). Having been expelled early and lacking the social competencies of smiling and gaze contact, the infant in the first few weeks relies on the irritating signal of fretting or crying. If the caregiver postpones (through early expulsion) her deepest attachment to the infant until after its greatest vulnerability to mortality from infection, the infant defends itself with this annoying signal. The infant may court an increased risk of predation, abuse, desertion, or infanticide, but the typical result seems to be that the caregiver's easiest route to reducing crying while optimizing reproductive success is through proximity maintenance and nurturance—more, perhaps, than is in the best interest of the caregiver.

Smiling and Mutual Gaze

Although development is continuous, there are two critical social transitions during infancy. The first, typically between ten and fourteen weeks, is a marked increase in social competence, including mutual gaze interactions, contingent-responsiveness in dyadic exchanges (Stern 1974, 1977; Stern and Gibbon 1977; Rochat 1999), and the easily elicited social smile, for which some cross-cultural data are available (Gewirtz 1965; Kilbride and Kilbride 1975; Kilbride 1980). In human adults, smiling in greeting is a cross-cultural universal. It has been filmed and measured in the same form and context in many cultures on all continents, some remote from mutual influence (Eibl-Eibesfeldt 1971a, 1971b, 1973a, 1988). Adults in widely disparate cultures also universally interpret pictures of smiles as signaling friendliness or happiness (Ekman 1973; Ekman and Rosenberg 1997). Young children exhibit the social smile in typical form and context and make the usual interpretation (Izard 1977).

Quantitative variation in the form and function of the smile may reflect the influence of learning—for example, in Bali, Ifaluk, and other Pacific cultures, smiling is considered essential public behavior, irrespective of underlying emotions (Lutz 1988; Wikan 1990), but such variation does not bear on the fundamental qualitative constancy of spontaneous smiling in greeting. It is as close as we are likely to come to a human species-typical social display. It is related to the "play face," an open-mouthed smile common during social play, and to the submissive, closed-mouth grin shown in greeting a dominant animal, both characteristic of Old World monkeys, apes, and humans (Andrew 1963; van Hooff 1972; Preuschoft 1995; Preuschoft and van Hooff 1995, 1997), but the relaxed friendly smile in social greeting is characteristically human (Blurton Jones 1971).

Well-formed *non*social smiles occur regularly in neonates during rapid eye movement (REM) sleep from at least thirty weeks of gestational age (Wolff 1963; Emde and Harmon 1972; Emde, Gaensbauer, et al. 1976; Messinger, Dondi, et al. 2002). But for practical purposes, *social* smiling is absent at birth and emerges during the first few months. The response cannot be clearly seen until some time in the second month (Spitz and Wolf 1946; Ambrose 1959; Emde and Harmon 1972; Sroufe and Waters 1976). The Duchenne smile, which includes high cheek raising, and the simple smile, which does not, follow a similar developmental trajectory (Messinger, Fogel, et al. 1999). Subsequently, beginning around nine months of age, the smile is not merely social but instrumentally so, becoming tied to intentional communication, shared attention, and means-end understanding independently measured (Jones and Hong 2001).

Cross-Cultural Universality

A classic test appraised the development of social smiling in early infancy in Israel by beginning with the infant in a quiet alert state and presenting a live but impassive adult in face-to-face juxtaposition with the infant (Gewirtz 1965). The number of smiles in one minute was recorded. Smiling was two orders of magnitude more frequent at four months than at term, whether in a normal middle-class family or in a foundling home. Variation is great after four months, but in early incidence and rate of emergence in samples in different environments, although statistically significant, it is minor (Gewirtz 1965). In naturalistic social contexts among the !Kung, the growth curve is similar to those in Israel. Infants of the Baganda of Uganda, the Samia of Kenya, and in China, Canada, and other cultures differ little from American infants in the age and context of social smiling (Kilbride and Kilbride 1975; Kilbride 1980; Kisilevsky, Hains, et al. 1998).

The social smile produces a marked change in parent-offspring relations. Mothers may report that they did not subjectively sense the existence of a relationship or even that they did not love the infant before the emergence of gaze fixation and social smiling (Robson 1967; Robson and Moss 1970). Two-month-olds may already smile differentially to the mother and a stranger (Farris 2000), and even "coy" smiling (with simultaneous gaze aversion and curving arm movements) may emerge during this period (Reddy 2000). By six months, four different kinds of smiles can be identified and associated with different types and stages of peek-a-boo and tickle games (Fogel, Nelson-Goens, et al. 2000). Also at that age infants discriminate between and respond differently to different facial expressions and intensities of smiling (Striano, Brennan, et al. 2002).

The absence of this endearing behavior at birth is an evolutionary puzzle; the solution lies in the constraint on gestation length and in the fact that newborns in premodern societies such as hunter-gatherers have death rates of 10 percent or more, mainly from infection. Since this is more a function of the immunological challenges of extrauterine life than developmental stage, it may be best for the mother to remain un-attached for this period, to spare her a grief reaction that could jeopar-dize future reproduction. But it is still to the infant's advantage to have all means to elicit care; the result is an adaptive contest and compro-mise.

Developmental Mechanisms

The neurobiology of the emergence of social smiling remains elusive. Anencephalic infants with midbrain-level functioning have reflex smiles (Monnier 1956), so these must require no higher a level than the teg-mental area of the pons (Luyendijk and Treffers 1992). Videotape studies of normal neonates show that highly coordinated and specific facial ex-pressions, involving sometimes widely separated facial muscles acting in concert, are present at birth (Oster 1978). Thus, with only lower brain functions mature, intricately timed facial muscle action patterns are al-ready under complex central control. Also at birth, gaze fixation and vi-sual following of a face can be elicited (Brazelton 1973; Als 1977).

Since blind infants develop reliable social smiling only a month or two later than sighted ones (Thompson 1941; Freedman 1964; Fraiberg 1977), a crucial role for visual perception can be ruled out. Since the mean age of onset of social smiling in samples of low-risk premature infants can be better predicted from postmenstrual age (forty-eight weeks) than postnatal age (Brachfeld, Goldberg, et al. 1980), a key role for conditioning seems unlikely, although it is prominent later in infancy (Ahrens 1954; Brackbill 1958; Ambrose 1959). Finally, since MZ twins are significantly more concordant in the emergence of social smiling than DZ twins (Freedman 1974), a genetic contribution to individual differ-ences is likely.

So a central connection between the social perception and the already well-formed motor output matures during the growth of this be-havior. Some relevant changes may not be explicitly social or emotional. For example, by two months of age, visual following of eyes and faces substantially improves (Haith, Bergman, et al. 1977), and the second month may be a watershed for the emergence of "reciprocal exchanges, affective mirroring, and mutual imitations with others" (Rochat 2007:8). Perhaps by two months, certainly by three, an infant is especially at-tracted to stimuli whose changes it can control. These "contingently re-sponsive stimuli" have been extensively studied (Rovee and Rovee 1969; Watson 1972; Watson and Ramey 1972; Rovee-Collier and Lipsitt 1982; Rovee-Collier, Earley, et al. 1989); a competent caregiver is such a stimu-lus. In playful face-to-face interactions the caregiver's signals change at a pace ideally suited to challenge infant attention at this age (Stern 1974; Stern and Gibbon 1977), but the requisite infant abilities are largely ab-sent earlier. Colwyn Trevarthen (1979; Trevarthen and Aitken 2001) has called the change *primary intersubjectivity*.

By four months visual pattern memory emerges (Super, Kagan, et al. 1972), as does "the smile of recognitory assimilation" (Zelazo 1971, 1972; Zelazo and Komer 1971), which reflects the pleasure of recognizing a fa-miliar stimulus after uncertainty, or of absorbing a mildly unfamiliar one. Playful mother-infant interaction makes good use of this function of smiling, and later, interaction with others does as well (Farris 2000). In a remarkable confirmation of the concept of intermediate discrep-ancy as most attractive to young infants (Kagan 1970), as well as the concept of optimal arousal level in three-month-olds' interactions with their mothers (Stern 1974; Stern and Gibbon 1977), there is an inverted U-shaped curve relating the degree of synchrony of mothers' vocaliza-tions with those of their four-month-old infants and maternal sensitiv-ity measured independently (Hane, Feldstein, et al. 2003). The most sen-sitive mothers show neither too much nor too little synchrony. But in addition to these changes, many other aspects of early-maturing com-petence are specifically social, in the ethological sense—"wired-in" for social functions.

Neural Maturation

A study by Tobias Grossman, Mark Johnson, and their colleagues (2008) opens a new window on the social development of infants at this age, particularly with respect to mutual gaze. Twelve four-month-old infants were shown photorealistic computer animations of faces with eyes turning toward them (mutual gaze) or away from them (averted gaze). A second later, in the context of the mutual gaze where it occurred, the expression on the face changed to a closed-mouth smile with eyebrows raised. Near-infrared spectroscopy (NIRS) and event-related EEG oscillations were used in separate experiments to detect regional brain activation.

Both methods showed activations of the superior posterior temporal cortex, known to be implicated in detecting biological motion, and the frontopolar cortex, which these researchers reasonably take to include the medial prefrontal cortex (mPFC), implicated in mentalizing functions later in development. Both were preferentially but not exclusively activated on the right. Moreover, there was separate activation temporally tied to the onset of the mutual gaze and the appearance of the eyebrow raise and smile. This worked whether the face was directly oriented to the infant's face or slightly averted (20 degrees off), if and only if the gaze contact was made.

Remarkably, this finding in four-month-olds is the same pattern of activation shown previously with fMRI in adults, indicating a very early appearance of brain activity specialized for face-to-face communication. It also contrasts with the decade-long, partly experienced-based developmental curve found for facial recognition (Cohen Kadosh and Johnson 2007). The right-leaning activations are also found in adults, suggesting that right-hemisphere specialization for the processing of emotion and social communication develops very early.

The authors do not discuss possible subcortical activations because their methods did not allow them to detect activation deep in the brain, but given the state of development of the cortex at age four months, it seems likely that basal ganglia and/or limbic areas are involved, since eye contact has activated them in other studies of adults (Whalen, Kagan, et al. 2004). One interpretation of the rapid development of activation for mutual gaze and response to smiling and eyebrow raising as opposed to the slow developmental course of face recognition is that the former is a human universal similar to a releasing mechanism (smiling in greeting) and may be wired subcortically and develop early.

Initial approaches to a developmental neurology of smiling must be indirect. Myelination of the motor roots of the fifth and seventh cranial nerves—which mediate facial sensation and control of facial muscles—is completed prenatally (Langworthy 1933; Rorke and Riggs 1969), consistent with nonsocial smiling even in prematures (Messinger, Dondi, et al. 2002). The motor nuclei of these nerves are close to the pontine neurons that control REM sleep (McCarley and Hobson 1975; Hobson and Pace-Schott 2002), a frequent context of early nonsocial smiles.

The basis of social smiling is more obscure. In adults, localized brain damage can lead to the separate loss of either voluntary smiling or emotional smiling (Monrad-Krohn 1924, 1927; Brain and Walton 1969; Rinn 1984; Weddell, Miller, et al. 1990). In corticospinal damage, voluntary retraction of the mouth corners is weak or absent, while smiling in appropriate social and emotional contexts is preserved. "Mimic" paralysis (the emotional form) is less clear in origin, but some cases result from basal ganglia strokes and injuries (Rinn 1984) and others from damage to the anterior cingulate cortex (Damasio 1994). The basal ganglia role receives support from the "mask face" of Parkinson's disease (PD) (Katsikitis and Pilowsky 1988)—facial unresponsiveness in emotion, with *voluntary* facial expression preserved. Since PD strongly affects the basal ganglia, the mask face suggests the possible mediation of social smiling by this subcortical circuitry (Rinn 1984).

An ingenious fMRI experiment supports the division of neural control into voluntary and spontaneous mechanisms. Adults looked at happy, sad, or neutral faces while trying to keep their faces immobile or while turning the corners of their mouths up (as in smiling) or down (as in frowning); dissonant pairings—say, smiling at a sad face—should require more voluntary effort. These dissonant pairings preferentially activated the inferior prefrontal cortex and the somatomotor cortex, but congruent pairings activated the amygdala, hippocampus, and parahippocampal region (Wild, Erb, et al. 2003). Together with the clinical evidence, this finding supports the subcortical (in this case limbic) mediation of social smiling.

With these results, some neuroethological findings that otherwise seem remote become relevant. MacLean and his colleagues maintained that the striatopallidal complex (most of the basal ganglia) plays a key role in fixed social displays (MacLean 1973, 1985); these structures or their homologues are prominent in birds and reptiles, which rely more than mammals on stereotyped displays in social behavior. In primates, lesions of the pars interna of the globus pallidus specifically abolish the species-typical fixed action pattern of genital presentation, a social display in the squirrel monkey, *Saimiri sciureus* (MacLean and Ploog 1978). No other deficits are seen in these monkeys, even in motor functions traditionally associated with striatopallidal circuits. This suggests that the striatopallidal complex is a governor of species-specific displays, with the globus pallidus as an essential way station.

These findings also suggest a tentative model of the growth of social smiling. In late prenatal life the smile appears in mature form peripherally, owing to maturation of cranial nerves V and VII, but not in mature context; its association with REM sleep perhaps depends on the proximity of the pontine reticular formation to those motor nuclei. After two postnatal months (forty-eight gestational weeks) the response is socially deployed. Regional brain growth changes likely to be involved are: (1) sensory changes, especially tectal; (2) motor changes, especially cerebellar; and (3) changes in the striatopallidal complex, especially the globus pallidus and its outgoing pathways, which show the most rapid changes at this age.

Although limbic myelination lags that of the striatopallidal circuits, it is likely that communication between these two subcortical systems is implicated in the growth of social smiling. In adults, deep brain stimulation can produce smiling and euphoria, and in some cases this involves stimulation of the ventral striatum, including the nucleus accumbens, a main reward center, and of some fibers of the internal capsule that project to the facial nerve (Okun, Bowers, et al. 2004). Although the smiles were often unilateral (after stimulation in one hemisphere), the association with giddiness led the researchers to call them limbic smiles. The ventral striatum projects to the globus pallidus, forming part of the striatopallidal complex, but it also links that complex to the limbic system and so could cause limbic smiles.

Finally, some maturation of the mirror-neuron system in the frontal premotor areas may play a role, together with the basal ganglia and the limbic system, in the emotional resonance between infants and others. An unusual pattern of mirroring has been found using fMRI in adults looking at and expressing pleasant facial affect, with the representation of the two functions overlapping greatly (Hennenlotter, Schroeder, et al. 2005). Frontal lobe regions roughly corresponding to the well-known MNS of the monkey frontal lobes were activated, but predominantly in the right hemisphere, which is more involved in emotion; the anterior insula, which monitors bodily sensations, and the amygdala in the limbic system itself are also involved.

The classical MNS would seem an obvious candidate to process face-to-face interaction, including mutual gaze, mutual smiling, mimicry, and other patterns. But is it active at birth? Arguments in favor are based mainly on mimicry in newborns and the fact that children age three years and older have fMRI and EEG evidence of MNS like that for adults (Lepage and Theoret 2007). Only one study looked at infants (six to seven months old), using near-infrared spectroscopy to show that watching a live model manipulating a toy activates in infants the same sensorimotor areas as in adults, suggesting possible MNS activity. However, the investigators "cannot confirm that the measured brain activity does not include any motor activity of infant movement" (Shimada and Hiraki 2006:937). This is tantalizing evidence that something like an MNS may be emerging at six months, but it does not speak to the much less mature newborn. The circular reasoning that neonatal mimicry is itself evidence of an MNS preempts the possibility that this behavioral resonance may be subcortical.

8

The Growth of Attachment and the Social Fears

The "fourth trimester" and the postnatal year of fetal brain growth are two of the five innovations in human ontogeny (the other three being overall lengthening of growth, the growth of language, and the additional lengthening of middle childhood). But although humans delay it until long after birth, attachment is no innovation; it is largely a legacy from our primate ancestors. In John Bowlby's (1969) model it is an adaptation to prevent predation, but its adaptive significance is far more complex, including temperature maintenance, nutrition, hydration, transfer of passive immunity, and the first steps in enculturation. It also influences future intimate relationships (Berlin, Cassidy, et al. 2008), although they have their own logic as partly distinct "affectional systems" (Harlow and Harlow 1965). Still, Bowlby's claim is consistent with the models of Janson and van Schaik (1993) and Sibly and Brown (2007, 2009; see also Chapter 2), which puts offspring mortality (and its prevention) at the crux of life history.

Infancy is a dangerous time. If the second and third months of life are the time of emergence of sociality—social smiling, mutual gaze, attraction to contingently responsive stimuli, and recognitory assimilation—then at the end of this phase transition sociality is established but undiscriminating. The emotion is positive but impersonal; anyone can elicit it and—despite its favoring primary caregivers—strong emotional bonds, which will prove vital to survival, do not yet exist. As we will see in Part III, it is a time when multiple caregiving, father involvement, and adoption are likely, because infants this age are attractive, and attracted, to everyone.

This changes markedly in the second half-year. Many infants become wary of strangers by five months of age, and increasingly so in subsequent months (Morgan and Ricciuti 1969; Tennes and Lampl 1964; Lewis and Rosenblum 1974). Crying when left by the mother in a strange situation, with or without a strange person, becomes more common, although it is far from universal (Bretherton and Ainsworth 1974; Ainsworth 1979). Vulnerability to the adverse effects of long separations from primary caregivers increases (Bowlby 1973). And attachment behaviors such as following, clinging, and cuddling are preferentially directed at the primary caregiver, especially in strange situations or with strangers (Ainsworth, Blehar, et al. 1978).

These are not, of course, all functions of the growth of fear, but they are changes in the emotional valence of the infant's interpersonal space that make certain individuals very attractive and the rest of the species less so, if not actually forbidding. Primary caregivers describe a deepening of their emotional bond with the infant, and theorists such as Bowlby (1969, 1980) and Mary Ainsworth (Ainsworth, Blehar, et al. 1978) saw these changes as the growth of attachment, a major phase in emotional and social development (Cassidy 2008).

Another hallmark of the period is peek-a-boo, which, following the infant's maturing cognitive competence, accomplishes the same thing as moderately arousing, moderately synchronized play between mothers and three-month-olds, but in a quite different way. Now arousal is achieved by toying with loss: mothers and others hide their faces briefly, bringing about a little fear, and reappear, to the baby's delight; this can go on for a long time. In a clever experiment, six-, seven-, and eight-month-olds played the game, but on some trials the adult who reappeared was different, or the same person in a new location (Parrott and Gleitman 1989). Evidently this was not so amusing, since it produced significantly fewer smiles. However, an age effect was also nearly significant; six-month-olds reduced their smiling less.

Universals of Human Attachment and Social Fear

Stranger reactions have been measured using variations of the following paradigm (Konner 1972; Ricciuti 1974; Chisholm 1983). The infant or toddler is sitting near the mother and is approached slowly by an unknown adult who may try to evoke a positive reaction. Positive and negative behaviors are both recorded, including smiling, approaching the stranger, gaze aversion, withdrawing, and crying. Responses are summed or weighted according to intensity, providing a score for a given infant at a given age in response to this stimulus. Inge Bretherton and Mary Ainsworth's (1974) more widely used test of attachment, the Strange Situation, measures a defined set of the infant's behaviors toward the mother or other primary caregiver in a novel setting with and without strangers (Solomon and George 2008).

The infant and mother are brought into a room with toys on the floor. After three minutes, a stranger enters; she chats with the mother and interacts with the infant for one minute each. The mother then leaves the room for three minutes, or until the infant cries for thirty seconds, and returns for three minutes with the stranger absent. Similar staged events, including leaving the infant alone, follow. The infant's behavior is recorded, with an emphasis on crying and other protest behavior when the mother leaves and on behavior at reunion. Reunion styles can be reliably used to classify infants and toddlers into three or four attachment types: secure, insecure, avoidant, and (in some studies) disorganized (Solomon and George 2008); these may be related to independent measures of caregiver sensitivity (Belsky 1999; Belsky and Fearon 2008). In a remarkable study hinting at the internal psychological dimension of these categories, securely attached infants ages twelve to sixteen months looked longer at a simple geometric animation of a mother leaving a child behind than did insecurely attached ones the same age, presumably reflecting different expectations (Johnson, Dweck, et al. 2007). Individual differences aside, there are consistent age changes; the tendency to cry when the mother leaves increases in the second half-year of life and (at least) well into the second year.

The initial changes are largely independent of social context; non-Western and Western cultures show similar patterns (Kagan 1976; Super 1981; van Ijzendoorn and Sagi-Schwartz 2008). The growth of social fears with the concomitant growth of attachment appears to be a universal feature of the second half-year, although individual expression is extremely varied. The ontogenetic emergence of this feature is species-specific. The percentage of infants who withdraw, fret, or cry when a stranger appears, who cry when left by the mother either alone or with a stranger, or who seek the mother rather than a stranger or secondary caregiver when mildly apprehensive (Kagan, Kearsley, et al. 1978) rises from the middle of the first year well into the second year, whether the sample is drawn from the !Kung San of Botswana (who have 24/7 mother-infant contact), traditional Navajo Indians (who bind their infants into cradleboards), a remote Guatemalan Indian village (where mother-infant contact is high but vocal interaction is low), an Israeli kibbutz (where the infant is separated from the mother in a nursery except on afternoons and weekends), or various subcultures and socioeconomic levels of the United States (Kagan 1976; Super 1981; Konner 1982; Chisholm 1983).

The effects of infant day care are instructive. In a study of Chinese-American and Caucasian-American subcultures in Boston, infants who were in day care for eight hours a day did not differ significantly from infants who had no such separation from the mother on measures of social fear and attachment at any age, despite the fact that the day care regimen began before four months of age (Kagan, Kearsley, et al. 1978). Similar findings emerged in other day care samples (Caldwell, Wright, et al. 1970; Ricciuti 1974; Brookhart and Hock 1976; Blanchard and Main 1979). Although day care can affect the security of attachment and is a source of individual differences (Belsky 2002; Maccoby and Lewis 2003; Watamura, Donzella, et al. 2003), there is no evidence that it affects the basic maturational pattern.

Animal Studies

Similarly organized behavioral patterns, with species-typical motor components and ontogenetic timing, are seen in higher primates (Blurton Jones 1972a; Rosenblum and Alpert 1974) and other mammals (Scott 1963). Analogous, although perhaps not homologous, events are seen in the first posthatching days of precocial birds (Bateson and Horn 1994; Bateson 2003). Observation in the environments of evolutionary adaptedness (Altmann 1979; Rosenblatt and Snowdon 1996; MacKinnon 2007), including ours (Konner 1972, 1977b, 2005), clearly suggests adaptive functions for attachment, including the prevention of predation and the intergenerational transfer of acquired knowledge. The ontogenetic association of independent locomotion and active imitation with the growth of social fear and attachment behavior in several species supports these two adaptive explanations. However, in mammals at least, we can add to these two the functions of nutrition, hydration, birth spacing, and the transfer of molecular and cellular immunity in milk (Chapter 15). In humans, the mother-infant bond forms part of the basis of cultural continuity (Part IV). Here we focus on the biological psychology of the growth of attachment itself, and on what cross-species comparisons may tell us about developmental mechanisms.

Imprinting in Birds

In precocial birds, hatchlings leave the nest almost immediately, but soon mainly follow the mother, forming an almost indelible preference. Genetic predispositions matter—head-and-neck shapes trigger an innate preference (Bolhuis 1999)—and prehatching experience influences the hatchling's choices of whom or what to follow (Sleigh and Lickliter 1998). Nevertheless, it can be imprinted on humans, members of other species, inanimate objects, or wall stripes, usually irreversibly, and this process influences adult mate choice. For several hours the hatchling will approach and follow any conspicuous object, and following itself intensifies the motivation to follow (Hess 1974). After a certain point punishment increases rather than decreases the behavior, contrary to conventional learning laws. Visual and auditory stimuli combine to intensify it (Bolhuis 1999). Within days the hatchling becomes exponentially less likely to approach other objects and finally avoids them (Collias 2000); both the attachment and the fear persist through much of growth (Bateson 1987). Imprinting is more flexible than originally thought, but it cannot be explained by conventional learning (Rauschecker and Marler 1987).

The neurobiology is increasingly understood (Bradley and Horn 1987; Horn 1991) and may turn out to be related to that of human attachment. In the bird's dorsocaudal neostriatal complex (dNc)—corresponding roughly to the association cortex of mammals—but not in other forebrain areas, dendritic spines are massively pruned during imprinting (Bock and Braun 1999b; Braun, Bock, et al. 1999), possibly serving to narrow the hatchling's perceptual focus. The pruning depends on activation of excitatory n-methyl-d-aspartate (NMDA) receptors (Bock and Braun 1999a), which in mammals are strongly implicated in memory. Blocking them prevents the normal pruning of imprinting and prolongs the sensitive period (Parsons and Rogers 2000). In this and another forebrain area, the neurotransmitters glutamate, norepinephrine, and dopamine are involved (Metzger, Jiang, et al. 2002; Gruss, Bock, et al. 2003). In addition, imprinting activates an immediate early gene, *fos*, in γ-amino butyric acid (GABA) neurons in these same circuits, indicating neural firing (Ambalavanar, McCabe, et al. 1999), so fear may be mobilized through excitatory glutamate receptors and inhibited by GABA.

Altricial, nest-bound birds like zebra finches develop slowly and imprint much later, as half-grown fledglings (Immelman 1966). They do not follow, but the consequences for later mate choice are similar to those in precocial birds that imprint right after hatching. A zebra finch male raised by a society finch female mates preferentially with that species, even if he has already mated with a member of his own species and fledged a clutch of eggs with her. As in the precocial species, the permanent preference is mediated by rapid pruning of dendritic spines (Bischof, Geissler, et al. 2002; Lieshoff, Grosse-Ophoff, et al. 2004). This could have mammalian parallels, but more understanding of comparative brain function is needed. We have seen (for example, in eye formation) that independently evolved processes may share genes (Pichaud and Desplan 2002); attachment in birds and mammals could share some mechanisms without having had a common origin in a reptilian ancestor.

Attachment in Higher Primates

The development of attachment in higher primates is much slower (MacKinnon 2007; Suomi 2008). Monkey and ape infants are carried by the mother from birth and never leave her immediate proximity for weeks to months (depending on species-typical growth rates). As we saw in the last chapter, the infant is born with reflexive (instinctive) clinging and develops other proximity-maintaining behavior. Although the consequences of deprivation and perturbation of attachments in monkeys and apes have been well studied (Chapter 14), the normal process is not as well understood. However, as with imprinting, the experimental substitution of inanimate objects for the mother revealed salient features of the attachment object (Harlow 1958b).

Rhesus monkey (*Macaca mulatta*) infants became attached to a sloping cylinder of wire covered with terrycloth and warmed; if the column also gave milk through a nipple, clinging and contact increased slightly. But given the choice between a milkless warm terrycloth surrogate and a wire milk-dispensing one, infants spent almost all their time on the cloth model, switching only to feed. When a frightening object—a windup teddy bear banging a drum—was introduced, infants always went to the milkless cloth model, undermining the claim that oral gratification is the basis of attachment (Freud 1940). Furthermore, equipping the cloth model with a device that gave the infant a periodic blast of cold air—a punishment, in conditioning terms—*increased* the infant's contact time (Rosenblum and Harlow 1963). As in imprinting birds, the infant sought comfort from the source of the punishment, something also seen in abused children.

This made it impossible to explain higher-primate attachment by reinforcement—unless (as in birds) the process is self-reinforcing (Harlow and Mears 1978). But this begs the question of genetic influence, and indeed it basically concedes that ethologists had the right model for this behavior: attachment, like imprinting, is wired into the infant's brain and is its own reinforcer. Essentially, attachment is an instinct—a complex, highly motivated, unlearned behavior pattern (Harlow and Mears 1978). Species differ. Pigtail macaques (*Macaca nemestrina*) react much more severely to removal of the mother than their sister species bonnet macaques (*M. radiata*), among other differences in social behavior (Rosenblum and Kaufman 1971). Bonnets show stronger preference for the mother over a stranger than pigtails do, and they dramatically increase stranger avoidance between the ages of one and six months, reversing that in the next six months; in pigtails avoidance remains low throughout the first year (Rosenblum and Alpert 1974).

Biological Mechanisms

Individual variation in the developmental timing and intensity of the social fears is partly genetic. In a number of human studies, fear of strangers in infancy is more correlated in MZ than DZ twin pairs (Abe, Oda, et al. 1984) and is stable over time (Andersson, Bohlin, et al. 1999), as seen in the longitudinal pattern of growth of fear (Freedman 1974) and in specific behaviors toward the stranger (Plomin and Rowe 1978). Attachment behaviors are stable within individuals across development in infancy (Bar-Haim, Sutton, et al. 2000). Rhesus monkeys too show heritability of specific fear- and anxiety-related behavior (Williamson, Coleman, et al. 2003), but temperamental predispositions toward fear and anxiety can be moderated by maternal care in both monkeys and humans (Gunnar and Cheatham 2003). In a striking study of 157 twin pairs age twelve to fourteen months, 52 percent of the variance in attachment type (based on reunion behavior) was attributable to shared environment, with almost all the rest due to nonshared environment or measurement error; almost no genetic effect was revealed (Bokhorst, Bakermans-Kranenburg, et al. 2003). This pattern is very unusual in behavior genetic research and strongly suggests a larger role for family environment in influencing attachment styles than for most other behavior.

It is equally clear from twin studies and experiments that large components of both fear and attachment are due to environmental influence, including conventional reinforcement contingencies (Gewirtz and Pelaez-Nogueras 2000). For the moment, however, we are interested in explaining not individual variation but universal features of maturation. In view of the known basis of fear and other strong emotions in the limbic system, we examine this system for maturational events that might help to dispel the developmental mystery.

The Physiology of Human Fear and Attachment

Still, we can gain insight from individual differences. As we saw in Chapter 2, timidity is a stable feature of temperament and is associated with a larger change in heart rate to novelty than in bolder children and with other physiological differences as well (Kagan 1994, 1996; Kagan, Snidman, et al. 1999, 2007). Timidity can be predicted from reactivity in four-month-olds, and it predicts shyness and hesitancy in adolescence. The amygdala has been proposed as a brain center that might be structured or set differently in these children (Schmidt, Fox, et al. 1997); it responds to higher brain centers and in turn affects physiology—including heart rate, blood pressure, cortisol, norepinephrine activity, pupil dilation, and facial skin temperature. On fMRI, amygdala response to novelty in drawings of faces (especially eyes) differentiates adults who have been timid from infancy (Schwartz, Wright, et al. 2003; Whalen, Kagan, et al. 2004). Research on house cats supports this: as in children, about one in seven is naturally timid, and these cats have excitable amygdalas, but no difference in other brain areas. Also, the growth of timidity in cats parallels amygdala development, as it does in human children (Adamec 1990, 1991).

Separation distress in human infants and children is also related to differential electroencephalography (EEG) activation of the cerebral hemispheres (Bell and Fox 1994; Davidson 2000). Higher-level control of fear is due in part to the right frontal lobe, which mediates negative emotions; it is more dominant in timid toddlers (Fox, Rubin, et al. 1995). The prefrontal cortex (PFC) projects to the amygdala, and glucose metabolism shows a reciprocal association between them. Overall, "the prefrontal cortex plays an important role in modulating activity in the amygdala" (Davidson 2000:1157), with the right and left PFCs playing different, balancing roles.

Other research has focused on the physiology of children's attachment (Fox 1998; Fox and Card 1999), although it is hard to separate temperament from attachment style. All older infants and toddlers tend to show physiological signs of stress, including increased cortisol and heart rate, in response to separation. Some physiological responses are stable from birth to age six and are associated with temperament, but maternal contact mitigates stress responsiveness (Spangler and Grossman 1999). In a similar model, inhibited, highly reactive infants used secure attachment as a physiological buffer against stress (Gunnar, Brodersen, et al. 1996; Gunnar 1998). Right frontal activation is associated with negative emotion as a trait or a state (that is, whether it is caused by a chronic reactive temperament or by a stressful situation), and left frontal activity with positive emotions and relationships.

The frontal lobe may be asymmetrical in its exchange of information with the vagus nerve and the autonomic nervous system generally, as well as with the subcortical limbic circuits that mediate fear (for example, the amygdala), affiliation, and pleasure. Through limbic regulation of the hypothalamic-pituitary-adrenal (HPA) axis, these higher centers also influence levels of cortisol and other aspects of the hormonal milieu (Gunnar and Donzella 2002). Finally, the quality of relationships in early life and the frequency and duration of routine separations from the primary caregiver may have lasting consequences for the child's HPA axis as measured by diurnal cortisol rhythms (Gunnar and Cheatham 2003; Watamura, Donzella, et al. 2003).

An Anatomical Model

Can we reconcile these findings with normal brain development to explain the species-wide developmental pattern? At three months of age there is little or no myelin in the striatum, the fornix, and the cingulum, but these approach the adult level by a year of age (Yakovlev and Lecours 1967; Brody, Kinney, et al. 1987; Kinney, Brody, et al. 1988). The mammillothalamic tract also myelinates during this period. Other marked changes during the second half-year of life relate to control of movement. With the absence of myelin in the fornix, the cingulum, and to a lesser extent the mammillothalamic tract, the level of functioning in the Papez emotion circuit is very poor at three months compared to the end of the first year, so it is not surprising to find that the emotional competence of the older infant is much greater. Furthermore, the gains in myelin in the striatum and in the fiber fields of Forel, connecting the thalamus with the globus pallidus and subthalamus, which facilitate communication within the basal ganglia, suggest the possibility that not only emotional competence but also the ability to express emotion in action are maturing.

The Klüver-Bucy syndrome of tameness, indiscriminate exploration of objects, hyperorality, and hypersexuality following temporal pole lesions has been mentioned. With the exception of hypersexuality, which probably requires reproductive maturity, the syndrome is in some respects reminiscent of the behavior of normal human infants in the four- to five-month age range. (They are rather emotionally promiscuous, perhaps an infant analogue of hypersexuality.) The absence of myelin in the limbic circuitry may give them a partial temporal lobe disconnection that in some respects mimics the Klüver-Bucy syndrome.

Separation in Monkey Models

The effects of prolonged separation and the long-term effects of repeated early separation are considered in Chapter 14, along with the more serious effects of isolation rearing. Here we consider what happens in response to a short-term separation, both behaviorally and physiologically. To the extent that attachment is a reluctance to separate and is measured in human infants in part by their protest against it, we have much to learn about it by studying brief separations, the other side of the coin of attachment.

As we will see in more detail in Chapter 14, removing the mother affects pigtail macaques (*Macaca nemestrina*) more than bonnet macaques (*M. radiata*) (Rosenblum and Kaufman 1971). Pigtail infants protest for hours, with increased activity, distress calling, and physiological changes (Reite, Short, et al. 1978); they played less with peers and the younger ones slouched, curled up, and looked sad. Physiological changes are also seen in the squirrel monkey (*Saimiri sciureus*) (Coe, Wiener, et al. 1985;

Coe and Erickson 1997) and marmosets (Dettling, Pryce, et al. 1998). Some of these effects include marked heart rate elevations, changes in body temperature, sleep disturbances, and activation of the HPA stress response. On brain imaging, rhesus monkey infants separated from their mothers showed higher activity in the right dorsolateral prefrontal cortex (dlPFC) and the right ventral temporal/occipital lobe but lowered activity in the left dlPFC, as in studies with human infants (Rilling, Winslow, et al. 2001). Activity in the dlPFC predicted cortisol levels, as did activity in limbic midbrain areas. Repeat separations may exhaust the stress response system, since they reduce cortisol in squirrel monkeys (Levine and Mody 2003) and marmosets (Dettling, Feldon, et al. 2002) as well as rodents. Short-term responses to separation can last if repeated or prolonged, but for now we simply note that the immediate physiological effects are unpleasant, and that part of the mechanism of proximity maintenance is the avoidance of them.

Neurochemistry in a Monkey Model of Fear of Strangers

The neurochemical mediators of infant fears have also been studied (Kalin and Shelton 1989; Kalin 1993, 2003). In these experiments, human strangers approach infant rhesus monkeys and either stare at them or make no eye contact. Both conditions induce fear, but with staring the monkeys tend to coo loudly—essentially, the separation call—and bark nervously, while with no eye contact they tend to freeze and crouch. Morphine, a centrally acting opiate, reduces cooing, and the antagonist naloxone increases it. In contrast, Valium, working via the inhibitory neurotransmitter GABA, specifically suppresses the other fear behaviors—barking, freezing, and crouching. This is consistent with the idea that separation protest is an opioid-receptor-dependent system biologically distinct from general fear (Panksepp 1998, 2003, 2005). Separation distress may be due more to depressed activity in brain opioid systems, while more general signs of fear are due more to lowered GABA function.

Freezing is also related to cortisol levels in both infants and mothers (Kalin, Shelton, et al. 1998). Cortisol is stable over time for a given individual, at least from ages one to three, or up to puberty in rhesus monkeys. Also, cortisol declines with birth order, a possible effect of maternal experience. Finally, both cortisol—as in the Rilling study—and fearfulness are correlated with excess activity in the right frontal lobe compared to the left; this is also a stable effect as the infant grows (Kalin, Larson, et al. 1998; Kalin 2003; Kalin and Shelton 2003). But what subcortical circuits mediate these effects?

Fear and Attachment in the Developing Brain

The human cortisol system is active from birth, and its activity during the first year is in part a function of the sensitivity of caregivers, interacting with temperament (Gunnar and Cheatham 2003). The system is mobilized in the separation phase of the Strange Situation in the second year; fear is correlated with cortisol level in that situation and with cortisol reactivity to routine inoculations during the first six months (Gunnar, Brodersen, et al. 1996). A working model might hold that the HPA axis is active from birth, but the system does not respond to amygdalar and hippocampal control until myelination of the fornix and stria terminalis (ST) later in the first year. Amygdalar output (via the ST) would report the fear itself to the hypothalamus, while the hippocampus (via the fornix) could report a mismatch between a new person or situation and well-known ones (Schaffer 1966; Kagan 1974).

We know from animal models that adult attachment depends on oxytocin (OT) operating on specific receptors in several limbic system structures (Carter, Williams, et al. 1992; Insel 2000). This may also be necessary for infant attachment; oxytocin receptors (OTRs) are first expressed in early infancy (Carter 2003), and OT knockout mice are less vocal in separations from the mother and less fearful in other situations (Winslow, Hearn, et al. 2000). OT injections markedly reduce rat pups' crying on separation (Insel and Winslow 1991), and OT antagonists eliminate their normal olfactory preference for the mother (Nelson and Panksepp 1996). Neonatal prairie voles treated with OT or OTR antagonists experience immediate and long-term behavioral consequences that influence adult pair-bonding, parental behavior, hypothalamic activity, and HPA axis activity (Carter 2003).

As in monkeys, opiate systems are also involved. In the first week, mouse pups with the μ-opioid receptor gene knocked out show far fewer distress calls when separated from the mother than controls do, although they show normal calling in response to cold temperatures and other stresses (Moles, Kieffer, et al. 2004). This is consistent with the finding that separation-calling (cooing) in infant monkeys can be soothed by opiates and increased by opiate receptor blockade. Both sets of experiments support Jaak Panksepp's model (Panksepp 1998). Whether or not these specific explanations relating infant emotionality to neuroanatomical and neurochemical studies are correct, no account of the changes in infant emotional competence during the second half-year will be satisfactory without an account of their relationship to the equally striking changes in limbic system structures known to underlie emotional behavior. We know from near-infrared spectroscopy studies that infants six to nine months old preferentially activate the right frontotemporal cortex when looking at their mothers' as opposed to others' faces (Carlsson, Lagercrantz, et al. 2008), and at a year of age, deeper into the process of attachment formation, the orbitofrontal cortex strongly predominates, especially when the mother is smiling (Minagawa-Kawai, Matsuoka, et al. 2008). The difference may reflect progress from recognition to love.

Models of infant attachment should include functional change in the human limbic system due to myelination from ages four to twelve months (Yakovlev and Lecours 1967; Brody, Kinney, et al. 1987; Kinney, Brody, et al. 1988; Utsunomiya, Takano, et al. 1999; Staudt, Krageloh-Mann, et al. 2000). In approximate order of development, these are the cingulum bundle (linking the frontal and cingulate cortex to the hippocampus); the stria terminalis (amygdala to the bed nucleus of the stria terminalis [BNST] in the hypothalamus); the mammillothalamic tract (hypothalamus to anterior nucleus of the thalamus); and the fornix (hippocampus to septal area, ventral striatum, and hypothalamus, especially the mammillary bodies). Other limbic tracts myelinating in this period are the stria medullaris (linking the hypothalamus and ventral pallidum with the midbrain raphe nuclei) and, to a lesser extent, the anterior commissure (connecting the temporal lobes) and the frontal cortical gyri. Infants in all cultures become capable of attachment during this period, and these maturational changes must in part underlie that growing capacity.

As we will see in Chapter 12, the resulting circuit connects structures that will later be involved in pair bonding and parental behavior, including the amygdala, the BNST, the prefrontal cortex, the ventral striatum, the ventral pallidum, and probably the periaqueductal gray (PAG) and ventral tegmental areas of the midbrain. The anterior cingulate cortex, mediodorsal thalamus, and PAG are activated in human sadness and in guinea pig separation distress; the PAG also involves the septal area, the BNST, and the dorsal preoptic area (POA) of the hypothalamus (Panksepp 2003, 2005). At a minimum, infant attachment shares mediation by three limbic system nuclei (BNST, POA, and PAG) with parental behavior and pair bonding. All attachment probably depends in part on OT and on the dopamine reward system.

So the delay of strong social bonds, due to the fourth trimester and the yearlong extension of the fetal brain growth rate, marks a departure from the timing of anthropoid infant attachments, but not from their form, their adaptive functions, or their foundations in biological maturation. The primary caregiver supplies the bridge as the infant's contribution grows, and it is not accidental that this happens just in time for the growth of independent locomotion, with all its added dangers. But as we will see in Chapter 16, the cooperative dimension of human child rearing is already in place, and it greatly intensifies with the growth of language.

9

The Growth of Language

As the primary attachment weakens and broadens, all higher-primate young are socialized: they widen their web of relationships and learn the prevailing patterns of social interaction. But unlike other anthropoid young, the human child is delivered into a stimulus world pervaded by culture—most impressively in language acquisition.

Apes can learn to communicate using signed natural languages (Gardner, Gardner, et al. 1989) or arbitrary lexigrams (Premack 1971; Rumbaugh, von Glaserfeld, et al. 1974; Rumbaugh, Gill, et al. 1975); bonobos *(Pan paniscus)* in particular can decode spoken English sentences (Savage-Rumbaugh, Murphy, et al. 1993). Evidence of ape intellectual capacities, including language, toolmaking, and emergent forms of consciousness, continues to grow (Savage-Rumbaugh and Rumbaugh 1998; Savage-Rumbaugh, Shanker, et al. 1998; Rumbaugh and Washburn 2003; Fields, Segerdahl, et al. 2007). Still, an average child will surpass the most advanced ape language abilities by the third birthday, and not surprisingly, children this age have social learning and social cognitive capabilities that apes do not have (Herrmann, Call, et al. 2007). It is one of the unique (derived) features of human ontogeny, and its timing may have facilitated our species' success by enabling earlier weaning in the context of cooperative breeding.

Decades of research have shown remarkable consistencies in the nature and timing of language acquisition, and some psycholinguists posit a "language instinct" (Pinker 1994). But how can behavior that is so variable, and so obviously influenced by the cultural milieu, seriously be thought of as instinctual? There are at least two answers. First, language is not as variable as it seems. The pairing of sounds to meanings varies enormously (Hockett 1960), although not randomly (Lakoff and Johnson

1999; Berlin 2006), but syntax shows more limited variability (Chomsky 1988, 1993). Second, in a roughly eighteen-month period between twelve and thirty-six months, children in every culture progress from a few words to hundreds and acquire most grammatical rules in their native languages (Lenneberg 1967); much of this actually takes less than a year. All children become members of their destined language community, a major step toward enculturation.

At first glance, it seems unlikely that biology could play a role. Although universal or at least very widespread features of infant-parent nonverbal communication speak to the unity of the human species and our recent evolutionary origin (Locke 1993), languages do not seem plausibly in this category. Their most obvious aspect, looking around the world, is mutual incomprehensibility. Despite the current wave of extinctions, there are thousands of languages—at least eight hundred in Africa and four hundred in New Guinea alone. If there were one argument for the flexibility of human behavior, it would seem to be this diversity of natural languages. Furthermore, if we exchanged any two normal neonates in any two cultures, each would acquire easily and perfectly the language of its adoptive home.

Yet the biological model holds that language is too complex to be learned, in any conventional sense, in the time in which it develops (Chomsky 1959; Lenneberg 1967; Gleitman and Newport 1995; Boeckx and Piattelli-Palmarini 2005; Gleitman 2006). In the world of the young child, the schedules of reinforcement for correct and incorrect speech are too inconsistent, and in environments in which deliberate instruction is rare and the sentences the child hears vary infinitely, direct imitation cannot suffice. The primary linguistic data cannot explain language acquisition without specific innate predispositions (Crain 1991). Some things the child says could never have been heard, and imitation is resisted in situations where the child has mastered a rule, as in this exchange with a twenty-seven-month-old boy who could name two or three letters:

> "That's a A."
> "You mean 'That's an A.'"
> "That's a A."
> "That's an A."
> "Hm?"

"That's an A."
"Hm?"
"That's an A."
"Hm?"

Utterances like "Hm?" are adaptive mechanisms for eliciting repetition, and they occur in many cultures. (In !Kung children the equivalent is a nasalized "Hih?" with the same effect.) But despite the repetition, he did not learn to say "an" for several more months, despite such complex utterances as "The cars . . . um . . . the cars gonin' up there and . . . um . . . gonin' to the store . . . and . . . um . . . buy some milk and . . . buy some milk and buy some chickens and mayonnaise and . . ." In another exchange, he was asked to identify a piece of a jigsaw puzzle representing a pair of feet:

"What's that?"
"Two foots."

This was a classic overgeneralization error; he had the rule for plurals and ignored exceptions.

"You mean, 'Two feet.'"
"Two feets."
"Two *feet*."
"That's a two foots."

His mental model did not permit a plural without an *s*. He was ready to change the sound of the noun, even to add another grammatical error, but not to break his pluralizing rule.

Persistent overgeneralization errors are among the observations that have led many to conclude that imitation is implausible as a sole strategy for acquiring language (Marcus, Pinker, et al. 1992; Pinker 1995). "Two foots" is not a random imperfect imitation; it results from an internalized rule. A learning theorist would say that the rule was learned, which is partly true, but that does not explain such insistence on it, or the child's ability to abstract the pluralizing rule from adult English. It also does not explain the developmental course; overregularization is preceded by a period of correct forms that are *later* overregularized (Marcus, Pinker, et al. 1992). So it can reasonably be claimed that the young child's brain has built-in biases (Crain 1991; Stromswold 2000).

These facilitate first the abstraction of rules from a bewildering array of inputs, then their tenacious application to the ordering and transformation of new words, mastered at the rate of about ten per day during most of early childhood.

A Language Acquisition Device

The common features of function and thought underlying the world's languages are known in the Chomsky-Lenneberg model as universal grammar (Chomsky 1957; Lidz, Gleitman, et al. 2003; Lidz and Gleitman 2004). Each language transforms the universal grammar to generate its particular rules and forms, the links being culture-specific transformational grammars. Associated with universal grammar is the language acquisition device (LAD), which all children theoretically share. This "organ" of the mind, according to this model, appears in the growing brain during early childhood and explains the linguistic interchangeability of newborns and the speed with which young children acquire their native languages, during the same age range, in all cultures. A related distinction is made between competence and performance, the former being cognitive and closer to the fundamental, neuropsychological LAD and the latter more behavioral and dependent on cultural input. Similarities in the acquisition of signed and spoken languages suggest that the neuropsychology of language is not tied to the vocal-auditory channel (Petitto 2000).

Genetically designed brain circuits would explain the "deep structure" of languages and the universal capacities for meaning, productivity of novel signals, and displacement in space and time from the thing talked about. The learning environment would set "switches" in circuits in the brain's language module, designed by evolution to use certain kinds of information—whether or not to distinguish between *l* and *r*, for example, whether to pluralize with or without gender markers, or whether to put tense markers before or after a verb root (Pinker 1994, 1995). In the model the environment is crucial but not all-powerful; its role is to take advantage of these evolved brain circuits (Pinker and Bloom 1990; Pinker 1997).

Three lines of research support this view. The evidence for language universals is overwhelming and does not depend on Chomsky's or any other specific theory but can be shown from wide empirical study of many of the roughly four thousand known languages (Greenberg 1963;

Greenberg, Ferguson, et al. 1978; Butterworth, Comrie, et al. 1984; Comrie 1989). Furthermore, generalities that fall far short of the universal point to biological constraints. There is virtually no feature of language, not even vocabulary (Lakoff and Johnson 1999; Berlin 2006), for which biases and departures from chance do not exist. The anthropological linguist Joseph Greenberg observed that *any nonrandom distribution* of features of sound, symbol, or syntax apparent in the world's languages requires explanation (Greenberg 1975; Greenberg, Ferguson, et al. 1978), and sometimes the explanation is biological.

A second basis for this view is the great variety of conditions that still allow language to flourish and grow. These include barriers to communication such as deafness in the parents or the child, many forms of mental retardation, abuse and neglect, and minimal stimulation of the child with adult speech (Lenneberg 1967; Pinker 1994). Infants exposed to two languages keep the codes separate in their minds by twelve months of age (Kovács and Mehler 2009), and they attain standard milestones of language development at typical ages in both languages, with modest loss in either. No one has claimed that learning is less than vital; they have merely called attention to the remarkable similarities—especially in grammar—in the attainments of children growing up in extremely different conditions, which suggest a large and specific role for the child's maturing brain.

The third basis is the direct evidence of genetic effects. In early childhood identical twins are more similar than non-identical twins in their mispronunciation errors (Locke and Mather 1989), vocabulary comprehension (Mather and Black 1984), and speech and language disorders (Lewis and Thompson 1992). One adoption study showed that one-year-olds' verbal performance can be predicted better from the general intelligence of their biological parents than by the adoptive home environment (Hardy-Brown, Plomin, et al. 1981). Importantly, two behaviors of adoptive mothers—vocal responding and imitation of the infant's sounds—also predict infant language in this study, a discovery made possible by controlling for genes. Similarly, four-year-old twins show genetic effects on all measures of language, but nonshared environment is another important influence (Kovas, Hayiou-Thomas, et al. 2005). In seven-year-olds, who have had plenty of time for environmental shaping, genes explained more than half the variance in measures of spontaneous conversation; shared family environment explained none (DeThorne, Petrill, et al. 2008).

Research on language deficits also supports a role for genes. The genetic effects are not precise—they do not uniquely knock out plurals or any other feature of grammar—but some familial patterns in specific language impairment (SLI) are classically Mendelian (Tallal, Ross, et al. 1989; Gopnik and Crago 1993) and have clear brain correlates (Mills and Neville 1997). As we saw in Chapter 5, in an SLI kindred, a *FOXP2* mutation is shared by affected relatives and an unrelated SLI individual (Lai, Fisher, et al. 2001). Some studies suggest subnormal activity in Broca's area linked to the mutation (Liegeois, Baldeweg, et al. 2003), but there are other motor system abnormalities in both adult and child brains (Takahashi, Liu, et al. 2003; Vargha-Khadem, Gadian, et al. 2005; Soriano-Mas, Pujol, et al. 2009).

Cross-Cultural and Other Evidence

Observations of !Kung San infants and children illustrate the general principles of language acquisition and support their universality. These children master a language that pervasively utilizes implosive consonants ("clicks") and, like Mandarin Chinese, semantically salient tones, but infants and toddlers acquire language in a way similar to Euro-American children. Babbling (meaningless positive vocalization) peaks at eight months in both cultures, followed within months by a steep, steady rise in the ratio of meaningful to nonmeaningful vocalizations.

In both cultures the first phase is dominated by one-word utterances that stand for whole ideas (holophrastic utterances). Common first words for !Kung in my research included *Aiyo* (Mommy), *Mba* (Daddy), *Na* (gimme), *Ihn* (take it), and *Ihn-Ihn* (no). An American boy's first words were *Mommy, Daddy, no* (a toggle-switch word meaning either yes or no), *uh-oh, ihn-ihn* (for refusal), *open,* and *woop* (for *whoops*). A girl's one-word utterances at twenty months included *Aiee* (hi), *Mama, Papa, oosh* (shoes), and *take-you,* a blend of *thank you* and *take it,* used to mark exchanges of food or objects in either direction (also a toggle switch). Like the !Kung, both American children proceeded to simple, rule-bound, grammatical sentences within a year. Studies of the acquisition of Russian, Italian, and other languages show that important features are universal (Slobin 1985–92; Bates and Marchman 1987; Caselli, Bates, et al. 1995).

It is a truism of cross-cultural research that children are bathed in an environment that is inherently and pervasively cultural (Boas 1938; Sapir 1994; Shore 1996), yet language *training* is difficult to identify in any culture. Anthropologists studying child language among the Kaluli of New Guinea (Schieffelin 1990) and in Samoa (Ochs 1988) failed to find adult responses that helpfully rephrased the child's speech or expanded it to enhance meaning. Such corrections and expansions of child speech pepper the parent-child conversations of well-educated Americans but are uncommon in many other cultures (Schieffelin and Ochs 1986). Such social "scaffolding" is considered essential by some psychologists, but !Kung, Kaluli, and Samoan children acquire their respective languages without it, and oratory is valued in all three cultures.

Caregivers in many cultures do speak to infants in "motherese" (Fernald, Taeschner, et al. 1989), which anthropologists have studied as "baby talk" (Ferguson 1964). They speak simply with exaggerated tonal variation, repeat near-words, shaping them toward meaning; if this is training, then it exists for infants and young toddlers in all cultures where it has been sought. But these tactics for communicating with infants reflect the young child's universal responsiveness to certain stimulus patterns (Fernald 1985). Parental speech to prelinguistic infants is an adaptive behavior rooted in the phylogeny of parenting (Falk 2004b), not a conscious or cultural tactic to aid language learning (Fernald 1992).

In many cultures corrections are elicited by the infant's repetitive utterance of a signal such as *Hm?* or *Hih?* and parents respond to an infant's pointing hand by naming something. Among the Manus of New Guinea, Margaret Mead observed up to sixty corrective repetitions by a parent of the same child utterance, the corrections alternating with the child's mispronunciation. I observed up to about twenty repetitions in the same kinds of exchanges among the !Kung (Konner 1972). However, the pattern is not common in non-Western cultures (Schieffelin and Ochs 1986). This simple repetitive questioning by the infant may be an unlearned behavior attached to an unlearned utterance and itself an adaptation for language acquisition, but the responses are not the same everywhere.

Pidgins and Creoles

An important source of cross-cultural evidence is the creation of creoles: languages that arise as blends between two others, usually during and after colonization, when a subject people must try to speak the language of their conquerors. The first idiom formed is not the creole itself, but a pidgin, a crude version of the language of subjugation that includes elements of one or more native languages (Bickerton 1990, 1998). Pidgins also commonly arise in slavery or indentured servitude; when people are transported thousands of miles and thrown together in a polyglot setting, they must somehow communicate with their masters and each other, and they create a code from the linguistic material at hand. These initial codes are crude and disordered, meeting symbolic but not syntactic criteria for language.

Yet in the next generation, when children grow up hearing the pidgin, they transform it as they acquire it. It becomes a true language, but one that resembles the language of children in key features. The resulting creoles share those features regardless of where in the world they occur or what languages they blend. The creole created from Japanese, English, and Hawaiian converges with the creole of Mauritius, which draws mainly on French. Children throughout the world have used their shared neurolinguistic circuits to fashion convergent creoles from disparate languages.

More remarkably, a crude signed code invented by deaf children has grown into a true signed language, much as creoles emerge from pidgins (Senghas, Kita, et al. 2004). In Nicaragua, when the Sandinista regime established the first schools for the deaf, children previously isolated in a hearing world came together with others like themselves for the first time. Like the polyglot oppressed, they developed a code by which they could understand each other, blending and regularizing their varied home sign systems. The first cohort, mostly age ten and older when they met, did not get past this pidginlike stage of signed communication. But as younger children joined them, some of the youngest began to transform the code into a true language as they acquired it, on the model of vocal-channel creolization throughout the world. A linguistic code with few rules became in less than a generation a true language with a childlike but valid syntactic order.

In a related, also carefully studied example, a group of Bedouin with a substantial proportion of deaf individuals has developed a new signed language over the course of three generations, during which it has evolved all the important properties of natural signed language (Sandler, Meir, et al. 2005; Senghas 2005). Cases like these reveal what have been aptly called the resilient properties of language (Goldin-Meadow 2005a, 2005b). These must in some sense be coded in the brain.

Children with Atypical Input

So language does not have to use the vocal-auditory channel. Deaf infants babble vocally in the second half-year, but these efforts lead nowhere and fade away (Stoel-Gammon and Otomo 1986). However, they also "babble" with their hands, and just as the babbling of hearing infants gradually leads to words, hand babbling, given appropriate responses, leads to signs (Petitto and Marentette 1991; Petitto 1993). Later, acquiring American Sign Language (ASL), deaf children show the classic rapid increases in vocabulary and in phrase and sentence length during the sensitive second and third years (Klima and Bellugi 1979). Although delayed, deaf children use the same strategies in acquiring sign language as hearing children do in the vocal channel, and both reflect similar underlying cognitive processes, and therefore similar neural machinery (Petitto 2000). Scaffolding does not seem to be needed. Signing parents of deaf children do exaggerate their signs when addressing young children—visual motherese—and may make signs in contact with the child's body to maximize attention and information transfer, but it is unlikely that this information matches the adult input provided to upper-middle-class hearing children of hearing parents.

A limiting case is deaf infants exposed to no signed input (Goldin-Meadow 2005a), a difficult but common situation affecting 90 percent of deaf children: they are deaf, their parents are not, and the parents often do not know a signed language. The infants are exposed to neither heard nor seen language, yet they sometimes invent one, a "home sign," with no resemblance to ASL. Such children, although unusual, structure their invented signs in ways that resemble natural languages (Goldin-Meadow and Feldman 1977; Goldin-Meadow and Mylander 1983, 1991; Goldin-Meadow 2005a). Structure is evident both at the level of early one-word signs and at the level of simple phrases and sentences. For example, "one child pointed at a tower, produced the HIT sign [fist swatting in air] and then the FALL sign [flat palm flops over in air] to comment on the fact that he had hit [act 1] the tower and that the tower had fallen [act 2]" (Goldin-Meadow and Mylander 1991:318).

These invented signs are conventionalized much as words are and combined as two-word phrases at the same age at which hearing children achieve that milestone. That such deprived children fall far short of the standard criteria of natural languages is not surprising; what they do achieve is noteworthy, and it "suggests that the human brain is strongly canalized to produce linguistic systems with hierarchical organization" (Goldin-Meadow and Mylander 1991:340). Eric Lenneberg (1967:155–58) inferred something similar from less formal observations.

Children with Developmental Disorders

In Down syndrome, language develops, albeit more slowly, in much the same way as in normal children; an IQ of 50 at age twelve or 30 at age twenty is compatible with grammatical mastery of English (Lenneberg 1964). People with Seckel syndrome (nanocephalic dwarfism) are far below normal in body weight, brain weight, brain to body weight ratio, and IQ, but they usually attain verbal skills like those of a normal five-year-old. Research on children with atypical brains shows that brain structure predicts much of their cognitive profile (Reiss, Eliez, et al. 2000) and confirms that some circuitry is dedicated to language acquisition (Bishop and Mogford 1993; Tager-Flusberg 1994). In Williams syndrome there is a partial disjunction: relatively competent (although not normal) language and social interaction against a background of global retardation (Bellugi, Marks, et al. 1993; Karmiloff-Smith, Klima, et al. 1995). Teenagers with this syndrome may be at a two-year-old level in visuospatial and other intelligence tasks, yet they are able to discuss what they can and cannot do. This suggests that language is not just a special application of general cognitive powers.

Biological Foundations

Like bipedal walking, language is a species-specific human behavior, and even some gross distortions of input cannot prevent its development. It cannot be explained by general properties of the brain; instead, it involves a dedicated set of circuits—the canalizing internal factors that constitute the LAD (Sakai 2005; Pujol, Soriano-Mas, et al. 2006). We know enough about brain development to doubt any model of first language acquisition that omits an account of the brain's maturing functions. Such an account should not simply give a nod toward the brain's increasing power, but recognize the development of specific circuits at different times and rates corresponding to domains of mental development.

For example, maturational changes in Broca's area follow a largely predetermined plan, on a timetable that roughly corresponds to the acquisition of speech. Fiber density in this area increases markedly during the first two years, including the period from fifteen to twenty-four months. It is likely that in the deaf child the region controlling the hand becomes responsive to Broca's speech area, a portion of which is involved in tool use. Signing shows significant plasticity, but it is still based on specific brain maturation (Sakai 2005). Language areas also participate in general processes of cortical development (Rakic 2000). New dendrites are formed and make new synaptic connections. Synapses broaden, neurons increase in size, and glia multiply greatly, making up most of the increase in brain weight; this paradoxically thins out neuronal density in the cortex, with unknown functional implications.

In addition to Broca's area, other myelination cycles are relevant to explaining the universals of language acquisition that share the relevant timing (Roch-Lecours 1975; Lecours, Lhermitte, et al. 1983). One is the years-long cycle in the radiation to the auditory cortex and adjacent areas. The much faster cycle for the visual cortex enables essentially mature levels of visual pattern perception at five postnatal months. But human auditory pattern perception includes language, and the much longer and slower myelination cycle of the auditory association cortex must contribute to the maturational part of language acquisition.

These relationships were confirmed in a study using three-dimensional MRI on one hundred children between birth and age three years (Pujol, Soriano-Mas, et al. 2006). Imaging focused on language-related areas, taken from an animation of their aggregated data at different ages and smoothing them. The criterion used was 10 percent myelinated fibers in an area, which accounts for the finding of earlier myelination compared with the classic postmortem studies. Even at this criterion, changes in myelin deposition are dramatic over the first fifteen months, and they continue until thirty-six months. Age is a strong predictor of both myelin deposition in these pathways and the exponential vocabulary increase after eighteen months of age. There appears to be a threshold of myelination above which the vocabulary increase is possible, but these are correlations, and causal inference should be made with caution.

Again, myelination, although it certainly improves function, is also correlated with other measures of brain development. In Broca's area, there is markedly more higher-order dendritic branching in the left hemisphere than in the mirror-image region on the right; this left-right difference in lateralization is much greater than the left-right difference in the nearby precentral gyrus, part of the primary motor cortex (Scheibel 1991). Many other structural and functional lateralization differences have been shown (Herschkowitz, Kagan, et al. 1999; Ors, Ryding, et al. 2005).

These circuits can be malleable. In cases of left-sided brain damage in childhood, the right hemisphere has a remarkable capacity to take over many language functions by reorganizing in ways still poorly understood (Muller, Behen, et al. 1999; Hertz-Pannier, Chiron, et al. 2002; Fernandez, Cardebat, et al. 2004). Congenitally deaf adults have wider bands of responsiveness in the visual cortex than hearing people owing to lifelong differences in input (Neville 1991, 1995). Also, language regions have among the most protracted myelination sequences, which continue well into adulthood (Sowell, Peterson, et al. 2003). But neither dramatic recovery from some severe early brain damage nor subtler changes in normal brains obviate the maturational plan. Discoveries about the learning process are clarifying the very large learned component in language (Chapter 24), but do not negate the genetically regulated maturational process.

However, there remains a fundamental disagreement over whether language ability requires a separate module of brain circuitry (Pinker 1994; Piattelli-Palmarini 2002; Boeckx and Piattelli-Palmarini 2005), or whether language ability and its development merely require our large general cognitive powers, applicable to language as well as to other spheres of cognition (Bates 2004). Over several decades some cognitive scientists (Gardner 2004) and evolutionary psychologists (Cosmides and Tooby 2000; Tooby and Cosmides 2000) have proposed modular models of brain function and mind, and there have been interesting attempts to reconcile them with cultural models (Karmiloff-Smith 1995, 2009). There is neurophysiological evidence that the brain's capacity to acquire syntax is not just localized, but localized differently from that for the acquisition of semantics (Neville 1995). However, there are surely some large, general cognitive powers that are shared by all cognitive circuits (Jensen 2000), as well as specific circuits that participate in many different functions (such as those in the dorsolateral frontal cortex involved in short-term memory, or the hippocampal memory-storage and retrieval circuits). As we saw in Chapter 4, increasing brain size poses a major packaging challenge as the white/gray matter ratio increases; ultimately a brain as large as ours must become more modular and, especially, more lateralized. But no one who understands brain function thinks that language is completely modular. Many of the motor circuits deployed for speech are also used for other functions, and word learning involves the hippocampus; grammatical functions, however, may be more classically modularized in language areas (Opitz and Friederici 2004).

Early Anatomical Preparedness

One argument for the LAD is the fact that the brain seems to prepare developmentally for language long before there can have been any significant linguistic input (Dubois, Hertz-Pannier, et al. 2009). It has been known for decades that the brain of the human newborn is already lateralized in language-related areas, with left-sided dominance in most brains (Witelson and Pallie 1973), that this asymmetry extends down to about thirty weeks of gestational age (Wada, Clarke, et al. 1975; Chi, Dooling, et al. 1977; Wada and Davis 1977), and that infants in the first few months of life are attentive to phonemic boundaries and other specific aspects of language (Eimas 1997). Now we know that lateralization is established in the human brain by about twelve weeks gestational age, with hemispheric differences in the expression of twenty-seven genes (Sun, Patoine, et al. 2005). The functional meaning of these differences will take time to determine, but they clearly show that lateralization is a genetically driven, early embryonic process.

Given these developmental foundations, it is not surprising to find very young infants beginning to process ambient language sounds. Lateralized auditory evoked responses using magnetoencephalography are seen in third-trimester fetuses (Schleussner, Schneider, et al. 2004). Infants of twelve weeks postnatal age imitate adult vowel sounds (Kuhl and Meltzoff 1996). An fMRI study of two- to three-month-old infants showed preferential activation of the left planum temporale area for speech in their native language (French), but not for an inversion of the same speech that had similar acoustic characteristics (Dehaene-Lambertz 2004). The left perisylvian cortex is activated as three-month-olds listen to stories (Dehaene-Lambertz, Hertz-Pannier, et al. 2006). Using event-related potentials (ERP) in a habituation-dishabituation paradigm to study brain activity in perceptual discrimination, the same investigators found that, at three and a half months and at six months, infants responded differently to a phonetic change *without* phonemic value in French (a change that would not affect the meaning of a word) than to a repetition, and still more differently to a phonetic change *with* phonemic value.

Practical and ethical issues limit fMRI studies in infants, but ERP and EEG have revealed a developmental sequence that matches infant competences. Summarizing decades of research and scores of studies in her laboratory, Angela Friederici (2005) constructed a summary chart that matches events in language development with the characteristic ERP types experimentally evoked by them (Table 9.1). The four categories of ERP were defined in language perception studies of adults, and the table shows the approximate ages at which they come on-line. To simplify, they reflect phoneme discrimination, prosody (intonation) awareness, word meaning, and grammar. The ERP changes are gradual, as are the perceptual and behavioral changes.

Are these patterns determined solely by preprogrammed brain maturation? We know that Dr. Seuss stories read to fetuses in utero influence their preferences as newborns, that newborns prefer the sound of their mother's voice, and that four-day-old infants in German-speaking and French-speaking homes each prefer their home languages to the other. All these are the results of environmental influence on the brain.

Table 9.1. Development of the Comprehension of Language in Relation to When Specific Event-Related Potentials Become Detectable

Age (months)	2	9	14	18	32 to 36
Function	Discriminate phonemes	Prosody awareness	Word meaning	Phrase structure	Sentence structure
ERP	MMN	CPS	N400		ELAN-P600

Source: Based on Friederici (2005) and other sources.

Theorists like Mark Johnson (2007; Johnson, Grossman, et al. 2009) and Annette Karmiloff-Smith (2009) consider the maturational view inappropriate and see brain development as a constant environmental remodeling process. In some sense this must be true, since there is no clear hardware-software distinction as in silicon. Brain software—learned routines—*remodels* synapses, higher-order dendritic branches, and other aspects of microstructure. The genes cannot build the brain on their own.

But the question remains: how much do the genes do? In the growth of language, a great deal. As Ghislaine Dehaene-Lambertz (2004:457–58) has written (my translation), "The acquisition of language is classically discussed by opposing two hypotheses.... Without totally resolving this debate ... our results show systems well adapted to the handling of language from the first months and would therefore favor genetically pre-constrained systems. These systems particularly adapted to speech permit the infant to find the adequate input in its auditory environment."

The Role of Learning

How do we reconcile the biological facts with the very large scope that must be assigned to learning? One school of thought uses a new approach to artificial intelligence (AI), connectionism, to model language acquisition. Old AI efforts programmed computers to solve problems that brains can solve, but they did not include any learning capacity, nor did they attempt to make the program's algorithm parallel the brain's; getting a computer to solve a puzzle would theoretically tell you something about how the brain does it. But the brain usually does it very differently.

Connectionist models, called neural networks, are quite different (Elman, Bates, et al. 1996). They are ordered webs of (somewhat) neuronlike elements programmed to change connection strength with experience. Thus they are capable of learning, including some language learning, and they have suggested to some investigators that no innate propensities are needed. A more moderate version concedes that the brain has mildly specific preparation for language learning but gives learning a much greater role.

But connectionism remains subject to the AI error: *if a machine can do something the brain does, the brain probably does it similarly.* Connectionist software elements are not *really* like neurons, nor have connectionist circuits learned anything resembling natural language, although they have amassed languagelike knowledge. In any case, no amount of elegant machine learning can prove that the brain, a pastiche made by evolution from old, inelegant parts, acquires language the way connectionist networks do. The central questions of language development must be answered by what the developing brain actually does.

These questions cannot even be answered by what brains can be *made* to do—the assumption that if the brain can do something in a laboratory, then that must be how it does it in nature. This might be called Skinner's error. Drawing on many elegant laboratory studies of verbal learning, B. F. Skinner (1957) argued that inborn elements must play little or no role in the child's acquisition of language. His logic was flawed and remains so. For example, a study of eight-month-old infants showed that they can abstract and remember something resembling a word from a two-minute stream of meaningless syllables, based on the frequency of occurrence of the string that becomes the "word" (Saffran, Aslin, et al. 1996). Moreover, this kind of learning has been shown to be relevant to the acquisition of Italian (Pelucchi, Hay, et al. 2009). But how much do infants rely on such mechanisms? The chaotic environments where they grow are not controlled laboratory settings. Also, we want to know how their brains may be specifically prepared for statistical language learning. Again, no one denies a role for learning; the question is this: how is language learning specifically and preferentially supported by brain adaptations?

Another prominent view is that early nonverbal interactions between mothers and children are essential for language to emerge (Scaife and Bruner 1975; Bruner 1981, 1983). Certain preverbal interaction patterns, such as jointly pointing at or playing with an object, seem to precede the first words, and nonverbal interactions, particularly shared attention to objects, seem to be necessary support (scaffolding) for language development. One leading theorist, Jerome Bruner, proposed a language acquisition scaffold (LAS) corresponding to the LAD. Most of the evidence for the LAS is from middle-class, college-educated, English-speaking families, in which adults have far more conversations with children than in most other cultures (Schieffelin and Ochs 1986). "Expanding children's utterances, using leading questions, announcing activities/

events for a child, and using a simplified lexicon and grammar to do so are cross-culturally variable" (Ochs 1986:6). Yet all children learn to talk. As for the joint attention of infants and caregivers to objects, considered essential scaffolding by some, analysis of !Kung infant interaction showed far less than in the American middle class (Bakeman, Adamson, et al. 1990). The need to explore developmental alternatives to gaze following and gaze alternation, rare in many cultural contexts, is being increasingly recognized (Akhtar and Gernsbacher 2008).

Second, most studies of scaffolding show it to be only correlational, not causal. The parent could be responding to a verbally precocious child, or the verbal skill of both could be the result of social status or shared genes. Also, these interpretations lack any reference to the actual facts of brain development. Some studies support a role for interactive learning, but others are more consistent with the maturational model. We saw that in middle-class, college-educated families the errors of young children are often followed by adult corrections (negative evidence) and that such corrections often stimulate "recasts" by the children. But other studies suggest that children are exposed to limited negative evidence and respond little to it when it occurs (Marcus 1993; Pinker 1995; Stromswold 2000).

Laboratory research motivated by this theory establishes the power of some forms of scaffolding under experimental conditions (Tomasello and Kruger 1992; Akhtar, Carpenter, et al. 1996). But the real question, which has not been addressed by the proponents of the various scaffolding models, is whether some form of scaffolding is necessary for language learning under natural conditions. Much other evidence suggests that it is not.

Language and Emotion

A revival of interest in the neurobiology of emotion has made it clear that models of mind that ignore it cannot adequately represent human cognitive functions (Damasio 1994, 1998; Dolan 2002). Some language development models stress the preparatory and contextual role of the infant's relationship to the primary caregiver, emphasizing preverbal vocal and facial interaction as fundamental bases of language acquisition (Locke 1993; Kuhl 2007). Although the cross-cultural variability of such interactions is less than that of the LAS components, it is still substantial (Klein, Lasky, et al. 1977; Konner 1977b) and thus creates a similar problem.

Still, any theory of environmental influence in language acquisition must include the infant's emotional investment in relationships with caregivers. We are bathed in communications from birth, and the child's ability to interpret and send signals is of the essence of relationships. This guarantees that the first efforts at semantically meaningful signaling is emotionally weighted, as is the reward value of being able to communicate wants and needs, which must quickly become autonomous from basic survival needs (if it is not autonomous to begin with) and be rewarding in and of itself. Language acquisition is emotionally and motivationally charged.

There are other cognitive changes during the second year of life that collectively show increased self-awareness, including self-descriptions, linguistic self-corrections, and signs of distress at an inability to imitate a task (Kagan 1981a), with close parallels in children that age in Fiji (Katz 1981) and probably in other cultures. These changes are not tightly yoked together, nor do they appear in a set order, but as they are achieved the child reaches a new level of consciousness. Michael Lewis (2003) has studied mirror self-recognition, the use of personal pronouns, and pretend play as indicators of self-awareness; these appear in the second year, with mirror self-recognition coming first. After controlling for age, the maturation of the left temporo-parietal junction, involved in language, uniquely predicted self-representation (Lewis and Carmody 2008).

In Daniel Stern's (1985) synthetic model of psychodynamic development, this is referred to as the emergence of a verbal self, the culmination of a series of changes beginning in early infancy. At two or three months, in connection with the emergence of sociality discussed in Chapter 7, a core or physical self emerges—that is, the subjective impression of being a physical entity and of the caregiver as a separate individual, also known as primary intersubjectivity. Around nine months the infant adds the sense of a subjective self—similar to the secondary intersubjectivity heralded by joint attention and social referencing. Finally, during the second half of the second year, a new sense of self emerges, based on or bound up with words, which allows the child to enter into a rudimentary conversation in which subjective states are discussed.

These stages are not passed *through* but are successively integrated, so that normal two-year-olds experience them in some sort of balance. The five-month-old has a core or physical self that can act on and be acted on by others, and the one-year-old can share attention to things. The two-year-old, however, has words, which "permit the old and persistent issues of attachment, autonomy, separation, intimacy, and so on to be reencountered . . . through shared meaning" (Stern 1985:173). With language, the child enters a new kind of discourse and a new variety and complexity of relationships. Language creates a certain distance from experience; words convey emotion less directly than gaze contact, smiling, hugging, or crying. But this means that the intimate primary relationships are no longer the child's only social option.

The emergence of language delivers the child into a wider world. It is one of the great changes in the evolution of childhood, and it may have helped our hunter-gatherer ancestors increase their offspring productivity by greatly reducing the ancestral ape weaning age against the background of human cooperative breeding. Weaning foods are the key to this transition, but language draws helpers to the child's and the mother's side. It is not a coincidence that the typical hunter-gatherer weaning age, between two and a half and three years, comes at the end of the child's greatest advance in linguistic capacity.

10

The Growth of Sex and Gender Differences

As with attachment and social fears, the growth of sex and gender differences is distinctively human, yet continuous with that in other anthropoids. In most cultures throughout history, psychological sex differences have been seen as intrinsic, with ample evolutionary justification. In *The Descent of Man, and Selection in Relation to Sex* (1871/1981), Darwin saw how they might have arisen: males compete for access to females, and female choice directs the divergent evolution of males. Seeing males as an evolutionary invention by asexual species that are basically female leads to the question of why males exist. The likeliest answer is that the variation due to sexual reproduction offered greater resistance against microbes and parasites and a richer substrate for further evolution. Given that males were invented, it is not surprising that they should diverge. But soon after Darwin some theorists began to take a very different approach to gender differences.

Gender Identity

Sigmund Freud (1905, 1924, 1933) proposed a psychodynamic model of gender differentiation in which anatomy is destiny but self-concept and family dynamics determine how anatomy shapes mind: (1) little boys fear the loss of their penises because, having masturbated, they are fond of them and imagine that girls have lost theirs; (2) boys desire their mothers and fear castration by their fathers; (3) girls envy boys their penises, think they themselves have had and lost them, and view pregnancy as a substitute; (4) girls fall in love with their fathers and see their mothers as rivals; and (5) both sexes must cope with these trials to maturely self-identify as male or female.

Many questioned this scheme, and it seems quaint in retrospect, but it influenced modern thought about how gender takes form in the mind. Although elements of it appear in clinical case studies and in anecdotes from various cultures, there is little research confirming the model; however, it challenged subsequent theorists to explain not just male-female differences in behavior but the concept of self as it becomes gendered during childhood. As we will see in Chapter 11, a new capacity for self-reflection is added in middle childhood, but self-identification as male or female begins much earlier, often by age two, almost always by three.

Also by age three, most children can readily and explicitly identify toys, games, clothing, tools, and work that in Western culture are stereotyped as masculine or feminine (Golombok, Rust, et al. 2007; Berenbaum, Martin, et al. 2008); in this first phase, children are more rigid than they will be later. They also correctly classify others, including the same-sex parent, as sharing (or not sharing) a gender label with them. Sometime between ages three and seven they decide that gender is permanent and not a choice. Paradoxically this subjective gender constancy goes along with decreased rigidity of stereotypes about how the sexes play, work, and dress (Serbin, Powlishta, et al. 1993; Katz and Ksansnak 1994); perhaps three-year-olds, uncertain about gender constancy, need stereotyping. Increasingly during this period they also prefer to identify with and imitate the same-sex parent or other same-sex adults (Maccoby 1998). However, exposure to models whose behavior runs counter to stereotypes helps reduce this preference.

But behavior begins to differentiate much earlier—neonatally and perhaps prenatally. Sex differences in behavior and responsiveness at birth—within hours or days of delivery—have long been known (Korner 1969, 1973). Girls have more sensitive skin and react more to a light touch; boys show more muscular strength and when laid face down make more effort to lift their heads. Girls smile more (nonsocially at this age) and show other signs of greater oral responsiveness; when a liquid they are sucking is changed from bland to sweet, girls increase their rate of sucking more and search more with their mouths. Girls respond faster to a flash of light. However, there are no differences in responsiveness to sounds or in overall bodily activity.

In other studies newborn boys have been shown to be more irritable than girls, to cry and grimace more, to have more widely fluctuating emotional states, and to smile and comfort themselves less (Osofsky and O'Connell 1977; Hittelman and Dickes 1979; Feldman, Brody, et al. 1980; Weinberg, Tronick, et al. 1999). Differences in social responses appear by two or three months of age (Moss 1967; Farris 2000), and perhaps at birth (Hittelman and Dickes 1979). One-year-old boys are more likely than one-year-old girls to physically attack a barrier (Goldberg and Lewis 1969); hitting and sex-typed toy choice show sex differences by the second year (Maccoby 1998). Preferences for trucks and soldiers versus dolls and baby bottles appear by eighteen months, and the difference widens for several years (Berenbaum, Martin, et al. 2008).

Same-sex playmate preference is consistent in Western cultures (Maccoby 1998). By age three, children are more likely to approach or play with a same-sex peer, and by four and a half years, in preschool situations where choice is possible, children show a threefold preference for their own sex. By six and a half years this preference has increased to elevenfold, and a wide gap is maintained until age eleven, independent of toy and game selection; it is quite pronounced even in gender-neutral play settings. Research on deliberate efforts to reduce sex-segregation shows that children resist such efforts. As Eleanor Maccoby writes (1998:30), "The separation of the two sexes is seen most strongly in situations where adult-imposed structure is absent, such as school corridors, lunchrooms, and playgrounds." Segregation is not related to how masculine or feminine their own behavior is within their own sex. However, Maccoby cogently argues that by the time adolescence brings them together again, boys and girls have essentially grown up in different cultures, largely of their own making.

Sex Differences in Aggression

Many sex differences are culturally generated, but a few appear so consistently across cultures as to be independent of even today's electronic media. Self-segregation is one; it appears wherever numbers are large enough to enable choice. Another is physical aggression, however measured. In most cultures, boys are rougher and less nurturing than girls by age three (Whiting and Edwards 1988). The sex difference is present at seventeen months and does not change at twenty-nine months in longitudinal follow-up (Baillargeon, Zoccolillo, et al. 2007). It persists throughout growth and is greatest in young adulthood (Archer 2004).

On phylogenetic grounds, some biologically encoded behavioral sex differences must be expected, and a likely domain would be physical aggression.

This is not because males were the hunters during human evolution. Predatory behavior in mammals has little to do with aggression, which involves different psychological states, uses different neural circuits, and recruits different behaviors. There is overlap between predation and aggression, but the main use of physical aggression is competition for resources, including mates. (One hypothesis of the invention of males is that females gained reproductive advantage by exporting risky competition to a nonreproducing sex.) Predation by carnivores and rodents resembles problem solving or play more than fighting, and human hunting involves little or none of the emotion that characterizes interpersonal violence. It is male predominance in fighting—phylogenetically much older than hominin hunting—that leads us to expect that biological sex differences underlie physical aggression.

In addition, past candidates for behavioral or psychological sex differences in children have included reliance on others for help; nurturance and affiliation; offering help and seeking companionship; and verbal and mathematical ability. In the first systematic review (Maccoby 1966), dependency was found to be higher in girls in sixteen studies but higher in boys in six, with the sexes equal in nine; later studies provided no evidence for a consistent sex difference in dependency (Maccoby 1998). Based on recent meta-analyses, the evidence for sex differences is similarly inconsistent for many variables (Hyde 2005); most psychological "sex differences" that the average person can think of are not real.

In other cases uncertainty persists. In a meta-analysis of sex differences in social smiling (162 research reports with a total of 448 effect sizes), there was a significant tendency for adolescent girls and women to smile more than males, but this difference was highly contingent on contextual variables (LaFrance, Hecht, et al. 2003); this finding extended a long-standing ambiguity in whether or not there are sex differences in sociability. A sex difference in functional brain laterality was clear in only five out of 112 experiments, although in all five, males had more (Hiscock, Perachio, et al. 2001).

But findings have always been more consistent for aggression, nurturance, and sexuality (which we consider in Chapter 12). In a review of all studies by psychologists until 1965 (Maccoby 1966), before either evolutionary psychology or second-wave feminism, subjects ranging from two-year-olds to adults were studied by many investigators with techniques including direct observation, parents' ratings, teachers' ratings, laboratory experiments, personality tests, self-descriptions, and fantasy during doll play. For nurturance and affiliation, forty-five of fifty-two studies found females showing more; two found the reverse, and five found no difference. For all types of aggression, forty-four of fifty-seven studies found more in males, four found the reverse, and nine found no difference. But for overt physical aggression, girls and women showed less in *all* studies. Reviews and meta-analyses of hundreds of studies in the more than four decades since strongly confirm these findings (Maccoby and Jacklin 1974, 1980; Eagly and Steffen 1986; Archer 2004; Hyde 2005; Tremblay, Hartup, et al. 2005).

Cross-Cultural Studies

Cross-cultural research reveals the same pattern, even while effectively challenging many other sex role stereotypes. In seven traditional societies in the Pacific, features of behavior and character that Western culture defines as male or female were not gender-bound (Mead 1949). Among the Mundugumor of New Guinea, women were "as assertive and vigorous as the men, . . . detest bearing and rearing children, and provide most of the food, leaving the men free to plot and fight" (Mead 1949:53–54). Among the Tchambuli of New Guinea, "women, brisk, unadorned, managing and industrious, fish and go to market; the men, decorative and adorned, carve and paint and practice dance-steps, their head-hunting tradition replaced by the simpler practice of buying victims to validate their manhood" (54).

Such reports negated claims that women everywhere would be more concerned about their appearance, men more industrious and managerial. Yet in both of these New Guinea groups, which fought often, men did the fighting and women the child care. This division of labor (unlike, for example, gardening) is universal in traditional societies. In 122 cultures, weapon making and fighting were invariably done by men (D'Andrade 1966). There has been no report of a culture in which men do nearly as much child care as women (Chapter 17), and in every culture violence—especially homicidal violence—is overwhelmingly accounted for by males, sometimes fighting over women. Even dreams show this difference. In 75 tribal societies men were more likely than women to report dreams with the subjects *weapon, grass, coitus, wife, animal, death, red, vehicle, hit,* and *ineffectual attempt,* while women reported more dreams including *husband, clothes, mother, father, child, home, female figure, cry,* and *male figure* (D'Andrade 1966).

Quantitative and qualitative anthropological studies of children have confirmed sex differences in dimensions that resemble physical aggression and nurturance. One study compared two- to six-year-old !Kung San children in Botswana with children in London (Blurton Jones and Konner 1973). Systematic observations were done under similar conditions. Six behaviors showed a sex difference in the same direction in both cultures: boys showed more aggressive faces, more aggressive acts, more frowns, and more interactions with other children; girls spent more time alone and showed more interactions with adults. But while London boys showed higher overall activity than girls, the !Kung showed no difference in the activity levels of girls and boys. Many similar findings cast doubt on the claim that sex differences in activity level are universal results of the sex difference in resting metabolic rates, but the sex difference in aggressive behavior is consistent.

In her comprehensive study of play among the !Ko of southwestern Botswana, another San group then living by hunting and gathering, Heide Sbrzesny (1976) found that of fifteen games or dances involving fighting or hunting, nine (of which four involved simulated combat) were played only by boys and the other six were played by both sexes together; rarely did girls play such games alone. However, she notes, girls exhibited "a fairly high level of aggression . . . in their constant nagging, teasing, and joking" (Sbrzesny 1976:307); this finding is consistent with many studies in Western cultures showing that verbal aggression and relational aggression are often higher in girls. Furthermore, "games of cooperation (melon-ball-dance) are predominant in girls and women. 11 dancing games were described with women in contrast to 3 with men. Only boys play hunting games and only girls play with dolls, which they construct from melons. The children thus identify themselves with their gender role without evident pressure from the side of the adults, who rarely intervene in the children's behavior" (307).

Beginning in the 1960s, teams of anthropologists led by Beatrice and

John Whiting conducted observational research on children (Whiting 1963; Whiting and Whiting 1975a; Whiting and Edwards 1988). The first wave, the Six Cultures Study, included a small town in New England and farming or herding villages in Mexico, Kenya, India, Japan, and the Philippines. In each, the behavior of children ages three to eleven was recorded by both a male and a female anthropologist, in hundreds of hours of observation, using uniform methods. Subjects were scored on twelve well-defined and reliably observed units of behavior, such as *seeks help, offers support, touches, reprimands,* and *assaults.* Two independent dimensions of behavior emerged from multidimensional scaling, one interpreted as egoistic versus nurturant, the other as aggressive versus intimate. Boys differed from girls in the two-dimensional space in each of the six cultures, showing greater egoism and/or greater aggressiveness than girls; that the magnitude of the sex difference varied among cultures; and that girls in some cultures were more aggressive or egoistic than boys in other cultures, although less aggressive and/or egoistic than the boys in their own. The analysis was later extended to five other quantitatively studied cultures—Kien-taa in Liberia, Kokwet, Ngeca, and Kisa-Kariobangi in Kenya, and Bhubaneswar in India—with similar conclusions (Whiting and Edwards 1988; Edwards 1993).

Since those studies, the greater physical aggressiveness of boys has remained a consistent finding in the anthropology of childhood (Fry 1987, 1990, 1998). Robert Munroe, one of the researchers in the Six Cultures Study, went on to lead field research in four other traditional cultures in Belize, Kenya, Nepal, and American Samoa (Munroe, Hulefeld, et al. 2000). A total of ninety-two boys and ninety-two girls ages three, five, seven, and nine were systematically and extensively observed in naturalistic settings. Boys on average exhibited aggression in 10 percent of their social interactions, girls in 6 percent, and the difference was present in all four cultures. Play-fighting, or rough-and-tumble (R&T) play, is easily distinguished from real fighting on objective observable criteria understood by children in all cultures (Costabile, Smith, et al. 1991; Smith, Smees, et al. 2004), but the sex difference prevails in both. Of twenty-seven cultures in a worldwide sample with relevant information, "21 descriptions link wrestling to boys, 6 to both girls and boys, and none solely to girls" (Fry 2005:65). The same is true of R&T play in great apes (Lewis 2005) and many other mammals (Fagen 1981). We consider cultural variation in later chapters, but here we try to explain the universals.

Neuroendocrine Foundations

Could neonatal differences be causally related to such complex behaviors as aggression and nurturance? Greater strength and lower skin sensitivity at birth might lead to more aggression; within each sex, infants with more sensitive skin at birth are less likely to attack an obstacle or barrier in a test at age five (Bell and Costello 1964; Bell, Weller, et al. 1971). Some social scientists still attempt to explain consistent sex differences solely by reference to child training practices, which could conceivably be consistent cross-culturally.

This is less plausible in adulthood; marked adolescent hormonal changes produce lasting sex differences. Males have much higher levels of testosterone (T); females have much higher levels of estradiol, the main active estrogen (E_2), and progesterone, which cycle monthly, steeply increase during pregnancy, and are relatively suppressed during lactation, when prolactin and oxytocin levels are high. Injection of T and related male hormones increases or enables aggression in humans and many other animals, and castration decreases it (Davidson, Camargo, et al. 1979; Dixon 1980; Breedlove 1992; Pope, Kouri, et al. 2000); T injection disrupts normal mothering in female mammals.

However, many species show sex differences in aggression *pre*pubertally, when sex hormone levels are very low (Mitchell 1979). In a classic study of rhesus monkeys, prepubertal males and females socially isolated from birth were placed with normal, unfamiliar monkey infants; males were more aggressive toward the infants, and females more nurturing (Chamove, Harlow, et al. 1967). As we saw in Chapter 6, some prepubertal behavioral differences are due to early androgens. In humans and other mammals, before and/or shortly after birth, males have high androgen levels (Hines 2006; Knickmeyer and Baron-Cohen 2006). During this phase androgens directly affect sexually dimorphic nuclei, including the medial preoptic area of the hypothalamus (SDN-POA), where male and female rats differ in neuronal connections (Raisman and Field 1973; Gorski 1996, 2000; Pak and Handa 2008). Males castrated at birth develop the female pattern, while perinatal injection of T gives females the male pattern (Reinisch 1974; McEwen 1978). Size differences in the SDN-POA are visible to the naked eye (Gorski, Gordon, et al. 1978; Gorski 1996), and parallel sex differences are seen in primates (Ayoub, Greenough, et al. 1983).

In addition, evidence has accumulated for structural sex differences in other areas of the brain (Gorski 2000), including the medial nucleus of the amygdala and the hypothalamic nucleus that the amygdala projects to—the bed nucleus of the stria terminalis (BNST) (Hines, Allen, et al. 1992). There are also sex differences in the region of the lumbar spinal cord containing motor neurons for the bulbocavernosus muscles, which control the penis in rats and humans. This difference appears prenatally in humans (Breedlove and Arnold 1980; Breedlove 1994; McCarthy and Arnold 2008).

Testosterone must mostly be converted to estrogen in brain cells before it can exert many of its effects (Naftolin and Butz 1981; Naftolin 1994), but female fetuses are not affected because serum E_2-binding proteins restrict it from crossing the blood-brain barrier; T does so easily and is converted to E_2 by an aromatase enzyme. The E_2 binds with its receptor and the resulting hormone-receptor complex binds to select DNA promoters, switching genes on. E_2-treated hypothalamic slices from newborn mice of both sexes greatly increase extension of axons and dendrites (neurites) compared with controls (Toran-Allerand 1976). Genes whose promoters are bound by the hormone-receptor complex make proteins for neural growth, including nerve growth factor (NGF); in addition, the actions of steroid hormones and growth promoters are cross-coupled and synergistic (Toran-Allerand 1996). Also, E_2 alters genetically guided neuronal movement patterns in the POA of hypothalamic slices (Knoll, Wolfe, et al. 2007).

Some studies have examined the human brain for corresponding sex differences, focusing on a group of four clusters of hypothalamic neurons, the interstitial nuclei of the anterior hypothalamus (INAH) (Swaab, Chung, et al. 2001). Investigators in the mid-1980s reported a sex difference in INAH 1, a finding that has been difficult to replicate, but several subsequent studies found large, consistent differences in INAH 3, with one reporting that it was three times larger in men and another also showing INAH 3 in the female range in homosexual men (LeVay 1991). Others have found large differences in INAH 3 and 4 (Allen, Hines, et al. 2007). Beyond the hypothalamus, brain imaging by multiple methods points to numerous possible differences (Cosgrove, Mazure, et al. 2007). Men have larger overall brain volume, women larger brain to body weight ratios. Corrected for total volume, the hypothalamus and amygdala are larger in men, and the hippocampus and caudate are larger in women. Sex differences have also been found in the distribution of receptors or transporters and in activity in several major neurotransmitter systems, including those using serotonin, dopamine, and γ-amino butyric acid (GABA).

Sex differences in cortical asymmetries have also frequently been reported (Wisniewski 1998; Ullman, Miranda, et al. 2008), although they remain controversial. In the corpus callosum the posterior part tends to be larger in women, suggesting that interhemispheric communication could be better in women in posterior cortical areas (de Lacoste-Utamsing and Holloway 1982; Allen, Richey, et al. 1991; Mitchell, Free, et al. 2003). Callosal sex differences are consistent with at least some findings from neuropsychological studies (Davatzikos and Resnick 1998; Hampson 2008; Ullman, Miranda, et al. 2008). A similar pattern in the rhesus monkey (Franklin, Kraemer, et al. 2000) suggests that it may have preceded the emergence of language. One hypothesis of the adaptive value of the difference is that males needed a different kind of spatial ability to range farther, as they do in many species for mating, and in humans for hunting as well (Gaulin 1992). Meanwhile, growing evidence suggests that the navigation and search tasks at which women perform better are adaptations for finding and gathering high-quality plant foods (Silverman and Eals 1992; New, Krasnow, et al. 2007). Sex differences in the brain outside the hypothalamus have also been studied in children (Giedd, Castellanos, et al. 1997) and may, like hypothalamic sex differences, be dependent on circulating levels of steroid hormones (Wisniewski 1998).

The difference in the SDN-POA apparently predisposes males in many mammal species to be more aggressive and less nurturing throughout life. Perinatally castrating or giving anti-androgens to males or injecting females with androgens reduces or abolishes the difference. In many species, including rats, mice, dogs, rabbits, and monkeys, females so treated are subsequently more aggressive and less nurturing than untreated females and show less lordosis and other female sexual behavior and more mounting and other male sexual behavior (Edwards 1968, 1969, 1970; Edwards and Burge 1971). In prairie voles (*Microtus ochro-*

gaster) virgin males show much less parental behavior than virgin females, but males castrated at birth respond to pups as females do (Lonstein, Rood, et al. 2002). Many such effects have been shown in nonhuman primates (Goy, Bercovitch, et al. 1988).

Clinical Models

For humans, patients in several categories are relevant (Federman 1967, 1982, 2006). The first comprises those who have experienced high prenatal androgen levels, owing either to a single-gene defect in an enzyme of steroid synthesis or to drug exposure (Ehrhardt and Meyer-Bahlburg 1981; Reinisch 1981; Breedlove 1994; Hines 2006). The classic example is congenital adrenal hyperplasia (CAH) in girls, resulting from a defect in either of two enzymes, steroid 21- or steroid 11-hydroxylase. The adrenal cortex cannot execute its normal pathway of synthesizing cortisol and aldosterone, and synthesis is shunted to excess androgen production.

Even if medication and surgery correct the condition, CAH girls will have already been exposed to circulating androgens from around nine weeks of gestational age. If treated from birth, they are quite normal in most measures, but they tend to be more conventionally masculine in their behavior and life goals than their sisters or unrelated controls (Hines and Kaufman 1994). The strongest differences are in childhood play patterns, which, as we have seen, show quantitative sex differences in normal preschool children (Hines 2003).

When subsequently compared—in interviews, mother ratings, and self-ratings—with their mothers, sisters, or individuals with other genetic (but not hormonal) disorders, affected girls show more intense activity and male-typical play, prefer playing with boys more, and are less interested in marriage, motherhood, doll play, and infant care than either their sisters or an unrelated matched control sample (Berenbaum and Hines 1992; Hines and Kaufman 1994; Iijima, Arisaka, et al. 2001). These are significant differences but are within the normal range of girls' behavior, which overlaps greatly with that for boys. Similar findings emerge from studies of girls and women exposed to androgenic progestins given to their mothers for pregnancy maintenance (Reinisch 1981; Reinisch and Sanders 1992). Treated CAH girls are also indistinguishable from unaffected boys in drawing tests (Iijima, Arisaka, et al. 2001), both groups being more likely to draw mobile objects and mechanical objects with dark or cold colors and to depict a bird's-eye view, while unaffected girls are more likely to draw people, flowers, and butterflies with light and warm colors and tend to arrange subjects in a row on the ground.

Some chromosomal (XX) females with CAH or other prenatal androgen exposure feel that they should be men, but most do not (Hines 2004). In a detailed study of four who did decide to become men through transgender surgery and hormone treatments, the transition was difficult (Meyer-Bahlburg, Gruen, et al. 1996). Three of the four had had their ambiguous genitalia shaped to female form in infancy, and all were designated as girls within a few weeks of birth, but their childhood gender identities were somewhat ambiguous. As they grew up and became sexually active, all four were exclusively interested in women, but unlike most women attracted to women, they gradually became determined to undergo sex change. Supportive romantic and sexual relationships with women were important in this, their last step toward male identity.

In the second clinical group are people with 5-α-reductase deficiency (Imperato-McGinley, Peterson, et al. 1979; Imperato-McGinley 1997), most of whom were found in three intermarrying rural villages in the Dominican Republic. Over four generations, thirty-eight individuals from twenty-three interrelated families were affected. Various single-gene mutations were identified there, in Mexico, Turkey, China, Italy, and India, and elsewhere (Canto, Vilchis, et al. 1997; Can, Zhu, et al. 1998; Lee, Lam, et al. 2003; Baldinotti, Majore, et al. 2008; Praveen, Desai, et al. 2008).

The Dominicans are the largest sample and have been carefully studied (Imperato-McGinley, Peterson, et al. 1979; Imperato-McGinley 1997, 2002). Nineteen infants appeared at birth to be female and were reared as normal girls. At puberty, they first failed to develop breasts, then had a mainly masculine pubertal transformation, including growth of a penis, descent of previously abdominal testes, deepening of the voice, broadening of the shoulders, and development of a muscular masculine physique. Physically and psychologically they became men, with normal or sometimes hypernormal sexual desire for women and with a surprisingly normal range of sexual functions except for being infertile, owing to abnormal ejaculation through a channel at the base of the penis. After many years of experience with such individuals, the villagers had identified them as a separate group, called *guevedoce* (penis at twelve) or *machihembra* (man-woman), but they often could not predict which girls would be in this group.

These children are chromosomally male (XY) but lack one enzyme, 5-α-reductase, which changes testosterone into dihydrotestosterone (DHT); they have almost no DHT, but normal T levels. The two hormones are respectively responsible for the formation of male external sex characteristics at birth (DHT) and at puberty (T). Lack of DHT leads to a female-looking newborn and prepubertal child, yet the ongoing developmental course of T yields a male puberty.

Almost all of these children become men of their culture. After twelve or more years of girlhood, with all the cultural influences shaping that gender role, they become almost typical men in terms of family, sexual, vocational, and avocational roles. Of the eighteen for whom data were available, seventeen made this transformation; the last retained a female role and gender identity. The transition for those seventeen individuals was not easy—some had years of confusion and distress—but they made the change without special treatment. Julianne Imperato-McGinley reasoned that prenatal T had masculinized their brains and "that in the absence of sociocultural factors that could interrupt the natural sequence of events, the effect of T predominates, over-riding the effect of rearing as girls" (Imperato-McGinley, Peterson, et al. 1979:1233).

Her view is supported by studies of the same syndrome elsewhere. Four of the five children in an Indian sample were raised as girls, and the three of those four who had gone through puberty made a transition to male identity (Praveen, Desai, et al. 2008). In a touching case in Pennsylvania, a twelve-year-old who had been raised as a girl wanted her ambiguous genitalia to be made unambiguously female; surgery was withheld, and counseling given. At fourteen, she seemed to be in transition to an increasingly male identity. Overall, a review of ninety-nine cases with "explicit psychosexual information" concluded that 63 percent had a gender role change in adolescence from female to male (Cohen-Kettenis 2005:400). Of eighteen others raised male, all remained so.

The Sambia of New Guinea view these individuals as a third sex and label them as "becoming male," or with the pidgin word "turnim man." This seems a stronger label than the Dominican ones and raises questions about whether even those labels influence gender change. But it remains clear that prenatal hormones contribute to male gender identity. Gilbert Herdt (1994a:443–45), the Sambia ethnographer, writes, "The 5-alpha reductase deficiency syndrome clearly creates extraordinary prenatal hormonal effects in gender development. . . . Gender identity is not entirely a social construction." Furthermore, "in some places and times, a third sex has emerged as a part of human nature; and in this way, it is not merely an illusion of culture, although cultures may go to extreme lengths to make this seem so."

A third category is cloacal exstrophy, also a rare condition in which XY individuals fail to develop male genitalia but have prenatal androgen exposure. Sixteen cases were studied with respect to the development of gender identity and gender-specific behavior (Reiner and Gearhart 2004). Two of the sixteen were assigned as males at birth and remained male by all psychological measures. The other fourteen, assigned as females, had varied developmental paths, but most had prominent features of psychological masculinity. At ages ranging from eight to twenty-one years, eight of the fourteen expressed a desire to be male, all fourteen had predominantly male toy and game preferences, nine had mainly male friends, and ten engaged in typically male levels of rough-and-tumble play. As with CAH, intrauterine exogenous androgenic treatments, and 5-α-reductase deficiency, cloacal exstrophy supports a role for prenatal androgen effects on the brain in the development of psychological aspects of gender. However, as with 5-α-reductase girls, some remain content with their female identity despite being chromosomal males (Mukherjee, McCauley, et al. 2007).

A fourth—fortunately very small—category is that of male neonates who lost their male genitalia owing to surgical error during circumcision and the surgical and hormonal "correction" of their status that followed. This was rationalized with the erroneous theory of psychosexual neutrality at birth. In one case an infant born in the 1960s (one of a pair of identical twins) was assigned as female shortly after the accident (Angier 1997; Diamond and Sigmundson 1997). Surgical construction of a vagina in childhood and female hormones in adolescence seemed to shape an identity-forming process that, along with family and cultural influences, would theoretically make this infant an infertile but psychologically normal woman.

But the woman this child became was very uncomfortable in her gender role, began to search for clues about her early development, and discovered her medical history. She said she experienced this news as a great relief, since it explained a lifetime of feeling like a misfit, and she took steps to reorient her gender identity again, voluntarily reversing the childhood process with male hormones and genital reconstruction. "He got himself a van, with a bar in it," one of his doctors said. "He wanted to lasso some ladies" (Angier 1997:A12). Tragically, this patient took his own life several years later. However, since his anatomically normal identical twin had also committed suicide earlier, and the patient said that he felt guilty and was depressed about his brother's death, we cannot assume that the gender changes were responsible (Black 2004).

A second botched circumcision followed into adulthood had a different course (Bradley, Oliver, et al. 1998). The penis was lost at two months, sex reassignment and corrective surgery were applied by seven months, and the child was raised as a girl. Female identity developed more securely in this child, although she chose "tomboyish" toys and activities. Interviewed in adulthood, she had had an unsuccessful relationship with a man and at age twenty-six was in a successful one with a woman. The patient described herself as lesbian. These and other relevant cases demonstrate that psychological gender assignment and gender-specific child rearing influence gender identity in important but not determinative ways (Hines 2004; Meyer-Bahlburg 2005). As Daniel Federman (2004:323) put it, "Taken together, such evidence points strongly to a hormonal role in the sexualization of the brain." Another physician-scientist is more emphatic: "Genetic males with male-typical prenatal androgen effects should be reared male" (Reiner 2005:549). This advice is highly controversial, but other reviews of the experimental and clinical literature also see a causal role for prenatal androgens (Cohen-Bendahan, van de Beek, et al. 2005; Hines 2005).

The psychological domain affected in these four categories is mainly social and emotional. A careful review of available studies of clinical syndromes and drug effects that could correspond to pre- and perinatal masculinization of the brain in nonhuman mammals concluded that "evidence is most consistent for a developmental influence of androgens on sex-typical play. There also is some evidence supporting a role for androgens in the development of tendencies toward aggression and ... of sexual orientation." But "suggestions that IQ, attainment of developmental milestones, and academic performance are enhanced after prenatal exposure to high levels of androgens or progestins have been refuted" (Collaer and Hines 1995:97).

Of course, prenatal androgens in the brain are only one force shaping gender. Overall, the most useful model of the development of gender posits the partial independence of its various psychological and biological dimensions (Federman 1967, 1982, 2006). Table 10.1 attempts to summarize these aspects of the differences between males and females, divided into mainly biological dimensions (sex) and mainly psychological ones (gender).

This model acknowledges that one can feel like a man but desire men, feel like a woman despite having male genitals, be a woman in almost every meaningful way despite having a Y chromosome, be a man in every way except for liking to wear women's clothing on weekends, be a hairdresser with an effeminate manner who is a heterosexual family man, be a "macho" gay man, be a "lipstick lesbian," or be one of countless other combinations. Some of these combinations result from genetic chances like androgen insensitivity, others from accidental hormone exposure, still others from divergent experience in childhood, adolescence, or adulthood. Increasingly flexible legal, medical, and psychological ideas about sex and gender will no doubt reveal that these variations are more common than is generally believed.

Nevertheless, evolution did not leave us without moorings in the winds of cultural influence. Sexual reproduction needed an anchor for behavior, and natural selection provided it, in the form of prenatal androgenization of the brain. The winds may be strong and the chain long, but we are not adrift. Whether the variants have adaptive functions or not, the core of predominant sexual identity is assured, at least probabilistically, and at least in the domains of aggression and sex. Heino Meyer-Bahlburg (2005:423), criticizing biological determinism, writes

Table 10.1.   A Simplified Model of the Partly Independent Components of Sex and Gender Identity

Sex	Gender
Chromosomal sex	Gender identity
Genetic sex	Sexual orientation
Hormonal sex	Gender-related behavior
Anatomical sex	

Source: Adapted from Federman (1967, 1982, 2006).

that current findings are "incompatible with the notion of a full determination of core gender identity by prenatal androgens." Wendy Wood
and Alice Eagly (2002:720), doing something similar, write that "the research findings are consistent with an interactive model that is only
partially captured by the influence of sex-typed hormones on behavior."
And Eleanor Maccoby (1998:9), the leading psychologist studying sex
differences in the modern era, writes that "the socialization account has
not proved adequate to the task of explaining gender differentiation,"
and that "whatever biological forces are powerful enough to have created the distinct, biological categories of male and female could, by implication, also be powerful enough to create binary distinctions with
respect to certain forms of behavior" (292). For someone who began his
career amid the outrage of psychologists and social scientists at the
mere suggestion of *any* such influence, this is a remarkable change, and
one consistent with evolutionary theory.

11

The Transition to Middle Childhood

The common claim that a child is most susceptible to influence during the first six years of life implies a belief that something unusual in development happens around age six. Many historical documents support the notion of an important transition at this time, suggesting that the "age of reason" begins at around seven. Sheldon White (1968, 1970, 1996) showed that this tradition is correct. In Western countries children begin school at age five, six, or seven, and this is an approximate lower boundary for the beginning of either formal education or the assignment of tasks and chores with significant responsibility in cultures throughout the world. At this age children attain a "more complex and subtle ability to understand and model the roles of their parents and of older children; they want to do this and want approbation at success" (Weisner 1996:320).

Cross-culturally as well as historically, this is a very widespread realization about children. The early-twentieth-century child psychologist G. Stanley Hall extended a long tradition of philosophical and educational observation when he wrote:

> After the critical transition age of six or seven . . . begins an unique stage of life marked by reduced growth and increased activity and power to resist both disease and fatigue which . . . suggests what was, in some just post-simian age of our race, its period of maturity. Here belong discipline in writing, reading, spelling, verbal memory, manual training, practice of instrumental technique, proper names, drawing drill in arithmetic, foreign languages by oral methods, the correct pronunciation of which is far harder if acquired later, etc. The hand is never

so near the brain. . . . Children comprehend much and very
rapidly if we can only refrain from explaining. (quoted by White
1996:20)

But, in modern terms, what emerges at this age is not "an absolute ability to reason; it is an ability to reason with others and to look reasonable
in the context of society's demands on the growing child to be cooperative and responsible" (White 1996:27). One developmental psychologist
calls this age "a sobering up" (Kruger 2009).

Many theories identify major changes at about this time. For Freud
(1905, 1920, 1924), it was the resolution of "Oedipal" feelings: the child
transcends an infatuation with the opposite-sex parent and anger toward the rival parent to enter a less sexual, less emotional phase he
called latency. For Piaget (1970), it was the transition to "operational"
thought—not just conservation of quantity through transformations
but a general ability to stand back from immediate sensations, not
through the imaginative exuberance of early childhood but through
new rational thinking skills (Flavell 1963). For Vygotsky, it was the time
when "interior speech" begins to be used to regulate action (Wertsch
and Sohmer 1995; Bruner 1999), and recent theory emphasizes the role
of language in children's new capacities for reflection on their own
memories and construction of autobiographical narratives (Nelson
1996a, 1996b). Interior speech and reflection on memories are parts of a
suite of abilities in metacognition, or thinking about thinking, often collectively called "theory of mind" or "mentalizing states." Although some
begin earlier than age five and others after age seven, there are steady
advances in these domains between early and middle childhood (Flavell
1999; Pillow 2008).

In learning theory it is a time when the simplest forms of classical
conditioning and reinforcement learning are augmented by mediating
mental processes, especially words, that help link the stimulus to the
child's own response, giving rise to a hierarchy of conventional learning
processes (White 1965). In Maccoby's (1998) model of the development
of gender-specific behavior and identity, middle childhood is a time of
increases in voluntary self-segregation by sex, even as sex role stereotypes weaken. Studies of self-concept from markedly different perspectives (Dunn 1996; Harter 1996) conclude that the potential for self-

criticism and comparison with others, a kind of social metacognition, increases markedly.

There have long been clues to a possible biological reality underlying these transitions. A cluster of growth changes sets the stage for cognitive development, facilitating enculturation and the acquisition of skills. White (1965, 1970, 1996) saw several reasons to suggest that this psychological shift is intrinsically neurobiological:

1. Electroencephalographic rhythms stabilize in a basically adult pattern.

2. Blindness beginning before the shift results in an absence of visual memories, but later-onset blindness preserves such memories.

3. Loss of a limb before the shift leaves the child with no sensation of the missing limb, but a later loss often leaves a phantom limb.

4. There is an inflection point in brain size growth such that brain growth slows markedly after the shift (Dekaban and Sadowsky 1978).

5. Myelination is largely complete in most major circuits, producing effectively adult levels of axon conduction, especially in longer pathways.

6. There are changes in the granule cells of the dentate gyrus of the hippocampus, a structure crucial to the encoding of new memories (Rose 1976).

7. The proliferation of synapses peaks in certain cortical regions prior to normal regression (Huttenlocher 1994).

8. Cortical energy consumption peaks, as measured by PET scanning (Chugani 1994).

Stepwise changes in observable behavior do not necessarily imply a stepwise underlying process. The five-to-seven shift may be an emergent phenomenon resulting from cumulative but gradual brain changes. Also, what look like stepwise changes in cognitive growth must (like evolution) be continuous at a finer level of resolution (Siegler 1995, 1996). But overall, there is evidence for relative discontinuity in some important measures of brain structure and function (Janowsky 1996).

There are also physical changes outside the brain, notably the eruption of the first molar tooth at around age six, considered a hallmark of human evolution (Anemone 2002; Skinner and Wood 2006) and linked to cognitive development in human and nonhuman primates (Parker 2002). Even more instructive is the midgrowth spurt (Meredith 1978; Bogin 2006), the sole prolonged acceleration of childhood physical growth velocity, which otherwise declines steadily from a prenatal peak to the pubertal growth spurt. Although it is not an acceleration in every child, it is at least a pause in an otherwise steady deceleration. This fact of purely physical growth supports a biological model of the shift. So does adrenarche, the onset of secretion of adrenal androgens in both sexes, detectable from around age six and gradually increasing until early adulthood (Havelock, Auchus, et al. 2004), due to developmental change in enzyme activities in the adrenal cortex (Rainey, Carr, et al. 2002). The adrenal androgen DHEA (dehydroepiandrosterone) is detectable at low levels before age five (Palmert, Hayden, et al. 2001), but steadily increases after that.

After several years of exposure, perhaps at a threshold level, the first pubic hair in both sexes appears. It used to be thought that adrenal androgens were involved in the initiation of puberty as defined by gonadal development, but these are separable; girls with adrenal insufficiency do not go through adrenarche but do go through puberty (Sklar, Kaplan, et al. 1980). As we will see in the next chapter, gonadal puberty results from maturational changes in the hypothalamic control of the gonads. However, the two processes share some genetic determinants (van den Berg, Setiawan, et al. 2006).

What are the adrenal androgens of middle childhood doing? They contribute to growth, but do not explain the midgrowth spurt, since they come too late to play a causal role in that event. A study followed nineteen children longitudinally with height measurements and assays for excreted metabolites of androgens (Remer and Manz 2001). Documenting the midgrowth spurt by synchronizing individual longitudinal height velocity curves to the peak of the spurt, the study showed a rise in adrenal androgens—after the peak. In a fascinating study, the age at which people of both sexes, including male homosexuals, remembered having their first crush was several years before puberty; this finding is consistent with the hypothesis that such early feelings may be due to adrenarche (McClintock and Herdt 1996), but they are separate from the sexual feelings of adolescence. It must be relevant that we share this feature of endocrine growth with chimpanzees but not with monkeys (Campbell 2006). Why should it have evolved before our last common ancestor with the apes?

An Evolutionary Approach

The midgrowth spurt has led to evolutionary speculation about the origins of middle childhood. Apes and monkeys have a pubertal growth spurt, although smaller than our own—few other species have it at all —but it comes much earlier, without a midgrowth spurt, and without as prolonged a phase of slow growth as occurs in human middle childhood. Human heterochronic evolution has changed the growth pattern of apes to produce a different adult form. Most phases of growth have been lengthened; for example, gorillas outpace us developmentally, even after our "fourth trimester." But no one process explains the evolution of human ontogeny, and there have also been a number of specific changes (Bogin 1994, 1999, 2006). In one model, the pubertal growth spurt was interrupted and most of it postponed from age seven, eight, or nine (when it occurs in apes) to the human timetable. The midgrowth spurt and the neuropsychological changes of the five-to-seven shift could be evolutionary vestiges of what was once the start of puberty (Bolk 1926; Gould 1977), but with sexual development—and the adolescent behavioral turmoil that often attends it—put off for years.

According to the theory, these years became middle childhood, an extension (after the brief midgrowth spurt) of the decelerating growth rate of earlier years. The extreme dependency, emotional turmoil, and intellectual confusion of early childhood have been surpassed (the "sobering up"), but the driven, sometimes chaotic, hormonally transformed mental and social life of adolescence has not yet begun. In these years the child is seen by societies throughout the world as a vessel into which knowledge, skill, and tradition can be steadily and reliably poured. It is also a time when susceptibility to communicable diseases is at a lifetime low (Worthman, 2010), coinciding with a lifetime high in the size of the lymphoid system; the high-risk life of the infant and weanling (Sibly and Brown 2007) has been successfully passed, and the risks of adulthood (Charnov 1993) have not begun. A phase of human growth that has apparently been inserted by evolution into a once almost uninterrupted

plunge toward puberty—a kind of partial braking process—made possible by our unique adaptation of postweaning provisioning (Kaplan,
Hill, et al. 2000), has in turn made possible what most distinguishes us
from apes: imbuing a child with a culture.

Cognition in Middle Childhood

Plainly, if evolution has designed a developmental plan that seeks to accelerate the acquisition of culture at this age, there should be universal
features of the growth of brain and mind to make this adaptation possible. Both classical studies of cognitive development at this age, which
focus on mental and moral challenges, and studies over the past quarter-
century, which focus on metacognition, social cognition, and theory of
mind, see many relevant changes.

Classic Studies: Cognition and Moral Development

Piaget described middle childhood as the period of "concrete operations," which he saw as a major advance in rational thought, although
rational thought in middle childhood still falls short of adolescent and
adult reasoning. One hallmark of the transition to middle childhood is
conservation—the idea that quantities of things often stay the same
through transformations that change their appearance (Flavell 1963;
Piaget 1970). In the conservation-of-number task, two straight rows of
coins, each with six pressed close together, lie beside each other. The
experimenter asks whether both have the same number or whether one
has more; the child says both have the same.

Then, as the child watches, the six coins of one line are spread out to
make it longer. If younger than five or so, she now says that the longer
line has more coins—a result that occurs even with several different
kinds of framing language, including "Can you buy more candy with
one of them, or would both buy the same amount of candy?" There are
many other similar tasks, such as pouring water from a narrower glass
to a wider one; children under five will say that there is less water in the
wider glass, having just watched the pouring from one to the other,
while children over seven get it right almost without thinking. There
are ways to set up the tasks differently, simplify them, or teach about
them that allow preschool children to show some understanding, but

they do not erase the difference between younger and older subjects: the latter do the task promptly, easily, under a wide variety of conditions, and without being taught.

This applies to many more characteristics of development across this age range. There is a marked advance in children's understanding of the roles played by members of the family (Watson and Amgott-Kwan 1983). At three, a child understands a mother's role enough to name two traits or behaviors appropriate to the role, but she cannot understand how a role change comes about, as when a daughter becomes a mother; at six or seven, she can grasp not only that but also the idea that one person can be a daughter, a mother, and a granddaughter simultaneously. Such understanding of basic kinship relations would clearly have great adaptive value and would probably have to precede a more complex process of enculturation.

Piaget's great contribution to one dimension of this process, moral development, was to study the understanding of the rules of the game of marbles achieved by boys of different ages playing in the streets of Geneva. He showed marked age changes in the nature and level of that understanding, which he believed were reflections of the stage of attainment of concrete operational thought. No one was giving lessons in the game of marbles; the boys were just playing, and they abstracted the rules as well as they could at a given age. If marbles can be taken as an instance of the complex dynamics of social and cultural life, this work may tell us a good deal about the adaptive value of middle childhood. We return to these questions later.

Recent Studies: Epistemological Development and Social Cognition

A great advance in the three decades since Piaget's time has been to study the development of epistemology in children—what they know about knowledge and mind, their own and others' (Flavell 1999; Pillow 2008). This includes perspective taking (the ability to appreciate that someone else's view of reality can be different from one's own) and metacognition (the ability to imagine or infer what others are thinking and feeling and to think about thinking itself). This suite of competences is often called "theory of mind," and it adds a vital new dimension to the child's relationships to others and to herself.

For example, five-year-olds understand that external events can cause sad or happy feelings, but not that memories or thoughts alone can cause them, while eight-year-olds understand all of this easily (Flavell, Flavell, et al. 2001); four-year-olds do not recognize that people who are silently reading or counting can be talking to themselves in their own heads, while six- and seven-year-olds readily recognize such inner speech (Flavell, Green, et al. 1997). This is not quite Vygotsky's intuition that inner speech begins at this age, but it shows that awareness of it does. Also in this age range, children begin to grasp the distinction between conscious and unconscious mental states (Flavell, Green, et al. 1999). Some abilities in perspective taking are achieved earlier. For example, in the false belief task, a child and a doll are looking at candy; the doll is taken out of the room, the child and the investigator move the candy to a hiding place, and the child is asked where the doll will look for the candy upon returning. All three-year-olds fail, saying the doll will look where the candy now is; most (but not all) five-year-olds give the right answer.

However, children's "understanding of mind" does not "begin and end with a grasp of the possibility of false beliefs . . . the altogether more complicated problem of appreciating the essentially *interpretive* or constructive nature of the knowing process does not occur until substantially later and is probably best understood as only beginning to get under way by 6 or 7 or 8" (Carpendale and Chandler 1996:1688). In fact, "understanding of many cognitive activities appears to begin between the ages of 5 and 7 years" (Pillow 2008:298), as exemplified by the following:

1. Four- to five-year-olds will state that they have always known a fact the experimenter has just taught them even though this statement contradicts what they said earlier (Taylor, Esbensen, et al. 1994); older children recount the actual learning process.

2. Two toys of different colors are shown to a child and a puppet, then hidden in separate cans. After the puppet looks in one can, the child is asked if the puppet now knows what color the other toy is. Six-year-olds say yes, and four-year-olds say no (Pillow 1999).

3. Children hear a story in which a character acts well or badly; kindergarteners and second-graders both say that seeing the behavior will

affect their peers' opinions of the character, but only second-graders think that gossip will make a difference (Hill and Pillow 2006).

4. A puppet is shown an ambiguous drawing, and the child is told that the puppet has a wrong belief about what the drawing is. Four- to five-year-olds say that the puppet will correctly identify the drawing, but six-year-olds say that the mistaken bias will mislead him (Pillow and Henrichon 1996).

5. The percentage of five-year-olds who say that people who are deeply asleep and not dreaming know that they are asleep was 61 percent in one study and declined to 39, 28, and 11 percent at age seven, age eight, and in adulthood (Flavell 1999:38).

We have here a constellation of new abilities that are vital, if not necessary, for the accelerated enculturation of middle childhood. As John Flavell (1999:37) put it, "Children discover that to acquire knowledge through perception of information, that information has to be adequate," and that "one's interpretation of an impoverished or ambiguous perceptual input may be influenced by one's biases." This new understanding of "the interpretive nature of knowledge" (Carpendale and Chandler 1996:1686) makes children more skeptical of the workings of their own and others' minds. But it also puts them into position to benefit from true imitation, teaching, and collaborative learning (Tomasello, Kruger, et al. 1993), which form the social-cognitive core of enculturation. We will see this in some detail in Chapter 26.

Cross-Cultural Constancy of the Shift

Cross-cultural studies confirm the universality of key changes at this age (Super 1991; Weisner 1996). Charles Super and Sarah Harkness gave the same cognitive tasks to children among the Kipsigis of the village of Kokwet, Kenya (165 boys and girls ages three to nine) and to 112 children in Duxbury, Massachusetts, a prosperous Boston suburb.

On the Bender Gestalt test, which involves copying a geometric drawing, the striking result was a precipitous and parallel drop in errors between ages three and six in both cultures. In the same age range there was a dramatic and cross-culturally parallel rise in the ability to cluster randomly spoken words, whether by function (fire-burn) or category (calf-cow). On a third test the increase was less steep but still parallel. "The first conclusion is an unusual demonstration of regularity in the behavioral development of our species" (Super 1991:244). The second conclusion was that the results seemed to present an example of exaptation; that is, since the Kipsigis children had no use for and no experience of these kinds of tasks, the abilities and their developmental transformations must have been by-products of an evolutionary process aimed at another adaptation in another context. Many other such cross-cultural tests have produced similar results (Mishra 1997).

Thomas Weisner (1996), in his review of the cross-cultural consistency of the transition, argues that it was part of the basic human adaptation for family life. In his ecocultural model, the transition to middle childhood arose to provide effective helpers in the care of younger children. As we saw in Part I, human success relied on postweaning provisioning and cooperative breeding, both of which can be facilitated by helpers in middle childhood, especially against the background of a heavy maternal workload. But children must be capable of learning child care and domestic tasks and of discharging them responsibly and effectively. At this age they have emerged from the sometimes disturbing emotions of early childhood but are not in a rebellious phase; they *want* to do the tasks assigned. Weisner (1996:300) draws on Roy D'Andrade's concept of cultural models (see Chapter 23) as intersubjectively shared schemas:

> "Buying something," for example, is a cultural activity that is schematically organized, involves shared knowledge, and includes concepts like money, seller, and price. "Taking care of an infant," "delivering messages to an uncle's house," "herding cattle," and "guarding the house and minding younger siblings" are practices embedded in cultural models. Children's abilities to use such cultural models and schemas appear to increase dramatically during the 5 to 7 period.

There are also cross-cultural confirmations of the established advances in epistemological understanding. John Flavell's (1986) experiments on appearance-reality distinctions in four-, five-, and six-year-olds in the United States and China show that performance on varied tests of the understanding of the distinction improves markedly in this age period in both cultures. Success on a basic false belief task, such as the doll scenario described earlier, is usually achieved by Western children before age five. Using a more naturalistic scenario, researchers tested children among the Baka of Cameroon in a hut with food and a cook instead of a doll; their developmental curve was similar to the Western one (Avis and Harris 1991). The same is true for Mofu children of Cameroon (Vinden 2002), Junin Quechua children of Peru (Vinden 1996), and children in two different cultures of Papua New Guinea (Vinden 1999), although in some of these studies more complex aspects of false belief (for example, how the doll or the person will feel upon discovering the mistake) are less cross-culturally constant and schooling makes a difference for some tasks.

A group led by Tara Callaghan and Phillipe Rochat conducted a basic false belief test systematically in Canada, India, Peru, Samoa, and Thailand, using a person, not a doll, as the target of the trickery (Callaghan, Rochat, et al. 2005). Although this most basic element of metacognition appeared in half the children by age five, it continued to progress at the same slope for six months or more. It remains to be seen how the more subtle aspects of the development of metacognition will hold up under cross-cultural testing. However, we do know that systematic efforts to teach younger children the subtleties of the false belief tasks and other aspects of metacognition are unsuccessful (Flavell 1986), suggesting that culture cannot shift these age curves very much.

Cross-Cultural Recognition of the Shift

Since key changes of the five-to-seven shift are cross-culturally universal, it is not surprising that most cultures implicitly recognize it. In a comparative study of cultures as described by ethnographers, Barbara Rogoff drew fifty communities from the Human Relations Area Files (HRAF), a database selected for the quality of the ethnographic accounts and the mutual independence of the cultures (Rogoff, Sellers, et al. 1975). The investigators rated "age of transitions in 27 categories involving cultural assumptions regarding responsibility or teachability in children or assignment of a more mature social, sexual, or cultural role . . . for 16 of the 27 categories there appeared to be a modal cultural assignment of social responsibility in the 5 to 7 year age range" (Rogoff 1996:274). In the other eleven categories the age of first chores or schooling was usually later, not earlier, suggesting that the shift forms a lower bound for many of the expectations imposed on children.

Among childhood chores in non-industrial cultures, care of younger children has a privileged evolutionary position, and it too is unlikely to be assigned formally until after the shift. In view of adolescents' uncertain reliability, one likely adaptive function of middle childhood was to ease the mother's burden in caring for infants and toddlers, allowing shortened birth spacing in some contexts. Since formal task assignment is more common in agricultural and pastoral cultures, the closer birth spacing and higher reproductive rates of post-hunter-gatherer populations—a faster lifestyle in the Sibly and Brown (2007) sense—depended in part on the five-to-seven shift and the calm cognition characterizing middle childhood.

!Kung children are relatively free of responsibility (Draper 1976), but among the Hadza, an east African hunter-gatherer group, children this age forage for a substantial portion of their own food (Blurton Jones 1993). !Kung children are responsible when necessary, as when a younger child is in trouble. Also, the concept of a latency period is harder to establish in the !Kung (Konner 1972) and many other cultures (Frayser 1994) where children play at sex throughout middle childhood. When they play "house," they simulate sexual intercourse; it is Western children who mask their curiosity by "playing doctor." Still, attachment is less emotionally intense in !Kung children after age five, and the sex play of the next few years does not have anything like the seriousness of sexual contact in puberty. In this sense even the many cultures with sex play in childhood may have a childhood phase that corresponds emotionally to latency. Like play-fighting, play sex has emotional and behavioral content quite apart from the real thing.

The Transition in Moral Reasoning

The five-to-seven shift is an ontogenetic evolutionary legacy that helps form the basis of cultural transmission through intersubjective teaching and learning, one of the defining features of the human species (Kruger and Tomasello 1996), so it is not surprising that it also forms the basis of rule-bound games and verbal traditions in middle childhood. But the shift also lays the foundations of relationships in later childhood, and of one regulator of relationships—the moral sense. Indeed, one of the most interesting findings about this age range is that some aspects of morality come not primarily from parental discipline or Sunday school lessons, but from social play.

Fairness, rules, oral contracts, sanctions against wrongdoers or just against those who are different are what many time-honored children's games, songs, jokes, and other traditions are about, and they require the mental facility granted by the five-to-seven shift. Piaget's research while playing marbles with children led to *The Moral Judgment of the Child* (1965), his seminal work on moral development. A game like marbles is part of the cultural sphere controlled by children, yet it is replete with rules and rests on a basis of fairness more crucial than skill. Declaring that "children's games are the most admirable social institutions" (Piaget 1965:13), Piaget mastered the game in many local variants among the poorer children of Geneva. He observed and interviewed children and spent many hours playing marbles, recording their reactions to his inadvertent or deliberate rule-breaking. He identified three developmental stages:

1. Before the five-to-seven shift, children may seem to play by rules, but have little grasp of them; they conform, but in an empty imitation of older children's rule-bound play, full of obvious errors reflecting lack of deeper understanding.

2. Between about seven and ten, the child displays greater knowledge but refers the rules to higher authority. (An eight-year-old girl said that the rules were invented by "a Gentleman" and that it was best to leave the game as it was.)

3. By ten or eleven, children accept that the game can change and that they themselves can make the rules. A new rule may be invented and adopted, provided it serves the game's two main goals of fun and fairness. Yet this flexibility with invented rules is combined with a stricter adherence to rules generally, perhaps owing to a deeper respect for their source. (A twelve-year-old, asked to try to invent some new rules, thought of playing in two squares, one inside the other. He said that his new game would be as fair as the old one if the players understood and agreed on the new rules.)

This account reflects the appreciation, in late middle childhood, that the rules come not from authority but from consensus and tradition— yet these older children obey the rules more strictly. In effect, they have absorbed the essence of morality—not the rules but the reciprocity and fairness the rules ensure. As Piaget (1965:196) put it, "Moral autonomy appears when the mind regards as necessary an ideal that is independent of all external pressure. . . . Autonomy therefore appears only with reciprocity, when mutual respect is strong enough to make the individual feel from within the desire to treat others as he himself would wish to be treated." This Golden Rule morality can replace received authority only when the cognitive faculties of perspective taking and metacognition have matured.

Theory of mind researchers have become more interested in moral development, and there are several ways the two overlap (Astington 2004). Second-order understanding ("He thinks that she thinks . . .") is not in place until age seven, and it appears to be involved in distinguishing irony or joking from lying. This kind of sophisticated perception is needed to establish intent, which is one of the core issues in moral judgment. Other investigators take a stronger stance, arguing that the years from four to seven bring about the transition from a passive understanding of false beliefs to an interpretive epistemology that includes attribution of *agency* to other actors (Sokol, Chandler, et al. 2004). This level of development is tantamount to understanding the thinking, not just the desires, of the other, and it is probably needed for both true intersubjectivity and appropriate moral judgment. Even independent of age, children with a more interpretive theory of mind measured by a different task were much more likely to judge harshly the puppet Punch when he deliberately attempted to kill Judy by throwing her in the trash; although she survived, they—and not the less "interpretive" children— judged him by his deliberate intent.

Evolutionary psychologists emphasize the need for a cognitive function, possibly modular and wired into the brain, for detecting cheaters, in view of the extreme dependency of hunter-gatherers on reciprocal relationships (Cosmides and Tooby 2000; Tooby and Cosmides 2000). This capacity may require developmental progress through the five-to-seven shift, and one of the purposes served by middle childhood may be to hone this cheater-detection function. It would be interesting to study the development of the social-cognitive tasks involved in cheater-detection. Similarly, the Prisoner's Dilemma, the Nash equilibrium, and other situations that depend on cooperation may require the level of intersubjective reasoning attained in middle childhood, and sophistication in these games may depend on learning between the five-to-seven shift and puberty. It is very unlikely that such effects could be seen at a much earlier age. The prominent role of transactions—reasoning about reasoning—in these studies exemplifies the metacognition of middle childhood. Just as memory changes with the reflection made possible by language (Nelson 1996a, 1996b), so moral reasoning does as well. We return to these matters in Part IV.

A Biological Model

In addition to early weaning due to postweaning provisioning (a feature of rearing conditions, not the ontogenetic plan), there are five basic changes in the trajectory of human growth as compared with great apes: (1) an overall lengthening of the life span, including development to maturity, by about 50 percent; (2) a relatively early expulsion of the fetus as gestation increased much less than other life-cycle phases, resulting in the "fourth trimester"; (3) fetal brain growth rates extended through the first postnatal year; (4) the emergence of language between two and three years, facilitating weaning; and (5) a relative prolongation of middle childhood, defined as the time between first molar eruption and puberty. Of these changes, the last may be the most critical.

Some of the evolutionary changes were described in James Tanner's classic *Growth at Adolescence* (1962). The rodent growth curve accelerates steadily from prenatal life through puberty. Weaning takes place at around three weeks, and puberty about two weeks later, at or after peak growth velocity. In higher primates the trajectory is dramatically different. There is quite a bit of variability among species (Shea 1983a, 1983b; Leigh 1996), but our appropriate reference point is the chimpanzee. Its life cycle is far slower and longer than that of rodents, and because chimps are highly precocial compared to them, growth rate peaks during gestation and slows until the adolescent growth spurt begins between ages three and four, around the time of first molar eruption (Robson, van Schaik, et al. 2006). Weaning takes place around age five (van

Schaik, Barrickman, et al. 2006), and menarche takes place shortly after females' peak height velocity (PHV), as in humans, but in chimps the chronological age is between eight and nine. Menarche today in the United States is at age 12.7, but among hunter-gatherers the age was two to three years later. Since hunter-gatherers wean at two and a half to three years of age (Konner 2005)—two to three years earlier than chimps and bonobos—the time between weaning and puberty onset in our early human ancestors was probably around a decade, but almost nil in those two closely related species. This is a huge difference.

The first scientific record of growth known was based on measurements made by the Count de Montbeillard of his son from birth to age eighteen, in the years 1759 to 1777. The midgrowth spurt is clearly present, and because it is an individual curve, the pubertal spurt is dramatic. The curve for an average boy raised in a good modern environment is similar, and the curve for an average girl is ahead by about a year at the midgrowth spurt and by about two years at PHV, which occurs shortly before menarche. Comparison with the chimpanzee curves supports the Bolk-Gould thesis that we inserted a period of slowing growth after the start of what was the ape pubertal growth spurt. This left the midgrowth spurt, followed by four to seven years between that change and the resumption of rapid growth. On the slower developmental plan in hunting-and-gathering cultures, it would have been at least six years.

Equally if not more important is the quiescence in sex hormones from late infancy to puberty. Levels of estradiol in girls and testosterone in boys, respectively, are extremely low during middle childhood. In a chronological and developmental epoch when a chimpanzee body and brain would be increasingly bathed in these hormones, human development suppresses them. Despite sexual play in childhood in many cultures, body and mind are not yet plunged into the emotions and risks of real sexual, and ultimately romantic, life. Gonadotropins are also at or close to their lifetime lows during this period.

Only adrenal androgens begin to come into play during middle childhood, and the slow rise in these hormones eventually produces the first pubic hair in both sexes, with psychological consequences. The emergence of pubic hair is the first sign of the end of childhood; 95 percent of children get it between ages nine and thirteen (Marshall and Tanner 1969, 1970; Parent, Teilmann, et al. 2003). Children with early adrenarche have an increase in anxiety and possibly depressive disorders, but it is not clear whether this is due to the indirect impact of body image or the direct impact of hormones (Dorn, Hitt, et al. 1999). In some girls it may be a harbinger of or marker for syndrome X—obesity, hypertension, dyslipidemia, and insulin resistance (Saenger and DiMartino-Nardi 2001)—and the increasing incidence of insulin resistance in children may prefigure a rise in premature adrenarche.

Benjamin Campbell and Zvi Hochberg have offered intriguing models of this phase of human development that give adrenal androgens a central role. In Campbell's model, suggestive evidence that dehydroepiandrosterone (DHEA) may have stimulatory effects on neural development (Suzuki, Wright, et al. 2004; Mellon 2007) leads to the hypothesis that the line leading to chimpanzees and humans evolved adrenarche to stimulate extended brain development during middle childhood and adolescence (Campbell 2006). However, brain development is about 95 percent complete by age seven in both species, chimpanzees and humans differ dramatically in the time elapsed between adrenarche and gonadarche (one to two years versus four to seven years), and the outcome of brain development in the two species differs by a factor of three in size and differs in many other ways as well. Still, if chimpanzees indeed parallel us in adrenarche, then adrenarche may explain some things we have in common and may have begun as a stage on the way to human status.

Campbell (2006:577) points to preliminary research suggesting that DHEA affects mood and cognition in several ways:

> 1) reduced fear and anxiety through changes in the amygdala, leading to increased social interaction with unfamiliar individuals; 2) increased ability to process the feedback from social interactions, related to changes in memory mediated by the hippocampus; and 3) increased plasticity in the form of synaptic connections that reflect the effects of social and environmental information.

These influences could be due to possible DHEA effects on the androgen receptor in the amygdala, the estrogen receptor in the hippocampus, and the $GABA_a$ receptor, which influences synaptic plasticity in many brain areas. However, all these influences are based on findings about the clinical effects of adrenal androgens on adults with various disorders (Arlt, Callies, et al. 2000; Hunt, Gurnell, et al. 2000; Wolkowitz, Brizendine, et al. 2000).

Hochberg (2008)* extends Campbell's model by emphasizing the relationship between adrenarche and increased fat deposition, especially in girls, suggesting a preparation for menarche, which we know depends in part on their percentage of body fat (Frisch and Revelle 1971; Frisch, Wyshak, et al. 1980; Frisch 2004). Evidence that low-birth-weight girls who gain weight rapidly in childhood have earlier adrenarche as well as earlier puberty (Ibanez and de Zegher 2006; van Weissenbruch and Delemarre-van de Waal 2006) could be consistent with the prediction of life history theory that early stress signals a need to grow up and begin reproducing sooner.

Both models deserve further investigation. However, it is far more likely that the great cognitive changes of the transition from early to middle childhood result from brain maturation than from adrenal androgens. This includes the maturation of EEG rhythms, the peak of synaptic proliferation in some cortical regions, the peak of cortical energy utilization, the maturation of the dentate gyrus of the hippocampus (possibly including neurogenesis of granule cells), advances in myelination, and other changes. The advent of some mentalizing ("theory of mind") functions by age four and their marked improvement by age eight has been attributed by some to the emergence of a "theory of mind module."

This is much less plausible than a series of maturational changes in brain circuits that serve different aspects of mentalizing, including basic functions (for example, gaze and voice processing, emotion recognition, and goal detection) as well as higher processes like working memory, recursion, executive functions, and metarepresentation (Stone and Gerrans 2006). Add to all this the more sophisticated and sober metacognition of middle childhood and the knowledge of the interpretive

---

* Hochberg follows Bogin (2006) in defining "childhood" as the period from weaning (age three) to the midgrowth spurt and first molar eruption (age six or seven), and "juvenility" as ages seven to ten in girls and seven to twelve in boys. I use the standard terms of psychology and pediatrics: early and middle childhood. Since weaning is not a milestone, I view basic language mastery (also by age three in all cultures) as the start of early childhood.

nature of knowledge, of the role of bias, and of the really deep differences between different people's perspective, and you have a suite of abilities whose brain representation will be challenging to delineate, but that is unlikely to constitute a module.

What is clear is that human development exhibits a long phase not shared by apes, characterized by advanced perspective taking, collaborative psychological states, metacognition, and heightened self-awareness and used by children in all human societies for an enormous amount of vital and detailed cultural learning. We return to this phase and its uses in Part IV. For now, suffice it to say that children in this phase help defray the costs of reproduction by tending younger children and by learning the skills of food gathering that make it possible for humans to wean at three or younger instead of at five or later. Our closest ape relatives are weaned almost as they begin puberty. We are weaned in half the time, despite our longer childhoods, and this may be tied to the maturational milestone of basic language acquisition. Whenever this happened in our evolution, it was a watershed for the apes, and the winner in the quest for reproductive success would be the species with the long, slow, strangely shaped childhood.

12

Reproductive Behavior and the Onset of Parenting

The evolution of human ontogeny left us with variable but late reproductive maturation. In terms of life history theory (Charnov 1993; Sibly
and Brown 2007), we have reduced mortality both during development
and in adulthood (Kaplan, Hill, et al. 2000), and this has allowed us to
postpone reproduction and invest more in individual offspring. As we
have seen about childhood generally, little in adolescence makes sense
except in the light of evolution, and thoughtful accounts have brought
to bear on this phase of life an evolutionary understanding of sex differences, risk taking, parent-offspring conflict, father-absence, parenting
styles, sexual experimentation, romantic love, and pair bonding (Weisfeld 1999; Weisfeld and Woodward 2004). Human adolescents, like the
near-adults of many species, must separate themselves from their parents and strike out on their own, competing for mates and other resources, a process fraught with risk. What we think of as problem behavior is largely the result of ancestral adaptations, without which we
would not be here.

Animals that reproduce by sex must become capable of the associated behaviors and, in many species, of parenting, but no higher animal
is born sexual. Sexual capacity develops according to a species-specific,
genetically controlled, maturational plan, with a particular pace of gonadal maturation and an age at first reproduction targeted to a limited
range (Sisk and Foster 2004). For small rodents such as mice and rats,
development fairly plunges from birth to sexual maturity, a mere six
weeks punctuated at around the halfway point by weaning. Growth velocity rises steadily from birth, starting to fall only shortly before sexual
maturation occurs, with full adult size reached at three to four months.
But in monkeys, apes, and humans there are two peaks of growth rate,

prenatally and pubertally (Tanner 1962; Bogin 1999). Humans superimposed on that pattern a lengthening of the life span and most of its components, with a special lengthening of middle childhood. After that, at about age eleven in females and twelve in males in the West (two to three years later in rural populations in the developing world), growth acceleration begins again; in both sexes this is one of the earliest signs that true puberty has begun.

Biological Changes in Puberty and Adolescence

The average timing and sequence of the principal pubertal events were first well demonstrated in a large British "standard" sample (Marshall and Tanner 1969, 1970). Although growth acceleration is the first sign in girls, breast buds—the nubbins of tissue under the nipples that are the first step in breast development—are found soon afterward, and the appearance of short, downy pubic hair soon follows. Peak height velocity (PHV) occurs about eighteen months after the spurt starts. Then growth rate drops, returning to prepubertal levels two years later. Pubic hair is fully adult by about this time, breasts within a year or so. Halfway between the peak and the end of the growth spurt, before age thirteen, the average girl in the United States has her first menstruation— *menarche*, from Greek roots for "moon" and "beginning." Girls complete growth at around sixteen.

Because of the sexual bimaturism that helps cause sexual dimorphism, boys in most European and American samples lag behind girls about two years in the growth spurt and to a lesser degree in other aspects of puberty. In boys the acceleration of growth begins only at age twelve and is not the first sign of puberty—the onset of growth in the testes is, at age eleven and a half. Next comes a light growth of pubic hair, followed by the lengthening of the flaccid penis. The growth spurt in boys peaks at fourteen and ends at sixteen, with slow growth until eighteen, each milestone lagging up to two years behind that in girls; the added growing time is part of what gives males their greater adult size.

No event in male puberty is as striking as first menstrual flow, but one marker is the first conscious ejaculation (*spermarche*, the result either of deliberate masturbation, which almost all boys eventually do, or a nocturnal emission the boy has noticed). It is less reliable than menarche, but it seems a logical first sign of emerging reproductive readiness. Alfred Kinsey and his colleagues, studying thousands of American men, found that this milestone occurred on average at age 13.5, although economically disadvantaged boys—who, in the 1930s and 1940s, went through puberty later—tended to have it at 14.5. It was 13.5 in a study in Israel (Laron, Arad, et al. 1980) and 13.2 in Belgium (Carlier and Steeno 1985). A large sample of Han Chinese boys (Ji 2001) reported it at 14.5 on average, with urban boys at 14.4 and rural boys at 14.6; the range among three socioeconomic classes was 14.3 (the highest class) to 14.7. First ejaculation is on the accelerating portion of the male growth velocity curve, while first menstruation is after the peak of the female curve.

Other secondary sex characteristics increase divergence. Males add muscle mass; females keep more of the adipose tissue of childhood, concentrated in the hips, buttocks, breasts, and upper arms. Androgens enlarge the slender muscles of the larynx just as they do other muscles, deepening the male voice, while the female voice changes more subtly. Shoulder width increases more in males, owing to bone as well as muscle growth. In both sexes the feet grow faster than the body, an ontogenetic allometry. There has been disagreement about the biological coherence of the various processes of puberty, with some insisting on frequent dissociation (Tanner 1962) and others on linkage (Harlan, Grillo, et al. 1979; Harlan, Harlan, et al. 1980). But more widely ranging cross-cultural studies show more dyssynchrony of pubertal events (Worthman 1993).

In addition, there has been a historical, or secular, trend toward earlier reproductive maturation (Tanner 1968) as well as larger size at any given age (Meredith 1978). This ongoing change (Karlberg 2002; Delemarre-van de Waal 2005) may be a facultative adaptation for adjusting the length of the reproductive period to environmental quality (Gluckman and Hanson 2006) and is considered in more detail in Chapter 20.

Is Individual Age at Puberty a Facultative Adaptation?

In a cross-cultural study in 1965, John Whiting (1965) found that cultures with customs that are stressful to infants—cold baths, ear or nose piercing, and the like—had on average earlier menarche. The study was inspired by animal models in which early stress accelerated growth.

Given the fact that good nutrition and control of infectious disease accelerate growth in humans, however, it seemed paradoxical for early stress to do the same. But in the light of evolutionary theory, there may be good reasons.

There is great individual variation in pubertal timing within contemporary populations, and efforts have been made to explain it in the light of life history theory and facultative adaptation (Draper and Harpending 1982; Belsky, Steinberg, et al. 1991; Ellis 2004). Both the original *K*-versus *r*-selection theory (Pianka 1970) and subsequent modifications (Sibly and Brown 2007) predict that species with higher mortality will have faster development in order to take reproductive advantage of what little time they have. It has been tempting to use this theory to explain within-species human differences, but it was not intended for this purpose, and until recently it has been difficult to find effects that are uniquely explained by evolutionary theory as opposed to other theory.

In an elegant (although retrospective) study, Jacqueline Tither and Bruce Ellis (2008) interviewed pairs of adult sisters who had grown up in either a biologically intact family (N = 93) or a disrupted family (N = 68) in urban New Zealand. The older sisters reported an average age of 11.8 years and the younger 5.4 years at the time of divorce or separation from their biological father. The main hypothesis, that younger sisters would have earlier menarche in disrupted families only, was confirmed by a significant interaction effect between family disruption and sibling age at the time of disruption. Although this effect was small, it was considerably greater in families where paternal dysfunction was high: "In disrupted families in which sisters were exposed to serious paternal dysfunction (e.g., substance abuse, criminal offending, violence), the younger sisters had substantially advanced menarche," amounting to a total of eleven months (Tither and Ellis 2008:1416).

This is a finding of great interest. Although the New Zealand study was not a controlled experiment, the design controlled for genes and for shared family environment, except for the different ages of paternal dysfunction and departure. Moreover, no other theory of environmental influence on age at puberty would have generated this hypothesis; it is thus a remarkable success for the application of life history theory and the concept of facultative adaptation to human growth. Other theories have argued that much of what we have long considered deprivation or disruption in children's environments may have consequences anticipated by evolutionary adaptation (Belsky, Steinberg, et al. 1991; Chisholm 1993). Disrupted early environments should lead to developmental paths that biologically and psychologically maximize reproductive success in alternative ways, such as earlier and less cautious sexual behavior. But other theories that have nothing to do with evolution (attachment, social learning, effects of poverty) make similar predictions. These are not necessarily contradictory, since they are different levels of explanation. Still, the prediction of earlier menarche from paternal absence and dysfunction (Draper and Harpending 1982) came uniquely from Darwinian theory and as such confirms its value in child development research.

Tither and Ellis review international adoption studies showing that the age of adoption of deprived girls in the developing world into comfortable Western families is inversely related to age at menarche; the earlier the adoption age, the later the menarche. This is counterintuitive, since we know that prolonged deprivation slows growth, and indeed the girls who remain home in the less developed countries (for example, India and Bangladesh) have later menarche. The pattern suggests something specific to the social transformation of adoption at a later age. But the pattern of later menarche in poor rural environments in the developing world has been confirmed many times (Eveleth and Tanner 1976; see also Chapter 20).

Ellis (2004) proposed a comprehensive model that sees a U-shaped relationship between stress in childhood and age at puberty, mediated by stress-reactivity. Subsequent studies have shown in longitudinal samples of children that a disrupted early social environment, especially harsh mothering, is followed by earlier menarche (Belsky, Steinberg, et al. 2007); that overall social deprivation, including absent or dysfunctional fathering, is associated with earlier adrenarche and the maternal component with earlier puberty (Ellis and Essex 2007); and that while five-year-olds currently living in low-stress environments have higher autonomic reactivity than those in higher-stress environments, seven-year-olds studied prospectively showed higher sympathetic and adrenocortical reactivity if raised under high-stress conditions (Ellis, Essex, et al. 2005). This seeming contradiction is resolved by the curvilinear model in which intermediate chronic stress levels produce the lowest stress-reactivity (Boyce and Ellis 2005), a sound adaptation to conditions in our environments of evolutionary adaptedness (see Chapter 21).

Control of the Onset of Puberty

Although the precise control of puberty onset remains elusive, we understand the alteration in hormone levels (Reiter 1987; Sisk and Foster 2004) and its anatomical, and to a lesser extent behavioral, consequences. The first relevant endocrine change is adrenarche, already discussed (Havelock, Auchus, et al. 2004); adrenal androgens produce pubic hair in most girls and (with the much larger amount from the testes) help govern puberty in boys. One monogenic form of premature puberty results from an enzymatic blockade that causes excess androgen secretion in childhood (Noordom, Dhir, et al. 2009). And as we saw, Martha McClintock and Gilbert Herdt (1996) proposed that adrenarche is responsible for the first prepubertal crushes.

But the main events of puberty are independent of these adrenal changes, as shown by their dissociation in clinical syndromes (Sklar, Kaplan, et al. 1980). In both sexes puberty involves three interdependent endocrine centers: the hypothalamus, the pituitary, and the ovary or testis. The hypothalamus—neural and neuroendocrine tissue—regulates the pituitary and responds to the rest of the brain. It is the brain area most demonstrably shaped by prenatal androgens, which prepare it for its pivotal role in puberty. The anterior pituitary—diverse endocrine cells—is joined to the hypothalamus by small blood vessels, through which the hypothalamus delivers gonadotropin releasing hormone (GnRH), signaling the pituitary to make gonadotropins and secrete them into the general circulation. These hormones, follicle stimulating hormone (FSH) and luteinizing hormone (LH), govern gonadal function, causing the testes to produce androgens and the ovaries estrogens and progesterone, collectively the elixirs of pubertal metamorphosis.

A central question is why puberty does not come earlier. The leading model holds that the hypothalamus of middle childhood is exceptionally sensitive to negative feedback from gonadal hormones. Although such feedback plays a role in most children, the suppression of GnRH secretion in children without gonadal function, such as those with Turner's syndrome, as well as evidence from animal models, suggests that CNS control and suppression of GnRH is also involved (Styne 1994). Damage to the posterior hypothalamus—or even pressure exerted on it, as in hydrocephalus—can result in precocious puberty, showing that a brain mechanism normally inhibits hypothalamic GnRH secretion. Although it remains unknown, it is one place to look for a key heterochrony in human evolution, the lengthening of middle childhood.

In any case, through a combination of central neural inhibition and gonadal negative feedback, GnRH and gonadotropin secretion are suppressed prepubertally. The gonads release only small amounts of testosterone (T), estradiol (E_2), and other gonadal steroids. These suffice to suppress the system, keeping all hormones of the reproductive axis at a low ebb known to endocrinologists as the juvenile pause. The hypothalamus gradually increases GnRH release and grows inured to negative feedback from gonadal hormones. This occurs through a slowly developing alteration of the central neural circuitry combined with a change in the GnRH cells themselves, such as receptor down-regulation or altered transcriptional control. The low levels of gonadal steroids that suppressed GnRH secretion no longer suffice, and the system escapes tight feedback control. The hypothalamus delivers more GnRH to the pituitary, which via LH and FSH stimulates the gonads to grow and accelerates the synthesis and release of gonadal hormones.

The changes are not subtle or small. The testosterone level rises about eighteenfold in boys and twofold in girls, leaving males at growth cessation with almost ten times the female level (Reiter 1987). Estradiol rises eightfold in girls and twofold in boys. The rate of change may be crucial, and a boy's more than tenfold increase in T may occur in a single year, yet girls, despite much lower levels, may in some ways be more sensitive to T; small increases may affect them as much as large ones affect boys. Gonadal hormones produce most anatomical changes. In boys, genital size, pubic hair, and total body growth are T-driven; in girls, estrogens, progesterone, and androgens all contribute, but other hormones are involved, and we do not have a complete model. Clinically, T or E_2 can accelerate puberty in delayed children. They induce a growth spurt, and in accelerating growth eventually stop it, first stimulating, then shutting down the growth plates near the ends of the long bones.

As mentioned, these events do not have to be lockstep. The Kikuyu in

Kenya had much later menarche—just under age sixteen—but also a different order of events (Worthman 1986, 1993). Kikuyu boys entered puberty earlier than girls but developed very slowly, so girls surpassed them and were taller for two years. Initiation ceremonies and separate housing for boys were tied to physical change, not age, making puberty a different subjective experience. Pubertal events may be discordant, but Kikuyu have a greater synchrony of culture and biology than Western cultures, where keen awareness of chronological age heightens anxiety over the pace of bodily change.

Still, social influences are not so important that biology can be ignored; the timing of puberty and the incidence of adolescent problem behavior are heavily influenced by genes (Plomin and Fulker 1987). For example, MZ twins have higher correlations than DZ twins for age at PHV (.80 versus .40), first sexual intercourse (.41 versus .18), and delinquency (.70 versus .48). Despite dramatic changes at puberty, correlations between MZ twins or between adopted children and their biological mothers hold steady or increase through adolescence (Burt and Neiderhiser 2009), and the pacing of puberty runs in families.

Growth and Change in the Adolescent Brain

One of the great achievements of the last decade in developmental research is the clear demonstration that significant brain changes are part of normal adolescence (Sisk and Foster 2004; Sisk and Zehr 2005; Steinberg 2009). Some are no surprise—for instance, the fact that the onset of puberty depends on hypothalamic maturation (Romeo 2003). It would also not be surprising to find that the hormones that transform muscle, fat, hair, skin, and bone, not to mention the lengthening of the larynx and the development of breasts, could directly influence the nervous system. Whether or not hormones are involved, it is clear from neuroimaging and histological studies that the adolescent brain continues to mature (Blakemore and Choudhury 2006).

The ratio of white to gray matter density increases in many brain regions in the teenage years, although there is less consensus on the explanation, which could be either expansion and myelination of axons within the cortex or synaptic pruning in the same regions (Paus 2005). There is ongoing myelination and/or pruning in the association cortex, which helps explain cognitive and social maturation (Baird and Fugelsang 2004; Blakemore 2008), and this may exculpate some adolescent defendants on the basis of brain immaturity (Steinberg and Scott 2003; Gur 2005; Scott and Steinberg 2008).

Overall, as sample sizes have increased and studies have multiplied, there has been considerable consistency in the finding that gray/white matter density declines from adolescence to adulthood (Steinberg 2009). Some studies beginning in childhood show a steady decline from age four to adulthood, and others show a curvilinear pattern with a peak in early adolescence (Lenroot and Giedd 2006). These disparate findings may be due to methodological differences, regional variation in brain areas, or the postnatal increases in synaptic density well demonstrated in postmortem studies (Huttenlocher 1979, 1994, 2003; Huttenlocher and Dabholkar 1997). What is clear is that the idea that brain development is basically complete at puberty, widely accepted a decade ago, is no longer tenable.

Is the change experience dependent or maturational? Probably both. While the striking developmental brain changes in the early years of life appear to be mainly extensions of embryonic processes, the adolescent changes, although important, are less dramatic, and known processes of brain plasticity loom larger. For now, we have impressive age changes that are difficult to tease apart from experience but easier to cite as explanations for some adolescent behavior, and a new science of adolescent psychology is rapidly forming. At least two important theoretical models have emerged from these studies.

Sarah-Jayne Blakemore (2008) has attempted to integrate findings about regional brain changes in adolescence and relate them to the neuropsychology of social behavior and emotion. Among the established relationships are that the cortex bordering the posterior part of the superior temporal sulcus (pSTS) is involved in the perception of faces, facial expressions, and biological movement, and that both the pSTS and the medial prefrontal cortex (mPFC) are involved in "mentalizing states"—efforts to think about what others are thinking, about their intentions, or even about one's own thinking. There are well-validated ways of measuring these functions, and some of them show changes in adolescence that are correlated with brain changes.

For example, classic studies showed that face recognition improves until age twelve in girls but then declines (Carey, Diamond, et al. 1980), and that girls in midpuberty do worse than prepubertal or postpubertal

girls, controlling for age (Diamond, Carey, et al. 1983). There is also a decline in reaction time to match facial expressions to emotion words between ages ten and fifteen (McGivern, Andersen, et al. 2002). Blakemore
cites studies showing an increase in prefrontal activity on fMRI in response to fearful faces from ages eight to fifteen and a decline in activation of the anterior cingulate cortex (ACC) and the left orbitofrontal
cortex (OFC) between adolescence and adulthood, also in response to
fearful faces. Adolescents and adults differ in activation of the mPFC in
relation to their success at such mentalizing tasks as thinking about intentions and understanding irony, and Blakemore believes that pSTS
differences across age are also involved in mentalizing. While the interpretation of these findings is not straightforward, the fact that the orbitofrontal and medial prefrontal cortex is undergoing change in these
years suggests that Damasio's model of its role in reasoning and judgment, achieved by mediating between emotion and thought and between body and mind, could help us understand the dynamics of declining and then improving judgment in adolescence.

Others, including Eric Nelson and his colleagues (Nelson, Leibenluft,
et al. 2005) and Laurence Steinberg (2007, 2008, 2009), have proposed a
more general two-factor model with simpler assumptions, especially to
explain adolescent risk taking and failure to plan for the future. In essence, the increase in risk taking between childhood and adolescence is
viewed as due to pubertal change in the social-emotional and reward-
seeking systems of the brain (Forbes and Dahl 2009; Wahlstrom, Collins, et al. 2009), while its decline from adolescence to adulthood is due
to maturation of the cognitive control systems. A disjunction between
at least partly hormone-dependent motivational and social-emotional
circuits, on the one hand, and more slowly maturing (and perhaps more
experience-dependent) gray/white matter density ratios, on the other,
explains the intense and often peer-associated behavior of puberty, resolved by maturing inhibitory circuits in later adolescence. There is
much to recommend this model.

Nelson and his colleagues divide the relevant circuitry, which they
call the Social Information Processing Network (SIPN), into three simplified parts: the *detection* node, the *affective* node, and the *cognitive-
regulatory* node, the first and third of which are similar to Blakemore's
two components. In the detection node they include (as she does) the
superior temporal sulcus (STS), but also the inferior occipital cortex, the

inferior temporal cortex, the fusiform face area, and the anterior temporal cortex, all involved in perceiving people and other animate objects and their movements and expressions. In the cognitive-regulatory node (mentalizing or theory of mind) they identify the prefrontal cortex, including the mPFC and the OFC, also emphasized by Blakemore. Very importantly, however, they add a major affective component, mainly limbic and reward system structures, including the amygdala, ventral striatum, septum, hypothalamus, and some of their inputs and outputs. No overall brain model of the psychological lives of teenagers is plausible without including this circuitry.

What develops in adolescence? First, it is clear that there are limbic system changes, and they may be hormone-dependent. For example, adolescents show greater activation of the amygdala, OFC, and ACC when viewing fearful faces, and when asked to direct their attention to a non-emotional property of the face (for example, nose width), they are less able than adults to disengage the OFC. Like Blakemore, Nelson and his colleagues also find changes in the detection and cognitive-regulatory circuits between adolescence and adulthood, with emphasis on inhibitory prefrontal circuits. In their overview, "the surge in gonadal hormones at puberty induces changes within the limbic system that alter the emotional attributions applied to social stimuli while the gradual maturation of the prefrontal cortex enables increasingly complex and controlled responses to social information" (Nelson, Leibenluft, et al. 2005:163). To the surge in gonadal hormones we can now add mounting evidence of marked pubertal changes in the dopamine circuits linking limbic and striatal structures to the prefrontal cortex (Wahlstrom, Collins, et al. 2009). Since these pathways are central to motivation generally, and especially to pleasure and reward, they may mediate between pubertal hormones and adolescent risk taking, or alternatively they may be maturing independently of hormones.

Steinberg (2007, 2008, 2009) and his colleagues have applied and extended this analysis with an emphasis on adolescents' risk taking and their difficulty in delaying gratification, theoretically due to their immature (largely dopaminergic) brain systems of motivation and reward (Chambers, Taylor, et al. 2003; Hardin and Ernst 2009). Studying 935 subjects between ten and thirty years old with standardized computer tasks (for example, stopping at or running a yellow traffic light), they found that sensation seeking increases from ages ten to fifteen and stabilizes or declines thereafter, while impulsivity declines steadily from age ten (Steinberg, Albert, et al. 2008). They attributed these age changes to pubertal stimulation of reward systems and maturing inhibitory capacity, respectively; at younger ages, sensation seeking was higher in boys than girls. In the same sample, future discounting—the tendency to choose a smaller reward sooner rather than a larger one later—declined across adolescence (consistent with self-descriptions), while future planning continued to increase well into young adulthood (Steinberg, Graham, et al. 2009).

Efforts have also begun to understand the neurobiology of resistance to peer influence, which in many studies and across ethnic groups is low in early adolescence but increases from age fourteen to adult levels at eighteen. Individual differences corrected for age are also important, with low resistance predicting delinquency. At age ten, before the action of pubertal hormones, low-resistance children watched videos of facial expressions and hand movements that did or did not show anger. While watching angry hand movements, low-resistance children had a weak pattern of correlated brain activation compared to high-resistance children, who showed coordination between activation in the fronto-parietal and temporo-occipital areas involved in observing movement and those in the prefrontal cortex involved in planning and control (Grosbras, Jansen, et al. 2007). Perhaps the latter children will be better able to invoke prefrontal functions later, when faced with pubertal hormone surges, but we do not know what makes their brains work differently.

One study with paradoxical findings may offer a challenge to the prevailing two-factor model (Berns, Moore, et al. 2009). DTI was used to delineate white/gray matter density ratios in the cortex of sixty adolescents between fourteen and eighteen years old, and an adolescent risk behavior questionnaire was administered to assess self-reported risky behavior. With this method, and contrary to expectations, self-reported risk taking increased with age, and after controlling for age, the white/gray matter density ratio in the frontal cortex predicted risk taking. This is puzzling because other studies show decreasing risk taking with age, while white/gray matter density ratios are normally increasing. The study bears following up, especially with younger and older adolescents included.

Although strong claims have been made on the basis of animal models (Romeo 2003; Sisk and Zehr 2005), we do not yet know how much of the pubertal brain change is due to pubertal hormones (Forbes and Dahl 2009). Girls precede boys in the peak volume of gray matter in the frontal and parietal lobes (Giedd, Clasen, et al. 2006), but this is consistent with their greater maturity from early childhood on. In a study of thirty-seven boys and forty-one girls ages ten to fifteen, it was found that, after correcting for age, global gray matter volume was predicted by T level in boys but negatively correlated with E_2 level in girls (Peper, Brouwer, et al. 2009), which could help explain why girls mature faster on this measure.

Sex differences aside, genes shape individual brains (Lenroot and Giedd 2008). Twin studies show high heritabilities for brain anatomy in adults, although there are interesting variations in heritability across brain regions. The National Institute of Mental Health (NIMH) Pediatric Twin Study has shown high levels of genetic influence for most brain regions in childhood and adolescence, as well as fascinating changes in heritability with age:

> Regions such as the primary motor and sensory cortices that were not highly heritable in adults were significantly heritable in younger children but became progressively less so with time. Conversely, areas that were strongly heritable in adults were less so earlier in development. Instead, heritability in these regions [including frontal and temporal regions] gradually increased over childhood and adolescence. (Lenroot and Giedd 2008:1168)

Declining heritability in sensory and motor areas is consistent with what we know about brain plasticity in those regions (Merzenich 1998; Bengtsson, Nagy, et al. 2005), while increasing heritability in others is consistent with increasing heritability of complex traits, as we emerge from the statistically messy environmental and hormonal "background noise" of puberty.

Studies by Sarah Whittle and her colleagues shed further light on individual variation in a large sample of early adolescents. First, differences in temperament are correlated with the size (not just activity) of limbic structures as well as orbitofrontal and anterior cingulate cortex (Whittle, Yücel, et al. 2008). Further, in behavioral interactions with their mothers, the teens' amygdala size predicted aggressiveness, and for boys there were relationships between volumes of prefrontal, in- cluding orbitofrontal, cortex and emotional behavior (Whittle, Yap, et al. 2008). But T decouples amygdala from orbitofrontal functioning, sug- gesting disinhibition (van Wingen, Mattern, et al. 2010). And young ado- lescents' brain structures are also correlated with the way their moth- ers behave (Whittle, Yap, et al. 2009). Mothers' negative responses to positive affect were associated with larger orbitofrontal volumes, among other differences, suggesting that some regional brain differences could result from parenting styles. We have much to learn.

The Psychological Impact of Body Changes

Of course, the look of the body, reflected both in the mirror and in adults' and peers' behavior, impresses the child's mind in ways that have nothing directly to do with hormones. In an elegant study of the impact of first menstruation, girls drew a person twice, six months apart, in three study groups of the same age but different pubertal stages (Koff, Rierdan, et al. 1978). One group had not yet menstruated for either draw- ing, the second did so between the drawings, and the third was postmen- archeal for both. Menstrual status, not age, was the best predictor. Girls who had menarche between the drawings were more likely to draw a female first, and when scored for sex development, girls past menarche tended to show the curving hips, narrow waist, and breasts of a woman. Neither the "pre-pre" nor the "post-post" group changed, but girls who first menstruated *between* the two drawings increased their sex devel- opment scores, strongly suggesting an impact of menarche on body im- age. They also felt more satisfied with female bodies after menarche, "a pivotal event for reorganization of the adolescent girl's body image and sexual identity" (Koff, Rierdan, et al. 1978:635).

Other prospective studies confirm that menarche has a large psycho- logical impact, but the impact is typically negative. Girls who have in one way or another been prepared for menarche, especially positively, have an easier time than girls who come upon it unprepared (Koff, Rier- dan, et al. 1981; Rierdan, Koff, et al. 1989a, 1989b). For most girls in our culture it is experienced and remembered unfavorably (Pillemer, Koff, et al. 1987); "menarche may have a positive personal significance" be- cause of its association with maturity, "but this positive impact is over- shadowed by a negative interpersonal significance as girls become increasingly self-conscious, embarrassed, and secretive" (Koff, Rierdan, et al. 1981:148). It is hard to separate the effects of menarche from those of the hormones that cause it, but self-image must play a role.

Then too many girls have bloating and painful cramps—even, some women report, as painful as labor—that send them to bed for at least a day a month. Primary dysmenorrhea is subject to genetic influence in adult women, as is the covariation between menstrual pain and psychological symptoms (Silberg, Martin, et al. 1987). Thus some mental effects of genes are mediated indirectly through dysmenorrhea, although it could be that psychological symptoms influence pain reporting.

Female pubertal changes can also be positive, including a sense of pride in the development of feminine body contours (Petersen and Crockett 1986), but if a girl values slimness, she may reject the change in shape, which results partly from laying down fat. In contrast to cultural preferences in the non-industrial world (Brown and Konner 1987), and those of the West until the twentieth century, physiological and expectable levels of subcutaneous fat are condemned by many girls in industrial countries. A pathological expression of this preference is bulimia, which affects as many as 10 percent of upper-middle-class women in late adolescence, or, in extreme cases, anorexia nervosa, which affects 1 percent. Boys tend to be happier with their anatomical changes; they wish for more muscles, and they get them.

Adolescent Hormones in Sexuality and Aggression

Neo-Darwinian theory makes testable predictions about specific aspects of mating strategy and behavior, jealousy, fidelity, pursuit, diffidence, number and frequency of partners, and sex differences in all these variables (Buss 1998; Okami and Shackelford 2001), many of which are apparent in adolescence. This is also true of adult patterns of intra- and intersexual aggression (Archer 2000, 2004), both of which are related to sexual competition, although they have other purposes. But how do we understand this transformative phase developmentally?

As with earlier phases, we begin with universals. Adolescence heralds readiness for mating in all cultures, although there is substantial variation in culturally determined events like the onset of sexual activity and the age at marriage (Chapter 22). Considerable research has been devoted to the emotions that may accompany these changes. In some studies, puberty increases depressive moods in both sexes, at least temporarily (Rutter 1990; Buchanan, Eccles, et al. 1992; Angold and Worthman 1993), but for girls self-esteem often declines through adolescence, while for boys it rises (McLean and Breen 2009); depression has been found more consistently in girls than in boys, although some studies find it increasing in both. But how do these changes relate to neuro-endocrine changes, as opposed to such factors as role change, learning, and self-concept in reaction to visible body changes? Unquestionably these factors interact with each other and with cultural frameworks for the pubertal transition (Petersen and Crockett 1986). Furthermore, it is now widely agreed that at least a large minority of adolescents either avoid significant emotional problems or encounter them only transiently in early puberty.

Still, the hormones of puberty also produce behavioral and psychological changes, perhaps through direct effects on the brain (Sisk and Zehr 2005), although they are psychosocially mediated (Petersen 1987; Richards and Petersen 1987; Steinberg 2005). There are effects in both directions. For instance, T facilitates aggression and dominance in many species, including us, perhaps by disrupting orbitofrontal inhibition of the amygdala (van Wingen, Mattern, et al. 2010). Clinical evidence shows that T stimulates sexual desire in both men and women, mediated by regulatory effects of T on neuronal genes in some brain circuits.

T also responds to behavioral interventions. It rises after an aggressive encounter or other stressful experience if it results in a "win," whether in a college wrestling match (Elias 1981), a tennis match (Mazur and Lamb 1980), or a !Kung San hunt (Worthman and Konner 1987). During basic training in officer candidate school, T levels fall, to recover in the senior phase of training (Kreuz, Rose, et al. 1972), while recruits in both basic training and special forces who anticipated imminent combat in Vietnam showed reduced excretion of T and three other androgens (Rose, Bourne, et al. 1969). Some successes that have no obvious physical component, such as medical school graduation, are also associated with T rises. So hormone-behavior correlations are difficult to interpret even in adults. But pubescent children experience large internally generated hormonal changes, and it makes sense to consider these hormones and their fluctuations and interactions to have causal roles in behavioral change.

Several studies link T to aggressive and oppositional behaviors in adolescent boys (Buchanan, Eccles, et al. 1992). Physical bullying in early adolescence is primarily a male phenomenon (Pepler, Jiang, et al. 2008), so it is not surprising to find T playing a role. In a longitudinal study of the total national population of bullies and victims in Norwegian schools, the overwhelming majority of aggressive and threatening children were male, mostly a small percentage of boys who sustained this behavior through their teens. There was a strong relationship between T level and bullying (Olweus, Mattsson, et al. 1980, 1988), consistent with earlier studies showing that in male prison inmates the age at first arrest, usually sometime during the teenage years, or the history of violent offenses (Kreuz and Rose 1972) was related to current T level, and more recent studies (Dabbs and Morris 1990; Dabbs, Carr, et al. 1995) showing a positive correlation between T level and violence in adult prison inmates. The Olweus study used path analysis to fold hormonal, social, and behavioral variables into a causal model. It is likely that the truth about biobehavioral growth during puberty will more closely resemble this than any unicausal model.

Another landmark study examined hormones, personality, and sex in two hundred adolescents in the eighth to tenth grades, using an Adjective Check List (Udry and Talbert 1988). Among other measures, subjects were asked to check which of three hundred adjectives applied to them. High-T boys more often checked *ambitious, cynical, dominant, original, persistent, pessimistic, robust, sarcastic, severe, showoff, spontaneous, stingy, temperamental,* and *uninhibited.* High-T girls checked a partly overlapping list: *charming, cynical, discreet, disorderly, dominant, enterprising, frivolous, initiative, original,* and *pleasant.* Combining sexes yielded more correlations with T: *outgoing, self-centered, sensitive, sexy,* and *understanding.* No single-dimensional label would summarize all these, but "intense and emotional" might capture some of them. In a stepwise regression, neither age nor pubertal development helped predict the adjectives, once T level had been accounted for. But for girls, progesterone level in the first half of the menstrual cycle was associated with *boastful, foresighted, obnoxious, quitting, showoff, stern, tough,* and *unrealistic.*

A related study (Smith, Udry, et al. 1985; Udry 1990) examined the effect of close friends on the sexual behavior of boys and girls between ages twelve and seventeen to explore the relative influence of social context and endocrine changes. Interviews with the aid of drawings determined pubertal stage. Subjects also ranked themselves on a scale of sexual activity from kissing to intercourse; a friend of each subject had a similar interview. For girls, a distinction was made between pubic hair, which reflects androgens, and breast and hip development, which reflects estrogen; for boys, overall pubertal development was viewed as reflecting androgens alone. The results differentiated the sexes with respect to the mechanism of action of biological and social forces. For boys, pubertal development had the biggest impact on sexual activity, but the sexual behavior of the close friend had additional explanatory power.

For girls, pubic hair stage, reflecting androgens, strongly predicted sexual activity, while breast development, reflecting estrogen, had a less strong and separate effect, consistent with the established finding that women's sex drive is influenced by androgens. As with the boys, pubertal development alone did not predict girls' sexual activity as well as when the influence of close friends was added, but the girls showed a significant interaction effect. Low-androgen girls resisted the influence of sexually active friends, while high-androgen girls were quite susceptible to it, suggesting that internal, hormonally guided motivation does not suffice but is responsive to social factors. However, all the measured factors taken together did not predict sexual activity very well for either sex. Others—parental attitudes, family structure, religious belief, exposure to media, history of sexual abuse, physical attractiveness, and popularity—must also influence sexual activity.

Hormones were more important than age or pubertal stage in a study of one hundred girls ages ten and a half to thirteen from privileged homes (Warren and Brooks-Gunn 1989). Five hormones were measured and three questionnaires administered: (1) the Youth Behavior Profile (YBP), which measures depressive behavior, aggressive behavior, delinquent behavior, and immature hyperactive behavior; (2) the Self-Image Questionnaire for Young Adolescents, made up of seventy-eight sentences like "I feel nervous most of the time" and "I am a leader in school" (subjects estimate how true each is); and (3) a scale of sports involvement. Mothers were interviewed about their daughters' moods, and a physician assessed pubertal development. E_2 level classified subjects into four categories ("hormonal stages"), which were significantly related to the YBP. All trends were U-shaped or inverted U-shaped, showing significant but transient impact. Impulse control first dropped as E_2 rose, but then increased; depression, aggression, and psychopathology all rose first, then fell. Visible pubertal change, including breast development, did not predict these scores. These patterns are consistent with an initial effect of rising or fluctuating estrogen levels, leading eventually to physiological tolerance and/or learning and mastery of cognitive and social tactics for managing new feelings and motives.

In later studies, including longitudinal and mixed longitudinal designs, T level predicted sexual activity in boys (Halpern, Udry, et al. 1998) and the initiation of sexual intercourse in girls (Halpern, Udry, et al. 1997), although other factors, such as religiosity, sexual behavior of friends, sports involvement, and stability of family background, are always found to be important. Overall, hormones explain a small proportion of the variation in adolescent sexual activity after controlling for other variables (Halpern 2003). However, controlling for age and physical development first removes much of the variation in hormone levels that may be developmentally important (Buchanan, Eccles, et al. 1992).

It is clear that T level is both a cause and a consequence of behavior (Worthman and Konner 1987; Mazur and Booth 1998) and that it interacts in complex ways with environments, other hormones, and neurotransmitters. There are conflicting results. For example, in the Great Smoky Mountains multicultural and longitudinal study of adolescents, rising T levels in boys nine to fifteen years old showed an effect on non-aggressive conduct disorder, but not on physical aggression (Rowe, Maughan, et al. 2004). Another group of ninety-six boys followed from ages twelve to twenty-one showed consistent correlations between T and aggressive behavior at a given age, but in the group as a whole aggression declined with age (the typical pattern), despite increases with age in T (van Bokhoven, van Goozen, et al. 2006); this result shows that adolescent brain maturation and accumulating experience ultimately modulate hormone effects. In another study, girls age fifteen to seventeen with aggressive conduct disorders had elevated T levels along with other hormonal changes (Pajer, Tabbah, et al. 2006). And in still another, both aggressive adolescent boys and girls had higher T levels than less aggressive matched controls, with high cortisol modulating the effect (Yu and Shi 2009).

Subtler approaches reveal a complex picture. Delinquent boys with a mean age of 13.7 showed a correlation between T and overt aggression, but only against the background of low cortisol (Popma, Vermeiren, et al. 2007). Twelve- and thirteen-year-old boys who have been bullied verbally or physically show an increase in T, while bullied girls that age show a decrease, consistent with the tendency of boys to externalize and girls to internalize their abuse (Vaillancourt, deCatanzaro, et al. 2009). Five minutes of casual interaction with a woman elevated T levels in aggressive dominant men more than in other men, an effect enhanced by being sexually inactive for a month and by not being in a committed romantic relationship (van der Meij, Buunk, et al. 2008). And the T response to winning or losing a competition predicts future aggression (which impairs performance) in men but not in women, while baseline T does not (Carre, Putnam, et al. 2009).

These patterns are more subtle than was suspected several decades ago. The genetics and neurobiology of human aggression are very complex and involve many genes, hormones, and neurotransmitters (Craig and Halton 2009). However, there is no doubt that T and other androgenic steroids, administered experimentally to animals or under conditions of abuse or enhancement in human adolescents, *causally* increase aggressive moods and behaviors (McGinnis 2004). Further research will delineate how, not whether, T and other pubertal hormones contribute to adolescent sexual and aggressive behavior, since they clearly do.

Cross-Cultural Regularities

A quantitative cross-cultural study of this phase of life comparing 186 different cultures (Schlegel and Barry 1991) showed that, despite much variety, there are consistencies. Girls in the majority of cultures stay closer to home than boys and have extensive contact with their mothers and other adult female role models. Boys do not have comparable contact with adults of either sex but gravitate after puberty to other boys and form all-male groups that take them farther from home and draw them into competition and aggression. This is a cross-cultural confirmation and an extension into adolescence of the pattern of self-selected gender segregation, which in Western cultures begins in early childhood and intensifies in middle childhood (Maccoby 1998; Rose, Rudolph et al. 2006).

Schlegel and Barry saw it as an evolutionary adaptation of our species designed to prevent incest, comparable to patterns seen in monkeys and apes. Lionel Tiger, in *Men in Groups* (1984), attributed to adolescent male bands a primarily collective-aggressive and political function, while Beatrice and John Whiting (1975) found that "fraternal associations" of adolescent males are more likely in societies that emphasize military glory —martial prowess, heroism, death in battle, and the like. In this model peer groups separate boys from women, the better to make them warriors. When nonbelligerent hunter-gatherers like the Eskimo and the !Kung become settled, at larger population densities, adolescent boys may aggregate into potentially destructive gangs.

We know that male excess in physical violence is a cross-cultural universal (Chapter 10). Margaret Mead first studied Samoan teenagers, who seemed to have sexual freedom and a minimum of the "storm and stress" believed common in Western societies. But Mead underestimated the conflict felt by Samoan girls and boys during puberty; subsequent studies showed that virginity is prized for girls and that they have to resist or compromise with importuning boys. A steep rise in delinquency occurs in Samoan boys after puberty, resembling the rise in industrial states; the Samoan curve for age at first arrest peaks in the mid-teens and is indistinguishable from those for Western societies (Freeman 1982).

Although these curves do not solely reflect physical aggression, violent delinquency shows a large sex difference throughout young adulthood, extending the sex difference in physical aggression that begins in infancy (Chapter 10). Playful and real aggressive acts show a male predominance in early childhood in all cultures studied. In middle childhood there is less overt physical aggression, but still a male preponderance (Whiting and Edwards 1973, 1988). At puberty fighting frequency decreases further but is much more serious. Antisocial behavior increases tenfold (Moffitt 1993). Knives and guns come more into play; crime, violent or not, becomes much more likely. Males predominate in aggressive conduct disorders, and their excess increases through adolescence (Zahn-Waxler 1993; Loeber and Stouthamer-Loeber 1998; Lahey, Schwab-Stone, et al. 2000; Zahn-Waxler, Shirtcliff, et al. 2008).

Most crime in all countries is committed by young males, "one of the few facts agreed on in criminology" (Hirschi and Gottfredson 1983:552). The rise in crime during the teenage years is steep and has a shape approaching invariance in nations and cultures around the world, as well as a high degree of historical constancy (Sanders 1970). Victims of violent crime are also mostly young males, except for rape, although it too has many male victims, especially in prisons. Young males are the perpetrators, and male violence against women (including but not limited to rape) is part of the environment that shapes girls in adolescence and beyond (Smuts 1992; Smuts and Smuts 1993). It may also lead them to use sex in a pragmatic, even cynical way, to increase their chances in a hostile and dangerous man's world.

With risks of homicide, suicide, sexually transmitted disease, unwanted pregnancy, and substance abuse widespread, parental fear of adolescence is understandable, yet most teenagers experience it as positive. Conflict with parents, impulsive behavior, depression, and aggression often increase during the pubertal peak itself, but soon begin to decline again. Only a minority of adolescents experience what popular wisdom defines as disorders of the age group: unwanted parenthood, sexually transmitted disease, suicide, homicide, drug or alcohol dependence, and accidental death. There is divergence not just between the sexes, but between problem adolescents and those who are happy or at least stable. A leading investigator (Petersen 1988:590) found that "about 11 percent of young adolescents have serious chronic difficulties, 32 percent have more intermittent and probably situational difficulties, while 57 percent have basically positive, healthy development during early adolescence." The high rates of such disorders are compatible with the statistically more common picture of a healthy, stable, only mildly conflicted adolescent phase. However, other studies suggest that by the late teens the majority have experienced at least some difficulty.

Genes contribute substantially to the individual variation in some problem behaviors. A meta-analysis showed that aggressive behavior has a heritability of at least 60 percent and that other rule-breaking behavior is much less heritable (Burt 2009). Moreover, while aggression remains stably heritable over the adolescent years, non-aggressive delinquency increases in heritability as adulthood approaches (Burt and Neiderhiser 2009). This confirms studies showing that adult delinquency is more heritable than teen delinquency (Lyons, True, et al. 1995) because of the social and cultural turmoil of adolescence, but this apparently does not apply to serious aggression, which is a more stable, and stably heritable, trait. Its inheritance is polygenic and very complex (Craig and Halton 2009). Risky sexual behavior and adolescent misconduct are correlated for mainly genetic reasons, but each also has an important environmental contribution (Verweij, Zietsch, et al. 2009). However, individual variation, whether genetic, environmental, or both, does not negate population universals, including those of populations of males and females. And we must bear in mind that in the distant as well as the recent past adolescent risk taking and even to some extent misconduct were adaptive, and in some ways still are. If we are still living with the partly genetic consequences, we came by them honestly, through evolutionary adaptation.

A Role for Romantic Love?

We saw that both heterosexual and homosexual adults assign their first "crush" to a modal age of ten, before the pubertal sexual drive but possibly tied to adrenarche (McClintock and Herdt 1996). Despite the fact that romantic love has not been involved in bringing about most marriages in human history, psychological studies often consider romantic attachments and couple relationships as if they were a unified phenomenon (Feeney 2008), but this is a largely modern and Western pattern. Yet, contrary to the claim that romantic love is a Western phenomenon, it is important in most cultures (Jankowiak 1995a; Lieberman and Hatfield 2006; Hatfield, Rapson, et al. 2007). It was found in 146 of 166 cultures in a worldwide sample (Jankowiak and Fischer 1992). What is Western and novel is the institutionalization of romantic love as a portal to marriage. Elsewhere it sometimes leads to marriage, but often it mediates premarital, extramarital, or postmarital relationships or resistance to an arranged marriage (Jankowiak 1995a)—*Romeo and Juliet* with a better ending. Lila Abu-Lughod (1990) calls romantic love a "discourse of defiance," and it occurs in cultures from the Bedouin to Communist China (Jankowiak 1995b).

In the West it has distinctive patterns in adolescence that are now under active study (Florsheim 2003; Crouter and Booth 2006). In a large national sample in the United States, 55 percent of adolescents age twelve to eighteen said they had had a romantic relationship during the previous eighteen months (Carver, Joyner, et al. 2003), and more than 70 percent of those who were sexually active described both their first and most recent sexual relationships as romantic (Manlove, Franzetta, et al. 2006). Moreover, most sexual activity ("hooking up") that is not so described occurs in the context of established friendships ("friends with benefits") and not with relative strangers (Manning, Giordano, et al. 2006); more than half of sexually active teens have had such relationships, often with the hope that a romantic relationship would result.

Many other romantic relationships do not involve sexual intercourse (although they may later), and many are one-sided, with large majorities of teens reporting having experienced romantic rejection (Fisher 2006). Prospective studies show that these relationships influence adolescent behavior in many realms, for better and for worse depending on the partner (Simon, Aikins, et al. 2008). Still, in the United States, romantic love is a major portal to a sexually active life, and much other adolescent sex occurs with friends. Given that encounters with strangers are rare in hunter-gatherers and many other non-industrial cultures, it is likely that the sexual activity common in adolescence in 60 percent of cultures in the ethnographic range (Broude 1975) occurred in the context of friendship or romance.

Romantic love has its own neurobiology, separate from sex and to some extent from pair bonding (Fisher 2006). Young adults who said they were deeply in love looked at pictures of their beloved as compared to other friends; on fMRI they showed preferential bilateral activation of the anterior cingulate cortex, medial insula, caudate, and putamen and deactivations in the posterior cingulate and in the amygdala bilaterally and in the right prefrontal, parietal, and middle temporal cortex (Bartels and Zeki 2000). Maternal love partly overlaps this pattern, and the shared areas resemble prairie vole brain regions dense with oxytocin receptors (Bartels and Zeki 2004). College students in the first months of romantic love preferentially activated the caudate, the septum, and the ventral tegmental area (Aron, Fisher, et al. 2005), involved in brain reward. The density of serotonin receptors in blood platelets was elevated in people in love *and* in untreated obsessive-compulsives (Marazziti, Akiskal, et al. 1999).

Young men in committed romantic relationships have lower T levels than controls (Burnham, Chapman, et al. 2003), as do new fathers, and they respond to casual interaction with a woman with smaller increases in T (van der Meij, Buunk, et al. 2008). This may reflect a post-adolescent down-regulation of sexual and aggressive drives, but it is probably associated with pair bonding more than with romance. Although falling in love seems subjectively a powerful biological phenomenon and has special neural substrates, it is not universal even in Western culture, which values it highly. It may have evolved as a facultative adaptation—a sort of tantrum—activated by parent-offspring or mating conflict. But with or without romance, the pair bond is a fundamental feature of human experience, and the existence of different marriage forms in different cultures does not negate its importance. One way or another, most humans who have reached the age of breeding have become involved in a pair bond. Despite obvious differences, the pair bond shares much in common with the primary caregiver–infant relationship, both phenomenologically and, probably, physiologically, and it builds on that relationship in the individual life course. Yet, as in most pair-bonding species, it is compatible with a spectrum of reproductive strategies (Zeifman and Hazan 2008). As we will see in Chapter 17, the pair bond also forms the context for, and shares its biological basis with, fathering.

Ideals and Abstractions

Romantic love, among other things, exemplifies the idealizing imagination of adolescence. In the more educated modern subcultures, cognitive and moral development after puberty enter a phase that Piaget called *formal operations,* which entails a new capacity for abstract reasoning (Inhelder and Piaget 1958). This capacity includes highly abstract reasoning, such as that required for understanding mathematics, physics, and probability, a kind of reasoning that does not occur in prepubertal children but is also beyond the capacity of teenagers who lack a good high school education; even many adults of normal intellect never attain this phase. Thus some tasks made possible by the capacity for highly abstract reasoning cannot meaningfully reflect universals of development.

Still, adolescent thinking *is* characterized by the ability to reason about possibilities that cannot be directly observed; by a capacity for a kind of thought experiment that mentally tries out different solutions to a problem; by thinking beyond conventional limits; and by second-order thinking—the ability to reflect on one's own thought processes. All of these kinds of thinking can be seen in younger children, but they are more characteristic of adolescents of all cultures.

Very young children can remember and briefly recount events, and much more organized stories about individual events emerge in middle childhood, but studies in Germany (Habermas and Bluck 2000; Habermas and de Silveira 2008) and Denmark (Bohn and Berntsen 2008) have shown that the ability to compose a coherent life narrative emerges only in adolescence, because such narratives (which are probably critical in identity formation and life planning) require the cognitive maturity of adolescence. College students in Utah narrated stories about themselves, the quality of which depended in part on whether they had good listeners (Pasupathi and Hoyt 2009). Adolescents are also highly subject to religious idealism and conversion experiences, with more than half in some studies saying that they have decided to live their lives for God, and many of these commitments prove lasting (Good and Willoughby 2008). Some become involved in political activities, imagining a better world and trying to bring it about. These patterns are a different way of coping with the "storm and stress" of adolescence, and the positive impact of adolescents' engagement in these activities probably more than outweighs the impact of the minority with aggressive, sexual, and substance abuse problems.

The Onset of Parenting—Maternal Care

Although not attained by all individuals (there are other ways to achieve reproductive success), the typical end point of reproductive development is parenthood. Almost all cultures in the anthropological and historical records have considered it the natural, if not universal, outcome of development, and in evolutionary terms it is the main route to success. Despite the undoubted role that education and culture play in shaping parental behavior, evolution left us with a physiological legacy that provides some largely fail-safe mechanisms.

Most of these mechanisms in humans and other primates are obscure, but rodent maternal behavior is better understood, and studies in voles have highlighted key neurobiological features of paternal care and the separate mechanisms of female and male pair attachment. While parental care occurs in invertebrates (Trumbo 1996), fish (Crawford and Balon 1996), amphibians (Crump 1996), and reptiles (Gans 1996), in birds and mammals it is vital for all species (Gubernick 1981; Gowaty 1996; Rosenblatt 2003). Our focus here is mainly on mammals.

The intensification of infant dependency and maternal care probably drove brain, and especially limbic system, evolution (MacLean 1985; Mega, Cummings, et al. 1997), so that hormones operating in and on that system affect parenting. Oxytocin (OT), prolactin, E_2, and progesterone (P) promote nest building and caregiving in female rats, mice, rabbits, hamsters, ring doves, and canaries. Prolactin causes mammary enlargement and milk synthesis in mammals and crop and brood patch formation—the equivalents of nursing—in birds; it was once considered the main maternal hormone, but the onset of mothering depends on OT, the hormone that also causes milk letdown, although E_2 and P are also implicated (Bridges 1990).

In a priming experiment, ovariectomized virgin female rats were exposed to pups in an effort to find out what might make them maternal. This could be done with a course of hormone treatments that mimic pregnancy, labor, and delivery: a gradual rise in E_2 and P, then an abrupt fall in both and a surge in prolactin (Moltz, Lubin, et al. 1971). But a virgin female could also be made maternal by linking her circulatory system to that of a recently parturient rat (Terkel and Rosenblatt 1972), which suggested that some unknown blood-borne factor, itself dependent on the hormonal changes of pregnancy and parturition, is involved. Remarkably, the effect of transfusing blood plasma from a recently parturient rat worked with thirty-day-old juveniles just nine days past weaning (Brunelli, Shindledecker, et al. 1987; Brunelli and Hofer 1990), and this worked better in females than in males, suggesting prepubertal preparedness of the brain for whatever blood-borne factor is operating.

OT was a candidate, but it cannot cross the blood-brain barrier. However, when introduced into the brains of virgin female rats, it made them fully maternal in thirty minutes (Pedersen, Ascher, et al. 1982). OT antagonists prevent maternal behavior only before it begins (Insel 2000); thus it is necessary for the onset but not the maintenance of mothering, and it must come from inside the brain. The hypothalamus is directly involved both in hormonal regulation and in brain circuits for retrieval and protection of the young (Numan 1994; Stern 1996; Numan and Sheehan 1997; Young and Insel 2001). The medial preoptic area of the hypothalamus, sexually dimorphic and implicated in sex, is essential for rat maternal behavior, and E_2 affects that behavior through this region (Numan, Rosenblatt, et al. 1977; Numan 1994; Numan and Sheehan 1997; Stern 1996). But the preoptic area is only part of a core neural circuit that mediates maternal behavior. The medial amygdala, influenced by olfactory cues, projects to the preoptic area, which in turn has OT projections to midbrain areas; these also function in infant attachment.

Feedback from the midbrain to higher centers occurs through dopamine (DA) neurons that project to the nucleus accumbens, one of the brain's main reward circuits. OT introduced into the midbrain tegmentum can initiate mothering through a DA circuit (Pedersen, Caldwell, et al. 1994). Some rat maternal behaviors (licking, grooming, the arch-backed posture, and nursing) depend partly on a tactile input (Stern 1996), through the lateral septal area and the trigeminal nerve, which innervates the face and whiskers and under hormonal influence becomes more sensitive to tactile stimuli from pups.

Still, olfactory input is central in many species (Lévy, Keller, et al. 2004). Human mothers can identify their own offspring by odor alone from postpartum day three, and mothers with higher cortisol levels can do it from day two (Fleming, Steiner, et al. 1997). In many species, infant odors are attractive; females eat the placenta and lick amniotic fluid off the young. In some species, mothers must recognize their own offspring. Rodent species vary in their degree of offspring preference, but herding ungulates like cows and sheep need maternal recognition to ensure that they care for their own young in a herd in calving season. Ewes respond to lambs by switching on the immediate early gene *c-fos* in neurons, showing activity in the piriform, orbitofrontal, entorhinal, and medial frontal cortex; these, along with the cortical/medial amygdala, establish lamb recognition, which does not occur when the sense of smell is blocked.

OT influences many social and reproductive behaviors (Carter 1998, 2003). As a hormone, it is made by neurons in the hypothalamus that project to the posterior pituitary and secrete it into the bloodstream. It causes milk letdown and is involved in uterine contractions. But it is also a brain neurotransmitter, binding to neurons that express OT receptors (OTRs), and marked sex differences in OTR distribution strongly influence behavior.

In the prairie vole (*Microtus ochrogaster*) OTRs lie in the limbic system, including reward centers, and stimulation of these reward circuits—with dopamine D_2-receptor stimulants in the nucleus accumbens, for example—promotes both partner preference in females and maternal behavior; thus OT makes attachment rewarding by influencing DA reward circuits (Gingrich, Liu, et al. 2000). The promoter region of the gene for the receptor mutates easily, facilitating evolutionary change (Young, Winslow, et al. 1997b). Attachment systems diverge even in closely related species, and mating systems are not highly constrained adaptations.

Knockout studies show that other genes are also involved. Mice lacking *fosB,* which activates transcription factors in certain neurons, selectively lose maternal behavior (Brown, Ye, et al. 1996); despite normal pregnancy, hormone changes, and lactation, their pups die within a day or two of birth. Randomly selected pups from the same knockout mothers, cross-fostered to normal ones, develop well. Normal virgin females primed by repeated pup exposure will nurture them, but knockout virgins show marked deficits. They perform normally on cognitive tests, have normal olfactory function, walk around, explore, eat, adapt to cold, copulate, become pregnant, and fight off intruders normally; their defect is specific to nurturing.

Species differ (Numan 1994). Virgin female mice are easily primed for mothering with a few hours of exposure to pups and no hormonal treatment (Noirot 1972), but in rats this takes six to seven days, during which time pups must be kept alive artificially (Rosenblatt 1967). Hamsters require still more exposure; often several litters must be replaced as each is killed in succession by the female (Richards 1965). Rhesus monkey mothers made abusive by their own abnormal rearing improve with exposure to infants, but in an individualized, genotype-dependent manner (Suomi 2003, 2008); temperament is crucial, independent of rearing condition. Free-ranging rhesus mothers on Cayo Santiago show natural variation in maternal behavior, and mothers who nursed and groomed their infants had higher blood OT levels, while more rejecting mothers had higher blood cortisol levels and more 5-HIAA, a serotonin metabolite, in their cerebrospinal fluid (Maestripieri, Hoffman, et al. 2009).

Research on sex differences in nurturing behavior (Chapter 8) suggests that male-female differences in parenting are in part due to testosterone. T administered to females reduces maternal behavior in many mammals (Zarrow, Denenberg, et al. 1972; Lonstein and De Vries 2000). In female rabbits (Fuller, Zarrow, et al. 1970) and rats (Quadagno and Rockwell 1972; Juarez, del Rio-Portilla, et al. 1998), T administered to neonates inhibits their future maternal behavior. In Wied's marmoset (*Callithrix kuhlii*) females' urinary testosterone excretion is inversely correlated with current maternal behavior (Fite, French, et al. 2005).

Research on activation of human mothers' brains while viewing their own infants (as compared with other infants) reveals circuits for maternal behavior consistent with animal studies. As mentioned earlier, maternal responses share brain circuits in common with those active in romantic love (Bartels and Zeki 2004), including limbic areas rich in oxytocin and vasopressin receptors. But mothers specifically activated the left orbitofrontal cortex (OFC) and the periaqueductal gray area of the midbrain, a pain-control region, more than did women in love. Both kinds of love activated the nucleus accumbens, ventral tegmental area and other reward regions, as well as the anterior cingulate, a center for social relationships, and the medial insula, known to be involved in pleasant bodily sensations ("limbic touch") coming to consciousness through the OFC (Damasio 1994).

Subsequent studies in other laboratories have found strikingly over-lapping activations in maternal love. In one study, activity in the moth-ers' brain reward areas depended on how positive the infants' facial ex-pressions were (Strathearn, Li, et al. 2008). In another, mothers imitated their infants' expressions, activating the mirror-neuron system, the in-sula, and the amygdala, showing right limbic activity to infants' smiles and engaging higher cortical functions when the expressions were am-biguous (Lenzi, Trentini, et al. 2009). An elegant study using near-infrared spectroscopy to measure brain activity in both mothers and their one-year-old infants showed that the anterior OFC was preferen-tially activated in both, and more so when the partner was smiling (Minagawa-Kawai, Matsuoka, et al. 2009). Finally, a remarkable study using magnetoencephalography not only to locate but to parse the tim-ing of regional brain responses showed that adults specifically and rap-idly activate the medial OFC in response to infant as opposed to adult faces, so much so that the investigators refer to the response as "a neural signature for parental instinct" (Kringelbach, Lehtonen, et al. 2008:e1664). These studies converge to vindicate the general limbic sys-tem model (MacLean 1990) and the specific role of the OFC in mediating between limbic and cortical structures in the processing of emotion and reward (Damasio 1994).

Paternal Care and the Pair Bond

Unlike the males of doves (Buntin 1996) and some other birds, which produce crop milk for regurgitation feeding, male mammals lack physi-ological adaptations for feeding young. But paternal care and pair bond-ing require brain adaptations. In prairie voles and pine voles, two spe-cies that are both pair-bonding and paternal, these behaviors occur in males mainly because of vasopressin (VP), a small peptide that differs from OT by two amino acids. Each evolved by one amino acid substitution from vasotocin (Hoyle 1999), which mediates nest building and egg-laying in some lower vertebrates (Goodson and Bass 2001; Woolley, Sakata, et al. 2004). Mesotocin, the avian equivalent, is important in both flocking and social bonding in finches (Goodson, Schrock, et al. 2009). OT may play a role in male behavior, as shown by knockout mice; when paired repeatedly with one female, they show specific amnesia for her without any other memory deficits; the effect is reversed by OT replacement, and is reproduced by OTR blockers in normal mice (Ferguson, Young, et al. 2000).

In the pair-bonding prairie vole (*M. ochrogaster*) OT enhances females' preference for their partners without affecting their sexual activity, but even at high doses infused into the brain it does not induce attachment in the closely related, multiple-mating montane vole (*M. montanus*) (Winslow, Hastings, et al. 1993). Stimulation of reward circuits, through DA-D$_2$ receptors in the nucleus accumbens, increases partner preference in normally monogamous female voles; as with maternal behavior, OT stimulates dopamine circuits to make attachment rewarding (Gingrich, Liu, et al. 2000). Male loyalty and fathering, on the other hand, are VP-dependent in prairie voles, so partner preference depends on OT in females but VP in males (Young, Wang, et al. 1998; Insel 2000). Yet the patterns are not determined by OT and VP levels (these are similar across species), but by the expression of their receptors on brain cells (Insel, Young, et al. 1997; Wang, Hulihan, et al. 1997). Injecting VP into male prairie vole brains enhances partner loyalty, while blocking VP decreases it, with no change in sexual activity (Winslow, Hastings, et al. 1993); no such effects occur in male montane voles, however, because they lack the appropriate receptors.

In prairie voles the VP receptor is strongly expressed in limbic system circuits, but in the montane vole much less so (Lim, Hammock, et al. 2004). As with the OTR, small changes in the gene promoter account for the species differences. Prairie and montane vole VPRs are 99 percent identical, but the prairie vole promoter contains a particular highly repetitive sequence upstream from where the gene begins transcription. This accounts for different VPR expression in the brain and for the different social behavior. Remarkably, the same difference obtains between the pine vole (*M. pinetorum*), which is pair-bonding and paternal like the prairie vole, and the meadow vole (*M. pennsylvanicus*), which is multiple-mating and maternal-only like the montane vole. That four species in one genus have such different mating systems is consistent with the promoter region being mutated easily, facilitating rapid evolution (Young, Winslow, et al. 1997b). Perhaps it is the case that some groups of mammals need flexible, species-specific social adaptations that can respond relatively quickly—at the species level—to environmental challenge, and that paternal behavior and pair bonding, while difficult to modify in individual life spans, are not the kinds of fundamental characteristics of organisms that are subject to long-term evolutionary constraint.

Remarkably, a transgenic (knock-in) *mouse,* supplied with a VPR gene containing a prairie vole promoter, expressed the receptor in cingulate cortex, olfactory bulbs, and parts of the thalamus, all abnormal (or perhaps supernormal?) for mice (Young, Nilsen, et al. 1999; Lim, Wang, et al. 2004). The genetically engineered mice, but not their littermates, showed increased social affiliation when treated with VP to their brains, but did not pair-bond. The pattern of receptor expression in the mice was not quite the same as that of truly monogamous voles, nor was their behavior.

Neural circuits active during mating and pair formation in prairie voles are all in the limbic system: the ventral pallidum, the nucleus accumbens, two hypothalamic nuclei (BNST and medial preoptic area), the medial amygdala, and the mediodorsal thalamus. Except for the accumbens, all are the same as the VPR-expressing regions of the prairie vole brain (Lim and Young 2004). Some of them help mediate sexual behavior in other species, but the ventral pallidum and nucleus accumbens involvement may be specific to pair bonding. Transgenically increasing VPR expression in the prairie vole, specifically in the ventral pallidum, accelerates pair formation and increases affiliation. Blocking the VPR in the same region blocks males' preference for their partners (Lim, Hammock, et al. 2004). In the biparental California mouse (*Peromyscus californicus*), new fathers show activity in the BNST and the dorsal raphe nucleus, and they specifically activate the medial preoptic area when presented with a pup (de Jong, Chauke, et al. 2009).

The OTRs in the nucleus accumbens are probably involved in the DA reward pathway; the VPRs in the ventral pallidum play an analogous role because, among other reasons, this structure receives important neural input from the accumbens. Both OTRs and VPRs are well placed, respectively, to make maternal behavior and partner preference in females (OT) and paternal behavior and partner preference in males (VP) rewarding and even addictive (Insel 2003). A well-known, simple pathway—from the olfactory bulb to the medial amygdala through the ventral pallidum to the nucleus accumbens—would then make the partner rewarding. In the common marmoset (*Callithrix jacchus*), a small, pairbonding South American monkey (Wang, Moody, et al. 1997; Wang, Toloczko, et al. 1997), the VPR is found in the nucleus accumbens, diagonal band, septum, amygdala, and hypothalamus. Although it differs somewhat from the prairie vole distribution, these limbic system structures would also affect social behavior in this species—the only primate studied.

Just as T interferes with maternal behavior in females, castration of males improves their paternal behavior (Clark and Galef 1999; Lonstein and De Vries 2000). Male T levels are lower in biparental vireos (a songbird) than in a closely related species that is exclusively maternal (Van Roo, Ketterson, et al. 2003). Compared to males of a solely maternal hamster species (*Phodopus sungorus*), males of a closely related biparental hamster (*P. campbelli*) show a correlation between low T, high prolactin, and paternal behavior, and they drop their T levels in the first postnatal days (Reburn and Wynne-Edwards 1999). Finally, the human male's T and cortisol levels are low during his wife's pregnancy (with T declining further after birth), while his E_2 rises (Berg and Wynne-Edwards 2001). No doubt future research will reveal many other patterns in the physiology of human parental behavior that overlap with those of other species, although humans will not simply mimic prairie vole adaptations. Is a pharmacological fix to make wayward men "commit" to women and children on the horizon? Not anytime soon, but it is a future possibility. Who decides when and how to use it is another matter.

Interlude 3

Thinking about Growing Up Gay

In the dark days when psychiatrists considered homosexuality a disorder, a close friend and former teacher of mine—let's call him Michael—was being psychoanalyzed by the most prominent doctor among those who thought they could "correct" it. Michael spent years on the couch, four days a week, trying to reorient his sexual and romantic feelings and change his life. In the course of this treatment he married a wonderful woman, who had been his friend for a long time. They seemed happy, but he died of a heart attack— probably a very late consequence of childhood rheumatic fever—at age forty-two, about six months after the wedding.

I am not a strong believer in psychosomatic illness, although psychologi- cal stress certainly matters (see Chapter 21), and we all have points of physi- cal vulnerability that can buckle under severe emotional burdens. I would never blame my friend's wife, but his "reorientation" was a stressful process, and it has crossed my mind to wonder whether it had just a little to do with his death. In any case, successive editions of the official diagnostic manual of the American Psychiatric Association have moved from calling homosexual- ity a disorder to calling homophobia one, while reserving a subcategory for psychological discomfort with one's sexual orientation, whatever it might be. Michael, I think, would have qualified for that diagnosis both before and after his psychoanalysis.

There is now ample evidence that sexual orientation as a stable trait—as opposed to situational patterns and experimentation—is largely genetic and neurobiological (LeVay 1997; Rahman 2005). No genes have been clearly identified, although some have been proposed; it is certainly multigenic and complex in determination. There is a possibility of X-linkage, which would help explain why, despite substantial differences in the overall percentage of people estimated to be same-sex-oriented in different surveys, there are usually about twice as many men as women found to be gay. There is evidence of a fraternal birth order effect such that later-born men are more likely to have same-sex orientations, and this suggests a maternal immunization process that changes the biochemical milieu of fetuses after a first male pregnancy.

Direct measurements of genital arousal in response to erotic pictures and literature show unequivocally that self-reported sexual orientation is reflected in bodily processes that we have little control over, although this effect is more specific in men than in women (Chivers and Bailey 2005; Bailey 2009). Neural correlates of arousal seen in fMRI studies largely confirm these effects; different activation in several brain regions of interest, including the amygdala, shows that sexual attraction is in the brain as well as the body (Safron, Barch, et al. 2007).

But in any case, twin, family, and adoption studies show clear patterns of inheritance explaining about half the variation in homosexual orientation (Pillard and Bailey 1998), and in some studies there are more relatives through the maternal than the paternal line who share this orientation. Prospective studies have found that boys with exceptionally high female identification and behavior ("sissy boys") strongly tend to grow up to be gay (Green 1987), and studies of childhood home videos of currently homosexual men and women confirm retrospective interviews: children of both sexes destined to be same-sex-oriented showed more gender nonconformity on the videos, beginning at an early age (Rieger, Linsenmeier, et al. 2008, 2009). Similarly, girls who between three and twelve years of age (mean age of nine) were diagnosed with gender identity disorder were followed into young adulthood: eight of twenty-five were bisexual or homosexual in fantasy, and six of twenty-five in behavior; these are far higher rates than in the general population (Drummond, Bradley, et al. 2008). There was some evidence that girls who were most boylike in their childhood behavior were most likely to have nonheterosexual behavior as adults.

Overall, "current theories have left little room for learning models of sexual orientation" (Rahman 2005:1057). In addition to the genetic evidence, hormonal models, and childhood videos, gay men and women remember having their first erotic feelings toward others of the same sex at thirteen for boys and sixteen for girls—both years before their first same-sex experience (Bell, Weinberg, et al. 1981)—and another phenomenon, the first "crush" on someone of the same sex, occurred around age ten, the same age as for first crushes in heterosexuals (McClintock and Herdt 1996). These and other similar findings suggest great stability and intrinsic disposition. Against this background, "coming out" as a developmental event can be ambiguous, liberating, or traumatic depending on the cultural and social context.

But unfortunately for simplicity, observers are not always talking about the same things. Despite biological factors, homosexuality can still be viewed as a social construct (Adriaens and De Block 2006), and debates continue as to how common true bisexuality (as opposed to experimenting with same-sex activity) really is. Is sexual orientation a continuous or dichotomous variable? Recent longitudinal research on young women who have sex with women, following them for a decade, found that their sexual orientation became more fluid—more bisexual—with time (Diamond 2008), at a minimum casting doubt on the notion that bisexual women are in a phase on their way to forming a *less* fluid orientation. Clearly, finding one's true sexual orientation can be a lifelong developmental process (Patterson 2008). Adrienne Rich's (1978) poems of lesbian love in middle age are among the best poems in the modern American canon. But what can we mean by "true sexual orientation"?

Up to this point we have been talking about sexual orientation as an enduring trait, influenced by genes and early hormones, prefigured in childhood, self-identified by adolescence, and enduring throughout life—a fundamental disposition, not a choice. This is important in political and religious terms, since many people expect themselves and others to refrain from homosexual activity, but if it is not a choice, such people should change their expectations. Of course, one can always try to choose celibacy, but for a male at least it may be physically very difficult to have sex with a person who is simply not enticing or arousing; for either sex, it can be an imposition that causes great suffering.

But of course, sex between men or between women is not always the result of a trait or lifelong sexual orientation. It can be a choice by predominantly heterosexual people to experiment, or a response to a transient or long-term cultural milieu. The culture of ancient Greece was in the latter category and was in part the pattern that the Judeo-Christian tradition rejected when it prohibited homosexuality. In the transient category are institutions like prisons, where homosexual *behavior* is common but is not continued by most inmates after release. Some male-on-male sex in the United States forms a pattern known as "the down low" or being "on the down low." These phrases refer to men who also or predominantly have sex with women, and in some cases their sex with men appears to be elective and condoned by local mores.

Looking beyond narrow cultural boundaries is, as always, illuminating. A cross-cultural sample showed that homosexuality was permissible in forty-nine of seventy-six societies in the ethnological range; another found it to be "not uncommon" in twenty-nine out of seventy. But the cultural response ranges from severe prohibition to active encouragement of situational same-sex behavior. Among cultures in New Guinea sometimes called "the semen belt," pubescent boys are expected to give oral sex to and imbibe the semen of older boys, a ritual process that is believed necessary to turn them into men (Herdt 1987), although the interpretation of the practice is less straight-forward than it once seemed (Knauft 1993). The boys almost always cease this activity in their twenties, marry, and become fathers.

However, in one case a young Sambia man, Kalutwo, persisted in his male-on-male activity in a way that was deemed inappropriate (Stoller and Herdt 1985). He had strong homoerotic attractions, entered four successive marriages to women, all without issue, and was dissatisfied with his sexual life. Unlike the other boys in his culture, he had homosexuality as a trait and would have had it anywhere. Anthropologists have long known that many cultures have specified roles for men and women with ambiguous or trans-gender inclinations, as described in the landmark volume *Third Sex, Third Gender* (Herdt 1994b).

Among some Native American groups, an institution known as *berdache* allowed certain men to permanently dress as and play the roles of women, including marrying men (Roscoe 1991, 1994); similar options exist for men in some South American and Polynesian cultures. The *hijras* of India are a religious community of transvestite men who often sell sex to men and usually elect castration to complete their transgendered status (Nanda 1999; Reddy 2005). "Boy-wives" and "female husbands" are found in various African cultures (Murray and Roscoe 1998). Of course, homosexual practice was common among the philosophers of Athens and the partnered warriors of Sparta; "sissy" boys may grow up to be gay, but many other men who have sex with men (including gay men) are highly masculine. And despite the serious stigma often attached to homosexuality, gay men and women have not just survived but created adaptive subcultures throughout history, in the hostile Judeo-Christian West and beyond it (Duberman, Vicinus, et al. 1989).

So same-sex erotic attraction and behavior manifest themselves in many ways in varied cultural contexts and have been aptly referred to as "homosexualities" rather than homosexuality (Bell and Weinberg 1978; Herdt 1994b). Genes, hormones, identity, permanence, inclination, behavior, and culture may vary independently along the dimension of sexual orientation (Federman 1967), and Western culture has seen a transformation of values and openness in our time (Clendinen and Nagourney 1999), with a growing effort to make research relevant to policy (Gonsiorek and Weinrich 1991; LeVay 1997).

Needless to say, homosexuality poses an evolutionary puzzle. If sex is for making babies, how did this nonreproductive sex escape the culling effects of natural selection? Related behaviors occur in many animals—for example, male-male mounting and thrusting in monkeys, a dominance signal, and female genital rubbing in bonobos, a central social behavior in one of our two closest relatives. But it is not clear how the numerous and varied same-sex contacts in animals relate to what we observe in our own species.

As to that, several hypotheses have been advanced. If gay men and lesbian women also had offspring heterosexually along the way, they might compensate for the time and energy they divert from reproduction. If their relatives were more fertile than average, that too could compensate. If they were needed helpers at the nest, they could advance the cause of siblings, nieces, and nephews enough to outweigh their reproductive disadvantage by playing a role in human cooperative breeding. And if they were exceptionally intelligent or creative, they could enter into arrangements yielding inclusive fitness benefits that would easily ensure the survival of their genes.

All of these hypotheses are reasonable; none has been shown to be true. Certainly helpers at the nest have been important in human survival (Hrdy 2009). Certainly some of the most outstanding creative people in history have been gay. There is some evidence for the fertile relatives hypothesis (Rahman, Collins, et al. 2008); support for the kin selection hypothesis is equivocal at best. But we are only at the outset of what should be a long future for such studies. We know that as gays and lesbians have fought for the opportunity to make formal long-term commitments and even to marry, many such couples have adopted or in one way or another conceived children. Studies show that the children of such couples grow up normally by all tests and standards; that if adopted, they are no more likely than the children of heterosexual couples to become gay or lesbian themselves; and that as they grow their only reliable difference from other children is that they are less homophobic (Wainright, Russell, et al. 2004; Wainright and Patterson 2009; Patterson 2009).

Freud, we have learned, was wrong about many things, but when it came to sexual orientation he got the basics right. In a 1935 letter to the mother of a gay man, he wrote that homosexuality "is nothing to be ashamed of, no vice, no degradation, it cannot be classified as an illness." Although he called it "a certain arrest of sexual development," he denied that treatment could change sexual orientation. "What analysis can do for your son runs in a different line. If he is unhappy, neurotic, torn by conflicts, inhibited in his social life, analysis may bring him harmony, peace of mind, full efficiency, whether he remains a homosexual or gets changed" (Anonymous/Freud 1951:786–87). The letter was sent to Alfred Kinsey in 1949 with a note attached: "Dear Dr. Kinsey: HEREWITH I enclose a letter from a Great and Good man which you may retain. From a Grateful Mother."

His disciples, alas, became convinced for a time that different sexual orientations are treatable illnesses. But the official *Diagnostic and Statistical Manual* has evolved in several editions from that position back to something like Freud's: whether gay or straight, a person can be at odds with his or her own sexual life, and psychotherapy may help, with or without a change in sexual orientation. So "homosexuality," whatever that is exactly, is no longer considered an illness, but ironically homophobia now is—an instance of cultural evolution with profound implications for psychosocial growth, and also for human happiness.

TRANSITION 2

Plasticity Evolving

"One of the oldest unresolved controversies in evolutionary biology
—and a source of many bitter arguments and failed revolutions—concerns the relation between nature and nurture in the evolution of adaptive design." So begins Mary Jane West-Eberhard's magisterial *Developmental Plasticity and Evolution* (2003:3), which removes any doubt that plasticity is one of life's fundamental processes and that neither survival nor evolution would be possible without it. There are many forms of plasticity in the living world, and they are not restricted to animals with significant nervous systems, nor, for that matter, to animals. Here we have to restrict ourselves to animals that learn and otherwise bear in their neural and neuroendocrine systems the imprint of lived experience. But we should bear in mind that these cases are only the tip of the iceberg of plasticity and that plasticity is not a footnote to genetics but is of the essence of adaptation and of (thoroughly Darwinian) evolution.

The controversies began long ago. At least since Jean Baptiste de Lamarck, some evolutionists have believed that the characteristics an organism acquires during its lifetime can somehow be transmitted to offspring; even Darwin succumbed to this temptation. It is so engaging a belief that we are inoculated against it in our first lessons about evolution, and contempt for Lamarck—a pioneering evolutionary scientist in an era chockablock with creationists—seems to be an integral part of the vaccine. Our first inclination is to hate the notion that randomness is at the heart of the history of life, so we incline strongly toward the alternative. But on the other hand, we hate the thought that all our hard-won life experience is wasted.

Of course it is not, since in our case at least there is a process of cultural evolution occurring in parallel and in interaction with genetic evolution. We will consider in Part IV some ways of looking at that. But there is an even more fundamental way in which experience is not wasted, whether or not an animal has culture. In fact, this resolution of a central philosophic question raised by Darwinian theory—*Is experience wasted?*—applies to any animal with a nervous system capable of changing its behavior. All such organisms have added to the information storage capacity of DNA the more rapidly malleable storage form of neuronal circuits.

In what sense can Ernst Mayr's (1988) claim that "behavior is the pacemaker of evolution" be true when it is clear that evolution is the ultimate cause of behavior? The paradox is resolved by the Baldwin effect, a strictly Darwinian process that highlights the role of plasticity in creating the conditions for evolution (Baldwin 1896; West-Eberhard 2003; Crispo 2007; Badyaev 2009). This effect has sometimes been dismissed as trivial or synonymous with natural selection (Mayr 1963), but its importance was acknowledged by Julian Huxley (1942), by G. G. Simpson (1953), and even eventually by Mayr himself (1988:408), who described a closely related phenomenon:

> In animals, almost invariably, a change in behavior is the crucial factor initiating evolutionary innovation. As has been stated so often, behavior is the pacemaker of evolution. . . . When an arboreal bird becomes more terrestrial, as did the mockingbird-like ancestor of the thrashers, this shift set up a selection pressure on strengthening and elongating the legs and strengthening the bill used for digging. . . . In birds the form of the bill is particularly plastic and apt to respond to shifts in behavior.

There is nothing strange or mystical here. Mayr certainly did not believe that modifications through use can be passed on to offspring. The thrashers' ancestors, as a population, became more terrestrial in habit, and small increments in terrestrial activity set up selection pressures for leg and beak changes that were strictly Darwinian. Indeed, by "plastic and apt to respond" he was referring, not to a change within an individual during a life course, but rather to the responsiveness of the genome transgenerationally, under selection pressure, as elegantly shown for the beaks of Galápagos finches (Grant and Grant 1989, 2008).

The initial behavioral change that pioneered the new niche *could* have resulted from random mutation and genetic drift, but all that is required for the Baldwin effect is that the ancestral thrasher be capable of modifying its behavior during an individual's life so as to spend more time on the ground than its parents or grandparents did. If this more terrestrial life succeeds, then after this initial response (through individual plasticity) selection pressures will have been altered. Genotypes randomly more disposed toward some degree of terrestriality *independent of learning* will subsequently be favored. Thus a genetically coded form of terrestrial behavior could emerge, and in parallel with or in consequence of it, anatomical and physiological changes such as longer legs and stronger bills. If behavior is the pacemaker of evolution, then any animal capable of learning (or otherwise modifying its phenotype) is a potential candidate for the Baldwin effect. This will almost always be a very slowly incremental process.

The term *phenocopy* is often used to describe a phenotype arrived at through environmental influence that mimics the result of genes that the phenocopy lacks. For example, familial hyperlipidemia in humans is the result of genetically coded alterations in cholesterol transport—a lipidemia genotype. But sufficient intake of cholesterol and saturated fats can in many people mimic key features of the phenotype of familial (genetic) hyperlipidemia, creating a phenocopy of the genotype. This can also occur in schizophrenia, which is often genetic in basis but can appear in phenocopies caused by viruses, cocaine or amphetamine overdoses, extreme deprivation during rearing, and other environmental processes.

Conversely, the Baldwin effect includes the emergence of a *genocopy,* a genetically determined adaptation—in behavior, physiology, or anatomy—that gradually arises in a population that has pioneered the change through individual plasticity. This is not the inheritance of an acquired characteristic; it is a completely conventional Darwinian process. Its only special aspect is that instead of beginning with an environmental change operating on randomly generated genetic variation in a basically passive population of organisms, the environmental change begins with individuals penetrating a novel environment as a result of plasticity during a single life course. The initial step is active, not passive; the organism's agency engenders selection pressure. In the next step the new environment thus penetrated operates on already existing, random genetic variation previously generated through mutation. This second step is simply and conventionally Darwinian.

However, there may be (also non-Lamarckian) ways for the first and second steps to be linked. In classic experiments on fruit flies, the British geneticist C. H. Waddington (1953), whose model of development we encountered in Chapter 6, also showed that certain environmental conditions can expose and exploit otherwise hidden genetic variation. Once the environment exposes the phenotype, selection can act on the underlying genes. For example, after heat-shock treatment (in which fly pupae spend four hours at 40°C), 40 percent of the emerging flies lack species-typical cross-veins on their wings. Interestingly, when these veinless flies are mated and their pupae are similarly heat-shocked, a higher percentage will be veinless at pupation. Repeating the cycle in subsequent generations increases the proportion. This is predictable, since we have selected for a facultative adaptation—the genetically determined trait of responding to heat shock during development with the disappearance of cross-veins.

What was new in Waddington's experiments was that after many repetitions of the cycle of selection, cross-veinless flies appeared *without* heat shock. The naive observer might mistake this for the inheritance of an acquired characteristic, but it too is perfectly Darwinian. Analyses of the genetics of cross-veinlessness, whether facultative or fixed, showed it to be multigenic. Heat shock in effect identified the 40 percent of flies that had *some* of the necessary genes in the wild condition. Repeated selection increased the proportion of the multiple relevant loci, and when a threshold was crossed in the proportion of those loci, the condition developed at normal temperatures. Waddington (1953) called this change *genetic assimilation,* while Mayr (1963) preferred the more precise (and more Darwinian) term *threshold selection.* Either way, it is a Darwinian process in which a facultative adaptation becomes innate, because the facultative adaptation has led the organism into a new environment with new selection pressures.

Note that this is different from the concept of *exaptation,* a term later introduced to refer to a preexisting unrelated characteristic that by chance becomes useful in a new environment (Gould and Vrba 1982). Genetic assimilation (threshold selection) is not random. Like the Baldwin effect (and in service to it), genetic assimilation should accelerate evolution by natural selection. Indeed, one might reasonably suggest that part of the selective advantage for individual plasticity and facultative adaptation generally is that it enables populations or lineages to test adaptations before evolving them in a full genetic commitment. There is no implication of progress here, at least not in any ultimate sense. Dinosaurs may well have availed themselves of genetic assimilation and the Baldwin effect repeatedly on their way to extinction. These are simply processes that may make the overall Darwinian dynamic more efficient in some—or, I suspect, many—local situations.

Another process that seems to skirt Lamarckian territory is epigenetics, not in Waddington's general sense but in the specific, molecular sense of DNA methylation and other effects that suppress or enhance genes on a short- or long-term basis (Keverne and Curley 2008). It is claimed that this burgeoning field requires a revolution in Darwin's theory (Jablonska and Lamb 2005). This claim has been subtly and effectively dealt with (Haig 2007), but not entirely disposed of. It has remarkable and widespread implications; DNA is epigenetically modified by drugs and toxins, but also by learning and even early experience. Some of the modifications are very durable, and some can be passed on to offspring. They allow maternally and paternally inherited genes to behave differently and may allow the two sexes to compete on epigenetic ground. In one remarkable system, a maternally imprinted gene is expressed in the rat mother's hypothalamus, activating milk letdown and maternal care, but also in the fetuses and placenta she is carrying, which indirectly has the same effect on her (Keverne and Curley 2008). Then, after their birth, the pups' suckling is promoted, further releasing maternal care. This may not be Lamarckian, but it is a new world of transgenerational effects, and we need to learn a lot more about it. Meanwhile, our best path to understanding how acquired behavior affects evolution remains the Baldwin effect.

Probably the best-studied instance of the Baldwin effect in a complex animal is the curious case of how the house finch (*Carpodacus mexicanus*) conquered America, studied over many years by Alexander Badyaev (2009). The house finch in 1850 was confined to the West Coast of North America, but a few were introduced here and there—Oahu, New York City (in a pet store)—and by the twenty-first century the house finch had spread throughout the contiguous states and the Hawaiian Islands. Huge amounts of variation were generated, including variations of two standard deviations in some body measurements in ten generations (about seventeen years). The finch arrived in northwestern Montana in the 1950s and had become a common breeding bird in Missoula by 1969.

This population experienced particularly strong selection on juveniles. In fact, the smaller but (in evolutionary terms) still rapid change in that population was shown to be due entirely to changes in the birth sequence of male and female offspring and the sensitivity of young males to environmental conditions during growth. Without this and other kinds of adaptability within individual lives, the population, instead of evolving rapidly, would have gone extinct. And this in turn depended on the adaptability of mothers. Due to a previously evolved facultative adaptation, female finches arriving in the northwest Montana cold experienced changes in prolactin, corticosterone, and androgen levels that adjusted the order in which they produced males and females, which can only be produced at the rate of one a day.

By producing males first and females later, the mothers exposed males to different environmental conditions during development, greatly increasing male variability. This in turn increased the rate of evolution among males and, through female choice, of the population. Thus the physiological adaptability of mothers placed their offspring in different situations such that their *developmental* adaptability ensured a more rapid rate of evolution and an avoidance of extinction. Genetic evolution is not involved until the last step, and it preserves the adaptability of both mothers and offspring so that they can live to evolve another day. Thus, in terms of geographic range and environmental variability conquered within a century, the house finch is the most successful of all extant birds.

The house finch is something, one might even say, like us. It is not difficult to come up with scenarios in which the Baldwin effect very likely operated in the long Darwinian history of our own species. Consider three examples.

First, in the evolution of the early vertebrate brain, the forebrain, the paired sense organs, and the whole front end of the head were added to the already existing midbrain, hindbrain, and spinal cord of the primitive chordates. Like all evolutionary transitions, this was achieved through changes in ontogeny. Many vertebrate brain changes appeared in the service of interpreting sensory information detected in the environment by newly evolved sense organs, especially for olfaction and light-detection by primitive retinas and a pineal eye. Thus information in the environment drove vertebrate brain evolution—not surprising since every organism's genome is a library of information about the environment encountered by its ancestors.

Compared with primitive chordates, the simplest vertebrates are more mobile and efficient predators. Natural selection for more effective predation (and evasion of predators) made vertebrates out of simpler chordates by altering neural ontogeny. One process involved was invasion, in which projections from the new sense organs penetrated the front end of the primitive chordate nervous system. These projections were probably mainly olfactory. If, as is likely, there was neural plasticity in the ancestral chordates, then within-life-course plasticity could have been pioneered in ecological niches that then produced parallel population-based changes in ontogeny through mutation and selection.

Second, in mammals the initial steps in the evolution of maternal care could also have been pioneered through plasticity within individual life cycles. Some parental care exists in many amphibians and reptiles (including some dinosaurs), but these would have been convergences with, not bases for, mammalian maternal care. Still, the enormously successful synapsid reptiles of 260 mya included the ancestor of mammals, and at some point in this line maternal care began to emerge. In any clutch or litter some offspring are weaker and more dependent than others. Under r-selected conditions, they are culled. But if early mammalian evolution increased the value of offspring through K-selection (or a slower lifestyle; Sibly and Brown 2007), then mothers who were able *facultatively* to adjust their further investment—with, say, increased egg-guarding—would have placed their heirs in a situation where genetic assimilation could produce genocopies. With further evolution of limbic system structures, genetically coded mammalian mothering could have evolved. In addition to parental instincts, the Baldwin effect would also have resulted in selective forces favoring new anatomical adaptations, the mammary organs, in the new niche pioneered by behavior. So a creature resembling in some ways the modern platypus, whose genome contains the same suite of milk-related proteins as ours (Warren, Hillier, et al. 2008), or the transitional fossil *Oligokyphus* may have emerged, with exquisite slowness, as the first mammalian parent.

Finally, consider the australopithecine ancestor of *Homo habilis,* whether *Australopithecus africanus, A. afarensis,* or another species. Given its apelike brain, it should have been capable of learning within the life course to make tools. In chimpanzees this occurs naturally with termite-fishing sticks, among many other instances (Wrangham, McGrew, et al. 1994; Whiten, Goodall, et al. 1999; Whiten and van Schaik 2007). The bonobo Kanzi acquired a refined stone-flake toolmaking process taught by human experts (Toth, Schick, et al. 1993; Savage-Rumbaugh 1994; Savage-Rumbaugh and Lewin 1994). Such observations highlight the capacity of an apelike brain to be modified in a way that enables skilled toolmaking.

So a new niche pioneered by inventive toolmakers would favor the evolution of brains more capable of mastering complex tasks. This differs from the first two examples, which proposed the genetic fixation of behavioral patterns previously achieved through individual plasticity; this would involve something more like the fixation of plasticity itself. But selection would also have favored greater manual skill, better eye-hand coordination, rapid sequential movement planning, and so on, involving parietal-frontal circuits and perhaps serving as an exaptation for later rudimentary speech (Stout, Toth, et al. 2008). As in the previous cases, brain modification that began as synaptogenesis and other individual changes leads indirectly—through genetic assimilation, genocopy, and the Baldwin effect—to brain expansion under genetic control, with consequences for the anatomy of the hand, jaw, teeth, and other structures (Wells 2005).

There are many other possible examples. Since juveniles are the most experimental members of a species, exhibiting much "useless" behavior (play), it may often be they who take the lead into a new niche (Bruner 1972; Pellegrini, Dupuis, et al. 2007). The first step in the evolution of language may indeed have been taken by a population of infants and mothers playfully babbling at each other (Falk 2004b). Alternatively, the fact that chimpanzees and bonobos can acquire some elements of language (Gardner, Gardner, et al. 1989; Rumbaugh and Washburn 2003) suggests that genetic assimilation may have been involved in the evolution of the human capacity for language, or at least the part of linguistic ability that is developmentally constant (Pinker and Bloom 1990). Finally, the human or protohuman invention of the infant-carrying sling may have paved the way for the last increase in human altriciality at birth (Zihlman 1981).

Note that I am not proposing a new process in evolution, but merely highlighting a neglected one. Against the assertion of the importance of random forces in evolution—which is quite valid in its place (Gould 1989)—we can counterpose a kind of directedness in Darwinian natural selection. This is not an *overall* directedness, and it is certainly not a push ever upward to something better, much less specifically human, but perhaps it constitutes a dim sightedness in our mainly blind excursion through evolution's darkness. At the leading edge of adaptation, experience during individual lives can establish a foothold for a new local dynamic of natural selection. Experience changes individuals, who enter new ecological niches, so that underlying genetic variation produces genocopies through genetic assimilation and/or the Baldwin effect; new features or capabilities of behavior, brain, and other aspects of functional morphology thus become innate. Experience, far from being wasted because of the independence of the genome from the rest of the organism, pioneers what may become fundamental genetic changes. Adaptability, a feature of phenotypes, leads to adaptation, a feature of genomes, populations, and species (West-Eberhard 2003, 2005).

Selection for Plasticity and Resilience

Given that plasticity can lead evolution, a problem remains: a plastic brain might also be vulnerable to insult, to jumping the track of tried-and-true development. Adaptation for finding food and water, self-protection, and mating are relatively fixed in most organisms because fixation enhanced survival. Perhaps selection produces a developing nervous system relatively impervious to insult, whose key behavioral functions emerge normally in spite of wide environmental fluctuations. To the extent that this is true, it can be explained by *canalization* in evolution and development (Chapter 3): each organism has a few possible developmental tracks, and departures are corrected (homeorhesis). But in evolutionary terms, canalization raises three problems.

First, while it would seem to be equally valuable for all organisms, species vary greatly in the environmental dependency of their developing nervous systems, and parts of nervous systems within the same or-

ganism are canalized to varying degrees, so factors other than a generalized necessity to protect vital functions must be operating.

Second, it is hard to see how selection could produce a nervous system with functions developing independently of features always present in the environment; selection would lose its cutting edge. The visual
system of kittens does not develop normally without an input of patterned light (Wiesel and Hubel 1965; Hirsch and Spinnelli 1970). Why,
we could ask, didn't natural selection give them a *really* good visual system, one that would grow reliably with or without patterned light? The
reason is that, since kittens have never been called upon to grow such a
system, there was no disadvantage for those that could not.

Third, the developing nervous systems of many organisms, including those with the least plastic behavioral functions, may exhibit highly
specific adaptive forms of environment-dependency. Salmon become
imprinted on the odor of the stream they are raised in, enabling return
after long migrations (Hasler 1960). Moths feed preferentially on the
species of leaf they lived on as caterpillars, despite metamorphosis
(Thorpe 1939), and a similar process appears to determine habitat selection in damselflies (Johnson 1966). As we have seen, the infantile attachment objects of many bird species determine mate choice in adulthood
(Hess 1974). This effect is modifiable, but at greater cost, and is readily
reestablished. Finally, in learning paradigms where rats are trained early
in life and the habits extinguished, those habits are much more easily
reinstated later in life than established anew in controls, despite the
same base-level responding (Campbell and Jaynes 1966).

This phenomenon may be widespread and should alert investigators
of early experience effects to the hazards of accepting the null hypothesis when we have studied only base-level adult responding. Even for
species whose behavioral repertoire is limited and relatively fixed, we
recognize *facultative* as well as obligatory adaptations (Chapter 3); they
evolve in a species that inhabits several different niches, and they require a switching mechanism to enable "choice" among several genomic
possibilities, the switch being responsive to cues in the environment.
If the choices vary along a continuum, they are known as a *range of reaction.*

Early experience triggers the switch in many animals, allowing,
among other things, within-species variation between fast and slow
lifestyles (Sibly and Brown 2007, 2009). Environments signal the infant

nervous system that certain facultative adaptations will be useful in adulthood and "tell" it to produce those rather than others. The mechanism by which rats stressed in infancy become larger and less fearful as adults (Levine and Mullins 1964; Denenberg 1964), probably via changes in maternal behavior after the separation and/or stress (Caldji, Tannenbaum, et al. 1998), could have resulted from such selection. Application of this concept to human development suggests that some early experiences usually considered detrimental may enhance later adaptations in certain environments (Belsky, Steinberg, et al. 1991; Chisholm 1993, 1999b; Chisholm, Burbank et al. 2005).

The view taken here is that plasticity, where present, is some combination of facultative adaptation, range of reaction, and generalized flexibility. It is not an off-road vehicle that can take behavior anywhere with impunity. Nor is it often quite as restricted in its travels as a railroad train—although that would still allow track switching. Rather, it is like a car that performs best on the expressway: leaving the high road is costly because it becomes slower, less efficient, and more likely to lose time getting lost. The odds are that most voyages in plasticity will keep to the major highways; still, some wanderers off the beaten path create new roads.

III

Socialization: *The Evolving Social Context of Ontogeny*

**wherein we discern the contributions of social life
to developing relationships and emotions**

That many kinds of mammals and birds . . . have a remarkably highly developed intelligence is no longer questioned by psychologists. But . . . many animals also show that they have the emotions of fear, happiness, caution, depression, and almost any other known human emotion. . . . Obviously these human characteristics could not have all originated by a big saltation when *Homo sapiens* was born. Naturally, we find the antecedents in many species of animals.

—Ernst Mayr, *What Evolution Is* (2001)

13

Paradigms in the Study of Socialization

We established in Part I that psychosocial growth evolved to serve survival and reproduction, both of which require social skills far too vital to be left solely to learning. But this evolved social and emotional equipment differs among species, in part because the social context evolves too. Even within a species, populations vary in social organization and dynamics and thus in the context of psychosocial growth, yet generalizations are often possible. For example, most reptiles take little or no care of their young after hatching or birth, but all mammals do; lemurs tend to have more linear dominance hierarchies, whereas monkeys have more coalitions; orangutans are more solitary, while other apes are gregarious; and humans invariably provision their young with food after weaning, while no other primates do so. Such generalizations enable us to sketch the evolution of social dynamics, and in particular the evolution of the social context of ontogeny.

But if the main processes of psychosocial adaptation are encoded in the genes and unfold in a maturational plan, what role remains for socialization? By definition, socialization is the process of bringing infants and juveniles "up" into functional social roles, *beyond* what maturation provides, in relation to the species-specific context and even the local social context. As we saw in Transition 2, the biological options range from rather fixed growth trajectories through facultative adaptations developed over a few canalized pathways to widely and subtly varying adjustments based on more generalized plasticity.

Field studies have characterized the life course, population dynamics, and social context; laboratory research elucidates socialization processes. Although it is not always easy to move between these levels, integration is possible. For analytic purposes we separate *socialization* (Part III) from *enculturation* (Part IV), although they form a continuum.

For our purposes socialization is widely shared by humans and other social organisms. Not all enculturation processes are unique to humans, but most are, and some others are shared with only a few species and in very limited ways.

Part III focuses on the many aspects of socialization that we share with other species. It discusses historical change within the human species since the hunting-and-gathering era, but the processes involved in this change are *emphatically* not those of our evolution from nonhuman ancestors. Cross-cultural and historical variation in human social arrangements results not from genetic change but from changes in ecological context, demography, and technology, as well as from disease, war, colonization, culture contact, and other historical processes. These variations in space and time within the human species, although they are more complex, are somewhat analogous to the differing social dynamics of chimpanzees in forest and savanna environments, but they are not at all analogous to the differing genetically based adaptations of chimpanzees and bonobos.

As with the subject matter of Part I, the evolution of ontogeny, and Part II, psychosocial growth itself, the subject matter of Part III, socialization, has been approached in several historically separate spheres, each with its mode of inquiry. The first two sections of Chapter 2 ("Neo-Darwinian Theory—The Adaptationist Paradigm" and "Life History Theory") are also essential to Part III. Other spheres are the subject of this chapter: (1) learning; (2) experience effects; (3) ethology, sociobiology, and field primatology; (4) ethnology and quantitative cross-cultural research; and (5) historiography and historical demography. The first two rest on laboratory research, while the third, along with paleontology and molecular evolution, enables us to model the biological evolution of social structures leading up to that of human hunter-gatherers. The last two, combined with archaeological evidence, enable us to model the nongenetic changes since the hunting-gathering era and to suggest how the social context of ontogeny has changed so rapidly that there may now be some genotype-environment discordance.

Laws of Learning

Despite their neglect in present-day psychology, laws of learning are well established (Domjan 1996, 1998; Hergenhahn and Olson 1996). More than a century of laboratory experiments in various species have established both the laws and the limits of their scope. New ways of looking at cognition in children offer many insights that learning psychology missed, but to ignore or belittle its achievements is not helpful. To some extent the theory's oversights were inevitable, given the exaggerated claims made by leading proponents (Skinner 1938), but the attempt to completely replace one successful paradigm (social learning) with another (social cognition) instead of attempting to integrate them is to give up a chance for more comprehensive models. Lawful learning processes include:

1. *Habituation.* A naturally occurring response wanes with repetition of the stimulus or becomes focused on a few appropriate stimuli instead of the many that initially evoked it. A young frog initially darts its tongue at any small dark object crossing its visual field but comes to ignore fragments of falling leaves or specks of soot in favor of edible insects. Human newborns habituate to repeated mildly noxious stimuli, and individual differences are of interest (Schanberg and Field 1987; Lundqvist and Sabel 2000).

2. *Associative learning.* Although this fundamental process can apply to any form of learning, it is most useful in describing perceptual learning. A set of stimuli is repeatedly presented, and the organism gradually represents them as a pattern in the brain (a schema in cognitive terms), as when the human infant mentally encodes its mother's face as a coherent visual pattern linked to a recognizable voice.

3. *Classical conditioning.* An inborn reflex such as salivation in response to food is extended when the ringing of a bell is repeatedly paired with the food's arrival, until salivation is caused by the bell alone (Pavlov 1927). It explains how a hungry infant comes to be soothed by its mother's voice or why the mother may wet her blouse with milk when hearing a baby cry. This process is close to the biological foundations of learning (Domjan 2005).

4. *Operant conditioning.* This begins not with a stimulus or reflex but as a naturally occurring act—an operant (Skinner 1938). A rat in an apparatus will walk around, sniff, occasionally rear up, or press a button or lever. Any operant can be changed in frequency by reinforcers (rewards) and punishments, and much is known about how various

schedules of reinforcement affect the rate and durability of learning (Domjan 1998). Some of these findings (more consistent reinforcement speeds learning) seem obvious; some (less consistent reinforcement produces a more lasting habit) do not (Skinner 1938).

Treatments and educational methods for retarded and autistic children rely on these laws for their success, as do behavioral therapies for various disorders of children and adults, including phobias for flying (Klein 1999) and spiders (Mineka, Mystkowski, et al. 1999), blushing (Mulkens, Boegels, et al. 1999), and aggression in mentally retarded children (Mineka, Mystkowski, et al. 1999). These laws also help explain normal behavioral change as children are socialized (Whiting 1941; Bandura and Walters 1963; Galef and Heyes 2004; Laland 2004). Habituation helps explain the infant's gradual cessation of smiling for a parent who never responds. Simple associative learning helps explain an infant's linking the mother's face, voice, and odor. Classical conditioning is part of what enables a toddler to use the toilet reliably. Operant conditioning is at least part of why children perform in school or do their chores.

Other learning phenomena are more subtle, and some play crucial roles in the biology of the emotions. In classic experiments, two rats in adjacent chambers are shocked simultaneously and identically, but one sees a light go on before the shock comes, and at this signal can turn the shock off for both rats (Weiss 1972; Coco and Weiss 2005). Both are shocked equally, but the rat in control learns avoidance, while the other learns helplessness—and develops more stomach ulcers, loses more weight, has higher serum glucocorticoid levels, and is more fearful afterward. Cells in the brain stem locus ceruleus undergo norepinephrine depletion, and they fire more rapidly, a pattern associated with depression (Weiss and Kilts 1995). Elsewhere in the brain there is a serotonin deficit, and in the mesolimbic reward circuits a depletion of dopamine blocks normal reward. The same circuits are more active in the rat that is in control (Coco and Weiss 2005).

In addition, there are forms of special learning—such as the "sauce béarnaise syndrome," the tendency to avoid a food after only one or two experiences with nausea even hours after eating it (Seligman and Hager 1972)—that are mode-specific; this type of learning, sometimes called the Garcia effect, is exemplified in, say, differences in the rate of learning to associate a sound with nausea as opposed to a taste (Garcia and

Koelling 1966; Garcia, Quick, et al. 1984; Garcia, Brett, et al. 1989). It seems evident that evolution should produce such patterns and that the brain would be wired for them, but they were certainly not predicted by standard learning theory. Other examples are imprinting processes—a term that includes the baby chick's rapid onset of following (Chapter 8), and a mother goat's learning the identity of her newborn kid within minutes, by smell (Klopfer and Klopfer 1968). These experiences differ greatly from what happens to a rat or pigeon in a Skinner box, but they qualify as learning.

Early Experience Effects and the Sensitive Period Question

Aside from the general or special laws of learning, certain interventions have stronger and more lasting effects, for good or ill, if they occur earlier in life. We cannot do most experiments on children, but animal models show how experience alters the growing brain, and they provide support for the idea of sensitive periods in limited domains, although they also prove the power of later intervention.

1. If one eye of a developing rhesus monkey is closed for a week before age six months, it will permanently have little or no depth vision (Hubel, Wiesel, et al. 1977; LeVay, Wiesel, et al. 1980; Hubel 1987). Depth vision depends on binocularly responsive cells in the visual cortex that normally respond to light stimulation in either eye (Parker, Cumming, et al. 2000); these cells receive inputs from both eyes during the first six months of postnatal life, when axons from the two eyes are in active competition for synaptic sites on those cells. Complete visual deprivation of one eye for even a few days during this sensitive period allows input from the other eye to take over most sites on the binocularly responsive cells.

2. Rhesus monkeys raised in social isolation for the first six months of life almost inevitably grow up to behave abnormally—they are socially withdrawn, sexually inept, and show inappropriate fear and anger (Harlow, Dodsworth, et al. 1965; Harlow 1971). Females may be impregnated and give birth, but they are often neglectful and abusive mothers (Champoux, Byrne, et al. 1992). Initial attempts to treat this syndrome by restoring social contact failed, but placing a previously isolated six- to twelve-month-old monkey with a younger monkey worked, provided the placement occurred early (Suomi, Harlow, et al. 1972; Novak and

Harlow 1975). Perhaps both the deprivation and the treatment must oc-cur in an early sensitive period. Less extreme forms of deprivation—partial isolation, rearing with peers only, rearing with the mother only, and isolation for shorter periods—also have lasting effects (Suomi 1997), especially against certain genetic backgrounds (Suomi 2003; Barr, New-man, et al. 2004). Even the relatively mild intervention of two six-day separations from the mother during early life has measurable effects on the monkey's emotional behavior at age two years (Hinde and Spencer-Booth 1971).

There are durable changes in stress hormone regulation in these ani-mals, and they have lasting imbalances in the norepinephrine (NE) (Kraemer, Ebert, et al. 1989; Kraemer 1992) and serotonin (5-HT) (Krae-mer and Clarke 1990) neurotransmitter systems as they grow; their greater vulnerability to depression, aggression, and self-injury appears to depend on these imbalances (Kraemer, Schmidt, et al. 1997). Other studies find lasting changes in the 5-HT metabolite 5-HIAA in cerebro-spinal fluid, the function of the 5-HT transporter gene, and brain glu-cose metabolism—all parallel to behavioral changes like increased ag-gression and alcohol consumption (Doudet, Hommer, et al. 1995; Higley, Suomi, et al. 1996; Heinz, Higley, et al. 1998).

The longer and more severe forms of early social deprivation also al-ter dendritic branching in the cerebellar cortex (Floeter and Greenough 1979; Pysh and Weiss 1979), although it is not clear that this is really due to emotional deprivation. The cerebellum was long seen as an organ of movement regulation, but it clearly plays a general role in learning (Ray-mond, Lisberger, et al. 1996). Also, rearing monkeys under social dep-rivation changes the corpus callosum in ways that are correlated with mental ability (Sánchez, Aguado, et al. 1998). Monkey social deprivation models are discussed in detail in Chapter 14.

3. Although this is not an effect restricted to early life, environmental enrichment or impoverishment, social or inanimate, changes the anat-omy of the rat cerebral cortex (Diamond, Krech, et al. 1964; Rosenzweig, Bennett, et al. 1972; Globus, Rosenzweig, et al. 1973; Diamond, Johnson, et al. 1985; Diamond 1991), especially the visual cortex, but paradoxi-cally even blind rats reared in enriched conditions show visual cor-tex changes, suggesting that parts of it may serve more global mental functions. Changes occur in cortical weight and thickness, synapse size and number, dendritic branching complexity and spine density (espe-

cially on pyramidal cells), and enzyme activity relating to acetylcholine, a widespread neurotransmitter in the forebrain. In mice studied with transcranial two-photon microscopy, learning a motor task (but not exercise alone) remodels the dendritic spines, both adding and subtracting, whether the learning occurs in the first month of life or in adulthood (Yang, Pan, et al. 2009). The remodeled spines enter into new neural circuits, and a large fraction of these last a lifetime, forming the basis of lifelong memories even from a time when the mice are very immature, and when synapse formation and pruning are developmentally dynamic.

4. Some related phenomena are probably widely characteristic of vertebrates, as shown, for example, by studies of the jewelfish (Coss and Globus 1978). Rearing in groups as opposed to social isolation changes the number, distribution, and shape of spines on the distal dendritic branches of pyramidal neurons in the tectum. These changes are traced to the stimulation that comes from social experience; repeated stimulation of the circuit by social experience makes the spine shorter and thicker, affording less resistance to incoming electrical signals (Valverde 1967). No doubt older structures of our own brains also change with social experience.

5. As we saw in Chapter 8, imprinting of chicks on the mother hen during the first few days of life rapidly and permanently changes not only behavior but also the pattern of protein produced in the forebrain roof, the neuronal structure in the hyperstriatum, a forebrain region; neuronal activity and dendritic spine density in other brain regions are also changed by imprinting. Imprinting exhibits a highly sensitive period, but it can be extended by neurochemical manipulations (Parsons and Rogers 2000). Thus the circuitry of attachment in birds is highly malleable with certain experiences, and the permanent behavioral impact of imprinting—not only the chick's following the mother throughout infancy but its restricted mate choice in adulthood—may be due to these brain changes.

Ethology, Field Primatology, and Sociobiology

Natural history has been a serious human endeavor at least since the dawn of culture, and hunter-gatherers have large and accurate bodies of knowledge about animal behavior, on which their livelihood partly depends (Lévi-Strauss 1968; Blurton Jones and Konner 1976). Since Aristotle, observations, systematic and otherwise, about animals in the wild have been published, often marred by anthropomorphizing, overdramatization, and anecdote. By the late eighteenth century some observations were scientific, and Darwin both engaged in observations and relied on those of others for his great works on sexual selection and the expression of the emotions, the foundations of behavioral biology (Darwin 1871/1981, 1998). In the twentieth century a new field emerged: devoted to the study of animal behavior, mainly under natural conditions, in comparative and evolutionary perspective, ethology was so different from American comparative psychology (as noted in Chapter 1) that there was almost no overlap in the methods and discoveries (Beach 1950; Lockhard 1971).

Ethology was driven by five essential concepts (Tinbergen 1963; Eibl-Eibesfeldt 1975; Lorenz 1981):

1. Behavior, like anatomy and physiology, is subject to evolution by natural selection and can best be understood as adaptation.

2. Each organism has a species-specific repertoire of behavior that is subject to within-species variability but nevertheless is partly innate (*Erbkoordinationen*, or legacy patterns, known as "fixed action patterns").

3. Each organism, owing to its unique perceptual apparatus, lives in a species-specific perceptual world, or stimulus envelope (*Umwelt*), and has at least some unique, instinctual responses to stimuli (innate releasing mechanisms).

4. Animal behavior must be studied first and foremost under natural conditions, in the species' environment of evolutionary adaptedness.

5. With this basic knowledge, experiments can test hypotheses about adaptation, innate versus learned in development, specifics of the stimulus envelope, and so on.

Rigorous standards of observation, experiment, and analysis steadily expanded the range of species under careful study. Ethologists did studies that became models for the discipline (Tinbergen 1960; von Frisch 1967; Lorenz 1970), and their work culminated in a Nobel Prize for Karl von Frisch, Konrad Lorenz, and Niko Tinbergen in 1972. The generation of scientists trained by them, and then by their students, transformed our understanding of behavior and ecology with more classic field studies (Geist 1971; Schaller 1972; Leyhausen 1979; Clutton-Brock, Guinness, et al. 1982; Caro 1994) and laboratory studies (Bateson 1966; Klopfer and Klopfer 1968; Hinde and Spencer-Booth 1971; Hinde and Davies 1972a, 1972b).

Meanwhile, growing interest in field biology and ecology was producing a convergence between ethologists and biological anthropologists studying primate behavior in the wild (Itani 1958; Schaller 1963; Carpenter 1965; DeVore 1965a, 1965b; Jolly 1966; Goodall 1968; Hrdy 1977). In addition to the general aims of ethology, these studies were strongly motivated by the attempt to shed light on the evolution of human behavior. Because of the greater complexity and plasticity of monkey and ape behavior and social organization, methods and concepts were somewhat different, with less explicit interest in innate behavior.

Women played a leading role in primate field studies, conducting now-classic research on species including chimpanzees (Goodall 1986), gorillas (Fossey 1983, 1984), orangutans (Galdikas 1985b), baboons (Altmann 1979; Smuts 1985), langurs (Hrdy 1977), and lemurs (Jolly 1966; Richard 1978), although men have also carried out major field studies (DeVore 1965a, 1965b; Kummer 1968; Kurland 1977; Kano 1992; Wrangham 1993; Mitani 1994; Sapolsky 1995). For our purposes, it is crucial that some primate field research has been focused on infants, juveniles, adolescents, and the social contexts in which they grow (DeVore 1963; Goodall 1967; Altmann 1979; Taub and Mehlman 1991; Fairbanks and Pereira 2002; Kappeler, Pereira, et al. 2003).

As we will see, early hopes for a simple phylogenetic analysis of primate social organization failed, because within-species as well as between-species variation showed that ecological factors were as important as species-specific behavior in determining group size and composition, and even in inhibiting or facilitating communication (Strier 2006). But if the larger social context of ontogeny proved more variable and puzzling than it had seemed at first, certain features of ontogeny and of the mother-offspring relationship did prove very characteristic of (at least) the Old World monkeys and apes, and we will consider these in some detail.

As we saw in Chapter 2, the new and powerful theoretical approaches of sociobiology and evolutionary psychology exerted increasing influence on ethology and field primatology; their history and the controversies surrounding them have been well summarized (Segerstråle 2000; Alcock 2001). An ethologist or primatologist who went to sleep in 1965 and woke up today would find both fields completely transformed by these ideas, which have also strongly influenced psychology and anthropology. For our purposes, the domains of special importance include possible evolved neurocognitive modules for social behavior (Barkow, Cosmides, et al. 1992; Cosmides and Tooby 2000; Tooby and Cosmides 2000) and the application of Darwinian thinking to social psychology (Buss 1989; Simpson and Kenrick 1997) and developmental psychology (Bjorklund and Pellegrini 2002; Bjorklund and Smith 2003).

Ethnology and Quantitative Cross-Cultural Comparison

Ethnology of a sort began in ancient times, when Herodotus and Thucydides in their otherwise sober histories repeated fantastic tales of bizarre humanoid savages beyond the civilized world. Not until the post-Renaissance European age of exploration did somewhat reliable reports begin to be published, mainly by missionaries. Serious ethnology began in the nineteenth century but did not become really professionalized until the early twentieth century (Harris 1968). As with field primatology, some of the greatest ethnographers have been women (Mead 1928/1961, 1930; Benedict 1934; DuBois 1944; Whiting 1963), but men were also among the pioneers (Boas 1889, 1940b; Kroeber 1901, 1909; Malinowski 1935; Evans-Pritchard 1940; Whiting 1941; Firth 1957). Early on, anthropologists wanted to study children (Mead 1930; Boas 1940a), whom they saw as the prime vehicles of cultural transmission. They went far afield to test what they saw as naive psychological views of developmental universals (Freud 1905; Werner 1948) and cast doubt on their universality (Mead 1932; Malinowski 1927/1961).

Franz Boas, one of the founders of American anthropology, had studied physical growth and encouraged his students (Margaret Mead, Ruth Benedict, Cora DuBois, and others) to describe childhood and child training in their ethnographies. Some anthropologists viewed the psychoanalytic study of childhood as the key to understanding culture and collaborated with psychoanalysts (Kardiner and Linton 1939, 1945; DuBois 1944). Others were skeptical but, agreeing that a focus on childhood would be an aid to the understanding of both culture and individual psychology, invoked other theories and more rigorous methods.

These alternative accounts included detailed and empirically oriented ethnographic description (Whiting 1941, 1963; Whiting, Child, et al. 1966), quantitative cross-cultural studies using large, carefully chosen, and representative samples of the world's cultures (Barry, Bacon, et al. 1957; Barry and Paxson 1971; Whiting and Whiting 1975b; Schlegel and Barry 1979, 1991; Munroe, Munroe, et al. 1981; Whiting 1981, 1990)—some of which tested psychoanalytic predictions (Whiting and Child 1953; Burton and Whiting 1961; Whiting 1990)—and finally, systematic, quantitative, consistent coding of infant and child behavior in ethnographic settings (Whiting and Whiting 1975a; Munroe and Munroe 1984; Whiting and Edwards 1988; Munroe, Munroe, et al. 1997; Munroe, Hulefeld, et al. 2000; Munroe 2004). Some of these studies converged with the methods of ethology (Konner 1972, 1977b; Blurton Jones and Konner 1973) and developmental psychology (Caudill and Plath 1966; Klein, Lasky, et al. 1977; LeVine, Miller, et al. 1988; LeVine, Dixon, et al. 1994). We have touched on some of their findings and will do so again.

Historiography and Historical Demography

Historical research is also cross-cultural. There are limits to what we can infer about childhood from archaeological excavations—although we learn about childhood disease, injury, and burials from fossils, and there is an archaeology of toys—but from the first emergence of written documents we have evidence of how people thought about children and their development.

Two once-influential notions about that history are now discredited. Phillipe Ariès (1962), a French social historian who helped found the subdiscipline, claimed that before the Renaissance it was not acknowledged that children were very different from adults. One line of evidence was pre-Renaissance paintings, where children are generally drawn as smaller in size but with adult proportions. According to this view, infants were recognized as different, but after early childhood children were seen as miniature adults. Anthropologists found Ariès's larger claim implausible because, although they could not directly study the families of medieval Europe, varied cultures throughout the world—including peasant societies like those of the European past—

revealed many folk views of childhood, but lack of awareness of the profound differences between children's and adults' capabilities and needs was not among them. Also skeptical of Ariès's claims were serious historians of medieval childhood, who have provided wonderfully rich accounts of real children in real families that decisively show that children were treated as children (Orme 2001), their lives complete with dolls, toys, schools, primers, rhymes and songs for children, and the eighty games depicted in Pieter Brueghel the Elder's 1560 painting *Kinderspiele*, most of which were very old at the time of the painting. Ariès's observations had more to do with representational tradition than concepts of childhood.

In the 1970s a second model dominated, that of Lloyd deMause, editor of *The History of Childhood Quarterly*. He held that in all past societies children were brutalized and that modern history has seen a steady improvement toward decent child care: "The history of childhood is a nightmare from which we have only recently begun to awaken. The further back in history one goes, the lower the level of child care, and the more likely children are to be killed, abandoned, beaten, terrorized, and sexually abused" (deMause 1974:1). But field research on parents, children, and the family in hundreds of cultures on every inhabited continent had produced only a few cases of child-rearing patterns that resembled the parenting that deMause and others ascribed to all past cultures (Lancy 2008). Ethnologists of childhood had described the affection and care lavished on children in many traditional cultures and the accommodations made in all known cultures to children's limited and developing capacities.

To be sure, life for children was not idyllic in most societies on record (Lancy 2008). Few cultures have been as child-centered as ours, and in many adult authority was absolute. In the pastoral and farming cultures that dominate the ethnological record, children were assigned chores from an early age and expected to learn and grow into more difficult ones—much as we expect with schoolwork. Physical punishment was common in many cultures, and abuse can occur anywhere. Painful initiation rituals occurred widely, and in strongly male-dominated cultures girls were severely restricted. But the world's cultures have always valued children and protected them from a dangerous world so that they could grow up and carry forward their parents' genetic and cultural legacies (Whiting 1963; LeVine, Dixon, et al. 1994). In evolutionary terms, it could not have been otherwise. The "nightmare" conjured by deMause did not exist.

Beginning in the 1980s, historians offered a more balanced account that demonstrated continuities as well as changes and gave attention to geographic and social class variations (Pollock 1984, 1987; Hawes and Hiner 1985, 1991; Hiner and Hawes 1985; Heywood 2001; Orme 2001; Mintz 2004). Without probing subtleties still in dispute, we can find consensus on key generalizations about a centuries-long, ongoing process that affects populations throughout the world. Some of these changes are tied to the demographic transition, which began at different times and proceeded at different rates in different populations; starting half a millennium ago, the transition is largely completed in some places while proceeding rapidly from a much more recent start in others. There have also been exceptions and temporary reversals; demographic transition is a process of cultural, not biological, evolution. Nevertheless, these broad trends have affected most populations and can be summarized as the impact of modernization on childhood:

1. *Mortality in infants, children, and mothers has declined steadily.* This change is not mainly due to improved medical care or vaccination but to improved nutrition and sanitation and decreased infectious disease (McKeown 1995). Farmers, grocers, plumbers, and garbage collectors have been more important agents of mortality reduction than doctors.

2. One to three generations after these decreasing death rates, *birth rates have declined in most populations*, even to some extent in African and Muslim populations, where they remain very high. World population is expected to stabilize at about 9 billion at midcentury, but that will suffice to destabilize ecologies globally and in many localities.

3. *Demographic transitions increase, then decrease, family size.* Societies as disparate as China, where family size was reduced coercively, and Italy or Thailand, where it declined naturally, are eliminating the traditional extended family; siblings, aunts, uncles, and cousins are rare in the lives of many children, intensifying some consequences of increased mobility with modernization. Professional and institutional child care, including school, is replacing the extended family.

4. *Decreases in infectious and nutrient deficiency diseases in childhood, along with increased nutritional quality and quantity, explain the first phase of the transition.* Yet owing to geographic disjunctions, hundreds of millions of children remain undernourished and/or infected with malaria, gastroenteritis, lung infections, and AIDS. Paradoxically, rates of childhood obesity and type II diabetes are simultaneously increasing throughout the world.

5. *Children are larger at a given age and grow faster, while the age at sexual maturity has decreased, mostly owing to health and nutritional factors.* This secular trend may have leveled off in the most modernized populations, with an average age at menarche of twelve or thirteen, but it continues in the developing world.

6. *Children's chores and labor increased during the change from hunting and gathering to agriculture, but decreased in the late-industrial and post-industrial world, while schooling and literacy, also a kind of work, have greatly increased.*

7. *With gradual urbanization of the human species for ten thousand years, a steadily increasing proportion of children have grown up in cities.* This continuous, millennial, worldwide social trend helps explain the replacement of child labor with schooling. Since physical and mental health are generally better in urban than rural populations, urbanization also helps explain declining mortality and secular trends in growth.

8. *Mass electronic communication is reaching an increasing proportion of children.* Throughout the developing world poverty coexists with television aerials and cell phones. Mass communication homogenizes some aspects of culture while raising material expectations that multiply the impact of population growth.

All these trends apply to the transformation of settled, agricultural, rural societies into mobile, industrial, urban ones; they are much of what we mean by modernization. But the deeper history of our species reaches down through scores of millennia into the hunting-and-gathering past. The transition from this basic human way of life to the peasant-based societies that dominate historiography was as momentous as the industrial revolution, and we will consider in the remainder of Parts III and IV the implications of this *first* demographic transition.

14

Early Social Experience

Psychiatrists and psychologists routinely assume that early experience figures in mental illness, but direct evidence for such effects in humans is surprisingly hard to obtain. Still, there have been at least five decades of active laboratory research in animals on the acute and late effects of early environment on behavior and physiology. This research has dealt extensively with representatives of at least four orders of mammals and a number of bird species, and some investigators speak of "programming life histories."

In higher primates, disorders of psychological withdrawal and depression can be created by early isolation and separation (Harlow, Dodsworth, et al. 1965; Akiskal and McKinney 1973; Gilmer and McKinney 2003). Since early social deprivation produces chronic, severely aberrant behavior in chimpanzee adults (Davenport 1979), such effects probably figure in at least some human mental disorders, despite the fact that genes are usually involved. In the realm of normal but variable adaptation, most social species of vertebrates achieve adult functions in part through experience during development, which influences behavior through long-term changes in neural and neuroendocrine systems. This chapter briefly reviews the evidence for that claim.

Early Handling, Stress, and Stimulation

By the midtwentieth century it was clear that daily stimulation or stress for the first three weeks of life permanently alters rat physiology and behavior (Denenberg 1964). Initially the attempt was to make a positive intervention, by giving rat pups several minutes of human handling and stroking, but other interventions soon suggested that the experience was probably stressful. Yet stimulated rats grow faster and as adults are longer, heavier, and less fearful in an open-field test (defecating less and exploring more); they have greater learning ability and survive longer under starvation, drowning, injected tumors, and other stresses.

It was suggested early on that the hippocampus and hypothalamic-pituitary-adrenal (HPA) axis might explain the effects (Levine and Mullins 1964); stimulated rats' adrenocortical function matured faster, and in adulthood their adrenal cortex released less glucocorticoid (GC—corticosterone in rats, cortisol in humans) during stress and weighed less afterward. But they responded *more* strongly to severe stresses like electric shock, suggesting that they were conserving adrenocortical resources (Levine, Haltmeyer, et al. 1967). They also had alterations in corticotropin-releasing factor (CRF), adrenocorticotropic hormone (ACTH) (Zarrow, Campbell, et al. 1972; Campbell, Zarrow, et al. 1973), and cerebral lateralization (Berrebi, Fitch, et al. 1988). Mice responded similarly when reared with rat "aunts" instead of with mouse mothers or foster mothers (Denenberg, Rosenberg, et al. 1969), suggesting a possible maternal-behavior basis for the effects.

We now know this is the case. Early stress (or separation, which always accompanies it) changes the mother's behavior, and that explains the long-term effects (Plotsky and Meaney 1993; Liu, Diorio, et al. 1997; Denenberg 1999). Rat mothers double their normal rate of licking and grooming after brief separations; also, there is natural variation, and pups whose mothers lick and groom more have the same long-term effects without being separated. Results with mice and rabbits are more equivocal regarding the maternal explanation of separation and stress effects (Denenberg 1999); even in rats the long-term physiological effects of infant stress differ from those of maternal separation (Meaney, Diorio, et al. 1996), and long (three-hour) separations result in decreased, not increased, maternal behavior (Boccia and Pedersen 2001), with different consequences for the HPA axis and hippocampal development (Huot, Plotsky, et al. 2002). But one way or another, altered maternal behavior accounts for many of the effects.

Lower stress-responsiveness could be due to reduced hippocampal GC receptor density, which causes greater sensitivity to feedback and so to HPA suppression by the hippocampus (Meaney, Aitken, et al. 1988; Sapolsky 1997). However, rat pups removed from their mothers for three hours repeatedly have more CRF gene expression and more CRF receptors in several brain regions as adults (Plotsky, Owens, et al. 1998;

Meaney 2001), an effect reversible by the serotonin reuptake blockers used to treat depression (Arborelius, Owens, et al. 1999). Finally, GC receptor gene expression, DNA methylation, and promoter binding differ in offspring according to how much licking, grooming, and arched-back nursing their mothers do (Weaver, Cervoni, et al. 2004). Maternal buffering of HPA activity also occurs in humans (Gunnar 2003), and the amount of nonmaternal care predicts aspects of children's daily GC cycle (Watamura, Donzella, et al. 2003), but these effects are not as large as those routinely seen in rodents.

Seymour Levine (2001), one of the founders of this line of research, summarized decades of work by noting that the period from four to fourteen days of age in rats is one of suppressed stress-responsiveness and that the mechanism of this suppression is regulation of the HPA axis by normal maternal behavior, with tactile stimulation, feeding, and passive contact playing somewhat different roles. But the pup's HPA system is functional, and after twenty-four hours of maternal separation it shows a robust ACTH and GC response to mild stress, along with *c-fos* messenger RNA (mRNA)—immediate early gene expression—in the paraventricular nucleus of the hypothalamus. The latter activation is consistent with vasopressin replacing CRF as the main regulator of ACTH in deprived pups. As we will see later in the chapter, early perturbations in this system have also proved important in human children (Heim, Nemeroff, et al. 2001; Heim, Plotsky, et al. 2004).

A new path has opened in the study of lasting effects of maternal behavior in the form of evidence that maternal licking and grooming (LG) widely influences the offspring's level of DNA methylation, affecting the genes for an estrogen receptor in the hypothalamus and a glucocorticoid receptor in the hippocampus, as well as a regulatory gene turned on by nerve growth factor (Champagne and Curley 2009). These changes persist into adulthood, influence behavior, including maternal behavior, for the next generation, and may be passed on to offspring genetically. This demonstration that maternal care can reach into the genome of the young has implications we have barely begun to understand.

Postweaning Isolation and Crowding

In addition to these very early stress and stimulation studies, social experience during development in rodents has been modified by varying litter size, postweaning grouping, age of weaning (with or without nutritional controls), and total maternal deprivation. Early studies showed that total maternal deprivation with incubator rearing produces higher adult basal and stress-induced GC in rats (Thoman, Levine, et al. 1967:75) and that male rats weaned early are more susceptible to conflict- or restraint-induced gastric ulcers compared to controls left with a non-lactating mother (Ader 1962).

Subsequent studies of similar and other manipulations have consistently showed a wide variety of effects consistent with lasting changes in behavior and its underlying neural and neuroendocrine functions (Sánchez, Ladd, et al. 2001). Mice weaned at fourteen instead of twenty-one days of age have a persistent increase in anxiety-like and aggressive behavior as adults, along with higher HPA axis activity in response to novelty (Kikusui and Mori 2009). Males weaned early have precocious myelination in the amygdala, decreased brain-derived neurotrophic factor (BDNF) protein levels in the hippocampus and prefrontal cortex, and other hippocampal changes.

Early social deprivation also damages male sexual behavior (Gerall, Ward, et al. 1967; Spevak, Quadagno, et al. 1973). Even reducing a rat mother's anogenital licking makes male pups less likely to copulate as adults (Moore 1992), probably because there are fewer neurons in the sexually dimorphic nucleus of the lower spinal cord (Moore, Dou, et al. 1992). Preweaning isolation (reducing a litter to a single pup) added to the effects of postweaning isolation alone: the earlier isolates were more reactive to handling, preferred to be near a single adult rat, and showed less heart rate increase in response to an intruder (Leigh and Hofer 1975).

But probably the most frequently used paradigm has continued to be postweaning social isolation. Three weeks of isolation after weaning (from three to six weeks of age) alter the level, turnover, production, utilization, removal, and enzyme activities of neurotransmitters (Essman 1968; Valzelli 1973; Tizabi, Massari, et al. 1980; Miachon, Rochet, et al. 1993). In mice the behavioral effects include prompt fighting in paired males, and this isolation-induced aggression has served for decades as a testing model for psychotropic drugs (Burkhalter and Balster 1979; Wilmot, Fico, et al. 1989; Haller, Makara, et al. 1996).

Rats reared in isolation have altered monoamine neurotransmitters, serum GC, and ACTH (Miachon, Rochet, et al. 1993); they also have greater amphetamine (AMP) sensitivity in the nucleus accumbens and are more sensitive to reward (Jones, Marsden, et al. 1990). Rats show be-havioral changes in response to postweaning isolation (Byrd and Briner 1999) or earlier brief maternal separation (McIntosh, Anisman, et al. 1999). Anatomical changes also occur: ten months of isolation after early weaning in rats increases the dendritic spines in the medial preoptic area of the hypothalamus (Sánchez-Toscano, Sánchez, et al. 1991).

Far away phylogenetically, jewelfish respond to exposure to social stimulation with a shortening of the stems of the spines on dendrites, an effect that probably makes it easier to bring about action potentials in the postsynaptic neurons (Coss and Globus 1978). In rats, environ-mental enrichment after weaning enhances rat brain structure and function (Diamond 1991; Bennett, Diamond, et al. 1996), while social iso-lation after weaning reduces the weight and thickness of the neocortex, the density of dendritic branching, and cholinesterase activity (Rosen-sweig 1966; Krech, Rosensweig, et al. 1966), shortens brain length (Walsh, Cummins, et al. 1973), affects the brain monoamine level and turnover (Stolk, Conner, et al. 1974; Jones, Marsden, et al. 1990), and increases sus-ceptibility to stereotyped behaviors caused by amphetamine and related drugs (Sahakian 1974). Isolation also affects gregariousness (Booth 1973), increases adult immobility and defecation in the open field (Warren and Ivinskis 1973), reduces problem-solving ability (Denenberg, Woodcock, et al. 1968), and, given later social experience, increases shock-induced fighting (Hutchinson, Ulrich, et al. 1965).

In mice, social isolation for three weeks reliably increases aggressive-ness in males (Welch and Welch 1971; Valzelli 1973; Terranova, Laviola, et al. 1998); increases activity (Essman 1966, 1968) and stress-induced gastric ulcers (Essman and Frisone 1966; Caputo, Essman, et al. 1968); and affects the level and turnover of neurotransmitters, including do-pamine (DA), norepinephrine, and serotonin, in some brain regions (Welch and Welch 1965; Valzelli 1973; Tizabi, Massari, et al. 1980; Had-field and Milio 1988). Increased sensitivity to amphetamine and to re-warding stimuli in isolation-reared rats apparently depends on changes in dopamine metabolism in the nucleus accumbens (Jones, Marsden, et al. 1990). In guinea pigs (*Cavia porcellus*) isolation-rearing causes a defi-cit in male sexual behavior (Valenstein, Riss, et al. 1955), and male Mon-golian gerbils (*Meriones unguiculatus*) isolated between ninety and two hundred days of age are more likely to attack a female when paired for mating (Ägren and Meyerson 1978).

Going beyond social rearing to overcrowding can have different effects; notably, it increases amphetamine toxicity (Thiessen 1964). All amphetamine effects, including adrenal secretion, catecholamine depletion, convulsion, and death, are increased by crowding after the drug is given; these effects are also increased by a rise in ambient temperature, by cortisol or ACTH, by shock, or by stress. But toxicity declines if the rats become accustomed to the more crowded conditions over a period of weeks before getting the drug (Welch and Welch 1966).

Social Deprivation in Monkeys

Rhesus monkeys (*Macaca mulatta*) raised in social isolation—with their physical needs being met but maternal care withheld—grow up behaviorally abnormal (Harlow, Dodsworth, et al. 1965; Suomi 2008). They are socially withdrawn when introduced to companions, and they often hug themselves and rock, bite themselves, bang their heads, and show other stereotyped, repetitive behaviors. They are fearful and may overreact with inappropriate aggression. Also, deprivation of contact with other juveniles, despite otherwise normal maternal rearing, still results in abnormalities, and peer-rearing (without mothers) produces behavioral and physiological abnormalities as well, especially in genetically vulnerable juveniles (Chamove 1973; Chamove, Rosenblum, et al. 1973; Higley, Suomi, et al. 1996; Novak and Sackett 1997; Worlein and Sackett 1997).

Numerous efforts at rehabilitating the most deprived monkeys failed or fell short, including placing them with experienced monkey mothers or same-age playmates, giving them human affection, and allowing years to pass. As adults they usually remained socially maladapted, fearful, prone to outbursts of aggression, and, if male, sexually inept (Harlow 1975). Some motherless females showed a particular ineptitude with their own infants, being not merely neglectful but abusive, clearly unable to adapt normally to the stress of new motherhood (Arling and Harlow 1967; Champoux, Byrne, et al. 1992).

Motherless monkeys given daily playtime with peers develop much more normally, but this has to start very early in life, not after months of isolation. Also, motherless mothers improve their maternal behavior with their second and third infants and vary in their susceptibility to isolation-induced damage to maternal care. Isolation-reared monkeys can become attached to gentle, long-haired dogs, clinging and riding under their bellies or on their backs; however, while this finding reveals persistent power to form attachments despite prior isolation, it does not address the question of long-term damage (Mason and Kenney 1974).

Finally, monkeys isolated until six months of age were caged together with normally reared three-month-olds (Harlow and Suomi 1971; Suomi, Harlow, et al. 1972; Suomi and Harlow 1972). The younger infants acted on the isolates as successive infants had acted on natural mothers: they were persistent, needy, and minimally threatening companions. The result was an almost complete elimination of the isolates' abnormal behavior, replaced with normal social behavior toward the younger monkey. Later, similar success was achieved with monkeys isolated for a full year. Isolation-reared monkeys may continue to have subtle abnormalities that appear later under stress. However, together with new evidence on isolation-reared children, these controlled experiments show how the young of higher-primate species can adapt to and survive deprivation.

In monkeys as in rodents, species differ. Pigtail macaques (*Macaca nemestrina*) react much more severely than bonnet macaques (*M. radiata*) to mother removal (Rosenblum and Kaufman 1971), yet bonnets show stronger preference for the mother over a stranger (Rosenblum and Alpert 1974). Separated pigtail infants showed a several-hour protest phase, including increased activity, distress calling, elevated heart rate and body temperature, and marked sleep disturbances on EEG (Reite, Kaufman, et al. 1974; Reite, Short, et al. 1978, 1981; Reite and Capitanio 1985). Nighttime temperature and heart rate declined in several sleep stages, reversing changes in the first postseparation hours. First-night measures predicted a depression phase the next morning, including depressed heart rate and temperature along with behavioral changes. Infants played less with peers and more with inanimate objects, and younger ones showed slouching, curling up, depressed activity, impaired coordination, and dejected faces.

The Neurobiology of Social Perturbation in Monkeys

In the squirrel monkey (*Saimiri sciureus*) separation causes HPA activation but reduces the immune-system hormone thymosin and white blood cell competence (Coe, Mendoza, et al. 1978; Coe, Wiener, et al.

1985; Coe and Hall 1996; Coe and Erickson 1997). In Goeldi's marmoset (*Callimico goeldii*) separation increases GCs and piloerection; both, along with proximity seeking at reunion, were greater in victims of parental aggression (Dettling, Pryce, et al. 1998). On positron-emission tomography (PET), rhesus monkey infant separation activated the right dorsolateral prefrontal cortex (dlPFC) and the right ventral temporal/occipital lobe, but deactivated the left dlPFC, consistent with the hemispheric bias in human studies. Activity here was correlated with GC levels, as was activity in brain-stem areas, including the raphe nuclei, locus ceruleus, and periaqueductal gray (Rilling, Winslow, et al. 2001). As in rodents, early repeat separations produced adult HPA hypofunction in marmosets (Dettling, Feldon, et al. 2002) and squirrel monkeys (Levine and Mody 2003), suggesting a kind of exhaustion of the system.

Nursery-rearing as opposed to mother-rearing reduces corpus callosum size and cognitive function in rhesus monkeys (Sánchez, Hearn, et al. 1998), even when nursery care is of the highest quality. A study of dopamine function in these two rearing conditions—important because DA disturbances are implicated in several psychopathologies—showed significant differences in brain measures that were associated with behavioral differences (Seraphin 2004). Responses to several different DA receptor agonists and antagonists were assessed in terms of behavior and changes in testosterone, cortisol, and prolactin. In another sample of rhesus infants in the two rearing conditions, postmortem brain study allowed the density of two DA receptors to be assessed in the prefrontal cortex and the basal ganglia. Nursery-rearing and its behavioral concomitants were associated with lower dopaminergic tone and levels of CSF DA metabolites, reduced prefrontal DA receptor densities, and altered sensitivities to the pharmacological manipulation of DA systems. This could mean a reduced ability to experience pleasure and reward.

In an ingenious intervention, one more reflective of the situations that might be encountered in the wild, bonnet macaque mother-infant pairs faced one of three conditions of foraging demand (Coplan, Rosenblum, et al. 1995; Rosenblum and Andrews 1995; Rosenblum 1998): low (food found predictably and easily), high (prolonged searching needed), or variable (high or low, in random alternation). All mothers were ensured adequate food intake, but infants of variable-demand mothers showed long-term anxiety and other symptoms linked to increases in corticotropin-releasing factor (CRF), somatostatin, serotonin, and dopamine in cerebrospinal fluid (Coplan, Andrews, et al. 1996; Smith, Coplan, et al. 1997; Coplan, Trost, et al. 1998; Rosenblum 1998).

Genotype-environment interactions are important: rhesus monkeys carrying the short allele of the serotonin transporter gene promoter are vulnerable to peer- instead of mother-rearing—showing decreases in the main serotonin metabolite but increased ethanol consumption, impulsivity, and aggression as adults—while those with the long promoter are resilient on all these measures (Suomi 2003). Considerable evidence now suggests that similar genotype-environment interactions are very important in children's psychosocial development and may predominate over main effects in determining many aspects of behavior, normal and abnormal. We return to this crucial matter in Interlude 7.

Human infants and toddlers show physiological stress—increased GCs and heart rate—in response to separation (Fox 1998; Fox and Card 1999), and maternal contact dampens stress-responsiveness (Spangler and Grossman 1999). In inhibited, highly reactive infants, secure attachment reduces physiological stress responses (Gunnar, Brodersen, et al. 1996; Gunnar 1998). In addition, separation distress in human infants and children is related to lateralized EEG activation (Bell and Fox 1994; Davidson 2000). Right frontal activation is tied to negative emotions (temperamental or reactive), and left frontal activity is tied to positive emotions and relationships. This may be due to asymmetrical exchange between the frontal lobes and the vagus nerve, the autonomic nervous system generally, and the limbic circuits that mediate fear, affiliation, and pleasure. These higher centers also influence GCs and other hormones (Gunnar and Donzella 2002); relationship quality and routine separations in early life may have lasting consequences for GC diurnal rhythms (Gunnar and Cheatham 2003; Watamura, Donzella, et al. 2003).

Experience in the Etiology of Psychopathology

There are a number of animal models of mental disorders (McKinney 1974; McKinney, Gardner, et al. 1994), including depression (Akiskal and McKinney 1973; Gilmer and McKinney 2003), psychosis-like illness (McKinney, Young, et al. 1973; Kraemer, Ebert, et al. 1984), and alcoholism (Kraemer and McKinney 1985; Higley, Suomi, et al. 1996), as well as many other forms of altered behavior. The applicability of such models to human behavior is always debatable. Nevertheless, since twin concordances in the major psychoses hover around 50 percent (Gottesman and Hanson 2005), nongenetic influences must play a major role, and this is true of many mental disorders. Although these need not be *behavioral* influences, they may be, and efforts have been mounted to develop animal models of the psychogenic factors in psychosis using isolation-rearing.

Another model, separation of infants from mothers or from peers, has been repeatedly described. In a classic study, John Bowlby and John Robertson followed infants who had to undergo prolonged hospitalization and identified four stages in prolonged separation or loss (Bowlby 1973). Stage 1 is protest, usually lasting a matter of days (Bowlby 1980). It is seen in the Ainsworth experiment as a reaction to a brief separation (Bretherton and Ainsworth 1974; Ainsworth, Blehar, et al. 1978; Gunnar, Brodersen, et al. 1996). Protest is the phase of resistance, active searching for the mother or primary caregiver, and refusal to accept the loss. Protest can be adaptive, bringing the parent back. There may be anger at the mother for leaving, at a substitute while she is gone, and at the mother again when she returns. Stage 2 is grief proper: the energy of protest is exhausted, and the futility of searching is accepted. Activity slows, and there is a dejected facial expression, quiet whimpering, and an inability to experience pleasure. Stage 3 is an emotionless adaptation. The dejected mood has passed, and there is evidence of recovery, but subtle experience with the child reveals a self-protective mode. Stage 4 is real recovery: the loss has been accepted, and the capacity for affection reemerges.

This outline, of course, simplifies: the phases are rarely so clear-cut, and overlap and variation are common. Prolonged separation in infant rhesus monkeys (*Macaca mulatta*) has been a model for these forms of grief and for treatment (Harlow 1959). The pattern is also seen in chimpanzees, orangutans, and gorillas (Codner and Nadler 1984), pigtail monkeys (*M. nemestrina*) (Wolfheim, Jensen, et al. 1970), Java monkeys (*M. irus*) (Schlottman and Seay 1972), and other species.

But infant monkey separation and loss is very different from isolation-rearing, which begins at birth; with isolation, the *absence* of social contact is being studied, and the result is a syndrome of social, communicative, and emotional incompetence, ranging from ineptitude to something like autism. Fortunately, such deprivation is rare in human life, and very few cases of human autism result from deprivation. In the separation studies the infant's capacity for affection and attachment has developed normally—the infant has been given an opportunity to form a strong emotional bond—and that capacity is then disrupted. The response in some species at least is very like that described for human infants: protest followed by a grief reaction—an animal model of depression (Akiskal and McKinney 1973; McKinney, Gardner, et al. 1994):

1. *Protest.* In this first stage, locomotor activity, vocalization, especially distress vocalization, and such indicators of "emotionality" as defecation or urination increase greatly. Rhesus monkey infants, still in the protest phase eight days after separation, were found to have approximately doubled levels of adrenal tyrosine hydroxylase, dopamine ß-hydroxylase, phenylethanolamine N-methyl transferase, and choline acetylase as compared with nonseparated controls, showing that all steps in epinephrine synthesis were induced, driven centrally by sympathetic arousal causing greater activity in the preganglionic cholinergic neuron (Breese, Smith, et al. 1973). These are characteristic responses to stress. The protest phase was also shown following mother-infant separation in the rat (Hofer 1975), the infant's locomotor activity and elimination being 60 to 70 percent higher than that of controls eighteen hours after separation, including controls who remained with a lactating mother and those with a nonlactating mother. Several other responses, including changes in sleep and cardiovascular and respiratory measures, were due to absence of lactation rather than separation.

2. *Despair.* The monkey data, which come from a number of different laboratories (Mineka and Suomi 1978; Rosenblum and Paully 1987), then show, as do the human and chimp data, a characteristic second phase of despair. This phase is like the depression produced in rhesus infants by reserpine, which depletes norepinephrine and dopamine: total depression of locomotor activity, huddling into a ball and self-clasping, withdrawal, and very sad-looking faces. It corresponds to the effect on adult monkeys of 6-hydroxydopamine (6-OHDA), which widely damages norepinephrine and dopamine neurons.

3. *Detachment.* In the third phase the depression is essentially completed and the infant appears to be relatively unconcerned as to

the reappearance of the separated individual. This can happen only when other individuals are present to replace the missing one. It has been difficult to demonstrate experimentally. We may think of the depression phase as resulting from exhaustion and collapse of the stress response system and perhaps of the sympathetic nervous system. The infant presumably has depleted brain monoamines and no longer has the levels of GCs and epinephrine reflecting vigorous activity. This phase may be produced by exhaustion of the stress response system.

If the separation is terminated, behavior returns to normal, after temporary proximity seeking. However, two or more years later these individuals are more fearful in a strange environment and, if separated again, become immediately depressed and continue to be depressed for a time after reunion. If the separation is not terminated, the eventual course for these infants is the same as that of infants isolated from birth—continued withdrawal, with an eventual appearance of stereotyped behavior.

Early Deprivation in Human Childhood

Freud likened early psychic stress to the effect of a pinprick on an embryo in then-current experiments, using the term *trauma,* Greek for wound. As the pinprick caused a divergence in the developmental path and a permanent anatomical abnormality, a psychological wound could theoretically divert the child's mind from normal growth, leaving emotional deformities.

Corresponding concepts arose among educators and cognitive psychologists, who posited certain deprivations in early mental growth that were irreversible. The "critical period" concept arose to denote the possibility that there are times in human development when certain experiences must occur or be avoided, memories laid down, feelings felt, and skills learned and mastered. Theoretically, failing timely intervention, some functions would be permanently impaired. A less limiting version, the "sensitive period," is more tenable, if the developmental track and the environmental inputs are well specified.

Unfortunately, there are many children with life experience relevant to this question: those raised in extreme poverty; those with handicaps such as blindness, deafness, or both; those in bad institutional settings; and those in severe social and cultural isolation (Skuse 1984, 1985). Children in this last category—some called "wild" or "wolf" children—have always been of unique interest because they suggest what is irreducibly innate. According to Herodotus, an Egyptian king in the seventh century BC had two children raised by shepherds who never spoke to them to determine what language they would speak; the result was not recorded. A thirteenth-century chronicler attributed a similar experiment to Holy Roman Emperor Frederick II, who reportedly thought the child would speak Hebrew.

A "Wild" Child

Tragically, some children have in fact been raised with severe deprivation of interaction. The "wild boy" of Aveyron was found naked and covered with scars in a forest in southern France in 1797, when he looked about twelve years old (Lane 1976). It was thought that he had raised himself on the acorns and roots of the forest; he could swim, climb trees, run well on all fours, find food and water, and burst out with shouts and laughter. But he was not able to speak or to comprehend, and so he seemed to offer an answer to the ancient question: the child not exposed to language speaks nothing at all. In the words of an early observer, "His affections are as limited as his knowledge; he loves no one; he is attached to no one; and if he shows some preference for his caretaker, it is an expression of need and not the sentiment of gratitude" (Lane 1976:39). The boy came under the care of Jean-Marc Itard, a young physician involved in teaching the deaf. From 1801 to 1806, Itard devoted himself to the boy; among other improvements,

> he has both a knowledge of the conventional value of the symbols of thought and the power of applying this knowledge by naming objects, their qualities, and their actions . . . an extension of [his] relations with the people around him . . . his ability to express his wants to them, to receive orders from them . . . a free and continual exchange of thoughts with them. . . . Victor is aware of the care taken of him, susceptible to fondling and affection, sensitive to the pleasure of doing things well, ashamed of his mistakes, and repentant of his outbursts. (160)

Victor's achievements were "the slow and laborious result of an intense training in which the most powerful methods are used to obtain the smallest effects," and his "emotional faculties, equally slow in emerging from their long torpor, are subordinated to a profound egoism" (159). In adolescence—on schedule biologically—his sexuality emerged as compulsive masturbation. Itard ended with something like a critical period hypothesis of Victor's limited learning capacity. However, even at the time debate highlighted the methodological problem with such cases: they have no history. The presumption of a normal infant somehow led to survive in the woods with no human contact is at the heart of Itard's interpretation, but Victor could have been impaired to begin with.

Modern Cases

In some modern cases, the history is better known: ten children from six different families were reviewed by David Skuse, who intensively studied two of them (Skuse 1993). Most had spent a few months after birth in normal environments and were known to be normal before being subjected to social isolation or other extreme psychological deprivation; several were nutritionally deprived as well. Found at ages ranging from two to thirteen years, they had spent their lives locked in tiny rooms or cellars, or tied to chairs or beds, with minimal to no human contact. On discovery, they were withdrawn, fearful, and profoundly abnormal in speech.

Most had impressive recoveries. Exceptions were the oldest, Genie, who was thirteen when found, had made little progress by age eighteen, and regressed thereafter; and another who died of liver disease at ten. But all eight of the others were in school by their teens, six in normal schools. Twin boys in Czechoslovakia grew up together but with no other contact from eighteen months to seven years. Their speech was very poor, they could barely walk, and they reacted with horror to normal objects. Yet, placed with foster parents, they developed normal attachments, showed rapid language learning, displayed above-average IQs, and at age twenty were both working steadily at maintaining office machinery; each had been dating and had fallen in love. Such cases contradict many predictions about the role of socialization in normal development, suggesting resilience even in severe deprivation, a powerful canalization of development into normal paths of neurobehavioral growth (Waddington 1942, 1971). Still, the effects of deprivation vary, and we need more than anecdotes to discern them.

Romanian Orphan Studies

The severe deprivation of many infants in Romania during the transition out of communism is a tragic natural "experiment." Of these, 111 were adopted into families in the United Kingdom after up to two years of both severe social deprivation and infectious illness (Rutter 1998; O'Connor, Bredenkamp, et al. 1999; O'Connor, Rutter, et al. 2000). "The catch-up in both physical growth and cognitive level appeared nearly complete at 4 years for those children who came to the U.K. before the age of 6 months. . . . The developmental catch-up was also impressive, but not complete, in those placed after 6 months of age" (Rutter 1998:465). Ongoing studies showed attachment disturbance, including indiscriminate friendliness (O'Connor, Marvin, et al. 2003), and suggested "some form of early biological programming or neural damage stemming from institutional deprivation," but the conclusion was also reached that "the heterogeneity in outcome indicates that the effects are not deterministic" (Rutter, O'Connor, et al. 2004:81).

Studies of institutionalized Romanian orphans adopted in the Netherlands and Canada have similar results. Age is critical, with children adopted before four months of age showing basically no difference from Canadian-born controls (Gunnar, Morison, et al. 2001; MacLean 2003). But children adopted after years of deprivation have not just attachment disorders but chronically elevated serum cortisol (Gunnar, Morison, et al. 2001) and reduced activity in limbic system brain regions (Chugani, Behen, et al. 2001), and they are at strongly elevated risk for autism, post-traumatic stress disorder, and inattention/overactivity (Kreppner, O'Connor, et al. 2001; Rutter, Kreppner, et al. 2001; Hoksbergen, ter Laak, et al. 2003, 2005).

Cortisol was not elevated in Romanian orphans adopted before eight months, but over one-fifth of the later-adopted sample had cortisol levels more than two standard deviations above the mean for early adoptees or Canadian-born children between six and twelve years old (Gunnar, Morison, et al. 2001). The researchers speculated that there might have been limbic system changes that influenced the HPA axis, and these do exist.

Harry Chugani and his colleagues used PET imaging to compare a sample of adopted Romanian orphans (averaging nine years old, having been an average of three years in the orphanage) with young adults and a sample of children (averaging eleven years) being evaluated for unilateral epilepsy but with normal contralateral hemispheres. Compared to either control group, the adopted orphans had "mild neurocognitive impairment, impulsivity, and attention and social deficits" and "significantly decreased metabolism bilaterally in the orbital frontal gyrus, the infralimbic prefrontal cortex, the medial temporal structures (amygdala and head of hippocampus), the lateral temporal cortex, and the brain stem" (Chugani, Behen, et al. 2001:1290). This is essentially the circuitry of judgment and psychosocial function (Damasio 1994; Nelson, Leibenluft, et al. 2005), and changes in that circuitry would help explain the elevated risk of impulsivity and autism spectrum disorders.

Following up on this study with diffusion tensor imaging (DTI), seven children averaging nine years of age who had been in Romanian orphanages in early life were compared with normally reared controls, with a focus on the cingulum, fornix, stria terminalis, and uncinate fasciculus—all major fiber tracts of the limbic system. The left uncinate fasciculus, which connects the amygdala and hippocampus with the orbital frontal cortex, was significantly smaller in the former orphans than in the controls, and all the other fiber tracts showed nonsignificant trends in the same direction (Eluvathingal, Chugani, et al. 2006). This finding adds an anatomical dimension to the functional differences seen in the PET study and strengthens the case for limbic system mediation of the psychosocial and emotional deficits resulting from social deprivation.

Finally, a study in Bucharest, Romania, began with institutionalized infants and children who were five to thirty-one months old and a never-institutionalized control group, but half the orphanage children were randomly assigned to foster care (Moulson, Fox, et al. 2009). Studies at baseline and thirty and forty-two months later included the recording of event-related potentials (ERP) in response to pictures of happy, angry, fearful, and sad facial expressions. At all three assessments the institutionalized group had occipital lobe ERPs of smaller magnitude and longer latency than did the family-reared children, but the children in foster care changed to become intermediate between the other two groups at the forty-two-month assessment—evidence suggesting both some degree of impairment of emotional processing in the brain and the potential for recovery.

Such institutional deprivation syndromes, as well as the potential for recovery, recall the monkey experimental data, including the neural and hormonal changes. Romanian orphans show a sensitive period for certain kinds of damage, with a threshold at four to eight months of age in different studies. But past experience with both human and nonhuman abnormal rearing suggests that further study will be needed to determine both the extent of the sensitive period and the pattern of residual deficits, especially under pressure.

Other Social Disturbances of Childhood

Apart from these extreme deprivations, there have been many studies, mostly retrospective, of the lasting behavioral and physiological effects of early psychosocial disturbances. Using some of the animal models discussed earlier as a starting point, Christine Heim and her colleagues have demonstrated changes in the HPA axis, and in CRF function in particular, in depressed patients with and without a history of childhood abuse (Heim, Nemeroff, et al. 2001; Heim, Plotsky, et al. 2004). In essence, they find different subtypes of mood and anxiety disorders depending on the presence or absence of childhood trauma, with CRF function better predicted by that history than by depression itself. Since CRF administration in the brain produces stress responses and anxiety-like behavior in animal models, it seems likely that the chronic CRF hyperreactivity found in survivors of childhood abuse could help account for specific symptoms in those adults.

Does every childhood stress leave a permanent scar? In a study by Glen Elder, the hardship of the economic Great Depression had long-term effects on the mood and behavior of children and adolescents, with differences depending on the father's behavior, the sex and age of the child, and (for adolescent girls) the child's physical attractiveness, which influenced how fathers treated them (Elder 1974). But these associations are modest; past thinking has underestimated "children as producers of their own socialization" (Elder, Nguyen, et al. 1985:373).

Also, such studies rarely consider prenatal exposure to alcohol, nicotine, cocaine, and heroin, pre- and postnatal malnutrition, or brain injury from accidents or abuse, much less genetic vulnerability. Depending on age at injury, the pre- and post-injury environment, and genetic resilience, the brain often recovers significantly from either biological or psychosocial trauma. Children may produce their own socialization, but their agency differs, owing to prior developmental pathways. We look again at stress and resilience in Chapter 21, but first we review the evolving social context of ontogeny.

15

The Evolution of the Mother-Infant Bond

As we have seen, parental care and offspring dependence occur widely in the animal world. They have evolved independently in insects (Trumbo 1996), many species of fish (Crawford and Balon 1996), tree frogs (Crump 1996), alligators (Gans 1996), birds (perhaps with a dinosaurian origin), and mammals (Rosenblatt and Snowdon 1996). In birds and mammals parental care is essential in all species (Gubernick 1981; Gowaty 1996; Rosenblatt 2003). The intensification of infant dependency and maternal care probably drove early mammalian brain evolution, especially in the extended limbic system (MacLean 1985; Mega, Cummings, et al. 1997).

Pair bonding—which in birds and mammals is usually associated with paternal care—is also widespread; it evolved independently in shrimp (Duffy, Morrison, et al. 2000), fish (Wilson, Ahnesjo, et al. 2003; Whiteman and Cote 2004), at least one reptile genus (O'Connor and Shine 2003), 90 percent (about eight thousand species) of birds (Tullberg, Ah-King, et al. 2002), and separate lines of mammals, including voles, foxes, lemurs, marmosets, gibbons, and humans (Fuentes 1999; Dupanloup, Pereira, et al. 2003; Rasoloharijaona, Rakotosamimanana, et al. 2003; Schuiling 2003). Kin alliances occur in all social species, and nonkin cooperation and reciprocity are occasionally found (Clutton-Brock 2009). As with the wings of insects, birds, and bats or the anatomical shape convergences of sharks, dolphins, and penguins that adapt them for rapid swimming, these convergences in the social behavior of widely disparate species show that under certain ecological conditions similar selective pressures produce similar outcomes, often from very different starting points.

Maternal Care in Mammals

In mammals in particular, maternal behavior was raised to a very high level. Since maternal care is well developed in the platypus (*Ornitho-rhynchus*) and other egg-laying mammals (monotremes), as well as in all marsupials, the mother-offspring relationship probably emerged with the mammals around 225 mya and continued to evolve with the roughly 100-million-year transition from egg-laying to internal gestation. "Mammal" (from *mammae,* or breasts) implies the nourishing of the young from modified sweat glands. In monotremes these are dispersed over the chest and abdomen rather than in pairs of distinctive organs, enabling the young to feed simply by licking the mother's skin. Thus the specific physiological adaptations for feeding in mammals long preceded live birth.

The same is true of retrieval in response to separation calls (MacLean 1985). From comparative evidence, we infer that these distress signals coevolved with the middle ear bones, which were derived from the reptilian jaw angle by 195 mya (Luo, Crompton, et al. 2001), a process that appears to have involved heterochrony, pedomorphosis, and specific regulatory gene changes (Ji, Luo, et al. 2009). The mammalian inner ear is also adapted for high-pitched sounds because of a specialized cochlea (Brownell, Bader, et al. 1985) that enables mammals to hear in a frequency range an order of magnitude higher than that of reptiles; the most likely predators on mammalian young, reptiles consequently cannot hear their high-pitched separation calls (Allman 1999). So by early in the dinosaur age mammals had evolved the reciprocal capacities of offspring and mother for attachment.

Mother and Infant Primates, Including Humans

The primate order, which originated about 70 mya, evolved singleton births, which require prolonged parental investment to maintain reproductive success. The parent-offspring relationship develops even more slowly and is longer-lasting in higher primates, especially catarrhines — Old World monkeys, apes, and humans (Martin 1995; Bolter and Zihlman 2007; MacKinnon 2007); close proximity from birth is formative for the relationship (Maestripieri 2001). The catarrhines evolved a suite of maternal and infant adaptations that can be called the catarrhine mother-infant complex: a hemochorial placenta; singleton birth with rare twinning; twenty-four-hour physical contact for weeks to months and close proximity until weaning; nursing at least three times per hour while awake and at least once at night; weaning no earlier than 25 to 30 percent of the age at female sexual maturity; gradual transition to a multi-aged juvenile playgroup; and variable but usually little direct male care.

None of these features is found in every individual, and a species may depart from one or more of them. Even in the macaque genus, as noted in Chapter 14, the response to separation varies from severe (rhesus, M. *rhesus,* and pigtails, M. *nemestrina*) to relatively mild (bonnets, M. *radiata*), depending on the extent of nonmaternal care (alloparenting or allocare) common in the species. This set of responses in turn varies substantially among the Old World monkeys and apes; in some species mothers allow no allocare, while in others it is common (Hrdy 1976). Allocare affords the possibility of earlier weaning, shorter birth spacing, and higher reproductive rates (Mitani and Watts 1997; Ross and MacLarnon 2000), and we will focus on it in the next chapter.

Male (not necessarily paternal) involvement varies even more. Gibbons (Hylobatidae) have extensive male-infant contact and carrying, and Barbary macaques (M. *sylvanus*) and Japanese macaques (M. *fuscata*) have a significant amount of male interaction with infants and juveniles. Orangutan (*Pongo pygmaeus*) infants and juveniles, in contrast, rarely encounter adult males except for the latter's occasional sexual visits with their mothers. Most monkey and ape infants have males around most of the time, sometimes affording protection, but without much direct interaction. As we will see in Chapter 17, to the extent that males contribute to the care and protection of the young, they are part of the spectrum of cooperative breeding; among catarrhines, males became very involved only in gibbons and humans. Overall, however, the list is useful, if only as a heuristic. The wide distribution of these features in monkeys and apes suggests that they are common to the catarrhines and may have been present in the common ancestor, between 30 and 40 mya (Martin 1990).

Primate groups include a core of genetically related individuals with associated nonrelatives. In most catarrhines, the core is a matriline, stable over the lives of individuals. In some species, including chimpanzees (Goodall 1986), bonobos, and several New World monkeys, the core is a patriline and the females are unrelated migrants (Strier 1996). Primates vary more than once thought in the structure of lineages within groups, and explanations require attention to ecological conditions within as well as across species.

Nevertheless, the mother-offspring bond remains central to socialization, and whatever the lineage structure, social support and generosity are preferentially distributed toward genetic relatives—yet not exclusively. Monkeys and apes may aid nonrelatives and can usually expect aid in return, as predicted by reciprocal altruism theory (Trivers 1971); selection favoring this process must have played a role in the evolution of social cognition and memory. Cooperation and competition are found everywhere, and one of the major purposes of cooperation is defense. Both sexes engage in conflict (Hrdy 1981); males are more aggressive, but female dominance hierarchies are important for many reasons, not least because they determine the ranks of their offspring.

For example, in stumptail macaques (*Macaca arctoides*) the birth of an infant to a high-ranking mother occasions exceptional interest (Gouzoules 1975), and a mother's rank strongly influences infant interactions with other individuals thereafter, with both male and female adults taking special interest in these infants (Gouzoules 1975b). As female langurs age their dominance increases through kin connections and their own skill, and their reproductive success also increases (Dolhinow, McKenna, et al. 1979). Female chimpanzees are strongly, sometimes brutally, subordinate to males, but in their sister species, the bonobo, female coalitions keep a firm lid on male aggression and dominance (Stumpf 2007), and the socialization of the young in these two species differs accordingly.

Still, social organization varies both between and within species (Cheney, Seyfarth, et al. 1986). Monogamy is present in some South American monkeys (including the marmosets, or Callitrichidae) and (among catarrhines) in gibbons (Hylobatidae), although the mated pair may not be exclusive. In most species larger group associations are the rule, including stable polygynous relations or temporary associations between individual males and females. In orangutans the usual social grouping is a female with her offspring, separate from solitary males. Thus one of the most cognitively advanced higher primates, an ape quite closely related to humans, has perhaps the most rudimentary social organization. The causes of this wide variation in social organization among the higher primates and its partial independence of cognitive capacity remain obscure.

In hunter-gatherer societies, fathers contribute substantial provisioning (Marlowe 2001, 2003) but limited direct care (Konner 2005). Among primates, only humans have postweaning provisioning, which accounts for our shortened birth interval and our success as a species (Lancaster and Lancaster 1987; Kaplan, Hill, et al. 2000). We also prolong female postreproductive survival, which is vital in a species in which postweaning provisioning and learning are uniquely important (Hawkes, O'Connell, et al. 1998; Hawkes 2006). Cooperative breeding benefits not only from paternal and grandmaternal investments in children (Hrdy 2009) but also from the quasi-parental role of child groups (Konner 1975). We will return to these other relationships, but first we consider the mother-offspring bond in one hunter-gatherer society.

Mother-Infant Relations among !Kung Hunter-Gatherers

As early as the 1950s, ethnographers had described !Kung infants as having extremely close physical relationships with their mothers and being indulged in every way (Marshall 1959). Childhood and adolescence were relatively free of assigned chores, and the playgroup figured importantly in socialization from toddlerhood on. Physical punishment was rare. Detailed quantitative observational research on infancy and childhood extended these generalizations.

The Nursing Pattern

!Kung infants were put to the breast whenever they cried or fussed and frequently at other times as well. In dawn-to-dusk observations, we saw that breast-feeding occurred for a few minutes at a time, several times an hour, throughout the day (Konner and Worthman 1980); using 15-minute observations divided into 5-second time blocks, we observed 45 infants and recorded nursing and other behavior. Throughout the first 80 postnatal weeks fewer than 25 percent of the 15-minute periods had no nursing, even though the observations were never begun during nursing.

In an independent set of observations, we monitored 17 mother-infant pairs (infants aged 12 to 139 weeks, with a mean age of 63.9, ±9.9) in three two-hour sessions, from 8:30 to 10:30 AM, 12:30 to 2:30 PM, and 4:30 to 6:30 PM, each on a different day, and recorded nursing bouts to the nearest 30 seconds (Konner and Worthman 1980). Over all ages from the neonatal period to 24 months, there were 4.06 nursing bouts per hour (±0.41), for a mean length of 1.92 min (±0.18), yielding 7.83 total minutes of nursing per hour (±1.27). The mean time elapsed between nursing bouts was 14 minutes, and the average infant's longest time between bouts was just over 55 minutes. Age did not predict average bout length or total nursing time, but it strongly predicted the average interval between nursing bouts ($r = .71$, two-tail p < .005).

Weaning and Birth Spacing

Traditionally, !Kung weaning occurred during the fourth year, after the mother was pregnant again (Konner 1977b). Nursing infants began supplemental food, often prechewed by the mother and sometimes "kiss-fed," when they were around six months old. Weaning was usually completed well before the end of the pregnancy. However, if there was no next sibling, breast-feeding could continue until after age five, especially for boys; one was breast-fed occasionally at age eight. There was no abrupt cessation of nursing and rarely any punishment, but weaning could involve protest and depressed behavior. Interviews of adult women revealed that some still had vivid memories of weaning and attendant sibling jealousy (Shostak 1981).

Sleeping Arrangements and Night Nursing

All !Kung infants slept beside their mothers at night on the same skin mat at least until weaning. In interviews of twenty-one mothers with children up to age three, twenty reported waking up to nurse at least once a night, and all twenty-one said that in addition their infants nursed without waking them, from two to "many" times or "all night." This pattern of highly frequent breast-feeding during the day with additional nursing several times a night was shown to be the most likely explanation for hormonal changes that reduced fertility (Konner and Worthman 1980; Stern, Konner, et al. 1986). It is also likely that this pattern played a role in lactation success; in infant sleep patterns, digestive problems, and blood glucose cycles; and in mothers' milk composition, attitude toward nursing, sexual arousal and receptiveness, and mood (Stern 1986).

Physical Contact

!Kung infants had very high levels of skin-to-skin physical contact in infancy, mainly with the mother (Konner 1972). Newborns were carried in a leather sling upright on the mother's side, usually with no intervening clothing. Spot observations (Konner 1976) revealed that physical contact with the mother declined from about 70 percent in the first months to around 30 percent at eighteen months, peaking at fifteen weeks. Physical contact with anyone peaked at about 90 percent of observations between ten and twenty weeks and declined to about 42 percent at eighteen months; fifteen-minute observations showed similar results (Konner 1975).

Nonphysical Interactions

If infants in traditional cultures have more physical contact and breast-feeding, perhaps industrial societies compensate with visual and vocal communication. This was true when Boston mother-infant pairs were compared with those in a Guatemalan Indian village: the total number of interactions was roughly equal, but in Boston about 80 percent were vocal, while in Guatemala 80 percent were physical (Klein, Lasky, et al. 1977). The infants of both professional and working-class Boston mothers were in physical contact 11 percent of the time at ten months. Comparing the Boston and Guatemalan data with the !Kung suggested a third pattern (Konner 1977b). Boston professional-class infants had more verbal interaction with their caregiver than working-class infants (Tulkin 1977), and Guatemalan infants had much less verbal interaction than in the Boston working class (4 versus 10 percent of five-second time blocks), but the !Kung infant, caregiver, and reciprocal vocalizations resembled those of the working class (Konner 1977b).

Overall Indulgence

!Kung are highly indulgent and responsive. One measure, response to crying, can be quantified. The likelihood of some nurturing response to spontaneous fretting of all durations was 78 percent during the eight- to twelve-month period (Konner 1972, 1977b). A recent analysis of response to infant crying among the !Kung confirms this picture (Kruger and Konner, forthcoming). Caregiver response was non-exclusive—several people often responded to a crying bout—but the mother was involved in the great majority of bouts. !Kung infants showed the "normal crying curve" with a peak in the first three months (see Chapter 7), and they also had as many crying bouts as infants in a Dutch sample, but their bouts were shorter and their total crying was only about half as much (Barr, Konner, et al. 1991), probably because of differences in care.

Mother-Infant Relations in Other Hunter-Gatherers

Systematic quantitative research has been done on infancy and childhood in at least five other hunter-gatherer groups—Hadza, Efe, Baka, Ache, and Agta. In this and other chapters in Part III, we consider the relevant results from these five cultures, looking for any possible generalizations. The accounts of maternal care and, in the next two chapters, nonmaternal care and male parental care are necessarily complementary, and the brief ethnographic background is given only once. In addition, we refer to more narrowly focused research on other hunter-gatherers, including the Bofi of the Central African Republic (Fouts, Hewlett, et al. 2001), the !Xun of North Central Namibia (Takada 2005), the Yora of Peruvian Amazonia (Sugiyama and Chacon 2005), the Martu of Australia's Western Desert (Bird and Bliege Bird 2005), the Meriam of Torres Strait (Bird and Bliege Bird 2000, 2002), and the Mikea of southwestern Madagascar (Tucker and Young 2005). Of special interest are studies that followed hunter-gatherers as they shifted toward horticultural or pastoral subsistence, such as the !Kung (Draper 1976; Draper and Cashdan 1988; Draper and Howell 2005) and the Baka of Cameroon (Hirasawa 2005), and controlled comparison of foragers and farmers in the same environments (Hewlett, Lamb, et al. 1998, 2000; Fouts, Hewlett, et al. 2005).

Hadza

The Hadza live in rocky, hilly terrain near Lake Eyasi in northern Tanzania, and when studied, they gained about 95 percent of their subsistence from hunting and gathering (Woodburn 1968; Blurton Jones 1993; Kaare and Woodburn 1999). The climate is milder than the !Kung's, and the environment has more game and plant foods. The Hadza had a slowly growing population. Some aspects of Hadza infancy and childhood resemble the !Kung (Blurton Jones, Hawkes, et al. 1989; Blurton Jones 1990):

> The Hadza child's first year of life appears not to differ greatly from that of the !Kung infant. The mother is the principal caretaker. The baby spends most of its time riding on the mother's side or back. Suckling is frequent, and often, but by no means always, "on demand.". . . The baby is likely to be surrounded by relatives, old, adult, and young, and receives attention from them and is carried by them. . . . Face-to-face interactions described in Western cultures (and in !Kung) can be seen between the Hadza mother and infant and other people and the infant. (Blurton Jones 1993:316)

These observations are confirmed in quantitative studies (Marlowe 2005a). Mothers provide the most interactions by far with infants and continue to predominate until age three. In thirty-minute focal-follow observations of the infants, they interacted with mothers in 78 percent of minutes, with fathers and older sisters in 18 percent, with maternal grandmothers in 9 percent, with older brothers in 8 percent, and with all others in 29 percent. The corresponding figures over the first four years were mothers 43 percent, fathers 17 percent, older sisters 17 percent, maternal grandmothers 10 percent, older brothers 9 percent, and 41 percent for all others.

Like the !Kung, the Hadza show maternal primacy in the context of multiple caregiving. During hourly scan observations, only about 30 percent of all holding of children (< 4 years old) is by someone other than the mother (Marlowe 2005a). Still, "Hadza are weaned a good deal younger than !Kung, at around 2.5 years old. Soon after they are 2 years old, Hadza children begin to be left behind when the mother gathers, although they may be suckled before the mother leaves camp and as soon as she returns" (Blurton Jones 1993:316).

How indulgent are Hadza parents? For Nicholas Blurton Jones (1993:316–18), "the Hadza are strikingly different" from the !Kung. "Mothers break off suckling bouts, evoking protest from the infant. One also often hears crying and observes parents ignoring a crying infant." Furthermore,

> We see Hadza parents use physical punishment, and we see and hear them shout prohibitions and commands at children. . . . This bleak picture should not be exaggerated, and quantitative analysis may also redress the balance. Hadza children are active and cheerful most of the time and are welcomed in their home. Even among these people, who seldom publicly show affection or warmth, parents can be heard to speak warmly of and to their child. But the overall picture is certainly not the developmental psychologist's dream presented by the !Kung. (Blurton Jones 1993:318)

Quantitative analysis did differ: "Hadza children are allowed to do as they like most of the time . . . children throw tantrums and pick up sticks and hit adults, who do little more than fend off the blows and laugh" (Marlowe 2005a:179). Frank Marlowe also cites two earlier ethnographers as portraying doting parents and an absence of punishment, which he confirms: "I too found children received considerable affection and were rarely punished. I only saw one spanking [by a stepfather] during a year of observation of men and children" (Marlowe 2005a:179).

Efe

The Efe are small-stature hunter-gatherers of the Ituri, part of the tropical rain forest of the northeastern Democratic Republic of the Congo (Bailey 1991; Peacock 1991). They also trade game with neighboring agricultural peoples, and "the majority of their caloric intake comes from cultivated foods acquired from the Balese" (Tronick, Morelli, et al. 1987:97). Still, they spend much of their time in the forest, are seminomadic, and live in small camps of six to fifty people from several extended families. Research on Efe infancy was initiated in the 1980s, and the results presented a direct challenge to what they called the continuous care and contact (CCC) model of !Kung infancy (Tronick, Morelli, et al. 1987). The contrast began with birth, which for the Efe was a group event. The newborn was not held by the mother first but passed among the women; some of them, lactating or not, would try to nurse. A lactating woman would in fact nurse the infant until the mother's true milk replaced the colostrum.

Still, the mother also nursed her infant; her involvement began within hours of the delivery:

> For the first few days of life the newborn is kept in or around the hut and is almost always in physical contact with the mother or another person. A mother does not resume her normal work schedule until four to five days postpartum. When regular tasks are resumed, the infant may accompany her mother on long out-of-camp trips. If this occurs, child-care responsibilities are generally shared by individuals at the work site. When the mother's work requires a short out-of-camp trip, she often leaves the infant in the care of another. Almost all females attempt to comfort a distressed or fussy infant. Comfort includes allowing the infant to suckle and often occurs in the mother's presence. But if unsuccessful the infant is returned to her mother. (Tronick, Morelli, et al. 1987:99)

As we will see, Efe multiple caregiving not only continues but increases over the first five or six months and resumes in the second year. Nonmaternal care, like maternal care, is highly indulgent: "Most interactions with infants appear positive and playful. But if infants do fuss or cry, they are responded to quickly" (Tronick, Morelli, et al. 1987:100). Fussing infants were responded to by *someone* within ten seconds 85 percent of the time for the first seven weeks, and 75 percent of the time at eighteen weeks. So the Efe are very like the !Kung in responsiveness, in part through multiple caregiving. However, the infants in this initial report were not yet in the main phase of attachment. Another paper (Tronick, Morelli, et al. 1992), also a challenge to the CCC model, focused on social contact—all the time when the infant was not alone, including when two or more individuals were in contact simultaneously. Infants were in social contact with their mothers about 50 percent of the time at five months (ranging from 36 to 70 percent) and about 21 percent at thirty-six months (ranging from 14 to 40 percent). But the highest level of contact with mothers is 63 percent, occurring at eight months of age.

There is an interesting difference between two figures in their paper.

In the first, the total time Efe infants are in social contact with mothers, *all* children, *all* adults, and fathers is displayed; except for fathers, non-maternal contact is high at each age. But the second chart compares time in contact with mother, father, the *average* child, and the *average* adult; this is meaningful because unless a specific individual spends an amount of time with the infant that rivals the time spent by the mother, the likelihood of a comparable attachment is low.

Their data show that the *average* adult is not in social contact with the infant more than a small fraction of the mother's time *at any age* and has less time in contact than the father at every age except eight months. The average *child* eventually approaches the mother's contact time, but not before age three. Most intriguing, however, is the pattern at eight months of age, when the ratio of nonmaternal to maternal contact is lowest in all three categories; at this key age for attachment formation, the mother occupies over an order of magnitude more of the infant's time than does the average child, the average adult, or the father.

Edward Tronick, Gilda Morelli, and their colleagues (1992:573) conclude that "the developmental course of the Efe infants' and toddlers' social relationships does not conform to the patterning of relationships predicted by CCC models." This is a reasonable inference at one level, but at a deeper level there are commonalities. They emphasize the density of social contact in which these infants are immersed: "Efe infants and toddlers spend almost all of their time in social contact with other individuals, and although the amount of social contact declines with age, 3-year-olds still spend most of their time in physical and social contact with other people. . . . Efe infants and toddlers are almost never alone in the sense of being out of sight or hearing of other people" (573–74). This account qualitatively resembles published accounts of the social context of !Kung infancy, but contact with mothers, although still very high, is lower for Efe.

Aka

The Aka are small-stature hunter-gatherers of the southwestern Central African Republic and northern People's Republic of the Congo (Bahuchet 1999). The tropical rain forest environment of the western

Congo basin is varied but rich, and when their infants were studied, the Aka mainly subsisted by foraging. They spent 56 percent of their time hunting, 27 percent gathering, and 17 percent working in nearby agricultural villages. However, their diet was mainly farm-produced food, for which they traded hunted game.

In contrast to the Efe, infertility was unusual, and the mean number of live births reported by postmenopausal women was 5.5 (versus 4.7 for the !Kung). The interbirth interval was about 3.6 years (versus about 4 for the !Kung). Infant mortality in both peoples was about 20 percent in the first year. Aka camps comprised twenty to thirty-five people (half under age fifteen) and varied from one to fifteen nuclear families; groups moved, aggregated, and dispersed, depending on food availability. Women were prominent in the key foraging strategy of net hunting (Noss and Hewlett 2001).

Barry Hewlett's classic monograph (1991b:32–33) describes the general pattern of infancy and childhood:

> Aka infancy is indulgent: Infants are held almost constantly, they have skin-to-skin contact most of the day . . . and they are nursed on demand and attended to immediately if they fuss or cry. Aka parents interact with and stimulate their infants throughout the day. They talk to, play with, show affection to, and transmit subsistence skills to their infants. . . . I was rather surprised to find parents teaching their eight-to-twelve-month-old infants how to use small pointed digging sticks, throw small spears, use miniature axes with sharp metal blades, and carry small baskets. . . . Unlike their village neighbors, Aka infants are carried in a sling on the side rather than on the back, which allows for more face-to-face interaction with the caregiver.

In addition, "besides being indulgent and intimate, Aka infancy also lacks negation and violence. . . . Seldom does one hear a parent tell an infant not to touch this or that or not to do something. . . . Violence or corporal punishment for an infant that misbehaves seldom occurs" (35). Aka say that either parent hitting an infant is grounds for divorce.

There are departures from the !Kung pattern. Since women, even nursing mothers, engage actively in net hunting, they may set a baby down on the forest floor and leave it to cry while the mother completes the kill. (The Ache, in contrast, although also forest foragers, rarely set their infants down on the forest floor.)

As we will see, Aka have high levels of nonmaternal care, especially paternal care—in fact, theirs is the greatest father involvement in the anthropological record. Still, mothers are by far the main caregivers: they hold infants 87 to 96 percent of the time during net hunts, and while in the forest camps but not hunting they hold infants 51 percent of the time in the first four months, 87.5 percent of the time from eight to twelve months, and 57.3 percent of the time from thirteen to eighteen months. Even more than in the Efe, the mother's role increases markedly just during the phase of attachment formation (Ainsworth 1979; Belsky 1999; van Ijzendoorn and Sagi-Schwartz 2008). Despite the role of fathers, infants show far more attachment behaviors during this age period toward mothers. Infants from eight to twelve months direct 15.5 percent of such behaviors toward fathers, 58.8 percent toward mothers, and 22.2 percent toward all others. So about one-fourth as many attachment behaviors at this age are directed at fathers as at mothers, although this rises to 58.4 percent in the second year.

There is further support for maternal primacy in the fact that at all ages the behavior "fusses for" (admittedly an inference) was recorded almost eight times as often for mothers as for fathers, even though fewer than one-fifth of the episodes of fussing for the mother ended in nursing. Holding by the mother declines until the key transition of weaning, which typically starts at age three or four when the mother becomes pregnant again.

Ache

Sometimes known as Guayaki, the Ache of eastern Paraguay hunted and gathered in a dense, subtropical, broadleaf, evergreen forest (Hill and Hurtado 1996, 1999). They were full-time hunters until the 1970s, when they were settled in a Catholic mission, and in the 1980s their demography, subsistence ecology, life history, and child development were studied. Foraging in the nearby forest still supplied 20 to 25 percent of their subsistence, and it was possible to reconstruct many of their presettlement social patterns. As for the context of infant care,

> women alternate between walking and carrying their young
> children, brief periods of vegetable, fruit, and larval food col-
> lection, and resting on the ground. Women spend very little
> time in direct food acquisition and in activities incompatible
> with childcare. Instead, they focus their attention on child su-
> pervision when not walking from one campsite to another. . . .
> Children younger than three years of age rarely venture more
> than a meter from their mother and spend some 80–100 per-
> cent of the time in tactile contact with them. (Kaplan and Dove
> 1987:191)

The Ache were traditionally highly nomadic, and when in the forest men provided 87 percent of subsistence calories and spent twice as
much time in the food quest as women did. So, unlike most hunter-
gatherers, the Ache approximated the common notion that women defer many subsistence activities in favor of child care. Indulgence is very
high: "Ache children of less than 4 years of age are spoiled by American standards (they are almost never chastised and win most conflicts
with parents simply by crying and whining)" (Kaplan and Dove 1987:197).
Older children are described as helpful and obedient. Subsequent quantitative ethnography confirmed and extended these descriptions (Hill
and Hurtado 1996:219–20): "Traditionally Ache infants spent the first
year of their life in close proximity to their mother, suckling at will and
sleeping in their mother's lap at night. Indeed, scan sampling and focal
infant follows suggest that in the forest, infants under one year of age
spend about 93 percent of daylight time in tactile contact with their
mother or father, and they are never set down on the ground or left
alone for more than a few seconds." This stricture is probably related to
the hostility of the environment, but in any case, "high-quality childcare
overrides other competing needs."

Mothers of infants and young children gather less food than other
women, despite having more dependents.

> After about one year of age Ache children still spend 40 per-
> cent of their daylight time in their mother's arms or lap, but
> they sit or stand on the ground next to their mothers 48 per-
> cent of the day. It is not until about three *years* of age that Ache
> children begin to spend significant amounts of time more than

one meter from their mother. Still, Ache children three to four years old spend 76 percent of their daylight time less than one meter away from their mother and are monitored almost constantly. (Hill and Hurtado 1996:220)

While traveling in the forest, infants are carried in a sling with skin-to-skin contact, their heads resting on the mother's chest. From about eighteen months they can ride on top of the mother's carrying basket, clinging to her head and ducking to avoid low-hanging branches and vines.

Ache nursing frequency resembles the !Kung pattern, but not weaning age or birth spacing. "Ache children generally continue nursing on demand until their mother is pregnant with her next child . . . although they may begin eating some solid foods such as armadillo fat or insect larvae . . . as early as 6–12 months. . . . Because Ache mothers wear little clothing and carry or sleep with children resting on their bare chest, nursing is frequent throughout the day and night" (Hill and Hurtado 1996:220).

Median age at weaning was twenty-five months in the settlement, but this may have been shorter than it was previously: "Unfortunately we have no way at present to determine the age at weaning in the precontact situation" (Hill and Hurtado 1996:221). But

> weaning is an extremely unpleasant experience . . . with children screaming, hitting, and throwing tantrums for several weeks . . . some mothers who became pregnant very soon after the birth of a child simply continued to breast-feed all the way through their next pregnancy, and then, if the interbirth interval was too short (i.e. less than two years), would simply kill the newborn child and continue nursing the first. (220–21)

Weaning from being carried, which occurs later, was equally unpleasant.

Nursing was frequent but less than for the !Kung. Bouts were about thirty minutes apart, and bout length declined from an average of more than ten minutes to about two minutes over the first eighteen months (Hill and Hurtado 1996). This pattern may have developed after reservation settlement and led to shorter birth spacing, although these measures did not predict birth interval in the small sample studied. Even in the presettlement period, birth spacing was shorter than for the !Kung, and some evidence suggests that Ache demography entailed rapid population growth followed by crashes (a more *r*-selected pattern), as opposed to the slow, steady population growth of the !Kung.

Agta

Although the Agta occupy diverse habitats, their main eco-niche is a somewhat seasonal tropical rain forest crossed by many streams, rivers, and waterfalls. They are widely distributed over the Sierra Madre Mountains, a rugged range paralleling the east coast of the main northern Philippines island (Griffin and Estioko-Griffin 1985; Griffin and Griffin 1999). Like the central African foragers, Agta rely on nearby agriculturalists, trading hunted meat for most of their plant foods and consumer goods. They are seminomadic and live in small camps. Although men hunt and fish full-time, Agta women are uniquely involved in hunting to an extent unknown among any other hunter-gatherers, bringing home up to half of the meat, including large game (Estioko-Griffin and Griffin 1981; Estioko-Griffin 1985; Griffin, Goodman, et al. 1992). Many anthropologists have thought of hunting as incompatible with extensive infant and child care, so it might be expected that this group would challenge any generalizations about maternal primacy.

During the main study years the population was in decline owing to high mortality, especially in the early years of life (Headland 1989). The Agta crude birth rate (CBR) was slightly above that of the !Kung; if the last child survived until the birth of the next sibling, birth spacing was slightly over three years, as retrospectively reported by women over forty-five (Goodman, Estioko-Griffin, et al. 1985). P. Bion Griffin and Marcus Griffin (1992:301) describe maternal care:

> The baby remains against the body of the mother nearly constantly in its first weeks, but is also in contact with the father, siblings if any, and other kin that may drop in to visit, nap, or play. . . . Babies sleep by mothers' breasts, between mother and father. . . . Grandparents may take in toddlers and older children on a "drop-in" basis or in the case of the parents' overnight departure for hunting and fishing. During the first 12 months an infant is usually carried in a sling at the mother's

> back, side, or front. . . . Nursed on demand, it is returned to the
> back for sleeping after suckling. Women are quite comfortable
> thus engaged in collection of forest materials, and some spo-
> radically hunt and kill game while transporting the baby. As
> the baby grows in its first year, it is increasingly handled by
> others, albeit in brief episodes . . . an infant under 1 year may be
> passed among several adults and youths, being returned to its
> mother if it becomes fussy.

The striking fact is that for some of the mothers hunting does not preclude continuous physical contact with their infants. Based on spot observations throughout the day, "infants are carried in a cloth sling much
of the time before exploratory crawling and first walking begins. Carrying does not cease then; usually a toddler is carried frequently by the
mother" (305).

Remove the references to hunting and this could be a description of
!Kung maternal care. Weaning occurs gradually, but weaning age can
be roughly estimated from the following: "Sometime when the child is
roughly between age 20 to 28 months, it nurses less and less. . . . This
gradual decrease in nursing seems to run about 3 to 6 months" (Thomas
Headland 2004, personal communication). For comparison with the
!Kung, we should try to estimate the earliest *completed* weaning age for
the Agta, which would be twenty-three months. Also, "small children
are almost never nursed after 28 to 30 months." From these accounts
we can reasonably estimate Agta weaning age as halfway between
twenty-three and twenty-nine months, or twenty-six months, although
this is lower than a previous published estimate, which stated that "at
about two and a quarter years, the nursing often continues but with less
intensity" (Early and Headland 1998:92–93).

Reconstructing Maternal Care: Phylogeny and History

We now attempt to reconstruct the history of the mother-infant relationship from the higher-primate condition to the post-industrial state,
based on good knowledge of higher primates (McKenna 1981, 1987;
Pereira and Fairbanks 1993; Martin 1995; Kappeler and Pereira 2003;
MacKinnon 2007), reasonably good knowledge of recent human hunter-
gatherers (Konner 2005), and good knowledge of both industrial and

non-industrial nonforaging societies (Whiting 1963; Whiting and Whiting 1975; Munroe, Munroe, et al. 1981; Whiting and Edwards 1988; LeVine, Dixon, et al. 1994; LeVine 1997; Munroe 2004). We use a comparative strategy to reconstruct evolutionary (to the hunting-gathering stage) and historical sequences.

For each domain of maternal care, probable events leading from our higher-primate ancestors to the condition found in human hunter-gatherers are suggested, followed by an account of the range seen in traditional farming and pastoral societies (the intermediate level), followed in turn by a sketch of the modern industrial adaptation, mainly as seen in the United States and Britain. The differences between hunter-gatherer adaptations and the others are due to cultural evolution.

It may be that the context of socialization in hunting-and-gathering societies is a "natural" one for development in our species, since these societies represent our environments of evolutionary adaptedness (EEAs). This argument for disjunction between our ancestral and current environments has been made for diet (Eaton and Konner 1985), chronic degenerative disease (Eaton, Konner, et al. 1988; Eaton and Eaton 1999), women's reproductive cancers (Eaton, Pike, et al. 1994), general cognitive processes (Barkow, Cosmides, et al. 1992; Cosmides and Tooby 2000; Tooby and Cosmides 2000), and emotional symptoms and syndromes such as depression and anxiety (Nesse and Lloyd 1992; Nesse 1999), in addition to long-standing hypotheses about socialization and childhood experience (Konner 1972, 1981, 1995; DeVore and Konner 1974; Konner and Shostak 1986).

However, several notes of caution are needed. First, the process of transformation from higher primates to human hunter-gatherers is evolutionary in the biological sense; the subsequent transitions are instances of sociohistorical change, such as may take place in a modern nation, a farming village, or an urban neighborhood in the course of a few decades, requiring no genetic change. There is no important biological difference between modern human hunter-gatherers and the inhabitants of modern industrial states.

Second, the brief characterization of each adaptation, as well as that of the "intermediate-level" societies, is a simplification. Hunting and gathering is defined to exclude equestrian hunting (a recent and divergent historical development), and greater emphasis is placed on warm-climate hunter-gatherers as being more representative of the human

EEAs. No attempt is made to exclude cases on the basis of other criteria, such as a history of contact with non-hunter-gatherers. Of the non-mounted hunter-gatherers, good information is available for the !Kung, the Efe, the Aka, the Agta, and the Ache, as detailed earlier. Generalizations beyond these to other hunter-gatherers, studied earlier and with more traditional methods, must be made with caution, but are strongly confirming.

Childbirth

The pregnant state alters the function of every major organ system and can bring out diseases toward which the mother has a genetic predisposition, such as diabetes or essential hypertension. But many of the half-million maternal deaths annually worldwide are due to childbirth itself—a major cause of mortality throughout history.

Higher-primate background. Evolutionary optimization is never perfect, but hominid evolution seems to have made birth especially suboptimal (Trevathan 1987; Rosenberg 1992; Rosenberg and Trevathan 2002). Great apes have an easy time at parturition (Chapter 5), as do most monkeys (Lindburg and Hazell 1972). This suggests that the last common ancestor (LCA), prior to the evolution of bipedal posture, had to endure perhaps two or three hours of discomfort. As hominins became adapted for upright walking, a stronger, thicker, less flexible pelvis made birth much more difficult; 4 million years later, an expanding brain confronted female hominins with cephalopelvic disproportion (CPD). The australopithecine pelvis was altered further to form the modern human pelvis (Berge 1998, 2002), but CPD worsened.

Human hunter-gatherers. In childbirth the !Kung were unusual in the cross-cultural range in that they preferred to give birth alone (Biesele 1997), citing an ideal of physical courage, although it was common only in higher parities (Konner and Shostak 1987). Of fifty-four births to fourteen women interviewed, the mother reported that she was entirely alone for the second stage of labor (the birth itself) in 57 percent. For the first two births, up to eight women other than the mother were involved. The favored position was squatting, and there were no specialized birth assistants, but the laboring woman's mother or older sister was likely to be there. Maternal mortality was comparable to that of other premodern populations.

Other hunter-gatherers do not share this ideal. Among the Ache, "all births have at least two major helpers . . . usually joined by a handful of secondary helpers" (Hill and Hurtado 1996:251); even the husband may be involved. For the Mbuti, most births were attended by the mother's closest friend, often a relationship formed at their initiation (Turnbull 1981). Among the Siriono of Bolivia, "childbirth normally takes place in the hut and is a public event" (Holmberg 1969:177); nevertheless, the mother commonly received no assistance. Of eight births directly observed by the ethnographer, "not a move was made by these onlookers to assist the parturient women, except in one case when twins were born" (178). The other births were unproblematic.

Intermediate-level societies. Births in gardening, farming, and herding societies were typically attended by nonspecialist helpers, often female relatives, and were characteristically conscious even where consciousness-altering substances were available (Newton and Newton 1972). "Birth in most societies is patterned not as an unaided solo act but as a social one in which others may help. . . . The emotional impact of the labor attendants is a key variable in the patterning of labor" (Newton and Newton 1972:162). Of sixty cultures for which information was available (Ford 1945), older women assisted in fifty-eight; the role of husbands varied. In a cross-cultural study of delivery position, of seventy-six non-European societies in the Human Relations Area Files, sixty-two used nonrecumbent postures—twenty-one kneeling, nineteen sitting, fifteen squatting, and five standing. All non-industrial societies had high maternal and neonatal mortality.

The modern United States. Although today in the United States one woman at most will die for every ten thousand births, a century ago the rate was perhaps one hundred times higher, and in the century before that higher still (Wertz and Wertz 1989). Widowers and stepmothers were common figures of life and lore, and the most likely cause of death in young women was death during childbirth or within one month after it. From the midnineteenth century to the midtwentieth, birth was removed steadily further from its social context. Disasters in the midnineteenth century notwithstanding—maternity hospital physicians were transferring deadly infections from mother to mother on their hands—childbirth became increasingly medicalized, with a transfer of control from home to hospital, from midwife to physician, from women to men. By the 1930s most American births took place in hospitals with physicians attending; the proportion rose from 5 to almost 100 percent in the first two-thirds of the century (Wertz and Wertz 1989), paralleling dramatic decreases in maternal and neonatal mortality. However, the association may not be causal, since most mortality was due to infection and probably responded more to public health measures than to medical interventions, as did infectious mortality generally (McKeown 1995).

Some obstetricians were receptive to "natural" childbirth (Goodrich and Thoms 1948), but overall death rates were still high, so medicalization intensified. By the 1950s, "twilight sleep"—a type of general anesthesia—removed the mother as a conscious presence, although she still had to be strapped down in some deliveries to prevent thrashing. Delivery rooms were surgical operating rooms; only medical personnel were admitted (Wertz and Wertz 1989; Mitford 1992). Yet the psychological aspects of childbirth received increasing attention (Macfarlane 1977). Some prospective parents demanded change, and hospitals adapted: allowing full maternal consciousness; keeping the surgical apparatus in the background, so that the room seemed familiar and calm; having a father or friend (*doula*) with the mother for emotional support, which in a randomized controlled trial shortened labor (Kennell, Klaus, et al. 1991); and encouraging interaction among infant, mother, and father immediately after birth (Seashore, Leifer, et al. 1973; Leiderman and Seashore 1975; Davis-Floyd 1992). Many births now involve midwives, but they are often certified nurse-midwives, practicing in hospitals under obstetric supervision (Mathews and Zadak 1991; Brody 1993). Other factors promoted by natural childbirth advocates, such as walking during labor, did not prove helpful (Bloom, McIntire, et al. 1998; Cefalo and Bowes 1998).

But natural childbirth did not replace technology (Mitford 1992). During the same period, Cesarean sections rose from less than 10 percent to more than 25 percent of births; the rate remains this high, despite efforts to lower it. Electronic monitoring during labor, using an electrode attached to the fetal scalp, is commonplace. Methods such as epidural injection replaced general anesthesia. Artificial induction and suppression of premature labor have become more common. These interventions spread in some population groups, while natural childbirth remained popular in others. Some women delivered at home against medical advice. But 10 to 20 percent of normal pregnancies classified as low-risk end with a serious delivery problem. Since doctors could not force people into the hospital, they had to lure them in or risk losing infants, mothers, and income.

As obstetricians say, "Childbirth may be natural for the species, but it's damn near pathological for the individual." Yet it has remained quite safe; none of the dire predictions made about these changes came to pass, while intervention has come under scrutiny. Some European nations—Ireland, for instance—achieved the same declines in mortality with one-fourth as many C-sections (O'Driscoll and Foley 1983; O'Driscoll, Foley, et al. 1984). Extensive research now seeks ways of reducing the rate (Belizan, Althabe, et al. 2007). Electronic monitoring spread in American hospitals, but a study in Ireland's leading obstetric hospital showed that a midwife checking the baby's heartbeat with a stethoscope on the abdominal wall was as safe as electronic monitoring; a study in the United States and Canada confirmed this (Shy, Luthy, et al. 1990). A leading obstetrician labeled it "a disappointing story" (Freeman 1990), and other studies have raised similar doubts (Leveno, Cunningham, et al. 1986; Sandmire 1990; Chen and Wang 2006; Devane, Lalor, et al. 2007; Freeman 2007).

A landmark study reported outcomes in almost twelve thousand women attempting to give birth outside of hospitals in eighty-four freestanding birth centers throughout the United States. Almost four out of five of the births were supervised by midwives. One in six of the women had to be transferred to a hospital, but the C-section rate was less than 5 percent. Complications and infant mortality were no higher than for comparably low-risk births taking place entirely in hospitals, and there were no maternal deaths, suggesting that birth centers are safe if located in or near hospitals (Rooks, Weatherby, et al. 1989). Other studies of free-standing birth centers have had similar results (Eakins 1989; David, von Schwarzenfeld, et al. 1999; Walsh and Downe 2004).

Although minimized intervention is called natural childbirth, it might better be called *cultural* (as opposed to *medical*) childbirth, since it restores childbirth as a rite of passage, a conscious family ritual (Davis-Floyd 1992). Pushing through the pain is part of the ritual, along with special breathing rhythms, talking, encouraging, coaching, hand-holding, hugging, even swearing and husband-blaming. Humans have difficult and perilous births. In the past, cultural rules and rites were introduced because nothing else could be done to confront the dangers;

now they return because the medical framework has made birth relatively safe.

Physical Contact with Mother and Others

Mammals vary in mother-infant physical contact (Blurton Jones 1972a). "Cache and carry" names two of the options, distinguishing mammals that hide their infants while mothers forage (such as some ungulates, rabbits, and tree shrews) from those whose infants cling to their mothers or are carried (marsupials, bats, and most primates). A third type, "following," involves low direct physical contact (except for nursing) but high proximity (zebras, wildebeest, and other herd animals). Nonhuman higher primates without exception maintain continuous physical contact with infants in the first weeks of life; except in marmosets, the mother accounts for most of this contact. Physical contact gradually wanes over the first weeks to months, depending on the pace of development in a given species.

Although the !Kung data came mainly from out-of-the-sling observations (underestimating total physical contact), carrying devices afford an approach to assessing physical contact that can be inferred from older ethnographies (Whiting 1971). Variations include (1) almost constant carrying in a sling at the mother's side, back, or front, with or without direct skin contact; (2) some carrying alternating with time in a crib, cradle, or hammock or on a blanket on the ground or floor; and (3) very little or no carrying, with infants tied in a cradleboard or tightly swaddled. (Cradleboarded or swaddled infants can also be carried.) These variations are related to ecological conditions, especially climate: forty of forty-eight cultures in the tropics had close and frequent physical contact, usually with carrying devices, whereas twenty-nine of thirty-seven nontropical cultures used heavy swaddling or cradleboards, regardless of continent.

Whiting argued that hunting and gathering is neither necessary nor sufficient for carrying in close physical contact. However, two exceptions to the rule about cold-climate societies are the Eskimo and the Yahgan of Patagonia, both resembling "classical" hunter-gatherers, except for inhabiting cold climates. The dominant hunter-gatherer pattern of infant care apparently persisted when these two peoples moved toward the poles. Hunting and gathering may thus be a sufficient condition for close contact but not a necessary one. But since most of human evolution took place in warm climates, the inference that early humans had close physical contact with infants, probably using a sling, remains sound, while intermediate-level societies cover the range from close contact in a sling to little or no direct contact except for nursing. It is likely that leaving behind the hunting-and-gathering mode of subsistence permitted, but did not cause, a decrease in direct contact using a sling as the main carrying method.

Until the recent reintroduction of the baby carrier, the Western infant was kept horizontal in a baby carriage or pram or reclined in a stroller. The widespread use of baby carriers, including sling- and pouchlike devices, beginning in the late twentieth century, may be a partial return to ancestral patterns, although without high levels of skin-to-skin contact.

Nursing Frequency

Mammals can also be divided by feeding type into "continual" versus "spaced" feeders (Ben Shaul 1962; Blurton Jones 1972b). Continual feeders cling (most primates, bats, and marsupials) or follow (the most precocial ungulates). Spaced feeders leave infants in nests (tree shrews and rabbits) or movable caches (eland and certain other ungulates). Milk composition (Ben Shaul 1962) and sucking rate (Wolff 1968) are related to the spacing of feeds: continual feeders have dilute milk, with lower fat and protein content, and suck slowly; spaced feeders have concentrated milk and suck rapidly. The lipid and protein composition and sucking rates of higher primates, including humans, are consistent with their classification as continual feeders. This is the case for most monkeys (Horwich 1974), chimpanzees (Clark 1977), and human hunter-gatherers (Konner 2005).

In intermediate-level (gardening, farming, and/or herding) societies the range of variation is great. While high feeding frequency occurs in some of these cultures, in others this pattern is precluded by the organization of subsistence, especially the mother's workload (Nerlove 1974; B. Whiting 1963, 1972; Whiting and Whiting 1975a; Whiting and Edwards 1988), which often results in daily mother-infant separations of several hours. Among the Kikuyu of Kenya, for example, the mother may work in the garden part of the day, leaving her infant with a sister or cousin in the home village compound (Leiderman and Leiderman 1977). This pattern is seen in many cultures (Lancy 2008).

Mothers in the West have long been spaced feeders. Differences between "demand" and "scheduled" feeding may be minor, since in midtwentieth-century American homes "demand" feeding sorted itself out to about six feeds a day, every four hours (Aldrich and Hewitt 1947); feeds about every three hours have become the norm in more recent decades. It would be useful to know more about the effects of feed patterns on infant feeding difficulty, colic, sleep-activity cycles, and blood glucose dynamics, as well as on maternal mood, milk production, and likelihood of conception.

Age at Weaning

In most higher primates, weaning is related to the onset of pregnancy. In many Old World monkeys the next pregnancy occurs when the infant is about a year old, although in baboons (which are larger and slower to develop) it is usually two years (DeVore 1963). In chimpanzees (Goodall 1967; Clark 1977) and bonobos, mothers become pregnant again at about five years, and in human hunter-gatherers it occurs at two and a half to three years (Konner 2005), thanks to the unique human trait of provisioning after weaning (Lancaster and Lancaster 1987; Kaplan, Hill, et al. 2000). In many traditional cultures, offering premasticated food ("kiss-feeding") ensures that infants share the mother's germs, so that the antibodies in milk and some delivered with the food (particularly immunoglobulin A) are specific to microbes both mother and infant have been exposed to. The advantage of such softened food in the weaning process, partly predigested by the mother's saliva, is probably significant; premastication and kiss-feeding are widely seen in intermediate-level societies as well (Eibl-Eibesfeldt 1983, 1988). Finally, the invention of cooking would have greatly facilitated earlier weaning (Wrangham 2009).

However, the weaning age for hunter-gatherers is very late by modern standards. As noted earlier, it can be difficult to determine, since in many traditional cultures there is high variability, a gradual decline, and no fixed limit on nursing. In Bofi, foragers in the Central African Republic, the process was driven mainly by the child, with the result that weaning distress as measured by fussing and crying was very low (Fouts, Hewlett, et al. 2001, 2005; Fouts and Lamb 2005). The children were all healthy, between eighteen and fifty-nine months old, and with two living parents (Fouts and Lamb 2005). They were rated as high-nursing, low-nursing, and weaned; the mean ages in months were 24.9 months for high-nursing (N = 9), 33.7 months for low-nursing (N = 5), and 49.1 months for weaned (N = 8). The low-nursing group was close to being weaned, so it is reasonable to assign the Bofi a weaning age of thirty-five months (Fouts and Lamb 2005). Thus the estimated weaning ages for seven carefully studied hunter-gatherer groups are: !Kung, 42 months; Hadza, 30 months; Efe, 30 months; Aka, 42 months; Ache, 25 months; Agta, 27 months; and Bofi, 35 months—for a mean of 37.3 months (Konner 2005).

Age at weaning in intermediate-level societies reportedly ranges from immediately after birth in the Marquesan Islands (Linton 1939) to a number of cultures that wean as late as the !Kung. As with direct-contact carrying, leaving the hunting-gathering subsistence mode behind appears to permit societies to wean earlier, depending on local constraints. The availability of cow's milk and cereal gruels may be critical. The decline of weaning age in !Kung hunter-gatherers as they became more settled and gained access to cow's milk exemplifies this possibility. Still, 83 percent of the 176 societies in the sample studied by Herbert Barry, Margaret Bacon, and their colleagues (1967) that were *not* hunter-gatherers had weaning ages of two years or older.

Daniel Sellen (2001) refined this cross-cultural analysis. In 112 non-industrial societies there is a normal distribution of weaning ages; 83 percent weaned at age two or later, 51 percent at two and a half or later, and 35 percent at three or later. Table 15.1 (from the same article) shows other descriptive statistics for the sample of societies, including the age at which liquid and solid foods were introduced. The data strongly suggest that the hunter-gatherer pattern of breast-feeding for at least two and a half years is preserved in many intermediate-level societies. So weaning is late in intermediate-level societies, though on average not as late as among hunter-gatherers. Most settled agricultural populations have in recent times had a birth interval of two to three years, with weaning in the second year (Morley 1973:306), part of the reason that agricultural societies have higher fertility than hunter-gatherers (Boone

Table 15.1. Transitions in Infant Feeding across Cultures

Transition (Number of Cultures)	Average Age Reported as Typical for the Transition		
	Mean ± SD	Median	Mode
Introduce liquids (18)	4.5 ± 6.0	2.0	6.0
Introduce solids (42)	5.0 ± 4.0	5.5	6.0
End breast-feeding (112)	29.0 ± 10.0	29.5	30.0

Note: Statistics refer to age in months.
Source: Sellen (2001:2712).

2002). As Sellen (2007) has pointed out, breast-feeding is to some extent a facultative adaptation in humans, and this flexibility sometimes leads to bad outcomes for the infant.

In the United States in the 1970s about 10 percent of infants were breast-fed at three months of age, and about 5 percent at six months (Fomon 1974:9), as opposed to 58 percent breast-fed at a year of age during the period 1911 to 1916 (2). Similar declines and similarly low late-twentieth-century levels were observed in Britain, Sweden, Poland, and other modern industrial countries. The developing world has now seen a similar decline in breast-feeding. For three decades the American Academy of Pediatrics (1978) has officially recommended breast-feeding, and this has increased breast-feeding in the United States, but it has not affected the worldwide decline, especially among the poor (Sellen 1998; Corbett 2000; Guttman and Zimmerman 2000).

Sleeping Distance

Mother-infant sleeping distance is a neglected feature of the infant's environment, even though bedtime protest and night waking are very common (Bernal 1973; Spock and Parker 1998; Goodlin-Jones, Burnham, et al. 2001). In all higher primates and human hunter-gatherers, mother and infant slept in immediate proximity if not direct physical contact. As noted earlier, hunter-gatherer mothers report that their infants wake up repeatedly during the night to nurse, and some additional nighttime nursing bouts take place while the mother sleeps; this has also been shown in co-sleeping American mother-infant pairs observed in the laboratory (McKenna, Mosko, et al. 1999).

Of ninety cultures in the sample of Barry, Bacon, and their colleagues

(1967)—the number for which information was available—mother and infant slept in the same bed in forty-one, in the same room with bed unspecified in thirty, and in the same room in separate beds in nineteen. In none of the ninety did mother and infant sleep in separate rooms, a pattern that probably did not precede the industrial state. Current Western culture derives from that of the agricultural peoples of northern Europe, who used cradles and swaddling. Americans often have infants sleeping in separate rooms from their parents. The syndromes of bedtime protest (Spock 1968; Spock and Parker 1998) and night waking (Bernal 1973) may be artifacts of this sleeping arrangement (Konner and Super 1987; McKenna, Thoman, et al. 1993; McKenna, Mosko, et al. 1999).

But the norm is often bypassed (Rosenfeld, Wenegrat, et al. 1982). A well-known pediatrician explained in a mass-circulation magazine how to get infants to sleep in a separate room; after receiving many letters from mothers who slept with their babies, he wrote again saying that this was acceptable (Terry B. Brazelton, personal communication, 1979). Systematic studies also show that some mothers in Western cultures have returned to the practice of co-sleeping (Hanks and Rebelsky 1977; Schacter, Fuchs, et al. 1989; Brazelton 1990; Sears and White 1999), with consequences for reproductive endocrinology as well as night waking (Elias, Teas, et al. 1986; Elias, Nicholson, et al. 1986; Stern, Konner, et al. 1986). But an eighteen-year longitudinal study of two hundred California families, many living a "countercultural" lifestyle, showed no long-term consequences of early bed-sharing for any psychological measure (Okami, Weisner, et al. 2002). Meanwhile, many parents in southern Europe and Japan never abandoned traditional co-sleeping to begin with (Caudill and Plath 1966; Gatti 1989).

Attachment Theory and the Mother-Infant Bond

John Bowlby (1969, 1973, 1980) brought together a wide range of research in a theoretical framework in the three volumes of *Attachment and Loss,* and research on behavior, development, and maternal care in non-human primates and human hunter-gatherers was its logical testing ground. For Bowlby, the emphasis in the first half-year was on visual and vocal-auditory mechanisms, such as visual-postural orientation, smiling, crying, the cessation of crying, and noncry vocalizations. He criticized "secondary drive" theory, which stressed the role of satiation of hunger and the pleasure of suckling as primary reinforcers for attachment (Freud 1920). Later in the first year, with the development of effective locomotion, proximity-maintaining mechanisms come into play—grasping, clinging, scrambling, and climbing on the mother—and later still, following behavior and the use of the mother as a base for exploration.

In lieu of learning and drive theories, Bowlby proposed an ethological unfolding of attachment behavior according to an unknown genetic program (for further discussion, see Chapter 8). This system "seeks" an object much as the neural mechanisms underlying imprinting in precocial birds do, although the higher-primate process is much longer and more gradual. The behaviors in question will fully emerge, change, and function in certain predictable ways only when an appropriate object is found. Because immature organisms must stay close to parents to prevent death by exposure or predation, these underlying neural mechanisms are under strong selection, and it is now essential to mental health for infants to pass through such an attachment.

But with current knowledge of human evolution, we know that the mother-infant bond is built on contact mechanisms from birth, not just in the second half-year. A possible role for reinforcement is suggested by the increasingly social complexion of hunter-gatherer nursing as the infant grows and by the web of attachment behaviors in the context of nursing. Habits in many animals are strengthened by reinforcing them with satiation of hunger. Since some components of attachment are habits, they probably can be strengthened by primary drive reinforcement, which is not to say that such reinforcement is necessary.

The hunter-gatherer evidence also suggests that the danger of death by predation and exposure is only one of the selective forces favoring attachment. If inadequate mothering produces sexually inept males and abusive mothers, this would result in selection as strong or stronger than any resulting from predation. The function of attachment as preparation for adult social behavior—essential to survival and reproduction—would give it as strong an edge in evolution as many advantageous physical characteristics. Finally, the hunter-gatherer data support the view that competent subsistence behavior (with its attendant emotions) is acquired beginning in infancy and is made possible by

Table 15.2.   Functions of the Mother-Infant Bond

Nutrition, hydration, and temperature homeostasis
Protection from predators, enemies, and rivals
Protection from insects, scorpions, snakes, etc.
Optimal birth spacing (nursing-frequency-dependent?)
Transfer of general immunity (nursing-dependent)
Transfer of specific immunity (nursing- and proximity-dependent)
Nongenetic transmission of behavior and psychological dispositions

proximity to adult models; this proximity, in turn, depends on attachment.

In the hunter-gatherer context infants are cared for primarily by their mothers, especially during the second half-year of life. Nutrition and protection from gastrointestinal disease and other illnesses (a danger even greater than that of predation) requires extended continual nursing and an even temperature and state. There is no nutritional or immunological substitute for milk. The adaptive functions of the mother-infant bond are shown in Table 15.2. However, as we will see in the next chapter, many cultures provide forms of multiple caregiving. As Margaret Mead (1962) pointed out in a critique of Bowlby, many studies of multiple caregiving, from traditional polygynous cultures to modern Israeli kibbutzim, have failed to show that multiple care has undesirable consequences, given a nurturing and uninterrupted human environment. Finally, where the risk of infant death is low, there is no evidence of ultimate biological advantage in leaving infant care exclusively to women; mounting evidence shows that men can be suitable caregivers and objects of infant attachment (Pruett 1998; Lamb and Lewis 2004).

Interlude 4

Thinking about Maternal Sentiment

There have been serious challenges to the notion of maternal instinct, and even maternal sentiment, and no treatment of the evolution of childhood can ignore them. They are taken to undermine a common concept of maternal care, which, according to critics, presents mothering as a compelling universal characteristic of women; it is not. But between that false claim of a universal instinct and the equally false claim that motherhood is something purely culturally constructed lies a great range of possibilities.

Clearly, maternal instinct can be maladaptive. One gorilla mother carried her dead infant for four days before letting it go (Angier 2008). Of course, any Darwinian worth her salt can come up with explanations that might make this seem adaptive. The mom might have a baby destined to recover; she might be protected from unwanted sex or aggression a little longer; she might be advertising what a devoted mother she is; and so on. Many would find this sort of reasoning unconvincing in the face of the risks: infectious disease, predation, and the burden of caring for an infant that is dead. Yet, although this case was observed in captivity, similar behavior has been seen in the wild in other higher-primate species (Sugiyama, Kurita, et al. 2009).

But if this behavior is not adaptive, why has it not been culled by natural selection? Well, for the same reason that low back pain and cephalopelvic disproportion stay with us: all three result from adaptive compromise. The Darwinian mill grinds slowly and in the end pretty finely, but it leaves a few coarse grains in every species' adaptation. Of course, species differ markedly. Hamsters eat their young in several different situations, while some nonhuman primates show maternal devotion after death. Human mothers, I have argued, are protected for two months from the most severe grief by the slow emergence of infant sociality (Chapter 7), and we considered arguments for predicting parental grief from reproductive value (Chapter 14). But humans, more than other higher primates, can accumulate experience (even across generations), anticipate consequences, and make choices. It seems that we should be able to use our cognitive powers and cultural traditions to match choices to circumstances, so that in our case "maternal instinct" would be small to nonexistent. Several challenges to the notion of maternal instinct or even maternal sentiment have been fielded; we will consider three of them.

Selective Maternal Neglect in Northeastern Brazil

In papers in the mid-1980s and in *Death without Weeping* (1992), the anthropologist Nancy Scheper-Hughes argued that some conditions lead to maternal withdrawal of affection in anticipation of death or even to infanticide. Maternal withdrawal is selective, so that the weak die more quickly, leaving resources for those more likely to survive. According to Scheper-Hughes, "maternal thinking and practices are *socially produced* rather than determined by a psychobiological script of innate or universal emotions, such as has been suggested in the biomedical literature on 'maternal bonding' and, more recently, in the new feminist scholarship on maternal sentiments" (Scheper-Hughes 1985:292 [emphasis in original]). Setting aside for the moment whether this is a fair description of psychobiological or biomedical claims, let us focus briefly on the evidence.

The setting is a hillside shantytown called Alto do Cruzeiro in the state of Pernambuco in northeastern Brazil. The data are impressive, the mothers' narratives moving, and the interpretation sympathetic to Brazilian women in abject poverty caring for children in situations that would challenge the sentiments of a saint. Scheper-Hughes interviewed 72 women who had 686 pregnancies, of which 251 resulted in death between birth and age five years. The average woman had 9.5 pregnancies, 8.1 live births, and 3.5 infant or child deaths; 70 percent of the deaths occurred before age six months, and 82 percent in the first year. Virtually all of this reproductive experience took place in the stress of extreme poverty and ever-present, potentially deadly, infectious childhood diseases, especially diarrheas.

Scheper-Hughes herself tried to save a child and was told, "If a baby *wants* to die, it *will* die," because of "the innate weakness of the child." Of such children, mothers often said something like "It is best to leave them die" (Scheper-Hughes 1985:294).

> It became painfully apparent that Alto mothers were often describing the symptoms of severe malnutrition and gastroenteric illness further complicated by their own selective inattention. Untreated diarrheas

and dehydration contributed to the baby's passivity, his or her disin-
terest in food, and developmental delays. . . . Because these hungry
and dehydrated babies are so passive and uncomplaining, their mothers can easily forget to attend to their needs, and can distance themselves emotionally from what comes to appear as an *unnatural* child,
an angel of death that was never meant to live. . . . A mother speaks of
having "pity" for such a child, but her grief is as attenuated as her attachment to a baby who never demonstrated more than a fragile hold
on life. (305–6)

Scheper-Hughes's compassion for these mothers is compelling; when she
challenged a woman about her inability to breast-feed, "she responded an-
grily, pointing to her breast, 'Look. They can suck and suck all they want, but
all they will get from me is blood'" (303).

We will return to whether this tragic nexus of suffering is relevant to "a
psychobiological script of innate or universal emotions," but first consider
another study of similarly impoverished mothers in northeastern Brazil (Nations and Rebhun 1988). Done at around the same time in the nearby state
of Ceará, this study involved three communities: Pacatuba, Guaiuba, and
Gonçalves Dias. The authors saw their investigation as a challenge to
Scheper-Hughes's formulation. One mother, Eveline, talked about her daughter Rosa, who had died at a year of age from "evil eye, teething sickness, and
doença de criança"—the same terms used by Scheper-Hughes to describe the
illnesses mothers give up on:

The kid was living dying. She was so thin. . . . We had faith in God and
faith in the doctor . . . but she died. . . . We have to be resigned (confor-
mar) to what God does . . . but we always had her clean, kept her
carefully. She never crawled on the floor or was in the hot sun. She
was always with us. But if God wants to take her, I have to bear my
cross. . . . Oh my God, we felt so much, we were dying living. . . . I
would lie there holding her, crying and trembling, seeing that in a
moment I would be without her, while her father and grandfather ran
here and there looking for medicines. . . . Now where I open the suit-
case and see her shoes, her little baptismal shoes, oh my God in
heaven, who has the courage to look? My father said, "You have to
accept it because there are other children." . . . I don't have the courage to look at Rosa's things. Because I was the one who fought for
her, day and night. . . . I couldn't eat. . . . Oh my sweet Jesus, when

my little girl got released from the hospital, I took her home. . . . I was almost dying of crying for happiness because I thought she would escape, but . . . she got worse again. (Nations and Rebhun 1988:143–44)

This quote is full of exceptions to all the generalizations made by Scheper-Hughes about mothers in a similar community in the same region of Brazil, with similar patterns of poverty, reproduction, childhood illness, and death.

But this mother's reactions were not exceptional. In contrast to the claim that mothers fail to seek medical help, almost all in this study did. In sixty-five terminal disease episodes, expert help (traditional or modern healers) was sought for the infant or child in all but three (95 percent), and in these three cases the parents said they would have done so if the illness had not become fatal so rapidly (Nations and Rebhun 1988:155). Physicians were consulted in 66 percent of the deaths. In contrast to the claim of grief avoidance, "informants confirmed again and again that losing a child is a terribly painful event" (59). Dacia of Pacatuba echoed Eveline of Guaiatuba, saying, "This is a horrible thing for us. Five children of mine have died. Nobody has desire to eat, nobody wants to do anything. It seems like life is over. All this lasts a long time, we are all knotted up inside . . . it seems like life is over." Another asked, "Who is the mother who does not cry when her baby cries?" (160).

Conformaçao—resignation and acceptance of God's will—is the goal, as it is in some sense for grievers everywhere, especially parents who have other children to care for. Moreover, in the communities studied by Nations and Rebhun, a premium is placed on not crying in public, which would lead some observers to conclude that there is little or no grieving. But with probing, mothers revealed folk beliefs that offer another reason to try to avoid tears.

Our informants believe that when a child dies, his consciousness is liberated from his body in the form of a little angel with wings (*anginho*), which then flies to heaven. . . . Our informants smiled as they described chubby *anginho* tumbling and playing happily in heaven. . . . Tears harm the *anginho*, for a mother's tears will wet the winding sheet or wings of her little dead angel, who will be knocked out of the sky and forced to wander on earth, harming relatives. . . . This lack of crying is a conscious effort; mothers describe turning their backs, biting their lips, balling their fists into their eyes to keep from crying. (Nations and Rebhun 1988:162)

Anginho is translated by Scheper-Hughes—Nations and Rebhun believe mistranslated—as "angel of death"; she says that some mothers' refusal to forget is seen as "inappropriate or even as symptomatic of a kind of insanity" (Scheper-Hughes 1985:312). But "our informants emphasize however that maternal grief is normal and expected, not insane at all. Mothers dry their tears specifically to allow their dead children to fly to heaven, where prayers can reach them" (Nations, Camino, et al. 1988:162). Grief *can* be maladaptive, and some cultures attempt to curtail it, but that does not mean it is absent.

Historical data on families that lost children in more favorable circumstances support this interpretation. Cotton Mather, the preacher-writer of colonial New England, lost thirteen of his fifteen children. His terse but clear diary shows that he mourned each loss. In 1694, when his four-year-old daughter Katherine—then his only child—was gravely ill, he wrote, "My little and my only, Katherine, was taken so dangerously sick, that small hope of her life was left unto us. In my distress, when I saw the Lord thus quenching the coal that was left unto me, and rending him out of my bosom one that had lived so long with me . . . I cast myself at the feet of his holy Sovereignty" (Pollock 1987:110).

Katherine survived, but later another child, age two and a half, did not: "I am again called unto the sacrifice of my dear, dear daughter Jerusha. I begged, that such a bitter cup, as the death of that lovely child, might pass from me. . . . Just before she died, she asked me to pray with her; which I did, with a distressed, but resigning soul. . . . Lord, I am oppressed; undertake for me!" (Pollock 1987:125). Mothers, especially of infants, were more affected. William Byrd, a planter, colonial official, and author, recorded his and his wife's reactions to the death of their six-month-old son in 1710:

> *On the day of the death:* . . . My wife was much afflicted but I submitted to His judgement better, notwithstanding I was very sensible of my loss. . . .
>
> *On the following day:* My wife had several fits of tears for our dear son but kept within the bounds of submission.
>
> *Three days later:* My wife continued to be exceedingly afflicted for the loss of her child, notwithstanding I comforted her as well as I could. (124–25)

In the context of high mortality, these parents bowed to what they saw as God's will, but grieved greatly and had neither a sense of inevitability nor anticipatory withdrawal of care or affection.

In one witchcraft trial, held at Springfield, Massachusetts, in 1651, part of the evidence against the accused, Hugh Parsons, was that he had allegedly not grieved over his infant's death; he tried to defend himself by proving that he had indeed grieved. John Demos (personal communication, September 18, 2009), a leading authority on witchcraft in early New England, wrote as follows:

> Several witnesses claimed that Hugh "shewed no kind of sorrow for ye death of his child." Supposedly, he remained overnight in a meadow at the moment of crisis, talked casually with neighbors about the event ("He came into my house and said 'I here my child is dead, but I will cutt a pipe of tobacco before I go home'"), and afterward declined to grieve "openly" with his wife. . . . He responded that, "he was very full of sorrow for the death of it in private, tho not in publik." . . . Indeed, with one of his friends "he could not speak to him for weeping"; moreover, . . . "he saith that when his child was sick and like to dye, he run barefoot and barelegged, and with teares, . . . to desire Goody Cooly [a local midwife] to come to his wife because his child was so ill." . . . His restraint in public, and in front of Mary, he explained this way: "he was loath to express any sorrow before his wife because of the weak condition she was in at that time."

Parsons was convicted. His wife, also accused of witchcraft, denied that charge, but confessed to killing the child by natural means; she was executed, but her confession led a higher court to reverse her husband's conviction. What the case clearly shows is that despite high mortality in colonial New England, grieving was expected and real—so much so that it was invoked, by both accuser and accused, in a trial of witchcraft, a capital offense.

Adoption for Minor Marriages in Traditional Taiwan

If some parents give up on dying children, others give up living ones, even quite willingly. Arthur Wolf (1993, 2003), who made great contributions to our understanding of incest avoidance (Interlude 6), has turned his attention to mothers. In a paper called "Maternal Sentiments: How Strong Are They?" he used a quantitative ethnographic-historical method to study adoption for minor marriage in Taiwan from the viewpoint of mothers giving up their infants. Like Nancy Scheper-Hughes, but with a very different phenomenon, Wolf conceptualized his study as undermining biological notions of motherhood.

The data, from eleven villages and two small towns in northern Taiwan, are striking. From 1906 to 1910, records show that of 1,094 live female births, the cumulative proportions given up for adoption were .547 by age one, .620 by two, .678 by three, .736 by five, and .807 by fifteen. The same figures for the 1,559 girls born from 1926 to 1930 were .440, .505, .543, .571, and .600. These data reflect Taiwanese traditions generally and also apply to traditional Japan and other Far East communities. A breakdown by months of age is also illuminating. Over the first two years of life, there are dramatic changes. Wolf (2003:S34) writes, "Adoption rates were very low during the first month following birth, peaked in the third, fourth, and fifth months, and then gradually declined. This means that the great majority of the Taiwanese women who gave away a child had been caring for her for several months before the adoption."

One inference is that the claim that infants are most easily given up in the first three days after birth and that breast-feeding activates maternal hormones that make this increasingly difficult are undermined for this culture. However, there is much that is comprehensible in these data. The low level of adoptions in the first month corresponds to the presocial phase, when death from infection is highest. The sharp rise in adoptions in the third month corresponds to the normal decline of neonatal crying and to the sociality heralded by social smiling and gaze contact, a maturational transformation owed mainly to brain development. The marked (not gradual) decline after five months—the rate in months 9 to 11 is less than one-third of that in months 3 to 5—corresponds to the maturating capacity for attachment, the period when attachment becomes well established in most human infants in all cultures.

There is much in these data to reassure a psychobiologist, if we assume that what is driving the age curve of adoption is the normal maturation of the infant's social brain through the changing mother-infant relationship, not the mother's hormonal state. The latter might help explain the low level of adoption in the first month, but the main reason for the delay is probably the need to stabilize the infant's biological state and to allow her to develop the stunning social competencies that elicit attention and care from everyone, not just mothers; this would help the infant survive in her new adoptive environment. But this window of opportunity closes two-thirds of the way with the development of attachment; the infant has singled the mother out, and she knows it. Just as we saw in the Efe, famous for multiple caregiving, and the Aka, famous for fathering, here too the age of attachment returns the infant strongly to the primary caregiver. The difficulty of adoption increases for both infant and mother, and the continuing decline in adoption with age suggests that it gets harder. The monthly risk of adoption in the second year of life is a little over one of every one hundred girls, or one of every two hundred infants. Calculated from the annual data in Wolf's Table 1, the monthly risk of adoption in the third year is .0039, and it drops to .0018 at ages three to five—a two-thirds drop between the second and third years and another 50 percent drop thereafter.

Still, adoption adds up over time, and despite the correspondence of the age curve with psychobiological predictions, this pattern of fosterage does present a challenge to the notion of maternal sentiments, resolvable only through the lens of culture. Wolf shows that wealth and poverty do not matter and that the major predictor of adoptions other than age is the composition of the sibship. Girls with no older siblings had an adoption risk of .213 by age one; this doubled for girls with an older sister and almost tripled for girls with an older brother. This is telling.

Recall that Wolf began by studying minor marriage, the pervasive custom of arranging marriage for daughters in infancy or childhood and sending them to live in their future husband's household. We do not know the history of this cultural pattern, but clearly, once a threshold was reached, it would be driven to fixation at a high level by the need to find spouses for your children of both sexes. If you wanted a good match, you had to make sure you were not left behind. In this severely patriarchal culture, daughters were devalued (although, as we will see, they can be devalued far more), but they were still useful pawns in the game of finding wives for sons. Daughters were not given away; they were traded.

This too has important sociobiological implications. For almost all parents everywhere, marriage is ultimately one of the main hopes cherished for children of both sexes. In traditional northern Taiwan, if you wanted the best for your sons *and* your daughters, you had to give your daughters away. Second, a high proportion of the mothers who gave away daughters (especially the wealthier ones) also *gained* adoptive daughters:

> How to find wives for their sons was an anxiety-laden problem for most Taiwanese families. Waiting until they reached marriageable age was dangerous because by then the great majority of the girls would have been claimed by other families. Besides, a marriage involving a young adult was an expensive undertaking, usually costing

more than most farm families earned in a year. Consequently, the most appealing strategy was to give away one's daughters and raise daughters-in-law in their place. Most men thought that any healthy child would do, but women preferred to adopt an infant, "because the girl you nurse yourself will grow up just like a daughter and listen to what you tell her." (Wolf 2003:S34)

This adoption preference is not a psychobiological surprise.

Anthropologists invited to comment on Wolf's article raise interesting questions. Min-Tao Hsu (2003:S43) writes, "I believe that he plays down . . . the women's emotional responses." Among other informants, she goes on, "I asked my maternal grandmother about her feelings about giving my mother away, and she spoke of sadness, fate, and 'the choice of no choice.' Crying and smiling at the same time, she said, 'You'll understand when you get married. . . . There are so many things we women need to do, whether it feels good or not.'"

Li Shuzhuo (2003:S44) writes about the "missing girls" phenomenon, the well-known skewing of sex ratios in China and some other Asian societies, due to infanticide, abandonment, and neglect. Li's research shows that two-thirds of the excess female deaths occur on the first postnatal day and that

> most cases were instigated by parents-in-law, acquiesced in by mothers, and undertaken by mothers-in-law. When we talked to several mothers of lost girls, they cried and said that they had no power to oppose their parents-in-law and husbands. . . . I argue that these women's natural maternal sentiments are "hidden" and that their behaviors need to be understood in the context of son preference.

And Sonia Ryang (2003:44) writes that Wolf seems to see maternal sentiments as

> almost synonymous with the maternal instinct to protect and raise the young, which, as far as I can see from his study, the Taiwanese women did as much as women elsewhere. Taiwanese gave away their own baby daughters and adopted and raised others' baby daughters, and they protected and raised their baby boys with no problem in any case. Therefore, the only thing that is confirmed by this study is that humans are capable of raising others' children along with their own—a fact that can be established without reference to minor marriage. The fact that Taiwanese women raised others' daughters and gave

away their own tells us more about their culture, a systemic effect of a particular form of patriarchy, and about gender ideology than about their maternal sentiments. Furthermore, if what is meant by maternal sentiments is emotion, then I find it troubling that the major source for the analysis is statistics.

Ryang's assessment aside, statistics can tell us something about "instinct" and emotion. But if they are used to tell us that a caricature of psychobiology, a straw man of maternal sentiments, is wrong, that is not helpful. Under the circumstances, a cultural context that clearly strained human nature, Taiwanese mothers did the best they could for their daughters and displayed maternal feeling in ample measure.

Another cultural context that seems to have strained maternal sentiments is that of the Israeli kibbutz, where for many years infants and children were expected to spend most of their time, including most nights, away from their parents and sleeping communally. Gradually, over the course of decades, every kibbutz ended the practice of communal sleeping, in response to demands by parents, especially mothers, to have their infants and children at home, at least at night and on weekends, turning the children's house into a day care system (Tiger and Shepher 1975; Spiro 1979; Maital and Bornstein 2003). Ideology trumped parental sentiments for many years, but not as a permanent cultural pattern.

The Fosterage-Abandonment-Infanticide Continuum

Sarah Blaffer Hrdy is to some extent a focus of Wolf's criticism and thus part of his straw man. This is odd, given that she has gone on record early and often as viewing maternal sentiments as facultative adaptations (Hrdy and Hausfater 1984; Hrdy 1992, 1999). She expresses this view clearly again in her comment on Wolf's article: "Human mothers have been known to abandon children, sell them, foster them out, give them to the church as oblates, drown them, strangle them, even eat them—and we can add to the list . . . that a large proportion of mothers in South China gave up infants to become child brides" (Hrdy 2003:S41–42).

Adoption for minor marriage was a relatively benign form of culturally systematic adoption. Parents in many societies give up children, sometimes into conditions that lead to a high risk of death. Some are abandoned in circumstances tantamount to infanticide: the modern Dumpster, for instance, or, in the ancient world, exposure in a wild place, leading to almost certain death; the infant Oedipus was abandoned in this way and saved only by the chance kindness of a shepherd. Some infants are deliberately killed not by unrelated individuals seeking competitive advantage for their offspring but by their own mothers and fathers, seeking relief from an untimely burden.

So in the realm of cultural practices that raise questions about maternal sentiments, there is a continuum from adoption through wet nursing and fosterage to abandonment, selective neglect, and infanticide. As we will see in Chapter 17, cooperative breeding in humans includes giving up infants and children for adoption, in some cultures very methodically. Joan Silk has carefully shown, in keeping with inclusive fitness models of cooperative breeding, that these adoptions distribute along lines of biological relatedness. The reasons vary, but the care and survival of the children is the goal. The same applies to the very different pattern considered in the previous section: you give a child up for adoption not to your blood relatives but to your future relatives by marriage—*through* the child in question—and those new relatives have strong incentives to take very good care of the child. If by maternal sentiments we mean wanting the best for one's children, those sentiments are operating in these adoption systems.

Wet nursing, or what Hrdy (1992) calls "delegated mothering," can be something else again. She cites the view that for much of European history it was a disguised form of infanticide, captured by the term "angelmaker" and its German equivalent, *Engelmacherin,* but she argues that this was only the end of a continuum. In Paris and Lyon in the late eighteenth century, wet nursing's heyday, 90 percent of infants might be nursed by women other than their mothers, and the intention clearly was not for most of these infants to die. The institution goes back to ancient Ur and Egypt and probably was the result of social stratification—a form of what William Durham (1991) calls "imposition" in cultural evolution (Chapter 28). Some women gained reproductive success at the expense of that of others. From the second century, wet nursing was a widespread commercial enterprise, although indentured servants and slaves continued to play a role. Rapid population growth in early modern Europe was fueled by this practice, and the upper classes had far more surviving children than the poor, who suppressed their fertility and withheld milk from their own infants while their well-to-do clients bred yearly.

Meanwhile, lower-class women were abandoning their infants in growing numbers. The Moscow Foundling Home in late czarist Russia is representative of much of Europe: women in urban and artisan groups, along with soldiers' wives and daughters, left infants at the home, but by far the largest and most rapidly increasing number was left by peasant women (Ransel 1988). Thousands of infants were left annually throughout the 1890s. The sex ratios of infants left in this and other Russian foundling homes were skewed toward girls until the late nineteenth century. The same class of women who were abandoning their infants were taking on foster children for pay. Many of the infants were left with identifying items, suggesting that mothers planned to reclaim them, and many were indeed reclaimed. For most of the century the sex ratio of reclaimed infants reversed that of abandoned children: more boys were reclaimed.

John Boswell's *The Kindness of Strangers* (1988) follows the foundling story in Europe from late antiquity to the Renaissance. In the fourth century St. Basil of Caesarea vividly imagined the plight of the starving father:

> How can I bring before your eyes the suffering of the poor man? . . .
> He turns his glance at length on his children: by selling them he might
> put off death. Imagine the struggle between the desperation of hun-
> ger and the bonds of parenthood. The former threatens him with a
> horrible death; nature pulls him back, persuading him to die with his
> children. Often he starts to do it; each time he stops himself; finally
> he is overcome, conquered by necessity and inexorable need. Then
> what are his thoughts? Which one shall I sell first? . . . If I hold on to
> them all, I will see all of them die of hunger, but if I sell one, how can I
> face the rest? . . . After a thousand tears [the father] comes to sell a
> beloved child. (Boswell 1988:165–66)

Basil is aware of and condemns parents who come to this without being so oppressed, but clearly sees this "choice of no choice" as the most common setting.

This account illuminates the spectrum of child abandonment, exposure, sale, and fosterage (Boswell 1988). Throughout the history of Europe monasteries and nunneries accepted abandoned children, some old enough to know that they were entering lives of celibacy and service, many not. This practice, oblation, waxed and waned periodically, as did overall abandonment. Foundling homes, invented in the Middle Ages, had infant mortality rates comparable to those outside them for a time, but experienced much higher rates with greater crowding. In most cases it is evident that these were difficult choices for parents crushed by grinding and unmitigated poverty, disease, and suffering.

There were no foundling homes in the world's "primitive" societies, and the majority resorted to infanticide, usually only at birth, in cases of deformity, birth spacing that made the infant a threat to its older siblings and unlikely to survive itself, one of a pair of twins, or, sometimes, the product of an extramarital union. In patriarchal societies infanticide affected girls disproportionately. These actions were typically taken within hours after birth and long before maternal sentiments were well established. The hormonal changes that help prepare mothers, and to some extent fathers, for parenthood (Chapter 12) are probably there to help the less-wanted child; the placenta—the child's genetic tissue—creates the great hormonal surges of pregnancy and the consequences of hormonal withdrawal at birth. However, those same hormonal upheavals can in some circumstances *cause* infanticide rather than prevent it, by precipitating psychosis or depression (Sichel 2003; Wisner, Gracious, et al. 2003).

But in theoretical perspective we should expect that mothers—consciously or not, with deliberation or in the midst of a psychotic depression—make "strategic" choices about their investment in offspring (Hrdy 1999). Up to a point, differential investment in offspring and even the harshness of parenting can be survived by the offspring, who themselves may adapt with different life history strategies (Draper and Harpending 1987; Belsky, Steinberg, et al. 1991; Chisholm 1993, 1999b; Ellis 2004; Ellis and Essex 2007).

Scheper-Hughes's idea of withdrawal of love in anticipation of loss seems logical, but the facts have been challenged by other anthropologists even for northeastern Brazil, and to the extent that they are valid they are a response to crushing poverty. Wolf's idea of mothers becoming accustomed to giving up their daughters under cultural constraint is important, but cultural constraints in Taiwan were severely oppressive to women; mothers did the best for their daughters under the circumstances, gave them away in an age pattern predictable from developmental attachment models, and often replaced their adopted-away daughters with adoptive ones. What little insight we have into what they felt does not suggest a lack of maternal sentiment. Mothers throughout European history abandoned and fostered infants, but did so from desperation and often with the hope of improving their lives. Infanticide is found in many non-industrial societies, but with serious constraints and immediately after birth; people in those cultures viewed it as most Americans today view abortion—regrettable, even tragic, but sometimes the lesser of evils.

My own experience among the !Kung helped me to understand much of this. Infanticides, all neonatal, at a rate of around 1 percent of births (Howell

1979), were usually controversial and saddening. Parents lost children to infectious disease at least as often as the Brazilians did, but they showed no
evident withholding of affection and typically had normal grief reactions after a death. Such reactions were considered culturally appropriate. In one
case a mother had obviously given up on a two-year-old child who was dehydrated and dying; she was criticized for her attitude and considered by some
to be mentally ill.

Parents, including mothers, are designed by evolution to withhold care,
distribute it differentially, give up on some infants and children, and give children away under certain economic and cultural constraints that disproportionately affect women and girls. To say that these facts undermine evolutionary or psychobiological views of motherhood is to caricature those views
in order to tear them down. This makes for good polemics but bad science.
Parental "instincts," like all motives and sentiments, can be modulated,
muted, or switched off, but that does not mean they do not exist; they do,
and when switched on they are very powerful.

16

Cooperative Breeding in the Extended Family

Although many species of fish, amphibians, and birds exhibit biparental or even mainly paternal care, the unique physiological adaptations of mammalian females suggest that the basal mammals were maternal in their care bias (Eisenberg 1981), as is the case for crocodilians (the only reptiles with parental care) and monotremes (the most primitive, egg-laying mammals). The core relationship in mammals is the mother-offspring unit, and other significant ties (if any) are added to it or build on it. The common ecological situation in which other caregivers could make an impact on the survival of the young was typically resolved either through the evolution of biparental care, as occurred in several rodents, canids, and primates, or through the addition of caregivers related to the mother—*alloparental care* or *allocare*. The addition of a father results in a nuclear family, while the addition of relatives of either parent makes an extended family. However, care by individuals unrelated to either parent is also seen, and it cannot be explained by kin selection.

The varieties of families have long been a central interest of anthropology, and both early and modern efforts have attempted to place the family in evolutionary perspective (Westermarck 1922; Gough 1975; Fox 1980). However, advances in the study of family structures in nonhuman species and their evolutionary bases (Fox 1980; Trivers 1985; Gouzoules and Gouzoules 1987) have forced rethinking of many aspects of kinship and family studies (Betzig, Mulder, et al. 1988; Betzig 1997; Chapais 2008). Stephen Emlen (1995), Sarah Blaffer Hrdy (2009), and others have helped set human infant and child care in a comparative context and emphasized the extent to which we are cooperative breeders.

Phenomenologically and psychologically, we can no longer believe, as many psychodynamic theorists did, that there is one basic relationship—that with the mother—from which all others are derived. Konrad Lorenz (1935/1970) saw in the bird's world several different varieties of "companions"—courtship partners, hatchlings, parents, and so on—that afford and demand different behavioral and perhaps neural capabilities. Similarly, Harry Harlow (1963, 1971; Harlow and Harlow 1965), studying monkeys, saw infant, maternal, peer, and other "affectional systems" and did not assume that they all depended on one mechanism or derived from one primary relationship. It is not only likely but demonstrated that the neural bases of different kinds of relationships overlap; that does not, however, make them varieties of one thing derived from one source.

Helpers at the Nest

Stephen Emlen's (1995) framework for applying Darwinian theory to families, especially in birds and mammals, defines the family as an association in which offspring maintain contact with the parents into adulthood. This definition seems too restrictive; for our purposes, prolonged contact with either parent will suffice. A dog surrounded by her pups, while the connection lasts, is a family, as is a mated prairie vole pair with their pups or a jacana male with his hatchlings. However, Emlen's criteria effectively define the *extended* family, and his model applies to them in particular. Extended families may be matrilineal or biparental, simple or complex, depending on whether one or more same-sex adults are breeding. Individuals called "helpers at the nest" may postpone or forgo their own reproduction in some situations.

Emlen (1995:8092) argues that "knowledge of four basic parameters, (i) genetic relatedness, (ii) social dominance, (iii) the benefits of group living, and (iv) the probable success of independent reproduction, can explain many aspects of family life." These lead to fifteen well-supported predictions about (extended) families (after Emlen 1995, Table 1):

1. Groups are unstable, dissolving when reproductive opportunities appear elsewhere.

2. Better resources lead to greater stability, sometimes to dynasties.

3. Help in rearing offspring is normative.

4. Help is greatest between the closest genetic relatives.

5. Incest avoidance suppresses sexually related aggression.

6. Breeding males invest less as paternity certainty decreases.

7. Loss of a breeder creates a vacancy leading to conflict over filling it.

8. Sexually related aggression increases as offspring may mate with stepparents.

9. Stepparents invest less in offspring than biological parents do; infanticide may occur.

10. Family members reduce their investment after replacement of a parent.

11. Stepfamilies are less stable than biologically intact ones.

12. Fewer ecological constraints lead to greater reciprocal sharing of reproduction.

13. Less asymmetry in dominance leads to more reciprocal sharing of reproduction.

14. Greater symmetry of kinship leads to more reciprocal sharing of reproduction.

15. Decreased genetic relatedness leads to more reciprocal sharing of reproduction; reproductive suppression is most likely among closest kin.

In brief, extended families, because they depend on advantages of staying or leaving, are unstable. Individuals that stay will help related young, and males will invest according to paternity certainty. After the loss of a breeder competition will surface, and stepparents will invest less in existing offspring, who, in turn, will invest less in future offspring. Where the environment is rich, the dominance hierarchy weak, or genetic relatedness low, reproduction will be less skewed and care reciprocally shared, with more equal results, but where subdominant relatives cannot go elsewhere, they will suppress their own reproduction.

Vertebrate Cooperative Breeding

Since these predictions, considerable further research on fish, birds, and mammals has supported many of them. As in predictions 1 and 2, richer opportunities outside the extended family make breakup more likely (Bergmuller, Heg, et al. 2005), while habitat saturation can stabilize such families (Schradin and Pillay 2004); sometimes, however, the main source of instability is inbreeding avoidance (Hatchwell, Russell, et al. 2000). A meta-analysis of fifteen cooperatively breeding bird and three mammal species showed that helping is biased toward kin and that, while this bias is variable, the greater the helping benefit, the greater the kin bias, confirming Emlen's predictions 3 and 4 (Griffin and West 2003). Female lions have minimal aggression, similar numbers of cubs, and mutual helping, confirming predictions 5 and 13 (Packer, Pusey, et al. 2001).

However, the selective mechanisms that maintain helping are quite diverse, including mating opportunities for helpers, the potential to inherit the territory, and inclusive fitness due to relatedness to breeders (Clutton-Brock 2002). Increased group size alone may favor helpers, given the safety from predation found in numbers (Kokko, Johnstone, et al. 2001). Ordinarily, helping enhances inclusive fitness if you stay where you were born (philopatry), but in some species—crows, for example—relatives may come to help from other territories (Baglione, Canestrari, et al. 2003). Helpers of different sexes may benefit differentially (Richardson, Burke, et al. 2002), and helpers may help in different ways depending on their relatedness to male versus female breeders (Stiver, Dierkes, et al. 2005). In sum, "no one factor by itself causes delayed dispersal and cooperative breeding" (Koenig, Pitelka, et al. 1992:111). However, "research over the last thirty years shows that cooperation in animal societies generally involves kin and is seldom highly developed in groups consisting of unrelated individuals" (Clutton-Brock 2009:55).

In vertebrates the most impressive example of cooperative breeding is a mammal—the naked mole rat, *Heterocephalus glaber* (Jarvis 1981; Sherman, Jarvis, et al. 1991). It lives underground in eastern Africa in colonies of thirteen to more than two hundred individuals, but typically seventy to eighty. They are nicknamed "saber-toothed sausages"; as adults they roughly resemble the hairless newborns of some mammals. There is never more than one breeding female, who mates with one to three large males. All other adults suppress reproduction and, along with subadults, serve as helpers at the nest, gathering plant food for the colony or defending it.

They are said to be *eusocial* (*truly* social), a term first applied to bees and some other social insects (Wilson 1971). The eusocial insects have genetic adaptations that make the level of relatedness of helpers to off-spring much higher than in most vertebrates. But in naked mole rats extreme inbreeding leads to similarly high levels of relatedness, making inclusive fitness (combined with ecological constraints) the best explanation for their exceptional level of helping. However, other species in the African mole rat family, such as the Damaraland mole rat, exhibit similar eusociality with much less inbreeding (Burda, Honeycutt, et al. 2000; Bishop, Jarvis, et al. 2004), probably owing to ecological constraints and limited reproductive opportunities for helpers.

Mammalian Allocare

Many other mammals have lower levels of allocare (Lewis and Pusey 1997; Solomon and French 1997). In African elephants, juvenile and adolescent females assist and comfort calves; they tend to be family members but not necessarily siblings (Lee 1987). Wild bottlenose dolphin (*Tursiops turcatus*) mothers guard their intensely attractive infants carefully in the first week but from week 2 allow them to swim tens of meters away with subadult females (Mann and Smuts 1998). Sperm whale calves (*Physeter catodon*) often surface with nonmaternal females while their mothers forage at some depth (Whitehead 1996). Male helpers among meerkats (*Suricata suricatta*) trade off helping and foraging; reducing their helping as they find greater success probably enables them to establish new territories, as in Emlen's model (Young, Carlson, et al. 2005). Also supporting the model is a population of urban foxes (*Vulpes vulpes*) in which dispersal causes high mortality, so offspring are more likely to stay in their natal group (Baker, Robertson, et al. 1998). Subdominants of both sexes provide food for the young of dominants, but subdominant females at least were in time able to bear young of their own.

Allocare in Nonhuman Primates

The first systematic review of allocare in primates was done by Sarah Blaffer Hrdy (1976), who has advanced a view of human evolution based on cooperative breeding (Hrdy 2009). From the outset Hrdy emphasized that it is not always clear where the help ends and exploitation of the young for the benefit of the "helper" begins. As with parent-offspring conflict, but more so because of lower degrees of relatedness, investment in the young is not always what it seems, yet it has probably made the difference between us and our last common ancestor with chimpanzees and bonobos. Still, a wide range of primates, like many mammals, benefit from helpers at the nest. Allocare is prominent in *Lemur catta* (Gould 1992) and in the ruffed lemur *(Varecia variegata)*, which typically has twins; it includes infant guarding and nonmaternal nursing and may mitigate the very high infant mortality in this species (Morland 1990). In the tarsier *(Tarsius spectrum)* subadult females predominate in allocare, but the levels shown are far below those of monkeys with comparably heavy infants relative to adult weight (Gursky 2000).

Among higher-primate species, helping is associated with faster infant growth and reproduction (Mitani and Watts 1997:213). A comparative analysis of anthropoid primate species (monkeys, apes, and humans) showed great variation among species in allocare and a clear correlation between the amount of allocare and female reproductive rates (Ross and MacLarnon 2000; Melo, Mendes Pontes, et al. 2003). Species with high levels of nonmaternal care have infants that grow faster and are weaned earlier (but at equivalent weights) than low-allocare species; this allows shorter birth spacing and more rapid reproduction, a pattern that foreshadows the human condition.

New World Monkeys

Like paternal care, nonparental helping is most developed in the marmoset family, where comparative analysis has been fruitful (Mitani and Watts 1997; Bales, Dietz, et al. 2000). In free-ranging groups in three species, the number of surviving infants is predicted by the number of nonpaternal helping males. Infant carrying is maintained longer in captive groups with helpers than in those without them, and in at least four species parents do less carrying and provisioning when helpers are available; helpers reduce parents' energy outlay, although fathers benefit more than mothers. Without enough helpers, environmental pressures may lead to infanticide and cannibalism (Melo, Mendes Pontes, et al. 2003).

Although the marmosets stand out, other New World monkeys have interesting patterns. The squirrel monkey (*Saimiri sciureus*) has little allocare, but as the infant rides on the mother's back (most of the time) it exchanges many more vocalizations with nonmaternal adults than with the mother (Biben 1992). The spider monkey (*Ateles geoffroyi*) exhibits little female allocare in the wild, probably because free-ranging groups are male-centered and import females, but both sexes may provide allocare in captivity (Watt 1994); in captive colonies where matrilines are preserved, not only do nonmaternal females provide care, but DNA shows that their care is predicted by genetic relatedness (Matisoo-Smith, Watt, et al. 1997). In free-ranging black howler monkeys (*Alouatta caraya*) care takes the form of carrying, comforting, or helping; although most of an infant's interactions are with adult and subadult females, males contribute as well (Calegaro-Marques and Bicca-Marques 1993). And in capuchin monkeys (*Cebus olivaceus*), where allocare is important, sibling females account for four times as much of it as do other females (O'Brien and Robinson 1991).

Old World Monkeys

Old World monkeys exhibit various allocare adaptations, but a few themes are apparent, and they are consistent with the expanded model considered earlier: allocare is especially beneficial to infants and mothers where there is ecological pressure; young females providing allocare benefit by developing future mothering skills; and allocare is preferentially directed toward kin.

Most colobines, including Hanuman langurs (*Presbytis entellus*), as well as some cercopithecines have newborns with flamboyant fur, which produces intense initial interest on the part of nonmaternal adults and juveniles that declines as the coat color becomes duller with growth (Hrdy 1976). In other primates the distinctiveness of the infant's coat is variable, although neonates can always be identified by size and facial and body shape. However, the contrast provided by the natal coat predicts the amount of allocare (Ross and Regan 2000), and it may, like the human "cute" response, be designed to elicit it.

As in most colobines, capped langur (*Presbytis pileata*) newborns are passed around to several experienced females, but over the first three months mother-infant pairs settle down with one allomother (Stanford 1992). After an observed birth in this species, the infant spent 9 percent of its time during the first fifteen days with four allomothers, three adults and one subadult, whose time with the infant was predicted by their age (Kumar, Solanki, et al. 2005). In free-ranging patas monkeys (*Erythrocebus patas*) older females predominate, but young ones participate, and their help is especially valuable owing to competition with rhesus macaques released on the same Caribbean island (Zucker and Kaplan 1981). In vervet monkeys (*Cercopithecus aethiops*) allocare has measurable benefits for the mother, and if the nonmaternal caregiver is a young female, she will later exhibit improved maternal skills (Fairbanks 1990; Meaney, Lozos, et al. 1990). Maternal grandmothers among vervets in particular increase the reproductive rate of their daughters and decrease their infant mortality (Fairbanks and McGuire 1986). In the blue monkey (*Cercopithecus mitis*) young females predominate and appear to benefit from practicing infant care (Forster and Cords 2005).

Japanese macaques (*Macaca fuscata*), which were the subjects of one of the first primate field studies testing kin selection theory (Kurland 1977), also show both nurturing and rough handling by nonparents, with the positive handling most likely to come from kin (Schino, Speranza, et al. 2003). Experiments with this species show that mothers and older sisters successfully aid young females in aggressive encounters, but aunts do not (Chapais, Gathier, et al. 1997). In a large captive colony of bonnet macaques (*M. radiata*), gentle handling of infants was common in females regardless of age; females doing more of it may simply be interested in infants (Silk 1999). As we will see in Chapter 18, Barbary macaques (*M. sylvana*) exhibit more male allocare than most monkeys, but females also provide substantial amounts (Small 1990).

Social context has also been shown in some species to influence reproductive success, although it is not clear whether indulgence of infants is part of the mechanism. In savanna baboons at Amboseli in Kenya, sixteen years of observational data showed that females with higher levels of social interaction had higher reproductive success, independent of rank and ecological conditions (Silk, Alberts, et al. 2003), and females commonly associated with close kin who collaborated with them in aggressive encounters.

Great Apes

In great apes, allocare is lowest in orangutans, with their usually isolated mother-offspring groups, but older siblings interact with younger ones, probably to the advantage of both. Bonobo adolescents, including males, may carry, groom, and otherwise provide substantial care to their infant siblings (Kano 1992). Among chimpanzees in the Mahale Mountains, nulliparous females are the main nonmaternal caregivers, while males do much less, and (in contrast to langurs) females who have given birth are indifferent to others' infants; mothers, for their part, regulate access to their infants according to sex and kinship (Nishida 1983). Given a choice in experiments, chimpanzees (despite being among the most prosocial primates) direct help almost exclusively toward kin (Silk, Brosnan, et al. 2005). However, also at Mahale, a 2.4-year-old male was cared for by two unrelated females in tandem for at least six days while separated from his mother, probably because she was ill (Uehara and Nyuno 1983). In a captive colony a mother who had lost her neonate but was still lactating adopted without difficulty a hand-reared five-month-old infant unrelated and unfamiliar to her (Palthe and Van Hooff 1975).

Overall, the rarely observed cases of adoption in monkeys and apes often, but not always, follow kinship, partly conforming to inclusive fitness theory (Thierry and Anderson 1986); other cases seem to involve reciprocal altruism or non-adaptive adoption. Lactating or late-pregnancy females are most likely to adopt neonates, while males and juveniles are more likely to adopt older infants. But all such cases are rare; in the wild the death of a mother usually leads to the death of her infant or juvenile offspring.

Does Helping Help?

We know that "help" is not always help (Hrdy 1976). Exploitative interactions between infants and unrelated individuals include males ingratiating themselves with the infants' mothers to develop "sexual friendships" (Smuts 1985), agonistic buffering—attempts to gain protection or rank by associating with the infants of dominant females (Gouzoules 1975; Deag 1980)—and competitive infanticide (Goodall 1977; Deag 1980; Hausfater and Hrdy 1984; Arcadi and Wrangham 1999; Soltis, Thomsen, et al. 2000). Another possibility is "aunting to death." An individual adopts an infant, not necessarily with any apparent intent to harm it, and keeps it neglectfully but decisively until it dies. Sublethal deleterious aunting has been observed in free-ranging Indian langurs (Hrdy 1976), and aunting to death has been described in Barbary macaques, a species where females in a group tend to be unrelated (Paul and Kuester

1996). Like genuine helping, it affords the nulliparous female a chance to practice mothering but, unlike real helping, also functions to reduce competition for scarce resources.

Patterns like these are the tip of an iceberg of behaviors that bring advantages other than inclusive fitness to older individuals that may or may not be advantageous to the young, and sometimes will do them great harm. Even the mother is not selfless with regard to her own off-spring, and abuse and neglect, up to and including infanticide, can be adaptive for her (Hrdy 1999). When seen as potentially adaptive for adults, such behaviors emerge as the end of a continuum of adaptation in which care is offered or withheld in varying degrees depending on the challenges posed by selection in different circumstances, and they may in some ecological settings have the effect of preparing infants for those circumstances (Chisholm 1993, 1999b).

Nonmaternal Care among !Kung Hunter-Gatherers

Relations with the father and with other children, also dimensions of cooperative breeding, are considered in the next two chapters; here we consider care by other adults, mainly women. An early descriptive account (Konner 1972:292) said of !Kung infants:

> From their position on the mother's hip they have available to them her entire social world. . . . When the mother is standing, the infant's face is just at the eye-level of desperately maternal 10-to-12-year-old girls who frequently approach and initiate brief, intense, face-to-face interactions, including mutual smiling and vocalization. When not in the sling they are passed from hand to hand around a fire for similar interactions with one adult or child after another. They are kissed on their faces, bellies, genitals, sung to, bounced, entertained, encouraged, even addressed at length in conversational tones long before they can understand words.

A later publication quantified nonmaternal care (Konner 1975): the total time in physical contact with anyone, and the percentage contribution of mothers, nonmothers, and children as a subset of nonmothers, compared with the corresponding results of Steven Tulkin's study of ten-month-old girls in Boston, using very similar methods (Tulkin

and Kagan 1972; Tulkin 1977). !Kung mothers accounted for 75 to 80 percent of all physical contact that infants received, with no significant
change over the first twenty months, but the percentage of physical
contact with anyone *other* than the mother (20 to 25 percent) was higher
than in either group of Boston infants. So despite the very close physical
relationship to the mother, !Kung infants also had more physical contact with others, even as a proportion of total physical contact. The distribution of face-to-face contact suggested a much higher proportion of
nonmaternal involvement, although the variability was high. Until three
months of age, mothers accounted for about 75 percent of face-to-face
contact; this contact declined to 50 percent thereafter and remained at
that level well into the second year.

The previously mentioned new analysis of response to crying shows
clearly that nonmothers play a prominent role (Kruger and Konner,
forthcoming). !Kung infants cried for about one minute per hour, mainly
in bouts of ten seconds or less. *Someone* responded to 88 percent of all
crying bouts, and almost all the others resolved within ten seconds.
Caregiver behaviors occurred at the rate of one for every three seconds
of crying, the most common being oral or tactile comforting. The mother
always responded to long crying bouts (more than thirty seconds), and
for half of these she was the sole responder. Only mothers nursed infants, and they provided significantly more comforting responses of
other kinds than did all other individuals combined.

Yet the !Kung mother was rarely alone with a crying baby. For nearly
half the bouts, other caregivers either were the sole responders or joined
the mother in responding. Even when others did not respond to the crying bout, they were nearly always present and offering care to the baby,
sometimes as the primary attendant, at some point during a ninety-
minute observation set.

The study of London and !Kung two- to five-year-olds mentioned in
Chapter 10 (Blurton Jones and Konner 1973) showed that !Kung children,
especially boys (p < .01; girls p < .10), ranged a greater maximum distance from mothers even though the London children were observed in
safe, familiar outdoor spaces in good weather, with well-known playmates available to play with. !Kung children were face-to-face with their
mothers less and were nurtured less by their mothers, or by anyone,
than London children were (all significant for both sexes at p < .02 or
better), suggesting that strong maternal dependency did not persist in

childhood. One possible explanation was that the dense social context made mothers more responsive to infants' dependent demands, thus lessening their children's proximity-seeking and other dependent demands at later ages (Konner 1977b).

Nonmaternal Care in Other Hunter-Gatherers

In the last chapter we compared maternal care in five hunter-gatherer groups. Here we consider the complementary roles of others, except for fathers, who are deferred to Chapter 17.

Hadza

Mothers predominate markedly in infants' care, interacting with them in 78 percent of minutes in the first year and 43 percent over the first four years (Marlowe 2005a). But nonmaternal care is substantial; the respective figures for other caregivers are: fathers, 18 and 17 percent; older sisters, 18 and 17 percent; maternal grandmothers, 9 and 10 percent; older brothers, 8 and 9 percent; and all others, 29 and 41 percent. (Several people could interact with the infant or child in any given minute, so the percentages exceed 100.) Although the involvement of fathers, siblings, and grandmothers does not change, the marked decline in mother involvement, combined with the increased interaction with all others, broadens the child's social world.

Efe

The Efe have much more nonmaternal care than the !Kung and are appropriately described as "the most extreme example of alloparenting in a foraging population" (Ivey 2000:857–58). Women other than the mother hold and breast-feed the baby first, and after several days the mother may begin to work and either receives help at the work site or leaves the baby for short periods with another woman. Even more than for the !Kung, dealing with a fussing baby is a group effort, and responses include nonmaternal nursing, although, as with the !Kung, intractable crying ends in the mother's arms.

We know that maternal primacy applies to the Efe despite their extensive nonmaternal care, but the alloparenting is remarkable. Nonmothers account for 39 percent of infants' physical contact at three weeks and 60 percent at eighteen weeks (Tronick, Morelli, et al. 1987). Infants are handed from caregiver to caregiver 3.7 times per hour at three weeks, 5.6 times per hour at seven weeks, and 8.3 times per hour at eighteen weeks. An infant is nurtured by 5 to 24 different people (mean = 14.2); 80 percent are breast-fed at times by nonmothers (five of seven infants at three weeks, two of eight at seven weeks, and six of nine at eighteen weeks). However, nonmaternal care is extremely variable; time spent with nonmothers ranges from zero to around 65 percent at three weeks and from 20 to 80 percent later in infancy.

Defining *social* contact as time not occupied with solitary activity, nonparent adults were in social (including physical) contact about 10 to 15 percent of the infant's time in the first year and about 20 to 25 percent in the second and third years, with fathers showing lower and children higher levels (Tronick, Morelli, et al. 1992). By age two, contact with mothers is surpassed by contact with children (in total), but not by contact with fathers or other adults. As predicted by kin selection theory, among those interacting with an infant there is a marked predominance of kin compared with nonrelatives matched for age, sex, and availability. Overall, care is indulgent: "Efe one-year-olds are in close proximity to a caregiver 100 percent of observed time and spend an average of 85 percent of time in direct care" (Ivey 2000:859).

Reproductive status is also important. Reproductive females provided about 11 percent of the female allocare, prereproductive girls 22 percent, postreproductive women 29 percent, and nonreproductive women the largest proportion, 38 percent. This last statistic matters because the average completed fertility of the Efe is 2.6 births, the lowest ever measured for a natural human population (Ellison, Peacock, et al. 1986). But the mean conceals a bimodal distribution: a high proportion of women are completely infertile owing to sexually transmitted pelvic inflammatory disease. A reasonable hypothesis would be that low fertility unevenly distributed makes the extremely high levels of allocare work (Hewlett 1991a). Nonmaternal breast-feeding may also be an inclusive fitness response to high infertility, since fertile women would reduce the likelihood of suppressing their own ovarian function with intensive nursing, thus shortening birth spacing.

This is a true instance of cooperative breeding in an extended family, and it illustrates how Emlen's four principles—genetic relatedness, social dominance, the benefits of group living, and the probable success of independent reproduction (low in the case of nonreproductive Efe women)—can explain many aspects of family life. At least four of Emlen's predictions are realized: (2) greater stability in a better resource base; (3) normative helping; (4) help greatest for closest relatives; and (15) reproductive suppression greatest in close relatives. Nonreproductive women closely related to the mother do most of the allocare.

Aka

Paternal prominence and maternal primacy notwithstanding, nonparent caregivers play a major part in Aka care. "While in the camp setting" (that is, not during transport), "Aka one-to-four-month-olds are held by their mothers less than 40 percent of the time, are transferred to other caregivers an average of 7.3 times per hour, and have seven different caregivers on average that hold the infant during the day" (Hewlett 1991b:34). Recall that on forest trips the mother holds the infant nearly 90 percent of the time, with transfers declining to two per hour. But as we will see, it is father involvement that most distinguishes the Aka. Together with other nonmaternal care, this certainly qualifies the Aka as cooperative breeders—perhaps not as much as the Efe, but more than the !Kung (Hewlett 1988, 1991b).

Ache

Multiple caregiving is less important among Ache than for most hunter-gatherers. However, between ages three and five, when children still cannot walk well and severely protest being set down on the forest floor, they are carried by other adults, including fathers and grandparents. After around age five, though, they are weaned from carrying too and either encouraged or forced to walk on forest trips; this "weaning from the back" was also a crisis for the !Kung (Draper 1976). Among the Ache,

> children scream, cry, hit their parents, and try everything they
> can think of to get adults to continue carrying them. Often,
> they simply sit and refuse to walk, prompting older band mem-
> bers to leave them behind. This tactic leads to a dangerous
> game of "chicken" in which parents and children both hope the

other will give in before the child is too far behind and may become lost. We observed one small boy to be lost for about half an hour during a parent-child transportation conflict. When the boy was finally located it was unclear whether he or his parents were more frightened. A small child cannot survive long in the Paraguayan forest, and if not found within one day is unlikely to survive. In any case, the boy's tactic paid off temporarily, since he was carried for the remainder of the day. (Hill and Hurtado 1996:222)

This episode underscores Bowlby's point about attachment as an adaptation to prevent death from predation; in this tantrum, the child's life was at stake.

Agta

The Agta are unique in the extent of women's hunting, which is surprisingly compatible with maternal primacy. Yet it could not be achieved without cooperative breeding:

Children are left in camp by mothers at increasingly frequent and lengthy intervals. Late in the first year an infant may be left for an hour or two; sisters or mother's mother or father are acceptable caregivers. . . . One sister sporadically nursed her sister's infant as well as her own, permitting the young mother time outside the camp. More frequently, however, babies are carried by mothers until the age of two or slightly more. (Griffin and Griffin 1992:302)

Multiple caregiving was described by ethnographers as substantial: "The infant is eagerly passed from person to person until all in attendance have had an opportunity to snuggle, nuzzle, sniff, and admire the newborn. . . . A child's first experience, then, involves a community of relatives and friends. Thereafter he enjoys constant cuddling, carrying, loving sniffing, and affectionate genital stimulation" (Peterson 1978:16, cited by Hewlett 1991a:13). But subsequent quantitative data from spot observations showed more modest nonmaternal involvement: "Within the residential cluster, mothers of children from age [0] to 8 years are caregivers slightly more than 50% of the time. Grandmothers and elder sisters come in at a modest 7.5 and 10.4%, and fathers follow with only 4.4%" (Griffin and Griffin 1992:303).

Cooperative Breeding in the Human Species

The application of models like Emlen's to human populations has confirmed some hypotheses arising from the models. However, three modifications have been suggested with respect to humans (Davis and Daly 1997). First, beyond the strains of environmental quality fluctuations and crowding, human groups apply additional pressure to each other in formal competition between groups, which is viewed by some as more important than conventional natural selection (Alexander 1979; Bowles 2009). Second, in human societies reciprocal altruism, like intergroup competition, is self-conscious, rule-bound, and probably more important. Third, our exceptional human life course affords a new category of helpers: postreproductive women. These helpers make an important contribution to our unique (among primates) trait of postweaning provisioning.

A comprehensive view of humans as cooperative breeders is needed and has been put forward by Sarah Blaffer Hrdy (2001, 2005, 2009). This model includes reciprocal helping by reproductive and nonreproductive women, especially grandmothers; helping by older siblings; and provisioning, defense, and care by fathers and other men. Citing Hilliard Kaplan, Hrdy estimates that it takes 10 to 13 million calories of energy expenditure to rear a human child to adulthood, and she plausibly argues that this cannot come from one person. Indeed, there is ample evidence that human mothers have always gotten help.

In hunter-gatherer societies, grandmothers, fathers, older siblings, and other relatives play varied but important roles (Konner 2005), as they do in horticultural, agricultural, and herding societies. For the more well-off, wet nurses and nannies have played key roles in the West, and in modern states day care centers and schools serve some of the same functions. One expectation of Emlen's model is that helping should be in part predictable from genetic relatedness (Hamilton's rule). But in an analysis of !Kung anthropometric and demographic data, Patricia Draper and Nancy Howell (2005:279) found that kinship measures could not predict the health and growth of children: "Our findings regarding . . . senior kin and sibling competition went contrary to our initial hypotheses. We expected children to benefit from having a full complement of parents and four grandparents, yet we found no such relationship. . . . Ju/'hoansi [!Kung] children ate and grew, apparently oblivious to the potential impact of various advantageous and disadvantageous family situations." Draper and Howell attributed this to the egalitarian nature of !Kung groups. Gift exchange (*hxaro*) networks among !Kung adults are predictable from relatedness (Wiessner 1982, 2002), but perhaps the "it takes a village" principle trumps Hamilton's rule when it comes to child welfare. Howell's (2010) further analysis showed that food is provided for children by !Kung adults independent of genetic relatedness. Among the Yora of the Peruvian Amazon the loss of a biological parent did not predict children's foraging, probably because "facultative cooperative breeding may buffer juveniles who have lost a parent" (Sugiyama and Chacon 2005:261).

If there is no society where mothers raise children without help, then theoretically, human reproductive success should be enhanced by helpers at the nest, whether adult or not (Kramer 2005). For example, a field study of the reproductive life histories of 794 Berber women in southern Morocco found "little doubt that human helping at the nest favors reproductive success and that the most obvious helpers are the socially and genetically closest individuals to the breeding pair, i.e. the offspring" (Crognier, Baali, et al. 2001:372). The same was true of a parallel study of 359 women age forty-five and older among traditional Aymara peasants of the Bolivian Altiplano (Crognier, Villena, et al. 2002). However, in neither case was helping limited to infant and child, nor was the burden carried disproportionately by older female siblings; a multicausal model including household chores, agricultural work, and child care seemed to apply.

The Grandmother Effect

But it is grandmothers who have stimulated the greatest interest and controversy. Humans are the only primates where females live for decades beyond their reproductive years, and grandmothers (when present) are almost always involved in child care. Great apes reproduce until they senesce and die. Although there are hints of menopause in individual cases (Gould, Flint, et al. 1981), even the Sumatran orangutan, with the greatest longevity in the wild, lacks menopause (Wich, Utami-

Atmoko, et al. 2004). Roughly speaking, the age at which women's reproductive capacity ends is similar to the age at which great apes naturally die, and so the evolution of the latter part of the human life cycle can be interpreted as a lengthening of the time until senescence without lengthening the reproductive span (Hawkes 2003).

Reproductive cessation occurs in some mammals (Cohen 2004), including a number of primate species, but in primates it does not occur long before death from old age, and few females live long enough to attain it (Fedigan and Pavelka 2007). However, Japanese macaque females with living mothers (postreproductive or not) have earlier ages of first reproduction and shorter interbirth intervals than those whose mothers are dead (Pavelka, Fedigan, et al. 2002). Similarly, in a captive group of vervet monkeys, females with surviving mothers were less frequent targets of aggression and had more pregnancies and more surviving offspring than those without them (Fairbanks and McGuire 1986).

Why did hominins find it advantageous to extend life so long after reproductive cessation? In a persuasive model developed by Kristen Hawkes (2003, 2004; Wich, Utami-Atmoko, et al. 2004), based originally on work by her and others on the Hadza (Blurton Jones, Hawkes, et al. 1997), the selective advantage of having a nonreproducing mother who for many years can help her daughters raise her grandchildren drives many aspects of human life history (O'Connell, Hawkes, et al. 1999). Considering how long it takes to raise a child to adulthood, as well as our unique postweaning provisioning of children, the survival of the postmenopausal woman's own last child or two was probably enhanced by postreproductive maternal care (O'Connell, Hawkes, et al. 1999; Peccei 2001). Mean life expectancy at the time of last birth in hunter-gatherer women is not very different from the maturation time of that last child.

There is evidence for both explanations, and it is neither necessary nor, probably, possible to choose. In addition, we know that earlier menopause is associated with lower risk of breast, ovarian, and uterine cancer and that hunter-gatherers had earlier menopause than we do, leading (along with later menarche, earlier first birth, and more time spent pregnant or nursing) to two-thirds fewer ovarian cycles, the probable protective factor (Eaton, Pike, et al. 1994). Prolonging fertility would have increased cancer risk, added to the increasing risk of pregnancy and childbirth with age. A combined causal model is needed to explain the evolution of this exceptional life history trait (Shanley and Kirkwood 2001). Hawkes (2003:380) concedes that "the data are scanty"—and for hunter-gatherers are likely to remain so.

Frank Marlowe's data suggest that among the Hadza, grandmothers are deployed in a particular situation. Mothers accounted for over 70 percent of the holding of infants and children under four, but when there was a genetic father in the household, the father accounted for one-fourth of nonmaternal holding and the mother's mother for only one-eighth. However, the Hadza have a relatively high divorce rate, and if there is no genetic father in the household, maternal grandmothers play a much bigger role, accounting for over 70 percent of nonmaternal holding. Among settled Martu aborigines in Australia, where men hunt with modern methods, grandmothers have a very prominent role and are the only nonmaternal caregivers who perform routine tasks such as feeding and cleaning, a unique role that has been referred to as "the grandmaternal niche" (Scelza 2009).

Other empirical studies reveal intriguing details. Among the Ache the fertility of men (but not women) is enhanced by having living parents or siblings (Hill and Hurtado 1996). On the Ifaluk Atoll in the South Pacific, mothers with living parents produce more children than those without them, as do mothers who have daughters first (Turke 1988). In rural Gambia between 1950 and 1974 a child's survival was improved by having a living mother, maternal grandmother, or older sisters, but not by having a living father, older brothers, paternal grandmother, or other paternal relatives (Sear, Steele, et al. 2002). The biggest impact was that of having a maternal grandmother, also shown to affect nutrition and growth (Sear, Mace, et al. 2000). In a noncontracepting, Bengali-speaking population in India, having a grandmother in the household shortened birth spacing, probably increasing reproductive success (Nath, Leonetti, et al. 2000).

Historical studies support the general hypothesis. For a village in central Japan between 1671 and 1871, records show that the presence of a maternal grandmother (but not of any other relative) exerted "a consistent negative effect on the likelihood of a child's death" (Jamison, Cornell, et al. 2002:67), although this finding fell short of statistical significance owing to the small number of grandmothers. Analysis of church registers in Bejsce parish in eighteenth- to twentieth-century Poland showed a positive effect of maternal grandmothers on fertility (Tymicki

2004). In a nineteenth-century polygynous population on the American frontier, infant mortality was lower when grandmothers, aunts, and uncles on the mother's side were nearby, but on the father's side only aunts had that effect (Heath 2003). However, in historical data for the British aristocracy, there was an *inverse* relationship between number of children and female longevity, suggesting a classic life history trade-off in this well-resourced population (Westendorp and Kirkwood 1998).

A compelling analysis was applied to two historical populations for whom there were excellent demographic records: 537 women born in Finland in rural communities from 1702 to 1823 and 3,290 women born in rural Quebec from 1850 to 1879 (Lahdenpera, Lummaa, et al. 2004). In each, women gained two extra grandchildren for every ten years they lived after age fifty. This worked through sons as well as daughters and applied to grandchildren surviving to breeding age as well as to grandchildren born. The finding is independent of genes, infectious disease, and the grandmother's own prior fertility. Also, men or women with a living mother began reproducing earlier, independent of the mother's age. Most interesting perhaps, the grandmothers' effect on their grandchildren's survival was not significant between birth and age two, but was highly significant between ages two and five, strongly suggesting that it is the weanling that benefits most from the presence of the grandmother. Not surprisingly, living grandmothers have grandchildren born after shorter interbirth intervals.

By contrast, the grandfathers in the Finnish database had almost no comparable effects (Lahdenpera, Russell, et al. 2007). The authors reasonably conclude that polygynous hunter-gatherers would be even less likely to show such effects, since older men have reproductive opportunities as fathers, and they interpret their data as strong support for the Hawkes hypothesis explaining prolonged postreproductive life span for women only.

Normative Adoption and Fosterage in Human Societies

A final aspect of human allocare is culturally sanctioned, systematic adoption or fosterage, which we now partly understand in its evolutionary context thanks to the work of Joan Silk. For example, in Oceania—often cited as a dramatic exception to kin selection theory—between 12 and 83 percent of households in various island cultures included at least one adopted child. Motives for adoption included childlessness (the most common reason), attempts to achieve optimal family size and gender balance, and substantial gifts from the natural parents. However, adoptions were by relatives in 73 to 100 percent of cases (Silk 1980), and the likelihood of adoption was highly predictable from genetic relatedness. This very strong correlation departs markedly from expectations based on greater numbers of kin of increasingly distant relatedness. Also, in ten of the fifteen Oceanic cultures for which information was available, biological offspring inherited larger shares of their parents' estates than did adopted offspring, and natural heirs sometimes even colluded to deprive adoptive heirs of their portion. Biological parents shared jurisdiction and could terminate the adoption and recover the child at any time. This pattern of adoption does not challenge kin selection theory, but confirms it in the context of cooperative breeding, in accord with Emlen's prediction 4; in humans, however, the cultural framework permits helpers to be at a distance from the nest.

In some West African cultures, more than half of all children spend years in a household other than that of their parents (Silk 1987b; Goody 1973, 1982). Virtually all are fostered to relatives, especially grandparents, aunts, and uncles. Fostering may occur at weaning or even birth, but typically children stay with their own parents for five to eight years. Foster parents take full responsibility and tend to be stricter than parents feel they could be, and the fostered children, especially girls, contribute to the foster household economy. A common motive is the death or divorce of the natural parents; among the Gonja, only 26 percent of foster children came from intact families. However, unless both parents have died, they maintain contact with their children and may see them daily if distance allows. Other reasons for fostering include education and moral training; providing a better environment for the child; conflict among co-wives over investment in a child; or even perceived danger from a co-wife. Advantages to foster parents include aid with household tasks, prestige, reinforcing kin networks, and in some cases reciprocal fosterage.

Among the Inuit (Eskimo) of the Arctic Circle, between 20 and 70 percent of children in various communities were adopted, most before a year of age, typically when parents had difficulty caring for the child or, for older children, after a parent's death (Silk 1987a; Guemple 1979). Inuit say adoption is an alternative to infanticide under conditions of privation or short birth spacing. Between 55 and 90 percent of adoptions involved kin, almost always genetically related. Close kin, including grandparents, predominated; ritual or fictive kin were rarely involved. Adoptions by more distant kin or nonkin involved large gifts from the biological parents. Adopted infants were treated very well, but older children were at risk for abuse. As in West Africa, Inuit parents gave up their children reluctantly, maintained contact with them, ensured the quality of their care, and if they were not well treated, might recover them at any time.

Overall, adoption is far more common in human than other higher-primate societies; as a normative institution it has important social and economic functions that would be irrelevant to nonhuman primates. However, kin selection comfortably predicts the predominant patterns in normative adoption, with reciprocal altruism explaining much of the rest. Yet as Silk (1987b:46–47) noted two decades ago, "It is clear that adoption and fosterage cannot yet be fully understood within the context of this model. It is not clear why adoption and fosterage are common in some societies and absent in others. . . . Other models that incorporate both biological and cultural factors may ultimately prove to be more useful." With the exception of the Inuit—who occupy an ecological setting that is not one of the main human environments of evolutionary adaptedness—all the cultures discussed in this chapter are horticultural or agricultural. Hunter-gatherers in warm and temperate climates do not have normative adoption or fosterage. (The special case of widespread adoption of girls in traditional China for "minor marriage" is taken up in Interlude 4, "Thinking about Maternal Sentiment.")

The Physiology of Alloparental Care

Little is known about the mechanisms by which the widespread behaviors associated with alloparental care are accomplished. In Mongolian gerbils some males naturally have low levels of testosterone; they show little interest in estrus females and spend 30 to 50 percent more time delivering allocare than other males (Clark and Galef 2000). In rodents generally, both across and within species, allocare in experiments is associated with oxytocin (OT) receptor density that is high in the nucleus accumbens and caudate-putamen but low in the lateral septum (Olazabal and Young 2006). In prairie voles, a pair-bonding but also cooperatively breeding species, treatment of female neonates with OT made them less likely to attack strange pups later in life, while males exposed neonatally to an OT antagonist were more likely to do this (Bales, Pfeifer, et al. 2004).

In common marmosets challenged with infants in an experiment, prolactin rose in those that showed allocare, in proportion to the duration of carrying, although the role of the physical effort is uncertain (Roberts, Jenkins, et al. 2001). The degree and nature of possible hormone changes in nonreproductive helpers among marmosets is controversial (Abbott, Keverne, et al. 2003; Mota, Franci, et al. 2006), but some studies have suggested that reproductive suppression may be physiological as well as behavioral.

Social Context and Mother-Infant Interactions

The hypothesis that social context may regulate maternal indulgence of infant dependent demands (or other aspects of mother-infant relationships) received early support from laboratory studies of several higher-primate species (Hinde and Spencer-Booth 1967; Kaplan 1972; Kaplan and Schusterman 1972; Wolfheim, Jensen, et al. 1970). A cross-species comparison would not support a simple inverse causal relationship between isolation of the mother-infant pair and maternal indulgence—orangutans show extremely indulgent mothering, for example, in isolated mother-offspring groups—but there is support for such a link within some species.

Robert Hinde and Yvette Spencer-Booth (1967) observed rhesus monkey mother-infant pairs longitudinally, with and without the long-term presence of other animals. In isolated pairs, mothers avoided and left infants more, while infants approached and followed more, than in group-living pairs. The proportion of time spent more than two feet from the mother was more for isolates for the first ten weeks, but more for group-living pairs from eleven weeks on. This may suggest that when infants are still motorically immature and have little control, isolated mothers keep their distance, but when locomotor facility emerges isolated infants subvert their mothers' attempts to escape them. Finally, isolate mothers tended to carry infants on their backs rather than in the ventral position during transport, and the infants climbed on them from behind, instead of from in front, as was more common in group-living infants.

In Joel Kaplan's (1972) similar study of squirrel monkeys (*Saimiri sciureus*), mothers in isolated pairs avoided and punished their infants more and retrieved them less than mothers in group-living pairs, and isolate-pair infants made more attempts to stay with their mothers and stayed closer to them, although (at least at some ages) group-living infants nursed more. Grooming and looking at infants, likely indicators of maternal concern, was higher in group-living mothers. The tendency for isolated pairs to have greater infant dependent demands with lower maternal indulgence resembles that found for rhesus monkeys. However, group-living mothers made more attempts to shake off their infants. A subsequent experiment with the same squirrel monkeys (Kaplan and Schusterman 1972) placed the infants in a choice situation: they could opt to be near their mothers, a strange adult female, a strange infant, or an empty cage. Group-reared infants showed a greater preference for their own mothers than did the isolate-pair infants. The authors interpret this as an indication of infants' preference for the mother and the closeness of the mother-infant relationship.

Similarly, in pigtail monkeys (*Macaca nemestrina*) (Wolfheim, Jensen, et al. 1970), mothers of fourteen- to fifteen-week-old infants in group-living pairs were more retentive, spent more time in ventral contact, and nursed more than those in isolated pairs. The authors interpreted this as protectiveness; infants in this species may come to harm from other group members. Hinde and Spencer-Booth interpret their data on rhesus macaques similarly. In view of the selfish or negative "allocare" seen in primates, this is a plausible hypothesis. However, Kaplan's explanation (based on experiments with squirrel monkeys) that the isolate mother is under so much steady pressure from the infant that she is less inclined to indulge it is also plausible.

Cooperative Breeding beyond Hunters and Gatherers

Intermediate-Level Societies

Multiple caregiving is very widespread cross-culturally, and it persisted after the hunting-gathering era. Anthropologists from Margaret Mead (1962) to Susan Seymour (2004) have shown the variety of multiple-caregiving arrangements. Seymour's study (1981) of twenty-four households in Bhubaneswar, India, Ruth and Robert Munroe's (1984) on the Logoli of Kenya, Herbert and Gloria Leiderman's (1974) of the Kikuyu of highland Kenya, Kathleen Barlow's (2001) of the Murik of New Guinea, and many others show that multiple caregiving is normative by long tradition in many societies where constraints of mothers' work and other demographic and economic pressures make it advantageous.

Often the extra "mothers" in polymatric cultures are young girls, the predominant sibling caregivers (Weisner, Gallimore, et al. 1977; Weisner 1987). Again, economic and demographic conditions are very important (Chapter 19); in Robert LeVine's research among the Gusii of Kenya, for example, rapidly rising family size in the midtwentieth century made households dense with children, and mothers relied on some of them to care for their younger kin (LeVine, Dixon, et al. 1994). To explain the consequences of household density for infant experience John Whiting (1961) looked at a sample of fifty-five cultures and found that infants were more indulged in those with large extended-family (87 percent indulgent) or polygynous households (83 percent) than in those with nuclear family (42 percent) or mother-child households (25 percent). Ruth and Robert Munroe (1971) tested the same hypothesis within the Logoli culture of Kenya; they found that overall indulgence was higher in extended-family households but the mother's role in the household was smaller. They also reanalyzed Mary Ainsworth's data on the Ganda of Uganda, which showed that more members in a household predicted less strong maternal attachment. Humans may react differently than monkeys to a denser social context: indulgence may increase in both, but for some humans this occurs through a diffusion of nurturance (allocare) rather than a closer mother-infant bond.

Modern Industrial States

In industrial societies, much depends on social class and specific cultural tradition. Carol Stack's classic study of cooperation among poor urban African-American women, *All Our Kin* (1974), showed extensive multiple caregiving, as did Vanna Axia's work on the "home cultural ecology" in Italian families (Axia and Weisner 2002); among other things, social support for caregivers predicted less intense reactions by infants to the stress of vaccination.

But on the whole, until the quite recent advent of infant day care, it was common for the middle class in Western countries to care for infants in an isolated nuclear family, with the isolation of the mother-child unit exacerbated by shrinking family size and/or divorce. As noted by cultural anthropologists (Seymour 2004), child psychologists (Lamb and Ahnert 2006), and biological anthropologists (Hrdy 2009), until the industrial revolution multiple care was more the rule than the exception, and it may be the isolated nuclear family that is anomalous.

Yet the importance of social support for mothers and, to a lesser extent, fathers has been of interest to psychologists (Cochran and Niego 2002), and the hypothesis that within Western societies such support improves the quality of mother-infant interaction and therefore of infant and child attachment has been tested (Belsky 1999). A meta-analysis of 66 studies (4,245 mothers of relatively homogeneous background— white, middle-class, and married) showed that both emotional and material support has a positive effect on maternal behavior (Andresen and Telleen 1992). A study of 62 low-income African-American mothers found that those with larger social support networks were more responsive to their infants and gave them more stimulating home environments than those with less support (Burchinal, Follmer, et al. 1996). The role of social support in lesbian families has also been shown, especially the involvement of relatives of the birth mother (Patterson, Hurt, et al. 1998). Finally, a meta-analysis of attachment security in modern settings of nonmaternal care showed that the sensitivity of the care setting predicts attachment security to both the mother and other caregivers (Ahnert, Pinquart, et al. 2006), although, as in many previous studies, attachment to the mother was stronger.

Questions are legitimately raised about the impact of day care on infants and children (Belsky 2001; Watamura, Donzella, et al. 2003), but the cross-cultural evidence demonstrates many precedents. These have almost always been based on kinship networks, as predicted by Emlen's model and by inclusive fitness theory. But as Seymour (2004:550) points out, Hrdy's conclusion "that the Western expectation of exclusive mothering is aberrant and that humans evolved as 'cooperative breeders,'" and Lamb's view that "multiple child care is 'a universal practice with a long history, not a dangerous innovation,'" deserve more attention and are finally getting it (Hrdy 2009). It is possible that the ways in which we in the West today do multiple care differ from how it was done in traditional kin-based and extended-family cultures and have different consequences. But it is clear that, cross-culturally and historically, children have survived and thrived in a wide array of multiple-caregiving arrangements.

17

Male Parental Care

Male parental investment resembles a component of cooperative breeding insofar as it is biologically optional help for the mother and evolved in some insects and amphibians (Trumbo 1996; Crump 1996), many fish (Gittleman 1981; Whiteman and Cote 2004), birds (Ketterson and Nolan 1994; Tullberg, Ah-King, et al. 2002), and some families of mammals (Kleiman and Malcolm 1981; Krasnegor and Bridges 1990; Moller 2003). It can include providing food for the female during the vulnerable breeding period, finding and defending a nesting place, building the nest, defending the female, brooding eggs, and feeding and protecting young. Since it can usually be selected for only when paternity certainty is high, it is typically associated with pair bonding (Reichard 2003). Despite the costs of parental care, males in primate species that have evolved it live longer relative to females (Allman, Rosin, et al. 1998); a positive effect of paternal involvement on fathers' health has also been found in humans (Bartlett 2004).

Species differences in parental behavior represent different strategies to maximize reproductive success (RS) (Clutton-Brock 1991); within-species variations by ecological circumstance may serve the same purpose. Some species have flexible programs for parental behavior that may optimize RS under different expectable conditions (Emlen 1995; Strier 1996). Fathering cannot explain the invention of males, which long preceded male parental investment, but it represents an advantage for many species that already have the burden of a large population that cannot themselves produce offspring.

Male Parental Investment and Reproductive Success

We can define parental investment (PI) as "any investment that en-hances the offspring's chance of surviving at the cost of the parent's ability to invest in another offspring" (Trivers 1972:139). The relative PI of males and females should correlate with sexual selection (courtship and mating patterns), sexual dimorphism, patterns of competition, adult sex ratio, and life span. In theory, the sex that invests more in offspring is the scarce resource competed for by the other, and initial investments invite future ones; other explanations are possible, but do not invalidate the correlations (Wade and Shuster 2002). In species with extensive, oc-casionally exclusive, male care, females tend to compete for males. And in some cooperatively breeding species, females have wider reproduc-tive opportunities, more competition, more variation in RS, and less sexual dimorphism than in related species without cooperative breed-ing (Rubenstein and Lovette 2009).

A continuum describing the interrelationships among the relevant variables is presented in Table 17.1. Not all aspects of this model are proven, but it pervades recent research. Data supporting different parts of the model have been collected for many species (Clutton-Brock and Godfray 1991; Allman, Rosin, et al. 1998; Reichard 2003). Examples are species at different points on the continuum, including humans, who fall between pair bonding and polygyny; in most traditional cultures po-lygyny is permitted but not very common. Even in monogamous socie-ties (Low 2003) pairs are not often permanent—aside from divorce and adultery, for most of history "till death do us part" meant fifteen or twenty years—and men are more likely than women to have secondary unmarried partners or engage in serial monogamy (Fisher 1989). Poly-andry is rare as an accepted marriage form. In a worldwide sample of non-industrial cultures (Murdock and White 1969), 1 percent were poly-androus, 17 percent monogamous, 51 percent occasionally polygynous, and 31 percent commonly polygynous. Human sexual dimorphism is consistent with this bias: men are moderately larger and heavier than women and have higher metabolic rates, higher mortality, greater mus-cle mass, less subcutaneous fat, wider shoulders, narrower hips, more facial and body hair, higher hairlines, and more prominent jaws.

Still, defining and measuring these variables is not easy. For exam-

Table 17.1.   Continuum of Parental Investment and Related Variables

Species	Elephant seal Red deer Baboon	Human	Ring dove Bat-eared fox Marmoset Gibbon	Jacana Phalarope
Parental investment (PI)	Male contribution negligible	Male PI less than female PI, but often substantial	Male and female PI about equal	Male PI greater than female PI
Mating system	Tournaments (leks), polygyny, promiscuity	Monogamy; some polygyny; polyandry rare	Monogamy/pair bonding	Polyandry
Courtship	Males compete for females; female choice	Variable	Low within-sex competition	Females compete for males; male choice
Sexual dimorphism	Males larger, higher metabolic rate, more brightly colored, and so on	Moderate (size, muscle, fat, face and body hair, and so on)	Minimal dimorphism	Females larger, more colorful, and so on
Variability in reproductive success	Males much greater than females	Males somewhat greater than females	Minimal sex difference	Females greater than males

Note: The distances from left to right are merely ordinal; equal distance between columns does not signify proportionate positions on the continuum.
Source: Adapted from Clutton-Brock and Godfray (1991), Allman, Rosin, et al. (1998), and Reichard (2003).

ple, does feeding the young represent more investment than defend-ing them, especially where the male risks serious injury? The model is easy to apply when the focus is on continuous or frequent care of indi-vidual offspring, especially carrying or feeding them, but defense and other activities also constitute parental investment. Many variants have been studied. Coyotes have a strong pair bond, and they and several other canid species simulate the provisioning and defense aspects of human fatherhood. This is also the pattern in thousands of species of monogamous birds; in some, like the ring dove, males are physiologi-cally adapted to feeding hatchlings (Lehrman 1955, 1965; Lott, Scholz, et al. 1967). Marmosets have twins, and males carry the infants 70 percent of the time, transferring them to mothers only to nurse (Epple 1975; Ma-son 1966). Gibbons are also sometimes pair-bonding, yet they do not ex-hibit such extremely high male parental care.

With DNA testing for paternity, it is possible to find out whether the offspring that males are caring for are genetically their own, as the model predicts. It is evident that in many purportedly monogamous species of birds (Ketterson and Nolan 1994) and primates (Garber 1994) males are caring for offspring fathered by other males. How can male parental care evolve and persist where cuckoldry is common? This is like the question of how cuckoo females can have their young reared by females of other species; the latter are simply losing in reproductive competition. However, birds that incubate eggs (a costly form of PI) have less cuckoldry than other species (Ketterson and Nolan 1994). In some marmosets and tamarins, males care for young in a multi-male group without regard to paternity, but in these species the males are usually brothers; inclusive fitness is optimized despite paternity uncertainty.

Males may also tolerate a certain amount of cuckoldry if the care they deliver offers access to an otherwise unavailable female, making infant care a kind of courtship. Extra-pair paternity produced less than 20 percent of offspring in almost 80 percent of 185 bird species and was completely absent in almost 30 percent (Moller 2003). No adaptation is perfect, and investment differs according to paternity in some species. Wild savanna baboons *(Papio cynocephalus)* have a promiscuous mating system, and DNA paternity testing shows that males protect their own more than other infants in aggressive encounters (Buchan, Alberts, et al. 2003). Preferential investment may occur in other species, but some studies fail to find it, and its role remains uncertain (Sheldon 2002). For females, the optimal number of fathers is not always one (Hrdy 2000); the arrangements in a species are a compromise between male and female goals.

Paternal Investment, Social Organization, and Ecology in Nonhuman Species

Mammals

Although some eight thousand bird species are pair-bonding and several are polyandrous, long-term pair bonding and significant male care of young are unusual in mammals (Kleiman and Malcolm 1981); 3 to 5 percent of mammal species and 14 percent of primate species are pair-bonding (Rutberg 1983; Moller 2003), and the higher rate in primates could reflect a need for prolonged parental care (van Schaik and Kappeler 2003). Female mammals have physiological adaptations for care; in most mammals parents associate only for sex and the male invests only semen, but there are important mammalian examples of high male parental investment, especially in rodents, carnivores, and primates (Kleiman and Malcolm 1981; Reichard and Boesch 2003). Voles have been essential to our understanding of the physiology of pair bonding and paternal care (Chapter 10); montane voles mate multiply, while prairie voles pair-bond, with males exhibiting mate guarding and pup retrieval, behaviors based on the brain distribution of vasopressin receptors in pair-bonding voles (Wang, Zhou, et al. 1996; Young, Lim, et al. 2001).

Experimental and field research in mammals shows that ecological factors are important. In the California mouse (*Peromyscus californicus*), which usually has isolated pairs and high levels of male parenting (Alberts and Gubernick 1990), experimental removal of the male significantly decreases pups' weight gain and likelihood of survival, especially if the mother is absent or weaning occurs earlier (Dudley 1974a, 1974b). In the wild, foraging requires that the mother be frequently absent, especially if resources are scarce. In hoary marmots (woodchucks; *Marmota caligata*) observed in two settings with different population density (Barash 1975), males lived in the same burrows as mothers and interacted with their young more in the less populated setting. In the higher-density setting males had separate burrows, interacted less with the young, and engaged in more defensive encounters. In three species of guinea pigs, males groomed and played with young, as predicted by paternity certainty (Adrian, Brockmann, et al. 2005).

Ecological factors have also been studied in relation to overall differences in social organization across groups of related species. It is more difficult to draw conclusions about specific ecological influences on male parental investment that would apply over a wide phylogenetic range, but associations are observed, and ecology is the framework within which paternal behavior or its absence must be understood.

Canids

In many wild canids both sexes feed the young by regurgitating after hunting, as some pair-bonding birds do (Gittleman 1986). Canids, like primates, include species that have mother-offspring units only, isolated pairs with their offspring, and larger groups. The most durable pair bonds occur where the pair lives separately, as in coyote or golden jackal pairs that outlast the breeding and rearing season (Ralls, Cypher, et al. 2007); bat-eared foxes (*Otocyon megalotis*) have especially strong pairs (Lamprecht 1979). In group-living canids, if the female is unable to hunt, she is provisioned, as are others unable to hunt (van Lawick and van Lawick-Goodall 1971; Fox 1975). In pack-living wolves the bond may also persist beyond rearing (Fox 1975). In the relatively asocial dingo (Australian wild dog; *Canis lupus dingo*) the male visits the den regularly without providing food (Corbett and Newsome 1975). But in the more isolated pairs that characterize coyotes the male provisions the female in late pregnancy (when she does not hunt) and the pups as they begin to take solid food (Gier 1975). Free-ranging domestic dogs show a pattern of male defense of pups and occasional direct feeding by males (Pal 2005).

In the cooperatively breeding African wild dog (*Lycaon pictus*) all pack members feed and protect the young (Estes and Goddard 1967), but male helpers have a greater impact on pup survival than females, who may be in competition with the mother (McNutt and Silk 2008). Canid packs are essentially extended-family units (Fox 1975), and their shared caregiving is mainly kin-directed altruism (Hamilton 1964; Emlen 1995). The three types of social organization may depend on ecological factors such as prey size and availability, dispersion, and predation pressure. Scarcity of resources, mate guarding, female choice, and protection from both predators and conspecifics may be important determinants of male proximity and parental investment.

Higher Primates

Although the pair bond exists in some prosimians, the typically low level of male parental care in monkeys and apes (MacKinnon 2007) suggests that in ancestral primates male care was minimal, but that it evolved independently several times (van Schaik and Kappeler 2003) in the marmoset family (Callitrichidae), a few other small New World monkeys, two Asian colobine species, and the gibbon family, all of which evolved some degree of pair bonding and paternal care (Wright 1990; Taub and Mehlman 1991; Moller 2003). Katharine MacKinnon (2007:582) prepared a table that gives a quick and helpful overview of male primate parental or parentlike behavior.

Literature reviews (Whitten 1987; van Schaik and Kappeler 2003) suggest that different forms of primate male parental investment—especially the "care" versus "protection" continuum—accompany different ecological conditions and patterns of social organization. Male care of infants does not follow phylogenetic trends but is more a function of ecology. In patrifocal species with female dispersal, such as the Brazilian muriqui (*Brachyteles arachnoids*) (Strier 1993) and the chimpanzee (Watts and Pusey 1993), relationships between male juveniles and adult males become increasingly important as the former approach adolescence; the younger males must be integrated into the patriline. This contrasts with male-dispersal species, common among Old World monkeys, where adolescent males may be driven out by older males, even their fathers, to face a risky and sometimes fatal group transfer.

The New World monkeys with the most male care (marmosets, tamarins, callimicos, night monkeys, and titi monkeys) tend to be arboreal and omnivorous and to live in isolated pairs or small extended families. Marmoset males may assist at birth, receiving and washing the infants, and then they may carry and groom them until independence. There is also some provisioning, with both males and females offering food (Snyder 1972). Extensive male care may enable twinning, but it also affords protection from infanticide (van Schaik and Kappeler 2003). Human pair bonding may have enabled meat provisioning, offspring protection (including from other humans), and long dependency (Lovejoy 1981, 2009; Tooby and DeVore 1987; Kaplan, Hill, et al. 2000).

In Old World monkeys and apes, males do not routinely care for infants but may retrieve and guard them as part of group defense. Males frequently respond to infants' distress cries, with protection if not nurturance. Play between males and young is seen in some species, and most show indirect interaction, such as orienting to or traveling beside an infant. Males occasionally "babysit" or adopt an infant, not necessarily for the offspring's benefit. There is significant positive interaction between adult males and infants in Japanese macaques (*M. fuscata*), Barbary macaques (*M. sylvanus*), olive baboons (*P. anubis*), and hamadryas baboons (*P. hamadryas*), although it may benefit the infant less than the male.

Males exploit the infants of other males in various ways. Caring for infants may be a way of courting their mothers (Moller 2003). Olive baboon males may show interest in an infant to befriend the infant's mother, increasing the chance of mating with her and easing their immigration into a new troop (Smuts 1985). Japanese monkey males use infants in agonistic buffering; less dominant males protect themselves from aggression by carrying infants, improving their position in large, hierarchical troops (Itani 1963). They tend to carry yearling females; some of these females are later likely to mate with the same males, who might otherwise have few chances. Hamadryas baboon males, who sequester prepubescent females in their harems, guard them against other males until they mature into breeding harem members (Kummer 1968). The male Barbary macaque interacts frequently with infants, a poorly understood pattern that may be a form of agonistic buffering (Burton 1972).

Males, especially subdominant ones, may need to guard females against other males or to guard infants that might be their own against infanticide by other males, and biparental care could be a by-product of this close association (Moller 2003). We know that in Hanuman langurs (*Presbytis entellus*) (Hrdy 1977) males taking over a troop systematically kill infants, whose mothers soon mate with the new males. This is the extreme case, but in many systems males exploit infants for their own gain, and so do some human males. Adult male interest in pubescent girls is legitimized in many human groups (Marshall 1959; Paige and

Paige 1981; Shostak 1981); Western baby care manuals warn against boys as babysitters because of risk of sexual abuse (Spock 1968).

In some species the dominant males, who, having most access to estrus females, are most likely to be the fathers, interact with and protect infants the most (Hall and DeVore 1965; Altmann, Hausfater, et al. 1988). As we have seen, DNA-proven paternity predicts male protection of infants in savanna baboons (Buchan, Alberts, et al. 2003). In hamadryas baboons, males "adopt" females to groom as future harem members (Kummer 1968). In Japanese macaques dominant males take over the care of one- and two-year-olds when the mothers are delivering new infants, but the fact that twenty of twenty-five infants adopted by males in one study were female (Itani 1959) points to a possible future-oriented courtship investment. Males clearly gain a status advantage by carrying infants, as in Barbary macaques, where males occupy about 8 percent of the infant's time (Burton 1972; Deag and Crook 1971; Deag 1980), but they also carry, retrieve, and protect infants very selectively, possibly depending on the likelihood of paternity.

Orangutans present the paradox of advanced cognitive ability combined with a rudimentary social order—mother-offspring units occasionally visited by males (Galdikas 1985a; Rodman and Mitani 1987). However, in settings with more predators and competitors, males do join the mother-offspring unit, probably protecting their past and future genetic investments (Knott and Kahlenberg 2007). Coercive copulation, males' defense of their access to females (much easier for fully adult males), and female choice have all been documented (Galdikas 1979; van Schaik and Van Hooff 1996). However, adult males need much more food and are disadvantaged when tied to females; this centrifugal tendency disrupts their association with the mother-offspring unit.

Table 17.2 compares selected pair-bonding mammalian species in relation to male parental investment and social organization (Katz and Konner 1981). Prairie and pine vole males guard and maintain high levels of physical contact with their mates and infants (Lonstein and De Vries 2000). Canid males provide protection and food resources. Marmoset males provide extensive direct care (almost continuous carrying), and human males provide protection, provisioning, and varying amounts of direct care.

Most male parental investment in human societies is indirect—defensive or protective, as in all higher primates—but the male economic

Table 17.2. Comparison of Selected Pair-Bonding Mammal Species

	Prairie Vole and Pine Vole	Coyote and Golden Jackal	Marmoset	Human
Number of young	Several	Several	Twins	Single
Location of young	Cached	Cached	Carried	Carried
Role of male in provisioning female and young	Low	High	Low	High
Importance of animal protein in diet	Low	High	Low	Medium
Female sexual receptivity	Noncontinuous	Noncontinuous	Noncontinuous	Continuous
Relationship of pair to other individuals	Isolated or extended family	Isolated or extended family	Isolated or extended family	Extended family in larger group

Source: Modified from Katz and Konner (1981).

investment (postweaning provisioning, and so on) is almost unique to humans among primates. So as we evolved from our ape ancestors, there was increased direct male care in the context of increased co-operative breeding. But the main male addition was provisioning with hunted meat, while maintaining the higher-primate male defensive-protective functions. The much lower mortality between weaning and reproductive maturity characteristic of human hunter-gatherers compared to other higher primates results from postweaning provisioning (Lancaster and Lancaster 1987; Kaplan, Hill, et al. 2000). We see the consequences of these changes in the basic human adaptation by looking at fathers among hunter-gatherers.

The Paternal Role among !Kung Hunter-Gatherers

The !Kung were classified as "high" on closeness of fathers to infants and young children in a sample of eighty independent non-industrial societies (Barry and Paxson 1971), and Nancy Howell's (2010) analysis of !Kung caloric exchange and demography suggests that fathers are part of a spectrum of helpers who enhance child survival. Often available near home, fathers held and fondled even the youngest infants, but returned them to the mother whenever they cried and for routine care. Young children often approached, touched, talked to, and requested food from their fathers and were rarely rebuffed. Boys did not begin following adult men on hunts until early adolescence or later, and they did not hunt independently for years. Information transfer on hunts was based more on observation than teaching. In traditional male initiation rites boys had to dance almost naked in the cold for several days. Adult men would frighten them in the dark and later make small cuts on their foreheads to signify their transition to manhood. Learning to fight was not considered important.

Fathers accounted for 10 percent of vocalizations to infants during the first three months (Konner 1976). A study in Boston in the 1960s (Rebelsky and Hanks 1971) kept a microphone in cribs around the clock and estimated the average amount of vocalization by fathers to three-month-olds to be thirty-seven seconds per twenty-four-hour period, so !Kung fathers, though much less involved than mothers, were more involved than Western fathers. The data on infants came from fifteen-minute observations marked in five-second time blocks, as described in Chapter 16. Father availability was not necessary for starting an observation session. Data on the number and percentage of observations of boys and girls in two age groups in which fathers interacted with the infant were analyzed. Overall father participation for the sexes and ages combined was 13.7 percent; fathers were more likely to interact with boys than girls at the older age only ($\chi^2 = 4.61$, p < .05).

Since observation onset required the mother's presence, the comparison was biased against fathers, but it showed that randomly distributed fifteen-minute time samples had about a 90 percent chance of finding interaction between mother and infant, compared with 13.7 percent for fathers. However, the density of interaction within the fifteen minutes was also much greater for mothers. In the study of two- to six-year-olds, !Kung fathers were present in 30 percent of observations, as opposed to 19 percent in London (Blurton Jones and Konner 1973). The Six Cultures Study of three- to eleven-year-olds showed father presence ranging from 3 to 14 percent, with 9 percent in the United States (Whiting and Whiting 1975a).

Paternal Roles in Other Hunter-Gatherers

Compared to the range of cultures in the anthropological universe, father involvement is high in hunter-gatherers generally, and highest in the Aka (Hewlett 1991b). Betsy Lozoff and Gary Brittenham's (1978) comparison of ten hunting-and-gathering cultures with other non-industrial cultures found more father-infant and father-child contact in the hunter-gatherers. We now consider male roles in five well-studied hunter-gatherer groups.

Hadza

Hadza infants and young children interact with fathers 17 or 18 percent of the time, just twice the rate for maternal grandmothers, despite the theoretical emphasis on grandmothers (Hawkes, O'Connell, et al. 1998). This difference is even greater when the data exclude stepfathers, a relatively common role due to a substantial divorce rate (Marlowe 2005a). Recall that kin selection applies; the only spanking ever observed was by a stepfather. Fathers' provisioning effort is critical when their wives are nursing infants under a year of age, a time when the mother's own foraging is less efficient than his, whether he brings in meat, honey, or plant foods (Marlowe 2003). However, yet again, this does not apply if the nursling is a stepchild. When no biological father is present, the maternal grandmother's role expands (Marlowe 2005a). In other analyses, fathers communicate with, play with, and nurture their biological offspring more than their stepchildren, and a father's involvement is inversely correlated with his mating opportunities, indicated by the number of unmarried younger women in the camp (Marlowe 1999).

Efe

Efe fathers interact with children much less than other adults or children during infancy, but more than the average adult at later ages. The father may be the second most important caregiver, but not a close second. In a study of ages one to three years (Morelli and Tronick 1992) the father was compared with other males. The *average* time a child was in social contact with men other than the father was more than half that spent with the father throughout the first three years and appeared to be mainly with one man at each age. One-year-olds spent more time with the average older boy than with the father, and the difference increased with age, so "the role Efe fathers played in the lives of their children relative to other males did not appear to be particularly special or unique" (Morelli and Tronick 1992:49).

Aka

The involvement and intimacy of fathers is what is best known and most distinctive about Aka infancy (Hewlett 1988); indeed, "Aka fathers do more infant caregiving than fathers in any other known society" (Hewlett 1991b:167). A policy report on fathers for the Society for Research in Child Development described the Aka as "the most nurturant fathers yet observed" (Engle and Breaux 1998:5). Still, the highest number reported for Aka father involvement is the percentage of all holding done by the father for infants up to four months of age (22 percent), a time when the mother does 51 percent and nonparents 27 (Hewlett 1991b:79).

As argued for maternal primacy, however, the presence of several or many nonparent "others" gives the father a good chance to take a strong second place in the infant's heart. On net hunts, infants this age are held by fathers 6.5 percent of the time, compared with mothers' 87.3 percent. On these and other trips to the bush, fathers hold infants only 8 percent as much as mothers do, a rate that declines to 3 percent over the first eighteen months. But the situation while in camp (which is most of the time) is very different: the paternal/maternal ratio of infant holding is .43 in the first four months, .25 from eight to twelve months, and .45 from thirteen to eighteen months.

As we saw in Chapter 15, from eight to twelve months, fathers receive about one-fourth as many attachment behaviors from their infants as mothers do, but this jumps to almost 60 percent in the second year. Over all of infancy and early childhood, neither any other individual nonmaternal caregiver nor all of them together rival the father as a secondary caregiver, so the opportunity for secondary attachment to him is great. And the interaction is different: "Mothers were most likely to provide nourishment and transport the infant, while fathers were more likely to hug, kiss, or clean the infant as they were holding" (Hewlett 1991b:94). In later childhood, "mothers and fathers contributed equally to the subsistence training of their children," but "fathers were more likely to educate sons and mothers were more likely to educate daughters. Fathers were more likely to transmit hunting, dancing, and singing skills while mothers were more likely to transmit gathering and child care skills" (Hewlett 1988:269). Although Aka men do not undermine the case for maternal primacy, their reputation for exceptional fathering is justified.

Ache

The Ache are even more nomadic than other well-studied foragers, and during their trips into the forest men supply 87 percent of the food (by caloric value) and spend twice as much time as women in foraging activity. Thus, in contrast to most hunter-gatherers, Ache mothers do defer subsistence activities in favor of child care. While in the forest, infants during their first year are in tactile contact with their mothers or fathers 93 percent of the time; this contact appears to be overwhelmingly with mothers, although Hill and Hurtado's (1996:219) mention of fathers at this point in their study suggests that fathers may be second in importance. Men approach fatherhood with great pride, and "new fathers are expected to hunt long hours and also pound and extract palm fiber for their wives" (Hill and Hurtado 1996:275). Children may spend part of their time with a man who helped create their "essence" by giving their mother meat during pregnancy.

Agta

Since Agta women hunt to a degree unknown in any other hunter-gatherer group, bringing in as much as half of the game, it might be surprising that mothers are the primary caregivers half the time. The corresponding figure for grandmothers is 7.5 percent, older sisters 10.4 percent, and fathers 4.4 percent (Griffin and Griffin 1992). Fathers are not often seen holding infants in camp, but they often carry toddlers and older children on subsistence trips and residential moves. In contrast to mothers, fathers were never observed in pacification play with a crying or fussing baby, but there was marked individual variation. In observations of seven fathers with infants under age two years, the ratio of paternal to maternal care was zero in two cases, more than half in another, and two-thirds in another. Still, of four infants under one year of age, father-to-mother care ratios were between one-to-six and zero. In summary, "the Agta fathers are not particularly active with children when compared with the !Kung, the Aka, and even the Ache" (Griffin and Griffin 1992:317). This limited role for fathers is all the more remarkable given the unique extent of Agta women's hunting.

Paternal Roles in Non-Hunter-Gatherers

For higher primates, humans have high levels of pair bonding and father involvement, but with much cross-cultural variation. The range of paternal roles in the ethnographic record is much wider than that for mothers. The following four cases illustrate the kinds of male parenting adaptations found in ethnographies, which form the basis of the quantitative cross-cultural comparisons to follow (Katz and Konner 1981).

Rwala Bedouin, North Arabian Desert, 1913 (Ethnographer: Musil 1928)

The Rwala Bedouin exemplified cultures that subsisted by large-animal husbandry; frequent warfare, strong male authority, and cultivated fierceness made fathers very distant. Camels were at the center of Bedouin life, and raids and war were common, motivated by spoils, territorial acquisition, and vendetta. Livestock management was a male province, and the culture was patrilineal, patrilocal, patriarchal, and polygynous. Fathers spent most of their time in a separate compartment of each tent, and children under seven talked to them only occasionally.

Sometime between ages three and seven, boys were dressed in black and circumcised by their fathers after feasting on a female camel sacrificed for the occasion. Then they spent more time with adult men, bringing them food and water and attending to livestock. In adolescence they learned to shoot and go on raids. Either parent could discipline a young child with a stick, but disobedient older boys were cut or stuck with a dagger or saber by their fathers, to harden them for their future lives.

Lesu Village in New Ireland, Melanesia, 1930
(Ethnographer: Powdermaker 1933)

Lesu was a typical Pacific island culture, with men (fishing) and women (taro gardening) making roughly equal contributions to subsistence. Most Lesu households were monogamous nuclear families, although some men had two wives in separate houses. Families were close, and men and women when together were affectionate and intimate, but they remained separate in public life; older boys, single men, and men with wives pregnant or nursing (most of women's reproductive years) slept in a separate men's house. Fathers often fondled and played with babies, even for hours at a time when women were gardening, and men and women were equally tender toward children. Groups of men passed or tossed infants around, dandling, kissing, and singing to them. Children were everywhere in the village, with their same-sex parent or older children, participating in communal food preparation, dance rehearsals, and rituals by watching and learning. Boys and girls increasingly participated in gender-appropriate activities, although boys tended to watch, and girls to participate. Between ages nine and eleven, dramatic initiation rites were staged, with fathers and mothers jeering and even throwing stones at one another, after which women danced while men circumcised the boys. For several weeks boys were kept away from women and taught fishing and hunting. As in most small island societies, warfare was rare, and boys were not trained as warriors.

Thonga of the Ronga Subtribe, East Coast of South Africa, 1895
(Ethnographer: Junod 1927)

Thonga subsistence was based on small-scale grain and vegetable agriculture (mainly by women) and cattle and goat herding (mainly by men).

The polygynous extended family occupied a compound where each wife had her own hut. Men tried to have several wives, many children, and large cattle herds, which attracted wives for their sons. Although men had ample free time, they had almost no contact with young infants, and they related to older infants and young children only in rituals. The mother and older siblings did the baby care, but after weaning toddlers lived with their grandparents for several years. Boys and girls respected and feared their fathers, and fathers were their sons' teachers. Older boys tended goats and were thrashed by their fathers when negligent. Thonga maintained an army with war costumes and weapons.

Koreans of Sondup'o Village, Kanghwa Island, 1950 (Ethnographer: Osgood 1951)

The agriculture of this community, part of a centralized state, differed from the small-scale farming of Lesu and Thonga in three ways: (1) men did it; (2) the use of plows enabled larger-scale planting while demanding greater strength; and (3) agricultural activities dominated other subsistence activities. Marriage was monogamous in an extended-family context, with patrilocal residence, patrilineal inheritance, and primogeniture. Fathers were stern and somewhat distant, although more tolerant and affectionate toward very young daughters. With sons especially, the father was the respected and feared disciplinarian; succor came from mothers.

Observable Patterns and Their Possible Significance

As these vignettes show, fathering patterns vary against the background of family structure and men's and women's work. The variable elements are: (1) the number of wives and children and the father's authority; (2) the father's contact with his wife (or wives) and children at different ages and the quality of those interactions; (3) the amount of direct paternal care; (4) paternal responsibility for imparting skills and values; (5) participation in rituals connected with children; (6) male and female subsistence contributions and relative workloads; and (7) how much men fight to defend or expand resources.

A wider cross-cultural sample shows that agricultural societies exhibit greater variation in father involvement than do other subsistence types, but on average they lie between the distant strictness of Thonga and Rwala fathers and the warmth of !Kung and Lesu fathers. In the two cultures with the most distant fathers (Rwala and Thonga), polygyny and warfare are common and male authority is stressed; the paternal role is to procreate, defend, command, and punish. In two of the three cultures in which women contribute most to subsistence activities (Lesu and !Kung), fathers are closer to their children. In the third culture (Thonga) fathers are distant and warfare and polygyny are present; statistically, distant fathering is associated with the need to train boys to be warriors (Whiting and Whiting 1975b). Another paternal role is to teach or model subsistence skills for boys. In some cultures the father plays with children and/or disciplines them, while the mother nurtures them.

This distinction often holds in American families (Yogman 1990; Cabrera, Tamis-LeMonda, et al. 2000), although there has been some convergence of roles with increasing father involvement (Roopnarine, Fouts, et al. 2005). Another aspect of the traditional Western father's role has been introducing the child to the world outside the home (Lamb 1975), a role that is more likely where the mother does not work outside the home. In many non-Western cultures the typical source of introduction to the world outside the home is the juvenile playgroup (Chapter 19).

Cross-Cultural Comparisons and Quantitative Ethnographies

Because of their different roles, exact mother-father comparison is difficult, but approximate ones have been made (Katz and Konner 1981). Herbert Barry III and Leonora Paxson (1971) used "regular close relationship" to denote the highest level of father involvement in their cross-cultural sample (for them, the !Kung), and their criteria suggest that this level of paternal involvement was comparable to involvement with infants "half or less of [the] time" on their maternal scale. Since there were few cultures with fathers in this highest category and few with mothers in this or the next-lowest category, paternal and maternal care distributions are almost non-overlapping in infancy, but the early childhood distributions overlap substantially (Table 17.3).

If this matchup of rankings is approximately correct, then fathers have a "regular close relationship" to infants in 4 percent of cultures, compared with 98 percent for mothers. In early childhood the corresponding figures are 9 and 66 percent. The marked discrepancy in in-

Table 17.3. Percentage of Cultures by Ranked Level of Parent-Child Proximity

Role of Father		Role of Mother	
		Infancy	
No close proximity	5%		
Rare proximity	15	0%	Practically all care is by others
Occasional proximity	37	0	Most care except nursing by others
Frequent proximity	39	2	The mother's role is significant but . . .
Regular close relationship	4	8	The mother provides half or less of care
		44	Principally the mother; others have important roles
		43	Principally the mother; others have minor roles
		3	Almost exclusively the mother
		Early childhood	
No close proximity	1		
Rare proximity	11		
Occasional proximity	19	2	Practically all time away from the mother
Frequent proximity	60	33	Majority of time away from the mother
Regular close relationship	9	39	Half or less time with the mother
		27	Principally the mother; others are important
		0	Almost exclusively the mother

Note: Charts for mothers and fathers have been adjusted to approximately match the respective definitions.
Source: Significantly modified from Katz and Konner (1981); data and definitions from Barry and Paxson (1971).

fancy lessens slightly in early childhood, but there is no human society in which males have primary or even equal responsibility for the care of offspring (Katz and Konner 1981; Geary 2000). These comparisons are mainly based on descriptive ethnographies, but the male-female discrepancy in infant and child care has been confirmed by quantitative studies in all patterns of subsistence. All hunter-gatherers, including the Aka, have it. Among the Ye'kwana, who subsist by hunting, gathering, fishing, and gardening in villages of the upper Orinoco basin in Venezuela, females account for about an order of magnitude more baby and child care than males do at every age (Hames 1988).

In the Six Cultures Study (five gardening or farming cultures and one New England town), despite a fivefold range of father involvement, time spent with children was consistently low (Whiting and Whiting 1975a): father *presence* ranged from 3 percent in the Rajputs of Khalapur in northern India and the farmers of Taira, Okinawa, to a maximum of 14 percent in Tarong, on the Philippines island of Luzon. In "Orchard Town," New England, the figure was 9 percent. Actual interactions were one to two orders of magnitude less frequent. Of the extended-family cultures, the percentage of all children's social acts directed to fathers was 0.1 in Taira, 0.2 in Nyansongo (a Gusii village in Kenya), and 0.3 in Khalapur; of the nuclear family cultures, the corresponding percentages were 1 for both Orchard Town and Juxtlahuaca (a Mixtecan Indian village in Oaxaca, Mexico) and 2 for Tarong—a difference probably due to the presence of older children and adult women in the extended-family cultures. As with the Aka-Efe comparison and some of the Hadza grandmother studies, this suggests a trade-off between paternal and other forms of nonmaternal care.

Cash and subsistence farmers in the village of Grande Anse on the northern coast of Trinidad show another interesting pattern (Flinn 1988). There is about a tenfold difference between mother- and father-offspring interactions up to age five (narrowing to about threefold between ages six and ten), but in the teenage years, when father involvement peaks, fathers have slightly more contact with daughters than mothers do, although they have only half as much contact with sons. This peak is associated with a much more than proportionate increase in angry interactions with daughters, often regarding prospective husbands.

None of these findings should be taken to imply that fathers are unimportant in psychological development; much evidence suggests the contrary, despite relatively low direct care (Lamb 1981; Feldman, Greenbaum, et al. 1997). New trends in the West are producing much greater father involvement (Cabrera, Tamis-LeMonda, et al. 2000; Marsiglio, Amato, et al. 2000) and, very significantly, the rise of the single-father family (Pruett 1998); studies unequivocally show that at least some males are capable of replacing almost all functions of mothers, at least in post-industrial cultures.

Subsistence Adaptation and Family Organization

John Whiting and Beatrice Whiting (1975b) showed that father involvement in infant care is part of a dimension they call husband-wife aloofness versus intimacy. Cultures with distant fathers also have distant husbands. These societies are unlikely to be hunter-gatherers but instead are those with important material resources requiring defense (for example, gardening cultures that are not on islands, or herding cultures with raiding neighbors) that draws the father into war and preparation for war. In addition, these cultures tend to have polygyny, which also draws the father away, at least in those cultures that have separate households for different wives.

Mary West Katz (now Mary Maxwell West) analyzed a world sample of eighty non-industrial cultures with regard to father-infant proximity, family organization, and subsistence (Katz and Konner 1981). The cultures were a subsample of George Peter Murdock and Douglas White's Standard Cross-Cultural Sample, chosen to be representative of the world range of thousands of cultures and for their mutual independence (Murdock and White 1969). Cultures selected had been rated by Barry and Paxson (1971) on their fathering scale at their highest confidence level for ethnographic quality.

The analysis showed that fathers are closest to their children where gathering or gardening (horticulture) is the main subsistence method and where polygyny, patrilocal residence, the extended family, and a male-dominant division of labor are absent (Katz and Konner 1981). These variables are related but not highly correlated; combinations of them predict father-child proximity. Important underlying factors are pressure for males to engage in warfare; the extended family, which provides nonpaternal allocare; and a heavy maternal workload, which draws fathers in. Earlier cross-cultural research related father absence to violent or hypermasculine behavior. In the Six Cultures Study, the two most father-distant cultures had more assault and homicide and a generally greater acceptance of violence and strife (Whiting 1963). In much larger cross-cultural samples, cultures with separate mother-child households had more theft and personal crime (Bacon, Child, et al. 1963) and greater pursuit of military glory (Whiting and Whiting 1975b; Broude 1990).

Beatrice Whiting (1965) invoked cross-sex identity conflict and protest masculinity to explain the relationship between father absence and interpersonal violence. In this model the absence of a male leads to excessive identification with the mother, which boys must overcome with hypermasculine behavior as they approach adulthood. Carol and Melvin Ember (1994) initially cast doubt on Whiting's model with a multiple regression analysis of cross-cultural data, which showed that the overwhelmingly important predictor of interpersonal violence is training boys for aggression in later childhood. But a subsequent study using new and more detailed codes did find that distant fathers predicted interpersonal violence, although socialization for aggression remained a stronger predictor (Ember and Ember 1994). In a combined model, cross-sex identification from exclusive maternal closeness would result in protest masculinity *if the boy later went through socialization for aggression by men* that separated them from their mothers, sisters, and younger siblings and rigorously prepared them for raiding and combat (Broude 1990; Chick and Loy 2001). In many such cultures adolescent boys undergo severe initiation rites that dramatically reorient their identification from mothers to fathers (Whiting, Kluckhohn, et al. 1958; Burton and Whiting 1961).

With cross-cultural correlations, other causal interpretations are always possible, but this anthropological view is more or less consistent with psychological and demographic research (Cabrera, Tamis-LeMonda, et al. 2000). Problems in children and adolescents in father-absent households include aggression and conduct disorder in boys, although it is not easy to separate the effects of father absence from those of divorce, economic hardship, and other variables (Sigle-Rushton and McLanahan 2004). A stable cultural arrangement that includes father absence (multigenerational, matrilineal African-American households or lesbian couples) would be likely to have different consequences from those of father absence that violates the cultural context. In the Katz study (Katz and Konner 1981) none of the three family-pattern variables alone predicted father-infant proximity, but any two or more of them did. For example, of the six societies with strictly monogamous and nuclear families, all had close fathers. Given an extended family, monogamy or infrequent polygyny predicted close fathering if the preferred residence was not patrilocal.

Warfare may underlie other mediating variables. In many patrilocal or polygynous cultures warfare and raiding are means of acquiring resources, including women, and both polygyny and patrilocal residence are associated with internal warfare and feuding (Van Velsen and Weterling 1960; Otterbein and Otterbein 1965; Otterbein 1968). Bride theft and raiding for wives are more common in polygynous cultures (Ayres 1974); even where women are not systematically kidnapped, warring groups often intermarry, exchanging women in peace agreements. Patrilocal residence keeps fathers and sons together when sons marry, facilitating war preparations; patrilocal clans act as military organizations. These clans or "fraternal interest groups" also assert paternity

rights (Paige and Paige 1973). The male reproductive strategy is to produce many children by having multiple wives, and a good deal of time is
spent in accumulating and defending the resources necessary to acquire
and maintain these wives, if not in actually kidnapping women (Betzig
1986, 1997; Mace and Pagel 1997).

These associations of aggressive activities with mate acquisition,
mate guarding, and polygyny somewhat resemble the cross-species
continuum, which links male competition to greater male size, strength,
and metabolic rate, as in human sexual dimorphism. Polygynous or
patrilocal cultures that must defend accumulated resources without
specialized armed forces represent the "distant" end of the continuum.
Most human EEAs were probably characterized by a relatively low level
of intergroup violence and the "close" end of the range of paternal behavior. But some had higher levels of intergroup violence, with all the
attendant consequences for fathering. The division of labor by sex is also
a factor; taken together, subsistence types where women make a high
contribution—gathering, horticulture, and shifting agriculture—are
significantly more likely to have close fathers (Katz and Konner 1981).

However, the reverse occurs in pastoral cultures, since large-animal
husbandry is never done by women. Raiding for livestock, population
growth, and land use may also be key ecological factors (Divale and Harris 1976); the complex of "male supremacist practices," including local
warfare, polygyny, and bride price, may promote population growth,
and these practices certainly promote the RS of some males. Also, in
simple field agriculture, where women are busy and important providers, polygyny tends to be common, fathers are distant, and siblings are
very involved in infant care; toddlers are often sent to live for a time
with their grandparents. This suggests an algorithm for nonmaternal
care: where the male's strategy for RS includes polygyny and/or military
activities, he contributes little to child care regardless of women's contribution to subsistence; when these are absent, if the mother's subsistence workload is high, fathers will do more, especially in nuclear family cultures; where the father is absent and the mother's workload is
high, grandparents and siblings will help with child care. These explanations are consistent with Emlen's (1995) adaptive explanations of allocare (Chapter 16).

So in some subsistence adaptations, intensified conflict and polygyny
were advantageous male strategies. As technology and political organization increased—as in plow-based advanced agriculture—women's

direct role in subsistence diminished, and the average men's role in defense gave way to specialized defense forces. Monogamy (in principle) often occurs in such cultures, and providing food becomes the main form of male investment. Father-child proximity varies somewhat in these cultures, probably contingent on details of the agricultural work (type of crop, distance of fields from home), but it tends to be in the middle of the range of father closeness (Whiting 1963; Whiting and Whiting 1975a).

The United States and Other Industrial Cultures

Modern industrial cultures have neither polygyny nor frequent local warfare, but the role of women's work is more complex. With the mass (re)entry of women into the labor force in the late twentieth century and a growing dependency on two incomes, father involvement has increased (Cabrera, Tamis-LeMonda, et al. 2000; Marsiglio, Amato, et al. 2000). While father absence has much less impact on psychological development than some psychodynamic theorists predicted, fathers serve a variety of important functions—as a more playful companion, as a part of the spectrum of cooperative care, and so on (Lamb and Lewis 2004). In studies in the United States, Germany, and other Western countries, supportive father involvement predicts more exploratory behavior throughout childhood (Grossmann, Grossmann, et al. 2008).

In some studies infants' attachment to fathers is less secure than to mothers in two-income families (Braungart-Rieker, Courtney, et al. 1999), but there is little doubt that fathers can be competent caregivers. Children of gay male couples grow up psychologically very similar to other children (Tasker 2005), even though they must deal with prejudice against their fathers and their families (Cameron and Cameron 2002). According to U.S. Census Bureau statistics for 2008, there are about 2.5 million single fathers in the United States, almost one-sixth of all single-parent households; apparently large numbers of men can nurture effectively, although differences between children raised by single fathers as opposed to single mothers may exist.

Dads and Cads

Patricia Draper and Henry Harpending advanced our understanding of fatherhood with a formulation based on sexual selection theory

(Draper and Harpending 1982, 1987). Like many other species (Trivers 1972; Parker and Schwagmeyer 2005), humans have facultative male parental investment. Under some circumstances, strong paternal investment is clearly selected for, or we would not be largely pair-bonding. But under other circumstances, males may find it more advantageous to withdraw investment in their offspring in favor of mating investment in many females, and the human pair bond does not rule out other sexual strategies (Marlowe 1999, 2000; Zeifman and Hazan 2008). Cads may be favored over dads if they can be confident that their offspring will be cared for by others (cuckolded males, mothers capable of rearing offspring alone, matrilines or other extended families); or that their offspring's chances of survival are unlikely to be much improved by their investment; or just that they have exceptional mating opportunities. This could be a facultative adaptation or an evolutionarily stable strategy, since female choice of one type or the other should be frequency-dependent. Dads should be selected to protect themselves against cads, while females should take advantage of both; they often mate with multiple males sequentially or simultaneously, and it is increasingly clear that the "optimal number of fathers" from the female viewpoint is often more than one (Hrdy 2000).

Frank Marlowe has contributed greatly to our understanding of how the dad-cad continuum could have evolved by studying men among Hadza hunter-gatherers. There is evidence that provisioning of the young and of nursing mothers by men (Marlowe 2000, 2001, 2003), including not only meat but honey and plant foods, plays an important role in survival and reproductive success. However, there is also evidence that Hadza men can be distracted from direct care of children if there are young women available (Marlowe 1999), suggesting that they may allocate energy between paternal investment and mating effort as opportunities arise.

Plasticity and Its Physiological Limits

Species differ in male parental investment, but within-species variability is considerable, suggesting environmental impact. Rhesus monkey males in the wild are indifferent to infants, but the pairing of adult males with infants in the laboratory leads to many attachment behaviors in both the males and their charges, including high levels of physical contact and other interaction (Redican and Mitchell 1973; Gomber and Mitchell 1974). In one experiment the male responded strongly to distress cries and to the infant's assuming a depressed posture in the cage.

Plasticity is also seen in natural and seminatural situations. In an olive baboon troop studied in the field (Ransom and Ransom 1971), one mother gave her infant very little care, but an adult male compensated by giving steadily more attention to the infant. Adult male langurs ignore infants in the wild, but inspected and nuzzled them in a captive colony, where one eight-month-old male whose mother was removed for two weeks sought and received care from an adult male (Dolhinow and Taff 1993). There can be marked individual differences within settings; for example, the amount of carrying done by male tamarins is contingent on helper availability (McGrew 1988; Tardif, Carson, et al. 1990). Beyond the primates, male marmots show different degrees of attention to the young in more densely populated as opposed to isolated settings (Barash 1975).

However, there is also evidence for neuroendocrine limits to plasticity (Chapter 12). Testosterone (T) administered to various mammalian females reduces their nurturing behavior. Male mice are more likely than females to kill strange pups presented to them, and T increases female killing dosage dependently (Davis and Gandleman 1972). Male Mongolian gerbils' parenting of neonates is inversely related to T levels, and T administration interferes with paternal responsiveness (Clark and Galef 2001). T given to rabbit females during the last twelve days of pregnancy causes depressed nest-building activity, postnatal scattering of offspring, failure to nurse, and eating of the young (Fuller, Zarrow, et al. 1970). In a semi-free-ranging rhesus monkey troop the only male showing substantial nurturance toward infants had been castrated (Breuggeman 1973).

But physiological differences go deeper than current hormone levels, owing to fetal brain androgenization (Chapter 10). Isolation-reared preadolescent male rhesus monkeys are less likely than females to nurture strange infants (Chamove, Harlow, et al. 1967). Female rats androgenized at four postnatal days show less nurturing behavior, approaching normal male levels (Quadagno and Rockwell 1972), while males castrated at birth show more nurturing behavior than intact males or males castrated at twenty-five days, suggesting a sensitive period effect. In some biparental birds males have elaborate neuroendocrine adaptations for these behaviors (Lehrman 1955). It is likely that fetal androgenization influences human nurturing as well (Collaer and Hines 1995).

In the Six Cultures Study (Whiting and Whiting 1975a), association with and nurturing toward infants (offers of help or support) were higher in girls than boys, increasingly so from ages three to eleven. However, virtually all cultures have socialization practices that tend to produce such sex differences—some more than others, depending on subsistence ecology (Barry, Bacon, et al. 1957; Barry, Child, et al. 1959). Carol Ember's classic study (1973) of Luo boys in Kenya showed that those assigned to nurture infants and children became good at it. The evidence of experientially induced paternal behavior in monkeys, along with the growing number of human single fathers, shows that the male nervous system can produce extensive nurturing behavior. Since several studies have shown a decline in T and other hormonal changes when human males become or are fathers (Berg and Wynne-Edwards 2001; Gray, Yang, et al. 2006), it is likely that exposure to infants and children has biological consequences, and this is strongly suggestive of plasticity.

Interlude 5

Thinking about "Oedipal" Conflicts

Freud attempted to describe universals of human thought, feeling, behavior, and development, but did not examine these matters in a wide range of cultures or in a documented phylogenetic context. From his experience with his own patients, Freud drew broad conclusions, and to some extent psychological anthropology grew up in reaction to these generalizations.

One of the most intricate was "the Oedipus complex"—named for the Theban king who, in Greek legend, returned after many years of wandering, unknowingly killed his own father on the road, and then, also unknowingly, married his mother. Freud held that every human child develops an attachment to the parent of the opposite sex and that this attachment is romantic or even sexual in nature, expressed in the common remark of children of this age, "I'm going to marry Mommy [or Daddy]." In theory, these feelings bring the child into conflict—usually only imagined—with the same-sex parent, who not only has the person the child wants to marry but also has authority, making the child feel weak and jealous. All these feelings must supposedly be resolved before further psychological growth can take place (Freud 1905, 1920).

The complex was thought to differentiate the genders. In the Oedipus complex, boys started out attached to their mothers and then gave them up, while girls had to first transfer their feelings of attachment from their mothers to their fathers before their "Electra complex" could form. A boy feared the father would castrate him as punishment for desiring his mother. A girl believed this mutilation had already happened to her, and she experienced a sense of loss along with penis envy. Because of the Oedipus complex, one of the worst things that could happen to a child was to see, even inadvertently, the parents having sexual intercourse; this "primal scene" would trigger the child's jealousy, castration fear, and sometimes permanent trauma.

Ethnographic research challenged this notion almost immediately. Bronislaw Malinowski (1927/1961) pointed out during the 1920s that among the Trobriand Islanders, a horticultural community of the South Seas, the person with disciplinary authority over the child is not the father, but—as is often the case in matrilineal societies—the mother's brother. A little boy might be jealous of his father, but would not fear him, because his father only plays with him and never punishes him; the boy would fear his uncle instead. Thus the components of Freud's Oedipus complex would be split. Margaret Mead (1935, 1949) posited "womb envy" as a counterpart to "penis envy"; if girls wished for penises, boys and men wished they could have babies. Evidence for womb envy comes from certain customs that are widespread in "primitive" societies—for example, the *couvade* practiced in parts of native South America and Oceania (Munroe, Munroe, et al. 1973; Munroe 1980; Broude 1988). When a woman gives birth, her husband pretends he is also going through labor, secluding himself while she is secluded and pretending to experience great labor pains if she is having a difficult birth, by this ritual asserting his role as the child's father.

Moreover, childhood sexuality is very different in many non-industrial societies than it is in our own, or was in the middle-class nineteenth-century world of Freud's Vienna. Children in rural communities frequently observe animals having sexual intercourse and ask about it, so they have ample opportunity to learn about sex. According to one study, the middle-class culture of Chicago in the 1940s more severely punished masturbation, sex play in childhood, and lack of modesty in dress than did the average non-industrial society (Whiting and Child 1953). Childhood sex play is very widespread in non-industrial societies (Frayser 1994), and punishment of pre-marital sex is weak or nonexistent in many of them (Broude 1975).

To Freud's credit, he promoted a certain degree of sexual liberation and specifically advocated a greater sexual education for children. The comfortable fin-de-siècle Viennese world he lived in, hypocritical in its patriarchal split between whores and mothers, was even more severe in punishing children's sexuality and in restricting sexual knowledge than middle-class Chicago in the 1940s. Freud correctly perceived that in his paternalistic time and place the authority of the father was important in the psychological growth of the child; his error lay in assuming that this pattern was universal.

We must also ask this question: since in many societies around the world children and parents sleep in the same room, and children have numerous opportunities to observe their parents having sexual intercourse, how important can the primal scene "trauma" be? In fact, the most common sequel of such an observation, so common in these societies, is that children play at sex, amid many giggles. These experiences, combined with opportunities to observe sexual intercourse in animals, amount to a natural sexual education for most children during ordinary childhood. We might conclude that if the primal scene is ever a trauma, it would only be so in cultures where children are almost always prevented from seeing it and where it is surrounded by a forbidding air of secrecy.

Freud also speculated on how the first Oedipal conflict might have played out during human evolution, using his narrative to explain the origins of the childhood version, by Haeckel's law: ontogeny recapitulates phylogeny (Freud 1913). In this scenario, the "primal horde" was a punitively patriarchal band of hunters in which, after much frustration, the brothers banded together, killed their father, and ate him. They thus took on his identity but were left with a deep sense of guilt, which must be played out forever in the Oedipal conflicts of every young child, much as the fetus must have its gills and tail and the child must crawl before it can walk. Although Freud made some effort in *Totem and Taboo* to cite ethnographic descriptions, he did not conduct real research, and the scenario outlined belongs squarely in the realm that he frankly labeled, in an unpublished essay, "A Phylogenetic Fantasy."

As for the developmental model, little of it has been supported with good evidence, although some children certainly have such thoughts and conflicts. Many children this age say they want to marry the opposite-sex parent, and some talk or fantasize about getting the same-sex parent out of the marital bed. But there is no evidence that all normal children must go through such a process, and cross-cultural studies since the 1920s have cast doubt on the hypothesis that the themes of this proposed family drama are universal—although, in the Trobriands, where it was first challenged and rejected, it may have some applicability after all (Spiro 1982).

Yet some related concepts must be valid. First, boys and girls differ anatomically, and most notice the difference by age three. Boys, through self-exploration, play, and accident, find out that their genitals can produce both strange pleasure and dreadful pain, and some do worry about injury and loss. Girls learn of their future procreative power and may discover clitoral masturbation. In preschool they sometimes envy boys their penises, although the typical reason given is that they too would like to pee standing up. Cultures and families where nakedness and masturbation bring reprimands or punishments—like the Austria where Freud came of age and practiced medicine—can certainly heighten children's conflicts. Among the !Kung San, fathers occasionally joked with boys about taking their penises, in much the way that

American fathers and uncles "steal" noses. In a lighthearted, loving context, such jokes might help boys handle their conflicts, especially in a culture where people are comfortable with such jokes; in a harsh or threatening context, the opposite could occur. Finally, there must be some psychological consequences of the fact that in order to reproduce, boys must mate with someone of the same sex as the overwhelmingly common first object of their attachment, while girls must almost always do the opposite.

What of the "phylogenetic fantasy"? Oddly enough, there are data from nonhuman primates that touch on this. First, there are many observations of young monkeys and apes (of both sexes) trying to interfere with their mothers' copulations, which they universally see. But in observations of stumptail macaques *(Macaca arctoides)* juveniles and subadults of both sexes interfered with copulations, and not exclusively those by their mothers (Gouzoules 1974a). Natural selection should favor young who could (safely) discourage their mothers from becoming pregnant again—a typical instance of parent-offspring conflict. In many species male violence against females, including rape, is observed (Smuts 1992; Smuts and Smuts 1993), and this must be highly salient for the offspring of those females. In fact, the stumptail juveniles appeared to be drawing some of the copulating male's aggressiveness away from the female (Gouzoules 1974a).

In his classic monograph on the bonobo, Takayoshi Kano (1992:147) published three vivid photos with these captions: "It is not uncommon for an immature to get between mating partners, and the behavior is tolerated by the adults; A juvenile comes between a male and a female during copulation; The juvenile screams while clinging to the copulating male." According to the accompanying text,

> juveniles show the greatest interest in the performance of copulation. They run up to watch, cling to the belly or back of either partner, and scream. Often, they are the children of the female partner, because a child is usually situated close to its mother. But, if copulation by a female other than the mother occurs nearby, the juveniles will hasten to that place in the same way. Their scream is not from fear or distress, but instead seems to come from sexual excitement; often after copulation, the child will enter into sexual intercourse (immature copulation) with one of the partners. (146, 149)

"Immature copulation" refers to the simulation that juveniles perform in play, without penetration or distress. But this is in a species where females have the upper hand. In orangutans, juveniles must watch as a male much larger than their mother emerges from the forest and rapes her (Galdikas 1985a; Knott and Kahlenberg 2007). This may be extreme, but coercive sexual harassment is seen in 60 percent of Old World monkey and ape species, including macaques, baboons, and chimpanzees (van Schaik, Pradhan, et al. 2004). Nor can we take much pride in our heritage; among primates, coercive sex is largely limited to catarrhines, including us.

Second, infanticide by males who take over a group is found in several higher primates (Hrdy 1977; Hausfater and Hrdy 1984; van Schaik and Janson 2000). In Hanuman langurs the ascendant males kill all infants and soon begin copulating with their mothers, despite the mothers' resistance (Hrdy 1979). Older juveniles observe all this. On the other hand, in some species, such as ours, fathers may have been selected to stay with their mates and offspring to prevent infanticide by other males (van Schaik and Kappeler 2003).

Third, male primates in adolescence must find a way to mate (van Noordwijk and van Schaik 2004). They may steal away with a willing female out of sight of a dominant older male—who in patrilineal groups like those of chimpanzees could be their father—but they are at risk of being caught and facing an attack. In the many Old World monkey groups that are matrilineal, males must change groups to mate. They may be driven out by their own fathers or other older males and stay solitary or roam in all-male clusters until they can try to join another group. This may involve fighting with resident males or subtler, peaceful strategies, but it invariably entails frustration and patience. Overall, this time of transfer is one of very high mortality for young males. Conflicts between older and younger females, though less severe, are also important (Hrdy 1981; Campbell 2004).

Some other attempts have been made to recast Freud's Oedipal theory in the light of modern Darwinian thinking, with varying degrees of success (see, for example, Badcock 1990; Erickson 1993), but such efforts, like the observations here, are far from *Totem and Taboo,* and there is little evidence of latent psychological dynamics in early childhood to match those that Freud posited. Nevertheless, the evidence does suggest that the general themes he touched upon are not completely removed from reality. Our monkey and ape ancestors grew up in a dangerous social world where they had to face and sometimes fight older and stronger individuals for access to reproductive opportunities. We do not have to accept psychoanalytic models, but the links between sex and violence are real.

18

Relations among Juveniles

Helpers at the nest include the young in many species (Emlen 1995), and from a developmental viewpoint the individual giving the care is also benefiting. Such relationships, especially among siblings, are of interest from the viewpoint of rivalry or conflict as well. But this subject is of special concern in our species because (1) to the extent that humans are cooperative breeders, reliance on children to offset the additional burden of caring for the young was crucial to our success; (2) in human societies the new burden of transmission of culture is borne in part by children; (3) the assignment of child care to children is a part of what made possible the replacement of hunting and gathering with food production; and (4) a focus on relations among juveniles forms a logical transition to the consideration of play.

Theoretical Considerations

Kin selection theory predicts that juveniles will enhance their inclusive fitness by aiding each other to the extent that they are related, limited by sibling conflict under strained resources—for example, if birth spacing is too close or sibships are too large. The phenomenon of siblicide in some frog and many bird species shows that even sibling aggression can sometimes be lethal, and theoretical expectations are complex (Mock 2004). As with infanticide by mothers, siblicide is the end of a continuum of cost-benefit ratios of help and rivalry.

Relatedness aside, there are other adaptive advantages to sibling care. Like nonmaternal adults, juveniles may gain improved status in the group by association with a higher-ranking female's offspring, through agonistic buffering (Hrdy 1976, 2001). Juvenile males may be priming unrelated younger females for future mating. Young females in any species and males in those with high male parental investment may improve their own reproductive success by practicing parenting (Fleming 1990; Fleming, Morgan, et al. 1996). In many primate species experience with infants improves later parental behavior (Pryce 1996), for males as well as females (Redican and Mitchell 1973). But relatedness aside, play also affords practice of the sexual and aggressive behaviors that promote reproductive success. Since selection could not create systems that develop independently of continuously present features of the environment, the neuroendocrine systems underlying sexual, agonistic, and parental behavior in higher primates in many species grow normally only when play with other juveniles is possible (Suomi 2008).

Juvenile playgroups are commonly called peer groups, but they often do not consist of same-age peers. Kinship is relevant, since relations among siblings, half-siblings, and cousins of various ages are partly embedded in these playgroups, which are very widespread in higher primates. Selection favoring them overlaps with the considerations outlined earlier, with two additions. First, since the juvenile is not explicitly burdened with the infant but is merely part of a playgroup that includes the infant, the cost-benefit ratio for the juvenile is improved, while the benefit to the infant, who is protected from harm, may still be substantial. So is any benefit the juvenile derives from practicing parental behavior. Second, the benefits of membership in a playgroup with older individuals should be greater than that in a peer group, since adaptive behaviors are learned more easily from those more advanced in development. Only the oldest members of the group would be selected, on this ground, to resist, and one might expect this to be balanced, especially for females, by the chance to practice parental behavior.

Juvenile Social Relations in Selected Mammals

Early social relations depend on ecology and phylogeny. Ecological factors affect population density, demographic structure, fertility, birth spacing, seasonality of births, time available for social interaction, and danger to infants and juveniles, among other variables. Phylogeny affects litter size, body and brain size at birth, speed of growth, and complexity and lability of the nervous system. Both kinds of effects may be illustrated by briefly considering the natural history of juvenile social contexts in some groups of mammals. In addition to those discussed in this chapter, social play has been studied in marsupials (Byers 1999b; Soderquist and Serena 2000), meerkats *(Suricata suricatta)* (Sharpe and Clutton-Brock 2002), dolphins (Bel'kovich, Ivanova, et al. 1991), elephants (Douglas-Hamilton and Douglas-Hamilton 1978), and other mammals (Burghardt 2005).

Rodents

Most rodents are born in litters and normally develop in a peer group of siblings. Social play is seen in many species (Pellis and Iwaniuk 1999). In spite of their rapid growth to sexual maturity (weeks to months), a substantial experimental literature shows rodent neural and endocrine systems to be labile with respect to early social contact and other forms of early stimulation, and postweaning isolation from peers has a broad spectrum of consequences for behavior and physiology, including brain structure and function (Chapter 14). In Belding's ground squirrels *(Spermophilus beldingi)*, play is mainly with littermates, with individuals tending to choose one partner, and the amount of play predicts later motor skills (Nunes, Muecke, et al. 2004).

Ungulates

Although these are the most precocial large mammals—most are ambulatory at birth, and their locomotor systems mature within a few days—play occurs in most species. Observations of domestic sheep and goats (Collias 1956; Orgeur 1994), domestic pigs (Chaloupkova, Illmann, et al. 2007), wild horses (Berger 1986), mountain sheep (Geist 1971), red deer (Clutton-Brock, Guinness, et al. 1982), pronghorn antelope (Miller and Byers 1998), and giraffes (Pratt and Anderson 1979), among other social ungulate species, suggest that normal adult sexual behavior and dominance behavior emerge gradually from the motor patterns of social play. Although these ungulates have singleton births, they are seasonally breeding, herding animals, as are most larger wild ungulates; in such species infants and juveniles always have peers available. Extensive juvenile play is observed in Siberian ibex, white-tailed deer, elk, bison, chamois, and boar (Fagen 1981; Prescott 1985; Miller and Byers 1998). The major exception are moose, which are unusual among large ungulates in being solitary in adulthood (Altmann 1958), like a number of small nonherding species such as duiker and steenbok. No play among juveniles is possible, although calves do play with their mothers; since adults are solitary, relations among juveniles may be less vital. Also, since juveniles in herding ungulate playgroups are more distantly related than siblings in litters, we might expect, on theoretical grounds (Popp and DeVore 1979), a higher incidence of aggressive play in the former; such play—head-butting, for example—is fairly common.

Terrestrial Carnivores

Of the land carnivores, the canids, felids, and bears are the best studied. Since they reproduce in litters, sibling peers are always present, and play among these siblings is observed in wolves, coyotes, and foxes (Bekoff 1974); domestic dogs, both in captivity (Scott and Fuller 1965; Bauer and Smuts 2007) and feral (Pal 2008); bears (Henry and Herrero 1974); hyenas (Pedersen, Glickman, et al. 1990); kittens (West 1974; Caro 1981); cheetahs (Caro 1994); and lions (Schenkel 1966; Schaller 1972). Such relations are necessary for normal social behavior development in dogs (Scott 1963; Scott and Fuller 1965). In lions (Schenkel 1966) the mother participates in and facilitates play-fighting among her offspring. This suggests that simple practice by trial, error, and exercise is not all that is involved; lionesses may facilitate their offspring's learning to hunt, as cheetahs clearly do (Chapter 19).

Aquatic Carnivores

In the Alaska fur seal (Bartholomew 1959), mothers separate from their pups about six days after seasonal birth, leaving the rookery to return to sea and coming back only once a week or so for the rest of the summer. After the mothers' departure, pups aggregate into "pods" of dozens or hundreds of age-mates that will persist into adulthood. Early social activities include play-fighting, grooming, and copulatory play, all of which gradually approach adult form. The extremes of aggressiveness and competition among males observed in the elephant seal (Le Boeuf and Peterson 1969) may depend on the pattern of socialization in very large peer playgroups of unrelated individuals; in these groups, selection should allow high levels of aggressive interaction. Social behavior in the Steller sea lion also appears to emerge gradually through practice in peer relations (Gentry 1974).

Primates

Seasonal breeding is characteristic of many but not all monkeys (Lancaster and Lee 1965; Campbell 2007). Higher primates normally have single births, with the exception of the marmoset family (Callithricidae). Most monkeys breed every year or two, and depending on group size and population density, a yearling playgroup comprising anywhere from a few to a large number of same-age peers will have been formed when their mothers are ready to next give birth. Such playgroups have been described for many species.

Whether patrilineal, matrilineal, or mixed (Strier 1994, 2006), the social group in most higher primates is a kind of extended family, so most juveniles in a playgroup will be cousins. However, as group size increases, the likelihood of associations among unrelated juveniles increases. For example, demographic features appear to determine variations in juvenile play behavior in squirrel monkeys (Baldwin and Baldwin 1974). Except in the Callithricidae and in rare cases of twinning in other species, precisely same-age sibling relations are absent. Also absent, except for certain populations of our own species, are the large groups of unrelated same-age peers that characterize large herds of ungulates or seal and sea lion rookeries.

In primates, three basic patterns are found—juvenile alloparental behavior, nonsibling playgroups, and multi-aged sibling groups. Juvenile alloparental behavior is typically female, although males do it in some species—particularly hamadryas baboons, Barbary macaques, and Japanese macaques (Chapter 17). Japanese macaques also show considerable juvenile alloparental behavior in females, who are typically younger than male caretakers.

The second pattern, the playgroup of variably related infants and juveniles, is the dominant form among monkeys. This form grades into the sibling group and need not exclude siblings. Among monkeys living in large groups, many playgroup members will be unrelated or distantly related; in addition, because of the seasonality of births and group size in some species, some will often be same-age peers. Predominantly same-age peer groups are found in rhesus monkeys and squirrel monkeys, but playgroups in most other cases are composed of one- to three-year-olds (from late infancy to puberty) and even occasional adolescents and adults.

The third pattern, the small sibling playgroup, is found in all the apes and in a number of monkeys, particularly those with dispersed populations. In the apes it is found to the exclusion of the second pattern, which is often demographically impossible; there are few same-age peers. The sibling group grades into juvenile alloparental behavior, typically in the playgroup context.

With the exception of hamadryas baboons, playgroups of both the second and third patterns are mixed-sex groups. Hamadryas playgroups are basically male, because yearling and two-year-old females are already being adopted by harem-building older males. The mixing of sexes in most species may be important from the viewpoint of function, especially in relation to the development of reproductive behavior; juvenile play-fighting (rough-and-tumble play) also includes mounting and other components of adult sexual behavior.

Relations among Juveniles in !Kung Hunter-Gatherers

Demographic factors made it unlikely that same-age peer groups in human groups would appear in the EEAs. The modal band size was thirty in twentieth-century hunter-gatherers in a wide range of environments (Lee and DeVore 1968). Taking the !Kung as a first example (Howell 1979), the proportion of women between the age of first birth (19.5) and the end of the reproductive period (about age 45) is 40 percent, or six women in each band of thirty. Infertility resulting from gonorrhea can be corrected for by subtracting the proportion of women who reported having had it (35/165) times their fertility relative to women who had not had it (2.4/5.1). The resulting mean number of fertile women per band of thirty will be about $x = 6 - 6(35/165 \times 2.4/5.1) = 5.4$.

The modal intersibling birth spacing is four years. If women are giving birth one year in four and if, when a child is born, a mean of 4.4 women in the band (other than the mother) are reproducing, then the likelihood of another giving birth in the same year is 4.4/4, or 1.1. Since mortality in the first year is 20 percent (most often in the first month), the chance of a given child having a peer in the band on her first birthday is 88 percent, which is quite high. The chance of having two peers is 4.4/16 x 80 percent, or 22 percent, and the chance of having three is one-fourth of this, or 5.5 percent. These figures decline as the child grows older, owing to continuing mortality. The chance of being a member of a peer group of three two-year-olds is less than 20 percent. These estimates define peers as infants born in the same year, so they may be as much as six months older or younger, a large difference in infancy. With the more restrictive definition of requiring peers to be no more than three months older or younger, the corresponding figures for the end of the first year are 44 percent for one peer, 11 percent for two, and 2.75 percent for three.

However, hunter-gatherer bands are characteristically nomadic and fluid. The band a child belongs to will probably split into subgroups, which will recoalesce to form new bands with other subgroups several times during the first few years of life. At some seasons the opportunity for peer play may be transiently greater, and at other seasons less. Although continuity of relationships will be preserved with some of the children in the band, a peer, if any, may not be among them. Further, the peers' mothers may dislike each other (as happened in one !Kung band with two infants of the same age), making interaction unlikely. Finally, other children are vying with the peer for an infant's or toddler's attention.

So from purely demographic considerations, peer groups were probably unusual during most of human evolution, owing to the carrying capacity of land for hunter-gatherers. The social context of infancy (Konner 1972, 1975, 1976, 1977b) and the social behavior of two- to five-year-olds (Blurton Jones and Konner 1973) and older children (Draper 1976, 1978; Draper and Cashdan 1988; Hames and Draper 2004) among the !Kung have been extensively described; of the three higher-primate patterns of relations among infants and juveniles, only the multi-aged playgroup occurs with any frequency in !Kung bands, and it is ubiquitous. It is made up of siblings and cousins and may consist, in a band, of six or eight children of both sexes ranging from late infancy to adolescence, or the playgroup may divide into smaller groups with smaller age ranges.

Infants begin to participate from the time they can walk, and a toddler may edge away from his mother to stumble into a pile of wrestling two- to five-year-olds. Better walkers will follow the group, and if they wander from the village-camp, these smallest children will be carried when they tire. The principal group activity is play, though this may include observation of adult subsistence activity; subsistence play, which, however playful, may produce food; "pretend" subsistence play, which does not produce food but assumes some semblance of adult subsistence forms; rough-and-tumble play; and sex play (Konner 1972; Blurton Jones and Konner 1973; Draper 1976; Shostak 1981). Playgroup activities also include protection, care, and teaching by older children. There is no child nurse comparable to that found in middle-level agricultural societies (Whiting 1963; Weisner, Gallimore, et al. 1977; Chapters 23 and 26); children are not routinely assigned to take care of infants, and mothers are almost always within shouting (or crying) distance. But there is informal care, protection, and teaching. Older children choose to participate since they too have few or no same-age peers.

Sibling rivalry is also evident. One three-year-old girl held her infant brother on her lap and played with him charmingly until she grew bored and stood up abruptly, allowing her brother to roll to the ground where she had been sitting. In spontaneous speech and in doll play she gave evidence of strong violent feelings toward him. But despite such fantasies, most caregiving by !Kung older siblings is nurturing, if informally so. Theoretically the typical four-year birth spacing in this culture reduces the chance of rivalry, since parent-offspring conflict over the birth and care of the next sibling should wane as the weanling's inclusive fitness interest in the younger sibling comes to match or exceed her own need for further care.

Heide Sbrzesny's (1976:307) superb study of play and games among the closely related !Ko San of southwestern Botswana also documented the importance of multi-aged child groups: "Socialization of the child, from approximately the fourth year on, mainly takes place in the children's play groups under the guidance of the elder children. These suppress and even punish the smaller ones' aggressive acts, and encourage friendly and cooperative behaviors, such as sharing and giving." However, she also documented the existence of competitive and even combative games, especially among boys (see Chapter 10). This seems in contrast to the situation described by Patricia Draper (1976:202) for the !Kung: "Competitiveness in games is almost entirely lacking. . . . To compete in a game or skill one needs one or preferably more close in age and perhaps sex with whom to compete, but the smallness of group size among !Kung usually ensures that several age-mates are not available."

Still, in both these cultures sharing and giving are emphasized from early childhood on. We return to the nature of play and games in the next chapter.

Relations among Juveniles in Other Hunter-Gatherers

As in previous chapters, we need to contextualize the !Kung pattern among those of other carefully studied recent hunter-gatherers. In the structure of child groups there is little variation, but there are some interesting differences in function.

Hadza

Although weaned earlier, young Hadza children made a gradual transition to a multi-aged child group and up to a point resembled the !Kung:

> Hadza children lack none of the charm and imagination of !Kung children. They have a robust humor and a pride in life that we find attractive and impressive. . . . Between the ages of 3 and 8, Hadza children seldom accompany their mothers on gathering excursions. Children over 8 may accompany their mothers, but do not always do so. . . . The children, usually in sizable mixed-age groups, may spend some hours out of camp. Sometimes they are at a favorite play site or at the water hole. (Blurton Jones 1993:316)

The most striking difference from the !Kung (Draper 1976) lies in the amount of real subsistence effort by Hadza children (Blurton Jones, Hawkes, et al. 1989, 1994, 1997). They play a lot, but "more often they are gathering food, independently of the women . . . [and] returns . . . are substantial. . . . In the foraging groups, even 3-year-olds try their hand at digging or picking up baobab pods and processing them." Furthermore,

> unlike !Kung children, Hadza children appear to be given many errands and to perform useful tasks, bidden and unbidden. Such tasks cost the children time and energy, and sometimes expose them to the hazards of the bush. Children of either sex may be asked to hold a protesting toddler when the mother leaves camp to forage. . . . Children commonly are sent

to fetch water and sometimes firewood. . . . Even toddlers are
sent to carry things from one house to another. (Blurton Jones
1993:316–17)

However, Hadza children had ample time to play as well as work, and
they did both in mixed-sex, multi-aged groups.

Efe

Efe infants' social contact with children increased threefold over the
first three years, reaching 29 percent of all social contacts at five months
and 62 percent at three years, even though infants' absolute contact
with adults did not significantly change (Tronick, Morelli, et al. 1992). At
six months, the *average* child was in contact for about 9 percent of the
infant's time, dropping to 5 percent at eight months, but rising again to
18 percent at three years, matching the mother's level and almost doubling the father's. Thus an individual child could be the most prominent
"other" in a three-year-old's life, in time spent if not in nurturance.
Five-month-old infants interacted with zero to four children per hour,
and three-year-olds with one to six.

Efe researchers use the words *child* and *peer* interchangeably, but a
high prevalence of infertility made same-age peers even less probable
than for the !Kung, and infants and toddlers interacted with a wide
range of ages. Efe girls helped their mothers gather plant foods and carry
out some tasks within the camp (Peacock 1985). Ituri foragers generally
(Turnbull 1962, 1965) are described as occasionally physically punishing
children but leaving them mainly to play, with little responsibility. For
boys, play was eventually interrupted by circumcision and initiation at
age eleven or twelve—a painful and, for some, traumatic experience.

Aka

After weaning at around age three or four, Aka children's lives changed
considerably. Generally,

the child . . . is not able to walk fast enough to keep up with the
net hunt . . . so the four to five year old frequently stays behind
in camp with one or two other children and an adult. . . . The
children play, explore and practice subsistence skills and sel-

dom venture more than fifty meters from camp.... In camp the majority of a child's time is spent within a multiage playgroup, but always in the company of adults. (Hewlett 1991b:36–37)

This transition from the intimacy of the parent-infant relationships to the multi-aged playgroup was very gradual (Barry Hewlett 2004, personal communication), but Aka playgroups might be same-sex. To the extent that they could keep up as they grew older, children might follow their parents on net hunts, in which women played vital roles; it was up to the children whether or not they tried to help. "Instruction is still primarily by observation and imitation, but verbal instructions are also given" (Hewlett 1991b:37). At around age eleven or twelve, the sexes segregated further. Aka girls of that age and older collected water, fruit, or nuts together, while boys hunted small game.

Ache

Weaning from the breast and from the back was sometimes fraught with conflict for Ache children. By this time they had already begun to learn about subsistence in the forest, but the process now accelerated. As they spent time foraging with women, both boys and girls acquired knowledge of edible fruits, vines with thorns, and stinging plants and animals, and they became skilled at collecting fruit, insect larvae, and small game. By age eight they learned another crucial, difficult, and subtle skill: that of tracking adults in the forest; then, unlike the boy left behind at an earlier age to teach him a lesson, they would not get lost. They spent a lot of time clambering in trees, "collecting fruits for themselves or knocking down fruits for the adult women to collect below. There is no segregation of play or foraging parties by sex, and children spend most of their time within 50 m of the adult women in mixed age-sex groups" (Hill and Hurtado 1996:223).

Both sexes became more independent around age ten; they might sleep at a relative's fire or even travel in the forest with another band. Relationships with adults who resembled godparents became important. Girls cared for infants, ran errands, and drew water; boys carried bows and arrows made for them by others. By age twelve girls might gather as much food as adult women, but they did not carry the aptly named burden-basket until after marriage; boys surpassed girls in food production by age sixteen, yet did not reach adult male levels themselves until their midtwenties.

Agta

Agta children made the transition to a similar multi-aged child group: "Playgroups are not age or gender segregated, but made up of most children in [the] local group. . . . Teen-aged girls bring toddlers on their hips to observe or join play" (Griffin and Griffin 1992:302). Like the !Kung, older children looked out for younger ones during play, but like the Hadza, Agta play contributed more to subsistence. Children began fishing in early childhood and were skilled by adolescence, when both boys and girls began hunting (Estioko-Griffin and Griffin 1981).

Relations among Juveniles since the Hunting-Gathering Era

We now consider generalizations about hunter-gatherers beyond these six cases and attempt to reconstruct what happened after hunter-gatherer life gave way to other subsistence ecologies, using cross-cultural samples (Konner 1975). The sample used here consisted of 182 of the 186 societies in the World Ethnographic Sample taken from the Human Relations Area Files (Murdock and White 1969). Four were omitted because they were mounted hunters. The sample was coded for societal complexity and subsistence variables (Murdock 1967; Murdock and White 1969; Murdock and Wilson 1972) and for infant and child care variables (Barry and Paxson 1971). The data were recorded and reanalyzed by Mary Maxwell West.

Nonmaternal care (NMC) in early childhood and the use of children as infant caregivers are both lower in hunters, gatherers, and fishers than in other subsistence types; this is consistent with an earlier finding that pressure for nurturance or parents' efforts to get children to care for infants are greater in societies whose economies involve more accumulation (Barry, Child, et al. 1959). Hunting, gathering, and fishing cultures favor the playgroup over the child nurse or adult nurse, while other cultures show no preference.

The question of the same-age peer composition of playgroups is more difficult to approach using available cross-cultural data. It is very common in horticultural and agricultural societies for playgroups to include children of different ages, although the majority of observations show sex-segregation, which increases with age, and girls are more likely to be with younger children than boys are; as in hunter-gatherer groups, "the 2- and 3-year-olds tag along after their older brothers and sisters, and even if the older children are not specifically asked to care for them, they are expected to see that their young siblings are safe" (Whiting and Edwards 1988:223–24). Of the nine cultures quantitatively observed and compared by Whiting and Edwards, all eight farming societies, regardless of level of political organization, fit this pattern. "It is only in Orchard Town [the New England suburban community] where the 4- and 5-year-olds who have 2- and 3-year-old siblings are never observed alone with these 2-year-olds" (224). Also, in American samples sex-segregation tends to become extreme in middle childhood.

In adolescence, cultures that hunt and gather, cultures without social stratification, and cultures with population units smaller than 1,500 are significantly less likely to have peer groups than more complex cultures (Harley 1963). However, adolescents segregate by sex in a wide variety of cultures (Schlegel and Barry 1991). But girls associate with women, while boys remain aloof from adults generally, and it has been suggested that young male peer groups have a primarily agonistic and political function (Tiger 1984). John Whiting found that fraternal associations of adolescent males are significantly more likely to occur in societies where a high emphasis is placed on military glory—martial prowess, heroism, death in battle, and the like. Such peer groups separate boys from girls and women and from their identification with women, the better to make them warriors.

Functional Considerations

We know that depriving developing monkeys of contact with other juveniles results in impaired adult male sexual behavior and inappropriate aggression. Under natural conditions, both rough-and-tumble play and sex play are universal components of juvenile play behavior in higher primates, including humans (Frayser 1994; Fry 2005), except for the low levels of childhood sex play in state-level human societies. Among higher primates, adult sexual behaviors are acquired gradually through open or clandestine observation of adults copulating (De Benedictis 1973; Gouzoules 1974a; Struhsaker 1967; Goodall 1967, 1986; Kano

1992) and practice—which is playful in childhood and more serious in adolescence (Goy and Goldfoot 1974; Tutin and McGrew 1973)—and the same is true of human hunter-gatherers and many other small-scale societies (Money, Cawte, et al. 1970; Shostak 1981; Konner and Shostak 1986; Frayser 1994; see Chapter 20). Deficits in caregiving behavior resulting from peer deprivation have also been shown (Chapter 14). Younger individuals are usually available, and in some species play-mothering (Lancaster 1971) and play-"fathering" (Burton 1972) are important.

Developmental Mechanisms

Consider the developmental process by which an infant comes to have relationships with other infants and children. Many studies in human infants and toddlers treat this subject as if it were only about same-age peer relationships (Lewis and Rosenblum 1975). From this perspective, cognitive developmental events during the second year of life make possible a quantum advance in the infant's ability to interact with peers. This is analogous to the process by which an infant comes to be able to control or manipulate the behavior of its mother at around three months of age, using various vocal and visual signals. As we saw in Chapter 7, one of the three-month-old's cognitive achievements is manipulating active objects, given that the object's actions are contingent upon some behavior of the infant's (Rovee and Rovee 1969; Watson 1972; Watson and Ramey 1972; Rovee-Collier and Capatides 1979). The ideal contingently responsive object is a caregiver who varies the level of stimulation and contingent-responsiveness to optimize the infant's arousal and maximize attention (Stern 1974; Stern and Gibbon 1977). Peers, however, exhibit mainly *non*contingent behaviors—egocentric, not tied to the infant's own actions, and, for the infant, largely random. They thus present a much less manageable and appealing set of stimuli. So the ability to manipulate them matures much later than for the most contingently responsive caretakers.

Despite early (Graziano, French, et al. 1976; Lougee, Grueneich, et al. 1977; Goldman 1981; Field 1982) and subsequent experiments demonstrating that relative age may matter, research and theory remain based mainly on the idea that relationships among infants and children are limited to same-age peers (Goin 1998; Hay, Payne, et al. 2004). Infant-only trios show evidence of some limited, basic group communication ability (Selby and Bradley 2003), and small groups of three same-sex two-year-olds can engage in triadic interactions (Ishikawa and Hay 2006). Evolutionary and cross-cultural considerations suggest that most children in most cultures have never had to develop social competence with same-age peers and that this is an unnatural situation for infants and toddlers (Konner 1975; Weisner 1987), yet this bias persists.

However, a few studies have compared the interactions and development of infants and toddlers exposed to same- as opposed to different-age partners. In one, toddlers age one and a half to nearly three played with same-age peers or with four- to six-year-olds (Howes and Farver 1987). Toddlers engaged in significantly more social pretend play with the older children than with age-mates, probably because the older children structured the play. A separate experiment asking the older children to teach or tutor the toddlers did not change the former's behavior, suggesting that teaching is a natural tendency. But the effect on the toddlers did not carry over to subsequent same-age pretend play, suggesting that the age difference only works in real time.

In a cross-cultural study, thirty-six two- to eight-year-old Indonesian children playing outdoors in Jakarta were compared with thirty-six age-matched American children playing outdoors in Los Angeles (Farver and Howes 1988). The Indonesian children tended to interact in mixed-age groups, but the American children did not; in this study, context mattered less than prior culture-specific experience. Other studies have suggested that mixed-age groupings provide cognitive benefits for both younger and older children (Christie, Stone, et al. 2002; Kowalski, Wyver, et al. 2004). There is also evidence that tutoring skills emerge naturally in young adolescents who interact with younger children and that the adolescents improve these and other skills in the interaction (Gray and Feldman 2004).

In a rare longitudinal random-assignment study of this question, fifty-nine children between twenty-one and sixty-seven months of age were repeatedly examined and observed over several years (Bailey, Burchinal, et al. 1993; Bailey, McWilliam, et al. 1993). In various domains of behavioral development, mixed-age children tended to score higher, but at younger ages only. In another study, twenty-four children were randomly assigned to same-age classrooms of one-, two-, three-, or four-year-olds while another twenty-four were assigned to classrooms consisting of one- and three-year-olds or two- and four-year-olds; sixteen of the children (eight in each group) had significant disabilities (Blasco, Bailey, et al. 1993). The disabled, but not the other children, showed significantly higher levels of purposeful play and social play in the mixed-age than in the same-age classrooms.

The bias also distorts theorizing about early social development by dichotomizing it between adult-infant interaction competence and same-age peer competence. This leads to a two-stage cognitive-maturational model of the development of social relations. But in the comparative context, this model is an artifact of the absence, in the studies to date, of juveniles of various ages. Infants in hunter-gatherer and other traditional cultures develop the ability to predict and adapt to the behavior of children of various ages—and degrees of contingent-responsiveness —throughout the first year, with increasing frequency, culminating in some ability to interact with same-age peers during the second year. These observations are best represented by a nonstage model.

The failure to study systematically relationships among children of different ages also creates confusion in the attachment literature (Thompson 2005; Waters, Corcoran, et al. 2005). It is not surprising to find that same-age peers are not readily seen as attachment figures for infants or toddlers, and the gap between these relationships and the relationship to parents and caregivers seems and is wide. But if we consider cultures in which siblings and many other children of varying ages are available to develop relationships with infants and toddlers, the gap shrinks considerably.

Same-age peer relations in human infancy and childhood are to a large degree an artifact of laboratory studies and of child care conditions in advanced industrial states. In this context we can begin to understand the forms of social behavior that psychologists have traditionally known in the laboratory and nursery as "parallel play" and "collective monologue." Infants may be inept in relating to each other because they were never called upon to do so during most of human evolution. However, same-age peer groups may serve purposes in many cultures—such as comparing children and fostering competition—that were less highlighted in the remote past.

19

Play, Social Learning, and Teaching

Whatever the age structure of juvenile groups, play is one of their main activities. Play is a biological puzzle, and the difficulty of defining it precisely is a clue to understanding it. It is generally defined as inefficient, partly repetitive movements in varied sequences with no apparent purpose. No definition of *play* avoids this inferential aspect of the term, and the otherwise inchoate nature of the definitions point to play's multiple functions and possibly multiple origins. This behavior, combining as it does great energy expenditure and risk with apparent pointlessness, is a central paradox of evolutionary biology (Lorenz 1965; Loizos 1966; Ewer 1968; Fagen 1977, 1981, 1993; MacLean 1985; Barber 1991). It seems to serve the functions of exercise, acquisition of information, and sharpening of subsistence, social, and reproductive skills (Lorenz 1965; Martin and Caro 1985; Byers and Walker 1995; Byers 1998). In some mammals the deprivation of early play opportunities has serious consequences for those skills.

Play is costly—it increases food requirements 5 to 10 percent (Siviy and Atrens 1992)—and risky, so it is unlikely to have evolved without significant adaptive value. Few now accept the hypothesis that play is a mere discharge of excess energy, but clearly energy matters and is part of the explanation for the age distribution; food provisioning increases juvenile play and scarcity decreases it (Sharpe, Clutton-Brock, et al. 2002). Risk also reduces it: cat and other predator odors decrease play in rats (Siviy, Harrison, et al. 2006).

It has long been noticed that the smartest mammals—primates, cetaceans, elephants, and carnivores—are the most playful (Ewer 1968), and the two traits have probably coevolved (Iwaniuk, Nelson, et al. 2001). If an animal is short-lived (usually implying smaller adult relative brain size), the young do not play much, because there is little time for them to gain from play's advantages. These associations are observed in birds as well as mammals (Spinka, Newberry, et al. 2001; Ricklefs 2004). Hypotheses relating play to life span, body size, type of selection, and other variables have been mathematically modeled and tested by Robert M. Fagen (1977, 1981, 1993). His model holds that (1) play is more likely in stable populations near the carrying capacity of their environments; (2) species and populations vary considerably in the scheduling of play over the life cycle, with several different adaptive optima; (3) younger animals play more; and (4) behavior patterns appear in play as soon as they appear in the general behavioral repertoire. Empirical data generally support the model.

Play can be locomotor activity, object play, or social play, and even closely related species differ in their patterns (Bekoff and Byers 1998). In studies comparing twelve rodent species (Pellis and Pellis 1998), eighteen equid species (McDonnell and Poulin 2002), and twenty-four monkey species (Masataka and Kohda 1988), consistencies as well as differences were found. Studies in many species have shown that, after an initial rise in early infancy, play declines with age (Byers 1998; Fairbanks 2000; Spinka, Newberry, et al. 2001).

The Evolution of Play

Various adaptive explanations of play include honing of innate motor sequences (Lorenz 1965); getting accustomed to destabilizing movements and unexpected events (Spinka, Newberry, et al. 2001); establishing or practicing social relationships; safe self-assessment with a view toward future risk-taking (Thompson 1998); and stimulating neuromotor development through neural and synaptic selection, either generally (Fairbanks 2000) or specifically through the synaptic proliferation of cerebellar purkinje fibers (which may have a sensitive period for permanent experience effects) and the differentiation of muscle fiber types (Byers 1998).

Rat pups whose entire neocortex was removed engaged in normal rough-and-tumble and other play with peers (Panksepp, Normansell, et al. 1994), showing that the patterns and rewards of this basic behavior are generated from older, more primitive parts of the brain. In Jaak Panksepp's model (1998) rough-and-tumble play evolved first, for the purpose of practicing social skills and other forms of positive emotion derived from it. His experiments confirm Paul MacLean's (1985) claim that play, being dependent on a well-developed limbic system but not on the neocortex, was part of the evolutionary achievement of early mammals.

Comparative evidence also sheds light on this hypothesis. Turtles show simple elements of nonsocial play, such as pushing repeatedly on floating objects with no apparent purpose (Burghardt 1988), and some investigators see playfulness in the behavior of lizards and fish (Burghardt 2005). It is clear that some birds play (Diamond and Bond 2003)—especially larger-brained species like ravens, parrots, and macaws—but bird brains are different from mammal brains, and we know little about how the behavior is generated (Ricklefs 2004). Since reptiles and amphibians show only rudimentary play, birds probably evolved the more elaborate forms independently of mammals.

A number of marsupials have full play capabilities (Watson 1998; Byers 1999b), and as Darwin himself observed, even primitive, egg-laying mammals like platypuses play socially to some degree. This places social play's origins near the dawn of mammals 200 million years ago (Fagen 1981, 1987). Still, larger-brained mammals play more; apparently the basic limbic system can generate the behavior, but an advanced cortex can elaborate on and benefit from play's moves, moods, and motives. This correlation may hold even within the Australian marsupials. Play—especially rough-and-tumble play—is prominent in larger-brained kangaroos, which exhibit both serious and playful boxing, but minimal in the relatively small-brained koala (Byers 1999b). In the chuditch, a cat-size Australian marsupial carnivore, littermates wrestle actively near the time when they must begin to disperse from their natal sites (Soderquist and Serena 2000). But although the relationship between brain size and play is strong *among* orders of mammals, it is inconsistent within them (Iwaniuk, Nelson, et al. 2001).

Specific functions of play have been addressed in some studies. Juveniles of Belding's ground squirrel (*Spermophilus beldingi*) develop greater motor skills if they play with more rather than fewer individuals (Nunes, Muecke, et al. 2004). Brown bear cubs (*Ursus arctos*) that play more in their first summer are more likely to survive to their second, after controlling for cub condition, salmon availability, and maternal characteristics (Fagen and Fagen 2004). In the Felidae, play appears to be central in learning to hunt (Caro 1981; Martin and Caro 1985). House cats toy with chipmunks or voles that are stunned and struggling—releasing and chasing them, batting them around with their paws, and playfully pouncing again and again (Leyhausen 1979). This occurs in most species in the cat family, demonstrates the role of practice in honing skills, and, as we will see, forms part of the basis of rudimentary teaching. Playing with live prey also is seen in young otters (Ewer 1968), fishers (Powell 1993), bottlenose dolphins (Bel'kovich, Ivanova, et al. 1991), and other carnivorous mammals.

Higher primates, among the most intelligent mammals, are also among the most playful (Fagen 1993; Lewis 2000). Comparative studies show that the macaques' play face is a true homologue of human laughter and smiling (Preuschoft 1995; Preuschoft and van Hooff 1997), and their facial and vocal expressions in behavioral context suggest that the emotions of social play, especially in the young, are very similar across higher primates (Griffiths 1997). In a study using tickling as a stimulus, the vocalizations of infant and juvenile bonobos, chimpanzees, gorillas, and orangutans were compared to those of human infants (Ross, Owren, et al. 2009). In all four species of apes the infants and juveniles had something resembling human infant laughter, and on acoustic analysis the phylogenetic distance of the apes' laughter from ours reflected the divergence estimated from genomics and paleontology, with bonobo and chimpanzee laughter closest to ours.

A study of twenty-four species of New World and Old World monkeys showed that play-specific vocalizations, possibly unique to primates, vary among species and occur most often in those that have allocare, which may function to provide supplementary food to defray the energy costs of play or to monitor play for risk (Masataka and Kohda 1988). The young of golden lion tamarins (*Leontopithecus rosalia*) spend about 4 percent of their time playing, mainly socially, but adults nearby are almost continuously vigilant for predators while the young are playing (de Oliveira, Ruiz-Miranda, et al. 2003). Juvenile chimpanzees use play signaling more when adults are nearby in an apparent effort to modulate play and avoid termination of play bouts by adults (Flack, Jeannotte, et al. 2004).

In general, the great apes have play patterns going beyond those of most monkeys; these include social play, object play, and (perhaps) rare instances of very rudimentary pretend play (Pellegrini and Smith 2004).

Their *social* play is predominantly rough-and-tumble but also includes play parenting and sexual or pseudo-sexual play, and there is some play in adults (Gómez and Martin-Andrade 2004). Their *object* play is difficult to parse separately from the larger domain of object manipulation, but it can be done; in keeping with differences in tool use and toolmaking in the wild, naive wild or captive chimpanzees do substantially more object play than gorillas do, suggesting a genetic foundation for the difference. Fantasy or pretend play is very difficult to identify in the absence of language and may occur only in highly trained captive apes enculturated to a human symbolic world (Gómez and Martin-Andrade 2004).

The Development of Human Play

It has long been observed that games of peek-a-boo or hide-and-seek motivate infants and children to re-create repeatedly the tension of what might be feared losses, presumably because of the pleasure involved in controlling and relieving the tension. Research on the "smile of recognitory assimilation" (Piaget 1952; Kagan 1971; Zelazo 1971, 1972; Zelazo and Komer 1971; see Chapter 7) suggests that tension is caused by a stimulus that is not well represented in the infant's or child's memory store and she smiles as the discrepancy is resolved. To evoke attention, a stimulus must stir something in memory, but the pattern must also fail to match the schema already stored. For the older child or adult solving a problem, the process is similar: if it is too unfamiliar, it will not be of interest or perhaps be frightening; if too easy, it will be boring. But if it is difficult but doable, it will evoke attention, arousal, and then, when solved, pleasure, often signaled by a smile.

In classic experiments mentioned briefly in Chapters 7 and 18, a three-month-old lay in a crib, gazing up at a mobile linked by a ribbon to her ankle, and she soon learned to control it (Rovee and Rovee 1969; Rovee-Collier and Lipsitt 1982; Rovee-Collier, Earley, et al. 1989). Smiling and cooing accompanied this apparent sense of control, but in the extinction phase frets of distress revealed disappointment. In other experiments (Watson 1972; Watson and Ramey 1972) three-month-olds learned to control a mobile by turning their heads on a pillow with hidden levers on both sides, and after several weeks they learned similar tasks more rapidly than infants who had watched the mobile move independently of their actions. In fact, once given control, the last group learned even more slowly than did subjects with *no* mobile experience. The "response-contingent stimulation" (Watson 1972:327) in the first experimental group was apparently enjoyable, and by definition rewarding, since these infants learned the task. The no-mobile babies took to the task and learned it well, albeit more slowly. But the worst off were those who had simply watched and who now expected to be passively entertained.

As mentioned in Chapter 7, humans lay the foundations of social play in the first relationship. In his paper "Mother and Infant at Play" and other studies, Daniel Stern analyzed video of face-to-face "play" in three-month-olds (Stern 1974, 1985, 1990). Mothers would naturally make expressions and sounds designed to arouse the infant by challenging his expectations, but within limits, modulating arousal at a level below fear. Loudness and pitch of voice, sudden brow raising and widening of the eyes, and vigorous physical stimulation were among the stimuli varied by mothers. When the babies gazed, smiled, laughed, and cooed, mothers judged the stimulation to be in the right range; averting a gaze, fussing, or crying meant the play had either lapsed into boredom or spilled over into distress. Stern distinguished between *satisfaction*, a state caused by meeting a need, and *joy*, a more complex rhythm of satisfying calm alternating with a desired level of arousal (Stern 1990).

In another study, there was a curvilinear function relating maternal sensitivity to the degree of synchrony in mother-infant interactions; the mothers rated as most sensitive by other measures had an intermediate degree of synchrony with their infants' vocalizations, while those with very low or very high levels of synchrony were rated as less sensitive (Hane, Feldstein, et al. 2003). The idea of optimal arousal is consistent with adaptation level theory (Kubovy 1999); intermediate stimulus patterns are a source of pleasure, which is enhanced by a certain degree of control. This is the zone of play, and the most sensitive parents lead infants into it. However, because adaptation level changes with long-term stimulation, the amount and type of stimulation that will produce optimal arousal also change—a key process in development.

It is possible that mutual gaze and playful gaze interactions are in some ways unique to humans and part of the foundation of human enculturation. Ann Cale Kruger (2010) has pointed out that the sclera, or whites of the eyes, have tripled in size since our ape ancestors just as brain size has, making human gaze direction much easier to determine and follow. The primary intersubjectivity of gaze interactions is succeeded in development by the secondary intersubjectivity of gaze following, in which older infants look where the mother is looking, or point and try to get her to look where they are looking. This "communion," as Kruger calls it, makes much of later enculturation possible.

The social playfulness seen in mother-infant dyads extends to other adults, then to older children, and finally to same-age peers. In the second half-year interactive play includes the peek-a-boo game (Parrott and Gleitman 1989); this more advanced version of the tension-modulation play seen at three months pointedly parallels the emergence of attachment. Now more is at stake as infants modulate and optimize their own arousal by toying with the temporary disappearance and reemergence of caregivers and strangers.

Although such games between adults and infants are commonly seen in Western cultures and even deemed essential for optimal cognitive development, many cultures have little or none of them. !Kung infants experience caregiver vocalizations and face-to-face interactions at levels comparable to or higher than those of working-class (but not as high as professional-class) infants in Boston; however, vocalizations were much fewer in a Guatemalan Indian village, despite very high levels of physical contact (Klein, Lasky, et al. 1977; Konner 1977b). Joint attention to objects by infants and caregivers is far less common among the !Kung than in the West (Bakeman, Adamson, et al. 1990; see Chapter 9), casting doubt on the idea that it is important to language development and also suggesting that joint play with objects is unusual in non-Western cultures. David Lancy (2007) reviewed cross-cultural variation in mother-child play, emphasizing intermediate-level societies—those whose subsistence base is horticulture, simple agriculture, or herding. Such play is far less common in these cultures than in the Western middle class (LeVine, Dixon, et al. 1994); mothers focus on nurturance and routine care, often including high levels of physical contact and breast-feeding, but are not as playful with their infants or toddlers. And contrary to the common finding in Western families that fathers are more playful with children than mothers are (see Chapter 17), this is not the case in many intermediate-level societies.

Developmentally, play soon becomes still more complex. A study comparing Puerto Rican and Anglo mother-infant play found large-body-motion games, clap-sing games, touch-turn-taking games, and hide-chase games as patterns in both cultures, although there were some cultural differences (Miller and Harwood 2002). Object play too follows a predictable trajectory (Piaget 1952): infants begin with exploratory manipulation, proceed to repetitive shaking or banging to produce systematic stimulation patterns, then go on, around ten months, to relate objects to each other; next comes playful use of the functional properties of objects, and finally the child recruits objects and pretend objects in fantasy play (Piaget 1951). By the end of the first year the play of boys and girls has begun to show differences (Goldberg and Lewis 1969; Lyytinen, Laakso, et al. 1999), expressed in toy preference and aggressiveness in relation to both objects and playmates. These differences persist (Maccoby 1998; Berenbaum, Martin, et al. 2008) and are found cross-culturally (Blurton Jones and Konner 1973; Whiting and Edwards 1988) and in some other primate species. As we saw in Chapter 10, monkey males and females differ in rough-and-tumble play owing to prenatal androgenization of the brain, and sex differences in the toy preferences of monkey juveniles parallel those of human children.

But at this stage uniquely human adaptations appear. Symbolic play emerges along with language as a central aspect of social play between mothers and infants in the second year. This generalization has been found to apply to the United States, Turkey, India, Argentina, Guatemala, Italy, Sweden, and some other (but not all) cultures, and there is some evidence relating the amount of such play to later cognitive development (Bornstein, Haynes, et al. 1999; Lyytinen, Laakso, et al. 1999; Bornstein, Venuti, et al. 2002). Object play, social (including rough-and-tumble) play with other children as well as adults, symbolic play, and various combinations of these are all characteristic of the early childhood years in all cultures studied (Konner 1972; Gosso, Otta, et al. 2004; Pellegrini and Gustafson 2004; Fry 2005; Smith 2005). Pretend play, which is central throughout this period, is highly revealing of children's emotional and cognitive functioning (Dunn and Hughes 2001; Lillard 2001). Although cultural variations are interesting and important—indeed, children have distinctive cultures of their own—the continuities are also remarkable.

Heide Sbrzesny's (1976) extensive study of play and games among the !Ko, a group of San hunter-gatherers closely related to the !Kung, illustrates the remarkable variety of play behaviors in human children even in a culture with limited technology and a small population. Much of their play is informal, consisting of exploring the village-camp and its environs in multi-aged child groups, just as with the !Kung (Draper 1976). But the uniquely human adaptations of symbolic play and ritualized games are common in both cultures (Marshall 1976, ch. 10).

Many of the games and much of the informal play among the !Ko and !Kung involve pretending to be various animals and the hunters pursuing them. An elegant melon-tossing game, played mainly by girls and women in both cultures, seems as much a dance as a game, and it hones physical skill while reinforcing ideals of cooperation. Melons are also used for making dolls; even a young girl can push a stick into a smaller melon at one end and a larger one at the other, and wrap, carry, and otherwise tend this very serviceable "baby." Toys such as miniature bows and arrows may be made by adults for children or by the children themselves; such artifacts are exclusively human and belong to the domain of enculturation.

Equally human are the ritualized schemas of even informal games. Both the !Ko and !Kung play a game of skill in which a feather tied to a weight is repeatedly tossed into the air with a stick and then caught by the end of the stick as it comes down. The children sharpen and display their skill, but they do not keep score or obviously compete—the players are too heterogeneous—and the almost ritual aspect of the play is elegant and enjoyable. The same is true of a game in which children (mainly boys) repeatedly throw a smoothed, long, thin stick forcefully across the sand in such a way as to make it slap against a rise in the sand, bounce, and sail rather elegantly away again—something like the American game of skipping flat stones on a lake.

The role of games and ritualized play in the Western world is perhaps best summed up by a glance—or rather, a long gaze—at Pieter Brueghel the Elder's 1560 painting *Kinderspiele* (Orme 2001). It shows over two hundred children engaged in at least seventy-five different games. "Some of the games are imitative (such as a wedding procession); others are skilful and involve toys or equipment. Objects depicted include dolls, dice, hoops, tops, windmills, a mask, and a hobby-horse. . . . There is riding on rails or on barrels, piggy-back riding, acrobatics, wrestling, swimming, and climbing a tree. The scene . . . is an encyclopedia of games . . . a celebration of childhood and its ingenuity" (Orme 2001:167). Some of these things can be seen in any ape or monkey, but the wedding procession, the dolls, the dice, and the mask, among others, are distinctively human. Even some purely physical games in childhood turn out to have explicitly stated rules. On the functions of play among the !Ko, Sbrzesny (1976:306) writes,

> The play behavior of children can be interpreted as a strategy to explore the environment, including its social aspects . . . the child perceives by manipulatory investigation the propensities of objects and investigates in which ways objects can be used as tools. It is in this latter aspect that a learning disposition for tool-using becomes evident. . . . Some patterns (poking and beating with sticks) are universal and link up to probable homologous patterns in chimpanzees.

In humans, "games can be interpreted as rituals of bonding and (or) aggression control" (307), and they frequently are resonant with symbolic meaning. In other species play also promotes bonding and aggression control, but not through symbols.

The Evolutionary Neurobiology of Play

If play is a pervasive adaptation, how is it wired in the brain? As mentioned, rats that were radically decorticated at birth showed remarkably intact play patterns as juveniles (Panksepp, Normansell, et al. 1994). Not surprisingly, specific lesions of the medial prefrontal cortex (mPFC) also left juvenile play behavior mostly intact, but adult social behavior *was* abnormal in these animals (Schneider and Koch 2005).

However, the role of limbic system structures has proved greater. For example, orbital frontal cortex lesions permit play-fighting in rats but damage judgment about appropriate play partners (Pellis, Hastings, et al. 2006), and primate species with larger amygdalas are more likely to engage in sexual play (Pellis and Iwaniuk 2002). Early lesions of the amygdala (but not the hippocampus) in rats decrease social play (Daenen, Wolterink, et al. 2002). Thalamic lesions associated with the limbic system (the dorsomedial and parafascicular nuclei) markedly change rough-and-tumble play (Siviy and Panksepp 1985), while lesions of some other thalamic nuclei do not (Siviy and Panksepp 1987). Activation of *c-fos* genes indicates neural activity in sensory and motor brain areas during juvenile play in rats, but also in the ventral striatum, which helps mediate reward in the brain, and the dorsal periaqueductal gray area, which controls pain (Gordon, Kollack-Walker, et al. 2002). Surprisingly, given its critical role in adult social behavior, removal of the olfactory bulb has little effect, but this underscores the intrinsic differences between play and other social behaviors.

The neurotransmitter systems involved in play have also been studied. Opioid systems have a key role: during social play, activity increases in the nucleus accumbens and other reward-mediating brain areas (Vanderschuren, Stein, et al. 1995). Opioid treatment enhances play, while opioid antagonists decrease it, probably through μ-opioid receptors (Vanderschuren, Niesink, et al. 1997). Catecholamines are also clearly involved; neonatal 6-hydroxydopamine (6-OHDA), which results in depletion of dopamine and norepinephrine in the nucleus accumbens and caudate-putamen, suppresses later responses to play initiation attempts. These neurochemical findings, like the anatomical ones, are consistent with the limbic system mediation of play, with a central role for reward systems.

Intelligent Players

As mentioned, the relative importance of the limbic system and unimportance of the neocortex is consistent with the MacLean (1985) model, in which the origin of mammals required limbic system expansion to generate play and attachment. However, the experimental findings seem inconsistent with the known relationship between play and brain size, which in mammals is mainly determined by neocortical size (Chapter 5). In a statistical comparison of fifteen mammalian orders, brain size, as measured by the encephalization quotient, explained at least half the variation in average play level (Iwaniuk, Nelson, et al. 2001). Primates, carnivores, cetaceans, and elephants played the most, and bats, insectivores, and monotremes the least. However, this relationship did not hold *within* mammalian orders, as tested on primates, muroid rodents, and marsupials, except for marsupials' near-significant correlation between play and neocortical size. Taxonomic grades have biologically different meaning, so regression analyses at different levels often produce different results, and they do not negate each other's possible significance (Pagel and Harvey 1988).

However, some small-brained species play a great deal and some large-brained ones do not, so there must be subtler relationships between the two dimensions. For example, adult-adult play behavior in primates is correlated with the size of the nonvisual neocortex, while sexual play levels are correlated with amygdala size (Pellis and Iwaniuk 2002). Within primates, there is a correlation between social play and neocortical size (Lewis 2000), as there is between group size and neocortical size. Also among primates, cerebellum size is correlated with social play (Lewis and Barton 2004), consistent with the role of play in smoothing motor sequences and learning to cope with postural imbalance, the role of physical self-handicapping in play, and the possibility that cerebellar purkinje circuits mediate play-responsive plasticity.

Developmental considerations also matter. It is reasonable to think that the allocation of brain development between pre- and postnatal phases is related to the prevalence of play, and indeed, the species in which there is more brain development postnatally have more play-fighting, a relationship that is true for both primates and muroid rodents (Pellis and Iwaniuk 2000). This correlation is also consistent with humans being the most playful primates.

Hormonal studies relating to play have mainly focused on sex differences (Chapter 10)—rhesus monkey females androgenized prenatally show more malelike patterns of rough-and-tumble play—but there are species differences. To take one example, juvenile play in free-living Belding's ground squirrels consists mainly of play sex and play-fighting. Males take masculine roles in sex play, but the sexes differ little in play-fighting; giving female neonates testosterone, however, makes them malelike in sex play without affecting their rough-and-tumble play (Nunes, Muecke, et al. 1999); parallel findings have been made in prenatally androgenized female lambs (Orgeur 1994). Generally, there is good evidence that girls exposed to androgenlike substances in utero have play and toy preferences that diverge from typical female patterns and toward typical male ones (Hines 2003).

Play, Learning, and Culture

Play is observable, and the underlying emotional states so far are not, but both phenomenological and neurobiological evidence strongly suggest that emotional states during play are positive, pleasurable, and rewarding. Positive emotions like joy, pleasure, satisfaction, and happiness grade into each other; they do not fully overlap, but they are difficult to define. Barbara Fredrickson (1998), viewing this as basic, developed a "broaden and build" model of positive emotions, acknowledging that positive emotions are less definable than negative ones. We have moderately clear distinctions between rage, fear, grief, disgust, and lust, but we struggle to differentiate such states as joy, satisfaction, contentment, pleasure, and happiness; perhaps this diffuseness is of their essence and relates to an evolutionary function of play. Research suggests that people in positive and playful moods are more open to experience and learn in better and in more varied ways (Isen 1993); if so, positive emotions may be in part an adaptation for expanding experience and knowledge.

This parallels the theory advanced by Brian Sutton-Smith (1997:229) in *The Ambiguity of Play*: "that play variability is analogous to adaptive variability; that play potential is analogous to neural potential; that play's psychological characteristics of unrealistic optimism, egocentricity, and reactivity are analogous to the normal behavior of the very young; and finally that play's engineered predicaments model the struggle for survival." The idea is that natural selection designed play to shape brain development, or as one researcher said of play-boxing kangaroo joeys, "Most likely they are directing their own brain assembly" (Byers 1999a:45).

The key role of play in human life and culture led one historian, who defined culture itself as an extension of play, to the designation *Homo ludens—ludens* meaning "playing" or "playful" (Huizinga 1949). Play is widespread among animals, yet *as adults*, humans are uniquely playful, and play is central to many aspects of culture (Caillois 1961; Geertz 1971; Bruner, Jolly, et al. 1976; Sutton-Smith 1997; Fromberg and Bergen 1998), a subject to which we return in Part IV.

Social Learning, Imitation, and Teaching

As we saw in Transition 2, in many animals substantial accumulation of information about the environment—and in some, information from conspecifics—occurs during development. Social play is a form of social learning. So is the ongoing reward and punishment received from other individuals; if an infant bites the nipple less often after a series of harsh

maternal vocalizations, or an adolescent male stops approaching estrus females after being threatened by adult males, social learning is under way.

A more advanced form of social learning is the enhancement of
behavior modification by watching other individuals. An increased rate
of learning by observation has been shown for a number of mammals
(Galef and Heyes 2004), including cetaceans (Rendell and Whitehead
2001) and primates (Caldwell and Whiten 2002), and for some birds
(Zentall 2003). In many mammals, especially those whose infants follow
or are carried, transmission of information about food sources and food
extraction is ongoing, one adaptive function of the mother-infant bond.
The mother is a more effective object for observational learning in kittens than is a strange individual (Chesler 1969). Even the acquisition of
complex skills, such as termite fishing in chimpanzees (Goodall 1967;
Lonsdorf 2006), can be facilitated in this way. Infants among human
hunter-gatherers acquire the basic movements of digging for roots,
cracking nuts, and pounding with a mortar and pestle by the end of the
second year, however ineptly they perform these actions, by observing
their mothers and copying them (Konner 1972).

However, it is not easy to define or demonstrate the various levels of
what seems to be observational learning. If I cough in a theater or yawn
in a meeting and then others do the same, they have not learned anything, except perhaps that it is all right to do what they felt like doing
anyway; this disinhibition is called *social facilitation* or *social contagion*. If
an animal follows others to a food source, watches them eat, and then
picks up or pecks at and eats some of the same food more readily, this is
a minor social assist to the usual individual processes of conditioning
and reinforcement learning and is called *stimulus enhancement* or *local
enhancement*. Similar enhancement processes have been demonstrated
in mate choice (Dugatkin 1992, 1996) and predator avoidance (Griffin
2004), and these forms of enhancement occur widely in fish, birds, and
mammals. Despite its simplicity, it is often vital to survival and reproduction, and theoretical models predict that because of the costs to innovators and the benefits to followers, the number of followers tends to
increase until there are not enough innovators, at which point a fluctuation around equilibrium may become an evolutionarily stable strategy (Laland 2004).

Finally, the most interesting domain of social learning is what is usually called imitation. Even after excluding social facilitation and enhancement, it is useful to break down processes of observational learning, and this has been done in different ways. According to a widely accepted view (Tomasello, Kruger, et al. 1993; Call, Carpenter, et al. 2005), *emulation* and true *imitation* differ based on goals or results and actions. Emulation occurs when an individual watches another perform a complex task and tries to duplicate the results, while true imitation entails achieving the same results by reproducing the model's actions. This is held to require an advanced form of perspective taking in which the imitator imagines being in the place of the model to the point of understanding the model's intentions, and the capacity to take this perspective may distinguish children from chimpanzees (Call, Carpenter, et al. 2005); emulation is in effect a higher form of stimulus enhancement or socially biased learning.

However, there is an important exception: chimpanzees that have been enculturated by human teachers do show true imitation (Tomasello and Carpenter 2005); so do orangutans that live among humans as they are rehabilitated for possible return to the forest (Russon and Galdikas 1993, 1995). These studies, along with many naturalistic observations indicating that apes may approximate true imitation, suggest that what is missing for apes in nature is a true *teacher* who is able to reciprocate their incipient perspective taking to create ideal situations for imitation—in other words, a person. However, as with ape language ability, a capacity for true imitation in the ape brain demands an explanation of why they have it and urges a search for its adaptive value in the wild (Whiten and van Schaik 2007). The true-imitation-capable brain may therefore have evolved before orangutans diverged from other apes, millions of years before the last common ancestor of chimpanzees and humans.

Of course, the label "imitation" has also been applied to much simpler processes. Echoing the actions of a model without understanding intentions or goals—usually limited to simple sequences—is a more primitive process sometimes called *mimicry* (Want and Harris 2002); this is a kind of motoric or emotional resonance that we may have in a primitive form from birth (Meltzoff and Moore 1977; Meltzoff 2002). Whether true imitation arises developmentally from mimicry, emulation, or both remains controversial, as does the question of just how much more advanced than emulation true imitation really is (Bard 2002; Want and Harris 2002; Whiten 2002).

Toward a Neurobiology of Social Learning

While the neurobiology of basic learning processes is increasingly well understood at both the cellular and nervous system levels (Bright and Kopelman 2001; Kandel 2001; Buzsaki and Draguhn 2004), social learning remains much more mysterious. However, as discussed in Transition 1, a major advance has come in the form of the discovery of the mirror-neuron system (MNS) (Rizzolatti and Craighero 2004), which is apparently unique to higher primates; perhaps we must begin to mirror action in premotor neurons to perceive it (Gallese, Rochat, et al. 2009). MNS neurons become active when the observed individual is *about* to perform an action, so it could serve the perception of others' intentions (Iacoboni 2005). These characteristics make it an ideal core substrate for imitation; it includes the ventral premotor area, the adjacent region of the inferior parietal lobule, and a visual-processing area of the superior temporal sulcus that relays the perception to the mirror neurons (Iacoboni 2005). When a person is actually engaged in *imitative learning*—trying to imitate an action not previously learned—the dorsolateral prefrontal cortex is also involved, and when facial expression of emotion is imitated, the limbic system and the insula are as well. There remains the question of whether nonhuman primates can understand the perspective or intentions of others (Hare, Call, et al. 2000; Tomasello, Call, et al. 2003; Flombaum and Santos 2005), but the fact that they have mirror neurons suggests that a continued search for evidence of this might be rewarding (Lyons, Santos, et al. 2006).

PET and fMRI studies show that the cerebellum, like the MNS, is active during the observation of movement (Petrosini, Graziano, et al. 2003:253), so "it appears that the cerebellar circuits are involved in the 'motor thought,' whether or not it is accompanied by an actual motor act." Further, these studies show that cerebellar circuits are especially active during the early phases of procedural or motor learning and decline in activity with repeated practice and mastery (Vaina, Belliveau, et al. 1998). The cerebellum also functions in word learning, and this major structure is disproportionately large in gibbons and humans, possibly reflecting the challenges of mastering brachiation or speech (Rilling and

Insel 1998). Classic studies showed that dendritic branching in the cerebellum is affected by different rearing conditions in monkeys (Floeter
and Greenough 1979). Finally, we have seen that cerebellar size is correlated with social play in primates, which may reflect its role in motor
stabilization, motor learning, and perhaps also observational learning.

Teaching: Uniquely Human?

As distinct from observational learning, teaching requires actual efforts
by adults or older juveniles to facilitate information transfer, and to
meet the strict definition, faster learning must be demonstrated (Galef,
Whiskin, et al. 2005). This, if it occurs at all (Caro and Hauser 1992), is
rare in the nonhuman world. Merely being available for observation is
not enough; teaching also requires deliberate modeling, active encouragement, exaggeration of the movements involved in the task to make
it more obvious, or simplification of the task to provide graded steps.
Such behavior is rarely observed in nonhuman primates, and efforts to
stretch features of their parental behavior to make it seem like teaching (Boesch 1991; Whiten 1999) really reveal how little teaching they do.
Careful studies of how chimpanzee juveniles learn the important skill
of termite fishing show that observation is clearly involved and that
mothers tolerate their offspring observing them, but that they do not
otherwise facilitate the learning process and certainly do not teach
(Lonsdorf, Eberly, et al. 2004; Lonsdorf 2006). Not surprisingly, some
experienced students of ape cognition have concluded that nonhuman
primates do not teach (Tomasello, Kruger, et al. 1993; Premack and
Premack 1996).

Felids, on the other hand, including lions (Schenkel 1966; Schaller
1972), tigers (Schaller 1967), and especially cheetahs (Caro 1994), commonly engage in behavior that functions to transmit hunting and prey-
catching skills. Most felid mothers lead cubs and kittens on expeditions
whose main purpose seems to be to acquaint the young with stalk-
ing (Schaller 1972), and they partially kill prey on the hunt, leaving the
young to finish the killing and intervening only if the prey is about to
escape. Cheetah mothers, however, bring back half-dead prey, which
their young then kill and eat (Caro 1994). Domestic cats also do this (Leyhausen 1979). These behaviors are closer to qualifying as teaching than
anything seen in nonhuman primates.

Subsistence Learning and Teaching in Hunter-Gatherers

Human hunter-gatherers have available a unique means of information transfer: language. Some teaching about hunting occurs in the form of storytelling and answering questions, means unavailable to hunting cats. So it is not surprising that some of the cat's teaching actions—such as furnishing the young with wounded prey to finish killing—occur very infrequently in human hunters and that linguistic information transfer, observational learning, and play are the main means of acquiring hunting skills. Fathers or the boys themselves sometimes make play bow-and-arrow sets. Boys also learn much about tracks, game movements, and local landscape from their mothers during gathering trips, and girls acquire much of the information they need about food gathering during the same expeditions, with little active instruction.

Task assignment among the !Kung San in this age group (three to eleven years) is minimal. This stands in contrast to the Hadza (Blurton Jones, Hawkes, et al. 1994), the Mikea of the Madagascar forest (Tucker and Young 2005), and the Meriam of Australia (Bird and Bliege Bird 2000), where children account for and may be held responsible for a substantial amount of food gathering. Still, a classic study of a large cross-cultural sample (Barry, Child, et al. 1959) showed that pressure on children to be responsible and obedient depends on subsistence economy and that hunter-gatherers exert less of this pressure than other non-industrial societies; perhaps the key variable here is task *assignment*. Finally, because of the multi-aged composition of hunter-gatherer playgroups, information acquisition and practice of skills are achieved in a context in which teaching, observational learning, and play are combined and, in effect, become one process. We address these questions further in Chapter 25.

20

The Contexts of Emerging Reproductive Behavior

Although New World monkeys are more variable, in Old World monkeys females tend to remain in their natal groups when they mature. Male transfer is often associated with high mortality (Kappeler and Pereira 2003; van Noordwijk and van Schaik 2004). In the Hanuman langurs of Jodhpur (*Presbytis entellus entellus*), which live in single-male harems, males leave or are driven out when their fathers are deposed by raiding males (Rajpurohit and Sommer 1993). There is a temporary survival advantage to males that continue associating with their fathers, but half die either way within nine months of emigration. In Tibetan macaques (*Macaca thibetana*), young adult males were the most active age group, in both mating and intergroup transfer, and they received the most wounds, especially serious wounds on the front of the body, along with subadult males (Zhao 1994). Over eighteen months of observation, five out of six of the young adult males changed groups, compared to five out of seventeen older males. Despite their high risk, younger males usually rose in rank after transfer, while older males fell. Subadult males emigrated rarely, but were likely to be seriously wounded; three died of severe attacks by resident males.

This may be why the great apes evolved a pattern of female instead of male transfer: with slower development and higher parental investment than monkeys, they could not afford to lose many males of peak reproductive value. Females transferring into a new group are welcomed by resident males. In any case one process or the other (or some combination) is necessary for incest avoidance (Moore 1983; Clutton-Brock 1989). Sexual contact by immigrant males is often not possible immediately after transfer, and the building of social ties may take time and effort (Smuts 1985).

The Development of Sexual Behavior in Monkeys and Apes

Among monkeys and apes, neither males nor females can rely entirely on instinct for sexual skills. Monkeys deprived of contact with other juveniles during development have impaired sexual and maternal behavior in adulthood. It is mainly males that lose sexual skills, but the impairments are clear. Harry Harlow described socially deprived monkey males attempting to copulate with the female's side or even her face (Harlow 1975; Suomi and Harlow 1975). (In lectures Harlow reportedly called these efforts "working at cross-purposes with reality" and "the head-start program.") Such sexual retardation was sometimes remediable with a patient, experienced female, but only after long practice. Major social deprivation has similar effects on the sexual behavior of male chimpanzees:

> Of the five males who reached sexual maturity and were given sufficient opportunity to engage in copulatory behavior, all but one have done so. For these animals, considerable learning appeared to be involved. Initial attempts at copulation were very poorly coordinated. For example, males with erections might mount the side or head end of the female and thrust against her, but with experience, particularly with the helpful tutelage of sexually proficient females who assisted with positioning and penetration, these animals have improved in frequency and style to the point that they are approaching normal species typical sexual behavior, except that they lack the usual signaling systems exhibited by wild born males. One (now fully adult) restricted male has neither attempted nor solicited copulation, and females rarely approach him. He masturbates frequently, sometimes to ejaculation, and occasionally uses a 55-gallon drum for thrusting. (Davenport 1979:356)

These outcomes recall those described by Harlow in rhesus monkeys. In the wild such males would have zero reproductive success, and even much milder deficits, under natural conditions of mating competition, would be a great disadvantage.

The social play of monkey and ape juveniles routinely includes mounting and thrusting. This may be same-sex activity and does not involve intromission, but juvenile monkeys do this instinctively or, having had many opportunities to observe adults having sexual intercourse, simulate it in play. Such play is part of the normal expectable environment, and even behaviors largely genetically coded and vital to reproduction depend on it in part.

This behavior is not as common as rough-and-tumble play, which predominates in juvenile interactions in higher primates and is probably necessary for the socialization of aggression and an aid to sex differentiation in social behavior (Chapter 10). Monkeys deprived of contact with other juveniles tend to be more, not less, aggressive as adults, and this inappropriate aggression generally translates into lower, not higher, dominance ranking. Given the self-handicapping behavior of mammal young at play, it is not surprising that extensive daily practice in inhibiting aggression, even during aggressive play, leads to appropriate inhibition in adulthood.

Adolescence among the !Kung Hunter-Gatherers

As previously noted, there has been excellent research on reproductive history and demography among the !Kung (Howell 1979, 2010). Prospective study of menarche (which is marked by a dramatic ritual) gave a mean age of 16.6 years and a median of 17.1, with the majority passing this milestone between 16 and 18, by which time about half were married. Careful retrospective study of women who were 45 or older in 1968 estimated the age at first birth at a mean of 18.8 and a median of 19.2 years, with all but a few mothers having had their first births between ages 17 and 22. (Completed fertility was 4.7 live births, with a mean age of last birth in the midthirties.)

As for the development of sexuality, playful experimentation with sex began in early childhood and continued through middle childhood (Shostak 1981; Konner and Shostak 1986). Since children did not assume serious responsibility for subsistence until the late teens and their playgroups were frequently out of sight of adults, sexual curiosity was not much restrained. Adults did not approve of sexual play, and when it became obvious they discouraged it by verbal chastisement, but interviews with adults suggested that they considered sexual experimentation in childhood and adolescence to be inevitable and normal. For adults, sexual activity was considered essential for mental health, and !Kung sometimes referred to mentally ill people (for example, a woman who ate grass) as deranged because of sexual deprivation.

Despite childhood sexual experimentation, the transition to the real sex of adulthood could be difficult, especially for girls. Half were married before menarche, typically to men about ten years older. Thus a teenage girl faced the sexual advances of a grown man after prior experience only with boys. These advances were supposed to be delayed until menarche, but the transition from sex play with other teens to adult sex was sometimes stressful (Shostak 1981). The years from 16.5, the mean age at first menstruation, to 19, the mean age at first birth—a delay due mainly to adolescent subfertility—were important. Maturation was in a sense suspended; the young woman was sexually mature but did not have to care for a family and made little contribution to subsistence. She could gradually adopt adult roles and adult sexuality without having to deal with the consequences of pregnancy.

Nisa, the subject of an extended life history (Shostak 1981), described frequent experience with childhood sexuality in the course of play with other children. In her early to midteens she had two unsuccessful arranged marriages to older men. Both were soon dissolved—the first because her husband had sex with a woman his own age in the marriage hut while his much younger wife was presumed to be sleeping, the second because Nisa's father disliked the young man. Both marital failures involved her resistance to her husband's sexual advances; despite extensive sex play in childhood, reportedly including sexual intercourse with penetration, her first experiences with adult men were awkward and unwelcome. Dissolution of these marriages was easily achieved, as was usual for couples without children (Howell 1979).

She was married for the third time—yet another arrangement by her parents—about a year before her first menstruation. She ran away from this husband several times before settling down with him. She finally agreed to have sex with him once, but then refused because of the pain it caused her, and it took years before their relationship assumed an adult cast. "We lived and lived, the two of us together, and after a while I started to really like him, and then to love him. I had finally grown up and learned how to love. I thought, 'A man has sex with you. Yes, that's what a man does. I had thought that perhaps he didn't.' ... I thought that, and gave myself to him, gave and gave. We lay with each other and my breasts were very large. I was becoming a woman" (Shostak 1981:166).

While a young woman's primary responsibility for feeding herself and her husband is deferred, it is borne by her mother and father. Even after her first birth, her need to be near her mother is recognized, and it is not until the second child that she may be expected to move to her husband's village-camp. At that time—when she is twenty-three or twenty-four years old—she is essentially on her own, with the full psychological and social responsibilities pertaining to motherhood. Still, she remains in a social and economic context dense with her own and/or her husband's relatives. It is rare for a marriage to be dissolved after it has produced surviving children.

Adolescence in Other Hunter-Gatherers

For the !Kung there was preparation for adult sexuality through learning and play, yet the transition was often stressful for girls, who usually married very young. What was the process for other hunter-gatherers?

Hadza

Hadza children did substantially more foraging for themselves (Blurton Jones, Hawkes, et al. 1994), and by age eighteen they could bring in as many calories as they consumed. One of the main Hadza rituals is the female initiation rite, in which "pubertal girls gather in a camp where they are covered with animal fat and adorned with beads, then chase boys and try to hit them with their fertility sticks" (Marlowe 2002:250). Thematically and behaviorally, the ceremonial dances have much in common with those of the !Kung, which we will discuss in Chapter 27, but the Hadza ceremony includes female circumcision, making it a drastically different experience (Power and Watts 1997); men are initiated in another ritual that occurs after a youth has killed his first large game and includes having his nose bloodied. The age at first marriage was about twenty for men and seventeen for women, with nineteen being a woman's median age at first birth (Marlowe 2005a). "Marriages are not arranged and there is usually no wedding ceremony. Marriage typically follows a brief sexual relationship and is evident when a couple begins living together. Polygyny is rare" (Marlowe 2005a:179).

Efe

At age eleven or twelve, Efe boys had their ceremonial circumcision, a painful and at times traumatic experience, but girls experiencing their first menstruation were feted with a long celebration. Few attempts were made to restrict sexual activity in adolescent girls unless they were involved with Bantu men in nearby villages, and sexual activity among Efe teenagers was not unusual (Peacock 1985).

Aka

Adolescents were permitted to eat and sleep with their parents but often did not, instead making trips to see relatives and explore the region. Initiation rites included circumcision for boys and painfully filing the incisor teeth to a point in both boys and girls. These rituals were not elaborate, but they were stressful, demanding resolve and courage, and they certainly gave the initiate a sense of having left childhood behind. There was considerable premarital sexual freedom, but apparently it was not taken advantage of to the same extent as among the !Kung (Barry Hewlett 2004, personal communication).

Ache

As in many hunter-gatherer groups, puberty among the Ache was often dramatic. When a girl passed menarche, at an average age of 15.3 years, she went through a rite of initiation and purification, along with "all men who have had sex with her. . . . Every woman we interviewed who had reached menarche before contact reported that she had engaged in sexual intercourse with at least one adult man prior to menarche. . . . 85% of the women . . . had also been married before menarche" (Hill and Hurtado 1996:224–25). Still, adolescence was a playful time, especially for boys:

> Both boys and girls begin experimenting with sex around twelve years of age . . . in a manner very similar to that described for the !Kung (Shostak 1981). Boys . . . spend most of their teen years visiting other camps and trying to form friendships and alliances with their same-sex age mates and older men. It is quite common to see these boys intimately joking, tickling, and touching each other or the adult men who have chosen to befriend them.

As in most human hunter-gatherers, as well as other higher primates where male alliances are important, such play coalitions are essential to survival and reproduction.

Girls, "despite their precocious sexual activity . . . are generally re-luctant and sexually reserved with most males most of the time. Indeed the best description of their behavior would be aggressively flirtatious but sexually coy to the point of causing frustration." Boys say girls are "stingy with their genitals," and "the major activity of girls at this time is walking around in small groups laughing and giggling and carrying on in any manner that will attract attention. They frequently spend much of the day visiting from hearth to hearth and are fed abundantly wher-ever they go" (Hill and Hurtado 1996:225).

Agta

Among the Agta, "premarital female chastity is not an ideal of much currency. . . . Although some data are difficult to collect concerning sex, almost certainly girls are able to engage in sexual activity with relative ease; promiscuity is not favored in any circumstance" (Estioko-Griffin and Griffin 1981:138). The mean and modal age of menarche was seven-teen, and often a girl's first sexual partner became her husband, but as among the !Kung, marriages were very unstable before the first live birth, around age twenty (Goodman, Estioko-Griffin, et al. 1985). Al-though "elopement by young lovers" was "not uncommon," it yielded less stable unions than the more typical arranged marriage, often be-tween a man of about twenty and a premenarcheal girl of fifteen (Estioko-Griffin and Griffin 1981:137). Women began hunting shortly af-ter puberty.

Broader Cross-Cultural Patterns of Premarital Sex

Contrary to some claims of culture historians, anthropologists find that liberal premarital sexual mores are not new for a large proportion of the cultures of the ethnological record and that liberal sexual mores and even active sexual lives among adolescents do not necessarily produce pregnancies. In fact, a great many cultures permit or at least tolerate sex play in childhood (Frayser 1994). Children in these cultures do not play "doctor" to satisfy their anatomical curiosity—they play "sex." They do play "house," as Western children do, but the game often includes pretend-sex, including simulated intercourse. Most children in non-industrial cultures have opportunities to see and hear adult sex, and they mimic and often mock it.

These general patterns of childhood sex play grade into the more serious adolescent mores of many cultures. Gwen Broude's research using the Standard Cross-Cultural Sample rated twenty sexual practices and attitudes (Broude 1975; Broude and Greene 1976). It was possible to rate 107 societies on frequency of premarital sex for males, and of these, premarital sex was rated as universal or almost universal in 59.5 percent, moderate or not uncommon in 17.8 percent, occasional in 10.3 percent, and uncommon in 12.1 percent. For females (in 114 societies), the corresponding percentages were 49.1 percent universal, 16.6 percent moderate, 14.0 occasional, and 20.2 uncommon.

She also rated 141 societies on attitudes toward premarital sex as explicitly expressed in the cultural norms. Premarital sex was rated as "expected, approved; virginity has no value" in 24.1 percent; "tolerated; accepted if discreet" in 20.6 percent; and "mildly disapproved; pressure toward chastity but transgressions are not punished and nonvirginity [is] ignored" in 17.0 percent. The cumulative percentage of these three categories is 61.7 percent, with the stricter, more disapproving categories distributed as follows: "moderately disapproved; virginity valued and token or slight punishment for nonvirginity" in 8.5 percent; "premarital sex disallowed except with bridegroom" in 4.3 percent; and "premarital sex strongly disapproved; virginity required or stated as required (virginity tests, severe reprisals for nonvirginity, e.g., divorce, loss of bride price)" in 25.5 percent. The distribution also appears to be bimodal: a minority of cultures had severe punishment of nonvirginity, while most of the others were quite tolerant of it; milder restrictions with token punishment applied to only 13 percent.

Parent-Offspring Conflict over Arranged Marriage

Descriptions of elopement among the Agta, flirtation among the Ache, and the pairing off of adolescents in other hunter-gatherers suggests something beyond sex. As discussed in Chapter 12, William Jankowiak and Edward Fischer (1992) identified romantic love (not necessarily at adolescence) in 146 out of 166 cultures, and Jankowiak (1995b) collected detailed descriptions from various cultures. For example, on the island of Mangaia in Polynesia, romantic love (*inangaro*) is common and valued (H. Harris 1995). Informants described seven elements of it that we would find familiar: desire for physical and emotional union, idealization, exclusivity, intrusive thinking about the beloved, emotional dependency, changing life priorities, and a powerful sense of empathy. Three adults recalled how they had felt when they fell in love with their spouse. A woman in her forties said, "I didn't want any other men, and I wanted him to look only at me, not at other girls" (H. Harris 1995:115); a man in his sixties said, "It's both feelings and sex. It's when both things come together; that is falling in love" (109); and a still older man recalled, "We felt together—close. When you look at some ladies, you know they are bright and good and beautiful, but someone else keeps coming into your mind" (111).

However, many cultures express a disjunction between romantic love and the marital bond, and it is likely that one purpose of romantic love as a facultative adaptation is to bring about extra-pair copulations or dissolution of the pair bond. The former function operated in the courtly love tradition of medieval Europe, where some historians mistakenly see the dawn of romantic love. Marriages in that subculture, as in most of the world throughout history, were usually "of convenience," but life could be nicer for a married lady when she was courted by a knight who was not her husband. Troubadour poetry and song underscore the power of such involvements, even at a distance. Yet the harsh reality of the iron chastity belts left behind by that culture shows that this kind of courtship was not reliably chaste.

For our purposes the Romeo and Juliet scenario is more relevant. In all cultures with arranged marriages, parent-offspring conflict consistently arose owing to differences in the interests of the parents, family, and lineage, on the one hand, and those of the young man and woman on the other. What Lila Abu-Lughod (1990) evocatively called the "discourse of defiance" sometimes entailed in romantic love might be described as the teenage equivalent of a temper tantrum. The annoyance caused by a young couple who are passionately in love may eventually exhaust the parents' resistance, and in fact many arranged marriages distasteful to the young have been derailed by this strange bit of biology.

Adolescent Sexuality in the Industrial World

Although traditional Western culture has been among the more restrictive with regard to premarital sex, there are many subtleties in the historical patterns. Evidence from historical demography sheds light on adolescent fertility and also takes into account changing frames of marriage and mores. Through church, family, and other records, the lateness of first childbearing in Europe, England, and the United States during the sixteenth to nineteenth centuries has been shown. The case of Shakespeare's Juliet, whose nurse chides her for not marrying and becoming pregnant by age thirteen (as Juliet's mother supposedly had done), was carefully considered by Peter Laslett (1965, 1977). His analysis shows that the nurse was a poor historical demographer, since such a case would have been exceedingly rare in either Juliet's Italy or Shakespeare's England. (Laslett also notes that Juliet's age in Shakespeare's source for the play was eighteen. The change gives the play more urgency and poignancy, but makes the nurse's impatience unrealistic.)

For the modern United States, teenage childbearing increased at least from 1940 to a peak in 1991, with a substantial decline thereafter, attributable in part to declining sexual activity and in part to contraception and elective abortion (Brindis 2006). But the problem persists and may be increasing again (Manlove, Ikramullah, et al. 2009). Attempts have been made to explain it in part by reference to the secular trend toward earlier menarche and to "the myth of an abstinent past" (Cutright 1972). This contrasting view holds that in the past teenagers were often active sexually, but did not become pregnant because of ovarian immaturity. Since the colonial period, premarital pregnancies *not* resulting in out-of-wedlock births ("shotgun marriages") have fluctuated rather than steadily increasing, and they appear to have declined during the early nineteenth century owing to more stringent sexual mores (Vinovskis 1988). Still, in the early United States pregnancies in the late teens were incorporated decisively into a framework of marriage and family, while pregnancies in the early teens were virtually unknown.

There is no doubt that there was a sexual revolution during the late twentieth century (Sulak 2003; American Academy of Pediatrics Committee on Adolescence, Blythe, et al. 2007). There have also been significant reversals since the 1990s. For example, the percentage of boys who became sexually active before age fifteen in the United States declined from 21 to 15 percent between 1995 and 2002, while the corresponding decline for girls was from 19 to 13 percent (Brindis 2006). However, this trend too may be reversing (Manlove, Ikramullah, et al. 2009), and a return to restrictive pre-1960s patterns is unlikely. Whether in the context of romantic relationships (highly conducive to the initiation of sexual activity), long-term friendships ("friends with benefits"), or (less frequently) "hooking up" with casual acquaintances, adolescents are sexually active at very high rates (Giordano, Manning, et al. 2006; Manlove, Franzetta, et al. 2006; Manning, Giordano, et al. 2006; Carver, Joyner, et al. 2003). But Western nations' liberal shift in mores regarding premarital sex has put them in a category that includes at least 60 percent of the traditional societies in the ethnographic record (Broude 1975).

Secular Trends in Growth and Maturation

Evidence that the age at menarche has declined in the United States and Europe for more than a century has been repeatedly summarized (Tanner 1975; Garn 1987; Karlberg 2002; Delemarre-van de Waal 2005; Kaplowitz 2006). There is disagreement about the magnitude of this change, its cross-national variability, and whether and where it may be continuing, but it is real. A very conservative estimate is that the age of menarche has declined two years since the early nineteenth century, and estimates as high as four or five years have been proposed.

This secular trend in growth has been well documented not only for age at menarche but for general growth rate (height and weight) and maximum body size. Such measures lead to an estimate of the rate of acceleration of growth that confirms the estimate derived from studies of menarche. The age of peak height velocity (PHV) in the pubertal height spurt, as indicated by changing height-for-age curves, has decreased at the same rate as the age at menarche: four months per decade (Meredith 1978; Garn 1987). This is of particular interest since studies of age at menarche are subject to methodological challenges that do not apply to studies of height growth.

The secular trend stopped by the midtwentieth century in some populations, notably in Oslo, Norway, where the earliest historical data come from; in London, England; and among old Americans—those whose families had been in the United States for generations—at Harvard and eastern women's colleges, but continued in some other populations, especially in the less developed countries (Zhen-Wang and Cheng-Ye 2005; Ayatollahi, Pourahmad, et al. 2006; Laska-Mierzejewska and Olszewska 2006). In northern Europe the trend appears to have stabilized at just about thirteen years and in New England at a slightly younger age. A large sample of girls in Newton, Massachusetts, menstruated for the first time at 12.65 years, no earlier than the menarcheal age of their mothers (Zacharias and Wurtman 1968; Zacharias, Rand, et al. 1976). In the wealthiest New England families the secular acceleration of growth apparently ended early in the twentieth century. However, there are recent indications that the trend may be resuming in the United States and among immigrants to Europe (Parent, Teilmann, et al. 2003; Herman-Giddens 2006), possibly owing to the new secular trend toward childhood obesity (Dunger, Ahmed, et al. 2005; Komlos and Breitfelder 2008) and/or a combination of low birth weight and subsequent rapid growth (Barker 2004).

Despite some exceptions, it now seems that the *mean* age at menarche may not drop much below twelve and a half to thirteen years in most of the world and that this may be an average biological lower limit in our species. But in the youngest estimates for the United States the mean was about fourteen during the early twentieth century. Norway, Germany, Finland, and Sweden, for which the data are much better, show a steady decline beginning in the midnineteenth century, when the average age at menarche for various populations was fifteen, sixteen, or perhaps even seventeen years. Conservative estimates thus leave room for a two- or three-year decline in northern Europe over 150 years.

There is considerable variation within populations, with the 95 percent confidence intervals bracketing an eight-year period. Thus reports of menarche at age twelve and pregnancy at thirteen in some girls occur even with the highest estimated mean menarcheal ages. In the U.S. population menarche at age eight or nine and pregnancy at ten or eleven are no longer considered to reflect hormonal pathology. Again, one can be skeptical about historical data on age at menarche, but the average age at which any given height is reached has also been dropping about four months per decade, and this too leveled off in populations with a long history of better resources.

The reasons for growth acceleration are not agreed upon, although improving nutrition and infectious disease control are clearly factors. Rose Frisch accumulated evidence that nutritional improvements led to increases in fatness at each age, which in turn led to earlier menarche through the production of physiologically potent estrogens in fat cells (Frisch and Revelle 1970, 1971; Frisch 2004). A delay in menarche due to intense and prolonged exercise was proven in young dancers and athletes (Frisch, Wyshak, et al. 1980), and this effect may work through the reduction of body fat or through hormonal stress responses. Other hypotheses of declining age at menarche include improved public health and medical care in controlling chronic childhood diseases; changes in environmental lighting; increased consumption of refined carbohydrates; increased stimulation during infancy; father absence or other stresses of childhood; genetic changes resulting from outbreeding or natural selection; and, most recently, hormones in industrial meats. Urban living is definitely associated with earlier maturation, an effect seen throughout the world (Eveleth and Tanner 1976), most likely because urbanization is accompanied by changes in other proposed causes.

We must also consider adolescent subfertility (Montagu 1957). The uneven quality of menstrual cycles during the first years after menarche has been repeatedly documented, as has a high proportion of anovulatory cycles (Adams Hillard 2006; American Academy of Pediatrics Committee on Adolescence, American College of Obstetricians and Gynecologists Committee on Adolescent Health, et al. 2006). Thus first fecundity lags behind menarche by one or more years in many populations, especially in less developed countries. In view of the effect of nutrition and exercise on fecundity, it is likely that this infertile period has decreased in recent generations, in parallel with declining age at menarche. If so, then mean age at first fecundity (or first consistent ovulation) would have dropped even more markedly and rapidly than age at menarche.

Secular Trends and Adolescent Behavior

Anthropological and historical-demographic literatures refute the notion that in primitive societies young people entered the childbearing phase of reproductive life in the early teenage years. Despite nonrestrictive rules and apparent high levels of adolescent sexual activity, early pregnancy and childbearing were uncommon. One explanation for this is probably that in most such societies puberty was followed by a significant period of adolescent subfertility.

A girl menstruating at fifteen and becoming a mother by nineteen could, 150 years ago, take her place as an adult in a relatively simple society that supported her through the firm institutions of marriage and family. Today a girl menstruating at twelve and becoming a mother by fourteen, usually unmarried and likely to remain so, is a child in a much more complex and unforgiving adult world. Has the growth of white matter in the cortex known to occur in adolescence lagged behind the physical and hormonal changes? We do not know. But even acceleration of all systems of maturation could not compensate for three fewer years of experience where maturity and knowledge are essential.

The secular trend is a large, nongenetic biological change in our species, with implications for psychology, education, and law. The secular trend and early reproductive activity do not need to be tightly coupled historically in order to be causally related. Cultures may resist change in sexual mores and behavior against the background of the long-term biological transformation, and then perhaps change quite rapidly. Nonlinear models based on catastrophe or emergence theory would accommodate that sequence. Other changes in behavior include rising school-age suicide, alcoholism, and substance abuse. Despite the complexity of these trends, they must be understood against the background of the prior secular change in the organization of human maturation. Evolutionary developmental psychologists rightly argue that there are environmental circumstances in which it is adaptive to grow up faster and begin reproducing earlier, but such arguments do not remove the need to make judgments about the kind of adolescence we want for them. In fact, they may help us to design it.

Interlude 6

Thinking about Incest Avoidance and Taboos

Theory notwithstanding, most men harbor no wish to sleep with their mothers, nor women with their fathers, nor (with instructive exceptions) brothers and sisters with each other. Seductions sometimes occur, of course, as do forcible encounters, but these are complex and often tragic outcomes. A very small number of cultures historically allowed brother-sister marriage, and they show that accepted incest is compatible with human nature, but these are rare cases in special circumstances. Yet the availability of family members as possible partners, the already existing intimacy and trust, and the fact that sex is private would seem to favor a higher incidence of incest than we see. There are multiple reasons why we do not, at multiple levels of explanation.

Classical geneticists predicted that incest would be avoided in most sexually reproducing species to prevent the birth of defective individuals; homozygous recessive genotypes occur more frequently in matings of close relatives (Wright 1922, 1932). This avoidance is seen far outside the human species (Hoogland 1982; Clutton-Brock 1989; Moore 1993; Chapais 2008). During courtship a great many types of animals recognize and rule out close kin, despite the individual's need to mate with its own species. In insects and some vertebrates, recognition is by pheromones (Greenberg 1979; Manning, Wakeland, et al. 1992); kin evidently do not smell sexy. But learning during development plays a role even in the ground squirrel (Holmes and Sherman 1982; Pfennig and Sherman 1995), and in humans the lack of evidence that pheromones are involved seems to require different explanations.

These have been sought for over a century; explaining the incest taboo has been a central challenge for anthropology since the field was founded. But the initial step was to see that incest avoidance and incest taboos are very different things. First, if incest avoidance were completely effective, taboos would be superfluous; cultures, codes, and religions prohibit things that some people tend to do (Fox 1980). Second, although all cultures share a core of forbidden matings—mother-son and father-daughter—prohibition of other matings with relatives is highly variable (Fox 1967; Lévi-Strauss 1969). Almost all cultures prohibit brother-sister mating, but among ancient Egyptian royalty, in traditional Peru and Hawaii, and in a few historical contexts in Europe, brother-sister marriage was allowed or even preferred (Fox 1962; Hopkins 1980; Bixler 1982a, 1982b; Goody 1983). In the absence of such acceptance, the psychological consequences of incest may be dire (Meiselman 1978; Lieberman 2009), but evidently such consequences can be prevented in some culturally determined situations.

Yet far from being limited to close relatives, where inbreeding is clearly a risk, human incest avoidance drives elaborate systems of sex and marriage rules that pervade the social orders of simple societies (Fox 1980; van den Berghe 1987; Thornhill 1991; Chapais 2008). Even the brother-sister taboo is not universal, and when we widen the circle of relatedness, we find that many cultures not only permit first-cousin marriage but permit or prefer some kinds (for example, a man and his father's sister's daughter) while prohibiting and punishing others as incestuous. Degree of relatedness alone cannot explain this cross-cultural variation, which has to do with the structure and alliances of lineages and clans. The rules force mates to move between bands or clans over generations, and the marital unions become the bases of ongoing economic exchange and mutual military aid—or at least a temporary peace (Lévi-Strauss 1969). However, psychologically these wider "incest" rules probably build on avoidance tendencies, even disgust, between close relatives (Lieberman 2009), and there may even be a reluctance that has to be overcome in order for first cousins to marry (Pastner 1986).

So cultures reinforce basic incest avoidance and extend it in very different ways. But what accounts for the basic avoidance if not pheromones? One view has been that early in human history cultures accumulated experience with homozygous recessive conditions and their defects and linked those to incestuous mating (Durham 1991). This may be part of the explanation, although even brother-sister marriage leads to obvious defects only some of the time, and cousin marriage does so infrequently enough to make this explanation less plausible in a small-scale society, especially since most birth defects would have other causes.

If early cultures did not consciously recognize the consequences of incest, then what led to its avoidance? A century ago Edward Westermarck (1922) hypothesized that incest avoidance results from proximity and intimacy during childhood, which he believed suppresses later sexual urges. This came to be known as the "familiarity breeds contempt" hypothesis (McCabe 1983; Spain 1987), although "familiarity breeds boredom" might be better. We now know this mechanism works.

Arthur Wolf (1970, 1995) studied the consequences of *sim-pua,* or "minor" marriage, in traditional China. Two children were betrothed to each other, and the infant girl went to live in the home of her future husband, where they were raised together. In a sample of more than twenty thousand marriages, sim-pua unions had lower fertility and a much higher rate of divorce, adultery, and resort to prostitutes than did "major" arranged marriages in which the wife went to the husband's household as an adult, with similar but smaller differences between sim-pua marriages and marriages by personal choice. In other words, it's not the arranging, it's the childhood contact. Furthermore, excellent records revealed a sensitive period for the effect: contact had to start before age five years but was more effective if it began before thirty months. Not surprisingly, given this framework, the effect is stronger for the younger of the two future spouses than for the elder (Lieberman 2009).

We might also expect to see an impact among peers raised together in traditional Israeli kibbutzim where communal rearing in a children's house was the norm (Talmon 1964; Shepher 1971, 1983). Personal choice governs most of their marriages, yet of 2,769 marriages of kibbutz children who grew up in children's houses, *none* was between two members of the same peer cohort, and no sexual activity was noted within a cohort despite liberal attitudes toward sexuality; these children seemed to see each other as siblings. Since friendships in Western cultures may morph into romance, it is remarkable that *no* kibbutz children became maritally, or even sexually, interested in their old childhood housemates.

Both lines of research support Westermarck's claim, and this is a paradigmatic instance of how natural selection shapes human behavior (Wilson 1978; Lumsden and Wilson 1981). All sexually reproducing species avoid the genetic consequences of close incestuous mating to some degree, since individuals who did not were selected against, but the avoidance relies on different mechanisms in different species. Some have off-putting pheromones that require no experience; genes produce kin-specific molecular signals that suppress sex. Others, like the ground squirrel, have somewhat weaker innate triggers supported by experience. Human incest avoidance, however, relies largely on that sensitive period for close social contact, supported by the uniquely human, conscious recognition of the adverse genetic consequences of close inbreeding and the corresponding intention to avoid them. This knowledge, passed on from parent to offspring, enters into a form of gene-culture coevolution (Lumsden and Wilson 1981; Durham 1991).

We humans may have a pheromonal dimension in our lives, but our incest avoidance depends mainly on experience. The Wolf sensitive period effect is a specific experience-dependent damping of sexual attraction and function in individuals who have been in close proximity from before age five, especially before age two and a half. This mechanism is clearly not in itself enough, since Chinese "minor" marriages often worked and produced offspring, so other factors must be in play. The most obvious explanation is also the most plausible: cultural prohibition of the strongest kind supports an already established psychological leaning. In Chinese minor marriage, however, the cultural imperative went the other way: since these were not really brother-sister matings, a strong cultural sanction worked against the sensitive period effect and partly overcame it.

In this analysis the purposes and methods of psychological anthropology and sociobiology are joined. The direct role of genes is obscure and often easily overcome by changes in the developmental niche, yet in the normal expectable environment brothers and sisters grow up together while unrelated boys and girls do not, so the demands of natural selection are usually satisfied.

It is difficult to overstate the importance of this example in understanding the evolution of human behavior. The selective forces on all sexually reproducing species are in this case roughly the same: close incestuous mating produces fewer viable offspring and is diminished by negative selection. However, invertebrates and some vertebrates with little learning ability and rapid development evolved a genetically coded pheromonal mechanism. Some mammals, such as ground squirrels, were selected to rely on pheromones with the support of developmental experience. And humans were selected to rely on developmental experience during a sensitive period, reinforced by culturally imposed, language-borne sanctions and taboos.

The selective pressure and the adaptive result are similar in all three kinds of animals, but—as with the flight of moths, bats, birds, and (with the aid of technology) humans—the intervening mechanism is different, and because of this, the human adaptation for incest avoidance can be and has been overridden, although not without cost. A vital, evolved behavioral pattern is not directly coded in human genes, although in other species, under the same selective pressures, it is. And for more distant relatives, humans have applied their usual creative approach to build on what has been selected for, shaping it to the service of other adaptations. The invocation of a selective force, an evolutionary process, and parallel adaptations in different species do not imply deterministic genetic control.

21

Stress and Resilience in the Changing Family

Life is stress, and so is evolution. It is stressful to be born, survive childhood, learn, grow, find and compete for food, water, and mates, be ill, age, and die. For hundreds of millions of years, stress was ubiquitous for all species ancestral to us; it is of the essence of evolution by natural selection and close to the essence of life itself. Since humans evolved under conditions far more stressful than those that most post-industrial people live in, we must be genetically adapted to cope. Stress responses "are normally adaptive and improve the chances of the individual for survival," and stress-responsive behavior includes positive as well as negative features, such as "increased arousal, alertness, and vigilance, improved cognition, and focused attention, as well as euphoria or dysphoria" (Chrousos 1998:312). Or as the psychiatrist George Vaillant (1977:374) said, "It is not stress that kills us, it is effective adaptation to stress that permits us to live."

Basic Stress Physiology

In mammalian biology, many physical and psychological stresses cause a similar response cascade, often called the general adaptation syndrome (GAS) (Selye 1976, 1998; Sapolsky 1992). One dimension is sympathetic nervous system (SNS) activation—the "fight or flight" response, known by the 1920s (Cannon 1915/1963, 1927). The SNS raises heart and breathing rates; mobilizes blood glucose; dilates arteries in the heart, lung, and voluntary muscles while constricting those of digestion and reproduction; and triggers sweating, pupil dilation, urination, and sensory heightening. SNS activation has two components: a faster one mediated by norepinephrine (NE) neurons, and a slower one mediated by adrenaline (epinephrine, E) secretion from the adrenal medulla. Adrenaline is a hormone, but the adrenal medulla originates in evolution and embryology from NE neurons; adrenaline is a modified form of NE that is hundreds of times more potent. NE and some adrenaline neurons within the brain are also activated under stress.

The second cascade of the stress response, added to Cannon's model by Hans Selye (1936, 1946, 1976), mobilizes the hypothalamic-pituitary-adrenal (HPA) axis. When stress activates the amygdala, hippocampus, and other limbic structures, they cause hypothalamic neurons to secrete corticotropin-releasing factor (CRF; or hormone, CRH) through pituitary portal vessels into the anterior pituitary, where it stimulates synthesis and release of corticotropin (adrenocorticotropic hormone, ACTH). ACTH in turn promotes synthesis and release of glucocorticoids like cortisol from the adrenal cortex. Like adrenaline, cortisol releases blood glucose to energize fight, flight, and other responses, with both positive and negative effects (Erickson, Drevets, et al. 2003). In chronic stress, glucocorticoids damage hippocampal and other neurons (Sapolsky 1992). But glucocorticoids can also accelerate healing; the HPA axis does not just activate coping, it also helps modulate and terminate stress responses.

The two main stress response cascades are synergistic. In vertebrate evolution the tissues we know as the adrenal cortex and the adrenal medulla moved anatomically (Norris 1997). They are juxtaposed in lower vertebrates, but in mammals the medulla ("pith" or "marrow") became surrounded by the cortex ("bark"). The cortex supplies high concentrations of cortisol to the medulla, where it activates the enzyme that converts NE to adrenaline, greatly augmenting the stress response. In addition, brain NE neurons cause the hypothalamus to release CRF.

Stress in Infancy and Childhood

The physiology underscores what is clear to all who study natural history: stress and coping are ubiquitous. For the fetus, infant, and child, stresses come from famine, natural disasters, infectious disease, predators, accidental injury, and the loss of loved ones, but also from routine events like noise, heat, cold, rain, insect bites, even the dark of night. Every change in the stimulus envelope is a kind of stress, or at least a challenge, to which the organism must respond physiologically and psychologically using innate programs shaped by individual experience. Differential activation of the complementary sympathetic (fight or flight) and parasympathetic branches of the autonomic nervous system is measurable in six-month-old infants and is stable across individuals from six to twelve months (Alkon, Lippert, et al. 2006).

Because of the competition entailed by natural (including sexual) selection, individuals frequently impose stress on each other beyond other environmental ones (Sapolsky, Romero, et al. 2000; Sapolsky 2001; Konner 2002). Fetuses must struggle with mothers who have lives of their own and might reproduce again; then the two must go through labor and delivery, the timing of which cannot be ideal for both. Parent-offspring conflict pervades this and later interactions; if the mother becomes pregnant again, the siblings are immediately in competition with each other. Siblicide in frogs, birds, and princes shows the extreme of this competition, as pedicide and abandonment show the extreme of parent-offspring conflict. Routine family conflict can lead to separation or abuse. Competition among nonkin is a basis of natural selection, and this too is a life-or-death matter, with fighting over mates and other resources taking center stage in and after adolescence. There are also innumerable stresses, risks, and minor hurts associated with children's subsistence, play, and learning.

As societies become more complex and stratified, competition becomes systematic imposition and exploitation, and some groups become slaves or servants to others. The deliberate imposition of stress is routine in war and peace. For children in complex societies, even in the best of times, the stresses of agricultural, household, or factory work—and in the modern context the stresses of formal schooling and team sports—are part of growing up.

Attention is the minimal organismal response to a stimulus change, followed by arousal and possibly frustration, fear, or pain (Davis and Whalen 2001; Ursin and Kaada 1960; Zald 2003). These reactions are mediated mainly subcortically (Liddell, Brown, et al. 2005; Ohman 2005). Parallel to this continuum of arousal is a continuum of action—or coping. Because so many stimuli cause at least mild and transient discomfort, coping is often rewarding and sometimes exhilarating—the smile of recognitory assimilation in the three-month-old being the simplest human example (Super, Kagan, et al. 1972; Zelazo and Komer 1971), the joy or relief after winning a game or passing a test being another. This is how we learn, grow psychologically, and become less tied to genetic influence, and how higher organisms transcend the instinctive behavior of some lower ones.

The notion that a goal for children should be the absence of stress contradicts all we know about evolution and the logic of our own experience. Selye used the term "general adaptation syndrome" (GAS) to suggest that far from being an abnormal response to an abnormal event, it is at the heart of adaptation itself. He recognized that some stress is undesirable and can impair future function (*dis*tress), while other stress is normal, brings about effective coping, and enhances long-term function (*eu*stress). It is not always clear where to draw the line (Selye 1975), but thinking of stress as potentially positive is crucial.

Stress in Early Life as a Signal for Facultative Adaptation

Evolutionary theorists are increasingly aware that stress in early life can be not just deprivation but a signal that *this* environment requires a different developmental trajectory (Belsky, Steinberg, et al. 1991; Chisholm 1993; Boyce and Ellis 2005). Evidence for such facultative adaptation comes from the Barker effect—certain outcomes of nutritional and other kinds of stress on the fetus in epidemiological and experimental studies (Barker, Osmond, et al. 1989; Barker 2004). Adults stressed in fetal life (as evidenced by low birth weight) but well nourished or overly nourished later in development were at greater risk for the metabolic syndrome and its consequences—diabetes, heart disease, and other chronic conditions. David Barker advanced the idea of a "survival phenotype"—a streamlined, smaller version of the human body designed to survive and adapt under sparse nutrition and other stresses.

Experiments show that the phenotype can be induced by undernutrition, deprivation of micronutrients, small maternal size, or cortisol crossing the placenta. Cortisol may signal the fetus to produce the efficient phenotype under many conditions of maternal stress—a physiological mechanism for a major prediction of life history theory (Worthman and Kuzara 2005). The small size but physical fitness of hunter-gatherers on all continents underscores the possible value of such a facultative adaptation (Eaton, Konner, et al. 1988). In this view, as in the older "thrifty genotype" hypothesis, the streamlined body is ideally adapted to sparse and stressful conditions but poorly prepared for abundant modern environments.

It is also increasingly clear that early postnatal stresses may act as signals for facultative adaptation: "A principal evolutionary function of early experience—the first 5–7 years of life—is to induce in the child an understanding of the availability and predictability of resources (broadly defined) in the environment, of the trustworthiness of others, and of the enduringness of close interpersonal relationships, all of which will affect how the developing person apportions reproductive effort" (Belsky, Steinberg, et al. 1991:650). This roughly corresponds, within the human species, to the concept of fast and slow life histories between species (Chapter 2) as successively advanced by Pianka (1970), Charnov (1993), and Sibly and Brown (2007).

According to this perspective, children growing up in unstable environments will mature faster, make fewer and weaker commitments, and themselves invest less in parenting as adults than children raised in stable and nurturing environments; both are appropriate adaptive strategies in their respective environments, and the developmental system is a facultative adaptation. Belsky and his colleagues specifically suggest that a girl growing up in an unstable environment—including father absence or erratic male presence—should take advantage of reproductive opportunities during her own adolescence, when she still has supportive female kin, since she knows that in her world a good man is hard to find.

Bruce Ellis (2004) reviewed the evidence pertaining to this and other models of pubertal timing and found the theory compelling but the evidence wanting. It is impossible with currently available data to tease early environment away from its correlations with genes, and the studies controlling for genes by using twin and adoption models are few in number. As for the more conventional environmental explanations of the timing of puberty (see Chapters 12 and 20), such as nutrition, illness, and continuous stress, these are *mechanisms* of facultative adaptation; they operate at a different level of explanation and do not contradict the evolutionary hypotheses. However, Ellis (2004) has put forward a model that promises to reconcile many seemingly contradictory claims and findings. According to this model, both low and high levels of stress result in higher stress-reactivity and later puberty, with heightened HPA-axis activity delaying puberty at both ends of the early-stress continuum (see also Boyce and Ellis 2005; Ellis and Essex 2007). Intermediate levels of early psychosocial stress are the path to low stress-reactivity and accelerated maturation. This model has the virtue of explaining apparent contradictions, since a study comparing children at different points in the declining part of the curve would come to the opposite conclusion of a study comparing children in the rising part. But as Ellis acknowledges, the model has yet to be properly tested.

Stress and Resilience on the Island of Dominica

The anthropologist Mark Flinn and his colleagues have carried out remarkable research on the Caribbean island of Dominica, a longitudinal field study of children's and parents' behavior, but with an emphasis on the effects of stress. The group's systematic behavioral observations and daily records of hundreds of children over decades would have been a major contribution to the anthropology of childhood, but with anthropometric measures, regular assessments of health and illness, and more than thirty thousand salivary cortisol measurements, the study is also a major contribution to our knowledge of stress and health in childhood —the causal arrow that goes *from* culture *to* biology.

These observations reveal, sometimes counterintuitively, how children subjectively perceive stress. Flinn (2006) has shown that strenuous work does indeed raise children's cortisol levels, but that they return to a baseline appropriate to the time of day within two hours. For example, repeated salivary cortisol measures for a twelve-year-old boy from early morning to evening showed an elevation of about 50 percent above the daily curve when he helped his father by carrying more than fifty pounds of wood at around 10:00 AM, but this was a transient increase. It was also highly adaptive, since the function of cortisol elevation is to mobilize blood glucose for action or work (Selye's GAS).

Cortisol elevations would also seem to be adaptive in response to childhood illnesses—the body must fight off invading organisms, rebuild damaged tissues, and prepare for catchup growth and reentry into normal life—and indeed, all of these responses do occur. However, there is also a correlation between children's average cortisol when *well* and the percentage of days that they are ill, usually with upper respiratory infections (URIs). This may be because the number of days ill reflects unstable family conditions, which appear to chronically elevate cortisol and suppress immune function (Flinn and England 1997). The chain of events from a particular psychosocial stress through elevated cortisol to a URI after several days was traced in some children, so cortisol elevation in response to this type of stress is not always adaptive.

Flinn studied a striking example of such a stress. A total of five full or half-siblings were living with their mother and father/stepfather when a serious marital quarrel occurred, causing a marked increase in the children's cortisol. The husband left home for several weeks, during which time the children's cortisol levels returned to baseline, only to spike again upon his return. This is a common reaction to psychosocial stress in Dominica. Traumatic family events, defined as change of residence of child or parent, serious fights and quarrels, punishment, or "shame," caused substantially elevated cortisol in 42 percent of cases; conversely, 19 percent of all marked cortisol elevations were temporally associated with such events (Flinn 2006).

The experience of this sibship can be generalized to a large sample of children in the community over many years. Step-offspring have higher mean cortisol levels and more days of illness than genetic offspring living in the same household (Flinn and England 1997). For days of illness, there is an interaction between genetic relatedness and socioeconomic status, such that the genetic versus step-offspring difference was greatest for the poorest one-third of the families. Also, children living with stepfathers and half-sibs, with distant relatives, or with single mothers and no other kin have higher cortisol levels than children living in families based on close genetic kin, and there were more days ill for a given cortisol level in children who had experienced significant growth disruption during the first two years of life.

Flinn and his colleagues went on to show that children who have experienced early trauma (ET) have higher average (baseline) levels of cortisol and approximately twice as much morbidity as children with no ET, and they reasoned (based on animal models) that diminished cortisol receptor function in the hippocampus blunts the negative feedback needed to keep normal children's cortisol level low. However, the kind of trauma matters; early social trauma (such as abuse or family instability) has a stronger effect than nonsocial trauma (such as a severe hurricane). Children with early trauma show a marked divergence between cortisol elevation under nonsocial versus social stress (almost double for the latter); other children do not, but both groups show a comparable return to baseline over an hour or two (Flinn 2006).

Mortality, Attachment, and Loss

Consider the stresses that human infants and children have faced. One obvious index of the health and quality of life of the young is of course mortality, which is indicative of untold suffering by the children and parents who survive, both because of the loss and because of the bouts of illness and trauma they also have endured or narrowly escaped.

High infant and child mortality was universal among not just hunter-gatherers but all human populations until about two centuries ago, when death rates began a decline that continues. Even leaving aside decimating epidemics, about half of live-born infants died by age fifteen, primarily owing to microbes and parasites. Some 10 to 20 percent died in the first year, half of those in the first few weeks. Even in the United States in the late nineteenth century at least 20 percent died before age five, and the rate was higher among the poor (Preston and Haines 1991). Throughout history in much of the world significant, though usually much smaller, numbers of neonates were killed to protect them from the inevitable suffering of prolonged deaths; to avoid compromising the survival of older siblings, who often died shortly after weaning; or even to manipulate sex ratios, usually by preferential care for males (Dickeman 1975; Scrimshaw 1984; Hrdy 1999).

Mothers are designed by evolution not to produce idealized infant and child care under all circumstances but adaptively to adjust their investment depending on the offspring and the circumstances (Interlude 4). They expel their newborns before they are really social, which cushions mothers against loss, and this sets the tone for parent-offspring conflict throughout the child's life. Parents withhold investment or distribute it unevenly in some situations, and in others they cut their losses by abandonment or infanticide. So it is not surprising that they may try to protect themselves against grief by withholding love in some circumstances.

Yet grief is common, even in nonhumans. Wild chacma baboons (*Papio hamadryas ursinus*) appear to grieve when they lose close kin; like humans, they experience an elevation of glucocorticoids and also exhibit an increase of mutual grooming with other survivors in the wake of such a loss (Engh, Beehner, et al. 2006). In an account by Jane Good-all (1971), when an old chimpanzee female lost a daughter, she and her son became closer. The young male returned to nursing, although his mother was now elderly and he was six years old. When he was eight, no longer nursing but still close to his mother, she became ill and died. He stayed with the body, inactive, looking dejected, and not doing anything else. When the body was taken for autopsy, he returned repeatedly to the spot, and he died a few days later. An autopsy showed peritonitis, but it seemed possible that his dejection had weakened his defenses.

Theory helps us to understand when and how grief occurs, beginning with Ronald Fisher's (1958:27) measure of reproductive value (RV): "We may ask . . . about persons of any chosen age, what is the present value of their future offspring[?] . . . To what extent will persons of this age, on the average, contribute to the ancestry of future generations?" Graphically, RV traces the amount already invested in the offspring and their reproductive potential, discounted by the risk that they will die without issue. RV is negligible in old age, but it is also low in infancy. With growth and development, investment by the parent accumulates and risk of death declines, so peak RV is reached at the onset of reproductive capability. The typical mortality curve is the inverse of the RV curve—lowest in adolescence and young adulthood.

Fisher's equation was first presented in the late 1920s; eight decades later the accumulated empirical evidence on parental grief matched the curve very well. As summarized by John Archer (1999:204), "Putting together the findings at specific ages indicates a steady overall increase in the intensity of grief from early pregnancy loss to loss of an adult. . . . An increase in grief with age is consistent with the increase in reproductive value of the offspring from conception to early adulthood, and the decline in parental reproductive value." Also as predicted by the RV curve, mothers grieve more than fathers in the early years, but the difference diminishes as the child grows and the parents' respective investments partly converge.

But why should there be grief? This potentially debilitating reaction detracts from the parent's investment in other offspring. But as with cephalopelvic disproportion, cancer, and many other aspects of our biology, grief results from a compromise: it is a cost of love, without which neither we nor our expensive, dependent offspring would have any value at all. Still, the equation predicts that grief will be minimal in circumstances where, like those of the Brazilian mothers studied by Nancy Scheper-Hughes, the chance of survival is minimal irrespective of the investment; it also predicts that the mother will withdraw when there is no hope.

Mortality and Loss in Aka and Ngandu Adolescents

Although untimely death is ubiquitous in all premodern cultures, few studies have focused on how children react. Bonnie Hewlett's (2005) research compared the experience of loss in two cultures in the Central African Republic, Aka hunter-gatherers and their farmer neighbors, the Ngandu. She interviewed twenty youngsters in each culture between the ages of ten and twenty, roughly balanced for sex and age. Because the Ngandu village had a larger and more densely settled population, it had at least one death a week during the study period. A twelve-year-old boy who had lost his mother and father in the same week four years before being interviewed was cared for by his mother's younger brother; the boy said, "He is like a father, he gives me food and I live with him and we go fishing." But when asked if he had someone like a mother, he started to cry, saying, "No one is like a mother to me, I miss my mother and still feel very sad because my mother and father are gone" (Hewlett 2005:326). The children often said that their grief was lessened by the care they got from others and that the spirit of the one they lost returned to watch over them. They also believed that children who died before age ten would be reborn, and they would look for signs of that in infants. The losses made them think, with varying degrees of emotion and acceptance, about their own deaths, which they thought could come at any time.

Aka adolescents remembered significantly fewer deaths, and their culture had less elaborate grieving rituals, but they spoke about their grief and consolation in terms very similar to those used by the Ngandu children. One girl felt the loss of her mother's older brother very deeply because he loved her a lot and was "happy and good with all the children." She said she was "sad and I had a feeling of love for the others and I was afraid that they would die too. I was sad for a long time and then I sang and danced again, but at first it was hard because my mother and father said that with death it is finished, it is goodbye for all my life" (Hewlett 2005:330). They all mentioned the family members who consoled them, and some mentioned ritual, singing, dancing, and explanations for the deaths as having helped them recover. There were minor cultural differences, but the prevailing sense was one of common experience.

It is difficult for us in the modern post-industrial world to fathom the experience of children who must grow up with such losses and yet carry on. As with parental sentiments over the loss of children, these youngsters grieved deeply and were not inured to death. Yet they recovered, and in similar ways, regardless of culture or gender. A fourteen-year-old Aka boy said that the sadness "stays and comes again and again." He also said that "when others get sick, I am afraid and I guard my brothers and sisters well" (Hewlett 2005:330). An older Aka boy concluded the interview: "Life is not always good with so much death, it is difficult sometimes, but it is good to be alive . . . but then there is death and life is finished" (333). These young people had understood the finality of death for years, but still struggled with the fact that it had to come to them—as which of us does not. Still another Aka boy concluded, "I can live in the forest, I eat well, I have family, life is good" (333).

Stress and Resilience in Exceptional Situations

At the margins of "normal" human life, which is stressful enough in itself, are situations that we view as completely abnormal, yet all too many children must adapt to them. Consider three.

Children on the Street

The number of children living on the street throughout the world is conservatively estimated in the tens of millions, but it is difficult to assess (Panter-Brick and Smith 2000). Equally difficult is the systematic study of these children, but Catherine Panter-Brick, a British biological anthropologist, has defied the odds and developed careful methods to assess their health, growth, and well-being (Panter-Brick, Todd, et al. 1996; Panter-Brick 2002; Worthman and Panter-Brick 2008). Among other things she has said in effect, *Homeless children, dreadful, yes, but compared to what? The Euro-American middle class?* That comparison is informative in its way, and of course quite dismal. But if you compare homeless children to children who are living the lives the homeless ones would otherwise be leading, you get some surprising results.

Studying four appropriately matched groups of boys in Nepal—one sampled from the street, the others from poor villages, poor urban settings, or poor schools—showed that the two non-school-based samples had higher frequencies of severe or moderate stunting and lower proportions of mild stunting or adequate growth than did the homeless boys. In fact, the latter were twice as likely to be in the "adequate" or "mild" categories than two samples of poor children, although half as likely to be in those categories as the school-based sample. Their smaller size might have been a facultative adaptation to their poor conditions, one they could make because of their independence. As Panter-Brick (2002:147) put it after many years of research, "Assessment that assigns street children to a category 'at risk' should not overshadow helpful analytical approaches focusing on children's resiliency and long-term career life prospects."

Ethnographic information reveals some remarkable psychosocial adaptations as well. A film account of a group of *abandonados*—eleven boys ranging in age from eight to twelve, living in the streets of Bogota, Colombia—documented a remarkable cohesiveness (Haines-Stiles and Montagnon 1992c). The younger boys were best at begging, while the older boys focused on group protection and defense. At night, in the *gallada,* or roost, the boys slept on dirty mattresses and blankets in the partial shelter of a building. On one occasion, when the youngest boy cried in his sleep, the older boy next to him reached out to touch his hair and stroked it until he stopped crying. Such tenderness and loyalty are impressive, but these street children turned a different face to outsiders. They stole and sold sex when begging and menial work did not meet their needs, and they behaved territorially with respect to other groups of street children, often exhibiting extreme bigotry in confrontations. Nevertheless, in their situation, all these responses may have been adaptive, and similar patterns are seen throughout the Americas and the world (Mickelson 2000).

Children at War

Extensive research on the experience and possible psychopathology of children exposed to extreme environments has concluded that they are both vulnerable and resilient (Williams 2006). The psychological damage is proportional to the amount, severity, and length of exposure, but there are controversies about whether some responses can be viewed as normal suffering and how much of the lingering effects are pathological —questions with serious implications for intervention, which can be unfounded and ill advised (Barenbaum, Ruchkin, et al. 2004). But the cognitive and cultural frameworks of the trauma and recovery are important, and "the amount of suffering has to do with . . . the child's psychobiological makeup, the disruption of the family unit, the breakdown of community, and the ameliorating effects of culture" (Elbedour, ten Bensel, et al. 1993:805).

Children are not just passive victims; they actively adapt, and sometimes the consequences are dire without being pathological. One review warned, "There is a real danger when order has disintegrated that frightened children will act to protect themselves by affiliating with terrorists, forming delinquent gangs, or emulating the violent behavior of adults. The undermining of civil society may be more of a threat to children's mental health in the long term than the 'distant trauma' itself" (Pine, Costello, et al. 2005:1789). This observation places the distant trauma in perspective and emphasizes the ongoing stress and suffering, but by depicting children as actors in their own lives it also calls attention to adaptation.

Researchers studying child survivors of the Holocaust and their children and grandchildren present a similar picture of vulnerability combined with resilience. It is clear that symptoms of post-traumatic stress are higher in child survivors than in carefully matched controls even half a century later. These women were four to fourteen years old during World War II, and they had suffered terrible trauma, including the loss of both parents (among other family members), before immigrating as orphans to Israel. Yet the differences vary between 0.2 and 0.9 standard deviations (the high-end differences being in "avoidance" and "unresolved state of mind"), meaning that the great majority were in the normal range. "Holocaust survivors (now grandmothers) showed more signs of traumatic stress and more lack of resolution of trauma than comparison subjects, but they were not impaired in general adaptation" (Sagi-Schwartz, van Ijzendoorn, et al. 2003:1086). Holocaust memoirs offer a similar picture of recovery and adaptation (Schneider 2007). As for "echo" effects on the children of Holocaust survivors, meta-analysis of thirty-one studies showed little or no evidence for such effects (van Ijzendoorn, Bakermans-Kranenburg, et al. 2003).

What about children who not only experience the trauma of war but are conscripted as child soldiers, a growing problem throughout the world? Some studies of adolescents who have survived this experience showed no specific association between it and post-traumatic stress (Bayer, Klasen, et al. 2007), but some careful studies have revealed lasting effects (Derluyn, Broekaert, et al. 2004). Individual and contextual variables are important. A study in Nepal comparing former child soldiers with matched controls who had suffered war trauma but had not been conscripted showed that the former child soldiers had higher levels of depression and (for girls) post-traumatic stress disorder (PTSD) than did the comparison group (Kohrt, Jordans, et al. 2008). But a study of elderly German men who had served as child soldiers in World War II found them to have less evidence of PTSD than counterparts who experienced other kinds of war trauma (Kuwert, Spitzer, et al. 2008).

Survivors of Childhood Cancer

Cancer in childhood involves prolonged pain, suffering, and debilitation from both the disease and the treatments; children are robbed for years of much of the joy, stimulation, and learning that their peers go right on having. Thus it is not surprising that there are lingering problems of psychosocial adjustment and sexual function in adulthood. However, these findings could be attributable in part to the physiological effects of disease and treatment rather than the psychological effects of stress, and studies have shown that educational achievement is impaired only in cancer survivors who had central nervous system (CNS) tumors (Koch, Kejs, et al. 2004) and that psychological adjustment problems are proportional to the amount of CNS radiation received during treatment (Vannatta, Gerhardt, et al. 2007). Childhood cancer is terrible, but survival and adaptation are possible. Resilience is real.

Child Abuse and Neglect in Western Industrial States

C. Henry Kempe, who pioneered the study of non-accidental injuries, coined the term "battered child syndrome" and showed that abuse and neglect are far more common in modern societies than was suspected (Kempe, Silverman, et al. 1962/1984). Pediatricians are now routinely suspicious of parental reports of falls and other accidents and alert for implausible explanations—for example, when partially healed wounds and fractures of different vintage are attributed to the same accident. Burn patterns may reflect an infant being held in a basin of scalding water rather than thrashing around; other stigmata of child abuse include cigarette burns, welts in a loop pattern made by a beating with folded electrical-cord wire, and odd-pattern burns like brands on buttocks made by holding a baby with a wet diaper down on a radiator or gas burner (Peck and Priolo-Kapel 2002). A subtler pattern, "shaken baby syndrome," occurs when an infant is held and shaken violently (Richards, Bertocci, et al. 2006). This leaves no obvious marks, but powerful forces tear blood vessels in the neck, brain, and eyes, sometimes causing blindness, mental retardation, or paralysis.

However severe an injury, if the evidence forces parents to drop the pretense of an accident, they often say something like "I had to teach him to mind." It has been argued that abuse in the West has its origins in a cultural framework that favors physical punishment, derived from Western religious traditions (Greven 1990). Proverbs 23:14 reads, "Thou shalt beat him with the rod, / And shall deliver his soul from hell." Although Jesus did not preach punishment, others did, including the philosopher John Locke; Susanna Wesley, whose sons founded the Methodist branch of Protestantism; some present-day Christian fundamentalists; and the folk view reflected in the saying "Spare the rod, spoil the child." Since many who have mutilated or killed a child cite discipline as justification, it is likely that the cultural framework produces a continuum from what many consider legitimate punishment through frank abuse.

Physical punishment was outlawed in Sweden in 1979—the law reads, "A child may not be subjected to physical punishment or other injurious or humiliating treatment"—but young Swedes are still civilized, considerate, and generous, suggesting that the folk saying could be wrong. In fact, there is evidence that corporal punishment increases aggressive behavior, sometimes in a lifelong pattern, including the kind of parenting that produces a third generation of aggressive children (Serbin and Karp 2004; Watson, Fischer, et al. 2004). A growing consensus among health professionals opposes such punishment (Elliman and Lynch 2000).

Serious child abuse is common. According to the American Academy of Pediatrics, one in twenty children are physically abused annually; of these children, numbering at least 2.5 million, 35 percent experience physical injury, 15 percent sexual abuse, and 50 percent neglect. One in four girls and one in eight boys are sexually abused before age eighteen. According to the Centers for Disease Control (CDC), 906,000 cases of maltreatment were *confirmed* by child protective services in 2002, of which 19 percent were physical, 10 percent sexual, 5 percent emotional or psychological, and 61 percent neglect, all conservatively defined. These figures are almost certainly underestimates. In the same year 1,500 children were confirmed to have died from abuse or neglect—28 percent from physical abuse, 36 percent from neglect, and the rest from combined forms of abuse.

The Children's Bureau of the U.S. Department of Health and Human Services analyzed data from 2007 and found that both being abused and being killed by an abuser decline with age; 76 percent of the abuse fatalities were children under age four. The cause-of-death distribution was similar to that described in 2002, but there was a decrease in the total number abused, to 794,000, and in annual rate per thousand children. The methodology involved screening calls known as referrals; two-thirds of these were accepted as reports, of which about one-third were substantiated on investigation. It is hard to estimate nonreferred abuse. The number of abuse-and-neglect fatalities was 1,760; four-fifths of perpetrators were parents (90 percent biological parents), and 42 percent were men.

Evolutionary Considerations in Abuse and Neglect

Evolutionary theory sheds light on these patterns, although that is small comfort to the victims. First, the age distribution is largely what would be predicted from the curve of reproductive value (Fisher 1958). Second, given the overwhelming predominance of males in homicide generally (Ghiglieri 1999; Konner 2006), the predominance of women at first appears puzzling, until it is realized that the 42 percent of child killings accounted for by men are about an order of magnitude higher than the proportion of time children spend with men.

Finally, although standard statistics do not consider this, there is strong evidence for another sociobiological prediction—namely, that unrelated males should pose greater risk. Although only 4 percent of child deaths were perpetrated by an unmarried partner of the mother, this figure too is not corrected for exposure, and there is often no way to identify stepfathers in these data. Such identification is possible in Canadian and British homicide records, and they show that the exposure-adjusted risk of fatal abuse by stepfathers is up to one hundred times higher than the risk from genetic fathers (Daly and Wilson 1996, 1998). Also, 28 percent of genetic fathers who killed their children then committed suicide, compared with fewer than 2 percent of stepfathers, and far more stepfathers killed with brutal beatings. Poverty, large family size, and maternal youth are all proven abuse risks, but the stepparent effect is independent of them. "Step-relationship remains the single most important risk factor for severe child maltreatment yet discovered" (Daly and Wilson 1996:79).

Reproducing Abuse

Child abuse is transmitted across generations. A classic study identified 267 new mothers at risk for abuse owing to poverty, unpreparedness for motherhood, and generally chaotic lives. (They lacked support from husbands, boyfriends, and family, and 88 percent of them moved—four times on average—during the child's first two and a half years; Egeland, Jacobvitz, et al. 1987.) Children were followed to age five by observation, questionnaire, and laboratory tests, and forty-four cases of child maltreatment occurred, including twenty-four cases of serious physical abuse. Questioned independently, forty-seven of the mothers reported a spectrum of serious abuse in their own childhood, including being burned with irons and hot water, thrown against walls and radiators, and beaten repeatedly with belts, switches, and electric cords; thirteen said they had been sexually abused by a relative. None had received psychological treatment. About 70 percent of the abused mothers abused their own children, compared to *one* of the thirty-five mothers whose parents had been emotionally supportive. The 30 percent of abused mothers who did not abuse their own children had a stable, intact home and family and the support of a husband who was the child's biological father.

Similar results have been found many times with many samples, although the estimate of the percentage of abused children who grow up to be abusive parents is typically lower, and important mediating variables have been identified (Kaufman and Zigler 1988; Oliver 1993). In one study of 4,351 families (Dixon, Browne, et al. 2005; Dixon, Hamilton-

Giachritsis, et al. 2005), 3.1 percent had at least one parent who reported having been abused as children themselves. Of these families, 6.1 percent abused their own infants before age thirteen months, compared with 0.4 percent of families without an abused parent. Three mediating risk factors (parents under twenty-one, history of depression or other mental illness, and residing with a violent adult) explained 53 percent of the transgenerational effect; adding poor parenting styles increases this explanatory power to 62 percent. Finally, this is not culture-bound; in a study of 150 Mexican mothers, having been abused as a child strongly predicted future abuse, mediated by such maternal variables as depression and antisocial behavior (Frias-Armenta 2002).

Inborn characteristics of the child, such as irritability, sleep and feeding difficulties, hyperactivity, physical disabilities, and, later, "acting out" or conduct-disorder behavior also contribute, by burdening or even provoking parents; some children are more difficult than others, making it hard to determine whether the transgenerational effect is a lasting effect of the abuse of the child or the result of partial genetic inheritance, the mediating maternal variables, or the child's characteristics. Abusive mothering is rare in nonhuman primates in the wild, but it occurs in captivity and is often associated with abnormal rearing, low rank, and other stresses (Nadler 1983; Maestripieri 1998; Brent, Koban, et al. 2002). The same problem of interpretation occurs in some studies showing that abusive behavior runs in families, both vertically and horizontally—for example, studies of sooty mangabeys (*Cercocebus atys*) and pigtail macaques (*Macaca nemestrina*) (Maestripieri, Wallen, et al. 1997a, 1997b).

However, a study of more than one thousand five-year-old human twin pairs and their families was able to apportion influence and found that the causes of both frank abuse and corporal punishment were mainly environmental, with a genetic contribution only to corporal punishment. This was mainly mediated by the child's antisocial behavior, which showed a large genotype-environment interaction (Jaffee, Caspi, et al. 2004). Physical maltreatment increased the chance of such behavior only 2 percent in children at low genetic risk, but 24 percent in those at high genetic risk (Jaffee, Caspi, et al. 2005). Still, laboratory primate studies prove that adverse early experiences can cause abusive mothering (Reite 1987; Champoux, Byrne, et al. 1992), which may echo in the next generation. Some genotypes are more vulnerable, but these experiments remove the genetic confound. However, it is important to emphasize that these monkey models of maternal abuse are artificially generated in captivity, and that abusive mothering has rarely been observed in the wild (Hrdy 1999).

Sexual Abuse

Although harder to measure, sexual abuse is very common. In retrospective surveys, from 20 to 35 percent of adult women recall having been forced as children, by an adult male, to have some kind of sexual experience, including exhibitionism without touching. Probably only a fraction of sexual abuse is ever reported to health or legal authorities. The best estimates are that there are more than 150,000 new cases of sexual abuse of children annually in the United States. Even victims who consult health professionals often do not report the crime; in a 1988 study of 156 medically documented cases of sexually abused children in Boston, only 96 were reported to the police. The majority of victims were age seven or older—35 percent were between thirteen and eighteen—and underreporting was much more common among non-white victims.

About 15 percent of adult women remember a sexual encounter in childhood—at an average of around ten years of age—that included physical contact with an adult. If exhibitionism, verbal pressure, and pornographic photography are included, the proportion rises to 25 percent. One-third to one-half of perpetrators are reported to be family members. Around 90 percent of sexual abusers of children are male —again, despite the fact that children spend most of their time with women. Ten percent of men report having experienced sexual abuse, but family members were rarely involved. Women are the perpetrators in perhaps one-tenth of all cases, more for boys than for girls. Parent-child incest is rare and is almost always father-daughter incest.

There are legitimate concerns that greatly heightened vigilance regarding sexual abuse has imposed abnormal limits on normal physical affection and that false as well as true memories of abuse have been acted upon by health and legal professionals (Wright 1994; Loftus 2003). The damage caused by false memories of abuse has in some cases been considerable, although not as great as that of ongoing and unrecognized abuse, which can include depression, anxiety, post-traumatic stress disorder, substance abuse, and sexual problems in adulthood.

Changing Family Structure in Western Industrial States

In addition to abuse and neglect, cultural and societal changes in family structure and child care arrangements that affect millions of children raise fundamental concerns about families and children (Moynihan, Smeeding, et al. 2004). The relevant changes, not all negative, include: parent-child separation and day care; divorce and single-parent families; a massive influx of women into the labor force; chronic poverty and economic depression; and growing numbers of unconventional families, such as those with two same-sex parents.

Change has been rapid. To take one index, the proportion of sixteen-year-olds not living with both parents was about one in four from 1900 to 1960 and rose to 43 percent by the 1990s. Taking all children under eighteen, 9 percent were living with a single parent in 1960 and 27 percent in 2000, although this proportion has been relatively stable since the 1980s (Sigle-Rushton and McLanahan 2004). The percentages are much higher for blacks than whites. The percentage of black children age zero to seventeen living with one parent has stabilized at around 53 percent, while the figure for whites hovers around 22 percent. The increase is due to a rise in both divorce rates and out-of-wedlock births; a large proportion of single-parent families have involved neither marriage nor even nonmarital cohabitation, which has also risen dramatically.

Divorce

Over one million children each year in the United States experience the divorce or separation of their parents, which is now a normative stress of childhood. It is not easy to separate the effects of divorce from other background variables. An influential study exemplifies the methodological problems (Wallerstein 1987; Wallerstein and Lewis 2004). Children followed from divorce to adulthood had adjustment problems right after the divorce and bitter and sad memories of it in adulthood, but there was no control group. The subjects were not studied before the divorce to determine whether their problems were new, nor were they compared to children who stayed in high-conflict marriages.

Better research suggests less dire effects. In a study led by Mavis Hetherington, seventy-two children of divorced couples (thirty-six boys and thirty-six girls) were compared to a matched sample of children whose parents did not divorce. Children of divorce had more adjustment problems—such as disobedience, especially in boys—than control subjects, but these differences had diminished at the two-year follow-up, and at six years they were less prominent still. Economic stress and other factors other than divorce were important:

> We did not encounter . . . a divorce in which at least one family member did not report distress or exhibit disrupted behavior. However, if these effects were not compounded by continued severe stress and adversity, most parents and children adapted to their new family situation within 2 years and certainly within 6 years. It also should be remembered that many households headed by mothers are exposed to excessive stresses and that without adequate support systems such families may show more sustained deleterious sequelae of divorce. (Hetherington and Cox 1982:285)

Other studies echo such complexities. One compared fifty-one young adolescents from recently divorced families to forty-six from intact families using structural equation modeling to find causal pathways (Fauber, Forehand, et al. 1990). In both groups conflict between the parents strongly predicted the children's problems such as depression and acting out, and the key variable was change in the parents' behavior—especially withdrawal from or rejection of their children.

Perhaps the first prospective data were collected by Jeanne and Jack Block, as part of a major longitudinal study (Block, Block, et al. 1986; Block and Block 2006). They followed 101 subjects from ages three to fifteen and looked back at the predivorce problems of children from families that later divorced. Up to eleven years before the divorce, children (especially boys) in that subsample had more behavioral problems than comparable children whose families were not destined for marital dissolution; the problems preceded the divorce and were very similar to the problems found *after* divorce in other studies. There was evidence that the predivorce problems stemmed from marital conflict, but other explanations are possible. Some problems, such as disruptive behavior, may begin with the child and help to cause marital conflict. Personality problems may run in certain families for genetic reasons, independently causing both marital conflict and children's behavior problems.

Later research has emphasized the variety of developmental pathways taken by children of divorce (Hetherington and Elmore 2003). However, "most eventually emerge as reasonably competent, well-functioning individuals" (Hetherington and Stanley-Hagan 1999:137). Children of high-conflict families, divorced or not, have externalizing and other disorders; children, especially girls, involved in divorces that *reduce* conflict are very similar to children from intact families. In addition to ongoing conflict, factors that predict vulnerability are a substantial drop in standard of living and loss of contact with one of the parents; clearly, these factors are unavoidable in many divorces. Overall, based on research on over 900 children and youths from 450 families, serious psychological problems arise in about 10 percent of children from intact families and in 20 to 25 percent of those from divorced families (Hetherington and Kelly 2002). This is a very substantial difference affecting millions of children, but most of the discrepancy is not intrinsic to divorce itself and can be reduced by interventions that control other risk factors.

In follow-up research with her original subjects after twenty-five years, Judith Wallerstein added a retrospective control group from the same neighborhoods and schools and concluded that the two groups differed in their ability to form intimate and lasting relationships and in the persistence of unpleasant memories. One conclusion expressed by subjects from divorced families was that "personal relationships are unreliable, and even the closest family relationships cannot be expected to hold firm." The investigators state that this realization "terrified" them (Wallerstein and Lewis 2004:359). One might argue, instead, that this skepticism is the adaptation that people need in a world of fluid and precarious relationships.

In fact, it is the world we evolved in. Not only did "till death do us part" mean fifteen to twenty years, but divorce, polygamy, adoption, father absence, and dynamic extended families were common stresses faced by children in most cultures. It is therefore not surprising that children are adapted for resilience in the face of family changes. Protecting children from stress is not a parent's sole priority, nor is it necessarily always good. The psychodynamic theories dominating psychiatry and much of Western culture up to 1960 would have predicted a pandemic of severe mental illness in response to the changes in the family in the half-century since. No such pandemic has materialized, because children are adapted to face a variety of family changes. Seen in the context of cross-cultural and historical variation, the current varieties of families do not seem so startling. Of greater importance are factors like poverty, prejudice, poor education, and lack of community support.

Abuse, Neglect, and Adolescent Aggression

Can abuse and neglect help explain transgenerational aggression *beyond* the impact of poverty? A prospective study (Widom 1989; Maxfield and Widom 1996) identified 667 children who had been abused or neglected from 1967 to 1971, according to the court records of one county in a midwestern city, and compared them to non-abused controls: for subjects abused before school age, the study matched sex, race, date of birth (within a week), and hospital of birth; for those abused during their school years the criteria were sex, race, date of birth (within six months), same class in same elementary school, and home address within a five-block radius.

After twenty years, 26 percent of the abused and neglected children and 17 percent of the controls had a juvenile delinquency record. The former also had significantly more offenses (2.4 per subject versus 1.4), were younger on average at first offense (sixteen versus seventeen), and were more chronic offenders (17 percent of the abused subjects versus 9 percent of the controls had five or more offenses). They also had significantly larger adult criminal records, and they had committed more violent crime (19.4 versus 13.5 percent). The crime rate, as in other studies, was much higher in males than females, especially for violent crime—there were at least five violent males for every violent female in both the abused/neglected group (19.4 versus 3.4 percent) and the control group (13.5 versus 2.4 percent), but the association with abuse or neglect, although much smaller than the sex difference, was consistent. *Violent* crimes were committed by 15.8 percent of those who had been physically abused, 12.5 percent of those who had been neglected, and 7.9 percent of the controls. (Sexual abuse alone did not raise violent crime.)

Abuse effects persisted at least six more years, to a mean age of thirty-two (Maxfield and Widom 1996).

Interpreting this study is not simple; if the children were abused or neglected by genetic relatives, then a biological confound could account for some of the association. If an abusive mother has a son who grows up to commit violent crime, it may be because they are both constitutionally violence-prone. But the twin studies and experimental animal studies strongly suggest that abuse in childhood can direct the developmental process toward aggressiveness, especially given particular genetic vulnerabilities. Moreover, studies using path analysis on large longitudinal samples find that environmental disadvantage predicts harsh and inconsistent parenting, "which predicts social and cognitive deficits, which predicts conduct problem behavior, which predicts elementary school social and academic failure, which predicts parental withdrawal from supervision and monitoring, which predicts deviant peer associations, which ultimately predicts adolescent violence" (Dodge, Greenberg, et al. 2008:1907).

The existence of genetic and hormonal predispositions to violence does not in any way negate the importance of such an environmental causal chain. As always, where experiments are not possible, the most convincing demonstrations of environmental effects come from studies that are designed to control for genetic effects. For example, in a study of 390 sibling pairs of varying degrees of relatedness (MZ and DZ twins, full siblings, half-siblings, and step-siblings), it was shown that having an aggressive sibling predicts externalizing problems in adolescence after controlling for shared genes and other environmental background variables (Natsuaki, Ge, et al. 2009). Many more studies like this one are needed to tease apart true environmental effects from correlations due to genetic relatedness.

Stress and Coping in Human Development

Both eustress and distress are ubiquitous in the environments of human evolutionary adaptedness, and so is resilience. Some people *seek* stresses, through extreme sports, dangerous occupations, or war (Haynes, Miles, et al. 2000), and adults vary on a continuum of thrill or sensation seeking that can be applied to different cultures (Neria, Solomon, et al. 2000; Wang, Wu, et al. 2000). Sensation seekers are physiologically different (Zuckerman 1984, 1990; Zuckerman, Buchsbaum, et al. 1980); not immune to trauma, they are nevertheless brave and resilient. The most resilient athletes are also the best (Holt and Dunn 2004; Martin-Krumm, Sarrazin, et al. 2003; Mummery, Schofield, et al. 2004; Schinke and Jerome 2002); decorated war veterans score higher on sensation seeking than other war veterans and are unlikely to develop psychiatric symptoms in response to war trauma (Neria, Solomon, et al. 2000).

Since athletic and martial abilities have been important in most cultures, such findings illuminate cross-cultural variation in stress-responsiveness and resilience. Without risk takers and sensation seekers—and a large reserve of stress-resilience—it is impossible to imagine the success of the human species. High-risk, high-gain strategies for reproductive success have been favored in many times and places, and resilience must have been pervasive.

A simplified and tentative model of human responses to acute stress (Konner 2007) suggests a continuum, consisting of three main types of individuals. The first, sensation seekers, pursue experiences that would be stressful to most people; they find novel and even dangerous experiences enjoyable and exhilarating. This group is characterized by two biological markers: low activity of the enzyme monoamine oxidase (MAO) and a particular polymorphism of the D4 dopamine receptor (Zuckerman and Kuhlman 2000). Those in the second—and largest—group are fairly resilient in the face of serious stress but have acute (dis)stress responses; when persistent, these responses might qualify as diagnosable psychiatric disorders, but few of them do persist. The third category is the minority who are vulnerable to psychiatric illness during and after acute stress. This group may develop post-traumatic stress disorder, depressive disorders, and/or generalized anxiety disorder, and these disorders may persist for years.

Some studies suggest that low hippocampal volume is a marker of vulnerability to PTSD, and that carriers and/or homozygotes for an allele for the shorter of two promoters of the gene for the serotonin transporter are vulnerable to depression and some other psychiatric problems following serious psychological stress (Caspi, Sugden, et al. 2003; Grabe, Lange, et al. 2005; Kendler, Kuhn, et al. 2005). However, meta-analysis has called this finding into question (Risch, Herrell, et al. 2009), and more needs to be done to identify genetic vulnerabilities. The proposed model is speculative, was advanced for heuristic purposes, and has many possible flaws. The first and third categories may not be independent of each other or mutually exclusive and therefore may not be ends of the proposed continuum. But it is clear that individual differences matter.

Sensation seekers aside, average people are resilient. A half-century longitudinal prospective study followed ninety-four men who were in college in the early 1940s (Vaillant 1977). At the thirty-five-year follow-up, in their midfifties, many were happy and successful by self-report and external criteria, and others were unhappy or failures. Developmental data suggested the advantages of a stable, nurturing early life; men who had been unhappy in childhood tended to remain so despite conventional success. Some functioned at a high level but could not form intimate relationships; one who was aware of this said that he could do nothing about it. This is consistent with other longitudinal studies, clinical evidence (Heim, Plotsky, et al. 2004), and experimental data from animal models (Sánchez, Ladd, et al. 2001; Bennett, Lesch, et al. 2002; Francis, Szegda, et al. 2003; Champagne, Francis, et al. 2003; Sánchez, Noble, et al. 2005; Suomi 2008).

But again, stress is not necessarily bad. One-third of the men spent at least ten days in continuous combat in World War II, and all ninety-four experienced major personal griefs and setbacks as adults, none of which predicted poor adjustment. The follow-up was extended another fifteen years in the 1990s and compared with longitudinal data from two other samples: forty women who had been intellectually gifted as children in the 1920s, and about three hundred men from poor families in Boston, followed from junior high into their seventies. In that sample, eleven men had had extremely bad childhoods and seemed at age twenty-five to be severely damaged psychologically; fifty years later, eight of the eleven were doing well (Vaillant 1993:287). Also, thirty men with "best outcomes" in later life were compared to thirty with "worst outcomes." Almost half of the "worst" had emotionally inadequate childhood environments, but one-sixth of the "best" did too; they compensated with a close marriage or exceptional business success.

Of another lifelong study, research director Jean MacFarlane wrote: "Many of the most outstanding mature adults in our entire group, many who are well integrated, highly competent and/or creative . . . are recruited from those who were confronted with very difficult situations and whose characteristic responses during childhood and adolescence seemed to us to compound their problems" (Vaillant 1977:299). Some children of schizophrenic mothers show deficiencies of social competence and attention problems in school. However, many others show few signs of pathology. They had abnormal mothers and separations when their mothers were hospitalized, compounding their genetic risk, yet they adapted well. Of six hundred children facing a different kind of deprivation, an inner-city childhood, many showed similar stress-resistance (Garmezy, Masten, et al. 1984; Garmezy and Tellegen 1984).

Finally, Emmy Werner followed 698 infants born in Kauai, Hawaii, in 1955 for more than forty years (Werner 1989, 1996; Werner and Smith 2001). About one-third were at risk, defined as having four or more of these risk factors: biological stress during or shortly after birth; poverty; a mother with little education; and a family environment marked by discord, desertion, divorce, alcoholism, or mental illness. Of this high-risk group, two-thirds—two-ninths of the larger original sample—developed serious problems with learning, behavior, delinquency, mental health, or pregnancy before age eighteen. But the last one-ninth (seventy-two high-risk children, thirty boys and forty-two girls) "developed instead into competent, confident, and caring young adults." In interviews these people cited two factors: a supportive adult outside their dysfunctional family and a strongly positive attitude toward life.

Still, two-ninths (two-thirds of the deprived one-third) were *not* resilient. Although not controlled for genes, this finding suggests that early disadvantage has serious, enduring consequences; it matters, and natural selection, for all its power, has not made children generally resistant to it. Even the resilient one-ninth had more difficulties at ages thirty and forty—including unstable marriages and stress-related medical complaints—than did their lower-risk counterparts. Still, for this subgroup there were few long-term results of early deprivation. Most remarkably, 44 percent of the resilient high-risk group, as opposed to only 10 percent of their low-risk peers, rated themselves as "happy" or "delighted" with their present lives, suggesting that overcoming adversity increases subjective well-being. As Helen Keller said, life is full of suffering, but it is also full of the overcoming of suffering.

22

Hunter-Gatherer Childhood—
The Cultural Baseline

It is somewhat arbitrary how we distinguish socialization from enculturation with regard to child rearing, but hunter-gatherer childhoods represent the culmination of the one and the foundation of the other. To whatever extent we grant nonhuman primates the transmission of limited behavioral traditions—protoculture—we should also grant them protoenculturation. Vast domains of human child rearing—attachment, play, sex differentiation, the development and control of aggression and sexuality, reward and punishment, early experience, and stress effects, to name a few—are largely shared in common with apes and monkeys and to a lesser degree with other species. What we think of as the essence of humanity is often not that at all, but a suite of capacities much older than we are.

Yet there is still much about humanness that lies in the domain of symbolically mediated culture and as such, for practical purposes, is just human. The hunter-gatherer phase of human history stands as the culmination of evolution and the beginning of history in the sense of rapid, transformative, nongenetic cultural change. Biologically and psychobiologically, "hunter-gatherers 'R' us," but in social order and culture we are very different. Combined with archaeological evidence, they are essential for our understanding of our origins (Lee and DeVore 1968; Dahlberg 1981; Marlowe 2005b). We have reviewed in Part III some basic facts of how child rearing changed after the hunting-gathering era. We will consider in Chapter 28 some mechanisms *by* which culture changes; these help explain how we have gotten here from there. They also account for the cultural patterns that are grist for the mills of ethnology and history. However, we first summarize the hunter-gatherer baseline of *both* socialization and enculturation.

Generalizations and Challenges

Some features of !Kung infancy and childhood appear to be representative of hunter-gatherers in general; collectively, these features have been termed the hunter-gatherer childhood (HGC) model. Viewed against its phylogenetic background, the HGC model is a species-specific human expression of the catarrhine socialization pattern. Present-day child care methods in industrialized countries are discordant with those in the human environments of evolutionary adaptedness (EEAs); this discordance has developmental implications.

Excellent research on hunter-gatherer infancy and childhood on the Agta, Efe, Hadza, Aka, and Ache seemed to call the HGC model into question. At the same time new theory in life history evolution strongly suggested that hunter-gatherer childhood should not follow a single pattern but should adjust to widely varying ecological conditions; infant and child care should be facultative adaptations. This has been called the childhood as facultative adaptation (CFA) model (Tronick, Morelli, et al. 1987); it challenges the HGC model and the discordance hypothesis and questions whether there is anything distinctive about hunter-gatherer childhood as a general adaptation.

The CFA model is highly consistent with recent advances in life history theory. All life involves compromises and trade-offs, some of which seem abhorrent to modern observers. Mothers in all species have competing demands on their energy that mitigate their investment in individual offspring, including the needs of other present and future offspring. Infanticide at birth is a choice sometimes exercised by mothers in traditional societies, including hunter-gatherers. Among the !Kung it was reported in about 1 percent of births (Howell 1979), with the stated goal of enhancing the care and survival of other children and avoiding efforts to care for defective ones, which were almost certain to fail. Some hunter-gatherers may have used it to bias the sex ratio toward males—a practice that is strongly correlated with the percentage contribution of males to the calorie content of the diet (Hewlett 1991a). Birth spacing and children's contribution to subsistence among hunter-gatherers are now clearly understood as partly facultative adaptations.

Nevertheless, there are two kinds of evolutionary causes of any character (Tinbergen 1963): (1) adaptation through natural selection in the EEAs and (2) constraints on solutions to adaptive challenges due to phylogenetic history (Maynard Smith, Burian, et al. 1985). If the monkey and ape background to hominin evolution entailed a consistent pattern of care of infants and juveniles, one would expect it to have coevolved with aspects of normal or optimal development that might be dependent on it. If so, this pattern would limit the facultative adaptations achieved in human evolution, or at least the range achieved without untoward developmental consequences.

The Hunter-Gatherer Childhood Model

The HGC model arose originally from research on the !Kung. I made an early attempt to set the !Kung findings in phylogenetic context, to evaluate evidence in the older literature of patterns in other hunter-gatherers, and to assess possible changes since the hunter-gatherer era that might be viewed as discordant (Konner 1981). My hypothesis that hunting-gathering societies resembled the !Kung in infant and child care more than did other non-industrial societies was supported. I defined hunting and gathering to exclude equestrian hunting, a recent historical novelty, and I emphasized warm-climate hunter-gatherers as representative of the human EEAs. No cultures were excluded for other reasons. Thirty years ago there were no quantitative studies of hunter-gatherer infancy and childhood besides the !Kung, but there were descriptive accounts that accorded with the !Kung findings (Turnbull 1965; Holmberg 1969; Balikci 1970). Of greater interest was cross-cultural survey research, much of which Robert Textor (1967) summarized; ethnology is a discipline where older studies can claim legitimacy.

Compilations of Older Studies

The measures used in older studies are not always currently relevant. Still, the data show a consistent tendency for societies that rely on food gathering to have more indulgent infant and child training practices than other non-industrial societies. Specifically, "pain inflicted on infant" was rated as lower (p < .05) and "overall indulgence of infant" (p < .05) and the ease and lateness of toilet training (p < .01) were rated as higher in societies that subsist primarily by food gathering (N = 19–40) than in those subsisting by other means (N = 22–34). In childhood, "anxiety over responsible behavior" (p < .01), "anxiety over obedient behavior" (p < .01), and "anxiety over self-reliant behavior" (p < .05) were rated as lower in food gathering societies (N = 30–35) than other societies (N = 37–42). Only at adolescence, when female initiation rites (p < .01) are more severe in foraging societies (N = 38) than in nonforaging ones (N = 27), was the pattern of indulgence reversed. Also, adolescent peer groups were less common (p < .05) in foraging (N = 23) than in nonforaging societies (N = 14), a finding that is consistent with the demographic unlikelihood of same-age peer groups.

Controlled Comparison

Betsy Lozoff and Gary Brittenham (1978) compared 10 warm-climate hunting-and-gathering societies with 176 other non-industrial societies on infant care practices as rated by Barry and Paxson (1971). The ten (!Kung San, Hadza, Mbuti, Semang, Vedda, Tiwi, Siriono, Botocudo, Shavante, and Chenchu) were the only societies that met the criteria of living between the latitudes of 22°30′ N and 22°30′ S and having less than 10 percent dependence on agriculture, animal husbandry, and fishing (Murdock and Morrow 1970). Close mother-infant contact, late weaning, and responsiveness to crying were characteristic of hunter-gatherers. Late weaning was also found in other non-industrial societies, and the mother was always the principal infant caregiver, but other aspects of the mother-infant bond were closer in hunter-gatherers. In addition, measured by body contact, sleeping distance, response to crying, or weaning age, mother-infant contact and maternal indulgence of infants were lower in the United States than in the average for 176 non-industrial cultures. This supported the finding that infant and child care and training in Chicago during the 1940s were substantially below the median in indulgence for a large representative sample of non-industrial cultures—except in aggressiveness training, where Chicago was more lenient. In "oral socialization," toilet training, sex and modesty training, and independence training, parents in Chicago were rated as more strict than the average non-industrial society (Whiting and Child 1953).

The variability in intermediate-level societies that rely on agriculture, animal husbandry, and/or fishing—which constitute the vast majority of the empirical base of cultural anthropology—was much greater than that for hunting-gathering or industrial societies (Lancy 2007). Just as variation in child care is not random with respect to mode of subsistence, it is not random with respect to societal complexity. Comparing large or small states with cultures in which the highest level of political integration is the minimal state, autonomous community, or family (Textor 1967) suggests a cultural-historical trend in child care. The hypothesis of decreasing indulgence with increasing political complexity was confirmed on six measures of infant and child care. In addition, three other variables relating to child life showed significant differences. Punishment of premarital sex was more severe in eighty-nine more complex societies than in ninety less complex societies ($p < .001$); exclusive mother-child households (with no extended family and the father sleeping elsewhere) were more likely in more complex societies ($p < .001$).

But paradoxically, desire for children was greater in those societies, despite lower indulgence. This is consistent with the higher birth rate of agricultural societies (Sellen and Mace 1997). Even small differences could account for Neolithic population expansion, but not for the substantial differences in infant and child care. However, Hewlett (1991a) found hunter-gatherers to be more likely to have multiple caregiving than horticultural and pastoral cultures, owing to the greater population density (compactness) of the immediate settlement, as Draper (1973) showed for the !Kung. Hewlett confirmed that hunter-gatherers' group size makes them more likely to have multi-aged child playgroups and cited other cross-cultural research showing that fathers are more involved with infants and children among hunter-gatherers and on islands, apparently because male-male competition is low in both ecological settings (Alcorta 1982; Hewlett 1991a).

Hunter-Gatherer Childhood in Evolutionary Context

Old World monkeys and apes had key features of the maternal relationship in common, the catarrhine mother-infant complex (Chapter 15). The HGC model proposes that many features of the catarrhine mother-infant complex were preserved in the human EEAs, but it goes substantially beyond infancy. Listed here are the main features of the HGC model as originally proposed in the context of !Kung infancy and child-hood (Konner 1972, 1975, 1976, 1977b, 1981; Konner and Worthman 1980; Katz and Konner 1981); the asterisks indicate features that have been subject to subsequent challenges.

1. Prolonged close physical contact with mother

2. High indulgence of dependent needs and demands

3. Frequent nursing (several times per hour) in waking hours

4. Mother and infant sleep on same bed or mat; night nursing

5. Weaning after age three and four-year birth spacing**

6. Strong separation and stranger protest until late stages*

7. Dense social context that seems to reduce pressure on mother

8. Nonmaternal care much less than maternal care until second year**

9. Paternal care much less than maternal care but more than in most cultures

10. Transition to multi-aged, mixed-gender child playgroup

11. Minimal childhood responsibility for subsistence or baby care**

12. Minimal restrictions on childhood or adolescent sexuality

These generalizations were presented as hypotheses for further study in the hope that others would do serious research on hunter-gatherer childhood; this hope was realized, and subsequent studies have raised important questions. Features 5, 8, and 11 (two asterisks) have been significantly challenged by this new research, and too little information has come forth in other studies to generalize about feature 6.

Summary of the Newer Evidence

Quantitative studies of infancy and childhood in the Hadza, Efe, Aka, Ache, and Agta have been described as departing from the HGC model, and some of the challenges have been considered in detail in Chapters 15 to 17. We can now place these departures in context.

Table 22.1 compares the findings of recent studies regarding key features of the HGC model. It suggests a high level of support for most of the original generalizations. Hunter-gatherers have frequent nursing, mother-infant co-sleeping, high physical contact, high overall indulgence, substantial to high nonmaternal care and father involvement, maternal primacy, transition to a multi-aged child playgroup, a relatively carefree childhood (except the Hadza), low restriction of premarital sex, and strong adolescent initiation rites. Only the Aka match the !Kung in age at weaning and interbirth interval, but the other three cultures have weaning ages over two years and interbirth intervals over three years. This is in the upper part of the range for preindustrial cultures and sustains the generalization that hunter-gatherers have relatively late weaning and long birth spacing.

To place these generalizations in phylogenetic context, the care of infants and juveniles in Old World monkeys and apes is characterized by (proportionately) somewhat later weaning age and longer interbirth intervals (especially in apes), the variable and species-specific importance of nonmaternal care, minimal direct male care except in gibbons, the near-absence of postweaning provisioning, and mixed-sex playgroups that can be same-age groups in strongly seasonal breeders. Other features of HGC are present in most catarrhines, including frequent nursing, late weaning, mother-infant co-sleeping, high overall indulgence, maternal primacy, and adolescent sexuality. The wide distribution of these features suggests that they may have been present in the common ancestor of the catarrhines, between 30 and 40 million years ago (Martin 1990:692).

Evaluating the Divergences

We now consider the implications of the divergences in four main areas: (1) weaning and birth spacing; (2) maternal primacy; (3) overall indulgence; and (4) responsibility in childhood.

Weaning and Birth Spacing

The strongest challenge to the original HGC model comes from data on weaning and birth spacing in the Hadza, Efe, Ache, and Agta. Respectively, they have weaning ages of thirty, thirty, twenty-five, and twenty-seven months, and interbirth intervals of thirty-eight, thirty-

Table 22.1.   The Hunter-Gatherer Childhood Model in Six Cultures

	Frequent Nursing	Weaning Age/Interbirth Interval (Months)	Sleeping with Mother	Physical Contact (All)	Overall Indulgence	Nonmaternal Care	Father Involvement	Maternal Primacy	Multi-aged Child Playgroup	Carefree Childhood	Premarital Sex
!Kung	+++	42/48	+++	+++	+++	++	++	+++	+++	+++	+++
Hadza	+++	30/38	+++	+++	+	++	++	++	+++	+	+++
Efe	+++	30/38	++	+++	+++	+++	+	++	++	++	++
Aka	+++	42/48	+++	+++	+++	+++	+++	++	+++	++	++
Ache	+++	25/37	+++	+++	++	+	++	+++	+++	+++	+++
Agta	+++	27/36	+++	+++	+++	++	+	+++	+++	+	+++

Note: The !Kung compared with five other recently studied hunter-gatherers on 11 aspects of infant and child care included in the HGC model.

Source: Konner 2005.

eight, thirty-seven, and thirty-six months. The process among the Aka is essentially superimposable on that of the !Kung, with about forty-two months of nursing and forty-eight months between births. The Hadza and Ache have shorter interbirth intervals, higher completed fertilities per female reproductive life, and faster-growing populations. The Agta population would grow faster but for its high mortality. The !Kung model for Paleolithic population growth—stability or gradual growth up to 0.5 percent per year (Howell 1979)—may have applied to some populations, but not all. Indeed, the Ache life tables have suggested a model of Paleolithic demography that entails increases alternating with crashes (Hill and Hurtado 1996). This model, if generalizable to the EEAs, presents a challenge not only to current ideas about hunter-gatherer demography but also to life history theory, which predicts greater population stability in long-lived, slowly developing species (Chapter 2).

However, prospective measures of Ache weaning were made during the reservation period, and it may have been later in the forest period. Given the forest-period birth spacing of just over three years, as well as the fact that "Ache children generally continue nursing on demand until their mother is pregnant with her next child" (Hill and Hurtado 1996:221), the traditional Ache weaning age was probably around thirty months. The observation that Ache mothers, "if the interbirth interval was too short (i.e. less than two years), would simply kill the newborn child and continue nursing the first" suggests a desired weaning age of more than two years. The Hadza weaning age of thirty months, in a setting far closer to their traditional way of life, is less likely to be an artifact of culture change. So far, however, it appears to be a lower limit of hunter-gatherer weaning age. The Efe interbirth interval of thirty-eight months occurred in a setting of exceptionally high infertility in the population. This could explain the shorter interbirth intervals in the fertile women, achieved in part through multiple caregiving, including nursing by other women.

Overall, however, these six populations have a lower limit on weaning age that is high by worldwide standards and extremely high by Western standards. Even if most hunter-gatherers were more like the Hadza, Efe, Ache, and Agta than like the !Kung and Aka, Paleolithic weaning would have been relatively late and interbirth interval relatively long. Still, against the background of ape patterns, this would represent a significant evolutionary shortening because we provision our young with food after weaning far more than other primates (Chapters 5 and 17).

Maternal Primacy

One of the most contested claims has been that hunter-gatherer maternal care supports the Bowlby model of the development of attachment, which includes a hypothesis of monotropy—attachment behaviors focused on a single caregiver (Bowlby 1969, 1973, 1980; Sroufe and Waters 1977; Bretherton 1992; Belsky 1999; Sroufe, Carlson, et al. 1999). Both the Efe and Aka studies have been cited as undermining this claim, but this challenge is easily met. First, the claim of maternal primacy in the !Kung literature was never as strong as it was made out to be; multiple caregiving and a supportive social context were always emphasized. Second, to the extent that maternal primacy was emphasized, there is little in subsequent research to undermine it.

The first of these statements is shown to be true in Chapter 15, so we focus here on the second, with an emphasis on maternal primacy, deferring discussion of the more abstract concept of monotropy. With the exception of weaning age (still very late by Western standards), the Ache experience of mothering is virtually the same as that of the !Kung. The Hadza wean earlier and separate more frequently than the !Kung, but nothing in the Hadza literature challenges maternal primacy; quantitative data strongly support it (Marlowe 2005a, Table 1).

The Aka have the highest level of father involvement not only among hunter-gatherers but throughout the cross-cultural range, an illuminating addition to our understanding of infant care. Still, even in forest camps where their involvement was highest, fathers held infants less than half as much as mothers at all ages, with a decline to 25 percent as much during the eight- to twelve-month period when attachment is forming. Nonparental involvement in care is very high by hunter-gatherer standards, yet mothers hold their infants 87 percent of the time on net hunts and more than half the time in the forest camps, where mother-infant contact rises to 87 percent between eight and twelve months, matching the contact time on hunts. Because there are numerous caregivers, the average nonparent does not approach the father in involvement, and his involvement is a distant second to the mother's. The nature of the attachment to the father will be interesting to study, but paternal attachment does not challenge maternal primacy.

The Efe have been cited as strongly undermining the HGC model, especially maternal primacy (Tronick, Morelli, et al. 1987). Multiple caregiving among the Efe substantially exceeds that of the !Kung, but no evidence suggests that any individual could rival the mother's involvement. Mothers account for about half the social contact with infants during the first half-year, a figure that rises to 63 percent at eight months. There is no time in infancy when the father, the average nonparent adult, or the average child accounts for more than a fraction of the mother's contact. Efe multiple caregiving is impressive, but also does not challenge maternal primacy.

Finally, all of these comparisons rely on observations made during the day. Proximity, nursing, and other aspects of parenting during the night have been repeatedly emphasized as important aspects of hunter-gatherer and other traditional child care (Konner 1977b, 1981; Konner and Super 1987; McKenna, Thoman, et al. 1993). Ache, Hadza, Aka, and Agta mothers sleep with their infants, giving them ample opportunity for night nursing. Efe infants reportedly sleep with others at times, but quantitative data are not presented, and descriptions suggest that maternal co-sleeping and nursing occur the great majority of nights throughout infancy. So the data we do have on co-sleeping also support maternal primacy.

The question of monotropy is far more difficult. Bowlby claimed that the infant tends to focus on one primary caregiver, even when multiple caregivers are available and beyond what would be predicted from the distribution of contact time and care. Despite the fact that this focus could be on an adoptive mother, a father, a grandparent, an aunt, or an orphanage nurse, the monotropy claim was viewed as an attempt to tie mothers to the home and was strongly challenged. In addition, the monotropy idea departed from prevailing psychological models, such as learning theory and social cognition, which predict a distribution of attachment proportional to contact and care.

However, many studies show that multiple caregiving does not prevent the development of attachment to the mother or other primary caregiver. This has been found repeatedly for high-quality day care (Caldwell, Wright, et al. 1970; Kagan, Kearsley, et al. 1978; National Institute of Child Health and Human Development 1997; McKim, Cramer, et al. 1999), in cross-cultural settings generally (van Ijzendoorn and Sagi-Schwartz 2008), in African multiple-care settings (Ainsworth 1967; Leiderman and Leiderman 1977), and even in the Israeli kibbutz, where nonmaternal care is very high in both amount and continuity (Sagi, van Ijzendoorn, et al. 1995). The quality of the attachment may be affected (Sagi, van Ijzendoorn, et al. 1994; Russell 1999), but it remains focused on the mother.

Monotropy would require special nervous system adaptations, but we have ample evidence from romantic attachments that an unreasonable focus on one individual, independent of objective behavioral input, is comfortably within the capability of the human nervous system. Brain research in the realms of imprinting in birds (Horn 1991; Bock and Braun 1999a, 1999b), pair bonding in mammals (Insel 1997, 2000), mother-infant separation in primates (Rilling, Winslow, et al. 2001), and romantic attachment in humans (Marazziti, Akiskal, et al. 1999; Bartels and Zeki 2000) is gradually taking the mystery out of monotropic attachments. Brain imaging, EEG, and evoked potentials will ultimately address the question of monotropy in human infant attachment. As for maternal primacy, it is found in all hunter-gatherers, including those seen as exceptions. Most notably, it is even clear in the Agta, where women do half the hunting. If maternal primacy were facultative, it seems likely that the Agta would depart from it. Exclusive maternal care is nonexistent and was never claimed, but maternal primacy is a general feature of hunter-gatherer childhood.

Responsibility in Childhood

Hadza children foraged for themselves very extensively, and the contrast with the !Kung case has been explicitly addressed (Blurton Jones, Hawkes, et al. 1994). Although !Kung children made a contribution to subsistence, it was very small in comparison. Since the Hadza lived in a rich environment that, if anything, was closer to those of our EEAs than that of the !Kung, the foraging of Hadza children suggests that in many such past environments children were expected to contribute substantially to subsistence. Investigators familiar with both cultures suggest that it is less safe for !Kung children to forage, because they can be more frequently out of the line of sight of their parents or the village-camp. The Hadza environment of relatively bare rocky hillsides makes it more difficult to get lost. The two cultures also differ in the amount of baby and child care assigned to older children. Agta children of both sexes fish during childhood. These facts suggest that the level of responsibility expected of children is a facultative adaptation among hunter-gatherers. As we will see in Chapter 25, this is supported by several other quantitative studies of children's subsistence effort.

Direct Comparisons in Fieldwork

One of the most persuasive sources of evidence about changes in child care since the hunter-gatherer era is direct comparison of these societies with neighboring horticultural or agricultural groups by the same investigators using the same methods or, better still, comparison within a hunter-gatherer group of more traditional and nomadic versus more sedentarized subgroups making a transition to other subsistence modes. The latter method was pioneered by Patricia Draper, who compared the behavior of !Kung children in more and less sedentary communities as some of their parents began to adopt horticultural and pastoral subsistence methods (Draper 1976; Draper and Cashdan 1988). There was no apparent conscious change in child-rearing goals or methods but rather an adaptation to new conditions.

Controlled comparisons of hunter-gatherers with nearby horticulturalists and agriculturalists are also very informative. Barry Hewlett and his colleagues compared Aka infant care with that of neighboring Ngandu farmers, !Kung hunter-gatherers, and Euro-Americans (Hewlett, Lamb, et al. 2000). These comparisons strongly confirm the results of broader cross-cultural comparisons (Chapter 15). The hunter-gatherers exceed the farmers in holding and touching, and the farmers in turn exceed the Euro-Americans. In responding to crying, the Aka (like the !Kung) are the least likely to have no response.

Ayako Hirasawa (2005), studying the Baka of southeastern Cameroon in several ecological contexts and comparing them with neighboring Bombong farmers, found that physical proximity with infants was greater among the Baka and did not vary among them whether they were engaged in gathering, fishing, farming, or doing nothing at all. Total time spent nursing was higher among Baka than Bombong (15 versus 10 percent of observations), as was nursing frequency (4.8 versus 2.9

bouts per hour). Compared with either Japanese or Euro-Americans, even the Bombong were frequent breast-feeders.

Finally, Hillary Fouts and Michael Lamb, studying Bofi foragers of the Congo rain forest in the southwest part of the Central African Republic, found typical hunter-gatherer patterns of nursing and weaning. In a direct quantitative comparison with neighboring farmers, also called Bofi, they found (together with Barry Hewlett) that "Bofi forager children are less distressed by weaning than Bofi farmer children in part because they continue to be held after weaning for substantial proportions of time, whereas for the Bofi farmer children weaning marks the end of intimate physical care" (Fouts, Hewlett, et al. 2005:47).

Conclusion: Facultative Adaptation, Discordance, or Both?

The CFA model has great merit—most human behavioral adaptations are facultative—but the HGC model remains robust. Facultative adaptation is always an option for natural selection, and it would be expected to apply to infant and child care in the human EEAs. But natural selection operating in any species must contend with the constraints of phylogenetic legacies, and in the case of the HGC model there are deep homologies with parallel patterns in Old World monkeys and apes, suggesting that the common ancestor may have already evolved these patterns between 30 and 40 million years ago.

The Ache keep their infants and toddlers off the forest floor and wean them at age two, but otherwise they bear a strong resemblance to the !Kung in their patterns of care. The Efe have more multiple caregiving, and the Aka have more paternal care, but the difference between them and the !Kung is less than has been suggested. Moreover, maternal primacy is high in both, making them only weak challenges to models of infant emotional development that center on attachment to a primary caregiver.

The Hadza have earlier weaning than the !Kung and by some reports are not as indulgent with infants, but some quantitative data suggest that Hadza parenting is also highly indulgent (Marlowe 2005a). The data on task assignment—where so much of human enculturation takes place—and specifically on foraging in childhood, show a marked contrast to the !Kung. Although !Kung children do forage to some extent, it is not expected of them, and their productivity is much lower than that of Hadza children. This difference, together with other evidence (Chapter 25), shows that child foraging is a facultative human adaptation. It is also likely that weaning age is partly facultative, varying between two and a half and three and a half years of age, probably in response to the richness and predictability of the foraging environment, the availability of suitable weaning foods, and the presence of infertile women who aid in infant care.

Although of poorer quality, the descriptive data in older ethnographies should not be discounted. There is no reason to believe that classical ethnologists had a bias that would have led them to find conformity to the HGC model, and their accounts support most aspects of it. But even without them, detailed recent studies allow some generalizations. Hunter-gatherer childhood was characterized by close physical contact, maternal primacy in a dense social context, indulgent and responsive infant care, frequent nursing, weaning between two and a half and three and a half years of age, high overall indulgence, multi-aged child playgroups, variable responsibility in childhood, strong initiation rites, and relatively liberal control of adolescent sexuality. These are durable features of the model. Given the risk, stress, and loss involved in hunter-gatherer life, they may have been necessary then but not relevant now. On the other hand, departures from them since the end of the hunting-gathering era may have consequences that merit further study.

TRANSITION 3

Does Nonhuman Culture Exist?

Many nonhuman animals have societies and socialization, and a developing individual would be maladapted if not socialized. But the question of whether culture and enculturation exist in animals is actively debated. To some it now seems well established, especially in chimpanzees; to others it is an almost absurd claim—culture is what critics and philosophers associate with the highest reaches of human capability, the subtlest kinds of awareness, the most sophisticated communications, the greatest attainments of civilizations, and the most sublime expressions of human creativity. What can those who see culture in animals be thinking of?

But the question is not about the most exalted habits and artifacts; it is about the simplest—the boundary between elementally cultural human activities and those animal activities that seem similar. Will a human father teaching his daughter to use a spoon be considered cultural but not a cheetah mother helping her son learn to hunt? Will the transgenerational transmission of human linguistic dialects qualify, but not the location-specific transfer of a song dialect to young white-crowned sparrows—even though that transmission depends on hearing their fathers sing that song in that dialect, as well as on subsequent practice while hearing themselves sing? Will we call it culture when a human child imitates her mother's use of a digging stick, but not when a juvenile chimpanzee matches her mother's fashioning of a termite-fishing tool?

No one denies that there are large quantitative differences between human culture and what is called culture in animals. The issue is whether a firm boundary can be drawn between culture and all nonhuman learned behavior. If a categorical distinction is to be maintained, it must be on a better basis than mere conviction and declaration, which have often prevailed in the past. Here I try to show that:

1. It is not difficult to define objective conditions that allow the word *culture* to be restricted to human groups.

2. It is somewhat more difficult, although not impossible, to deny the designation to certain behaviors of animals, especially apes, that are acquired from human teachers.

3. Nevertheless, the attempt to identify processes in nonhuman animals that plausibly have a rudimentary similarity to human culture has great value.

4. Animal behaviors to which we deny the term may teach us much about what culture is and how our capacity for it evolved.

5. If the word is used in its restricted archaeological sense in reference to the toolmaking traditions of early hominins, then there is no basis for denying culture to chimpanzees.

6. But if it is used in a broader sense, we are led to a different conclusion.

Efforts by skeptics to test claims for animal culture are valuable, but some scoffing has been defensive and shrill. The bar can always be raised and definitions changed so that the latest and most culturelike animal capability remains just shy of a mark, revised ad hoc to protect human uniqueness. Skeptics as well as proponents of animal culture can miss the point: this is a search for the *rudiments* of culture, for something that natural selection might have started with, in the brain function and social life of a nonhuman species, to build a true capacity for culture. Whether we call these phenomena rudimentary or (as I prefer) protoculture is of little consequence as long as we understand them.

Defining the Extremes

If we locate the ends of a long continuum of definitions for the word *culture,* these are clearly incompatible and readily explain some unfruitful debates. At one extreme, cultural anthropologists have defined culture as "that complex whole which includes knowledge, belief, art, morals, law, custom, and any other capabilities and habits acquired by man as a member of society" (Tylor 1871:1). Some definitions have relied on a linguistic context. For example: "The concept of culture I espouse . . . is essentially a semiotic one. Believing, with Max Weber, that man is an animal suspended in webs of significance he himself has spun, I take culture to be those webs" (Geertz 1973:5).

To say that such definitions make it difficult for nonhuman species to compete is not to deny their validity, but we have to parse such definitions pretty finely to come up with a comparative method that can even logically, much less empirically or technically, extend beyond the human. However, the exercise is not hopeless, and it is an interesting irony that centering the definition of culture on symbolic communication may facilitate comparison and in some ways favor claims for nonhuman culture. But it serves little purpose to truncate debate by fiat—defining culture as inherently human.

At the other extreme, John T. Bonner begins *The Evolution of Culture in Animals* (1980:9) by defining culture very broadly as "the transfer of information by behavioral means, most particularly by the process of teaching and learning." As we will see, there is a great gulf between these last two processes. The primatologist Kinji Imanishi (quoted by Nishida 1987:462) defines culture more simply as "socially transmitted adjustable behavior," by which standard almost all social animals have it (for example, in following more experienced group members into different feeding locales).

Andrew Whiten and his colleagues (Whiten, Goodall, et al. 1999:682), introducing an impressive analysis of social learning in wild chimpanzees, define "a cultural behaviour" as "one that is transmitted repeatedly through social or observational learning to become a population-level characteristic." Again, without narrowing the scope of the behavior in question, this would apply to many social animals. Not surprisingly, these more inclusive, weaker definitions tend to come from students of animal behavior, who sometimes apply terms like *culture* and *cultural transmission* not merely to complicated animals like chimpanzees and bonobos (McGrew 1992; Wrangham, McGrew, et al. 1994; Whiten, Goodall, et al. 1999) but to species like the indigo bunting (Payne, Thompson, et al. 1981) and the fathead minnow (Chivers and Smith 1995). Kathleen Gibson (2002) has instead referred to these patterns as "customs."

But in fairness to the animal behaviorists, Paleolithic archaeologists pointed the way to legitimizing the study of cultural transmission, if not in the fathead minnow, then certainly in the chimpanzee, by in effect defining it in terms of *material* culture, the only culture left behind by early hominins. Since the emergence of culture was one of their main research questions, it is not surprising that they defined it differently from anthropologists who lived in and with the cultures they studied.

Cultural psychology and cultural anthropology tend to define culture, cultural transmission, and cultural learning in ways that exclude non-human species by definition—on the basis of language or other concepts that are almost meaningless in the absence of language, such as intentionality of teaching and learning. Since we know that the most sophisticated language competence in a nonhuman species is roughly equivalent to that of a two-year-old human child in productive ability (Brown 1973; Greenfield and Savage-Rumbaugh 1991) or a two-and-a-half-year-old in receptive ability (Savage-Rumbaugh, Murphy, et al. 1993), and only when taught by humans, such definitions exclude naturally living nonhumans and impede the effort to identify elements in the social or mental life of animals that could have given rise to human culture.

In contrast, efforts to understand the dawn of culture focus on material artifacts, since that is what is available for study in the fossil record. All agree that the artistic creations of the people who lived, say, in southwestern Europe 30,000 years ago qualify as culture. But since these humans could have been living 300,000 years after the evolution of *Homo sapiens,* that conclusion does not help with the question of emergence or the problem of nonhuman-human boundaries.

The Approach from Material Culture

Since for most of evolution the only preserved artifacts are stone tools, these have long borne the burden of the cultural in archaeology. Mousterian culture, for example, refers to the stone tool assemblies associated with the remains of some Neanderthals, Acheulian culture to the hand axes made throughout the world by *Homo erectus* for around one million years, and Olduwan or pebble tool culture to the crude stones with one or two flakes chipped off associated with *Homo habilis* as long as two million years ago. Although named in each case for the type site where they first came to scientific attention, these artifacts occurred more widely in space and time; indeed, it is their lack of variation that allows some to question their cultural status, since a hallmark of culture in *Homo sapiens* is variety in space and progress in time.

We will return to that argument—consider how much of today's cultural variety would be missed if we had only hammers and knives to study—but for the moment we will accept the archaeologists' definition. The culture of the hominins at Olduvai consists, as far as we know, of stones with two or three chunks chipped off to make a sharp edge. They were simple enough to be mistaken for natural products—say, of a rockfall—until microscopic examination showed that they were used for butchering prey, cutting wood, and other tasks of interest to members of *Homo habilis,* who thus earned the designation "handy man."

This was not trivial: archaeologists took toolmaking seriously enough that even older texts sometimes referred to tool use by animals as an evolutionary precursor of culture. References to the apparently deliberate dropping of shellfish on rocks by vultures or to otters swimming belly up while they smashed mussels on stones held on their chests called vivid attention to the fact that animals use tools. In the 1800s anthropologists, interested in cultural evolution, studied these cases; Lewis Henry Morgan, known for his Iroquois ethnology and his evolutionary view of culture, also wrote a monograph on beavers and their works.

It was said at the time that animal artifacts are made by instinct, but that is not very satisfying. So when some of our closest relatives were found to be not only using tools but making them, this was of great interest. In the Gombe Stream Reserve in Tanzania, chimps of various ages sit in groups at the side of termite mounds inserting twigs into holes in the mounds, pulling termites out, and eating them off the twigs (Goodall 1986). Further, they strip small branches off the twigs so that they fit more easily into the holes. Finally, they are sometimes seen to prepare these stripped twigs at a distance from the termite mounds, on the way there. Unlike, say, the use of twigs by certain birds, there is no possibility that chimpanzee preparation and use of twigs for termite fishing is specifically hardwired. Field studies strongly suggest that juveniles learn it from adults by observation and imitation (Lonsdorf,

Eberly, et al. 2004; Lonsdorf 2006); significantly, baboons of various ages sit nearby and make the same observations, but do not reproduce the behavior.

To address the question of whether this and similar behaviors qualify as culture, formal criteria have been developed (Wallis 1992). Six of the criteria are taken from a classic anthropological source (Kroeber 1928): innovation, dissemination, standardization, durability, diffusion, and tradition. Two additional criteria, nonsubsistence and naturalness, are included in this formulation (Wallis 1992); language and symbol play no role.

Japanese macaques (*Macaca fuscata*) seem to satisfy Kroeber's criteria (the first six) with their well-known potato-washing innovation (Itani 1958), along with other socially learned, subsistence-related behaviors. Their new and socially learned behaviors of bathing in hot springs and in the sea may even satisfy the seventh criterion. None of these behaviors, however, meet the eighth criterion of naturalness, since all are directly or indirectly dependent on human intervention.

Chimpanzees come much closer. Chimp innovation under natural conditions is very rare, but the remaining criteria are collectively met by termite fishing, diver ant fishing, leaf sponging of water in inaccessible places, the grooming hand clasp, and other behaviors both instrumental and social. As a species, chimps probably meet criteria for material culture as well as *Homo habilis,* whose cultural behaviors can only be known from fossils and associated artifacts. Bonobos (*Pan paniscus),* as closely related to us as chimps are, use tools rarely in the wild and have not been seen to make them (Stumpf 2007). However, one bonobo, Kanzi, learned from a human teacher to make and obtain food with remarkably effective stone blades of the core-flake type (Toth, Schick, et al. 1993). If the hominins that made similar tools had culture, why not this bonobo? We know that *Homo habilis* passed their skills on from generation to generation, and that even if Kanzi did that the skill would still have had a human starting point, but chimpanzee toolmaking—equally complex but not set in stone—is both natural and transgenerational.

The Approach from Socially Learned Local Variation

Since local variation in traditions is a hallmark of human culture, the existence of nonhuman analogues is noteworthy. For example, as we saw in Interlude 1 on birdsong, in the white-crowned sparrow (*Zonotri-chia leucophrys*) in the vicinity of the San Francisco Bay, the male's spring song shows geographic variation that depends on a two-stage learning process: the infant male must hear the adult sing, and the adolescent must practice what he heard earlier. The learned local variants, stable over generations, are fairly called dialects (Marler 1993; Kroodsma 2004).

In oystercatchers (*Haematopus ostralegus*), two methods of mussel opening coexist (Norton-Griffiths 1969). The birds either hammer the shell with their beaks against a hard surface until it breaks or stab and pry into the opening and cut the adductor muscle so that the shell opens. Both techniques have been shown to depend on social learning and to require years of observation and practice. Individuals use one or the other method and tend to mate with individuals that use the same method.

In chimpanzees there is local variation in tool use, toolmaking, nest-building techniques, and gestural communication, all of which qualify as socially transmitted local tradition. Chimpanzees use tools in all thirty-six wild populations studied (McGrew 1992; Wrangham, McGrew, et al. 1994; Whiten, Goodall, et al. 1999; Whiten 2007). Gombe chimps alone show eleven different habitual tool-use patterns; as in other populations, termite fishing is the most frequent, but it is only one among many. Chimps in the wild use makeshift pestles, wave leafy twigs as fly-fans, aim and throw various objects, and remove marrow from bone with sticks. Research from seven field sites, totaling 151 person-years of observation, "turned up no fewer than 39 chimpanzee patterns of behavior that should be labeled as cultural variations, including numerous forms of tool use, grooming techniques and courtship gambits. . . . This cultural richness is far in excess of anything known for any other species of animal" (Whiten and Boesch 2001:62).

Except the human animal. The claim that a group of nut-cracking chimpanzees in the Tai forest "in many ways . . . could indeed be a family of foraging people" (Whiten and Boesch 2001:61) is problematic, to say the least. Wild chimpanzees are devoid of language and symbol, almost completely lack true teaching, and give no evidence of metacognition, or self-awareness of mental process; these are universal properties of human culture. Moreover, it is difficult to show consistent differences among localities in habitual chimpanzee patterns that serve similar purposes, so the claims remain open to the challenge that local "cultures" may just be idiosyncratic group responses to local conditions. Chimpanzee tool using, toolmaking, and the more variable aspects of communication may be a universal response to common environmental and social situations, manifesting itself differently in different locations (McGrew 1992, 1994). The local varieties could also result from genetic variation, which is much greater in chimps than in humans (Woodruff 1999). It is surprising to find true cultural variation in chimpanzee tools, since it is minimal in hominin evolution until early *Homo sapiens*, but chimps do show such variation.

The Approach from Teaching and Cultural Learning

Some of the most interesting attempts to define the boundary between human culture and nonhuman social learning involve observations and experiments in teaching and cultural learning (Tomasello, Kruger, et al. 1993; Kruger and Tomasello 1996). The opening sentence of a landmark paper, "Many animal species live in complex social groups; only humans live in cultures" (Tomasello, Kruger, et al. 1993:495), seems to limit culture to humans by definition, but in what follows it we are given some criteria: "Cultures are most clearly distinguished from other forms of social organization by the nature of their products—for example, material artifacts, social institutions, behavioral traditions, and languages. These cultural products share, among other things, the characteristic that they accumulate modifications over time." But the essence of cultural learning is that, "simply put, human beings learn from one another in ways that nonhuman animals do not." This is a well-defined and testable claim.

But first, cultural learning must be distinguished from social learning, which we will do more carefully in Chapter 26. Briefly, cultural learning involves a very advanced level of perspective taking or, if mutual, the intersubjectivity of teacher and learner; the relevant studies of apes have failed to show that they are capable of it (Lonsdorf 2006), with the possible exception of apes that are partly enculturated, with great effort, to a human symbolic world. In addition, a human teacher not only demonstrates and guides but displays the *intention* to teach by constantly readjusting the teaching behavior and materials ("scaffolding") to take account of the dynamic status of the learner in relation to the criteria of mastery. Finally, at least some human teachers keep at it until the learner has met the criteria. None of these standards has been met by any nonhuman primate.

So, despite the probable existence of true instructed learning in the cat family (see Chapter 19), it is unlikely that even the most advanced nonhuman primates have this ability more than trivially. Not only have humans "greatly increased the art of true teaching," but "teaching is inevitably linked to a rise in the complexity of language" (Bonner 1980:179). Since our wisdom has always been open to question, our species may merit the designation *Homo docens,* or "teaching person" (Barnett 1968), more than it does that of *Homo sapiens.*

The Approach from Language and Symbol

Still, crucial definitions of human culture rest on neither artifacts nor teaching but on language and symbol (Shore 1996). These should be separated, for at least two reasons. First, if language proves to be cognitively and neurologically modular, then to define culture by language rules out too many other mental functions—for example, image formation —that we might call cultural. Second, unless the definition of language is stretched greatly, there are countless phenomena in art and ritual that do not entail language per se, and some even exclude it. Still, language and symbol are central to the understanding of culture, especially when meaning itself is viewed as central. So what is the evidence that nonhuman primates have language?

Comparative Studies

As we saw in Chapter 5, there has been interest in the evolution of language for over a century, but it is very difficult to study. Endocranial casts of fossil hominids have been useful, but more indirectly, during the 1970s reports began to be published of the mastery by apes of some productive language. Such studies have continued, using not the vocal-auditory channel, which apes are ill adapted to use, but hand-signals from a signed language of the deaf, American Sign Language (ASL) (Gardner, Gardner, et al. 1989), arbitrary printed symbols for words (Premack 1971), or touch-sensitive arbitrary symbols on a computerized keyboard (Rumbaugh, von Glaserfeld, et al. 1974). When apes had mastered at least dozens of arbitrary symbols, some prominent psycholinguists (Brown 1973) conceded that they had progressed to a stage of human language typically reached by two-year-olds: the apes had gone beyond one-word utterances to two-word phrases.

One chimpanzee study (Terrace, Petitto, et al. 1979) seemed to show a complete absence of the simple word-order grammar that governs two-word phrases in human toddlers, as well as a lack of such elements of toddler language behavior as turn-taking with a conversational partner and naming things in the environment for the sheer fun of doing so. (Most—although not all—chimpanzees' acquired symbols are used instrumentally to obtain a desired object or activity.) Still, the apes mastered a large number of arbitrary symbols and used them in a manner displaced from the objects they signified, an achievement long considered far beyond their capacities.

By the late 1980s, however, some studies had suggested that the bonobo was superior to the chimpanzee in learning, especially language learning (Savage-Rumbaugh, Murphy, et al. 1993). Interest in this possibility was heightened by anatomical, physiological, biochemical, and behavioral evidence suggesting that we are more closely related to the bonobo. One bonobo, Kanzi, seemed more keyed in to human interaction and speech than any chimp had been, and investigators working with him saw a species difference (Savage-Rumbaugh 1991; Savage-Rumbaugh and Lewin 1994). He had a large vocabulary in the keyboard language "Yerkish," his utterances were rule-bound, and he had about as much grammar as a two-year-old (Greenfield and Savage-Rumbaugh 1991, 1994). This was not just subjective impression, nor was Kanzi unique. At least one other bonobo showed similar abilities.

Kanzi also appeared to understand much of what was said to him in complex English sentences, so this was tested. In a carefully controlled series of experiments, accepted as valid by most critics of ape language research, Kanzi displayed comprehension of English comparable to that of a two-and-a-half-year-old child (Savage-Rumbaugh, Murphy, et al. 1993); any doubts about Kanzi's ability now applied equally to the ability of any human toddler.

Other features of human language frequently called distinctive or unique include internalization of speech, using symbols in communication, passing them on to one's offspring, and inventiveness, metaphor, and meaning; all have been displayed by apes to some degree. Washoe was observed talking to herself in ASL (Gardner and Gardner 1974, 1975;

Gardner, Gardner, et al. 1989); two other chimpanzees, Austin and Sherman, communicated with each other in an experiment using a visual-symbolic language (Savage-Rumbaugh, Rumbaugh, et al. 1978); and Washoe communicated in ASL with her infant, who learned some signs in this way (Gardner, Gardner, et al. 1989).

Washoe invented the name "water bird" to describe a duck and spontaneously called a Brazil nut "rock berry" (Fouts 1974). Lana, learning the iconographic language Yerkish, knew the symbols for apple and the color orange and produced, after a few false starts, the request "Tim give apple which-is orange" to get an orange (Rumbaugh, von Glaserfeld, et al. 1974; Rumbaugh and Gill 1974). And the chimpanzee Lucy, while learning ASL, called radishes by the generic term "food" for three days, after which she spontaneously started calling them "cry hurt food" (Fouts 1974:379). It can be argued that most of these symbolic inventions had instrumental value, but so do most human ones; if produced by a small child, we would view these inventions not only as having meaning but (probably delightedly) as expressions of creativity.

Meaning has to do with the complexity of cognitive and emotional resonance around a stimulus pattern. Cross, crescent, Star of David, hammer and sickle, stars and stripes, the five Olympic rings, and even the Nike logo have resonances that ramify through brain and mind, often with substantial physiological consequences. So do family crests, team mascots, clinking glasses, blown kisses, and obscene hand gestures. None of these relies on language, but all draw on the human capacity for meaning; this, unlike some other aspects of language, probably resides in the frontal lobe, which facilitates almost endless associations (Deacon 1997a, 1997b) and, through the orbital frontal cortex, also gives emotional meaning to symbols (Damasio 1994). The frontal lobe, already large in the great apes, is larger, richer in intracortical connections, and more convoluted in humans (Rilling 2006, 2008; Sherwood, Subiaul, et al. 2008). Finally, the arcuate fasciculus, which connects the meaning-making frontal lobe to the grammar-making superior temporal sulcus, is uniquely prominent in humans (Rilling, Glasser, et al. 2008).

The Approach from History

The last claim for human culture is that, over time, it is cumulative. The closest instances to natural culture in animals are of two types. In the first, an invention is made by one individual in a species and spreads to others: One or a few blue tits discover that aluminum foil caps on milk bottles can be pecked through, and soon their comrades throughout the British Isles are visiting the sunny porches of pensioners and breakfasting on cream. Imo the Japanese monkey "genius" invents sweet-potato washing and the sifting of wheat from sand, and many others follow her example. A captive monkey learns to drink from a water fountain, and the ability soon spreads through his group. Yet these discoveries are never built upon; they remain isolated instances, neither leading to nor joining even the most slender stream of culture history.

The same appears to be true of protohuman traditions. The pebble tool "culture" of *Homo habilis* did not change notably until it was extinct; the Acheulian hand-ax tradition of *H. erectus,* one of its successors, remained essentially unchanged for a million years. In contrast, even the simplest and most static of human cultures is an engine of inventive mutual influence and change.

Furthermore, at least orally, human cultures preserve *historical* record, imaginative or real, couched in a human language. The past pervades human consciousness to some degree even in the simplest societies, and discussions of past events—narrating, sometimes dramatically, commenting on the narration, challenging points of fact or logic, and co-constructing a suite of stories—occupied many an evening for perhaps 300,000 years, but not for millions of years before that. And while our ancestors were arguing, any ape communities not far away in the forest were making their—yes, traditional—nests and drifting off to sleep. The only modern apes that have learned language learned it from human teachers, and none of their wild counterparts has anything like it. Even if their individual minds preserve some private history, it is difficult to see how they could have a collective one without being able to tell it to each other and to their young. All human cultures can, do, and probably must.

Steps have been made by nonhuman species in the direction of culture as defined by some of these approaches. If we are to dignify the Olduwan tool assemblage with the name *culture,* then certainly some of the achievements of chimpanzees in the wild deserve the same name. But no chimpanzee has been seen to make anything as complicated as an Acheulian hand ax, which members of *Homo erectus* made a million years before our own species. As for teaching, there is very slight evidence of it in chimpanzees, and none in other nonhuman primates. Felids evolved a pattern in which adults facilitate the learning of prey-catching skills by young, and in the cheetah at least this has reached a level that probably deserves the name teaching. But there is no other teaching in the cat family, and nothing like it in apes.

Considering language and symbol, all four species of great apes have shown some ability to master human symbols, but none has been seen to exhibit any comparable ability in the wild. Vervet monkeys show a limited range of semantically significant utterances in nature, so parallel abilities in apes may reward further study. But in any case, the predator-specific calls of vervets do not constitute being "suspended in webs of significance" that they themselves have spun. As for historicity, there is no evidence for its existence in any nonhuman species.

Even the most generous appraisal of the achievements of nonhumans leaves them just a short way down three or four of these five paths to culture. For humans to have evolved it, our ancestors had to have greater attainments in all these areas, and without assuming that they evolved strictly in concert, they must have influenced each other in a pattern of circular and cumulative causation—a doubly meaningful self-organization in the cultural realm. By the "more is different" rule of complex systems, even if the human distinction on each of the five dimensions was merely quantitative, the integration of the human advantages in each of them would have systemic effects that amounted to an emergent qualitative difference.

While research on animal protoculture, by whatever definition, is of great interest, we are studying rudiments—roots of culture, not the tree itself. Only humans, among primates, teach. Only humans create and build culture cumulatively. Only humans are suspended in webs of significance we ourselves have spun. And each of these claims applies in full to the simplest extant human cultures ever visited. By any reasonable criteria not restricted to one or two domains, the answer to the question "Do animals have culture?" is "No."

But as we arrogate to ourselves this admirable uniqueness, we should be careful not to exaggerate the importance of culture. It is not the same as, nor is it necessary for, social life, relationships, family, love, grief, cooperation, competition, conflict, altruism, sacrifice, heroism, loyalty, jealousy, pride, shame, status, dignity, awe, insight, creativity, memory, or thought. It pervasively tinges but does not decisively influence all human behavior. Many nonhuman species survive and breed through complex socialization, without culture, and that involves modifying behavior in the young to fit the demands of life among others over a long course of dependency. As a result of this necessity, many qualities and capabilities that we consider essentially human emerge and mature. They are not cultural products, although they are modified by culture. They are products of a brain with certain properties developing in a socializing context, with or without enculturation. If we deny culture to animals, we should be careful not to encompass in that denial certain "distinctively" human capacities that are not primarily cultural and that are not in fact distinctively human.

IV

Enculturation: *The Transmission and Evolution of Culture*

---

wherein we come to understand what culture changes

The development of human speech represents a quantum jump in evolution comparable to the assembly of the eukaryotic cell.
—Edward O. Wilson, *Sociobiology* (1975)

23

Paradigms in the Study of Enculturation

Despite the strong role played by phylogenetically primitive brain mechanisms in human social behavior, and despite the power of the genes, human beings have evolved a unique role for cultural experience, based on unique features of human brains. As the cultural anthropologist Claudia Strauss (1992:4) has aptly said, "The fact that a human baby born anywhere in the world can acquire language and culture anywhere in the world while a chimp baby cannot (or cannot in the same way humans do) indicates that all humans have a built-in receptiveness to the form human cultures take, and all human cultures probably share some bedrock commonalities because of these co-evolved features of human neurophysiology and morphology." And Roy D'Andrade (2001) has cogently argued that in a deep sense there is only one human culture.

Still, some fundamental processes in enculturation at the individual level are the same as those in socialization and are shared in common between humans and many other animals. The basic paradigms of learning psychology and early experience effects were introduced in Chapters 13 and 14. To understand how the processes of psychosocial ontogeny, generated by evolution and canalized through neural and neuroendocrine growth, go on to produce the remarkable variety we see in human culture (at a more superficial level than D'Andrade's), we must extend our treatment of learning and integrate individual and cultural experience. Accordingly, this chapter briefly introduces four additional paradigms: (1) laws of learning, expanded; (2) culture and personality; (3) the Whiting model; and (4) culture and mind.

Laws of Learning, Expanded

The basic laws of learning described in Chapter 13—habituation, associative learning, classical conditioning, operant conditioning, avoidance learning, emotional conditioning, and some forms of "special learning"—apply to humans and rats alike, although they do not of course explain all human learning. Still, the first even modestly successful treatments for autistic and retarded children have been based on such knowledge, and behavioral psychotherapy for some psychological conditions is widely accepted (Goldstein and Foa 1979); it helps in fear of flying (Klein 1999), fear of spiders (Mineka, Mystkowski, et al. 1999), fear of blushing (Mulkens, Boegels, et al. 1999), aggression in some children (Mineka, Mystkowski, et al. 1999), and auditory hallucinations (Shergill, Murray, et al. 1998).

But current learning theory goes beyond these processes. Duane Rumbaugh, who began as a conventional animal learning psychologist, has attempted to reconcile learning theory with advances in cognitive science, showing that higher-primate brain evolution produced an increase in cognitive processing power that not only overshadows simple learning laws but results in something qualitatively different (Rumbaugh, Washburn, et al. 1998; Rumbaugh and Washburn 2003). Extending Harlow's concept of the learning set—the ability to apply what has been learned to new and different tasks—he developed a novel method, the transfer index, to assess nonhuman cognitive competence. It showed a strong correlation between relative brain size and the ability to generalize over different learning tasks.

Primates with simpler brains, including prosimians and even some New World monkeys, do *worse* in a new learning challenge if previously trained in another, related task; the habit laid down by past training is too persistent. But for most monkeys and all apes, a point is reached when the prior experience makes them step back from the specific task and begin learning again with a different mind-set. At the level of the transfer-index task at least, they have learned to learn, a first step in metacognition—thinking about thinking. The transfer index measures an evolutionary increase in mental power and reflects the fact that in brain evolution as in many other natural phenomena, more is different (Anderson 1972).

This nonlinearity is the fundamental insight of complexity theory, and it applies in many branches of science. Quantitative change eventually produces a qualitative improvement, as in the learning experience of apes. Applying complexity theory, Rumbaugh used the term *emergent* to designate the insight in which an ape appreciates that although the game is similar, the rules have changed; this qualitative cognitive advance helped lay the evolutionary groundwork for human thought. This advance also facilitates apes' enculturation by human investigators into a world of symbols and meaning (Rumbaugh and Washburn 2003).

But the ability measured by Rumbaugh's transfer index is only one of several ways in which an internal cognitive process intervenes between stimulus and response to enhance conventional learning. In observational learning, common in mammals and birds, the chance to observe a response speeds subsequent trial-and-error learning. It is one basis of formal training in many cultures, and a basis of informal training in all of them (Mead and Wolfenstein 1955; Mead 1964; Kruger and Tomasello 1996). But as we saw in Chapter 19, imitation does not always involve learning (Zentall and Galef 1988), and it can be hard to distinguish from social facilitation—the wave of yawns in a theater. Some well-studied cases, such as the social enhancement of overeating (Wyrwicka 1976) or the tendency of children to pummel a large doll after seeing an adult on film do the same (Bandura and Walters 1963), are probably intermediate points on a continuum between social facilitation and imitation.

Farther along the same continuum is true imitation (Kruger and Tomasello 1996); the learner does not just watch the activity—say, weaving a shawl—but tries to take the model's perspective, "seeing" and "feeling" the warp and woof as the model sees and feels them. In the still more complex interaction of true teaching, the teacher not only sees and feels as if with the child's hands and eyes, but then adjusts the task to make the next, observable step in learning attainable, often by making it progressively harder. This is scaffolding, and it requires a kind of intersubjectivity that evolved along with the human brain.

Observational learning at its most advanced is possible only in humans, but many other animals use it. As mentioned in Transition 3, the oystercatcher depends on it for subsistence (Norton-Griffiths 1969). For two years the young watch as adults open oysters and then, gradually, the young improve their skills, but it is unlikely that they do this through true intersubjectivity. Being with the parent for two years during feeding affords the young the opportunity to learn by trial and error, with an assist from observation—socially facilitated learning. At a higher level would be chimpanzee termite fishing, in which the young not only are afforded the chance to practice but watch the adults carefully, apparently studying the process (Lonsdorf, Eberly, et al. 2004; Lonsdorf 2006).

Early experience effects (Chapter 14) are a type of special learning, in this case because evolution has shaped a maturation process in which sensitive periods make the brain susceptible to certain kinds of input. In human cultural terms, this is expressed in the saying, variously attributed to Lenin, Kierkegaard, Jesuit educators, and many others, "Give me a child until he's seven and he'll be mine forever" (Dru 1959:170). This belief, which is based on the notion that the young child's brain faithfully records sense impressions in a way that the more mature brain does not, has shaped theories of psychotherapy and education and, to the extent it is true, has major implications for cultural transmission and enculturation.

Some theorists shrink this period to three years, attempting to ground the process in brain development—the "infant-determinist premise" (Bruer 1999:206). What happens in this stage of development is important, but so are later experiences (Kagan 1984). Small interventions at age eight can produce large changes in children's behavior (Kruger 1992), and the environment during adolescence shapes adult behavior as much as or more than anything that happens earlier (Hetherington 1993). However, given strong evidence for early sensitive periods in animals (Chapter 14) and because the first six years of human brain development entail extensive shaping of neural circuits through selective, activity-dependent pruning of axons and synapses, there are probably important early sensitive period effects in humans as well.

Culture and Personality

The idea that cross-cultural differences in perception, thought, and action go deeper than consciously chosen customs is ancient, and the implication is that ethnic groups have different psyches. The culture and personality paradigm of psychological anthropology addressed these questions, starting from culture rather than biology. Prior to this development, racial theories prevailed; in the United States in the early nineteenth century, "anthropology" was largely synonymous with scientific racism, and it figured importantly in the "intellectual" basis of slavery (Stanton 1960; Chorover 1979; Gould 1981). A century later Nazi racial theories led to the murder of six million Jews and hundreds of thousands of Gypsies, among other minorities, based on alleged psychological and moral defects associated with ethnicity or race.

Early modern notions of national character were also a pseudoscience of ethnic bias:

> The Franks are simple, blockish and furious; the Bavarians sumptuous, gluttons, and brazen-faced; the Swedes light, blabbers and boasters; . . . the Belgians good horsemen, tender, docible, and delicate; the Italians proud, revengeful and ingenious; the Spaniards disdainful, cautious and greedy; the Gauls proper, intemperate, rash-headed; . . . the Bohemians cruel, lovers of novelties, filtchers; the Greeks miserable. (Mercator, quoted in Harris 1968:400)

Franz Boas (1940b, 1940c), a founder of American anthropology, and other twentieth-century anthropologists demolished racial theory, replacing it with the concept of culture (Stocking 1968). But some, including students of Boas like Ruth Benedict and Margaret Mead, found a quite different version of national-character theory acceptable: Races by any criteria are not cultural units; each embraces many cultures, and culturally convergent groups embrace different races. But it seemed more plausible that a nation could be a single cultural unit, a large-scale version of the much smaller cultures that ethnologists typically study; nations, like tribal units, might have a cultural style or pattern leading to strong commonalities in behavior, cognition, and emotion.

Measuring Personality Cross-Culturally

The first scientific attempts to discover culturally determined personality differences were made in studies of small-scale, traditional societies. During the 1930s and 1940s, psychoanalytic theory seemed to many the most promising domain in psychiatry, and anthropologists viewed it as a tool for the study of culture (Spindler 1979; LeVine 1982, 2007). Substantial interaction between psychoanalysts and anthropologists was focused on processes by which culture might influence personality development and by which personality, in turn, might sustain cultural differences (Kardiner and Linton 1939, 1945). But as George Peter Murdock (1945), a pioneer of systematic cross-cultural study, put it, there is "a common denominator of cultures." And as Alfred Kroeber and Margaret Mead, leaders of cultural anthropology, recognized by the 1950s, the delineation of cross-cultural variety also inevitably reveals constant features of human behavior and mental life, and many invariants of language, culture, cognition, facial expression, parent-offspring interaction, and other aspects of human behavior are now apparent (Brown 1991).

This is in some ways a return to the field's origins. Darwin's 1872 book on facial expression cited the occurrence of certain expressions in primitive societies as evidence for their biological basis, and he attempted to relate them to facial expressions in nonhuman animals. Edward Westermarck (1922), whose theory of incest aversion ("familiarity breeds contempt") was discussed in Interlude 6, appealed to ethnography to illuminate a proposed universal of psychodynamic development. In the early twentieth century, ethnographers tested members of "primitive" societies for proposed perceptual universals. It was also common for anthropologists to array non-industrial societies in an evolutionary sequence of social complexity, religion, or culture.

Anthropologists came to reject the evolutionary sequencing of cultures, replacing it with respectful description of cultures as independent human achievements. Proposed universals were met with derision, and ethnographers loved to say, "Not among my people they don't"—the anthropological veto. With the rise of postmodernist anthropology in the 1980s and 1990s, not only the search for universals but even any effort to define the features of an individual culture came to be viewed as "essentialist" fallacies stemming from the ethnocentric arrogance of scientists (Rosaldo 1993). For most of the twentieth century, however, it was assumed that ethnology could scientifically challenge psychological truisms. Early claims of Piaget and others were tested cross-culturally, especially those of psychoanalysis (Spindler 1979). We saw in Interlude 5 that Bronislaw Malinowski attempted to demolish the universality of the Oedipus complex by describing a separation of male authority (vested in the mother's brother) from the object of male jealousy (the biological father) in the Trobriand Islands, a claim long debated (Spiro 1982). Some American anthropologists undermined psychoanalytic claims with cross-cultural data (Mead 1928/1961), but at the same time used psychoanalysis to try to explain culture (Kardiner and Linton 1939, 1945).

The fundamental theorem of this school was that cultures differ because of different child-rearing patterns and that psychoanalysis and cultural anthropology could unite to explain cultural variety and elucidate laws of psychological development. In *Patterns of Culture,* Ruth Benedict (1934) contrasted the Kwakiutl Indians of the Northwest coast of the United States, whom she described as "Dionysian" and inclined to uncontrolled emotion, with the "Apollonian" Zuni of the southeastern states, who were supposedly reserved and controlled. Similar speculations were made about national character, especially during and after the Second World War (Benedict 1946; Gorer 1962; Mead 1965).

Later, samples of societies and cultures from the Human Relations Area Files were used to test hypotheses about the relationship of child-rearing practices to other aspects of culture. In an influential study, Whiting and Child (1953) showed that childhood experiences of interest to psychoanalysts were correlated with themes in religion, folklore, and other cultural expressions in a large cross-cultural sample. However, they knew that such correlations might arise from causes other than psychodynamic ones and that childhood experience had to be measured more rigorously.

Basic, Modal, or Variable?

By the 1950s the paradigm had produced results that undermined some of its own propositions. Attempts were made to measure adult psychological disposition and child training through personality testing and direct behavioral observation.

Cora DuBois (1944) challenged the paradigm in her study of the horticultural and fishing people of the Pacific island of Alor. She did a standard ethnography, including detailed descriptions of child care and long psychological interviews with a sample of Alorese adults, and gave the then-popular Rorschach inkblot test to thirty-seven people. Both the Rorschach results, which were assessed by an expert on the test, and the eight life histories, analyzed by a psychiatrist, showed great variation. One man was described as passive and inhibited, another as a strong character without inhibitions, and a third as a prophet who attempted to launch a religious revival. The psychiatrist, Abram Kardiner (in DuBois 1944), one of the originators of the basic personality theory of culture, tried to show how these three and others with unique personalities all fit what he saw as the basic personality of Alor, but what stands out about Alorese personality is the marked variation in a small-scale society.

An alternative, *modal* personality—patterns shared by a plurality of a group—was exemplified by a Rorschach study of Native Americans of the Tuscarora and Ojibwa ethnic groups (Wallace 1952). Averaging responses over each sample population yielded no meaningful patterns, but 37 percent of the Tuscarora shared a similar pattern of responses, while only 5 percent of the Ojibwa exhibited that pattern. This was interpreted to mean that the Tuscarora were "more able to undertake a wide range of social relationships," while the Ojibwa tended more to avoid any form of dependency; this difference was statistically significant, but the 63 percent of Tuscarora who did not share the modal personality pattern varied widely—as in Alor, and as we now know in all societies.

In fact, using far more sophisticated measures of personality, the same five factors emerge from each of fifty cultures (McCrae, Terracciano, et al. 2005). However, this is not to say that there are no differences in the distribution of personalities within cultures. There could still be some consistent differences, owing to gene-culture coevolution, child-rearing practices, or other factors, along the lines of modal patterns, and different personalities could play different roles within or outside the modal pattern. In this view, a culture resembles a symphony orchestra; strings may predominate in one symphony, brass in another, but more of the difference is due to the score than to the distribution of instruments. Culture is then not just a mechanism for psychodynamic self-reproduction through child rearing, but an algorithm for articulating the inevitable variation in personalities. We will return to this notion of culture as an articulating algorithm.

But it is worth noting here the similarity between this model and the evolutionarily stable strategy (ESS) of behavioral ecology. Just as Hawks and Doves or Cooperators and Defectors can reach an equilibrium in the quest for reproductive success, so the natural variants of human personality—extrovert and introvert, anxious and confident, conscientious and laid-back, and others—can reach an equilibrium within a culture. But in human groups culture itself is a major player; it sets parameters that alter the balance between and among types. This is in fact a major process in gene-culture coevolution and, as we will see, helps give cultures their distinctive qualities.

Kinesics and Proxemics

Some anthropologists, critical of the personality construct, sought cultural differences in basic behavioral tendencies. Research on cultural variation in body movement and interpersonal spacing led, respectively, to *kinesics* and *proxemics* (Hall 1959; Birdwhistell 1970). For example, southern Europeans move their hands while speaking more than northern Europeans do, and studies began to explore how the typical postures and movement patterns of a human group are tied to its style of dancing, singing, and work positions (Lomax 1968). Such studies, sometimes called *choreometrics,* used filmed performances and showed, for example, that separate trunk and pelvic movements are common in the dance and movement of African cultures but very uncommon in Native Americans. Physical spacing patterns also differ among cultures (Hall 1963); in a classic example, Arabs traditionally stand or sit closer together during business conversations than Americans do, making both uncomfortable in an interaction across the cultural boundary. Human ethologists have found many universals of expression, movement, and spacing (Konner 1972; Eibl-Eibesfeldt 1988; Ekman and Rosenberg 1997), but universality does not imply the same frequency, intensity, or contexts.

Measuring Child Behavior

Boas (1940a) encouraged students like Margaret Mead to study children, and this led to influential descriptive ethnographies of childhood (Mead 1928/1961, 1930), but descriptive ethnography was later supplanted by systematic quantitative research. This approach, pioneered by Beatrice Whiting and John Whiting, represented a major advance in methodological precision, and we encountered it several times in Part III. The Whitings' contributions included a method for the reliable measurement of children's social behaviors in different cultures and a causal theory relating child training, personality, and behavior to ecology, society, and culture. This work was carried forward by Robert LeVine, Robert Munroe, Ruth Munroe, Susan Seymour, Carolyn Edwards, Thomas Weisner, and others.

The Whitings recognized that neither personality testing, with its difficulties of interpretation, nor traditional ethnography was adequate for psychological description, yet they sought stable cultural differences in behavior and emotional expression. Their standardized system for observing and recording social behavior assumed that group differences in personality would be revealed in measurable day-to-day behavior. They observed children rather than adults, reasoning that children's outward behavior would reflect inner processes more honestly than adult behavior would, and hoping to trace the course of development that would explain cultural differences.

Their code had a standard, defined set of behaviors ("offers help," "insults," "assaults," and so on) that was applied in a uniform way to children of many cultures. By counting occurrences in a specified time period, quantitative ethnographers could compare, say, nurturance or aggressiveness cross-culturally. Multivariate statistical methods such as multidimensional scaling were used in analysis. The Whitings devoted thirty-five years to such measurement in societies around the world (Whiting and Whiting 1975a; Whiting and Edwards 1988), and many other investigators have followed (Munroe, Munroe, et al. 1981; Edwards 1993; LeVine, Dixon, et al. 1994; Seymour 2001; Munroe 2004). Cultural differences in children's social behavior were shown and led to interesting generalizations. For example, rates of aggression are lower in communities with large extended-family households than in those composed of individual nuclear families, probably because aggressiveness cannot be tolerated as well in a dense collection of relatives. As in the personality studies, however, there was not uniformity within cultures but organized diversity; sex and age differences within cultures were of the essence of social structure and function.

Beatrice Whiting and Carolyn Edwards extended the Six Cultures Study to six *more* cultures and analyzed the data for all twelve in a landmark book, *Children of Different Worlds* (1988). With the exception of Orchard Town, a New England community, all of them are non-industrial, non-hunter-gatherer cultures—intermediate-level societies. This analysis has been referred to above and will come up again, but to summarize briefly:

1. There are impressive cross-cultural consistencies specific to certain dyads, such as mother and infant, mother and school-age child, or school-age child and toddler. Infants "have the power to elicit nur-

turance and responsiveness from children as young as 2 years of age. The universality of this pattern . . . suggests that infants are born equipped with physical features and behavioral systems that evoke positive behavior and nurturance from both adults and children" (Whiting and Edwards 1988:270).

2. Of four major categories of children's behavior—"nurturance, dependency, prosocial dominance, egoistic dominance—two emerge as associated with sex differences." In early and middle childhood, "girls on average are more nurturant than boys in all dyad types . . . while boys are more egoistically dominant than girls" (Whiting and Edwards 1988:270). But girls spend proportionately more time with infants, often as assigned caregivers, and boys with peers, so it is difficult to separate biological differences from the influence of setting and learning. Boys are also found farther from home cross-culturally.

3. With the exception of the Americans (and unlike most hunter-gatherers), children are assigned many chores in these cultures. Girls do more work than boys except among the Kenya pastoralists, where boys herd livestock and the sexes do about equal amounts of work. Child nurses are important in all the non-Western cultures and after age seven they are almost always girls.

4. Despite similarities, cultural differences are apparent in the rank order of some maternal behaviors, for which the cultures tend to fall into three groups: "training" is emphasized in all the sub-Saharan African cultures, "sociability" in Orchard Town, and "control" in the others. Training involves guiding and teaching children in assigned chores, while control includes reprimanding, threatening, or punishing. However, these are differences in emphasis and frequency, not absolute patterns, and are explained in relation to the mothers' workloads; African mothers have the heaviest farming duties, so they assign and teach more tasks to their children.

The Whiting Model

Based on the original Six Cultures Study and other data, Whiting and Whiting (1975a) developed an influential model of the causal roles of fundamental features of society—such as ecology, economy, and vulnerability to external attack—on child training practices, which in turn might give rise to certain consistent adult predispositions. Although it was rarely possible to establish such causality, the model had the virtues of clarity and falsifiability.

The widely recognized version of the model posits that (1) some aspects of society and culture, called *maintenance systems,* are likely to determine major features of childhood experience; (2) such childhood experience markedly influences the adult personality of the typical member of the society; and (3) some other aspects of society and culture, called *projective systems,* are more likely to be consequences of the typical adult personality and so of the childhood experience. The rationale for indicating the flow of causation in one direction was not to rule out feedback but to encourage testable predictions.

It is also, without explicit mention, partly Darwinian in intent. That is, the subsistence environment was considered primary, and the society and culture a response to it. In today's terms, the mothers with heavier workloads have intensified parent-offspring conflict; however, the Whitings would have said that parents raise their children to get the adults their culture needs, consistent with what we would now call facultative adaptation. The evolutionary models in no way contradict the developmental ones. Adding local history does not change the basic concept but allows for divergence due to, say, culture contact or exceptional leaders. Later in the book we consider modifications to the model based on evolutionary (including life history) theory and phylogenetic constraint.

Broader Cross-Cultural Analyses

Later cross-cultural applications of psychoanalytic theory continued to have at least heuristic value (Levy 1973; Paul 1989), but most recent studies rely on measurements of behavior and development. Some use the Human Relations Area Files (HRAF), encountered in Part III (Barry and Paxson 1971; Ember 1994). This data set consists of nested samples of societies and cultures throughout the world, especially a core sample of 60 and a larger sample of 186. Criteria for inclusion include (1) the quality of data (the training and language competence of the ethnographers, the person-years of study at the field site, the number of published pages, and other measures); (2) geographic and cultural representativeness of the entire known range of several thousand societies and cultures; and (3) freedom from mutual influence (Galton's problem) in order to maximize the likelihood that each society entered in the world sample functions as a statistically independent unit.

Consider a cross-cultural study that exemplifies both the appeal and limitations of the approach (Whiting and Whiting 1975b). The analysis targeted a dimension labeled "husband-wife intimacy," measured by three independently rated but intercorrelated measures: whether husband and wife ate together; whether they slept together; and father involvement in the direct care of the children. All three vary markedly in the world cultural sample, but with high covariance. Also, all three relate to how much a society is involved in preparations for war. At one extreme are bellicose cultures like those of highland New Guinea; there men eat and sleep in collective houses, separate from women and young children, and teenage boys begin their martial training in these separate houses for men. At the other extreme are non-warlike societies like those of the Pacific islands, which have high husband-wife intimacy and more contact of men with infants and young children.

Applying the model to these correlations, some societies are thrust into warfare because of demographic and geographic conditions, while others (for example, those on small islands) are protected. Distinct maintenance systems arise in each type of society, with consequences for child care. Among these are bringing boys into an all-male world to train for war. Projective systems, such as distinctive male dress and hairstyles, religious beliefs, and beliefs about male superiority, would be seen as epiphenomena of the typical male personality.

This study is appealing because it involves primarily commonsense assumptions, but it remains an interpretation of correlations, not a causal demonstration. It might be argued that in the warlike societies the first event was a historical accident—say, a very belligerent leader. This individual could have invented men's houses and trained boys for war, which eventually led to a state of chronic belligerency with neighboring societies. In this kind of model, accepted by some historians and anthropologists, ideology and individual predisposition precede fundamental environmental adaptations. Recent historical ethnography of highland New Guinea, for example, casts doubt on the hypothesis that tribal belligerence was caused by increasing population density (see Wiessner and Tumu 1998:143–50 for discussion). Although this may be plausible for a given society at a given moment, it is difficult to argue convincingly that the regularities observed in broad cross-cultural analyses, involving scores or even hundreds of societies, could emerge from a collection of such ideological or historical accidents.

Extensions and Modifications of the Model

Carol Worthman (2010b) has reviewed the Whiting model and its heirs and set them in historical context. Three extensions are of special interest; their differences are mostly in emphasis. First, Thomas Weisner's (1997, 2002) ecocultural version emphasizes the starting point of the model—environment, history, and maintenance systems and how they shape the child's learning environment. The Whitings and Carolyn Edwards emphasized such variables as the mother's workload, the need for child nurses, the social structure of the family and the family compound, and the threat of external attack; Weisner has tried to understand how these variables translate into the day-to-day ecology of child life—not just their preparation for adulthood, but the simple constraints on their lives that have little to do with parental goals. His initial fieldwork was on the Abaluyia of western Kenya, but he has applied the same approach to how sibling relationships respond to different ecological pressures in different cultures, as well as to the very different adaptations of poor urban and counterculture families in California. These different family lifestyles have afforded Weisner and his colleagues the opportunity to carefully follow infants and children over an eighteen-year period with a variety of psychological measures. Among other results, they find negligible or mildly positive consequences of exposure to family nudity (Okami, Olmstead, et al. 1998) and no detectable positive or negative consequences of bed-sharing in infancy and childhood (Okami, Weisner, et al. 2002).

Robert LeVine and his colleagues (LeVine, Dixon, et al. 1994), especially in their quantitative ethnography of the Gusii (see Chapter 26), emphasize a different part of the flowchart, as do Sarah Harkness and Charles Super in their longitudinal multi-nation study of parental beliefs and practices. They focus on the child-care-related beliefs and goals of caregivers ("caretakers" in the older studies), which are only partly attributable to ecological causes. LeVine and his colleagues contrast the "pediatric" model of parenting among the Gusii—having protection as the goal and soothing as the means, decreasing interaction after infancy, and using the main cultural "scripts" of responding to distress, modulating excitement, and giving commands—with the "pedagogical" model of Americans—having active engagement and social exchange as the goals and stimulation and protoconversation as the means, increasing interaction after infancy, and using scripts of eliciting excitement, asking questions, and giving praise. These categories roughly correspond to Whiting and Edwards's controlling and sociable mothers, but they viewed sub-Saharan African cultures as emphasizing training rather than control. In any case, the LeVine group attributes the contrast at least as much to cultural goals as to ecology.

This is equally or more true of the Harkness and Super "developmental niche" model, which is based on collaborative research with coordinated teams in seven Western industrial states (Harkness, Moscardino, et al. 2006). By restricting the range to one modern ecological subsistence type, these researchers can highlight differences in parental beliefs and practices that are unlikely to reflect ecological pressures. They apply a common code of descriptors to parents' discourse about infants and children, a child temperament questionnaire, a more open-ended exploration of parents' ideas about co-sleeping, and an attempt to extract "cultural themes" from various kinds of data, including data from child care and school settings as well as the home. The fact that these cultures, all in modern ecological settings, have significantly different beliefs, practices, and results in infant and child care shows that cultural history and ideology exert subtle and important influences.

These might originate in the history part of the model, but would also require feedback from cultural ideology to the child's learning environment—which would not surprise a culture-and-personality theorist or anyone for whom culture is self-reproducing. Parenting practices are often implied or explicit in religion and ritual, the arts, games and play, attitudes, and many other aspects of culture not directly determined by either biology or ecology. However, having begun with research on infants and children among the Kipsigis of western Kenya (Super 1976; Harkness, Edwards, et al. 1981), Harkness and Super would not deny a role for ecology, especially in explaining differences in parenting in dramatically different ecological settings.

Finally, Worthman's own "bioecocultural" model draws on decades of research, including multi-year studies of adolescents in a Kikuyu community in Kenya (Worthman 1986; Worthman and Whiting 1987) and in three subcultures of North Carolina (Costello, Angold, et al. 1996; Angold, Costello, et al. 1999), among other cultural settings. Her model resembles the others but emphasizes biological dimensions. It focuses on the "microniche," the child's immediate environment, anticipated by biology but shaped by both ecology and culture. "The developmental microniche comprises the zone of proximal development representing the lived experience and actual operating conditions of the child *in relation to* the child's characteristics and capacities. . . . It is through the microniche that culture gets under the skin" (Worthman 2010b). The Vygotskian concept of zone of proximal development is important; the child can only respond to the immediate stimulus envelope, and if it is chaotic, unduly stressful, or meaningless, the result may be unfavorable, but if it is appropriate to the child's needs and developmental level, both adaptation and development are enhanced. But Worthman—a biological anthropologist and endocrinologist—differs from typical Vygotskians in that, for her, development is thoroughly embodied.

Challenges to the Role of Early Experience

Some challenges to the model lie not in the relationship between maintenance systems and ideology, nor in that between parental beliefs and child care practices, but, surprisingly, in the relationship between childhood experience and adult predisposition. In the 1970s Jerome Kagan began studying cognitive development in Guatemala (Kagan and Klein 1973; Kagan 1977; Kagan, Klein, et al. 1979), and this work led him to challenge the fundamental assumption that early childhood experience has a special and lasting effect on the formation of adult characteristics. A study of cognition in a remote Guatemalan Indian village found that lack of stimulation and substantial deficits in infancy did not lead to deficits in later childhood, and that within the United States day care attendance for eight hours a day throughout infancy did not affect available measures of cognition, attachment, or other dimensions of behavior (Kagan, Kearsley, et al. 1978). This and other research before and since (Caldwell 1964; Caldwell, Honig, et al. 1970; J. Harris 1995; Bruer 1999) seemed to support the counterclaim that little that happens in early life is irreversible.

While psychodynamic clinicians routinely accept retrospective interviews as evidence, they are not persuasive. Many developmentalists do prospective studies using excellent measures that seem to show consequences of early experience for later development, but without random assignment of subjects or genetic controls, the correlations can be otherwise explained. Reviews of research on the lasting effects of early deprivations or interventions have often raised doubts. This includes studies of low Apgar scores, lack of stimulation in infancy, breast-feeding versus bottle-feeding, day care in infancy, Head Start preschool interventions, and a host of other wide-ranging variables. In general, the more rigorous the study and the longer the follow-up, the less detectable the effect (Kagan 1984; Bruer 1999).

Others have specifically challenged the role of parental nurturance and place more emphasis on the role of peers in socialization (J. Harris 1995, 1998). Such an emphasis is compatible with the cross-cultural model, which does not specify that developmental influences must be parental, but it still flies in the face of standard beliefs among clinicians, educators, and parents. In fact, we now know that about half the normal variation in personality is due to genetic variance (Bouchard and McGue 2003; Caspi, Roberts, et al. 2005), with stability across the life span (Shiner, Masten, et al. 2003), especially in adulthood (Johnson, McGue, et al. 2005; Terracciano, Costa, et al. 2006; Hampson and Goldberg 2006), and that child-rearing patterns have limited effects (Rowe 1994; Loehlin, McCrae, et al. 1998); this generalization may also apply to chimpanzees (Martin 2005). Personality disorders are probably ends of continua of normal variation in personality (Trull and Durrett 2005).

Counterarguments are numerous: measures in childhood or adulthood may be inappropriate, behavior under stress rather than baseline behavior may be the right outcome measure, random assignment of subjects is unethical, and so on. But the fact remains that developmental psychologists, psychodynamic psychiatrists and other psychotherapists, and educational researchers of many theoretical and methodological persuasions have failed to show to the satisfaction of reasonable skeptics that decisive, lasting effects of early experience in the normal range exist, much less how such effects might work.

The most decisive evidence comes from studies with animal models (Chapter 14), which remain a touchstone. Animal studies using random assignment and rigorous control of other independent variables have repeatedly shown that early experience can have a lasting effect not only on behavior and psychological predisposition but also on neural and neuroendocrine structure and function. Only an assumption of the most unlikely discontinuity between the nature of human and animal brain and behavioral development could support a denial of similar effects in humans. But to hold strong specific beliefs about how early experience and cultural variations in child care influence adult personality without better evidence will probably impede discovery.

Culture and Mind

A persistent stereotype—more respectable than the racial and national-character theories described earlier, but almost equally patronizing—was the concept of "primitive mind." Mental capacities have grown more complex during human evolution, over millions of years, but in the early twentieth century there was a belief among psychologists that some contemporary human cultures represented earlier stages of mental evolution (Werner 1948). Language barriers and cultural differences prevented some psychologists from appreciating the quality of mental functioning in traditional cultures.

Proper studies of "primitive" language and thought revealed them to be complex and sophisticated (Shore 1996). Despite genetic differences and selection for particular genes, there is no evidence that different human groups have evolved at different rates, or that any are at earlier stages of physical or mental evolution. In fact, because of the flow of genes among human populations around the world, there has been little opportunity for different human groups to evolve separately (Mayr 1963; Lewontin 1972; Cavalli-Sforza, Menozzi, et al. 1994; Brown and Armelagos 2001; Kidd, Pakstis, et al. 2004). But it took sophisticated methods of cross-cultural cognitive study to recognize the complexity of "primitive" thought.

Testing subtle mental functions requires that the tester give elaborate verbal instructions and completely comprehend the answer, which was often impossible when testers worked through interpreters or with their own limited command of the native language. Ethnocentrism suggested that some languages were too simple for mental subtleties; we now know that all languages are complex, although often in different ways. The fact that they can be translated into one another and that all perform the same basic functions shows that this aspect of mental life is in crucial ways universal (Chomsky 1988, 1993; Pinker 1994). Another obstacle was researchers' ignorance of things that their study subjects knew best. The languages of non-industrial societies lack technical words for objects that their culture has no knowledge of, but they have larger vocabularies in some domains. For example, !Kung San has more words than English does for different ways of carrying objects and children (Lee 1979). And although a dictionary of English records far more words than exist in any nonwritten language, the number used by the average person is similar cross-culturally.

Vocabulary often reflects deeper knowledge. Claude Lévi-Strauss (1968) showed that in some "primitive" societies the knowledge of botany is extensive and precise, but that knowledge had to be studied by trained botanists before it could be appreciated. The same is true of Eskimo knowledge of anatomy and physiology. !Kung knowledge of animal behavior could not be evaluated until an ethologist collaborated with an anthropologist conversant in the language (Blurton Jones and Konner 1976).

Primitive mind theorists claimed that the thought of adults in non-industrial societies resembles that of children in our own, a supposed instance of ontogeny recapitulating phylogeny. The psychologist Heinz Werner viewed animism—the tendency to explain the movements of objects and other natural events by reference to spirits, ghosts, and other supernatural forces inhabiting the objects—as a common tendency in both children's and "primitive" thought (Werner 1948). Animism is in fact found widely in the religious systems of non-industrial cultures, and children do pass through a stage in which they rely on such explanations rather than more scientific or commonsense ones (Piaget 1951).

However, Margaret Mead (1932) studied animism developmentally among the Manus of New Guinea, comparing the explanations about the movements of objects given to her by Manus children of different ages and by adults. In spontaneous speech Manus adults frequently gave supernatural explanations for ordinary events; a canoe that drifted away, for instance, was explained by reference to the bad intentions of the canoe. Mead questioned children on such examples. But even when prompted with an animistic explanation, three- to six-year-old children rejected it; in a typical response, one girl said, "No, the canoe wasn't fastened right." When glass chimes jingled in her house, Mead prompted adults and children with an explanation consistent with Manus beliefs: the chimes were her charm and were crying for things that she herself wanted. Adults accepted this explanation and wanted to know what the chimes were asking for. Children rejected it; at the age of five or six they typically said, "The wind blows and knocks the pieces of glass together, so they make sounds"; this response closely resembled Piaget's findings in Swiss children. Other experiments confirmed that "Manus children not only show no tendency towards spontaneous animistic thought" but even harbored "a negativism towards explanations couched in animistic rather than practical cause and effect terms. . . . The contention that a tendency to spontaneous animistic thought is a function of immature mental development must . . . be dismissed" (Mead 1932:233). In fact, the animistic thought of Manus adults was complex and subtle, and children had to learn it.

So two truths about culture and mind emerged from early-twentieth-century anthropology: (1) there is no evolutionary sequence of thought from traditional to modern cultures; and (2) some basic patterns of human cognition and its development are universal. In addition to language universals, Boas (1938), Lévi-Strauss, and others posited a psychic unity of the human species and identified some of its features. Lévi-Strauss (1963, 1968) found a universal dualistic tendency in human thought, which had already been studied by Robert Hertz (1960) in the early twentieth century and was well demonstrated from subsequent ethnological materials (Almagor 1989). Lévi-Strauss believed that this dualistic tendency is related to the dichotomies that give phonemes their meaning in language (Jakobson and Halle 1956/1971) and he tied it to basic brain function.

In the face of postmodernist skepticism about human universals, this line of research languished until cultural anthropologists like Roy D'Andrade (1987, 1990, 1992), Claudia Strauss (1992), Naomi Quinn (Quinn and Holland 1987; Strauss and Quinn 1997; Quinn 2006a), and Bradd Shore (1996) began to try to reconcile cognitive psychology with ethnology by showing the value of cognitive concepts in understanding culture and by rejecting the notion that there are no moorings in the mind to which all cultural input must fasten. D'Andrade used the cognitive concept of *schema* to try to understand what he called "folk models" of the mind, on the assumption that all human minds work similarly but that culture provides specific schemas or cultural models: "A cultural model is a cognitive schema that is *intersubjectively shared* by a social group" (D'Andrade 1990:99).

As we will see in Chapter 26, intersubjective sharing of knowledge and perspectives is at the heart of human culture. Claudia Strauss (1992) relied more on connectionist models of the mind, with their interesting compromise between hard-wiring and plasticity, and with Quinn (Strauss and Quinn 1997) applied them to the establishment of cultural meaning in the mind. Quinn (2006b) went on to incorporate them into a universalist theory of human child rearing (to which we will return). In her cultural theory of the self (Quinn 2006a), she invoked brain as much as mind models (as seems appropriate for the twenty-first century), bringing both Michael Gazzaniga's (2008) model of brain lateralization and Joseph LeDoux's (2003) model of an experience-responsive *synaptic self* to articulate an "extended self" able to use universal cognitive elements but also respond deeply to cultural influence. I suspect that all these theorists would accept Geertz's claim (Transition 3) that we are all suspended in webs of cultural meaning, as apes are not, but that they would say that meaning must be reconciled to universal properties of brain and mind.

Bradd Shore (1996:35, 37) set his model of culture and mind squarely on a foundation of cognitive science, attempting frankly to recover the role of universal features in the human mind; we return to that model in Chapter 29. But as Shore notes, "a kind of cognitive romanticism . . . has begun to crystallize" in cultural anthropology in recent years, founded on the fallacy that "if cultural practices or beliefs are not fully determined or universally shared . . . then they must be arbitrary and thus infinitely variable." This fallacy, itself ironically an example of our universal dualizing tendency, has been espoused by some ethnologists and cultural psychologists. Richard Schweder (as quoted by Shore 1996:36) summarizes their view: "The mind, according to cultural psychology, is content-driven, domain specific, and constructively stimulus bound; and it cannot be extricated from the historically variable and cross-culturally diverse intentional worlds in which it plays a constitutive part."

The strong version of this claim is clearly false: much of the mind can be so extricated, as shown by Darwin, Eibl-Eibesfeldt, Ekman, Tooby,

Cosmides, and many others in evolutionary psychology and by Boas, Lévi-Strauss, Almagor, D'Andrade, Quinn, Shore, and other cultural anthropologists who accept in some form a universal human cognitive apparatus. In the weak version of the statement, it is trivial and everyone accepts it. Except in extreme pathological instances, no human child can develop without culture, and so all human behavior, thought, and feeling is tinged or brushed or saturated with cultural color, and this color is different in different traditions. In that sense, and that sense only, the human mind is "cultural all the way down."

Some psychologists too have fallen prey to cognitive romanticism, claiming that cultural psychology has overturned the theories of Jean Piaget and others about universal processes in human mental function and its development. As mentioned in the prologue, many of these claims refer to the theories of the Russian psychologist Lev Vygotsky, but they have been in the literature of cultural anthropology for over a century. Again, there is a weak form of the claim that cannot be denied, but the strong form—that there are no universals of mental function— is demonstrably false.

To the extent that the weaker version was valid, it was the common coin of anthropological writing for generations before Vygotsky (1962) was translated into English. Boas, Mead, Benedict, Linton, Hallowell, and others, despite their defense of psychic unity, emphatically described the pervasive role of culture in shaping the mind of the developing child and constructed a variety of theories to account for this influence. Benjamin Whorf (1988) and Edward Sapir (1994) took an even stronger stand in the linguistic domain, claiming that language differences shape perception itself. And as we have seen, definitions of culture have always included mental life (Tylor 1871:1).

By the time Ruth Benedict (1934:3–4) wrote her influential *Patterns of Culture* it was clear that

> the life history of the individual is first and foremost an accommodation to the patterns and standards traditionally handed down in his community. From the moment of his birth the customs into which he is born shape his experience and behavior. By the time he can talk, he is a little creature of his culture, and by the time he is grown and able to take part in its

activities, its habits are his habits, its beliefs his beliefs, its impossibilities his impossibilities.

And Alfred Kroeber (1948:8) wrote in his textbook of anthropology—the
model for all subsequent ones—"The mass of learned and transmitted
motor reactions, habits, techniques, ideas, and values—is what constitutes *culture.*"

Vygotsky and his disciples did add to these generalizations. Such concepts as self-directed (private), then internalized speech; intersubjectivity; scaffolding; and the zone of proximal development served to analyze
the role of social and cultural contexts in mental development (Wertsch
and Sohmer 1995; Fernyhough 1997). These psychological concepts
specified more clearly some of what anthropologists called socialization
and enculturation. Also, cultural psychology appeared at the right time
in the history of anthropology. Many cultural anthropologists were influenced by postmodernism, and some rejected the very goal of scientific research on culture (D'Andrade 2000; Greenfield 2000a). Psychological anthropology faltered, and the banner of scientific method was
dropped, just when increased adoption of psychological measurements
could have had the greatest impact. It is fortunate that cultural psychology picked up the banner (Kitayama and Cohen 2007).

But Vygotsky's view was extreme, and probably a product of his own
environment. He wrote in the 1920s in the Soviet Union and took a position in psychology analogous to that of Lysenko in genetics. This was
a political world hostile to strong individuality and to any features of
the human mind that might be resistant to change. Extreme plasticity
in every realm of life, even plant genetics, was an article of faith. Vygotsky's belief that mind is merely an expression of the culture held in
common by the members of a community is not surprising in this context.

But a main thrust of this book has been to propose a structure for the
evolutionarily generated, genetically guided maturational plan that all
biologists and most psychologists today accept as given, although it cannot yet be fully described. Psychodynamic theorists spent a century trying to build a plausible maturational plan for the emotions, and Piaget
and others pursued a similar goal in cognitive development. This book
attempts a synthesis of what survives from those efforts, combined with

neural and neuroendocrine development in the light of evolution, to sketch our species-specific pattern of biosocial growth. Part IV deals with the patterns and mechanisms by which culture completes the behavioral development of every human child, to realize the full potential of human biological endowment.

Interlude 7

Thinking about the Question "How?"

Atherosclerosis causes most deaths in Western countries and, barring some pandemic emerging virus, will soon cause most deaths throughout the world. It is the disease underlying the vast majority of myocardial infarctions (heart attacks) and the preponderance of strokes, which are, respectively, the first and third most deadly illnesses. If you do not die of cancer, chances are you will die of athero. A friend of mine in graduate school was killed by it at age twenty-seven; such an early death was and is rare, although it may not always be. His genetic luck was bad: he came from a family with hereditary hypercholesterolemia. This was before the era of statins—cholesterol-lowering drugs—and people with these genes had uncontrollable cholesterol levels regardless of what they ate or how they exercised. Their arteries clogged with fatty plaques and many died before age thirty.

Most of us do not have such bad genes, but atherosclerosis kills millions of us anyway. It was not always so. From the countless and endless harangues by health authorities, we all know by now that diets high in saturated fat and low in fiber combined with sedentary lifestyles make us much more vulnerable to athero's depredations. But why would evolution by natural selection have left us with such a high and wide gap in our defenses?

Basically, it had no choice. We evolved in environments where high-fat, low-fiber diets and inactive lifestyles were virtually impossible to come by. Try as they might, our ancestors had trouble getting animal fats in large doses and could not get enough calories without willy-nilly ingesting what for us would be enormous amounts of fiber. As for activity, they were almost always on the move; food was much harder to find, prepare, eat, and digest than it is for us. All told, they could hardly get their cholesterol levels over 150. Even at age sixty they had few signs of hardened arteries, while in our population twenty-year-old men killed in battle in Korea or Vietnam showed fatty streaks on their arteries at autopsy; they were already on the way to atherosclerosis.

So we have a discordance between the environment we evolved in and the one we are in now, with consequences that are very well established (Eaton and Konner 1985; Eaton, Konner, et al. 1988; Eaton, Eaton, et al. 1997). Natural selection cannot have protected us against dangers that did not exist in our environments of evolutionary adaptedness, the crucible in which our genomes were formed. Multiple genes control the response to those dangers. If, like my friend who died so young, you have a really bad set of genes, you will likely develop athero regardless of your environment. With a more ordinary shuffle of the genetic deck, you will not—unless of course your diet is high in saturated fat and low in fiber and your lifestyle is characterized by low levels of exertion. Since most people today grow up in such environments, atherosclerosis is endemic; it pervades our population, and at earlier ages than ever. Children dying of heart attacks in grade school is still rare, but no longer unheard-of. Of course, not everyone will develop atherosclerosis even in this new and dangerous environment. Just as some shuffles of the genetic deck are very unlucky, others are lucky indeed. Nothing in biology is 100 percent.

So what does this have to do with psychosocial development?

First, our understanding of how genes and environment interact to produce this deadly physiology is a model for the determination of any complex human trait. A bell-shaped curve of vulnerability describes the distribution of human genomes very well. At the left-most tail lie a handful of individuals whose genes make them atherosclerotic early and in almost any environment. These genes might confer different conformations of transport proteins for cholesterol, a different susceptibility of arterial walls to microscopic injury, different features of the clotting cascade, and so on. In the middle we find the bulk of human genomes: no atherosclerosis until old age, if then, in the normal expectable environment of the species as it evolved—low exposure to saturated fat, high fiber intake, and almost daily vigorous physical activity. And in the right-hand tail we find a handful of people whose robust collection of relevant genes makes them nearly impervious to atherosclerosis, regardless of their lifestyle. They are the genetic opposites of my friend with familial hypercholesterolemia, although they are probably protected multigenically, while he may have been undone by a single gene.

So here is the question: is atherosclerosis genetic or environmental, evolved or acquired, biological or cultural? Well, the question is not stupid, because great advances in both our knowledge of it and our ability to intervene, now and in the future, depend on the responses (plural) to this question.

The short answer is "All of the above," but this does not lead medical scientists to wave their hands and make pronouncements about the meaninglessness of asking about genetic versus environmental effects—to say, in effect, that it is all one inextricably interwoven developing system. It leads them, on the contrary, to roll up their sleeves and answer, for every point on the continuum, a smarter question: "How?" In the psychological realm this question was famously posed by Anne Anastasi (1958), who properly pointed out half a century ago that the questions "Whether?" and "How much?"—as interesting as they may temporarily be—are of no use unless they ultimately give way to the question "How?"

The difficulty for developmental psychology is that until quite recently the question "How?" has been answered mainly in one way: detailed description of the phenomenology of development. Freud and Piaget alike, although both originally biologists, said as little as possible about just how innate factors interact with the environment, even while making it clear that both are critical; learning theorists ignored the innate altogether. Descriptions got more detailed and led to new theories (Siegler and Crowley 1991; Thelen and Smith 1994), but they did not answer the question.

As with atherosclerosis, biological causal analysis does more than even the most detailed description or the most elegant phenomenological theorizing. Biological analysis consists first of description of phenomena at the cellular and molecular level, which is crucial in elucidating causal chains, and then the analysis moves to genetic contributions. Both biological components—not to mention the evolutionary dimension—have been ignored by most developmental psychologists.

This is changing. As it has become increasingly clear that these approaches need not lead to interventional nihilism or simple-minded determinism, new paths have been opened. Sandra Scarr and her colleagues pointed out that innate factors lead children to choose key aspects of their own environments (Scarr and McCartney 1983; Scarr 1992, 1993), a kind of genotype-environment correlation that confounds some conventional heritability analyses and leads to optimistic approaches to intervention. Robert Plomin and others have repeatedly shown that parsing out genetic influences directs attention precisely to the most powerful features of the psychological environment, again leading to focused and potentially powerful interventions

(Reiss, Hetherington, et al. 1995). Jerome Kagan and his colleagues have found a continuum from timidity to boldness that is biologically based and can be stable throughout life (Kagan, Snidman, et al. 2007), but they have also shown that certain experiences can shift children along the scale.

Meanwhile, neuroimaging, often with a molecular dimension, and related technologies for the study of infants and children have transformed our understanding of behavioral and psychological development (Part III)—not because the neural phenomena explain the behavioral ones (they often do not), but because the question "How?" will never be answered without reference to them. Among many other findings, research on the adolescent brain has led to fundamental revisions in our understanding of behavior, emotion, and even culpability in this crucial life phase (Spear 2000; Chapter 12). Nor will the "How?" question be answered without a sophisticated use of animal models, in association with human studies, to elucidate early experience effects on the neurohormonal bases of emotion and psychopathology (Heim, Plotsky, et al. 2004).

One recent approach to the question "How?" is in the molecular genetics of genotype-environment interaction. The risk for depression is highly predicted from accumulating stressful life events, but this appeared to be more true of individuals who have a gene for the short form of the promoter just upstream from the serotonin transporter than for those with the long form (Caspi, Sugden, et al. 2003; Kendler, Kuhn, et al. 2005), and rhesus monkeys displayed a similar interaction (Suomi 2003). Analysis of the neurophysiological mediation of this interaction effect has been under way (Pezawas, Meyer-Lindenberg, et al. 2005; Taylor, Steffens, et al. 2005). It seemed one of the most promising avenues in psychiatric genetics, and a vindication of the long-standing idea of interaction effects.

Alas, meta-analysis has cast doubt on this particular association (Risch, Herrell, et al. 2009), but similar studies of other polymorphisms—genes for an adrenergic receptor (Maestu, Allik, et al. 2008), a dopamine receptor (van Ijzendoorn and Bakermans-Kranenburg 2006), and the neurotransmitter-removing enzyme monoamine oxidase (Kim-Cohen, Caspi, et al. 2006)—have shown interaction effects with environmental stresses to shape behavior. The success or failure of one hypothesis or another does not change the fact that this is one of the most important research paradigms in child development today: for the first time molecular and physiological substance is being given to a century of generalizations about the importance of genotype-environment interaction. Such interactions may turn out to be the most common form of genetic influence in child development (Rutter 2006, 2007; Belsky, Bakermans-Kranenburg, et al. 2007), as they also are in the development of behavior in animal models (Cairns 1993; Suomi 2003).

Genotype-environment interaction may also work on a societal scale. People of low socioeconomic status have different serotonin sensitivity than do those of higher status, and this difference, if true, could help explain higher levels of psychopathology in deprived communities (Manuck, Bleil, et al. 2005). Getting back to the discordance model of atherosclerosis, the poor suffer even more discordance against the hunter-gatherer background than our society as a whole does. Since hunter-gatherer societies are not hierarchical, this evolutionary novelty has implications for mental as well as physical health.

But that's not all. As with the lifestyle factors that lead to atherosclerosis, our culture as a whole has departed markedly from the hunter-gatherer baseline in social arrangements and in particular in child care. We saw in Part III what hunter-gatherer cultures share in common in their child-rearing processes. Could our marked departure from these patterns be producing a discordance in the realm of child rearing comparable to the one for diet and lifestyle? And could this discordance have deleterious effects that parallel those that have been well demonstrated for atherosclerosis? We do not know.

But to summarize what we can learn from the science of clogged arteries, which is far more advanced than ours: The genes behind it are many and various; some collaborate in hundreds to produce the disease, some interact with particular environments, and some rare genes almost guarantee the disease with any environmental input. But environments, specifically diet and exercise, are enormously powerful, and in some cases override genes. Personal habits matter, public health programs matter, poverty matters—and so does the drastic departure we as a species have made from the environments in which our genomes were formed. Each of these statements could have a parallel in psychosocial development and its disorders.

24

The Culture of Infancy and Early Childhood

Every human group must transmit to children an enormous amount of information about its culture and way of life, including language, knowledge of the physical and social environment, rules of social interaction, subsistence skills, music, dance, plastic arts, dress and body decoration, modesty, humor, attitudes, explanations, ways of thinking and feeling, and religious ritual and belief. Children must learn to regulate their impulses, including hunger and thirst, fatigue, verbal and non-verbal expressiveness, aggression, and sexuality, so that these are expressed in a time and manner appropriate to the culture and in full knowledge of its preferences, strictures, and taboos. Some transfers occur predominantly at one stage of development—language in early childhood, sexuality in adolescence—but each is transmitted to some extent throughout life. This book cannot include a comprehensive account of the transmission of culture to children, but we can highlight its various forms.

Culture in Utero?

Developmental scientists have long considered the possibility of cultural influence before birth. In some senses this is well established—for example, through cultural influences on the substances ingested by the mother and on the mother's state of nutrition and health. Because embryonic development follows genetically guided patterns, the influence of nutrients and toxins depends on the stage of pregnancy. Learning is not the only kind of nurture, and there are critical or sensitive periods in which nurture makes a massive incursion. Foods, drugs, and toxins are powerfully formative environmental forces, the first strong winds of nurture blowing over the natural landscape of development.

Some environmental influences can be evenly distributed over development, while for others there is a more vulnerable phase. For example, thalidomide, used in the late 1950s to treat the nausea and anxiety of the first trimester, was a devastating teratogen, or embryonic poison (Newman 1986; Juchau 1993); around eight thousand children were born with severe birth defects, including the absence of arms and legs and major defects of vision and hearing. Yet thalidomide had little effect later in pregnancy, when the major organs were already formed.

A partly parallel account can be given of the most commonly abused nonprescription drug, ethanol. An estimated 2 percent of all potential mothers are alcoholics, and many more drink moderately. Fetal alcohol syndrome (FAS), a deforming consequence of heavy drinking during pregnancy (Spadoni, McGee, et al. 2007), includes underdeveloped heads and brains, malformed eyes, heart defects, and a facial anatomy distorted by widely set eyes and a broad, flat nose. Severe mental retardation is common in these children, especially those heavily exposed in the first trimester. But exposure later in pregnancy and early, lower-level exposure can also be toxic, especially to the brain, with its long course of development. These subtler injuries, fetal alcohol effects (FAE), are found at birth and for years afterward (Coles, Platzman, et al. 1997; Calhoun and Warren 2007). The recommendation of zero tolerance for the consumption of alcoholic beverages in pregnancy followed these findings—a cultural change.

The placenta is permeable to almost everything in the mother's blood, in addition to indirect effects of altered maternal physiology. Smoking reduces umbilical blood flow, and newborns of pregnant smokers weigh at least 5 percent less, on average, than those of nonsmokers. Pregnant users of heroin, methadone, or cocaine have infants who are themselves addicted and go through withdrawal at birth. With cocaine, as with alcohol, there can be permanent brain damage, including hyperactivity and reduced intelligence throughout childhood (Offidani, Pomini, et al. 1995). Pollutants, certain infections such as toxoplasmosis, and several kinds of radiation can cause irreversible damage during sensitive stages of fetal development.

But the main distorting influence on the prenatal environment throughout history has been inadequacy of the mother's diet. Low birth weight—2,500 grams (five and a half pounds) or less in a full-term neonate—is generally considered a result of poor nutrition during pregnancy (Kramer 2003). According to United Nations data, at least 18 percent of neonates in developing countries are low-birth-weight. On the Indian subcontinent, the proportion is 25 to 30 percent, and in the United States and Britain 7 percent, but some developed countries, including Japan, France, and Spain, do better. Substance abuse contributes to low birth weight.

Inadequate caloric intake is only one syndrome. Kwashiorkor, or protein-calorie malnutrition, is widespread; specific deficiencies of vitamins and minerals like iodine and iron can affect the fetus profoundly. Anemia—mainly iron-deficiency anemia—affects about 60 percent of pregnant women in the developing world and about 15 percent in the developed world. We tend not to view these variations in prenatal environment as cultural, but often they are, and they may be a large part of the environmental variance in quantitative genetic studies. Customs regarding wine and beer drinking during pregnancy and lactation are very different in Italy and Egypt, and women chewing coca leaves in Bolivia or betel nuts in Fiji may be shaping their babies' brains.

The normal fetal brain grows under a constraining genetic plan. However, as we saw in Chapter 6, neurons are generated in excessive numbers, and half or more die a preprogrammed death; trillions of connections forming networks among the brain cells that survive are also pruned. But regression is not random. The fourteen-week fetus shows spontaneous activity—slow rhythmic squirming of the body and limbs, combined with occasional jerks and kicks—and responds to stroking on the palm of the hand or the sole of the foot by clenching the fingers or toes. A month later other reflexes appear, including sucking and swallowing amniotic fluid. Thumb sucking, also seen at this time, requires the coordination of two neuromuscular systems and has a less reflexive, almost voluntary quality—unlike the response to foot stroking, which can be coordinated purely through the lower spinal cord. By the seventh month the nervous system generates diaphragm movements as the fetus in effect breathes amniotic fluid. This is the paradox that governs all of development: genes guide neuronal movement and build connections, yet the final circuitry depends on transactions with the environment.

What does the fetus know about the outside world? Studies of prematurely born infants show that all five senses are functioning to some degree by seven months of pregnancy. The most advanced sense prenatally is proprioception, which matures months before the other senses, and the foremost intrauterine experience is motion, both passive and active. Premature infants do better when they are kept in an incubator with a gently rocking waterbed (Korner, Kraemer, et al. 1975; Korner 1979) or on the rising and falling chest of a mother or father in the kangaroo care method (Feldman and Eidelman 2003). Newborn infants are especially responsive to being held and rocked partly because of the maturity of the system, but also because of long experience with vestibular stimulation and prenatal learning. Cultures where women are very active must produce neonates more accustomed to movement.

Studies of mice and other animal models indicate that psychological stresses on the mother during pregnancy can change postnatal behavior and that cross-fostering in utero produces changes in rat infant behavior independent of genetic and postnatal influences (Francis, Szegda, et al. 2003). The human fetus can probably detect bright light or darkness through the mother's abdomen, the uterine wall, the amniotic fluid, and its own translucent eyelids. Clearly it responds to sound; fetal heart rate and movement respond to tones systematically delivered, even when the mother cannot hear the tones herself. Fetal heart rate accelerates in response to sudden loud noises; such external events stimulate activity and may leave a memory trace. Studies confirm the impression of pregnant women that their fetuses respond directly to music and other sounds.

Ducklings learn the sound of their species-specific contact calls—or those of a substituted species such as domestic chickens—through the shell and fluid of their eggs, and they attend to those calls preferentially after hatching (Gottlieb 1980, 1991a, 1991b). Human fetuses after thirty weeks' gestational age habituate to a stimulus against the mother's abdomen that vibrates and makes a sound, and they remember the experience for at least ten minutes (Dirix, Nijhuis, et al. 2009). Newborns prefer their mother's voice, having heard it in the womb; they suck more to produce their mother's voice than another female voice (DeCasper and Fifer 1980). But closer to being cultural are studies showing that reading to a baby through the mother's abdominal wall during the last trimester leaves traces measurable by different reinforcement potential (De-

Casper and Spence 1986) and attention, including fetal heart rate de-celeration (DeCasper, Lecanuet, et al. 1994). The effect may depend on autonomic nervous system maturation in late pregnancy (Krueger, Holditch-Davis, et al. 2004). Whatever the mechanism, fetuses discriminate familiar from unfamiliar rhymes. Since in different cultures speech has different rhythms, pacing, musical tones, and sound patterns, intrauterine influence could affect language through habituation and associative learning. Specific cultural forms, such as the Lord's Prayer, the Moonlight Sonata, "I Can't Get No Satisfaction," Mexican *mariaches,* or the yodeling cadences of !Kung trance-dance music, could enter fetal consciousness as enduring schemas.

Cross-Cultural Variation in Infant Care

Although variation in infant care practices among hunter-gatherers is limited, there is wider scope for variation in the intermediate level—the horticultural, pastoral, and agricultural societies that comprise most cultures studied by anthropologists. Despite cross-cultural regularities in the way mothers interact with infants (Lewis and Ban 1977), there are differences of great interest. With the industrial revolution there were major changes, but a relative narrowing of variation around central tendencies quite different from those of hunter-gatherers. Ethnographic descriptions are informative and rich, and in the end there is no substitute for reading them, but it is fortunate that quantitatively oriented anthropologists—Herbert Barry III, Margaret Bacon, Irwin Child, John Whiting, and Leona Paxson, among others—carefully coded many infant and child care practices using standardized cross-cultural samples (Whiting and Child 1953; Barry, Bacon, et al. 1967; Barry and Paxson 1971).

Carrying Devices and Practices

Consider just one aspect of infant care, carrying methods. Carrying their infants in slings in bodily contact with the mother, hunter-gatherers are within the range for other higher primates on this dimension. But for human cultures generally, variation is great; as John Whiting (1981) showed, Native Americans of the United States and Canada often used cradleboards, many European and Asian cultures used swaddling and/or cradles, and many cultures in Africa, Oceania, and South America used slings or carried infants in arms. The best predictor of whether cultures used cradles (including cradleboards) was the mean temperature in the coldest month of the year. The 10°C isotherm effectively separated these two groups of cultures, with cradles, swaddling, and/or cradleboards present in 66 percent of 100 colder-climate cultures and absent in 92 percent of 149 warmer-climate cultures. Differences were significant for each of three separate north-south comparisons (Africa and Europe, East Eurasia and Oceania, North and South America). The use of slings or carrying in arms was strongly characteristic of warmer climates, but these were common practices even in colder ones.

Hunter-gatherers in cold climates, such as several Eskimo groups and the Yahghan at the extreme southern tip of South America, are among the cold-climate cultures that use slings or carry infants in arms. The Eskimo carried their infants in skin-to-skin contact in a sling under a parka. Thus it is still reasonable to claim, as we did in Part III, that slings and arm-carrying were used for most of the hunting-gathering phase of our history, which was mostly in warm climates, and that a majority of cultures began using other methods when they moved into colder climates. So while these data support the first part of Whiting's psychocultural model (see Chapter 24), suggesting a causal relationship between environment and child care practices, they also reflect phylogenetic constraint and cultural evolution.

Dramatically different from the sling was the cradleboard, a carrying device used by many North American native groups. The baby was wrapped snugly if not tightly in swaddling clothes and tied into the cradleboard to make a tidy package. Movement for the infant was essentially impossible, except for isometric exercise. James Chisholm (1983), studying the traditional community of Navajo Mountain in southeastern Utah, found it to be an effective method of care from the viewpoint not only of the mother but of the infant, who when in distress was comforted by and slept better in the cradleboard—effects that have been experimentally proven for swaddling (Lipton, Steinschneider, et al. 1965; van Sleuwen, Engelberts, et al. 2007). Some Navajo infants seemed to fuss and cry for the opportunity, and some had to be gradually weaned from the device as they outgrew it. They also fussed and cried when they awoke, wanting to get out of the cradleboard to nurse or play, and mothers responded.

In the four quarters of the infant's first year in the most traditional Navajo sample during the 1970s, cradleboard use accounted for fifteen to eighteen, twelve to fifteen, nine to twelve, and six to nine hours per day of the infant's time. Studies by Clyde Kluckhohn and others suggest that the durations may have been higher in the past. Meticulous observations, despite showing a flurry of changes immediately before and after being tied into or released from the cradleboard, reveal surprisingly little overall impact of the extent of cradleboard use on common measures of mother-infant interaction.

As Chisholm points out, citing Barry and Paxson (1971), this is not rare cross-culturally: "Of the 139 societies for which information is fully available, 20.9% use '. . . a cradleboard or other mechanism for restricting movement of the limbs as well as the body' for at least 'much of the time' during the first 9 months of life (1971:468). When the category defined as 'body movement is limited by swaddling, heavy blankets, or a small cradle' is added to the 'pure' cradleboard figure, a surprising 51.8% of all societies reported were found to practice some form of infant restraint" (Chisholm 1983:72). Cradleboards and swaddling keep infants safe and comforted, especially while they sleep, and are an effective cultural adaptation in infant care.

Possible Mechanisms of Influence

Psychodynamic culture-and-personality models (Spindler 1979; LeVine 1982) held that patterns of infant care produce deep and enduring emotional consequences. Russians were supposedly reserved and depressed because as infants they were tightly swaddled (Gorer 1962). However, from time to time the Russian infant would be unwrapped, flail its arms and legs, and cry. According to this post-hoc interpretation, Russians were usually reserved but had periods of sudden violence in their history owing to this infantile experience. This and similar claims arose largely from Freudian theory writ large.

But few studies confirmed psychoanalytic claims, and a growing consensus held that such factors were greatly exaggerated. Reviews of family influence in behavior genetic paradigms (Plomin and Daniels 1987;

Rowe 1994) and of the role of early nurturance generally (J. Harris 1995, 1998; Bruer 1999) are at variance with common convictions about its importance. As for the swaddled Russians, there is no evidence that tight bundling in infancy had any lasting effect on them, or for that matter on the culturally and psychologically very different Navajo and other Native Americans who wrapped infants tightly onto cradleboards. While there may be an effect in later infancy (Chisholm 1983), there is no evidence of deep psychological effects persisting into adulthood. It seems that there should be long-term effects of slings, cradleboards, or strollers, but we lack the research to say "whether" or "how."

Of course, animal models have shown important lasting effects of early experiences on mind and body (Chapter 14). We cannot do comparable studies on humans, so cross-cultural studies serve as natural experiments. They have suggested that early stress or stimulation may have an effect on body size in humans similar to this effect in rats (Landauer and Whiting 1981). Cultures with infant care practices that include deliberate stressing of infants, such as circumcision or repeated cold baths, have earlier first menstruation (Whiting 1965) and larger adult size (Landauer and Whiting 1964). Cultures with systematically imposed mother-infant separations show similar patterns (Whiting 1965), and the introduction of the mild-to-moderate stress of infant vaccination led to increased adult stature (Whiting, Landauer, et al. 1968), apparently independent of disease rates.

None of these studies completely rule out other variables that might explain the effects, but because of the parallel rat experiments, the inferences cannot simply be dismissed. On somewhat shakier ground are cross-cultural correlations between infant and child care and characteristics of the cultures as practiced by adults (Whiting and Child 1953; Rohner 1975; Rohner, Khaleque, et al. 2005), since cultural themes or patterns could shape both child care practices and adult behavior or beliefs. However, some intervention experiments manipulating human maternal care suggest that such effects are possible (van den Boom 1995).

But most enculturation in infancy and early childhood is of a more conventional, yet still remarkable, kind. Key cultural themes and symbols are rarely outside an infant's or toddler's world; indeed, the "simpler" the culture, the more likely it is that the most important events of life will take place on a single stage with players of all ages present. But human cultures go far beyond allowing participation; they inculcate the important themes from the earliest ages. !Kung infants and toddlers are deliberately taught the elements of what will be their most important religious and healing ritual, the trance-dance. Matters this central are not left to chance or to a simple reliance on imitation and passive learning.

Similar enculturation processes begin in infancy and early childhood in all cultures, and since psychologists rarely study religious development, they underestimate its importance, which is large in cultures at all levels of political complexity. The medical anthropologist Ronald Barrett (2008) photographed a man feeding his grandson the ashes of wood from a crematory pyre beside a sacred pool at an Aghori ashram in Banaras, India. This is done to protect children from common and deadly illnesses. However little this boy may have understood, he had a dramatic learning experience that put him on the path to uniquely human emotions of lifelong power.

Language Acquisition and Language Learning

The innatist view of language acquisition, summarized in Chapter 9, has of course not gone unchallenged. One critique is that parents provide more supports for language learning than the language acquisition device (LAD) model assumes and that language acquisition constantly demands and uses social contexts; the child is learning new social behavior—the *pragmatics* of language—not just labels and rules. Children growing up in rich social contexts with many and frequent conversations—around the dinner table with their parents, for instance—learn language better and sooner than less advantaged children (Pan, Perlman, et al. 2000). In many cultures corrections are elicited by the infant's repetitive utterance of a signal such as "Hm?" or "Hih?" Among the Manus of New Guinea, Margaret Mead observed up to sixty repetitions by a parent of the same child utterance, the repetitions alternating with the child's mispronunciation, and I saw up to about twenty repetitions in the same kinds of exchanges among the !Kung.

So there is a great deal of learning, consistent with Vygotskian models (Bruner 1997, 1999). Certain preverbal patterns, such as pointing at or playing with an object together, precede the first words and theoretically could prepare the child's mind for language learning (Bruner 1983).

Nonverbal interactions, such as joint attention to objects, could provide a "scaffolding" to assist the child in developing language—an LAS, or language acquisition scaffold, to complement the neural LAD. Most of the evidence comes from observational studies of middle-class, college-educated, English-speaking families, and most correlations in such studies are low. Still, children in the human species have learned thousands of different languages, and as far as we know any human newborn placed in any linguistic environment will grow up speaking *that* language and not any other—nor, certainly, will the infant speak any generic language belonging to the species as a whole, which does not exist. Therefore, somehow, infants and children in each culture are led by their elders into the linguistic and paralinguistic forms and rules that characterize that culture. But no one has a good idea of the precise processes involved.

A few laboratory studies have contributed to our understanding, supporting a role for interactive learning. Eight-month-old infants can remember something resembling a word from a two-minute stream of meaningless syllables, based on the frequency of occurrence of the string that becomes the "word" (Saffran, Aslin, et al. 1996). However, natural settings are far more chaotic. It has also been shown that in middle-class, college-educated families, correct sentences spoken by two-and-a-half-year-olds are more likely to be repeated verbatim by parents, and in addition a substantial proportion of young children's errors (34 percent) are followed by adult corrections (Bohannon and Stanowicz 1988). Such corrections often stimulate attempts by the children to get it right. However, whether such results prove the *dependence* of language acquisition on such corrections is controversial (Bohannon, MacWhinney, et al. 1990; Gordon 1990).

Still, one study that controlled for genetic influence by including adopted children showed that while toddlers' verbal performance can be predicted moderately well from the general intelligence of their biological parents but not their adoptive parents, a measure of adoptive parents' responses to their children's linguistic efforts was also correlated, independently of genes, with measures of the rate of language development (Hardy-Brown, Plomin, et al. 1981; Hardy-Brown and Plomin 1985). Once again a convincing demonstration of environmental influence came from a design that controlled for genetic effects.

Other evidence for potential parental roles comes from elegant experiments. For example, in teaching seventeen-month-old toddlers the names of unfamiliar objects, such as "gauge," "clip," "bow," and "wrench," parents could best ensure learning by "following-in" to the child's focus of attention (Tomasello and Farrar 1986), another developmental instance of what Ann Cale Kruger (2010) has called "communion." Parents might be told to say a sentence emphasizing one of four nouns (say, "gauge") only when the child was already playing with the gauge. Another noun (say, "wrench") was used with equal emphasis, but only when the child was not playing with anything. Two weeks later, the children understood "gauge" better than "wrench." In another experiment an invented verb, "plunk," was highlighted to fifteen-month-olds in a sentence either before, during, or after pushing a button that made a doll slide down a ramp (Tomasello and Kruger 1992). Learning was facilitated most by using the verb for an impending action—"Look, Jason, I'm going to plunk it." The label apparently needed to come in advance to maximize the chance that the child would link it with the event.

Such experiments prove the power of some forms of joint attention by adult and toddler in promoting language learning. Observational studies had called attention to the mechanism of learning (or scaffolding), which was then tested experimentally. As a result, we now know not only that middle-class American children's environments include such scaffolding, but also that these tactics can promote language learning. However, this does not prove that they are needed in language learning outside the laboratory.

Connectionism has also been used to help explain language acquisition (Karmiloff-Smith 1995; Elman, Bates, et al. 1996). Connectionist networks are ordered webs of (somewhat) neuronlike elements that change the strength of their connections with experience. They are capable of learning, including some language learning, and this has led some to conclude that no innate abilities are necessary. Others accept that the brain is specifically prepared for language learning to a mild extent, but give learning a far larger role than genetic preparedness (Elman, Bates, et al. 1996). Connectionist models are valuable, but their software elements are not really like neurons, nor have connectionist circuits learned anything resembling language. But in any case, no amount of machine learning can prove that the brain learns language the same way. The central questions of language acquisition must be answered by what the brain actually does, not by what machines can be made to do (see Chapter 9).

That said, the world's thousands of languages are different because of enculturation, not genes. It is likely that the great bulk of this enculturation is due to a combination of ordinary association learning, operant conditioning (in which some of the rewards are cognitive), observational learning, and the rule-bound organizing activities of the LAD—with an assist, in some cultures more than in others, from teaching.

25

The Culture of Subsistence

Higher primates do little or no teaching; carnivores do something like it in relation to hunting, but human hunters have language as a crucial additional teaching tool, partly in the form of telling stories and answering questions. !Kung boys go on hunts with their fathers and uncles only during adolescence—earlier, they would be too much of a liability—but by that time they have already absorbed much in the form of observational learning, linguistic information transfer, and play. Hadza children are expected to forage for much of their own food, but even this responsibility does not rival the chore assignments that characterize farming societies. Whether for play or work, middle childhood in hunting-and-gathering societies is characterized by an increase in self-reliant, but not in obedient or responsible, behavior.

Obedient and responsible behavior is in greater demand in middle childhood in agricultural and herding societies (Barry, Child, et al. 1959; Weisner 1996; Lancy 2007). A seven-year-old girl may carry and tend her infant sibling for hours each day while her mother works in the gardens, and a boy that age may be assigned full responsibility for guarding a herd of cattle that represent his family's entire (and substantial) wealth. In a hunting-and-gathering society, such chores are unnecessary in middle childhood, but as some !Kung became more sedentary, herded more goats, and foraged less, they began to demand more obedience and responsibility of their children, while the more laissez-faire outlook of most hunter-gatherers persisted in other !Kung communities that had not shifted their subsistence (Draper 1976).

In simple terms, then, there have been two major transformations in middle childhood experience. In the first, beginning ten thousand years ago with the spread of agriculture, birth intervals shrank, reproductive success increased, and work—including infant and child care—replaced a large proportion of play in many cultures. The second transformation came very recently, with modernization, and replaced most subsistence work with school.

Work, Play, and Cultural Transmission

All societies have explicit and implicit theories about how cultural knowledge is transmitted that may or may not correspond to psychological models. For example, the !Kung have an essentially Piagetian folk theory of child development: behavior is appropriate to stage of development and changes naturally with growth (Konner 1972). Children do not need to be beaten out of sinfulness, carefully taught every behavior, inclination, and schema they must one day have, nor even consistently rewarded for desirable behavior. What they mainly need, according to this theory, is to grow, engage with the world, and learn on their own. Infants are said to teach themselves as they play with an unwieldy mortar and pestle or a heavy set of nut-cracking stones. The word for teaching and the word for learning are the same—*n!garo*—and "She's teaching/learning herself" is a common phrase. As for undesirable behaviors such as crying, one mother said, "It's a baby, and has no sense—that's why it cries. Later, it will be older and have sense, it won't cry so much."

And it is Piagetian in a deeper and more specific sense: a child has a behavior or an idea about the world that leads to a challenge in which the event encountered may be absorbed into an existing cognitive schema—this is a different kind of tuber, harder or more delicate, with a less conspicuous dry brown tendril trailing up a brown branch the only evidence of its presence, but I can get out my digging stick and extract it. But if the tuber is very deep in hard ground, other tools and tactics may be called for; the child must *accommodate* to the new challenge with new schemas, not just iterate old ones. Accommodation forces cognitive growth, while assimilation processes similar experiences with well-established routines.

Culturally appropriate features of behavior, emotion, and mind are transmitted throughout life, but the years between the five-to-seven shift and puberty are a human adaptation for a large amount of cultural acquisition (Chapter 5). After the turmoil of early childhood and before the sexual and maturational challenges of puberty, but with the developmental advantages of metacognition, concrete operations, and other cognitive advances in the five-to-seven shift (Sameroff and Haith 1996),

middle childhood is a phase of calm and active acquisition of many domains of culture, from subsistence and rules of emotional expression to
ritual and faith.

As they grow, hunter-gatherer children spend more time observing
subsistence—mainly with mothers—going on gathering expeditions,
and gradually mastering the location, detection, and extraction of plant
foods in a complex environment. Girls have more exposure to this than
boys, but it is essential knowledge for both; men must gather plant foods
during hunts, especially less successful hunts. Acquiring this knowledge
involves little or no formal teaching, although teaching often occurs
during puberty rites. In a classic paper on the !Kung, Lorna Marshall
(1960:341) wrote, "They play in very engaging ways. The older children
have toys but the young ones play mostly with the family's utensils, often imitating what the parents do. Parents are likely to pause to show
three-year-olds how to twirl a drill or dig with a digging stick or pound
berries in the mortar."

Later work with the !Kung confirmed this, and Marshall's vivid account (1976, Chapter 10) of !Kung children's play and learning provides
many rich and revealing details. Boys began to acquire hunting skills
through a process in which play was highly significant. When, as teenagers, they began to accompany their fathers on hunting expeditions,
they learned by direct observation. When girls and younger boys followed women on gathering trips, they gradually gained knowledge but
did not work very hard. They also gathered their own food in a desultory
but playful way while roaming the bush in multi-aged groups. Skills
were socially facilitated, emulated, and imitated, but not drilled or
taught, and indeed, play and learning merged. Nancy Howell's (2010)
analysis of !Kung life histories shows that by the time they are grown up
and producing more food than they consume, !Kung have in effect accumulated a large caloric debt to the community, not to mention the
skills they have acquired under the very forgiving conditions of their
long childhood and adolescence.

A classic account of the life of a !Kung woman (Shostak 1981:99) includes her memories of a happy middle childhood and illustrates what
life could be like at its best for children before the Neolithic revolution:

> My heart was so happy I moved about like a little dog, wagging
> my tail and running around. . . . "The rainy season has come to-

day! Yea! Yea!" . . . My heart was bursting and I ate lots of food and my tail kept wagging, wagging about like a little dog. And I'd laugh with my little tail, laugh a little donkey's laugh . . . shouting, "today I'm going to eat caterpillars . . . cat—er— pillars!"

Another excerpt, in which she remembers killing a small antelope, illustrates the difficulty of separating subsistence and learning from play in such cultures.

> We started to follow the tracks and walked and walked and after a while, we saw the little kudu lying quietly in the grass, dead asleep. I jumped up and tried to grab it. It cried out, "Ehnnn . . . ehnnn. . . ." I hadn't really caught it well and it freed itself and ran away. We all ran, chasing after it, and we ran and ran. But I ran so fast that they all dropped behind and then I was alone, chasing it, running as fast as I could. Then I picked it up by the legs and carried it back on my shoulders. I was breathing very hard, "Whew . . . whew . . . whew!" . . . When I came to where the rest of them were, my older cousin said, "My cousin, my little cousin . . . she killed a kudu!" (101–2)

Killing an antelope calf was not assigned or expected—it was one activity in the continuum of leisure, play, and subsistence that occupied !Kung children. In fact, a young girl would not be *expected* to hunt, ever, although adult women opportunistically captured and killed small game. But she was forming the attitudes and acquiring the cultural framework in which hunting was highly valued (Marshall 1976; Lee 1979; Biesele 1993), and she would pass these on to her own children.

Older ethnographic descriptions of the process in other hunting- and-gathering societies substantially agree. Among the Walbiri of Australia, for example (Meggitt 1965:116),

> After the boy is aged five or six, he roams the bush with other lads, and his father sees little of him by day. Men spend much time cutting up damaged boomerangs to make throwing-sticks for their sons to use on these jaunts; and the boys display remarkable accuracy in killing small birds and lizards with the weapons. They now learn which flora and smaller fauna provide the best foods, they develop their tracking skills and ac-

quire an intimate knowledge of the bush for ten miles or so around the camp. During this period, men take little part in educating or disciplining their sons, whose behavior towards them often reflects this lack of control.

Studies and films of the Baka, hunter-gatherers of Cameroon, contain similar observations (Agland 1990; Haines-Stiles and Montegnon 1992a, 1992b). A four-year-old will follow women cutting plantains in a banana grove and heft a heavy machete with difficulty, trying to chop fungus off a fallen tree. An eight-year-old goes along with his father in a group of men and boys to gather honey, courting multiple stings. He hacks at a tree trunk with an ax, while grown men do the same nearby; they break out the honey, and the children eat it in the comb. Boys age four and eight watch their father smoke termites out of underground mounds; then, to their delight, they get to eat them. As children forage in groups for snails or honey, they clearly enjoy themselves, but they also find food and learn the facts and skills on which their lives and reproductive success will depend. It is remarkable that critical information transfer and skill development should be left to informal processes, but apparently they work.

Hadza

Nicholas Blurton Jones and his colleagues have focused attention on Hadza children's foraging because it was evident early on that they do a lot of it, and his previous experience with the !Kung made the contrast unmistakable. It is true that "Hadza children have fun while they forage" and "engage in foraging with a spirit of adventure, joy, fun, and achievement. Their foraging is interrupted by chat, joking and gentle teasing, resting, grooming, singing, and rushing about, all accompanied with smiles and laughter. Their foraging seldom appears to be a response to instructions from adults" (Blurton Jones, Hawkes, et al. 1997:282). Still, amid the fun, they feed themselves, each other, and even their families. The caloric value of food—mainly tubers and baobab fruit—brought back to camp by Hadza children and teenagers beyond what they ate in the bush showed an inflection point around the five-to-seven shift, but even the children in that age range could collect hundreds of calories an hour and could easily feed themselves if they could live on just tubers and baobab fruit.

Blurton Jones and his colleagues reject the idea that Hadza children forage to acquire skill, because collecting these particular items is too easy; this ignores the emotional and motivational aspects of enculturation, but their argument that children do this for their own and each other's sake is persuasive. Boys slack off in adolescence, while girls produce more than they need, probably reflecting the emerging division of labor whereby boys must learn to hunt while girls must improve on what they already do well. Two observations illustrate the trade-off for Hadza boys between manly pursuits and tending to kin. A twelve-year-old became temporarily lame from a large thorn in his foot, and then spent much of his time tending his two-year-old cousin. After healing, he returned to the bush with other boys or guarded the women on foraging trips. Juma, a boy about fifteen, was orphaned and lived with his grandmother. He and his uncle, an age-mate and close friend, spent time on most days in the bush, collecting honey and shooting at things, sometimes killing a small bird.

But toward the end of the author's stay Juma's three-year-old half-brother came to stay. Juma spent all of each day collecting and processing baobab fruit and feeding it to his little brother. He gave up whatever boys get from rushing about in the bush to tend his orphaned little brother (Burton Jones, Hawkes, et al. 1997:298).

In our cooperatively breeding species, kin selection often takes precedence over optimal foraging, and not just for children; we are always optimizing inclusive fitness first, with foraging efficiency a handmaiden of that process.

Mikea

Among the Mikea of southeastern Madagascar studied by Bram Tucker and Alyson Young, conditions were even more favorable to children's foraging than among the Hadza. Their forest was virtually free of predators and poisonous snakes, and certain tubers were easy to extract:

> Within this forest, Mikea forage for wild tubers, small game, and honey in addition to cultivating maize and manioc, and

participating in market-oriented activities. Mikea children tuber foragers have exceptionally high net acquisition rates, averaging between 536 net kilocalories/hour for girls and 504 net kilocalories/hour for boys. As such, the Mikea example offers a potentially interesting test case for theories of juvenile dependency and foraging strategies. . . . We find that while Mikea children experience over twice as much leisure time as adolescents and adults, they allocate similar time to foraging, especially for the wild tuber called *ovy* (*Dioscorea acuminata*). Mikea children make a significant contribution to the household food procurement effort. (Tucker and Young 2005:150)

The distribution of activities across different age classes among the Mikea was based on 6,637 observations of 46 individuals (infants excluded).

Adolescents resembled adults in their activities, but children spent their time differently. While leisure was the leading category for all groups, children had about twice the leisure time of adults. All made important contributions, with foraging, and tuber foraging in particular, predominating among all except adult men, who did more herding and less housework than women. In a longitudinal study, the net acquisition rate (controlled for seasonal and annual variation) increased markedly in three out of four children as they moved through adolescence, and in four out of four adolescents as they became adults, but all groups continued to preserve leisure time.

Mikea children were not bringing in all the calories they could have, nor were they foraging as efficiently as they might have done, but they were easily meeting all their caloric (although not all their nutritional) needs. "We argue that Mikea children are not trying to be 'efficient' at all. . . . Because parents provision children from their surplus, children are not energy-limited. . . . Children forage for the physical and mental challenge, and because it is an enjoyable social activity. [At one point there was] a 'food fight' between the boys and the girls. Several kilograms of edible tubers were destroyed in the ensuing volley. For children, foraging is an extension of play that occurs outside camp" (Tucker and Young 2005:168).

It is difficult to imagine a food fight among !Kung children, who must not waste food. "Are you a person who ruins food?" was a question often put to a child—or to an anthropologist making an inept attempt at nut cracking. In both the Mikea and the !Kung, play seems to be the root of subsistence learning. Among the !Kung, foraging is more difficult for children, yet it is a natural extension of their play-expeditions, despite low yield; among the Mikea, the abundance of tubers and ease of extraction allow a playful process to produce over five hundred calories a day. In both cases one consequence is learning, but as Tucker and Young (2005:169) point out, "they learn at their own leisurely pace."

The Okavongo Delta of Northern Botswana

John Bock's (2005) research on the development of foraging in four ethnic groups (including two San or Bushmen groups related to the !Kung) living near the Okavongo swamps has shed light on the process in this special environment. His data on lifelong success at cracking mongongo nuts are also relevant to the !Kung, for whom these nuts are a staple food. Unlike crushing a fallen baobab fruit under foot and eating the succulent, vitamin C–dense pulp, mongongo-nut cracking is a challenging skill, even for those who start learning it in infancy. Age-specific nut-processing returns across the life span in Bock's sample form an inverted U. Bock found arm strength to be less important than age, but he did not consider finger strength, hand size, arthritis, and other physical variables that might have affected the youngest or oldest subjects. Nevertheless, skill is central, since processing success is negligible until the late teens and continues to improve until midlife.

Meriam

Thanks to Douglas Bird and Rebecca Bliege Bird, we have many new insights into shellfish collecting (or is it fishing? or hunting?) by children among the Meriam of Australia. Although shellfish collecting lacks some of the drama of life on the African savanna or in tropical forests, it may be more important for our understanding of the evolution of modern *Homo sapiens*. We have seen that our ancestors were collecting shellfish extensively 164,000 years ago in South Africa (Chapter 4), while also using advanced stone tools and colored pigments (Marean, Bar-Matthews, et al. 2007). Since hominins much more ancient than that lived near lakes and rivers, the use of shellfish could be much older, and since the ancestors of all humans outside Africa may have left the Horn of Africa as recently as 60,000 years ago and reached Australia only 20,000 years later, it is likely that this rapid spread was along the southern coastline of Asia, where similar subsistence adaptations may have served all along the way.

Other than as reproductive vehicles, could children have played an important role in this process? For the Meriam, who inhabit some of the easternmost islands of the Torres Strait between Australia and New Guinea, two clam species and one conch made up 90 percent of the children's collecting, while adults collected quite different species, disdaining the ones that children could easily get (Bird and Bliege Bird 2000).

Bird and Bliege Bird (2002) calculate that both children and adults forage efficiently given how far they go from camp, how fast they walk, and so on, but they emphasize the opportunity costs for children, who need to put so much energy into growth. They argue that since adults seek different species of shellfish, which are extracted and prepared differently, children's learning of adult skills is of minor importance. But this seems implausible given the great overlap in motivation, attitude, emotion, and skill in a broader sense between adult and child shellfish collecting. Bird and Bliege Bird (2000) argue more cogently that the common archaeological finding of shell piles or middens needs to be reconsidered in terms of the range of species collected by children and child-adult differences in where and how the shells were discarded. While there seems little likelihood of deciding the relative importance of learning and consumption for children in these cultures, it is clear that these patterns shed important light on human evolution and history.

Martu

In contrast to the Meriam, the Martu live inland in Australia's Western Desert. Here too "children above the age of five often search for and pursue game animals. But they focus their efforts in different resource patches than adults," just as with Meriam shellfish collecting. One woman's recollections—those of Yupa Nakanangka, in Melanesian pidgin, quoted by Bird and Bliege Bird (2005:129)—are reminiscent of Nisa's:

> Mothers and fathers gone out hunting and leave us kids in camp. When we got hungry we go hunting for little lizard, get

him and cook it and eat him up. Me little bit big now, I go hunting myself, tracking goanna and kill him. . . . Soon as mother leave him, little ones go hunting, kill animals, blue tongue, mountain devil, take them home before mother and father come back, cook and eat it. . . . Mother come back and feed all them kids. . . . Morning again, father one he go hunting. All little kids go hunting self. . . . Morningtime, father one bin for hunting long way way. He been get and kill an emu, bring and cook him. Everyone happy, they bin say he good hunter.

Her narrative makes contrasting points. First, she remembers her pride in successfully hunting and cooking for herself, along with her friends. Second, the statements "Mother come back and feed all them kids" and "Everyone happy, they bin say he good hunter" show that children did not really fend for themselves. They were expected to help collect plant foods and grubs, but were not encouraged to accompany adults on hunting trips until their midteens.

Yora

Lawrence Sugiyama and Richard Chacon (2005) studied children's contributions to subsistence among the Yora of Peruvian Amazonia, with 6,448 observations of 70 individuals. As among the Mikea, Yora children spent less time in subsistence work and more in leisure than adolescents or adults. Virtually no child care was assigned to children, although it seems likely that Yora had multi-aged child groups that would have doubled as a child care setting (see Chapter 18). The amount of both fishing and hunting increased with age, but for hunting the sexes diverged in childhood and markedly differed in adulthood. Children used simpler fishing methods and their hunting was limited by physical strength, so that "independent hunting by juveniles was almost exclusively for small prey (e.g. lizards, birds) around the village or in the company of adult males" (Sugiyama and Chacon 2005:258).

Ache and Hiwi

The Ache of eastern Paraguay were extensively discussed in several chapters of Part III. As noted in Chapter 15, Kim Hill and Magdalena Hurtado (1996) were able to do some work on Ache foraging patterns even though the Ache had been settled on a Catholic mission and got only 20 to 25 percent of their subsistence from hunting and gathering. They also studied the Hiwi, otherwise known as Cuiva, who lived in grass- and swamplands of Venezuela (Hurtado, Hill, et al. 1992; Hurtado and Hill 1992). Hiwi had exceptionally high precontact mortality for hunter-gatherers, largely owing to warfare and other violence (Hill, Hurtado, et al. 2007). With the exception of larvae collected by Ache girls, children under ten did almost no foraging of any kind in these two groups. There was a sharp rise during the teenage years, especially in hunting, in both groups. However, there were also increases in gathering, with Ache boys and Hiwi girls achieving about half their modest lifetime peak in honey and palm fiber gathering, respectively. Peak hunting did not occur until the late thirties in either group, and peak gathering for Hiwi of both sexes occurred much later still.

Tsimane

Michael Gurven focused on the question "How long does it take to become a proficient hunter?" to shed light on whether one purpose of our long childhood is to acquire subsistence skills that cannot be mastered in less time (Gurven, Kaplan, et al. 2006). Measuring hunting returns and using specific tests of skill among the Tsimane forager-agriculturalists of Bolivian Amazonia shows that it in fact takes years *longer* than the long childhood to reach peak performance. To become skilled hunters, boys and men had to learn to recognize the smells, sounds, spoor (tracks), and scat (feces) of the seven most important prey species (paca, coati, tapir, howler and capuchin monkeys, brocket deer, and collared peccary), but that was only the beginning. They had to progress to the far less frequent *direct* encounters—sightings, pursuits (89 percent of sightings), and attempted kills. These peaked at age thirty-seven—about ten years later than indirect encounters.

The third step—without which there was nothing—was the kill. In this study, "over half of all kills were made with shotgun and rifle combined, 16% with bow and arrow, and the remainder with machetes, dogs, and other techniques" (Gurven, Kaplan, et al. 2006:461). In addition to almost half the kills being made with traditional methods, a high proportion of the kills made by rifle and shotgun today would also have been achieved in pre-gun times. With guns 57 percent of encounters resulted in kills, and without guns 48 percent—not a large difference. "Overall kill rate for the seven prey species shows a peak at age 39" (461). Age, but not strength, independently predicted hunting success, and this effect was greater with more reliance on traditional hunting methods. Indirect encounters tracked strength—both correlated with age—until adulthood, after which strength declined faster. Direct encounters, kill rate, and caloric return—in that order—increasingly lagged behind strength. This strongly suggests that the easy part—encountering smells, sounds, spoor, and scat—depended on having the maturing strength to get out there and trek, but that the next step, and the next, and the next, depended not on strength but on experience and skill, learned in hard-fought win-or-lose encounters.

In arrow-shooting contests, Tsimane boys improved dramatically through the teenage years and on average not at all thereafter; adult men's differences in accuracy were due mainly to individual differences, with body size an important influence. So bow-and-arrow marksmanship was learned or matured (with size) in adolescence, but the ability to recognize prey, encounter it, and make the kill increased steadily throughout adolescence and early adulthood until age forty—a relatively advanced age for a hunter-gatherer. A twenty-year-old who had not spent his childhood and adolescence trying would already be at least ten years behind.

Children's Work in Farming Cultures

Moving from hunter-gatherer to horticultural, agricultural, and pastoral cultures, we see that observational learning and play continue to be important and pervasive, while formal teaching efforts remain secondary in the acquisition of much subsistence and cultural information (Pelissier 1991). Teaching occurs widely, but usually falls into the context of punishment for misbehavior, scaffolding, and occasional deliberate modeling, or in a more formal sense is confined to rare occasions, such as initiation rituals or particular specialties such as drumming and navigating.

As hunting and gathering were increasingly replaced by pastoral and horticultural subsistence methods, !Kung children were assigned and did more work, wandered farther from home, interacted more with peers, and showed more male-female differences in behavior (Draper

1976; Draper and Cashdan 1988). This real-time study of the transition in subsistence patterns and its consequences for child care is invaluable. But we also know a great deal about child life in intermediate-level societies from classical descriptive and more recent quantitative ethnographies. The major new development at this level is the large increase in task assignment. Chores are a central feature of child life (Whiting and Whiting 1975a:84; Lancy 2007) and are essential to the ongoing subsistence strategy (Whiting 1972).

For example, among the Kpelle of Liberia, Gerald Erchak (1980:46) observed that "girls begin to help in infant care, to fetch water and other things, and to sweep the yard and hut at about age 5 or 6, and imitatively 'rehearse' these tasks much earlier; boys do not have many serious responsibilities until they are about 10, when they begin to drive birds and other pests from the rice farm, and perhaps bring home some firewood." Erchak quantitatively analyzed commands and prohibitions directed at young children. Between ages one and three, parents are most concerned with child safety and survival, but by ages four to six directives having to do with economic activities (for example, "Bring water for cooking") constitute the largest category of commands (31.2 percent).

In the Six Cultures Study child rearing and behavior were measured among five farming and herding societies (in Kenya, the Philippines, Japan, India, and Mexico) and a New England town (Whiting 1963; Whiting and Whiting 1975a). In New England the percentages for work, play, and casual social interaction were 2, 30, and 52. But situations involving formal learning occupied 16 percent of observations, while the figure for the five other societies averaged 5.2 (0 through 9 percent). There is almost a reversal of the relative proportions of work and learning when the industrial and intermediate-level societies are compared, with not much difference in the other measures (Whiting and Whiting 1975a:48). The highest proportion of work for the five agricultural societies was 41 percent. Early industrialization produced exploitative child factory labor and even child miners, but their work burden was eventually returned to a level found in farming communities and then to a much lower level, to be replaced by schooling.

The difference between 17 and 2 percent for work and between 5 and 16 percent for formal learning is substantial, but by most accounts this work is no more oppressive than school. Chores too entail learning and can be a main mechanism of cultural transmission—and also of ex-

panding horizons. Many chores are outdoors, interesting, even fun—
consider an eight-year-old driving a cattle herd to pasture. As anthropologists have pointed out, the extremely negative perception of child
labor common in Western cultures is based on exploitative factory and
mine labor during the nineteenth century, which set a moral standard
inappropriate for extension to children's farm labor worldwide (Niewenhuys 1996). Children's work in farming and herding cultures is often
vital to family survival, and traditionally there has been no conflict between work and learning, since work *was* education for adult roles. And
school is certainly a chore for most children.

Mark Flinn (2006), whose longitudinal research on the Caribbean island of Dominica has shed so much light on stress biology in childhood
(see Chapter 22), showed that strenuous work does indeed raise children's cortisol levels, but that they return to baseline within two hours,
as with the twelve-year-old boy who showed a transient elevation of
about 50 percent when he helped his father carrying wood. This was
highly adaptive, since the function of cortisol elevation is to mobilize
blood glucose for action or work, but the response is of questionable
value under many forms of psychosocial stress.

Studies of U.S. farm families support the positive role of work. Research in rural Iowa (Elder, King, et al. 1996) showed that youth experience work on the family farm as an apprenticeship that gives them a
sense of competence. A study in Dodge County, Georgia, by Peggy Barlett (1993) found that, while difficult, chores give children skills they are
proud of throughout life and often bring families closer together. Work
gives children a sense of worth, since their help is truly needed, and it
also transmits values. A mother said, "It's important to make them feel a
part of the farm," and when she went on to say, "You have to start them
at birth, practically, to raise 'em to be what you are"—a clear articula-
tion of deliberate cultural transmission—she might have been speaking
for parents in any traditional culture.

School tends children so that parents can work, while preparing them
for work in a post-industrial world. In these two senses it exploits children just as chores do, while giving many a sense of failure. In the developing world, school often prepares children for jobs that do not exist.
And much of children's play in modern society also becomes work. Team
sports teach ideals and behaviors like practice, discipline, skill, team-
work, fair play, and sportsmanship, but they may also cause pain and

humiliation. When parents shout from the sidelines, publicly showing disappointment at children's failures, team sports can be as coercive as child labor. Furthermore, children who work on their family farms are preparing for a career that they can have, while for most child athletes a career in professional sports is a vain dream.

Compared with hunter-gatherers, information transfer by deliberate teaching is much greater in modern states, but information transfer is overwhelmingly adult to child, not older to younger child. Of course, a larger and more complex body of information is transferred. But it would be of interest to know the results of trends toward mixed-age grouping, individual pacing, and "open" classrooms in some school settings.

26

The Culture of Middle Childhood

All cultures believe that a child must be socialized and enculturated in part by training and teaching, and we have seen many details of the acquisition of subsistence knowledge, motivation, and skill. But there is more to culture than subsistence. Reproduction is the goal of subsistence, and in our long-lived, relationship-dense species enculturation for successful reproduction is if anything subtler and more complex than enculturation for subsistence. It evokes all the richness of culture, including the spiritual and aesthetic dimensions that give survival and reproduction meaning—a meaning that, far from being unrelated to adaptation, is what makes adaptation possible for our species.

Infancy is part of this process, but later childhood and adolescence are more important in it. Although the culture acquisition device (CAD) to be proposed in Chapter 29 is more complex, in the traditional view socialization and enculturation together subsume at least four processes: (1) reward of behaviors that adults like; (2) punishment of behaviors that adults dislike; (3) imitation by the child of the behavior of adults and older children, sometimes enhanced by instruction; and (4) observation of rituals, stories, educational experiences, and the natural environment. Experiments show that each of these processes can change children's behavior.

A first attempt to use these principles to explain psychosocial development in a small-scale society was *Becoming a Kwoma* by John Whiting (1941). Beginning with the theory, supported by laboratory evidence, that frustration can lead to aggression or fighting (Dollard, Doob, et al. 1939; Killeen 1994; Felson 2000), Whiting made observations in the context of that theory. Instead of discouraging the reaction of frustration, Kwoma society encouraged it and guided it in certain directions throughout childhood and adolescence, with the result that Kwoma adults responded to the frustrations of everyday life with anger directed outward, toward enemy groups, in head-hunting expeditions. Many experiments show that a child's aggression toward others is influenced by reward, punishment, observation, and imitation. In a classic experiment, children who watched a film of an adult beating up a large doll were more likely to beat up a similar doll themselves when placed in a room with the doll after the film (Bandura, Ross, et al. 1961; Berkowitz 1993). Adult social expectations and cultural style constantly reinforce habits learned in childhood, but disruptions can occur when enough individuals become unresponsive to social expectations and cultural style, as when cultures are changing rapidly.

B. F. Skinner (1948) proposed that human behavior can be completely controlled by reward and punishment in combination with observation and imitation and that these processes can produce an efficient and happy society. Most behavioral scientists find this naive, but even those who see a strong biological basis for human behavior accept the fundamental importance of socialization and enculturation in producing culture-specific behavior (Barkow, Cosmides, et al. 1992). The child is designed by natural selection to develop with socialization and enculturation in the normal expectable environment and needs them. Reviewing the spectrum of enculturation for subsistence, we found much that is done through very informal modes of social learning and cultural transmission. Yet other aspects of culture require something different.

Enculturation among the Gusii of Kenya

The Gusii inhabit the highlands of southwestern Kenya and traditionally lived by cattle husbandry and agriculture. In their study of Gusii infant and child care, Robert LeVine and his colleagues (LeVine, Dixon, et al. 1994:81) defined three parallel life trajectories, which they called the reproductive, economic, and spiritual careers:

> The reproductive career is clearest in the Gusii context, for there are gradations of adult status that require specific levels of reproductive performance. To reach the age grade of elder, for example, a man or woman must have married children, and steps toward that goal (such as the circumcision of one's first

son) are publicly celebrated. Falling behind one's local age peers in such steps is inevitably a source of concern. The reproductive career includes not only one's own role in procreation, but also that of one's children.

The Gusii themselves define the stages of the life course for both males and females. "There is a life plan for each sex . . . and a trajectory of purposive behavior. Men and women want children, worldly goods, and spiritual safety and immortality. There is a timetable for the attainment of these goals" (LeVine, Dixon, et al. 1994:81). Infancy ends at weaning, circumcisions for both sexes occur during puberty, and elder status depends on having married children. That circumcised boys are designated as warriors and only women's status changes at marriage reflects the marked difference in sex roles in this culture.

Following the "situated learning" model (Lave and Wenger 1991), LeVine and his colleagues emphasize the role of "legitimate peripheral participation" in children's acquisition of subsistence skills. Tantamount to a kind of soft apprenticeship, and like the knowledge acquired by hunter-gatherer children when they follow their parents on gathering and hunting expeditions, situated learning is a formalization of the ways in which ethnographers have long described learning in traditional societies, and it bears some resemblance to Barbara Rogoff's (1990) psychological concept of "apprenticeship in thinking."

Excluding the initiation ceremonies, LeVine and his colleagues summarize Gusii childhood educational experience as "learning without teaching" in the setting of the household and homestead,

> the omnibus location for virtually all the activities in Gusii life: food production and processing, child care, social interaction, ritual, dispute settlement, and so forth. Children had observational access to these activities and participated in many of them, often as part of a sibling group. Participants were age-heterogeneous, and activities were hierarchically organized, with older participants delegating certain tasks to younger ones and often giving them commands. (LeVine, Dixon, et al. 1994:88)

As we will see, these last patterns do qualify as teaching, but for now just follow this rich description:

As the child became older, he or she was expected to approximate culturally valued forms of virtue and competence. Skills and moral virtues were not sharply distinguished . . . as skilled performance was considered to indicate moral maturity, and it was inconceivable that one could be morally virtuous without the willingness to acquire and perform skills of value to the home. . . . Children often did attain the expected skills, sometimes at remarkably young ages: Two 6-year-olds were observed in 1956 to have cultivated a small field of maize by themselves, from sowing to harvesting; most girls by age 8 could perform most of the household tasks done by married women. Boys were slower. (88)

We know that some students of hunter-gatherer subsistence reject the idea that we need the long human childhood to acquire subsistence skills because they are too easily acquired. Not all are easily acquired, but there is a far more important kind of enculturation, as revealed by this description: children must wholly engage the moral and motivational aspects of subsistence, including the mutual dependence and reciprocity entailed in all human survival and reproduction.

Gusii children were not at first expected to contribute. Boys "tagged along as their older brothers herded cattle, holding switches with which they would imitate their brothers' motions in beating the cows' rumps; soon they would be expected to herd the sheep and goats in this way, and by age 8 or 9 to herd the cattle themselves. If a boy were the eldest . . . the father would intervene to provide instruction. . . . Normally, however, participatory learning involved little parental intervention." However, "Gusii children learned by observing, imitating, and receiving corrective feedback, some of it harsh" (LeVine, Dixon, et al. 1994:90). Parents believed that praise undermined the social hierarchy, and they acted as "managers, protectors, and disciplinarians, rarely as teachers"; yet feedback, harsh or otherwise, is a form of teaching, and at a minimum parents acted as a fail-safe mechanism for situated learning when skills and values absolutely had to be learned.

Enculturation Processes beyond Conventional Learning

Revived interest in the concept of culture has created a new field of psychology devoted to understanding how psychological processes differ across cultures, complementary to psychological anthropology (Kitayama and Cohen 2007). Among other approaches, cultural psychologists have tried to find cognitive mechanisms for cultural learning that go beyond association, classical conditioning, instrumental conditioning, and other phenomena that have been well established in the laboratory. Three lines of research illustrate these contributions.

Guided Participation

Barbara Rogoff's *Apprenticeship in Thinking* (1990) proposes that culture is overridingly important; culture does not "have a role" because *every* psychological process is cultural, and to be human means to be a part of a shared culture from which individual minds are inseparable. In apprenticeship, (1) children actively recruit others for social guidance in learning; (2) many routine arrangements promote children's participation in skilled cultural activities, enabling learning in the absence of formal instruction; and (3) cultures vary in the goals of development and "the means by which children achieve a shared understanding with those who serve as their guides" (Rogoff 1990:8). This *guided participation,* which may or may not be explicit, stems from intersubjectivity, "a sharing of focus and purpose between children and their more skilled partners and their challenging and exploring peers."

This view closely resembles ethnographic accounts of childhood in traditional societies throughout the twentieth century. One summary of this literature states:

> Learning how to be a Tikopian, or Samoan, or Talensi happens, for the most part, in the course of everyday activity. People do not learn how to build canoes, for instance, in a course on canoe building in which they are lectured about the principles of canoe construction; rather, they learn it experientially, by helping out with the building of a canoe intended for use, not for purposes of instruction. Skills, then, as well as norms and roles, are learned in the doing. (Pelissier 1991:88)

Rogoff's approach to similar phenomena is psychologically informed, and such concepts as guided participation and scaffolding are valuable, but additional psychological processes are needed to explain apprenticeship systems in which complex skills are truly taught.

Mayan Weaving

The best-known such system is Zinacantec Maya weaving, studied by Patricia Greenfield and her colleagues over three generations of weavers in Nabenchauk, Chiapas, Mexico, beginning in 1969. This work uses a conceptual framework for cultural transmission with "two cultural pathways through development: the pathways of independence and interdependence" (Greenfield, Keller, et al. 2003:461), which correspond to psychological processes in individual and collective cultures (Greenfield and Suzuki 1997), a central distinction in cultural psychology (Kitayama, Markus, et al. 2000; Kitayama and Cohen 2007).

Mayan weaving is a four-thousand-year-old tradition that has used similar technology for most of that time, and Greenfield (1999, 2005) has studied it extensively. Videotape studies showed exceptionally long attention spans in girls as young as seven, spent mainly in observation (joint attention). In addition, there was the play (not pretend) weaving on toy looms that 86 percent of mothers and 98 percent of girls reported had been part of their own childhood experience, beginning by age three—earlier than involvement in real weaving (Greenfield 2005:37). Unlike toy ones, real looms require that the warp be prewound on a separate apparatus, which might require (in Piaget's terms) concrete-operational thought, so it is not surprising that real weaving starts with the five-to-seven shift. Experimental comparison of four-year-olds with seven-year-olds confirmed this hypothesis (Greenfield 2005:46–50). But the earlier phase begins the whole-body physical and postural preparation for weaving many hours at a time (*technique du corps*) that gives native learners a great advantage (Maynard, Greenfield, et al. 1999).

Greenfield's colleague Ashley Maynard (2002) went on to study the development of teaching in children, which is also central to cultural transmission and its evolution. She did extensive videotaping of sibling pairs at home, with thirty-six younger siblings (eighteen boys) age twenty to thirty-six months (mean age, two years) and seventy-two older siblings (twenty-five boys) age three to eleven years (mean age, 6.8 years), focusing not on weaving but on household tasks—washing, cooking, taking care of baby dolls, making tortillas. The definition of teaching was "any activity attended to by the younger child that had the possible effect of transmitting cultural knowledge," including "intentional and unintentional teaching." As we will see, teaching can be defined more rigorously, but for present purposes, "play became synonymous with sibling teaching in this study because there was so much teaching in the multiage play" (Maynard 2002:973). Like the !Kung (see Chapter 25), Zinacantecans have an essentially Piagetian conception of cognitive development, not a culturally determinist or Vygotskian one (Maynard and Greenfield 2003).

The discourse variables coded included nonverbal task simplification, talk with a demonstration, commands, explanations, feedback, and guiding the learner's body. The frequency of each tactic increased significantly with age, but the last three were done overwhelmingly by children eight and older. The eight- to eleven-year-olds essentially "demonstrated the skills of adult scaffolding" (Maynard 2002:978) in support of the learning process. None of the three- to five-year-olds used task simplification in their interactions with younger siblings, while 45 percent of eight- to eleven-year-olds did, consistent with an increase in perspective taking and other aspects of metacognition across the five-to-seven shift. Similar observations were made in a qualitative comparison study of sibling caregivers among the Wolof of Senegal (Rabain-Jamin, Maynard, et al. 2003).

Cultural Learning

These findings are consistent with the most systematic approach to cultural learning in a social cognitive framework—that of Ann Cale Kruger and Michael Tomasello (Tomasello, Kruger, et al. 1993; Kruger and Tomasello 1996). They defined three types of cultural learning (as distinct from the broader category of culture acquisition): (1) *imitation,* an advanced type of observational learning in which the learner understands and adopts the intention of the model, who need not take any action toward the learner or even be aware of the process; (2) *instructed learning,* in which the more proficient individual, the teacher, makes adjustments to the task in response to the learner's progress with the intention of helping the learner increasingly understand the task as the teacher does; and (3) *collaborative learning,* in which the difference in proficiency between two individuals is minimal, but by intentionally communicating and comparing their respective understandings of the task they can co-construct new skills or concepts.

In this framework all three types of cultural learning require perspective taking—unidirectional in the first and bidirectional (intersubjective) in the other two. They are more mature versions of what Ann

Kruger (2010) has elsewhere called "communion" between infants and parents. Also, instructed and collaborative learning are considered uniquely human. Even imitation, by Kruger and Tomasello's definition, is almost uniquely human (except for enculturated apes), since it requires an understanding of the model's intention, as contrasted with the two simpler types of observational learning: *emulation,* in which the learner attempts to reach the observed goal without reproducing the means; and *mimicry,* in which movements or sounds are reproduced without understanding of or attempts to reach the goal.

In a subsequent analysis, Kruger and Tomasello considered cultural learning from the viewpoint of the instructor rather than the learner and applied this analysis to a range of different cultures. The analysis defined three distinct educational styles favored in various situations by various cultures, again in terms of adult intentions, beliefs, and activities (Table 26.1).

1. *Expected learning.* The first educational style, expected learning, rests on the belief that a behavior or skill will emerge or be acquired more or less automatically with development as the child interacts with the affordances of the environment. This belief is well characterized by the !Kung statement that the child "learns/teaches herself" and by the Tallensi saying "Heaven teaches them" (Fortes, quoted by Kruger and Tomasello 1996:378). It is also exemplified by Margaret Mead's (1930:26) description of how Manus children in New Guinea learn to swim: "Swimming is not taught: the small waders imitate their slightly older brothers and sisters, and after floundering about in waist-deep water begin to strike out for themselves. Surefootedness on land and swimming come almost together."

Table 26.1. Three Educational Styles and Their Associated Adult Beliefs, Activities, and Practices

	Adult Beliefs	Adult Activities	Types of Practice	Example
Expected learning	Self-guided development	Laissez-faire	Informal	Manus swimming
Guided learning	Learning needs assistance	Scaffolding	Semiformal	Mayan weaving
Designed learning	Learning needs direct instruction	Teaching	Formal	Pulawat navigation

Source: Modified from Kruger and Tomasello (1996:377).

2. *Guided learning.* In guided learning, which is more focused and precisely defined than Rogoff's guided participation, adults believe that the acquisition of the behavior or knowledge requires assistance and they take steps to make it easier for the child to learn. This is scaffolding, and it may involve doing as little as adjusting the activity to enable the child to see it better or as much as guiding the child's hands in attempts at performing a skill. Guided learning includes assigning simplified aspects of a task to the child, offering corrections, making toy versions of a bow and arrow, canoe, or loom, and many other interventions. As the child learns, the scaffolding is gradually and intentionally removed. However, it does not involve a formally designated environment for learning. Unlike the case of swimming, the acquisition of canoeing in Manus involves a combination of expected and guided learning. The learning of weaving by the Zinacantecan Maya described earlier is mainly if not completely in the guided learning category, and the developmental study of sibling teaching showed that eight- to eleven-year-olds are quite capable of this kind of instruction (Maynard 2002). (Interestingly, children who had been to school used significantly more discourse methods in teaching their younger siblings, and in particular they used more explanations [Maynard 2004]. Neither method is necessary for guided learning or typical of teaching in small-scale traditional cultures.)

3. *Designed learning.* The most formal style of instruction, designed learning, entails at a minimum very deliberate teaching in relation to criteria of performance that adults feel must be met, including moral strictures, taboos, manners, and appropriate rules of adult behavior. In every culture, some subset of these expectations is not left to expected or even guided learning but is actively taught, sometimes in initiation rituals, but also on an ongoing basis throughout childhood and adolescence, often with techniques of physical or verbal punishment or humiliation. There are also many positive instances of designed learning; for example, the !Kung value sharing very highly, and from the time their infants are six months of age mothers and other adults frequently say, "Na," meaning "Give," when a bit of food is in the infant's hand and on the way to its mouth. The criterion is that they should inhibit the very strong impulse to eat and reliably turn the morsel over to the adult making the demand.

Interestingly, designed learning is sometimes invoked where it is not needed. The !Kung and many other African cultures believe that infants must be taught to sit and walk, and they apply deliberate strategies to make this happen (Konner 1972; Super 1976). However, the most common form of designed learning is Western-type schooling; only a minority of nonliterate cultures have had traditional equivalents. One classic example is the Dagomba drum school of Dagbon in northern Ghana, which involves not only drumming but committing to memory traditional genealogies and oral narratives that are the heart of social memory in this culture:

> The unbroken historical narrative and royal genealogy which they remember and recite is the charter of the political structure of the kingdom, and the story of the origins of the people, and as such is vital to the continuity of the traditional system. The task of learning this massive oral body of historical material is an arduous and painstaking one, requiring long hours of patient application and practice on the part of the teacher as well as the pupil. . . . Not only must the young pupils learn to drum but they must begin to learn songs and later the narratives and genealogies. (Oppong 1973:54)

It may take the boys anywhere from two to ten years to learn the repertoire, and some boys are known to be especially skilled in memory, at singing, and so on:

> If a boy begins to learn as early as six or so he may be able to sing and play a number of songs by the time he is about eight and will be able to practice playing in public at funerals, naming ceremonies or festivals. Learning is thought to be a very difficult process accompanied by ear-pinching and pulling and beating from the teacher. Wisdom is considered very hard to acquire. A typical teaching situation is for the teacher to recline in his entrance hall after supper in the evening, while the children gather round him, sitting and kneeling upon the floor. (55)

Other specialized roles in the Dagomba kingdom are also taught through apprenticeships, including those of butcher, blacksmith (weapons maker), barber (also a healer and circumciser), lute players to royal patrons, and diviners. A boy of five or six may begin drawing his bow back and forth across the single string of the lute while the teacher does the fingering; the boy learns words and music simultaneously and mouths the words silently if he does not sing them aloud.

Another example is the teaching of navigation among the people of Pulawat Atoll in the South Pacific, some of whom have the remarkable ability to traverse great expanses of ocean in light boats, out of sight of land, reckoning by the stars (Gladwin 1970). Highly selective and competitive, this apprenticeship may take twenty years, and much of it is verbal instruction carried out on land, aided by drawing maps in the sand of rising and setting constellations. Apprentices must memorize these maps as well as a voluminous amount of information about waves, currents, reefs, weather, and sea animals and plants. Years of didactic exercises on land are followed by years of practice at sea, resembling the pattern of Western medical education. Criteria are set and must be met at every stage of the process before the learner moves on to the next stage.

Enculturation by Children

Children socialize and enculturate each other, and this process may be just as important as parental nurturance and family environment in shaping personality, if not more so (J. Harris 1995). Clearly enculturation by children helps shape gender identity (Maccoby 1998). Since many human and animal societies have child playgroups that are not peer groups and age variation affects the interaction and mutual socialization, the vertical transmission of culture from older to younger children parallels, complements, or sometimes undermines adult-child transmission. Knowledge of subsistence, relationships, aggressive encounters, nurturance, modesty, dress, language, religious beliefs, and many other essential features of culture can be carried from child to child. All the processes detailed in the culture acquisition device (Chapter 29) can occur in multi-aged child groups, and most occur even in same-age peer groups. But it is not just culturally mandated behavior and knowledge that is transmitted: child groups may transmit what adults see as undesirable behavior, and they can become agents of cultural change.

Child Culture

There is another, independent kind of childhood tradition that lies in a domain inaccessible to the young of any other species. Formal games, jokes, superstitions, chants, songs and ditties, spoken formulas, symbolic hand-signals, and homemade toys form substantial and deep-rooted child cultures that span the globe (Opie and Opie 1987). The spoken formulas include riddles, jeers, curses, oaths, pranks, tricks, and pacts. The patterns differ in space and time, responding to opportunities of technology and history, but they can be stable in purpose and function, and their continuity even in form is sometimes remarkable. In Europe some such verbal traditions can be traced back to similar forms in the medieval period, which had many more of its own (Orme 2001).

Although elements of it may have originated with adults, children control this culture, which pervades the behavioral space around families, schools, and media, and it inculcates values. "Liar, liar, pants on fire," "Cross my heart and hope to die," and "Finders keepers, losers weepers" are value-laden formulaic remarks. Others, such as "Made you look!," "Crybaby," and "Baby, baby, stick your head in gravy," are exercises of power through humiliation, like the nonverbal threat gestures of monkey and ape juveniles. But the symbolic domain is pervasive in children and absent in other juveniles. Some examples of symbolic play and ritualized or formal games among !Ko and !Kung San hunter-gatherers were discussed in Chapter 19; they are a central aspect of human socialization. But because they involve language, symbol, ritual, rules, and toys fashioned for the purpose, this play becomes uniquely human, a vehicle for enculturation as well.

Moral Learning in Middle Childhood

We saw in Chapter 11 that Piaget (1965) identified the developmental stages of moral thinking associated with games such as marbles. By age seven children understand the rules but consider them immutable, while at age ten they understand that the rules are under their control and can be changed. Such a change could occur through collaborative learning, if the rule change is co-constructed as a product of discourse between approximate equals.

An elegant, random-assignment, controlled experiment tested and confirmed some of these observations (Kruger and Tomasello 1986; Kruger 1992, 1993). The forty-eight subjects were eight-year-old girls who viewed pictures with set stories relating to questions of fairness—problems in distributive justice. For example, in one story a class of children working together on a project earned money that had to be distributed according to some combination of work, skill, and need.

An experimenter sat with each girl and discussed one of the stories (randomly chosen), asking the child to comment on a fair solution but making it clear that there was no right answer—the goal being to assess the girls' process of reasoning. A randomly chosen half of the girls were paired with their friends, the other half with their mothers, for a similar discussion of two more problems while the experimenter was out of the room. After these conversations the experimenter sat alone with each girl and talked about the fourth, remaining problem. All conversations were recorded. An index of moral reasoning (Damon 1980) was devised for the child's contributions to the "before" and "after" discussions, and another coding system scored the conversations that the girls had with their peers or mothers (Berkowitz and Gibbs 1983). The scoring focused on *transactions*—remarks or questions that involved "reasoning about reasoning," that is, children's comments on their own or the other's ideas.

The results were that: (1) girls paired with friends showed more sophisticated moral reasoning afterward than girls paired with their mothers; (2) girls paired with peers used more active transactions than those paired with mothers; and (3) among the girls paired with mothers, those who used the most active transactions showed the greatest improvement in the index of moral reasoning. In effect, the more *peer*like the mother's role in the intervening conversation, the greater the child's gains in moral reasoning (Kruger 1992, 1993).

This countered the claim that moral thinking comes solely from adult guidance. "The peers' equal status allowed a critical reciprocity that was infrequent in adult-child dyads. Thus, Piaget's contention that symmetry of power leads to greater moral reasoning development is supported, and ... active reasoning is the essential element in the process" (Kruger 1992:210). Further analysis showed that reasoning should be not only active but spontaneous; mothers who used the Socratic method of questions leading toward a preconceived answer did not improve their daughters' thinking, but those who encouraged autonomous and spontaneous comment did. In a third analysis, each turn in the peer conversation was judged as undermining or supporting previously stated ideas or creating new ones. It was interactions "tearing apart solutions they find inferior" (Kruger 1993:178), followed by co-creation of an agreed-upon solution, that yielded the largest gains in moral reasoning. Important aspects of moral thinking in middle childhood are discovered by children through co-construction and collaborative learning.

But are these elements of moral reasoning culturally specific or simply part of the maturational plan in the normal expectable social environment? We know that core aspects of moral development are by-products of cognitive (and therefore brain) development (Piaget 1965; Kohlberg 1963; Colby, Kohlberg, et al. 1983; Levine, Kohlberg, et al. 1985; Snarey and Samuelson 2008). At the most basic level, a test of distributive justice given to children in seven different cultures showed a significantly greater tendency to share rather than hoard at age five as compared to age three, although the change was greater in rural and collectivist cultures than in others (Rochat, Dias, et al. 2009).

There are also clear cross-cultural regularities in the first four stages of Kohlberg's six-stage scheme: an obedience and punishment orientation (stage 1); instrumental purpose and exchange—"you scratch my back, I'll scratch yours" (stage 2); mutual interpersonal expectations—"do unto others as you would have them do unto you" (stage 3); and social system and conscience maintenance—good citizenship to keep society functioning (stage 4). Beyond these stages there is considerable variation (Miller 2007). Different cultures have different approaches to stage 5—a prior rights and social contract orientation (Edwards 1981, 1994; Snarey 1985), and the hypothesized highest stage—universal ethical principles—is expressed only in some cultures.

In Kenya, level of education and modernization influences progress beyond stage 3 (Edwards 1975, 1978), while in the Israeli kibbutz a strong emphasis on communal values leads to higher levels of moral reasoning (Snarey, Reimer, et al. 1985). A review of forty-five studies in non-Western cultures suggests the need for a less culture-bound framework that emphasizes the values associated with *gemeinschaft*, or community (Tönnies 1957; Snarey and Keljo 1989), consistent with the difference in cultural psychology between individualistic and collectivist cultures (Kitayama and Cohen 2007). In addition, the balance between justice and care is influenced by gender (Gilligan 1982), and the morality of care concept may apply more to some cultures than to the West (Miller 2007). Finally, the interactions that Piaget described are only one route to moral understanding; others include moral exemplars and membership in a moral community (Snarey and Samuelson 2008).

Inculcating Morality?

But what of the role of punishment? Despite the assertion of Piaget, Kohlberg, and other psychologists that moral judgment emerges naturally through play and conversation with other children, as well as growing evidence that it has nonhuman counterparts and an evolutionary history (Hauser 2006), many adults view children as basically wild or even evil creatures who have to be socialized and made moral through corrective training. These principles, when implemented, can result in harsh treatment.

Corporal Punishment across Cultures

In a cross-cultural study of corporal punishment, Carol and Melvin Ember (2005:609) examined the 80 percent of 186 societies in the Standard Cross-Cultural Sample that were accessible in the Human Relations Area Files; they defined corporal punishment as "hitting, striking, wounding, or bruising a dependent child for the purpose of punishing, disciplining, or showing disapproval."

As always, the quotations from original descriptive ethnographies illuminate the spectrum of variation. For the Rwala Bedouin studied by Alois Musil in 1910, corporal punishment was idealized in the belief that the rod came from Paradise and could lead people back there. "If they deserve it they [the children] are spanked with a stick, not only by their mother or father, but by the slaves both male and female" (Musil, quoted by Ember and Ember 2005:609). Adolescent boys were also punished with a dagger or saber to harden them for their adult lives.

The Mapuche (Araucanians), Native Americans of Chile, represent the middle range of frequency and severity of corporal punishment. Children with language comprehension who misbehaved were "coerced . . . to conform to behavioral standards . . . by corrections in stern tones, by scolding in angry tones, by deprivations, by slapping the child's hands, and by spankings or whippings" (Hilger 1957, quoted by Ember and Ember 2005:610), although other sources suggest that spankings and whippings were used infrequently.

This pattern of escalation with troublesome children is common among societies in the world range, but there are some in which corporal punishment is extremely rare. The Embers use the example of the

Copper Inuit (Eskimo), studied from 1913 to 1916 by Diamond Jenness and later by Richard Condon. In Ember and Ember's (2005:610) summary, "A child who misbehaves might be teased or briefly chastised, but adults will rarely use verbal threats or physical punishment. In fact, if a child ignores initial reprimands, parents will just ignore the child even if the child hits or swears at the parents." It is not a coincidence that the Inuit are hunter-gatherers; this description matches what I and others observed among the !Kung and many other hunter-gatherers.

However, the intermediate-level societies show a range of punishment types (Laney 2008). Ember and Ember did multiple regression analyses of carefully measured societal background variables and found that social stratification—inequality within a society—and alien currency, a marker for colonization, both predicted higher levels of corporal punishment, but the strongest predictor was nonrelative caretakers. In a separate analysis, local democracy predicted less corporal punishment, and warfare predicted more.

In sum,

> Our results suggest that corporal punishment is likely in societies that are marked by power inequality . . . [since] societies with exclusively parental caretakers are lowest in the employment of corporal punishment and societies with Non-Relative Caretakers are the highest, it seems that biological distance is more likely to increase . . . corporal punishment. . . . More-than-rare warfare, which may reflect a culture of violence, also seems to make corporal punishment more likely. (Ember and Ember 2005:617)

At first glance, the fact that nonparental care in general does not reduce corporal punishment seems to cast doubt on some of our reasoning in Chapter 16, but as soon as relatedness is introduced the picture becomes consistent again, in keeping with Emlen's (1995) model of cooperative breeding (Chapter 16), and with much other evidence on the impact of nonrelatedness on child abuse and pedicide (Daly and Wilson 1988a; Chapter 17).

Punishment in History

With this background, it is not surprising that highly stratified, undemocratic societies characterized by foreign dominance and/or frequent warfare—such as those that gave rise to all the major religions—favor corporal punishment, which they justify with a theory about change-worthy behavior quite different from the "she'll grow out of it" theories found among hunter-gatherers. This includes ideas about inborn nega-tive tendencies that have to be expunged by culturally guided parental practice. For fundamentalist Christians this tendency is the devil in the child, or "original sin," and for orthodox Jews it is the *yaytzer ha'ra,* or "evil inclination"—the tendency to be selfish, harm others, and break rules. The religious view also sees good in children: in Jewish tradition children have a good inclination working against the evil one, and Chris-tians believe that Satan is in the child's heart, but that Christ is also in the child.

Still, religiously motivated child training has at times in history—in the Puritans' New England, in traditional Catholic Ireland, among fun-damentalist Muslims, and in some ultraorthodox Jewish sects—seen the evil inclination as much stronger than the good. In view of the evi-dence that children's moral development is largely inherent in normal social-cognitive development, it seems odd that "Spare the rod, spoil the child" became the watchword of any method of child rearing.

The psychological burdens imposed on children by some cultures were deliberate and determined. A Methodist homemaker in England wrote in 1738, "I have been endeavouring to my utmost to convince my children of their natural sinful state, and the necessity of a saviour . . . while others are mourning over the sins and follies of their children, I have the pleasure to hear mine mourn in secret over their own sins" (Pollock 1987:217–18). A letter from a twelve-year-old boy to his father, composed in New England in 1638, shows how such guilt is internalized. "I feel also daily great unwillingness to good duties, and the great ruling of sin in my heart; and that God is angry with me and gives me no an-swer to my prayers; but many times he even throws them down as dust in my face; and he does not grant my continued request for the *spiritual blessing of the softening of my hard heart.*" He believes God has rejected him: "I think that the reason of it is most like to be because I belong not unto *the election of grace*" (215). This meant that he was already destined for eternity in hell, the terrors of which were impressed on him daily.

Parents in England and colonial America used whippings as well as hellfire, and even the doubts some expressed show how common they were. James Erskine (Lord Grange), an English judge, in his 1712 diary dissented from the methods of his son's tutor:

> As to the perverseness of the poor young child, it rarely is un-
> conquerable in a boy so very young, if proper methods be taken.
> I know the boy had a wantonness, as such of his age use to have,
> and is more pliable by persuasion than by rough treatment.
> But Cumming's crabbed peevish temper made him use the last
> method, and often to beat him severely for trifles, and some-
> times when the boy was more in the right than he, till I put a
> stop to it, and now he says himself that the boy does well. (Pol-
> lock 1987:194)

There were also fathers like late-seventeenth-century New Englander
Samuel Sewall, who often whipped his sons "pretty smartly"—for instance, when his ten-year-old was punished "for breach of the ninth
commandment, saying he had been at the writing school, when he had
not" (183). But this parenting style was also resisted. Margaret Woods, a
middle-class fifty-one-year-old in England, wrote in her 1799 journal:

> The love of liberty and independence is strongly implanted in
> the human mind. How far it should be indulged in the educa-
> tion and conduct of youth, will, by many people, be differently
> determined. . . . Either extreme, I believe, is prejudicial. Too
> tight a curb sometimes makes young people fret under it, and
> produces an impatience to be entirely free, when more gentle
> discipline might have produced submission. Little benefit can
> arise from mere compulsion.

And in a letter to a friend sixteen years later she wrote:

> I have always wished that they should be afraid of doing wrong,
> but not afraid of me. I would encourage them to lay open their
> little hearts, and speak their thoughts freely; considering that
> by doing so, I have the best means of correcting their ideas, and
> rectifying whatever may be amiss. I am, from judgment, no
> great disciplinarian; if I err, I had rather it should be on the le-
> nient side. Fear and force will, no doubt, govern children while
> little, but having a strong hold on their affections will have
> most influence over them in their progress through life. Obe-
> dience I do consider as an indispensable thing in education;
> but perhaps it would be imprudent to call it forth too fre-
> quently on trivial occasions. (187)

In 1815, at age sixty-seven, Woods sounds almost modern. To hunter-gatherers and many other traditional peoples, the common disciplinary tactics of eighteenth-century Anglo-Saxon parents would simply be deemed abuse. Recall that even middle-class parents in Chicago in the 1940s, who were far less harsh in their child training than these seventeenth-century predecessors, were scored as harsher in parental discipline than the spectrum of traditional cultures in every domain except aggression, where Chicago parents were lenient (Whiting and Child 1953).

Some "evil" tendencies—idleness, fantasy, risk taking, sexuality, selfishness, and violence—are vital evolutionary adaptations, but can be maladaptive as well. Children differ greatly in how and how much they show these tendencies, and the idea that one method of child rearing is best for all children is illusory. Children have different natures, so they should have different goals and different paths to those goals. But as for corporal punishment, there is a growing consensus against it (Pinheiro 2006).

Children and Religion

As with the "natural" growth of moral reasoning, there may be a natural emergence of religious ideas in an appropriate cultural environment. Anthropologists find concepts of gods and spirits in every culture (Swanson 1964), and psychologists have long been interested in the psychological underpinnings of these beliefs (James 1902; Freud 1961a, 1961b), most recently with an experimental cognitive approach (Tremlin 2005). For Darwin, religious devotion consists "of love, complete submission to an exalted and mysterious superior, a strong sense of dependence, fear, reverence, gratitude, hope for the future, and perhaps other elements. No being could experience so complex an emotion until advanced in his intellectual and moral faculties to at least a moderately high level" (Darwin 1871/1981:68).

Faith Development

This applies to development as well as evolution. Many if not most children have spiritual or religious inner lives that are important to them (Coles 1990), intertwined with the development of relationships, emotions, and mind. There is evidence of common patterns of growth of some religious ideas, and attempts have been made to define their sequential development. In the model developed by James Fowler (1995), the stages include primal, intuitive-projective, mythic-literal, synthetic-conventional, individuative-reflective, conjunctive, and universalizing faith (Fowler 1991; Fowler and Dell 2004, 2006), some of which are usually not reached until adulthood, if at all. This model needs to be validated and tested cross-culturally, but it has been accepted by a variety of religious traditions. Attempts have been made to integrate the emergence of faith in early childhood with psychodynamic models of the development of attachment and the self (Granqvist and Kirkpatrick 2008), suggesting that religious faith comes initially out of the process of individuation in the context of dependency on and love for parents (Rizzuto 1981; Fowler 1989). God is the "imaginary friend" your parents want you to believe in, with more superpowers than you can fathom, but also the parent, unseen and all-seeing, who knows if you've been bad or good and reacts accordingly.

Understanding Death

The severity with which the Puritans imbued their children with the fear of God was in part a response to an ever-present threat of mortality that made heaven and hell seem only a breath away. Families were constantly adjusting to deaths, especially the deaths of children. Death was equally common in hundreds of primitive cultures that did not burden children with guilt and fear, but given the Puritans' beliefs, they had to try to ensure their children's salvation; even their harshest punishments would be kind compared to damnation.

Still, children must deal with death even in a society where it affects mainly the old, and it occupies a changing place in mental life during the transition to middle childhood. Studies of children's concepts of death (Stambrook and Parker 1987; Cassem 1999) show that before age five children see it as reversible, much like a sleep, illness, or journey, although they are already thinking about the possibility of their parents dying. After age seven they begin to view it as inevitable and irreversible, but it remains fairly remote until ten or eleven, when they grasp that death is universal and that it applies to them. Individual variation is substantial, however, and there have been no systematic studies of cross-cultural differences, particularly in relation to the varying frequency of childhood death.

God's Presence and Companionship

Religion is largely about relationship (Granqvist and Kirkpatrick 2008). Interviews by Robert Coles (1990) have addressed the question of how children—most between ages eight and twelve, from various religions and cultures—use what they are taught about God, heaven, the devil, spirits, life after death, and other concepts of faith to explore and deal with their problems. Many feel that God is always with them, and this mitigates life's pain and loss.

One child described the then-recent experience of being one of the first African-American children to integrate an elementary school in 1962. She had had to walk through a mob of bigoted whites: "I was all alone, and those people were screaming, and suddenly I saw God smiling, and I smiled." At the school door, "a woman was standing there and she shouted at me, 'Hey, you little nigger, what you smiling at?' I looked right at her face, and I said, 'At God.' Then she looked up at the sky, and then she looked at me, and she didn't call me any more names" (Coles 1990:119–20).

Although religious differences underlie many group conflicts, as religious emotions inflame ethnic ones, culture-specific religious beliefs are not always negative, as seen in Coles's interviews of Hopi children in the American Southwest. Natalie, at ten, was described by her teachers as neither especially bright nor likely to give a good interview. Coles (1990:25–26) noticed her casting a long steady gaze at a thundercloud. "The home of the noise," she said after some time, pointing. She contrasted her beliefs with Anglo ones:

> The sky watches us and listens to us. It talks to us, and it hopes we are ready to talk back. The sky is where the God of the Anglos lives, a teacher told us. She asked where our God lives. I said, "I don't know." I was telling the truth! Our God is the sky, and lives wherever the sky is. Our God is the sun and the moon, too; and our God is our people, if we remember to stay here. This is where we're supposed to be, and if we leave, we lose God.

She said she prayed for the Anglos, and said of the Hopi-Navajo land dispute, "They want the land, and we believe it has been here for us, and it would miss us . . . the land can feel the difference." She felt sorry for the Navajo and the Anglos, who "don't feel at home near their mesa, so they want all the mesas in the world! . . . they'll never be able to choose which one to call their home! Their ancestors must go from one mesa to the next, and they must cry, because they don't know where they can stay and be together, and they don't know if they'll ever be seen by the people" (Coles 1990:157–58). She believed that hawks in flight were the transformed souls of the ancestors, hovering over a particular spot on the earth's landscape, benignly watching her. She mentioned "praying to our ancestors," and Coles asked if she prayed on her knees. "No. I sit. I close my eyes. I let myself join our people. I ask them to include me in their number. I talk to them. Then I see them" (152).

This conception of God is about relationships. A thirteen-year-old Tunisian boy said, "Allah watches over us, and He's there [watching] at night too. When you wake up and think of Him, it's because He has visited you and told you something" (Coles 1990:239). A nine-year-old girl in rural Tennessee kept having the same dream, fleeing from a frightening house and seeing a flash like lightning: "I think it was Jesus. He'd come to help me. . . . I was walking with Jesus, maybe following Him, and I think He was telling me something; I think He was saying you should be careful where you go" (131). A Jewish boy the same age heard the *shofar*—the ram's-horn trumpet—blown in his synagogue; a week later, "I was riding on my bike, and I could hear the sound of the *shofar* in my head. Maybe God is nearest you when you think of Him, and talk with Him, and remember the sound of the *shofar*" (145). Many children have such relationships and feel comforted.

But as with Puritan Protestantism, there is a dark side to religions. In the Kachina Society initiation rituals of Pueblo groups like the Hopi, children, especially boys six to ten years old, were sequestered in a sanctuary, the *kiva*, and systematically terrorized by adults costumed as threatening supernatural spirits wielding yucca whips (Eggan 1943; Goldfrank 1945; Aberle 1967). For example, the boys were subjected to "a We'e'e Kachina in a blue mask and carrying a long black and white ringed pole . . . [and] two Natackas with fierce bulging eyes and huge black bills fashioned from large gourds, each with a bow in the left hand

and a saw or a large knife in the right hand" (Goldfrank 1945:514). The stated purpose was scaring children into good behavior.

> The children tremble and some begin to cry and to scream. The Ho Kachinas keep up their grunting, howling, rattling, trampling and brandishing of their yucca whips. All at once someone places a candidate on the sand mosaic, holds his (or her) hands upward and one of the Ho Kachinas whips the little victim quite severely. . . . Some of the children go through the process with set teeth and without flinching, others squirm, try to jump away and scream. . . . It was also noticed on several occasions that some of the boys, probably as a result of fear and pain, involuntarily micturated and in one or two cases even defecated (Goldfrank 1945:517–18).

This too was a key part of Hopi children's traditional religious experience. It was said to be for their benefit—even to save their lives—as well as to make them obey. Eventually children learned of the dramatization and fakery and reacted with anger and disappointment, but as one grown woman said, people had decided it was for the best and the only way to teach the children.

Finally, consider the acquisition of faith in trance-dancing, which is both the central religious and central healing ritual among the !Kung. We have seen how infants and toddlers are brought into the motivational and performative aspects of the ritual; even they can feel awed by the spectacle. But it is only later in childhood that they begin to understand its spiritual meaning. Women sit in a circle around a large fire, singing in yodel-like tones and clapping in complex syncopation; men dance around behind them steadily and monotonously until some go into trance and fall down as their souls leave their bodies (Lee 1968; Marshall 1981, 1999; Katz 1982). They are rubbed and partly revived but stay in trance, lay hands on the person in need of healing, and shriek as the particles of sickness leave the body of the person healed and shoot out through the healer's neck and shoulders, back into the world of the spirits. The gods are not companions; healers yell at and berate them when they cause harm. The entire community comes together on behalf of one who is ill, and the ritual serves myriad functions that are clearly adaptive but would be very difficult to measure.

For many children religious commitment intensifies in adolescence, when brain changes bring about a new capacity for abstraction, spirituality, self-examination, autobiographical narrative, contemplation of the future, and the management of emotion; this is also a time when all cultures force the young to begin to see themselves in and assume adult roles in the community. Community for most humans is intricately bound up with religious commitment, and shared beliefs about the supernatural are no less real to them than fire or rain. But before adolescence, faith is already established in most children's emotional lives.

27

The Culture of Gender in
Childhood and Adolescence

For Freud, looking at nineteenth-century Vienna, penis envy was central to gender identity. Anthropologically, there seemed more evidence and logic in Margaret Mead's concept of womb envy; giving life to a child is more impressive than having a penis, and in some cultures this is made explicit. So cultures clearly differ in how they construe gender. But we have seen that behavioral sex differences—in toy preference, voluntary sex-segregation, aggressiveness, nurturance, and other measures—precede the gender concepts. Young children resist efforts to socialize or enculturate the sexes identically, and hormonal and brain differences account for some sex differences, but culture accounts for many others.

Culture Stretches Biology

Darwinian considerations generally require that the sexes have somewhat different roles. But for reasons we do not completely understand, most if not all cultures have chosen to exaggerate the biological differences. Partly, perhaps, because of the human mind's dichotomizing tendency, and partly because too much is at stake for a blurred distinction, the culture of gender intensifies biological tendencies.

Differences between the sexes at birth are small, and boys and girls overlap greatly; it is doubtful that the degree of difference between boy and girl infants at birth could drive the behavior of mothers and fathers. By contrast, there are great differences between boys and girls in the ways their parents think about and treat them. The first question asked about the birth of a baby, the first information offered, is about its sex.

We seem to require that information to know how to act even toward a neonate. Among many higher-primate species the first reaction of adults to a birth is to inspect the infant's genitals (Lancaster 1971; Hrdy 1974; Pereira and Fairbanks 1993), suggesting that immediate perception of gender shapes adults' behavior toward infants.

A human infant elicits very different responses from adults depending on whether it is dressed in pink or blue (Will, Self, et al. 1976; Smith and Lloyd 1978), and an audiotape of a child's mischievous utterances draws amusement and encouragement from adults who are told that the child is a boy, but negative judgments from matched adults who think it is a girl (Rothbart and Maccoby 1966). Similar findings have been made many times in many settings (Stern and Karraker 1989), and there is a systematic rating scale for parents' gender stereotypes about neonates, the Parental Sex Typing of Newborns Scale, or PASTON Scale (Leeb and Rejskind 1998). Fathers stereotype more than mothers (Rubin, Provenzano, et al. 1974; Karraker, Vogel, et al. 1995), but both do it, with their own and others' infants.

In a study of three-week-old infants and their mothers in the United States (Moss 1967), boys cried more, were more active, looked at mothers more, and slept and vocalized less than girls, but maternal behavior also differed in several ways: mothers of male infants held, burped, rocked, stimulated, aroused, stressed, looked at, talked to, and smiled at them more than did mothers of girls. Such differences could have affected the infants over the first three weeks or even during the observation period itself; for example, boys may have slept less because their mothers aroused them more. In a study at six months of age, boys showed more emotional distress than girls in face-to-face interactions with their mothers (Weinberg, Tronick, et al. 1999), but the mothers had had six months to shape theirs sons' and daughters' tendencies.

There is a large body of evidence for parental and other adult gender stereotypes (Golombok and Fivush 1994), and since children are socialized by friends and playmates, not just (perhaps not even mainly) by parents and teachers, the power of voluntary sex-segregation must be considered. Once exposed mainly to same-sex playmates, a child is influenced by their behaviors. These "two cultures of childhood" (Maccoby 1998) could maintain sex-specific behavior patterns even if adult models were neutral, which they are not.

Among other environmental influences, television delivers a frequently sex-stereotyped message to young children (Klinger, Hamilton, et al. 2001; Larson 2003; Aubrey and Harrison 2004), and it reaches more of the world's children every year. Cartoons, commercials, and other programming shape choices from early ages. In the United States, television viewing begins around age two and a half, occupies twenty-seven hours per week in the preschool years, twenty-three hours between ages six and eleven, and twenty-one hours in the teenage years (Leung, Fagan, et al. 1994), augmented by other media with similar messages (Jordan 2004).

Certain gender differences appear in so many different cultures that they clearly are independent of electronic media. As shown in Chapter 10, the social behavior of boys and girls does differ consistently across cultures, in measures related to aggressive behavior and nurturance (Whiting and Edwards 1973, 1988; Whiting and Whiting 1975a). But what if parents the world over were trying to *produce* the same sex differences? Cross-cultural studies of socialization suggest that such efforts are indeed made (Barry, Bacon, et al. 1957). Table 27.1 shows the number of cultures for which information was available on a particular child training practice. Parents in 82 percent of the cultures expect more nurturance from girls, *none* expect more from boys, and 18 percent expect no difference. For achievement and self-reliance, parents' expectations are also consistent, in the opposite direction: 87 percent expect more achievement from boys, and only 3 percent expect more from girls, with 10 percent expecting gender-equal achievement.

A reanalysis of cross-cultural data used more sophisticated statistical techniques, including factor analysis, to arrive at composite child-rearing variables (Hendrix 1985); despite differences in details, the con-

Table 27.1. Cross-Cultural Variation of Socialization Practices

Number of Cultures	Desired Behavior of Child	Percentage of Cultures: Parental Training Efforts		
		Girls More	Boys More	No Difference
33	Nurturance	82	0	18
69	Obedience	35	3	62
84	Responsibility	61	11	28
31	Achievement	3	87	10
82	Self-reliance	0	85	15

Source: Adapted from Barry, Bacon et al. (1957).

clusion that the sexes are socialized differently was confirmed. So although some sex differences are universal, parents in most cultures are trying to produce such differences. In the realm of nurturance at least these widespread child-rearing practices reinforce what is probably a biological tendency, yet there is some evidence that it could be substantially muted through experience and training.

In a study of Luo children in Kenya (Ember 1973) boys and girls in a rural community were assigned very different sorts of chores. Girls typically fetched water and firewood, made the fire, cooked, served food, and tended children, but some families had too few girls to manage these chores, which were then assigned to boys instead. Social behavior was examined using methods like those of the Six Cultures Study. In aggregate, boys showed more aggression and dominance, less altruistic and positive social behavior, and more dependency than girls, but the social behavior of boys who had done girls' work shifted in a female direction. This shift was not statistically significant, but when the boys were further subdivided into those who did girls' work *outside* versus *inside* the home, the latter were more girl-like than other boys on all measures. In particular, boys who did a lot of baby tending showed very little aggression.

In an observational study it is difficult to know how much the boys were changed by the tasks, since those tasks may not have been assigned to random boys. But it is likely that the boys who did girls' work inside the home actually changed and that training and experience can make boys more like girls in nurturance and aggressiveness, biological predispositions notwithstanding. This would be consistent with marked changes in father involvement with children in the West in the past few decades (Pleck and Masciadrelli 2004), the demonstrated effectiveness of some programs for improving paternal nurturance (McBride and Lutz 2004), the influence of changing workplace practices on fathering (Russell and Hwang 2004), and experimental evidence that substantial nurturing toward infants can be induced in male macaques (Mitchell and Brandt 1972; Redican and Mitchell 1973; Gomber and Mitchell 1974).

As we saw in Chapter 10, biological predispositions before puberty cause boys to be more aggressive and girls more nurturing, but these predispositions are subject to influence by training, experience, and parental expectations. Different cultures have different expectations and provide different experiences, which explains why the sexes are very different in some cultures and less different in others, and also why the girls in some cultures are more aggressive than the boys in others, even though they are less aggressive than the boys in their own culture. But parental training and cultural expectations in many cultures do widen the gender gap. We have seen that cultures that have to be prepared for combat separate fathers from wives and children until their sons are old enough to join the men in separate living and/or eating arrangements (Whiting and Whiting 1975b). The boys then enter a phase of deliberate training for aggressiveness, and many reviews of the psychological literature and cross-cultural comparative analyses show that this training does indeed produce hypermasculine men (Broude 1990; Chick and Loy 2001).

But parental and cultural expectations are not the only agents of gender socialization; same-sex friendships and sex-segregated groups play a major role as well (Maccoby 1998). By adolescence they are effectively separate cultures, and this happens in the great majority of societies, if not all (Schlegel and Barry 1991). As noted in Chapter 10, Heide Sbrzesny's (1976) research on play among the !Ko, a group of San hunter-gatherers in Botswana, showed that self-segregation by sex and largely sex-specific games and dances produce a different socialization and enculturation experience for boys and girls, without parental intervention, even in a hunting-and-gathering culture.

As we have seen, hunter-gatherers cannot always constitute same-sex playgroups, but given sufficient numbers, even in the least stereotyped environments, self-segregation is very strong. Parents and teachers can make boys and girls play together, but only firm control of children's time and choices can effectively prevent voluntary sex-separation. Anticipating Maccoby's two-culture concept, Alice Schlegel and Herbert Barry III (1991) examined adolescence throughout the cross-cultural range and devised their own model of gender divergence. They show that human gender roles fall within the higher-primate spectrum, with distinctive human features. But they consider that spectrum applicable to a wide enough range of cultures that they call it "an ethogram of human social organization"; however, it is really an ethogram of the social order of gender.

The trajectory begins with the family of origin, where some divergence occurs in early and middle childhood. Thereafter, Schlegel and Barry consider it universal for boys and almost universal for girls to have a period they call social adolescence between reproductive maturation and marriage. Girls have a stronger tendency to associate with adult women, and a weaker tendency to form isolated adolescent groups. For boys, the opposite is the prevailing pattern: they tend more strongly to form peer groups and less strongly to associate with adult males. Edgar Degas's painting *Young Spartan Girls Challenging Some Boys* appears on the cover of their book, and it illustrates several of their claims.

1. Boys and girls are separate and to some extent competitive.

2. The girls at least understand that they must in due course come together again.

3. The prevailing ethos of a culture affects both sexes, without abolishing sex differences.

In this case, Spartan culture produced competitive, even aggressive, girls. This pattern was confirmed by the Six Cultures Study of younger children, which showed that while boys and girls in all cultures differ in the same direction on the two dimensions of egoism and aggression, girls in some cultures can be more "masculine" on these dimensions than boys in others.

When lecturing on the development of sex differences, Irven DeVore used to depict the influence of culture against the background of biology by drawing what he called the "Toonerville Trolley" model. In this model males and females start life with the tendency to become somewhat different (parallel tracks). However, there is always substantial overlap between the distributions for any behavior, and any behavior can be influenced by culture. Males and females can both be made more feminine, more different from each other in conventional ways, more masculine, or more similar (converging tracks). Although there is no known culture that has tried to reduce sex differences to zero, it is theoretically possible to do this, or even to reverse the differences on some measures. But we now have the answer to another question. Margaret Mead used to say that we do not know what would happen if boys and girls were to be raised exactly alike, because it has never been tried. Well, now it is being tried, and although it has not been a systematic experiment, we know enough about sex differentiation of brain and behavior to predict the result: if raised identically, on parallel tracks, boys and girls would stay somewhat different.

Cultural Tradition in Adolescent Development

Some aspects of adolescent behavioral change are due to programmed hormonal dynamics, and others to ongoing brain development (Chapter 12; Giedd, Clasen, et al. 2006). For some the results are at least transiently negative, but there is clearly a continuum of vulnerability, with determinants that mostly remain obscure. Genes certainly play a role (Plomin and Fulker 1987; O'Connor, Hetherington, et al. 1995; Lenroot and Giedd 2008), as do insults such as exposure to alcohol or drugs and head trauma. But at least equally important are the cultural factors: in the United States young teens continually exposed to sex, violence, alcohol, and drug abuse will be much more likely to have a developmental crisis. For them the conflicts of puberty can initiate enduring problems. But in many cultures, tradition guides adolescent development much more than it does in the West, and if that means less freedom, it can also mean less anxiety. All such cultural influences take advantage of the fact that the adolescent brain is in a sensitive period of its development (Steinberg 2005; Blakemore 2008).

We say, often explicitly: *in a few years, you will take care of yourself.* This impending withdrawal of support is more disturbing to some adolescents than others, but few miss the message, and part of teenagers' oppositional behavior must be resentment of it. The psychoanalyst Erik Erikson (1959, 1968) characterized adolescence as a time of "identity crisis" and that term has become a cliché about not just youth but some other phases of the Western life cycle. In traditional cultures specific expectations are greater and options are fewer, but young teens know that they will have a place in the world they grew up in, which in some ways will always take care of them. Most cultures in fact have rituals to demonstrate this commitment and at the same time to preempt any identity crisis.

Rites of Passage

Such cultures make demands as well, some through rituals that mark life's major transitions—birth, marriage, parenthood, death—with symbolic practices designed to reclaim these events from mere biology and signal that there are proper ways to go through them. These rituals are called rites of passage, and many cultures have dramatic ones at or around puberty. Of societies in the World Cultural Sample, 68 percent have adolescent initiation rites for boys, and 79 percent have them for girls (Schlegel and Barry 1991); these rituals invariably emphasize not just adulthood but manhood and womanhood; they are powerful cultural devices for further differentiating the sexes.

Girls' initiation ceremonies are typically associated with menarche and are more likely in hunting-and-gathering cultures (Kitihara 1984). The average !Kung girl's first menstruation is at age sixteen and a half. Wherever she is, day or night, on seeing a trace of blood she must sit on the ground silently and wait. This can be dangerous if she is out gathering alone, since lions, hyenas, or wild dogs could get wind of her, especially given the smell of blood. But the women in her village—mostly relatives—know that she has been approaching menarche, notice her absence, and search for her. In this and throughout the ritual no men are allowed. The women track her easily, pick her up—her feet may not touch the ground—and carry her to a place not far from the village. They quickly build a small grass hut and set her inside it. A dance begins. All adult women in the village take part, and as the day, the night, and the next day pass the dance becomes raucous and bawdy. Women wear only a small leather pubic apron, tossed around haphazardly as they dance, or they pull it up to flash their genitals, provoking much laughter, but meanwhile they sing, dance, and clap without ceasing. No men dare approach. Throughout the ritual the menstruating girl sits alone in the seclusion hut, expressionless, never laughing or speaking, but getting the message: a ritualized yet uninhibited celebration of womanhood (Shostak 1981; Marshall 1999).

Rituals of manhood, for their part, have been among the most dramatic standard experiences of puberty and adolescence throughout the world (Gilmore 1990). !Kung teenage boys had to kill their first antelope and were proudly incised with small tattoos on their foreheads, but they also underwent "an intense and rigorous ritual," called *choma*, in which "the initiates experience hunger, cold, thirst, and the extreme fatigue that comes from continuous dancing" over a six-week sacred period; during this time much cultural knowledge is transmitted (Shostak 1981:239).

Among the Baka, small-stature hunter-gatherers of Cameroon, teeth filed to a point are considered handsome, and boys past eleven or twelve may elect to have their teeth painfully chipped and filed, proving their endurance and maturity. The boy lies motionless on the ground biting down on a piece of wood while a skilled relative chisels away at his front teeth. If he whimpers, another man may apply warm plantain skin to soothe him. Other children watch this group, walk in a circle around them, and tease the boy, sometimes leaning down to squeeze his upper lip. There is also an elaborate ritual in which boys are intensely hazed, frightened, and poked with sticks in order to be brought into the service of Jengi, the chief spirit of the forest (Higgens 1985).

Like the Baka, the Nuer, cattle-herding warriors of Sudan, respect the boy's sense of his own readiness for the *gar* ritual; he tells his father he wants the ceremonial scars that will make him a man. If his father agrees, one of his sisters shaves his head, and he lies naked and perfectly still, arms tightly folded across his chest, hands gripping his shoulders, surrounded by singing, dancing, spear-brandishing warriors. A male relative takes a sharp blade to his forehead and makes cuts so deep that they may etch his skull, extending from behind the forehead over one ear, straight across the brow to a point above the other ear. Four parallel lines like this are made and allowed to bleed and form scars by being left open. The resulting bands of scar tissue are thick, striking, and permanent—the unmistakable mark of a Nuer warrior, earned by a boy who has shown great courage, not speaking, weeping, or crying out throughout this painful, public, unanesthetized surgery. Finally he is given a new name, one only assigned to a grown man (Gardner and Harris 1971).

The Native American vision quest imposed a different form of stress. In the Great Plains bison-hunting cultures a boy had to purify himself in the sweat lodge and paint his body with white clay. Then he was sent out naked and alone for several days to an isolated place deemed nearer to the spirits. He ate nothing during this time and focused on one goal: attaining and keeping an altered state of consciousness that would generate his vision. Essentially, he had to become delirious from hunger, thirst, exposure, solitude, and the fear of mountain lions, bears, and enemy war parties. The vision he sought was private and came personally to him—an ancestor, an animal, a plant, even a storm. This vision became his guardian spirit and gave him a new identity, beyond boyhood (Lowie 1954).

In the very different Jewish bar mitzvah (meaning "son of the commandment"), a boy becomes a man in religious terms at age thirteen; he, not his parents, is now responsible for his sins. Originally the boy said a blessing before and after a portion of Torah (the first five books of the Bible) was read. The ceremony is more elaborate among American

Jews today. The child usually must master the complex melody for the Torah portion, chant many lines of it in Hebrew, and often lead the congregation in a substantial part of the service in Hebrew, in addition to delivering a speech in English of his own devising, usually a Torah interpretation. In the past century girls have increasingly had an analogous ritual, the bat mitzvah, except among Orthodox Jews. It is a major effort that often takes years of preparation, and the event itself is stressful. As is typical of initiation rites, its strong symbolic power makes learning very intense (Goldberg 2003).

Many cultures have pubertal rites explicitly directed at sexual organs. Throughout the Sudan and Sahel in Africa the genitals of millions of adolescent girls are operated on, including full or partial removal of the clitoris, in a ritual designed to make them less sexual but without which, it is believed, they cannot become women (Cook, Dickens, et al. 2002). They may also undergo infibulation, in which their labia are partly cut and then sewn together, an attempt to prevent premarital penetration. Considered mutilating by most health authorities, even those in the countries where this ritual is practiced, it remains in use by women in traditional North African cultures, who see it as essential for their daughters and granddaughters to become real women.

Among the Australian aborigines (Elkin 1964) pubescent boys were initiated by both circumcision and a second, more mutilating operation, subincision. The boy was laid on a rock, held down by his male relatives, and required to be stoical. For the first ritual, one of them removed the boy's foreskin with a sharp stone. In the second stage, a few years later, the underside of the penis was slit open from base to tip, also with a stone blade. This deep cut formed a thick scar, which could interfere with urination, but was said to enhance women's pleasure during sex—the opposite of the goal of clitoridectomy. During healing, the boy was taught by adult men for several weeks, a chance for the culture to make a clear impression.

Functions of Initiation

All these rites of passage serve at least three purposes, and each takes advantage of the sensitive period of brain development that we have come to understand adolescence is (Steinberg 2005; Blakemore 2008). First, the child is subjected to stress—often severe pain, usually fear—intended to give him or her a stronger adult identity: "I can take it as the adults once did, so I am no longer a child." In the emotional logic of symbolic ritual it is persuasive; it resembles the logic of Friedrich Nietzsche's epigram "Whatever doesn't kill me makes me stronger."

Second, like the hazing that leads to college fraternity membership, pubertal initiates are made to feel a part of the culture, the community, and the domain of either adult women or adult men. The initiates are *in*, and those who are not yet old enough or who belong to other groups are not. This is invariably a chauvinistic process, and those left behind are pitied or disdained.

Third, the time before and after the ritual provides a teachable moment for the transmission of culture. Deliberate, formal instruction ("designed learning") is usually part of adolescent initiation rites, even in cultures where there is little formal instruction in any other context (Kruger and Tomasello 1996). The child's eyes are wide with fear, then with stress or pain, and finally with exhilaration. This is the moment when your elders say explicitly, *This is what it means to be a Sioux man, a !Kung woman, a Jewish adult. These are our customs. These are your obligations and privileges.* The ritual not only makes such lessons vividly memorable but encodes them at a deeper cognitive and emotional level than is possible in normal learning.

It is relevant here that the sensitive period for Japanese children who live in the United States to consider themselves American is between ages nine and fifteen, when initiation rituals typically occur (Minoura 1992). Given that puberty occurred later in most traditional societies than recently in Japan or the West, one would expect the completion of puberty rather than chronological age to set the upper limit, and it does. Age eleven is the approximate lower bound of formal operations in Piaget's progression; crossing this boundary would help enable the child to form an abstract concept of the transition: *I am a child, but if I go through this, I will be an adult; I am a girl (boy), but if I do this ritual, I will be a woman (man).* Such if-then abstract propositions characterize formal operations. As for the upper bound, it could be a kind of affective closure to experience or the advent of a rebellious resistance to being taught.

Ideals and Commitments beyond Gender

In many cultures there is a paradox of divergence in adolescence: both bad and good behavior increase, although not to the same extent in the same individuals. The bad behaviors justify the idea of adolescence as a problematic phase of life. But it is also often a time of high ideals—charity, patriotism, religiosity, music, scouting, athletics, conservation, pacifism, racial integration, military discipline, holy war, and martyrdom, along with specific ambitions that set teens on the path to becoming carpenters, doctors, ministers, dancers, or soldiers, sometimes a complex negotiation of an identity crisis (Erikson 1959, 1968). In modern culture these choices have less to do with gender identity than they once did. Of course, individuals diverge at puberty owing to unique blends of biology, experience, and chance, but in many, problem behavior and idealism increase simultaneously.

Religiosity, including a conversion or "born again" experience, is central for many. As many as 55 percent of American adolescents in a survey of households said that they had decided to devote their lives to God through one of many religious paths, and about half of all teens said that they had had a powerful experience of worship and had had a prayer answered or received guidance from God (Smith and Denton 2006). Their affiliations with religious groups can have lifelong consequences for marital choice and other relationships. This surge in religiosity has been plausibly attributed to characteristics of the adolescent brain and mind (Good and Willoughby 2008), but other industrial countries are far less religious than the United States, so it would be interesting to see whether adolescents elsewhere find other ways to achieve parallel purposes.

The adolescent's acceptance of ideas and ideals beyond the self may not be universal, but it occurs in many cultures. Young people have played a key role in rebellions and revolutions in many nations, not just because they embrace naive political claims, but because adolescence makes them both idealistic and oppositional. They can imagine a better future, but they lack the experience that tempers enthusiasm, and their aggressive impulses lead them to take strong action to bring about change. In any case, the emotional dynamics of adolescence produce an ongoing opportunity, beyond initiation rites, to be enculturated in a strongly social context (Collins and Steinberg 2006). Friends, romantic partners, siblings, parents, schools, teachers, ministers, other adults, and media are the routes by which culture is transmitted in this phase.

28

Evolutionary Culture Theory

Cultural evolution is quite different from biological evolution, and one of its meanings is the succession from hunting and gathering to horticulture and/or pastoralism, to full-scale agriculture, to industrial and post-industrial modes of subsistence—a succession that entails changes in social and political complexity and results in changes in child care. Transmitting culture to children is more complex and much less well understood than genetic inheritance.

But aside from the grand long sweep of history, another meaning of cultural evolution is a kind of microevolution, the mechanism of cultural change. In the past few decades several models of cultural evolution have emerged based on analogies to and interaction with genetic microevolution. Some have introduced neologisms or idiosyncratic use of previously established terms, as well as mathematical reasoning that does not always improve on verbal argument in ways that are nontrivial and empirically testable. Some also dismiss or ignore large bodies of anthropological and sociological knowledge and theory ("verbal models") about cultural evolution and culture change. Nevertheless, they make meaningful contributions to our understanding of how human culture changes. One of them, William Durham's *Coevolution: Genes, Culture, and Human Diversity* (1991), is explicitly synthetic, cognizant of the others, and mindful of the contributions of the traditional sciences of culture; it is emphasized here.

Cultural Macroevolution

Independently of Darwin's influence, social, cultural, and linguistic anthropologists have thought about cultural evolution since the dawn of the discipline (Harris 1968). Studies of ancient languages, religions, and societies and the search for rules governing their historical change led to the use of archaeology and cultural anthropology to model earlier stages of language, society, and culture than were accessible through ancient texts, documents, stelae, and other artifacts. Theories of an evolutionary succession of stages motivated major works of nineteenth-century anthropology, including Lewis Henry Morgan's *Ancient Society* (1877), which was drawn upon by Marx and Engels in their accounts of economic evolution and the evolution of the family, and Edward Tylor's *Primitive Culture* (1871). These models had a restrictive and tendentious ordering of cultures, societies, or religions in stages in a frank attempt to see later ones as inevitable and better, and cultural evolution as progress.

Despite the demise of both Marxist stage theory and the nineteenth-century ideology of continuous progress, there is validity in stage models (Johnson and Earle 1987). In historical and anthropological linguistics, there was less interest in universal stages of evolution than in the construction of language phylogenies and the discovery of mechanisms of linguistic change, and thus these lines of language research became forerunners of present-day evolutionary culture theory (Mace and Pagel 1994, 1997). They also gave rise to efforts in the phylogeny of language diversification and mechanisms of language change (Swadesh 2006).

The Meme Model and the Question of Coherence

In 1976 Richard Dawkins introduced a useful term that has been adopted by many investigators: the *meme,* an elementary unit of cultural transmission. The term is cognate to *mimesis,* Greek for "imitation," and is appropriate to replication and transmission in the cultural sphere. By analogy to the gene, it was hoped that the meme would have a deep and clarifying effect in theories of cultural evolution, but there were problems with the analogy.

Although they began as abstractions, in the midtwentieth century genes took on a substantive biochemical reality. Memes, which are mental or behavioral schemas, have no comparable biological reality. Even though they are instantiated in neurons, as genes are in DNA, memes lack a comparably coherent, straightforward, and generative biological basis. Genes are discrete, making evolutionary transmission largely digital. This discreteness is mitigated by transposable elements, micro-RNAs, triplet repeats, alternative splicing, and other significant features and processes, but these features do not obviate the basic identity of genes. Genes have start-and-stop signals for transcription, and their products are individualized as functional RNAs (including micro-RNAs), proteins, or precursors of proteins. Memes have no natural boundaries, can be infinitely expanded or contracted, and grade into each other continuously.

Because of this continuity, cultural transmission is basically not digital, but analog, and the frequent treatment of it as digital is distorting. Also, imitation notwithstanding, there is no replication process for memes that approaches in simplicity or predictability the one we know for genes, despite replication errors. The culture acquisition device (CAD) proposed in Chapter 29 includes many different mechanisms, all of which are involved in cultural transmission. Finding anything in this process that corresponds to the meme is a challenge.

Yet despite these disanalogies, the meme has proved useful in understanding cultural stability and change; at least four theories of cultural evolution have been based on it or some equivalent. As with genes, the faithfulness of replication and its associated repair mechanisms can ensure stability over time. And also as with genes, the replication process is imperfect and errors have a creative function, serving as grist for the mill of culture change, much as mutations offer the genetic variation that natural selection acts upon. But memes are largely independent of genes and enter a process of cultural evolution to some extent independent of its biological counterpart.

Because it is susceptible to this confusion, the term "cultural evolution" is not ideal, but it has become widely accepted. Terms such as "sociohistorical change," "culture change," and "culture history" have different connotations; "cultural evolution" implies systematic, scientifically comprehensible processes at a microcultural level of resolution, some of which can be usefully modeled by equations and tested by simulation, observation, and experiment. But the microcultural process of meme transmission, faithful or not, runs into a difficulty that does have a significant partial analogy in genetic evolution: the problem of cultural coherence.

Except for Durham's, the main recent models of cultural evolution ignore or skirt this problem and instead assume that memes are independent of one another. This is basically the inverse of the meme boundary problem mentioned earlier. Just as a lack of internal structure and weak boundaries cause the meme to lose its discreteness, it loses its independence owing to cultural coherence, but this does have an analogy in the coherence of a genome.

In a first recognition of a limit on the independence of genes, Mendel's law of independent assortment of loci had to be modified in deference to linkage, which causes some genes to segregate together simply because of their physical proximity on a chromosome. But this was only the first example of genomic coherence. Because large proportions of the genome consist of regulatory elements (genes that turn other genes on and off) and because of the multigenic determination of traits and pleiotropic effects of genes, genomes cannot be moved with equal ease in any direction.

As we noted in Chapter 2, the position of adapted genomes can be visualized as resting at saddle points in graphs in a multidimensional space (Levins 1964). In three dimensions the corresponding phenomenon would be a basin in a volcano. A species may be at a local minimum, or trough, in the adaptive landscape. In theory it could be far better adapted if it could escape the trough and roll downhill under the pull of natural selection to maximal adaptedness. But the genome would first have to become less adapted to make the transition to the better state. Natural selection of a gradual type could not take the species over the maladaptive hump regardless of the greater adaptedness beyond.

Traditional anthropologists have always at least tacitly assigned a similar coherence to cultures. The reasons offered for such coherence have included a basic or modal personality arising from child-rearing practices, an overarching cognitive model that brings many of the culture's lesser models together, and a we-they dichotomizing of self and other that simultaneously heightens identity within and exaggerates and denigrates differences without. Any of these psychological mechanisms and others as well would act as constraints on cultural evolution. That is, cultural selection might have unexpected difficulty detaching a meme from the rest of the culture to maximize its adaptedness. Culture would resist piecemeal change.

Of the four theories of cultural evolution considered here, only one (the Durham model) gives serious attention to cultural coherence. Two of the theories, that of Cavalli-Sforza and Feldman and especially that of

Boyd and Richerson, posit that cultural evolution is highly independent of biological evolution. The other two, Lumsden and Wilson's and Durham's, begin with more empirically sustainable propositions about human psychology that result in a more strongly coevolutionary view of the relationship between cultural and genetic change.

Cavalli-Sforza and Feldman

Beginning in the 1970s, Luca Cavalli-Sforza and Marcus Feldman developed a pioneering quantitative approach to cultural transmission and evolution (Cavalli-Sforza and Feldman 1981; Feldman, Cavalli-Sforza, et al. 1994), but it gave insufficient attention to genetic factors in the traits studied. For example, political conservatism, religiosity, and preference for adventure movies were assumed to be transmitted only culturally, while we now know them to be both culturally and genetically transmitted (Benjamin, Li, et al. 1996; Bouchard 1994; Tellegen, Lykken, et al. 1988; Zuckerman, Buchsbaum, et al. 1980). In addition, their analysis was restricted to Western cultures—non-Western customs were only briefly mentioned—and to individual traits, or memes, considered as independent. They criticized the meme concept for artificial discreteness, but even the more continuous traits they dealt with, like religiosity, were treated as independent of other cultural traits. Finally, they underestimated the possible impact of cultural evolution on fertility. Of such culture elements as Coca-Cola, Frisbee, volleyball, and yo-yos, they write, "It is obvious that in none of these examples does participation appreciably alter the probability of surviving or having children" (Cavalli-Sforza and Feldman 1981:15). In fact, it is likely that in the developing world such habits profoundly affect status, especially during childhood, and status very often affects survival and reproduction. This is even truer of religious or political traits.

Still, Cavalli-Sforza and Feldman attempted to deal with real-world data. They classified cultural transmission as: *vertical* (transfer from parent to child); *horizontal* (transfer between peers); and *oblique* (transfer from adults to other people's biological offspring). They used linguistic change as a touchstone for cultural evolution, since this is by far the most systematically studied of patterns of cultural change. For example, Morris Swadesh (2006) proposed that in a core list of highly used words the frequency of cognates between two languages decreases exponentially with time, and that there is a 34 percent decrease in the probability of cognates in one thousand years of separation. He proposed the simple equation $p = 100 - rt$, where p is the final percentage, t is the number of years, and the constant r is .00042.

Not surprisingly, this equation has proven overly precise, and its validity declines with language separation time, but it gives some sense of the goals of quantification in cultural evolution. As for the core transmission process in linguistic stability or divergence, the equation suggests that vertical transmission of core words has an imitation error rate on the order of 10^{-3} to 10^{-4} per generation, or at least an order of magnitude higher than the genetic mutation rate. Some other cultural innovations are much faster than changes in the core vocabularies of languages.

Empirically, individual cultural innovations are frequently found to spread through a population according to a logistic, or S-shaped, curve, in which the slope or derivative—representing the rate of spread of the innovation—increases to a maximum and then decreases to zero. This suggests that some mechanism makes an almost exponential increase in converts possible in the early stages and that the reverse happens as the population of susceptible converts reaches saturation. Alternatively, the curve may reflect the distribution of different degrees of susceptibility among psychological types in the population.

On the question of coevolution, or interaction between cultural and natural selection, Cavalli-Sforza and Feldman at first considered only the possibility that they might be opposed to each other and appeared to assume that cultural selection always prevails, often leaving behind behavior that is maladaptive in Darwinian terms. For example, they viewed the clitoridectomies and related operations ("female circumcision") that are widespread in northern Africa as maladaptive in a Darwinian sense, but they presented no evidence regarding differential fertility. In fact, the peoples of northern Africa often create conditions that interfere with reproduction in young women who do not have the operation by making them unmarriageable. In this cultural context, women who have been operated on almost certainly have higher fertility, despite the risk of infection and other sequelae of crude surgery. On the other hand, Cavalli-Sforza and Feldman's example of the cultural spread of birth control and the associated small family ideal is a clear case in which cultural selection decreases fertility in the short run, although perhaps not in the long run (Myrskyla, Kohler, et al. 2009). Still, this example creates difficulty for those who believe that Darwinian theory can explain all of human behavior (Vining 1986).

Cavalli-Sforza and Feldman later tried to derive formal models for gene-culture coevolution (Feldman, Cavalli-Sforza, et al. 1994). Their model for the coevolution of milk production and lactose tolerance was unsuccessful by their own criteria. Their model of the coevolution of deafness (genetic) and sign language (cultural) assumes that only deaf people sign and that deaf children learn only from their parents; in fact, since most deaf children have hearing parents, most transmission of sign language is oblique or horizontal, not vertical, and may be from hearing as well as deaf signers. In their account of the coevolution of cultural bias in favor of males and genetically determined sex ratios, Cavalli-Sforza and Feldman fail to consider that the effect of pro-male bias on reproductive success (RS) may be contingent on environmental quality (Trivers and Willard 1973). Unrealistic assumptions weaken formal models.

Still, some nontrivial, testable results are suggested by their equations. For example, where vertical transmission is weak, a cultural trait is predicted to be more persistent with oblique transmission than with horizontal, while the reverse holds where vertical transmission is strong; that is, adults as models trump peers when parents are weak or absent, but peers follow parents in importance in normal conditions. Other results of their modeling are more predictable; for example, cultural homogeneity is favored by teacher-type transmission (a strong form of oblique transmission), and a greater contribution from earlier generations slows the rate of cultural evolution. According to a Jewish folk proverb, a mother is worth a hundred teachers, but for some forms of cultural transmission the reverse is clearly true.

Lumsden and Wilson

The formal model developed by Charles Lumsden and E. O. Wilson (1981) focused on gene-culture coevolution and generated a very different result from the classic argument that culture always blunts the effect of natural selection (Lumsden 1989). Here the assumption of similar learnability for all sorts of memes—which they called "culturgens," although Wilson (1998) later gave qualified acceptance to "meme"—is replaced by the more realistic assumption of strong differences in learning curves. Learnability is determined by the "epigenetic rules" by which culture and genes interact to guide development. The Lumsden-Wilson model is well informed by developmental psychology and does not make naive assumptions about the heritability of "cultural" traits, so it produces more realistic results. Innate biases shared by all infants and children restrict the cultural variation through meme modification, yielding strong pan-cultural tendencies. The model is exemplified by the following four cases:

1. Cross-culturally, color-naming systems follow predictable rules (Berlin and Kay 1969). If a culture names only two colors, they will be black and white; if three, red will be added; if four, then either green or yellow; if five, then the other; if six, blue is the addition. Something about the physiology of color saliency, genetically coded in all humans except the color-blind, makes it unknown for a culture to name, say, red and blue, omitting black or white. Although the naming of emotions cannot be studied with comparable precision, we know that emotion naming too builds from a limited to a wider range according to common principles (Heider 1991).

2. The discrimination of sounds within their physical continua was long thought to be arbitrary, with culture imposing contrasts to make meaningful distinctions (Jakobson and Halle 1956/1971). While culturally salient contrasts remain central to linguistic theory, we now know that the way the continuum between, say, *pa* and *ba* is broken for a meaningful distinction is not arbitrary. Human newborns attend preferentially to the transition through a particular part of this continuum, which shapes the distinctions that languages will make (Eimas 1997, 2000). Such innate categorical boundaries, although only probabilistic, do exist and have a similar shaping effect in many realms of perception. This does not mean that cultures do not vary, just that innate universal perceptual bias constrains variation.

3. Childhood fears and phobias also do not vary nearly as much as predicted from an assumption that they arise through culturally patterned learning. Strangers, snakes, heights, darkness, and a few other stimuli are widely found frightening. Cultures build on these through

conditioning or reduce them through habituation and modeling, but fears do not simply follow cultural rules, because innate biases matter (Bronson 1972; Hebb 1946; Konner 1972; Lewis and Rosenblum 1974; Mineka and Cook 1988; Muris, Merckelbach, et al. 1997).

4. Incest avoidance between brothers and sisters is widespread as a strong tendency in almost all cultures (Interlude 5). Its adaptive function (inbreeding reduction) is like that in other species, but in humans it does not depend on genetically coded kin recognition, such as pheromones, found in many species. Human mating avoidance is a function of contact during childhood, particularly early to middle childhood, as demonstrated by traditional China—where adoption of future brides was common and fertility was significantly lower in marriages where extensive contact occurred during the sensitive period—and by the avoidance of marriage between men and women raised together in Israeli kibbutzim.

In these and many other cases, no lockstep genetic adaptation determines the human mental or behavioral functions, yet neither are these functions free to vary independently of the shared human genome. The genome provides learning biases that culture responds to through the epigenetic rules (Waddington 1942) by which culture shapes ultimate function (with limited, often predictable cross-cultural variation). Equations modeling this process have some logico-deductive power and yield some interesting results.

For instance, Lumsden and Wilson show that only a small innate learning bias can yield a large shift in the "ethnographic curve" for cross-cultural variation. That is why brother-sister incest avoidance, despite its adaptive advantages, has not evolved as a fully determined instinctive process. In fact, assortative mating would make it *more* likely for a brother and sister to be sexually attracted to each other in the rare situation in which they have had no prior contact, and brother-sister marriages have been preferred in a few cultures. Also, "false positives" occur in Chinese adoptive marriage and kibbutz communal rearing, where fertility is reduced or mating prevented without genetic relatedness. Natural selection could afford to be imprecise about this, because in the great majority of societies, including the EEAs, brothers and sisters have contact during childhood, and cultural evolution produces rules to prevent brother-sister mating in most other cases.

Thus a nearly universal human behavioral pattern can emerge from a small or indirect innate bias through epigenetic rules. This result has the important implication that universal patterns may be less resistant to change within an individual life course than is suggested by their cross-cultural invariance. While this may be of no practical interest in the case of brother-sister incest, it would be of great interest in such cases as ethnocentrism or the division of labor by sex. If these universals result from weak innate biases, they should be relatively easy to change, despite their initial universality, given appropriate changes in the epigenetic rules.

In addition, not all learning biases result in the fixation or near-fixation of a cultural trait. For example, classic studies of fashion in European women's formal dresses over the past four centuries (Richardson and Kroeber 1940; Curran 1999) suggest a cyclical fluctuation in actually measured skirt length, waist height, and décolletage (shoulder and chest exposure). These cycles can be explained by innate biases toward, on the one side, competition for status among women and income among couturiers and, on the other side, waistlines and limits to exposure. As in color-naming systems, carrying the iteration of a given epigenetic rule farther from the simplest initial choices can generate considerable cross-cultural variety.

Another, more controversial, consequence of Lumsden and Wilson's argument is that genetic evolution can be greatly speeded by cultural bias. This result departs strongly from the conventional idea that cultural evolution reduces the rate of genetic evolution, and here the genes in question are not necessarily the universally shared ones of the general human learning biases.

To begin with, they show that John Locke's tabula rasa is an unstable situation for a species and cannot long persist even if it has somehow evolved. In nonmathematical terms, a tabula rasa population consists of individuals who are equally likely to learn all possible memes; that is, "the developmental field is flat" (Lumsden and Wilson 1981:13), a reference to C. H. Waddington (1942). In reality, the developmental field is contoured and sloped, so that the "ball" of behavioral or biological development rolls into and down favored paths. A population of Lockean individuals would roll aimlessly on a flat plane, directed only by the winds of experience.

Such a population is not in an evolutionarily stable state, since it is vulnerable to the invasion (through mutation, migration, and natural selection) of genetic adaptations that make individuals more likely to acquire adaptive memes. In other words, there is a great advantage in having learning biases if no one else in your population has any. Lumsden and Wilson, using a different metaphor, refer to this as the "leash principle," since culture is kept "on a leash" by natural selection. It could as easily be called the "guidance principle"; few if any domains of human behavior and mind are devoid of learning biases, which are guidance biases for cultural evolution.

On the other hand, as guidance biases become more stringent, a point of diminishing returns is reached at which individuals are closed to parts of the range of available memes that might enhance adaptation. This works for short-lived, high-reproductive-rate organisms whose populations can respond promptly to environmental changes—they can basically discard large numbers of overcommitted nervous systems with inflexible guidance biases and replace them with others that have different biases—but it is not evolutionarily stable in a large-brained, long-lived mammal, and especially not in humans.

It is true—and confirmed by formal deductions in this model—that culture can slow the rate of genetic evolution, but under other circumstances genetic evolution can be very rapid. Even assuming guidance biases much less influenced by genes than has already been demonstrated for many developmental processes, some genetic change can take place under gene-culture coevolution that is rapid enough "to achieve partial replacement of one allele by another within as few as ten generations, or about two hundred to three hundred years." This, Lumsden and Wilson (1981:295–96) believe, would require "extreme conditions . . . but substitutions can occur under a wide range of conditions in a thousand years," or fifty generations. Under gene-culture coevolution an autocatalytic process may greatly accelerate change, and the relatively rapid evolution of the human brain may have been such a process.

Boyd and Richerson

Like that of Cavalli-Sforza and Feldman, the model proposed by Robert Boyd and Peter Richerson (1985, 1988) assumes that coevolution and gene-culture interaction are minimal and that culture evolves with an extreme degree of independence. They hold that it is too costly to evolve new neurophysiological machinery to bias cultural learning. This view would correspond to the tabula rasa shown by Lumsden and Wilson to be an unstable strategy, but it does not apply to the real world of human evolution.

A second unrealistic assumption in this model is that cultural parents and naive offspring associate randomly with regard to their respective genotypes. As we saw with respect to religious and political conservatism, an example from within-culture variation in industrial states, this is simply not true; genotype-environment correlation is well demonstrated (Scarr and McCartney 1983; Rutter 2007). Such correlation also clearly affects the genetic and cultural contributions to general intelligence (Layzer 1974; Plomin, Petrill, et al. 1996), antisocial behavior (O'Connor, Deater-Deckard, et al. 1998; O'Connor, Neiderhiser, et al. 1998), and many other partly heritable traits.

Despite these limitations, there are interesting derivations within the model. For example, Boyd and Richerson provide a plausible explanation of why cumulative cultural evolution is rare in nature: frequency dependence. Under several different assumptions, the psychological capacities required for cultural evolution are much less advantageous and spread much less rapidly when they are infrequent than when they are common. Thus the evolution of cumulative cultural change (one requirement of true culture) is a threshold phenomenon that is inherently unlikely. Also, while Boyd and Richerson largely ignore specific genetic influences on cultural transmission, they fully accept cultural effects on genetic evolution. For example, new cultural conditions may affect sexual selection patterns, according arbitrary advantage to, say, women or men with a certain body shape or particular speech style.

The Durham Model

The most comprehensive treatment of gene-culture coevolution is that of William Durham. In *Coevolution: Genes, Culture, and Human Diversity* and other writings, Durham (1990, 1991, 1992) synthesized elements of previous models and added new features. Durham acknowledges a *dual inheritance system* in which the mutual dependence of cultural and biological evolution is strong but not complete. In agreement with Lumsden and Wilson, he finds them to be "two distinct but interacting systems of information inheritance within human populations" (Durham

1991:419–20). Culture is a paragenetic transmission system with many structural parallels to genetic transmission. "Theoretical formulations with the equivalent of a single fitness measure—including sociobiology on the one hand, and much of cultural anthropology on the other— are simply incomplete and logically insufficient for the task. Even as first approximations they may be misleading and ultimately counter-productive" (426).

The analogies, as far as they go, are presented in Table 28.1, modified from Durham's (1991) Table 8.1. The table is largely self-explanatory, but several points deserve emphasis.

1. Regarding the elements, or memes, Durham (1991:421) assumes "that the units of transmission vary widely with respect to scale and complexity . . . from single units of meaning in culture (like the morphemes of language), to more complex ideas, beliefs, and values, and on to entire languages, ideologies, symbol systems, and culture pools." He sees the disanalogies between genes and memes—levels of inclusiveness, discreteness, transmission mechanisms, and so on.

Table 28.1. Genetic versus Cultural Evolution: Structural Processes

	Genes	Memes
Units of transmission		
Variant element	Allele	Allomeme
Individual set	Genome	Memome
Population	Gene pool	Meme pool
Sources of variation	Assumed random	Random or purposive
Primary	Mutation	Invention
	Replication error	Mimesis error
Secondary	Recombination	Synthesis
	Migration	Migration
	Viral vectors	Diffusion
Transmission by:	Reproduction	Communication
Processes	Meiosis	Learning
	Syngamy	Teaching
Mode	Vertical	Vertical, oblique, horizontal
Ratio	Fixed, 1:2	Variable, 1:?
Transformation process	Gene-frequency change	Meme-frequency change
Non-adaptational	Mutation	Innovation
	Migration	Synthesis
	Genetic drift	Migration
Adaptational	Genetic selection	Cultural selection
	Genetic selection	Natural selection
	Genetic imprinting	Transmission bias
	Mitochondrial DNA	Imposition
Constraints	Genomic coherence	Cultural coherence
	Organic constraints	Constraints on learning

Source: Adapted from Durham (1991:426).

Still, memes have fitness independent of that of genes, and the fitness of memes, like that of genes, is their success in replicating and spreading in time and space.

2. Teaching and other intersubjective learning processes are central to human cultural transmission. Although all the models take the definition of teaching for granted, we know (Chapter 26) that it involves well-defined processes such as true imitation, scaffolding, and collaborative learning, some of which are uniquely human (Kruger and Tomasello 1996). Cultures differentially employ human potential for intersubjective learning, although all make use of it. Those that institutionalize teaching have different susceptibilities to cultural stability and change.

3. Central to Durham's model, but omitted from the others, is *imposition*, an extremely important type of transmission bias or sociological constraint that belongs to the general adaptive phenomenon of deception. Individuals routinely deceive others, even in their own lineage or other local grouping, allowing some to gain RS at others' expense. This is exploitation; it is used by kindreds, lineages, or collections of lineages such as castes and social classes as well as by individuals. Contrary to claims for the adaptedness of whole cultures, some individuals or groups in every culture are doing better than others, and what we describe as "a culture" or "a society" is a cross-section of unresolved internal conflicts. Imposition refers to the different degrees of choice that individuals have in accepting or rejecting transmitted memes, based on their differential power to create adaptive advantages. Colonizers may opt to learn a native language, but the colonized have no choice.

Processes 2 and 3 are examples of "secondary value selection," by which some carriers of culture direct or control the acquisition of memes by others and thus define the memes' fitness. Primary values correspond to guidance biases, the genetically determined contours in the developmental field. Secondary values are carried not in the genes but in the culture, and they spread through cultural transmission, just as do the memes whose fitness they help determine. Secondary values are thus the feedback of culture on itself; this can be negative feedback, which would stabilize cultures as coherent collections of memes, or positive feedback, which—whether intrinsic or imposed—would promote change. Secondary values (cultural selection) can either reinforce or oppose the tendencies provided by primary values or guidance biases.

Overall in the theory, a meme pool or culture pool changes over time, but not completely smoothly. The important features include: (1) *innovation,* the introduction of new variants; (2) *cultural drift,* the influence of chance events; (3) *diffusion,* the social transmission between populations; (4) *migration,* the movement of carriers between populations; (5) *differential reproduction,* the natural selection of cultural variants; (6) *choice,* or preservation through free decisions; and (7) *imposition,* or coerced preservation. Cultural constraints include the limits imposed by technology, mental habit, and other inertial factors that correspond to stabilizing cultural selection, the default condition of cultural transmission. Values, imposition, and cultural constraints, among other factors, affect the flow of memes, so that different ones have different degrees of likelihood of being transmitted to the next generation's culture pool.

Durham does not explicitly integrate psychological models of cultural transmission, but this is easily done. For example, in Bradd Shore's (1996) analogical schematization model, a culture's central schema would be a main gatekeeper that allows lower-level schemas to pass more easily if they are cognitively consistent with the central schema. In the classic culture and personality models, child-rearing practices would create modal personalities that act similarly, privileging for the transmission of cultural beliefs and practices consistent with the modal personality (LeVine 1982; Wallace 1961). In Kruger and Tomasello's (1996) model of intersubjective learning, established and prevailing patterns of imitation, teaching, and collaborative learning within dyads could under different circumstances foster transmission or innovation.

Defining Culture

But what is it that is transmitted or changed? Following LeVine, Durham cites five features that are central to what we mean by culture:

1. Culture has *conceptual reality,* including values, ideas, and beliefs shared by human beings. In a classic distinction, Clifford Geertz (quoted by Durham 1991:6) wrote, "Culture is the fabric of meaning in terms of which human beings interpret their experience and guide their action; social structure is the form that action takes, the

actually existing network of social relations." In Roy D'Andrade's (quoted by Durham 1991:6) phrase, cultural information "creates realities."

2. Culture undergoes *social transmission* through social learning and other extragenetic, interpersonal means. If an element of culture is not learned from other individuals, it is learned through the conditions created, often to some extent intentionally, by other individuals.

3. Culture depends on *symbolic encoding,* or "learned systems of meaning, communicated by means of natural language and other symbols" (D'Andrade, quoted by Durham 1991:6).

4. Culture has *systemic organization* and tends "to form an integrative whole," with a "strain toward consistency" (Murdock, quoted by Durham 1991:7). In Clyde Kluckhohn's conception, "every culture is a structure—not just a haphazard collection of all the different physically possible and functionally effective patterns of belief and action, but an interdependent *system* with its forms segregated and arranged in a manner which is *felt* as appropriate" (quoted by Durham 1991:7).

5. Culture has a *social history,* since the shared ideas, values, and beliefs in question have been handed down from previous generations. "Cultural standards have a history (they are 'historically created designs for living') and it is this aging process which gives to culture its unique quality" (Freilich, quoted by Durham 1991:8), even if it is not recorded or remembered explicitly by the people who carry it forward. The culture itself is the record.

Such a definition as this makes it difficult to sustain a case for non-human culture even without including intentional teaching as a defining feature. It also raises doubts about those formal models of cultural evolution that view culture as a collection of small independent elements and fail to accord it symbolic and historical coherence.

Applying the Model

Durham carefully takes up specific instances of gene-culture coevolution, advances competing hypotheses about the nature of the interaction in each case, and ranks the hypotheses on the basis of these tests. In addressing these examples, he recognizes five types of interaction between genetic and cultural evolution:

1. *Genetic mediation* corresponds to what Lumsden and Wilson call the "leash principle" and what I have called the "guidance principle," in which innate learning biases encourage the acquisition of some available memes rather than others. Durham, like Lumsden and Wilson, accepts the cultural evolution of color-naming systems as an example.

2. In *cultural mediation,* "the configuration of cultural instructions in a population directly affects the differential reproduction of genes and genotypes" (Durham 1991:226). An example is the selection pressure on sickle-cell anemia genes stemming from agriculture in West Africa. Sickling heterozygotes are more resistant to malaria than homozygous "normals," so malaria and sickle-cell anemia are geographically correlated. But West African ethnic groups with longer histories of slash-and-burn agriculture, which fosters the *Anopheles* mosquito, have more sickling genes, since these are indirectly favored by cultural adoption of this agricultural method.

3. *Enhancement* is a two-way coevolutionary process in which genetic and cultural trends encourage each other. An example is the effect of dairying on lactose tolerance. If it were a case of cultural mediation, the emergence of dairying would simply have set up selection pressure in favor of genes for lactase persistence after weaning, allowing for postweaning milk consumption. Substantial evidence supports this instance of the Baldwin effect and of Mayr's dictum that behavior is the pacemaker of evolution. But there is also evidence of positive feedback between the cultural evolution of dairying and the genetic evolution of lactose tolerance in dairying populations; increasing lactose tolerance also reinforced dependence on dairying.

4. *Neutrality* is the common situation in which trends in cultural evolution seem to have no effect on genetic fitness. An example is the largely arbitrary sound patterns emerging in the divergence of languages. While there are a few sound patterns that arise in relation to physical or emotional experience (Lakoff 1987; Lakoff and Johnson 1999; Berlin 2006), much phonetic variation is arbitrary.

5. *Opposition* is the process, considered common by Cavalli-Sforza and Feldman and by Boyd and Richerson, in which cultural evolution and genetic evolution tend in opposite directions. For the former, birth control is in this category, for the latter, war. Durham analyzes the practice of funerary cannibalism among the Fore of New Guinea, which, through a misfolded protein, or prion, spreads the deadly brain disease kuru. No aspect of the ritual enhances genetic fitness, but its cultural fitness—the tendency for it to be imitated and to fit into other cultural schemas—is evident. Needle sharing among drug users at risk for deadly diseases might be another example.

Other case studies, more directly relevant to social behavior, reveal the sophistication as well as some limitations of the approach. Among the Mundurucú, head-hunters in the Amazon valley, there was apparently an opposition of cultural and genetic fitness, since the man who succeeded in killing the victim on a head-hunting expedition was consigned to a sacred celibate status for two years or more. However, most men returned from head-hunting expeditions without heads, yet they benefited by eliminating competitors for scarce game in the region. Meat was readily converted into RS through what the anthropologist Janet Siskind (1973) calls an "economy of sex." So cultural opposition to genetic fitness for the successful head-hunter may be outweighed by fitness advantages for others on the expedition, and perhaps also for his relatives, who may find more game.

Second, Durham's analysis of incest taboos departs from Lumsden and Wilson's view. He accepts that the aversion theory is well established—children growing up together are sexually uninterested in each other—but incest rules apply much more widely than can be explained by aversion alone; moreover, cultural variation in incest rules cannot be for genetic reasons. He argues that since almost all cultures recognize the dangers of close inbreeding, this accurate cultural belief (meme) enhances the aversion, while independent cultural evolution results in marriage rules for much less related individuals. This misses the possibility that clan exogamy could result largely from growing up with cousins. In any case, partly independent cultural evolution is fully compatible with an innate guidance bias. Sexual aversion could result from childhood association (see Interlude 6), and this could generalize to wider sex avoidance rules, which would in each cultural setting depend mainly on patterns of exchange, alliance, and lineage structure.

Third, Durham analyzes polyandry among serfs *(thongpa)* in traditional Tibet, based on field research by Melvyn Goldstein and others. Although the marriage of several men to one woman coexists with other marriage forms in this group, it is much more common there than in the vast majority of cultures. Durham's (1991:97) analysis, including some interesting computer simulations, suggests that *thongpa* polyandry is due to "the monomarital principle"—a cultural mandate for a plot of land to pass to only one family of heirs, preventing its fragmentation and the family's downward social mobility: "Under the assumption of constant land constraints, a sustained belief in the monomarital principle soon produces more surviving descendants per household than any of its alternatives." Although the simulation uses RS as an end point, cultural choices based on ecological constraints (including imposition) are the mediators, and the result does not contradict the prediction that in human cultures polyandry will be rare and, where it occurs, brothers should marry together to protect inclusive fitness in case of nonpaternity. Maternity uncertainty is so rare that there are no corresponding concerns for women. Sociobiological predictions are confirmed by a worldwide sample (Daly and Wilson 1983; Murdock 1949): the great majority of over eight hundred cultures permit polygyny, but only four permit polyandry, and in all four it is fraternal. Neo-Darwinian theory predicts species-wide tendencies; ecological and cultural theory helps explain variation around them.

Finally, Durham did field research on ecological factors in the "soccer wars" in El Salvador and Honduras. Although he agreed with previous investigators that the peasant-occupied rural regions of these countries were "an ecologist's nightmare" (Durham 1991:362), this was not a simple case of cultural opposition to biological adaptation. The peasants' subsistence patterns seemed to be maladaptive cultural choices, but historical investigation showed that "it was the self-interested manipulation of national land tenure policies by the landed elite in the nineteenth century" (362) that drove peasant subsistence down ecologically devastating (and ultimately war-provoking) paths. Thus *imposition* through land expropriation, not simple cultural *opposition* to adaptation, explained the peasants' behavior. They were adapting as best they could, which was not very well, to conditions designed by the landed

Table 28.2. Comparison of Key Elements of Some Influential Models of Cultural Evolution

Authority	Units	Main Mechanism	Concepts of Fitness	Special Features	Gene-Culture Interaction	Comments
Dawkins	Memes: independent, any size	Imitation (mimesis)	Cultural fitness = chance of imitation	Memes as parasites	Uncoupled	No cultural coherence
Alexander	Not given	Social learning	Culture serves inclusive fitness	Strong learning biases	No independent cultural fitness	No separable cultural evolution
Lumsden and Wilson	Culturgens (memes) = mentifacts and artifacts	Social learning and teaching, with innate biases	Distinct but interacting fitnesses	Epigenetic rules; multiplier effect	May slow or accelerate each other	Culture can create strong selection on genes
Feldman and Cavalli-Sforza	Cultural traits	Social learning and teaching	Cultural fitness only	Horizontal, vertical, and oblique transmission	Gene-culture coevolution or competition	No cultural coherence
Boyd and Richerson	Behavioral traits	Social enhancement, imitation	Culture influences group fitness	Biased transmission in diffusion	Largely group selection model	Very abstract models
Tooby and Cosmides	Highly modular neural circuits	Prepared module-based learning	Culture serves inclusive fitness (or not)	Discordance hypothesis	Strongly prepared learning	Limited role for culture
Durham	Memes	Imitation, teaching, imposition	Distinct but interacting fitnesses	Conveyance forces	Five distinct types of mutual influence	Cultural coherence; comprehensive model

elite to enhance their own biological and cultural fitness at the peasants' expense. It is a virtue of Durham's theory that he does not just see maladaptation as runaway cultural evolution harmful to RS, but often as one group's disadvantage at the hands of another.

Some Models Compared

These different theories of cultural evolution, and several others, all have merit but take only partly overlapping approaches. Table 28.2 presents a simplified comparison designed to highlight key features of each approach and their similarities and differences.

Each of these approaches assumes a central role for childhood in the process of cultural transmission and evolution. What is missing from all these models is any understanding of the mechanisms by which infants, children, and adolescents function as agents of cultural stability and change. The next chapter does not propose a new model of cultural evolution, but it does attempt a general account of the universals of human behavior and culture and proposes a comprehensive culture acquisition device to account for some cultural universals and all cultural variation in the light of gene-culture coevolution.

Interlude 8

Thinking about Boys at War

One of the less fortunate but probably crucial contributors to cultural coherence is that members of a group feel superior to, and even actively denigrate, members of another group—especially those just over the next hill, who may be ridiculed and despised as filthy, hypersexual, violent, incestuous cannibals. This tendency, marshaling slightly modernized epithets, is at the heart of some racial, ethnic, national, and religious divisions in the world today. It can be seen to some extent at the level of hunter-gatherer bands. I often heard !Kung San speak disparagingly and sarcastically of the people living at a waterhole twenty miles away; although they spoke the same language, there were slight differences in accent or slang that could be made fun of. One woman from the other group who had married into the band in question and stayed for decades never felt quite at home. And throughout the history of every continent, ethnic, religious, and national identities have been a source of pride in both the good and bad senses of the word and have spawned no little amount of violence (Konner 2006).

It has often been said that all wars are boyish. A classic experiment in social psychology, of a sort that could perhaps not be done today, illuminated the process of group formation and the causes and consequences of the us-versus-them distinction, using boys as experimental subjects. In this study, known as the Robbers Cave Experiment, Muzafer Sherif led a group from the University of Oklahoma that studied twenty-two average, normal eleven-year-old boys, all middle-class Protestants with similar educational backgrounds (Sherif, Harvey, et al. 1961). During the summer between their fifth- and sixth-grade school years, the boys were taken to a two-hundred-acre camp in the Robbers Cave State Park, a densely wooded section of the San Bois Mountains of southeastern Oklahoma.

In the first stage of the study, which lasted a week, the boys were randomly assigned to one of two matched groups that did not differ in any measurable way. In this phase competition between the two groups was discouraged by the adults, and there were many noncompetitive joint activities. Still, boys in both groups began to display competitive attitudes—they named themselves Eagles and Rattlers, made negative remarks about one another, and began to react with hostility to each other's territorial "incursions."

In the second stage, competition was encouraged. The two groups were pitted against each other in contests including baseball, tug-of-war, tent-pitching, skits, treasure hunts, and cabin inspections, with trophies, medals, and four-bladed knives as prizes. "After the second day of the tournament, the 'good sportsmanship' stated in specific words during the initial period and exhibited after the first contests . . . gave way, as event followed event, to increased name-calling, hurling invectives, and derogation of the out-group to the point that the groups became more and more reluctant to have anything to do with one another." In a matter of days, "derogatory stereotypes and negative attitudes toward the out-group were crystallized."

In the third and final stage, the two groups of boys were blended together again and assigned set goals they had to attain cooperatively; for example, told that a water tank had been damaged by vandals, they were instructed to fix the tank so that they all would have drinking water. This process greatly reduced prejudice and conflict within a few more days. As stage 2 ended, the boys were asked whom they considered their friends; there was practically no crossover between the two groups in the answers. But by the same measure there was considerable healing of the split by the end of stage 3. The boys were returned to their families with, presumably, their positive attitudes toward all and their faith in cooperation restored.

Many studies with adults have produced similar findings under a variety of more controlled experimental conditions (Tajfel 1982; Robinson and Tajfel 1997), strongly confirming the ease with which prejudice against arbitrarily created and designated out-groups arises. Such prejudices are easily exacerbated by experimentally giving the in-group members frustrating experiences or experimentally lowering their self-esteem. Further studies also confirm that such artificially created prejudice can be reversed fairly readily when in-group and out-group members are brought together again and given appropriate interventions.

However, in the real world they are usually not brought together again, except under the unusual circumstances of formal peacemaking or imposition from above, often after prolonged and costly conflict. In small-scale societies, alliances between enemies can be formed by intermarriage, as expressed in the saying "We marry the people we fight"—a statement that contains as much of the history of Europe's royal families as it does of primitive cultures' alliances. Hunter-gatherer bands may fission temporarily after internal conflict and recoalesce after months of separation (Lee 1979). But as human populations grow, groups may split and remain separated, allowing the emergence of prejudices that are not counteracted by subsequent mingling. Over years and decades separation inevitably produces divergence in language and custom proportionate to the length of separation.

It is easy to see how the process of group identity formation observed in the Robbers Cave Experiment over the course of days might operate over generations. By building on fear of strangers in infancy and other childhood fears, adults would socialize and enculturate children into xenophobic attitudes; prejudice would be imbibed with mother's milk. If identity, bigotry, and animosity can rapidly emerge within days in children at a summer camp, beginning with no differences between the groups and no initial reason to resent or dislike one another, how much more must this be true over a period of generations during which genuine grievances—including loss of land and violent deaths—have accumulated?

But the most remarkable thing about the experiment is that group identity itself forms and coheres largely in contradistinction to the out-group. Even where there is no fundamental or even noticeable distinction, "they" must be defined as different and inferior, and "we" must be defined not just by what we are, but by what we are not—which bears a strong resemblance to what *they* are. If this model is correct, cultures cohere for very strong natural reasons, and the power of conformity within them comes partly from the need to show unquestioned loyalty in an ongoing or iterating competition or conflict. This would be a source of strong resistance to piecemeal cultural change and highlights the need for models of cultural evolution that allow for such resistance (Durham 1991).

That conformity even in the absence of group identity is a very powerful motivator in experimental as well as natural situations—even to the point of altering basic perception—has been shown in both classic (Asch 1955, 1956) and later studies (Nail 1986; Levy and Nail 1993; Nail and Helton 1999; MacDonald, Nail, et al. 2004). Mathematical models and empirical evidence strongly suggest that conformity and other factors amount to biased cultural transmission that helps groups to cohere and that ethnic markers, which tend to be stronger at ethnic boundaries, can arise and be reinforced through pair-wise norms of mutual aid (Soltis, Boyd, et al. 1995; Henrich 2001; McElreath, Boyd, et al. 2003; Bowles 2009). These findings confirm the predictions of Durham (1991), Shore (1996), and others that cultural selection and cultural coherence make it difficult for some individual traits to change in a trait-by-trait-independent manner. However, most models do consider the power that negative assessment of out-groups may have in making cultures coherent and stable.

A powerful if somewhat discouraging model begins with evidence from the fossil record that ancestral populations of hunter-gatherers may have engaged in frequent intergroup conflict and goes on to show that such a process could have produced the human tendency to cooperate, which is strong compared to that in chimpanzees, through a process of group selection (Bowles 2009). This model requires a lot of assumptions, but if group selection occurs, this could be one of its prime instances. Also, the archaeological evidence is growing in support of the claim of violence in our past (Keeley 1996; Leblanc and Register 2003; Konner 2006) and in some of our closest relatives (Wrangham 1999; Ghiglieri 1999).

Sociologists like Georg Simmel (1903) and Lewis Coser (1956) developed theories of social conflict pointing to its integral nature in society and to its positive functions. Some of these are seen in the third stage of the Robbers Cave study, when the end of conflict brought the boys together in a way that was probably more meaningful than it would have been before the conflict. But the most powerful social functions are those that operate within each group. Even at the outset, when conflict was discouraged, the boys wanted identities and named themselves "the Rattlers" and "the Eagles." In the second stage flags became important objects of affection or ridicule. And as stereotypes formed about the opposing group, coherence, loyalty, organization, motivation, and affection were all strongly fostered in the in-group. Such responses to conflict with an external competitor or enemy have contributed to the coherence of every culture on record.

Evidence pointing to the involvement of the limbic system and its cortical connections has emerged in some studies. For example, in a complex social task involving cooperation and trust, adult subjects were given a chance to perform it with in-group versus out-group partners (Rilling, Dagenais, et al. 2008). Those who said they felt differently with the two types of partners showed stronger activation in the dorsomedial prefrontal cortex (dmPFC), which plays a role in mentalizing tasks (see Chapter 12), and they showed stronger connectivity between the dmPFC and limbic reward centers (like the ventral striatum) after bilateral cooperation with in-group members than with out-group members. Moreover, subjects who discriminated against out-group members showed stronger activation of the frontoinsular cortex during those interactions, suggesting arousal and aversion. Finally, subjects who did not have strongly negative feelings about out-group members showed stronger activation in the dorsolateral prefrontal cortex (dlPFC) in such interactions, suggesting a cognitive or executive effort to overcome discriminatory tendencies. Studies like this do not prove biological causality, but they show preparedness in the human brain to do the things that behavioral studies show us, for good and ill.

The good news is that, if the boys at Robbers Cave could self-organize group conflict in a setting of essentially zero real group difference, maybe these variables are separable in practice. That is, in the real world, where differences between groups are far from zero, perhaps a certain amount of mingling can mute conflict and generate mutual respect. If you do not need difference to generate conflict, you may not need similarity to make peace. But recent examples like Rwanda, the former Yugoslavia, and Iraq suggest that contact, mingling, and even extensive intermarriage do not suffice to prevent the most vicious conflict across what seem for decades softened ethnic boundaries.

Of course, at age eleven, the Robbers Cave boys had no doubt had ample experience defining others downward—in sports, national rivalries, religion, even race. Children become conscious of these distinctions at very early ages, and given the history of our species, wariness may be adaptive. Pride and a sense of belonging in one's own culture may in the short run be a more secure path to survival than the lofty ideal of identification with all humanity. We can hope that it is only a matter of time before the latter replaces the former, and we want our children to be a part of that future and not an obstacle to it. But we also want them to be members of our culture, and the balance is likely to remain a difficult one to strike.

29

Universals, Adaptation, Enculturation, and Culture

Although the main goal of cultural anthropology in general and of psychological anthropology in particular has been the description and analysis of cross-cultural variation, that enterprise has always had an inevitable, if tacit, complement: characterizing the features of human behavior that vary relatively little or not at all.

Universals of Human Behavior and Culture

The concept of universals has at least five different meanings:

1. Behaviors, such as coordinated bipedal walking or smiling in social greeting, that are exhibited by all normal members of every known society

2. Behaviors that are universal within an age or sex class, such as the Moro reflex in all normal neonates or the muscle contraction patterns of orgasm in postpubertal males

3. Statistical regularities that apply to all populations but not to all individuals, such as the sex difference in physical aggressiveness or the suitability of the same five factors to explain personality variation in widely disparate cultures

4. Universal features of culture rather than of behavior, such as taboos against incest and homicide, some form of marriage, or the social construction of illness and attempts at healing

5. Characteristics that, although unusual or even rare, are found at some level in every population, such as homicidal violence, thought disorder, major depression, suicide, and incest

The list of characteristics that fill these five categories is a very long one, much longer than prominent anthropologists of the classic phase of ethnography might have predicted. It includes remarkable constancies in nonverbal communication (Eibl-Eibesfeldt 1971a, 1971b, 1988), their interpretation (Ekman, Friesen, et al. 1987; Ekman and Rosenberg 1997), and the semantic domains of the words used to describe basic emotions (Heider 1991) as well as social arrangements and many features of culture (Brown 1991). Despite claims to the contrary, the scientific search for societies that have no violence, no gender differences that go beyond childbearing, or no mental illness, or that lack weapons or fire, has been a vain one. Although there is convincing documentation of variation in the incidence or context of expression of most human behaviors, the existence of a large core of expected features is a demonstration of the reality of human nature and its validity as a scientific construct. These universals are even more deeply fundamental to the nature of our species than the features that are found in human hunter-gatherers and that have been departed from by later forms of society, since they persist in the face of historical and ecological transformation and are likely to be intrinsic to human nature.

Traditional cultural anthropologists frequently have shown little or no interest in such universals, viewing them as trivial or outside their subject matter, but the elucidation of universal features of human behavior and culture is increasingly being recognized as a central task of the discipline, and one likely to increase the sophistication of research on cultural variation. Some cultural anthropologists—most prominently Claude Lévi-Strauss (1963, 1968)—have attempted to delineate such universals as symbol systems and mental structures whose common underlying characteristics link widely disparate surface manifestations in art, language, and ritual (Shore 1996). This intellectual strategy owes much to linguists, who find common functional features of all languages that transcend their specific manifestations (Brown and Witkowski 1981; Chomsky 1988; Comrie 1989; Greenberg 1963, 1975; Greenberg, Ferguson, et al. 1978).

We saw in Part II the importance of the universal behavior of developing individuals, but we have only touched on the universals of culture. Many of these universals have been extensively studied (LeVine, Miller, et al. 1988; Comrie 1989; Eibl-Eibesfeldt 1988; Brown 1991; Konner 1991; Greenfield 2000b), and thinking about how to discover and evaluate putative universals has advanced (Valdés-Pérez and Pericliev 1999; Cournoyer and Malcolm 2004; Wierzbicka 2005). Table 29.1 lists the five major categories of behaviors or features of social and cultural life that may be called universal. Again, it is clear that this term has been and should be applied to different kinds of things. Some, like bipedal walking, are "hardwired" products of genetically controlled neural maturation in all normal humans. Others, like strong attachment to a primary caregiver, are characteristic of certain life-cycle stages. And still others, like the prohibition of in-group homicide or rape, are probably not wired in but are universal group responses to tendencies that occur in some individuals in all populations.

Some features of human culture, we must acknowledge—such as a pervasive division of labor by sex, having girls marry at menarche or earlier, or slavery in cultures above a certain population size—would once have been universals, yet they prove to be historical contingencies. They offer insight into human tendencies but are thoroughly misleading if thought of as permanent features of human life. Some of the universals we are about to consider—weapons? ethical dualism?—may turn out to be similarly impermanent. With these qualifications, and using the fivefold definition of universals shown in Table 29.1, many characteristics of the human adaptation qualify. Table 29.2, based first on the work of Donald Brown but modified and augmented by that of

Table 29.1. Five Major Categories of Universals

Category of Universal	Example
1. Characteristic of all normal adults	Bipedal walking; smiling in greeting
2. Characteristic of all members of a specified age/sex class	Moro reflex in neonates; nocturnal emissions in adolescent males
3. Characteristic of all populations	Males on average more physically aggressive than females; factorial models of personality variation
4. Characteristic of some members of all populations	Homicide; rape; depression; altered states of consciousness
5. Characteristic of all cultures	Rules against homicide and incest; religion

Sources: Compiled from LeVine, Miller, et al. (1988); Comrie (1989); Eibl-Eibesfeldt (1988); Brown (1991); Konner (1991); Greenfield (2000b); Valdés-Pérez and Pericliev (1999); Cournoyer and Malcolm (2004); Wierzbicka (2005).

Table 29.2.   Cultural Universals

Language, including:
   Broadcast transmission/directional reception
   Rapid fading of signal in basic usage
   Interchangeability of speakers
   Extragenetic transmission
   Specialization for communication
   Semanticity/symbolic capacity
   Arbitrariness of link between signal and object
   Displacement in time and space; abstraction
   Discreteness of elements
   Productivity at phonemic and morphemic levels
   Duality of patterning (meat versus team; man bites dog)
   Functions of communication and deception
   Phonetic contrasts (vocalic/nonvocalic, stops/nonstops)
   Phoneme number between 10 and 70
   Nouns, verbs, and possessives
   Other grammatical features; mutual translatability
   Synonyms and antonyms
   Color-naming sequence patterns
   Words for certain concepts (for example, dark, light, face, hand)
   Binary conceptual discriminations (good/bad, old/young)
   Certain kin terms (mother, father, son, daughter)
   Age/sex terminology (child, adult, male, female)
   Pronouns (at least three persons and two numbers)
   Proper names for individuals
   Conversational turn-taking and other pragmatics
   High-pitch, high-inflection speech to infants
   Special forms for special occasions
   Poetic speech (broken by pauses about three seconds apart)
   Consequences for prestige or status
   Narrative: autobiographic; event-specific; fictional/mythic
   Historical linguistic change
Cognitive/perceptual processes, including:
   Tendency to dichotomize from underlying continua
   Hierarchically arranged color taxonomy priorities
Nonverbal communication, including:
   Some equivalent facial expressions including those for:
   Joy, sadness, anger, fear, surprise, disgust, and contempt
   Smile with brow flash in greeting
   Coy/flirtatious behavior sequence
   Laughter, squeals of delight, weeping in joy and grief
   Childhood physical aggression (hit, kick, shove, bite; boys more)
   Play-fighting (chase, flee, wrestle, play face; boys more)
Social psychology/social cognition, including:
   Sense of self as subject and object
   Recognition of individuals by faces
   Intersubjective folk psychology; social metacognition
   Sense of distinctive peoplehood
Emotions, including:
   Childhood fears: separation, strangers, snakes, loud noise, dark
   Joy/happiness
   Disgust
   Generalized anxiety
   Sadness; grief in loss

Anger/rage
Attachment, from infancy; parental devotion; romantic love
Sexual attraction; sexual jealousy
Material culture, including:
Systematic, varied toolmaking from stone, bone, wood, metal
Tools to cut, pound, pierce, gouge, throw, and so on
Containers (such as hollowed wood, skulls, ostrich eggs)
Devices for tying things together (string, vine, sinew, wire)
Shelter from the elements
Draping, painting, or otherwise enhancing/covering the body
Weapons
Controlled, productive use (not necessarily making) of fire
Knowledge and use of medicinals or alleged medicinals
Consciousness-altering substances
Social organization, including:
Mother-child unit, usually with associated male(s)
Marriage (legitimation of sex and offspring)
Pattern of socialization/enculturation of children
Grandparents' participation in child care
Relatives distinguished from nonrelatives
Division of labor by sex and age
Male attraction to nubile females
Female attraction to powerful males
Male abuse of power
Reciprocity: exchanges and sharing; cheating
Property, at least personal property
Conflict at individual and group levels
Most organized and spontaneous violence due to men
Ascribed and achieved status/role beyond age, sex, and kinship
Hierarchically arranged kin naming priorities
Cultural patterns, including:
Etiquette, customary greetings, hospitality
Rituals, including rites of passage and mourning the dead
Dietary prescriptions and proscriptions; taboos
Folk narratives, music, poetry, song, dance
Decorative/plastic arts beyond body adornment
Sexual regulation, including extended incest rules
Standards of sexual modesty
Proscription of in-group violence, rape, murder
Redress of grievances; mediation; punishment
Distinguishing right from wrong
Recognizing intentionality and responsibility
Ethical dualism: different rules for in-group versus out-group
Belief system, including:
Supernatural entities beyond the visible and palpable
Theories of illness and healing, life after death
Worldview; the world's structure and the place of people in it
Dreams and their interpretation
Historical and origin narratives
Aberrant behavior at some low level, including:
Crimes of rape, assault, theft, homicide, murder
Mental illness: anxiety, thought and conduct disorders, depression, suicide

---

Source: Compiled from Darwin (1998); Murdock (1945); Hockett (1960, 1973); Count (1973); Tiger and Fox (1971); Eibl-Eibesfeldt (1979); Brown (1991); Ekman and Rosenberg (1997).

others, is a partial listing of presently known universals of behavior and culture in the human species. Given the claims of ethnographers in the heyday of twentieth-century anthropology about the extent and power of cultural variation, it is remarkable that there are so many constancies.

Cultural variation is powerful. It includes obligatory ritual homosexuality in adolescence in some cultures and punishment of homosexual activity by death in others; a high value placed on male fierceness in some cultures and a preference for restraint and docility in others; female genital surgery to ensure virginity in some cultures and sexual freedom for teenage girls in others; cannibalism of relatives in some cultures, of enemies in others, and abhorrence of cannibalism in yet others; routine corporal punishment of children in some cultures and adamant rejection of it in others; rigorous monotheism in some cultures and a varied pantheon in others.

These differences are scarcely trivial and certainly need explaining. Cultural relativism, which means not ethical relativism but open-minded understanding followed by interpretation and scientific analysis, is anthropology's most important contribution to human discourse, and the key to making sense of such differences. No biological theory is helpful in explaining most of them.

Yet underlying even such drastic contrasts as these are universal tendencies every culture deals with, and neither within-culture nor between-culture variation is random; laws of natural selection, often operating through facultative adaptation, govern both (Betzig 1997; Chisholm 1999b; Hrdy 1999), and within-culture variation is subject to proven regularities, as in the consistent applicability of similar factor-analytic models to personality variation (Eaves, Eysenck, et al. 1989; McCrae and Costa 1997; Hofstede and McCrae 2004). While there may be items in Table 29.2 that will prove to be as malleable as the once-universal division of labor by sex, I believe that most will prove more durable and that all of them (including the now-greatly-diminished division of labor by sex) reveal *some* enduring and biologically informative facts about our species. Finally, recall the wise observation made by the anthropological linguist Joseph Greenberg (Chapter 9): whenever the distribution of a linguistic (or, I would add, behavioral or cultural) trait departs from random expectations, and especially when the nonrandom bias is strong, something is revealed about the nature of the human mind.

All these universals have been subject to gene-culture coevolution and involve complex interactions between the human genetic endowments and the circumstances of human life. If they did not exist, meaningful communication across cultures would be impossible. But each is also the base for elaboration, in enormously varied ways, of its manifestations in individual cultures. We now turn to the mechanism by which both the acquired component of the universals and the culturally determined variation beyond them are transmitted and maintained.

A Culture Acquisition Device

Despite human universals and biologically guided patterns of adaptation, and in part because of them, the particulars of culture invade and affect the developing mind. Even some universals of culture, such as the proscription of in-group homicide, have to be transmitted. But how?

The answer requires a much broader range of psychological processes than is conventionally considered. The culture acquisition device (CAD) proposed here is meant to illuminate the normal enculturation of developing children—vertical or oblique transmission in cultural evolution theory—with the proviso that older children as well as adults may be doing it. The acquisition of culture by children, like their acquisition of language, is a human adaptation, the protocultural achievements of other animals notwithstanding. But if there is a language acquisition device (LAD), as proposed by Noam Chomsky, Eric Lenneberg, and others, it remains mainly unknown. The CAD consists of elements that are better understood. A proposed version is presented in Table 29.3. It consists of twenty processes divided into four categories: (1) reactive processes; (2) facilitative processes; (3) emotional processes; and (4) symbolic processes. The processes that are italicized in the table are probably unique to the human species.

Reactive Processes (Cultural Habitus)

Reactive processes are the most basic mechanisms of behavior modification that occur inevitably while growing up in a certain context.

Table 29.3. A Proposed Culture Acquisition Device (CAD)

Reactive processes (cultural habitus)
  Habituation (response decrement)
  Classical conditioning (including emotions)
  Associative conditioning (perceptual learning)
  Instrumental conditioning (cultural selection of actions)
  Social facilitation (relaxation of inhibition)
Facilitative processes (social learning)
  Local enhancement (ad hoc/unintentional scaffolding)
  Mimicry (mirroring without goal)
  Emulation (goal attained without mirroring)
  Imitation (goal-directed mirroring)
  Instruction (intentional/intersubjective scaffolding)
  Collaborative learning (intersubjective co-construction)
Emotional processes (psychodynamic recruitment)
  Attachment (recruitment for socialization)
  Positive identification (focused modeling)
  Fear of strangers (the comfort of the familiar)
  Negative identification (us/them polarization)
  Emotion management (the comfort of ritual)
Symbolic processes (cognitive enculturation)
  Cultural construction of perception (collective assimilation)
  Cultural schematization (collective accommodation)
  Narrative construction (narrative meaning)
  Cultural coherence (overarching themes)

Source: Adapted from Brown (1991); Whiting (1941); LeVine (1982); Tomasello, Kruger et al. (1993); Shore (1996).

Habituation. Habituation, or response decrement, is the adaptive waning of an inborn response that is repeatedly elicited without serving a function. For maturational reasons, eight- to ten-month-olds show wariness of strangers, but in American culture, with its high frequency of encounters with strangers, the response wanes, while in !Kung culture, where strangers are uncommon, it lasts much longer.

Classical conditioning. Infants react initially to a restricted group of unconditioned stimuli, linked through innate neural circuits to reflexive responses. Suppose that the calming effect of being held by a calm mother is an unconditioned response. Then if being in church calms the mother, this setting will gain the power to calm the child, as the mother's calm mood becomes tied to conditioned stimuli like liturgical music.

Associative conditioning. This ubiquitous process ingrains expectations for the spatial and temporal contiguity of stimuli and so establishes patterns or schemas even in the absence of action. In Samoa a welcoming ritual places an orderly set of sights, sounds, actions, and utterances before a child, who does not need to participate in order to learn the pattern. In highland New Guinea, watching men prepare for war has a similar impact.

Instrumental conditioning. In all societies some naturally occurring behaviors are rewarded and others are punished. From infancy !Kung children are socially rewarded for sharing, a fundamental cultural value and pattern, but reprimanded for approaching a fire; in some American subcultures boys are praised for physically fighting to defend themselves and reprimanded for shrinking from a fight. Within limits, the behaviors wax or wane accordingly.

Social facilitation. A child observes others performing an action for which a tendency is present but which is released from inhibition by observation, the classic example being yawning. This is one process involved in the acquisition of culturally specific dietary patterns and patterns of emotional expression. Public expression of grief is socially facilitated in rural Greek villages and among Plains Indians but strongly inhibited in Bali, with predictable results.

Facilitative Processes (Social Learning)

Facilitative processes include all forms of *specifically* social learning, including those that rely on the ability to take the perspective of the other in one or both directions.

Local enhancement. A kind of (often unintentional) scaffolding, or enhancing of the child's opportunity to learn, in which the ordinary course of events in the lives of parents or others brings the child into situations that make learning by trial and error easier. A berry-picking expedition, for example, brings a child into a setting where the reward of eating berries and the punishment of encountering thorns shape her behavior.

Mimicry. The action of another is mirrored by the observer, but without regard to goals and without apparent understanding. In infants and children it probably relies on a combination of limbic system structures and mirror neurons in the cortex, although in some species (such as parrots) it has a different neural basis. Neonatal imitation of facial movements and repetition of words and phrases without comprehension are examples.

Emulation. Here the observer understands the goal attained by the model and desires it, but finds a way to get it without using the same movement sequence. It implies perspective taking with regard to the reward value of the goal—vicarious experience of the reward—but not with regard to the means. It is developmentally the next stage after mimicry and a transition to true imitation, which is necessary for human cultural learning.

Imitation. In this more advanced form of social learning the child matches both the actions and goals of the model. It does not require intention on the part of the model, who does not have to take the child's perspective, but it does require the child to take the model's perspective. Like mimicry and emulation, imitation differs from social facilitation in that the behavior being learned is not an internally driven response under initial inhibition. Imitation relies on limbic circuits, mirror neurons, and vicarious experiences of reward, but also on the higher cortical processes of perspective taking.

Instruction. Almost unique to humans, instruction (true teaching) involves bidirectional perspective taking—an advanced form of intersubjectivity (Chapter 26). The teacher must also engage in deliberate scaffolding that tailors the learning process to the child's changing ability. Although much social learning takes place in human cultures without it, teaching is seen in all cultures. In addition to its role in the cultural transmission of complex skills such as Mayan weaving (Greenfield 2005), !Kung trance-dancing (Katz 1982), Dagbon drumming (Oppong 1973), and Pulawat navigation (Gladwin 1970), it functions widely in many kinds of cultural transmission (Kruger and Tomasello 1996).

Collaborative learning. Two learners of approximately equal skill, knowledge, and cognitive capacity address a challenge and co-construct a solution by learning interactively (Tomasello, Kruger, et al. 1993). As with instruction, collaborative learning requires two-way perspective taking. The solution is found collaboratively as two learners, often children, observe and comment on each other's skill and understanding. It is a central process in the cumulative growth of culture, an outcome uniquely human.

Emotional Processes (Psychodynamic Recruitment)

Emotional processes include all transmission based on developing emotions designed by evolution to shape behavior in relationships.

Attachment. Attachment to a primary caregiver and to others, in both human and nonhuman groups, lays an emotional foundation for ongoing social learning and socialization. Social skills emerge and are practiced in this emotionally rich context, reinforced by affective rewards and punishments. In human cultures this process does not just *socialize* the young but forms a major part of their enculturation. Acceptance, affection, and nurturance are children's rewards for mastering social and cultural codes.

Positive identification. This emotional process organizes imitative tendencies by focusing them on one or more individuals the child likes, admires, and feels similar to. It motivates the child to study and imitate the model, beyond instrumental, goal-oriented perspective taking. It is involved in the formation of gender-specific behavior in all cultures, as well as behavior specific to a family, a clan, a craft or profession, a class, or the culture itself.

Fear of strangers. Xenophobia, or fear of strangers, sets a limit very early in life on the outward reach of identification and imitation. It drives infants and children toward primary caregivers, enhances attachment, and strengthens within-family identification; it sets a boundary on the domain of people who are likely to be imitated. It begins the rejection of unsuitable models but does not yet necessarily involve denigration and negative identification that drives cultures apart.

Negative identification. Fear of strangers may, and does in most cultures, develop by middle childhood into prejudice against one or more out-groups. It is no longer just fear and the flight to a parental protector. The protector has become one's group of origin, and the negative emotion toward the outsider has become more complex. It may entail hatred, disgust, and/or fear, and these emotions form the basis of a desire to differ from the out-group, reinforcing in-group cultural identity. It becomes rewarding not only to do, think, and feel as certain people do, but to avoid doing, thinking, and feeling as certain others do. Whether or not they actually do the negative things, they serve as a negative model.

Emotion management. Life is full of unsettling experiences, which often generate unpleasant, awkward, or unfamiliar inner states, including anxiety, disgust, fear, grief, rage, hunger, lust, and romantic love, among others, so cultures provide rules and rituals that calm uncertainties and manage or channel the associated behavior. Culture transforms a highly intense or even disturbing individual experience into a comprehensible collective one, an experience that feels idiosyncratic and frightening into one that has been shared and worked through by others.

Symbolic Processes (Cognitive Enculturation)

Symbolic processes rest on uniquely human capacities for the formation of symbols and their integration into culturally significant schemas.

Cultural construction of perception. Because the brain sets the receptivity of sense organs, selecting and shaping input, cultures can shape perception. People whose cultures make no phonemic distinction between *l* and *r* have permanent difficulty attending to the distinction, a trivial matter for two-year-olds in other cultures. Also, susceptibility to specific and apparently simple optical illusions depends to some extent on the visual environments in different cultures; people growing up in small-scale rural cultures are not surrounded by straight lines and right angles and are less subject to some standard illusions (Segall, Campbell, et al. 1966).

Cultural schematization. Not only do cultures filter perception at the sense organs, but they actively build *schemas* to represent the world and structure thought. In highland New Guinea cultures, goods and services flow toward one respected man, who redistributes them (and other goods and services) in appropriate ways. People in these cultures tend to see all of economic and social life in terms of this schema. Cultures that recognize witchcraft perceive some categories of people as dangerous, and their mental lives may revolve around this schema.

Narrative construction. The human gift for creating narratives orders life in a time dimension much like that of the life course, with a beginning, a middle, and an end. Stories can be true or invented, funny or sad, instructive or baffling, cynical or edifying, and they may end happily or tragically, but each story told by humans helps to order life events in a way that is meaningful and comprehensible. As the narrative of one's own life, accurate or not, gives it meaning and orders future events, so the narrative of a culture's history—fighting against oppression for the Irish, suffering and overcoming it for the Jews, pioneering for Americans—functions for a people.

Cultural coherence. Cultures can have themes, and these may organize many schemas into a linked web of cognitive function. Warlike toughness may have been such a theme for the Plains Indians and the Yanomamo, as might cooperation for the Japanese, independence for Americans, submission to God for Muslims, sharing for the !Kung, and a positive attitude for the Balinese. While at best gross oversimplifications, such themes are simplifications made by the cultures themselves, and they may make the large array of symbols and schemas in any culture cohere to some extent, enable people to learn them more easily, and give them emotional weight and resonance.

A Model of Culture in Biological Context

The first three parts of this book considered three approaches to the biological explanation of psychosocial development: the evolution of ontogeny, the maturation of neural and endocrine systems, and the comparative sociobiology of socialization. In Chapter 22 we considered what may be distinctive about psychosocial development in hunter-gatherers in order to establish a baseline of adaptations characteristic of our species in its EEAs—at once the biological end point of past phylogenetic changes and the paradigm of fully and uniquely human culture.

Although these features of hunter-gatherer childhood are certainly older and arguably more "natural" than those in other cultural adaptations, and though departures from them could have important consequences, they do not have the status of cross-cultural universals, which reflect stronger underlying tendencies. In addition, as human subsistence ecologies have diversified, cultures have taken advantage of evolved facultative adaptations to create coherent solutions. These include features of social and economic organization, culture, and child rearing that vary predictably in relation to subsistence ecology, demography, disease, technology, and war.

Finally, there are highly variable features of culture that seem to have no lawful determination at all. For example, whether we say *dog, hund, chien, perro, kelev, Canis familiaris,* or any one of thousands of other possible names does not appear to have any relationship to the value of that species nor any particular consistency, despite the fact that the human association with dogs has lasted at least twenty thousand years. Such variety is the product of freely functioning human inventiveness, and it applies to many variable and intriguing features of culture.

But here we are interested in the constraints. A new model of the role of phylogeny and its consequences for human nature and cross-cultural differences can be built from the Whiting model (Chapter 24), which did not include phylogeny but did recognize a universal course of psychological development at the heart of the psychocultural dynamic. That model also privileged the ecological situation in shaping psychosocial realities. It took a materialist and functionalist stance: people and cultures must adapt to ecological circumstances, and cultural psychology is best understood as an adaptation. In this model history and culture are not driven by ideas; ideas emerge as human responses to situations and needs.

The modified, evolutionary version suggests that universals of human behavior are key to understanding the role of phylogeny. It directly affects the developing individual—not only the "innate needs, capacities and temperament," as in the Whiting model, but also core elements of cognitive and psychosocial growth. But natural selection created not only individuals with needs, capacities, temperaments, and canalized maturational plans but also equations—"if-then" statements—relating the environment to the social system, the social system to the child-rearing pattern, and so on. These are the facultative adaptations and norms of reaction, which set limits on plasticity by guiding the human organism's responses to environmental variation.

Referring again to the example of husband-wife intimacy (Whiting and Whiting 1975b), phylogeny left a legacy such that separating men from women and small children enhances their effectiveness as warriors—not that men must be warriors or must be aloof from their wives, but that choosing aloofness may increase effective martial readiness. The universal feature is not a trait, but an underlying mechanism relating two behavioral continua. This also shows how a Darwinian explanation of a social adaptation can be fully compatible with a psychological explanation of its development, the latter serving as the means to achieve the adaptation in our long-lived, slowly growing, cultural species.

To imagine that these putative causes are mutually contradictory is to confound the levels of explanation proposed by Niko Tinbergen (1963) and expanded in Chapter 1. When asked why a teenage boy punches another, we can say that he does so because of

- the secretion of a neurotransmitter in the amygdala,
- in a neural circuit primed by testosterone,
- in response to a verbal insult,
- after a lifetime of frustration and observation of violence,
- given a fetal brain hurt by alcohol,

- and shaped by prenatal androgenization,

- against a background of maleness and individual aggressiveness,

- caused by natural selection favoring male status and self-defense,

- on a phylogenetic foundation of reproductive competition.

None of these explanations contradicts any of the others; in fact, we do not *have* an explanation until we have all nine levels.

The application of neo-Darwinian or sociobiological theory to ethnological materials has produced some findings that at first glance seem to bypass complex questions about the relationships among society, culture, and individual development. For example, cultures in which men inherit land from their mother's brothers are less controlling of women's sexuality than those in which men inherit from their fathers (Hartung 1982); in societies that allow polygyny, wealthier men tend to have more wives, the extreme being despots who have had hundreds (Betzig 1986, 1992), and in small-scale societies in which adoption of children is common it largely tracks genetic relatedness (Silk 1980, 1987a, 1987b). Those making such findings usually do not claim direct genetic causes of these cultural practices, and indeed great confusion comes from a failure to distinguish between neo-Darwinian theory and behavioral genetics.

In the red-winged blackbird, males singing on richer territories mate with several females instead of one. But the mechanism of this flexible adaptive system—a classic facultative adaptation—may be quite different in blackbirds than the parallel effect in human beings. The wings of insects come from thorax, of birds from forearms, of bats from fingers, and of humans from technology. These four paths to the adaptive advantage of flight reach similar ends with different points of origin. The same is true of adaptations in social behavior.

To return to another classic example (Interlude 5), incest is avoided in most species, so adults seeking mates must rule out close kin. In many species this depends on pheromones, but we achieve inbreeding avoidance through a psychological mechanism dependent on cultural choice, even though the genetic results are the same. In analyses such as these the purposes and methods of cultural anthropology and evolutionary psychology are joined, and the study of childhood is much better served than by debates about who is right. Both are.

V

Conclusion

wherein we see, as through a glass darkly, how human
relationships and emotions may actually emerge

Thus, from the war of nature, from famine and death, the most
exalted object which we are capable of conceiving, namely, the
production of the higher animals, directly follows . . . whilst this
planet has gone cycling on according to the fixed law of gravity,
from so simple a beginning endless forms most beautiful and
most wonderful have been, and are being, evolved.

—Charles Darwin, *The Origin of Species* (1859)

30

The Ultimate Epigenetic Enterprise

The great German poet, scientist, and philosopher Johann Wolfgang von Goethe believed two centuries ago that he had found the secret of life. He called it the *Urpflanz*: the ultimate, original plant, a fundamental and endlessly generative form that was the core and basis of all living things. To get an orchid, a giraffe, or a poet, some mysterious process would transform the Urpflanz so that this latent core form blossomed into something new and marvelous. Living things were developmental variations on this theme. He meant this conception as an analogy to or substitute for the ordering of the universe described in the laws of his rival Isaac Newton; life, Goethe believed, had laws of its own, and something like the Urpflanz was at the heart of it.

As a scientist Goethe was no match for Newton, but as poetic insight the Urpflanz had a certain charm. It also had a certain prescience. Goethe would be pleased to see that the Urpflanz has been found; it is DNA, a powerful and generative form as elegant in its way as Newton's laws and, in the realm of life, as comprehensive. For over three billion years the universe has been tweaking it, and as Darwin put it decades after Goethe, "endless forms most beautiful" were and are being evolved. Darwin discerned in natural selection a true biological equivalent of a physical law, and the great secret behind the tweaking. The structure of DNA and the way it codes RNA and protein was the second great secret of life. But the mechanism by which our molecular Urpflanz generates the endless forms is still a mystery. We know a little; homeobox and other regulatory genes with marvelous names like *sonic hedgehog, ultrabithorax, wingless,* and *dick-kopf*—German for "thickhead"—boss each other around in a vast cascade of commands that are at the core of development.

Now all we have to do is spend a couple of centuries working out the cascade. We will need a very big piece of paper—it won't fit on a laptop screen—but we will eventually draw it, and it will explain everything. Up to a point. Beyond that, there is chance, chaos, and countless outside influences, especially in the flexible realm of behavior. These may make our elegant diagram look like a Jackson Pollock painting. Still, these outside influences are partly lawful, just as Pollock's drips and splatters, a mess to the untrained eye, are strangely ordered and intentional in their provenance. These external forces, like the cascade itself, are powerful, and understanding them better will give us a certain measure of control.

Nevertheless, we now know that it is first and foremost the cascade that builds brain and behavior, not just in the embryo but throughout development and life; the cascade proposes, the environment disposes. The cascade is the key creative element in the story. So we behavioral scientists might now show our respect for it—and break decisively with a century of disdain—by enunciating a law of psychogenetic inertia: developmental plans in motion will stay in motion according to predetermined guidance unless diverted by outside forces. It is perhaps just another way of saying canalization, and it is hardly as elegant as Newton's first law, but it may serve to remind us that all creatures, children included, come into the world with a plan. Having said that, we can now turn our attention to the vicissitudes of the Urpflanz and see whether we can find any general principles for transforming it in evolutionary and developmental time.

A General Theory?

Can there be any general theory that applies to the various levels of explanation? This has been the hope of theoreticians throughout the history of biological and behavioral science. Attempts at generalizations across levels have led to (usually transient) enthusiasm for general systems theory, information theory, entropy theory, catastrophe theory, chaos theory, complexity theory, and other kinds of overarching mathematical models. Aside from the hope of wide applicability, these have appealed to life and behavioral scientists because they have in common that they are nonlinear dynamical systems models and thus hold out the possibility for modeling living systems, which paradoxically join predictable and unpredictable processes.

Traditionally, the most accessible processes in the physical world, at least above the level of the atom, have been those that can be modeled by continuous linear equations. Since physics, if anything, lagged somewhat behind the life and behavioral sciences in statistical modeling, and since statistical models eventually transformed physics, it was thought until recently that linear models tempered with probabilistic generalizations would suffice to describe such phenomena as evolution, development, and learning. This no longer appears to be the case.

Chaos, Self-Organization, and Complexity

With the advent of modern computing capacity, it became apparent that good predictions would not be possible at all steps in all processes that were fully deterministic—not even such superficially simple ones as turbulence in fluids. This was not the famed indeterminacy of quantum mechanics, which occurs at size levels many orders of magnitude smaller. It was, rather, unpredictability in processes and among objects large enough to be well approximated by Newton's laws, processes that are deterministic in spite of their unpredictability. This paradox has generated enormous interest in widely disparate fields of scientific activity, including the biological, behavioral, and social sciences.

Very general theory evokes reasonable skepticism on the grounds that something that explains everything may really explain nothing. This criticism has been leveled at neo-Darwinian theory, general system theory, and complexity theory, among others. Generalizations about processes at different levels are risky even if limited to the living world. Still, there appear to be parallels among levels, noted by many investigators, including some with little interest in general theory who compared only two levels and pointed out commonalities.

This has occurred especially in comparisons of natural selection in biological evolution with at least superficially similar processes in the realm of learning (Skinner 1966), neuroembryology (Purves and Lichtman 1980; Purves 1988), circuit activity (Hebb 1949; Edelman 1987), postnatal neural development (Dawkins 1971; Huttenlocher, de Courten, et al. 1982; Huttenlocher and Dabholkar 1997), individual learning (Skinner 1966), language acquisition (Piattelli-Palmarini 1989), child training (LeVine 1982), cultural models (Shore 1996), and cultural change (Durham 1991). Another example would be proposed roles for self-assembly or self-organization in both evolution and embryology (Kauffman 1993), as well as in behavioral development (Thelen and Smith 1994).

There is a difference between finding or claiming an applicability of general theory at different levels within the biological sciences and suggesting one general theory for all complex phenomena. It may well be that chaos or complexity models are useful for the study of such disparate phenomena as turbulence in viscous fluids, the formation of coastlines, and the vicissitudes of market economies. However, as even physicists have freely conceded (Ruelle 1991), chaos and complexity theory has been best at explaining why certain things—like the weather on your street two weeks from now—are very hard to predict. Despite decades of promising and varied applications, complexity theory has predicted very little, except to say, "Our models show it *could* have happened this way." This has sometimes been very interesting—as in Stuart Kauffman's (1993) model of the origin of life as a daisy chain of linked enzymatic reactions among proteins instead of as a plodding extension of RNA bases, or in the model of brain and behavioral development conceived by Jeffrey Elman and his colleagues as a self-stabilizing, environmentally responsive neural network (Elman, Bates, et al. 1996; Elman 2005). However, these are models. They show what computer simulations can do, not what happens in life.

But consider the possibility that analogous models might be broadly applicable at very different levels of living systems—including behavioral and cultural ones—not just because they are complex but because they are living. That is, because of their evolutionary history and the nature of the continuity they exhibit through that history, they may preserve in their current organization some processual features of their provenance. In coastlines, tree branches, and other examples of the fractal geometry of nature, there is a self-similarity of structure at many levels from macro- to microscopic. Is it conceivable that there could also be a self-similarity of *process* at different levels of the dynamics of living systems, with different resolutions in time and space?

This self-similarity would be a kind of recapitulation of evolutionary process rather than of phylogenetic stages and would be seen in various processes of ontogeny and in ongoing nongenetic adaptive change, whether behavioral, cognitive, connectionist, or cultural. The initial process of evolution built complex adaptive systems out of simpler ones, and so did subsequent phylogeny in many lines. A parallel construction of complexity must occur in the individual life course of each multicellular organism. It would not be surprising to find that the processes that the living world discovered during the phylogenetic generation of complex adaptive systems were somehow made use of in the generation of such systems ontogenetically. Likewise, if life had found a set of algorithms for the production of change in genetic materials in response to environmental challenges, one might expect parts of this algorithm to reappear in the process of change in living systems above the level of the gene.

A Theory of Generative Variation

What follows is a speculative attempt to synthesize various observations about such parallels. I propose that the algorithm *variation, self-organization, challenge, selection* is a general one in living systems, applicable to the levels of evolution, neuroembryology, neural connectivity, learning, socialization, enculturation, and culture change.

Evolution

Darwin's evolutionary process had variation at its core: if there was variation within a population with regard to a certain trait, if the variation was partly inherited, and if the partly inherited variants had different reproductive success, evolution by natural selection would occur. He did not have to know the ultimate source of variation, which is now no mystery. His model gave little consideration to random processes, like genetic drift—the luck of the draw in a given generation in the distribution of heritable variants, particularly in small or isolated populations. We now know that such stochastic processes have played a role in phylogeny. Still, against the long-term background of such genetic and phylogenetic noise, the signal of adaptation through natural selection remains highly detectable. Through revolutions of thinking—mathematical genetics, molecular genetics, punctuated equilibrium, and nonlinear dynamical models of chaos and complexity—Darwin's theory remains our best explanation for the success of life in all its variety and for the history of changes that have made that success possible.

Darwin and his followers may have underestimated the significance of self-organization, but contrary to the claims of some of its proponents (Kauffman 1993; Goodwin 1994a, 1994b), it poses no problem for Darwin's theory. He was a great and avid student of the structure and function of organisms, and it would not have amazed him if there were laws favoring the assembly of some structures rather than others for intrinsic, rather than only for extrinsic, reasons—a fact universally accepted by present-day evolutionists (Maynard Smith, Burian, et al. 1985). The expectation of such laws, however poorly understood, may help us rise to such challenges as explaining the stasis common in the fossil record, the coherence of the genome, and the pervasiveness of canalization in development. Darwin's theory does not stand or fall on the degree of difficulty in dislodging a genome from its coherence or a species from a long-standing pattern of adaptation; as long as variation remains after the self-organization of the hereditary materials and environmental challenges yield different reproductive success among the self-organized variants, evolution by natural selection occurs. That this happens in nature is very well established, in real time (Endler 1986; Grant and Grant 1989; Hereford, Hansen, et al. 2004).

So the insertion of a self-organization step in the evolutionary algorithm, between the initial generation of hereditary variation and the variants' encounter with environmental challenges, is a valuable modification that may help smooth certain wrinkles in the modern evolutionary synthesis. Natural selection must operate on materials subject to laws—gravity, chemical bonding, energy storage, electrochemistry, and many more. It does not get around them, but rather uses them as a starting point; it can do the same with laws of self-assembly.

In a computer model of self-organization, the challenge of the surrounding world may be left out, and the idealized result might be a system that seems to be governed only by laws of internal organization. But in the real world, self-organized systems must face environmental challenge. We could claim self-assembly itself as a kind of natural selection: some systems are better adapted to the internal demands of their interacting elements; these stabilize while others dissolve. But even conceding that this is something more than what Darwin meant, there will be variation among the self-assembled structures that can become grist for a classic Darwinian mill.

Maturation

Deciphering the genetic code and sequencing genomes seemed to be great scientific challenges in the mid- to late twentieth century, but these are trivial endeavors compared with the central puzzle of embryogenesis: how does one genome, shared by almost all cells, produce multiple cell types? Distribution of cytoplasmic contents before division, physical position, and chemical influences from other cells—some in systematic gradients—are all involved in making cells different and turning some genes on and others off in myriad temporal patterns.

Some of the processes involved are understood and were discussed in Part II. One of the striking ones is that roughly twice as many neurons are generated during embryogenesis as will survive, and survival is not random. It is a process that depends on trophic connections with other cells and then on activity in circuits. The genes in all these cells are formally identical, but their patterns of expression make them functionally distinct epigenetic entities, competing for survival. Similar activity-dependent competition may take place among neuronal groups, even those in which the local micro-circuitry is more genetically determined.

There is also a huge overproduction of dendritic branches, axon terminals, and synapses, many if not most of which die back, also in an activity- and stimulation-dependent way. Both the generation and pruning of connections continue throughout life (to a much greater extent than neurogenesis), but the peak number of synapses is reached at various times in different brain regions and systems. Some occur months to years postnatally and are therefore subject to influence in ways that concern parents and educators.

Neurons continue to die in large numbers normally and continuously throughout life, and activity plays a role in the survival of those that remain. This may help to explain why the most overlearned or extensively practiced activities survive longest in dementia. This optimistic view of neuronal loss suggests that it is part of a process of lifelong shaping of neural circuits by experience. Finally, neurons are generated throughout life (at least some neurons, in some brain regions), and some of these new neurons are subject to the influence of activity in a brief critical phase after they are born that helps to determine their survival and function.

So in several different developmental processes there is competition within the nervous system for survival and adaptation, and this competition depends to some extent on function, analogous to natural selection but on a cellular scale within the brain. There cannot be true natural selection among genetically identical cells or among the connections between those cells. But the cells are epigenetically different; they differ in some transposable elements, in DNA methylation, in X-chromosome inactivation, and in gene expression, and both the neurons and their extensions differ in adaptive value in their local extracellular environments. The differences give rise to different rates of survival, and the end result is, if not optimal, better.

Socialization

Whether in societies of humans or other social vertebrates, the young behave differently from adults; many juvenile behaviors disappear while others survive, and some new ones are generated. Some behaviors are strengthened, habituated, or extinguished during trial-and-error interaction with the environment, including the social group. Human adults consciously treat some child behaviors—those that are improper, immature, inadequate, bad—as change-worthy, but nonhuman adults are also annoyed by many juvenile behaviors. Both human and nonhuman parents, other adults, and older juveniles, intentionally or not, distribute rewards and punishments, and the laws of learning favor the selection of desirable behaviors and the elimination or reduction of others. Infant and juvenile needs and demands in the context of parent-offspring conflict generate behavior that may be punished, while local enhancement, social facilitation, and direct reward facilitate others.

Play in particular generates energetic, varied, seemingly random behavior and draws the young into a shaping process that narrows behavior into mainly functional paths. Other natural processes, such as weaning or the food quest, can be deadly serious mechanisms that favor some behaviors over others. This process of selection begins the first time the infant is put to the breast and continues through the jarring first experiences of mate competition and beyond.

Once again in analogy with evolution by natural selection, the immature brain generates behavior that is more exuberant, random, and variable than can be sustained, in this case over the course of development. New behaviors appear and function as mutations do in transgenerational evolution. Babbling becomes speech, stumbles turn into strides, smashing nuts to bits is transformed into the smooth extraction of precious food. Much of this is maturation, but much is also selection. Whether they emerge as random variants or imperfect copies, behaviors that are adaptive and functional within a given developmental context survive and displace those that are not.

Enculturation and Culture Change

Human societies carry out this process of social selection intentionally and extend it greatly. Modesty, amount and type of speech, time and place of play, posture, movement style, interpersonal distance, and the time and context of the expression of emotions are subject to cultural selection that varies impressively from one human group to another and goes far beyond the socialization we share with nonhuman primates. Much of the selection in humans is deliberate and aimed at modifying explicitly defined change-worthy behavior. Other uniquely human behavioral selection processes are less deliberate. All human infants while babbling produce the entire range of sounds used in all human languages, but only those used in the language spoken around them are selected for and survive further development.

But of course, human adults do not merely select among naturally occurring child behaviors, favoring some over others; we lead our young to generate new behaviors that would never occur unless imitated or even taught. In addition, humans, including children, learn collaboratively, co-constructing not only behaviors that have not occurred before in the lives of the children in question but on some occasions never before, period. Thinking about and commenting on one's own and another's thoughts introduces a new dimension of selection never before seen in the history of life.

Some new behaviors, whether co-constructed or invented by individuals, survive the process of cultural evolution; others do not. So at the level of the individual life cycle and at that of the evolution of culture by nongenetic and coevolutionary means, some culturally instantiated behaviors survive better than others because they are more adaptive than others. But unlike neuronal competition within developing brains and far more than with socialization, selection-by-enculturation within an individual life course can result in new shared behaviors and psychological states that are transgenerationally transmitted outside the genome. While there are protocultural traditions in some other species, the human species is uniquely adapted for cultural evolution, with modes of transmission and fitnesses unique to its own domain, in interaction with genetic evolution.

Selection, Epigenetics, and Development

It appears that there are analogous processes occurring at each of the four levels of observation discussed in the four parts of this book. In each, *variation* is generated, *self-organization* stabilizes some varieties, *challenge* makes them more or less adaptive, and *selection* favors some over others. This may not be satisfying as a theory, or even as what I have called self-similarity in process, but it may be an insight into the nature of living things. Genome, nervous system, society, and culture are all information-storage devices and as such are life's answers to the second law of thermodynamics; that is, they use information against entropy, the universal tendency of all things toward disorder and sameness. It works because the information is about the environment, and it works to the extent that the information is right. It is therefore of the essence of life to interact with, know about, and change in response to the environment. Individual change through such interactions—epigenetics in the broad sense—is therefore what life, and especially childhood, are all about.

But childhood is also the outcome of an even grander interaction with the environment, a vastly wide and glacially slow gleaning of information that has gone on for billions of years. We do well by our children to recognize the costly and valuable information they bring into the world as we help them shape and organize themselves in response to the equally vital information they gather as they grow.

Reprise

What are the take-home lessons of this very ambitious book?

1. Nothing in biology makes sense except in the light of evolution.

2. Life is inherently improbable. It resists entropic disintegration only by storing information about the surround so that it can dodge some of the thousand natural shocks that it is heir to. Information is stored in four forms: molecular strings of nucleic acids; networks of excitable cells that stably change with use; social organization, which biases those individual patterns of excitation; and culture, a transgenerational, symbolic store of information that evolves nongenetically in interaction with the genome.

3. Adult phenotypes do not evolve, life cycles do; life *is* development, and modifications of development are of the essence of evolution. Prominent among them are the modification of regulatory genes, differential acceleration or slowing of organ-specific, system-specific, reproductive, and bodily growth rates (heterochrony), and the modification of both specific and general plasticity.

4. Functionally irrelevant factors have figured in the history of life. Macroscopically, these factors have included meteorite showers, volcanic activity, and oscillations in the earth's magnetic field. Microscopic factors have included cosmic radiation, chemical mutagens, and errors in DNA synthesis and repair. An important developmental factor has been extreme sensitivity to initial conditions, or formal chaos. At a population level, there is genetic drift and the founder effect. Such factors matter, but they do not negate selection; they are the background noise of the history of life.

5. Adaptation through natural selection is a signal we detect against that background noise, which provides the opportunity for selection.

Selection was faster and bigger after mass extinctions than before them, and without the variation due to mutation and the reshuffling of genes in sex, selection loses its force. The signature of selection can be difficult to detect and easy to misinterpret. But since the dawn of life, the signal has always been there.

6. Evolution's goal is reproduction, survival being a path to it. No understanding of biology, behavior, or development can be achieved without full appreciation of this simple fact. With sexual reproduction, two functionally distinct forms emerge. This distinction is more plastic in some species than others, but in no sexual species can biology, behavior, or development be understood without understanding gender.

7. Behavior is the pacemaker of evolution. Any action by an organism on or in response to its environment modifies that environment or draws it into a different one. In doing so, the environment creates new conditions of selection that act on the organism's progeny. The eventual result may be a genocopy of the acquired behavior, but it will often be another anatomical or behavioral change.

8. Mammals have lawful life histories, with two continua: one spans from "be small, live fast, die young" to "be large, grow slow, live long"; the other is a matter of lifestyle at a given size and ranges from having a rich resource base and breeding prolifically in the face of high mortality to being protected from predators and reproducing slowly and carefully. We come from primates, protected and slow in the treetops. As apes, we got bigger and slowed more, but cooperative breeding, post-weaning provisioning, hunting, tools, and a richer resource base made us the best ape. In a sense we ended with the best of both lifestyles.

9. Neural networks evolve capabilities for detecting information, storing it, evading danger, finding and ingesting energy-rich objects, maintaining body temperature, growing, and reproducing. Neural networks have achieved these functions in many evolutionary lines. Nothing ensures a trend, but increases in relative and/or absolute brain size are common in evolution and decreases are not. Many aspects of network design are encoded in the genome, but others are very responsive to input.

10. Major advances in brain size and complexity are often achieved by relatively small, highly mobile predators. This was true of the early vertebrates, amniotes, mammal-like reptiles, mammals, primates, and hominins. Larger brains tend to go with slower development, more effi-

cient energy processing, larger body size (after the initial brain expansion), and longer life span. Among mammals, these traits are associated with play and parental care.

11. Our very large brain was not intelligently designed but expanded by accrual. It is a supercomputer built around an adding machine built around an abacus, all steeped in hormones and charged with emotion. Emotion has been neglected in cognitive neuroscience, focused as it tends to be on the supercomputer. Essential in understanding psychosocial growth are the limbic system circuitry, the orbitofrontal cortex and its connections, the lateralization of language and emotion, the mirror-neuron system (expanded to include subcortical circuits), and autonomic function. Also important are the social circuits in the superior temporal and medial prefrontal cortex. These and all brain functions are end points of phylogenies.

12. If information is stored in genes, much of it is in those that direct the brain's assembly. We are starting to see how neurons are born, differentiate, migrate, extend axons and dendrites, form connections, and persist in circuits. Most of these processes are genetically guided. Cell death and synaptic pruning involve activity and stimulation in circuit survival, but even cell death is partly genetically programmed. Prenatal environment effects, including learning, occur in many species. Postnatal brain development extends prenatal processes and is also under substantial genetic guidance.

13. Although genes generate individual differences, many gene effects are species-wide. Universals of human development encoded in the genome are answers to adaptation's past challenges. Among the features of human psychosocial development grounded in genetically guided neural maturation are the rise and fall of early infant crying, the emergence of social smiling and mutual gaze, the development of attachment and social fears, the growth of sex-differentiated physical aggression, the emergence of language, the five-to-seven shift, prolonged middle childhood, and an increase in sexuality and other changes at puberty.

14. In our species, the phylogenetic process produced a unique developmental plan. Compared to apes, we have had all phases of life lengthened by roughly 50 percent, with three exceptions: gestation is only a bit longer, leaving a more helpless neonate; middle childhood, a time for intensive enculturation, has been greatly lengthened; and there has been little or no lengthening of female reproductive life, hence wonder-

ful grandmothers. Pair bonding, however imperfect, expanded the role of fathers. Postweaning provisioning and cooperative breeding shortened breast-feeding and birth spacing.

15. These changes, along with the evolution of upright posture, parent-offspring conflict, and, later, language and culture, required or enabled five great innovations in human ontogeny: the overall lengthening of development; the presocial phase of neonatal life, or the "fourth trimester"; the maintenance of prenatal brain growth rates for a year after birth; the emergence of language and its concomitant appreciation of other minds; and the additional, specific lengthening of middle childhood. These innovations, along with key features of the developmental plan that are legacies from our higher-primate ancestors, make up human childhood.

16. Because the life span has been stretched by one-half, but gestation by only one-eighth, pregnancy should be twelve months, not nine. Getting the infant out early may also be an instance of parent-offspring conflict. In response to high neonatal mortality, adaptation called for a slower growth of maternal love. Ontogeny's answer was to postpone smiling and mutual gaze to the third month, probably achieved through the growth of the basal ganglia; the infant's answer to maternal detachment is crying, which peaks at six weeks, a second instance of conflict.

17. Six million years ago bipedal walking evolved, and two million years ago body and brain size began to increase; brain size, which eventually tripled, did so much more quickly. Gestation in mammals is more tied to brain growth than body growth; by brain size standards our gestation should be only a little shorter than the elephant's, but then birth would be impossible. So instead of twenty-one months' gestation, we have a year of postnatal embryonic brain growth; this explains much about human infancy.

18. Despite truncated gestation, birth is much more difficult in us than in other apes, hence birth attendants, a component of cooperative breeding. Another step was postweaning cooperative provisioning, partly by men (with hunted meat), partly by grandmothers (with plant foods and time and energy after menopause), and partly by older children (with the same during postponed puberty). Tools and fire afforded protection from predators and facilitated hunting, plant food extraction, and cooking, but among the first tools were probably infant slings.

19. Infant predation being one of the greatest primate adaptive challenges, intense, bilateral mother-infant attachment arose some 40 million years ago. Stretched human development postpones infant attachment and fear of strangers to late in the first year, a cross-cultural universal due to the growth of the limbic circuitry and the right prefrontal cortex. In hunter-gatherers, grandmothers, other women, fathers, and children help with baby care, but in all hunter-gatherer cultures the mother is primary during attachment formation and weaning is later than age two.

20. The evolution of language facilitated cooperative breeding, post-weaning provisioning, and teaching, on which enculturation depends. It is not an accident that language emerges as lactation wanes in hunter-gatherers, but its development is universal to all cultures, owing to the growth of the auditory projection to the cortex and of the language circuits of (usually) the left temporal, parietal, and frontal lobes. Evolution designed this genetically guided process to deliver the child into a wider social world. Despite our slower lives, a human child is three or four years old when her next sibling is born, not five or six as in other apes, a huge advantage.

21. Females being a scarce resource for reproduction, males compete for them; many are called but fewer are chosen. This results in sex differences in size, aggression, and sexual behavior, realized in part through the masculinization of the brain by fetal androgens. Against a range of species, sex differences are mild to moderate in humans, but they emerge in early childhood and are evident in greater male physical aggression. In hunter-gatherers the behavioral differences are softened by mixed-sex playgroups throughout childhood, but they intensify in cultures with dense populations and segregation or self-segregation by sex.

22. As we evolved, our ape growth trajectory, with little time between weaning and puberty, was interrupted by a time of quiescent physical growth. After acquiring language and perhaps a younger sibling, the child begins a transition to middle childhood. The midgrowth spurt, once the start of puberty, and the initial rise of adrenal androgens, detachable from puberty, come between ages five and seven, as do great changes in reasoning and thinking about minds. This enabled enculturation and moral development and made children helpers at the nest,

another huge gain over apes. The mechanism is brain changes, including hippocampal maturation and peak synapse number in association areas, and possibly adrenal androgen effects.

23. Reduced juvenile mortality allowed us to postpone reproduction, but reproduction remains the goal, and adjustable but genetically guided puberty the means. Four or five years after the five-to-seven shift (six or eight years after it in hunter-gatherers), faster growth resumes. Hypothalamic control of the gonads becomes less sensitive to feedback, and gonadal hormone production rises steeply, transforming body and brain systems, producing sexual behavior, risk taking, and serious aggression. Synaptic pruning and myelination of cortical axons allow the attainment of adult inhibition and reasoning, but not before years of hormonal surges. The relative timing of these two processes is a facultative adaptation responsive to nutrition, disease, and social environment.

24. Adolescent sexuality, romantic involvement, and pair bonding have overlapping but separate networks in the brain, all including limbic circuits and all maturing and changing in this phase of life. Contrary to old views, there may be organizational effects of hormones on the brain at adolescence, and it may be a sensitive period for the effects of hormones and experience. If parenthood occurs, neural and neuroendocrine gene expression facilitate pair bonding and parenting, although they do not guarantee either.

25. Plasticity has evolved many times in many forms and is key to adaptation and evolution. Information input during development is privileged in early life in some species; during sensitive periods it leaves traces more easily and durably. Lifelong facultative adaptations may result—reproductive advantages gained through different evolved developmental paths in different external conditions detected in early life. Others are maladaptive. But the young are resilient, and some damage is less likely to last if it comes early. It is far from clear why the observed specific patterns of vulnerability and resilience have evolved.

26. Socialization—the chance during development to acquire and store information about social life—occurs in all social vertebrates, including those in which the only others encountered by the young are one or both parents. However, socialization is crucial in species with slow development, prolonged parent-offspring relationships, and social groups larger than the family. Larger brains tend to occur in species with larger social groups, provided they are networks of relationships,

not herds. With some exceptions, larger brains are broadly associated with more complex social groups, slower development, more socialization, and more play.

27. In mammals social experience during development has profound effects on the brain and behavior. Such experiences include early separation, stress, and maternal licking and grooming (or lack of it) in rats; postweaning social isolation in mice and rats; the social experience and other enrichment or deprivation even late in life that has an impact on cortical structure and function in rats and monkeys; social isolation in early life in monkeys, apes, and humans; and, with milder effects, early separations in those species. Resilience is also impressive, and some of these impacts are reversible.

28. Old World monkeys, apes, and humans, especially in our environments of evolutionary adaptedness, share features of parental care, including prolonged close contact with mothers, frequent nursing, late weaning, co-sleeping, high indulgence, variable male involvement, playgroups (often multi-aged), and sexual play during development. Human hunter-gatherers have additional features: more nonmaternal care by fathers, grandmothers, siblings, and others; postweaning provisioning, which shortens nursing to two or three years and birth spacing to three or four; and, most important, enculturation.

29. Stress is ubiquitous. Life is stress; evolution is stress. Children cannot and should not avoid it, but they should have more positive stress (*eustress*) than *dis*tress. The normal stress response mobilizes neural and hormonal activations that in turn mobilize energy and oxygen; it is great for a burst of effort in physical work, fight, or flight. But abuse, family conflict, poverty, and war may impose chronic stresses in which the same physiology becomes maladaptive and even damages the brain. We must learn more about resilience and its limits.

30. Enculturation is far more rare in nature than socialization. Culture in the weak sense (transgenerational, stable, dialectally varying, extragenetically transmitted patterns of behavior; customs; protoculture) occurs, at a minimum, in some songbirds, apes, and hominins. Culture in the strong sense (based on symbol systems and intersubjective learning, including teaching) has occurred only in hominins.

31. Cumulative cultural evolution has also occurred only in humans, beginning perhaps as recently as 200,000 years ago. It has accelerated steadily for the last 10,000 years, finally reaching a rate of change that

limits traditional cultural transmission and partly reverses the direction of teaching and learning across generations. However, the latter effect is consistent with the ancient evolutionary tendency for the playful, curious behavior of juveniles to draw themselves and their elders into new ecological niches.

32. As a result of subsistence ecology, cumulative cultural evolution is largely directional. Improved hunting and gathering increased population densities, as did horticulture, the domestication of animals, plow agriculture, irrigation, urbanization, and industrialization. Each had predictable consequences for social organization and child rearing. In the first demographic transition, agriculture shortened birth spacing and increased family size; in the second, family size was reduced by industrialization. Family dynamics changed greatly, and some population pyramids now look like population columns.

33. Accelerating cultural evolution changed socialization, so there are major differences between modern patterns and those in our environments of evolutionary adaptedness. Since much in psychosocial development, including the channels and boundaries of learning, is under genetic guidance, there may now be some discordance between our socialization patterns and children's biological preparedness, analogous to the proven discordance between our diets and activity and what is healthy given the genomes we evolved.

34. Or cultures may have the socialization that produces the adults they need. This is not true of abuse and neglect, and not likely true of all cultural practices. Socialization responds to subsistence ecology; it is designed not just to produce certain kinds of adults but for the convenience of adults, the efficiency of economic activity, and the enhancement of parental reproductive success. These purposes may depart from the best interest of the child. Yet children are resilient and adapt to many circumstances both biological and cultural. Again, we need to know more about resilience and its limits.

35. Socialization, shared with many nonhumans, involves reactive processes (habituation; classical, associative, and instrumental conditioning; and social facilitation), some social learning processes (local enhancement, mimicry, emulation, and some imitation), and some emotional processes (attachment, fear of strangers, and positive and negative identification). Enculturation, unique to humans, involves in addition social learning processes that require full intersubjectivity (advanced

imitation, instruction, and collaborative learning), the psychodynamic process of emotion management through ritual, and all symbolic processes, including the cultural construction of perception, schematization, narrative construction, and cultural coherence.

36. Despite cultural variety, many aspects of behavior in our species are universal. Some universals (for example, extreme xenophobia or strict division of labor by sex) proved to be products of once-universal conditions, not inherent biological tendencies. Others (like smiling in greeting or the statistical sex difference in physical violence) are far more rooted in biology. New methods in quantitative and molecular behavioral genetics and in brain imaging, especially applied to infants and children, will help resolve many controversies.

37. The second law of thermodynamics prescribes entropy, but life, childhood, society, and culture respond by gathering, ordering, and storing information about the environment on one scale or another. The process of generating variation, organizing it, challenging it, and selecting among more or less adaptive self-organized alternatives operates at each level—in evolution, development, socialization, and culture change. Generative variation in interaction with a selective environment is of the essence of life, childhood, society, and culture.

38. Most of that interaction took place during the several billion years leading to the formation of each living genome, so clarifying the role of the genes and of neural and neuroendocrine maturation in human development is central to our understanding of childhood. It is also central to all efforts to intervene on behalf of children, including those efforts that rely on plasticity. Gaining this understanding is the ultimate epigenetic enterprise.

39. Nothing in childhood makes sense except in the light of evolution.

40. Genetic determinism died long ago, and now environmental determinism is dead too. Long live complex, measurable, and mainly deterministic interactionism.

Epilogue

Finishing this book three decades after beginning it, I find myself in a wildlife sanctuary on the Gulf coast of southwest Florida, barefoot on the beach at dusk. White ibis make straight tracks across my wobbly one. A pelican pair passes, one with a fish in its mouth that it drops back into the water, to breed another day. A fisherman about my age lands, with difficulty, a twenty-pound pregnant snook on a ten-pound line. He measures her against his rod—she is almost three feet long—removes the hook, and eases her back into the sea.

Thanks to the steep slope of the beach, I am eye to eye with a great blue heron, slender, regal, a few white wisps of her crown plumage disarrayed by the breeze. I try to close in on her, but she prances a bit to keep her distance—a wise move, and no recent legacy. Finally, just to be sure, she takes to the air, where neither I nor any other possible predator can follow.

The beach is dotted with sea turtle nests, dents in the sand that in some places are just yards apart. They are staked, ribboned, and placarded with statutory warnings. We must block off all our lights at night lest the hatchlings take a wrong and fatal turn—they come prepared to crawl toward moonlight or even starlight glancing off the sea. One night we help the experts release more than sixty we have rescued from raccoons onto the beach at dusk; they must crawl before they swim, and they must imprint on this stretch of coast so that, if they survive, the females can return in thirty years to lay new nests of eggs. Slowly they rouse themselves, take a few wrong turns, get their bearings, and wobble into the surf. Children watch and cheer them on in three languages. Our species, after eons of the struggle for life, of battling nature, of proper exploitation of countless other species for our survival and that of our children, must now reverse that history.

One hundred sixty-four thousand years ago, give or take a few millennia, a group of quite human beings sat around a fire in a cave just above a South African beach. They were probably not the first of their kind—our kind—and certainly not the last. As the moon shone on the sea, they chatted—perhaps someone lapsed into oration—about the events of that day and the next, about this one's health and that one's baby, about this one's hunting blunder and that one's alleged adultery. The pigments they used show us that they had an idea of beauty; the tools they left tell us that they were masters of their world. Probably they had gods; our kind sees agency everywhere, and we do not just try to read other minds—we imagine them. That is about all we know about this group of humans—except that they deeply loved and wisely cared for their children. If they had not, we would not be here.

During the next thousand centuries they occupied many African coastlines and countless inland lakes and waterways. Then some of them—perhaps as few as hundreds—left Africa for Asia and ranged along the southern coasts of that continent, reaching Australia within a few thousand years. As the crab-eating fox left other canines behind in productivity of offspring because the seashore provided a rich resource base, so these humans may have outstripped other hominins by exploiting the same littoral environment—and not just crabs but all manner of fish, sea turtles, and the myriad birds and mammals that rely on the sea for food.

Not only would all this have fueled offspring productivity, but fish would have supplied the fatty acids needed for those children's exceptionally large and rapidly growing brains. Probably only modest modifications in subsistence technology would have sufficed to move them from one stretch of beach to the next, and the next. Whom or what they displaced, other than crabs and crab-eating foxes, as they covered those thousands of miles in as many years, we do not know, but what a journey it must have been.

Did children play a role? I suspect so. I think of the Meriam children of the Torres Strait islands, casually gathering clams and conchs, feeding themselves while playing. Children no doubt collected clams and conchs—and crabs too—all the way from Africa to those very islands.

Could *their* idle wanderings have sometimes led the way? Perhaps. But their parents were the cleverest and most dangerous predators those coastlines had ever seen, and their most advanced methods had to be taught. It was this ability as much as toolmaking, language, helpers at the nest, and postweaning feeding that made them and their children what they were.

But do not underestimate what the children were doing on their own. In my lifetime adolescents and even children have led the way in cultural change. Sexual revolution, popular music and films that swept the world, antiwar and integration politics, fashion, electronic media, and one transformation after another in the social deployment of communications technology—these and many more are the ways in which the young have taught us as we have taught them. Their huge, fast-growing, thoroughly dynamic brains transform them—why not us as well? And so it may have been for scores of thousands of years.

All three of my children love nature; in this they take after their mother and me. Or do they? Biologically, each is a marvelous but unexpected reshuffling of our genes; fortunately, none is quite like me. Culturally, they are amalgams of many influences—peers, teachers, television, music, and that cresting wave they were among the first to catch: the World Wide Web. When I assert my beliefs, they often find them foreign.

You hold your child in your arms for the first time and you imagine a world of influence that you do not really have. The sleek intertwining of heredity and environment, by which their mother and I were going to produce three, by our lights, almost perfect people yielded instead three marvelous works-in-progress, approaching perfection by their own lights. I will never know quite why they turned out the way they did. Life is not a controlled experiment.

Things happen. When their mother died thirteen years ago, our eldest was four days past her eighteenth birthday, our son was fourteen, and our younger daughter was nine. Their resilience during their mother's long illness and for the first year or two after her death was remarkable. Since then, there have been ups and downs that have made me realize that resilience has its limits, and yet they are still moving, growing, leaning into life.

From the day my first child was born I forgave my parents everything. How could I have known how very hard it was? At least I could hear. My parents were deaf, poor high school dropouts—but they had grit, grace, common sense, and dreams. Still, the fact that my brother and I grew to maturity, succeeded, and raised children of our own now seems almost miraculous.

And yet it is a miracle that happens every day. We are here because our ancestors were able to meet the challenge, and because they could, we can. Parenthood may seem at times a comedy of errors and kids a bunch of wacky creatures on the precipice, but there are underlying, ancient, biological forces that guide their stumbling, as well as ours.

Being human, I have other ancestors, other heirs. I once knew a descendant of Darwin's. She was well bred and well read, and we had much to discuss in English literature and philosophy—but not her great-great-grandfather's work, which she had not gotten around to. I teach and write about his ideas, passing them on to hundreds every year. Who is his real descendant? We both are.

Our children bring us the legacy of eons of evolution. *There is grandeur in this view of life,* and in this view of childhood. We do put our stamp on our children—the stamp of 100,000 years of human culture and centuries or millennia of particular traditions. But we must share our children with many others. In consolation, we may confer some of our likes, dislikes, skills, ideas, passions, and values on the genetic offspring of other parents. No other species can really make this claim.

Children, it is said, are messages we send into a future we cannot visit, even in our dreams. But we do visit the future in our dreams, often if we are lucky, and we see our children in it. By the strangely mingled and refracted light of both biological and cultural evolution, through the sometimes unwilling auspices of the young, we ourselves may join them in that future, if we tread lightly enough beside them on their way.

References

Acknowledgments

Index

References

Abad, F., N. M. Diaz-Gomez, et al. 2001. "Oral sucrose compares favourably with lidocaine-prilocaine cream for pain relief during venepuncture in neonates." *Acta Paediatrica* 90(2): 160–65.

Abbott, D. H., E. B. Keverne, et al. 2003. "Are subordinates always stressed? A comparative analysis of rank differences in cortisol levels among primates." *Hormones and Behavior* 43(1): 67–82.

Abe, K., N. Oda, et al. 1984. "Behavioural genetics of early childhood: Fears, restlessness, motion sickness and enuresis." *Acta Geneticae Medicae et Gemellologiae* 33(2): 303–6.

Aberle, D. F. 1967. The psychosocial analysis of a Hopi life-history. In *Personalities and cultures: Readings in psychological anthropology,* ed. R. Hunt. Garden City, Natural History Press: 79–138.

Abu-Lughod, L. 1990. The romance of resistance: Tracing transformations of power through Bedouin women. In *Beyond the second sex: New directions in the anthropology of gender,* ed. P. R. Sanday and R. G. Goodenough. Philadelphia, Univ. of Pennsylvania Press: 313–37.

Adamec, R. E. 1990. "Role of the amygdala and medial hypothalamus in spontaneous feline aggression and defense." *Aggressive Behavior* 16: 207–22.

———. 1991. Anxious personality in the cat. In *Psychopathology and the brain,* ed. B. J. Carroll and J. E. Barrett. New York, Raven.

Adams Hillard, P. J. 2006. "Adolescent menstrual health." *Pediatric Endocrinology Reviews* 3(Suppl. 1): 138–45.

Ader, R. 1962. "Social factors affecting emotionality and resistance to disease in animals: III. Early weaning and susceptibility to gastric ulcers in the rat. A control for nutritional factors." *Journal of Comparative Physiological Psychology* 55(4): 600–2.

Adolph, E. F. 1968. *Onset of physiological regulations.* New York, Academic.

Adolph, K. E., B. Vereijken, et al. 2003. "What changes in infant walking and why." *Child Development* 74(2): 475–97.

Adriaens, P. R., and A. De Block. 2006. "The evolution of a social construction: The case of male homosexuality." *Perspectives in Biology and Medicine* 49(4): 570–85.

Adrian, O., I. Brockmann, et al. 2005. "Paternal behaviour in wild guinea pigs: A comparative study in three closely related species with different social and mating systems." *Journal of Zoology (London)* 265(Part 1): 97–105.

Agland, P. 1990. Baka: People of the forest. United States, National Geographic/Questar Video.

Ägren, G., and B. J. Meyerson. 1978. "Long-term effects of social deprivation during early adulthood in the Mongolian gerbil *(Meriones unguiculatus).*" *Zeitschrift für Tierpsychologie* 47: 422–31.

Ahlgren, S., P. Vogt, et al. 2003. "Excess FoxG1 causes overgrowth of the neural tube." *Journal of Neurobiology* 57(3): 337–49.

Ahnert, L., M. Pinquart, et al. 2006. "Security of children's relationships with nonparental care providers: A meta-analysis." *Child Development* 77(3): 664–79.

Ahrens, R. 1954. "Beitrag zur Entwicklung des Physiognomie und Mimmerkennens." *Zeitschrift für Experimentelle und Angewandte Psychologie* 2: 412–54, 599–633.

Aiello, L. C., and P. Wheeler. 1995. "The expensive-tissue hypothesis: The brain and the digestive system in human and primate evolution." *Current Anthropology* 36: 199–221.

Ainsworth, M. D. S. 1967. *Infancy in Uganda: Infant care and the growth of attachment.* Baltimore, Johns Hopkins Univ. Press.

Ainsworth, M. D. S., S. Bell, et al. 1974. Infant-mother attachment and social development: Socialisation as a product of reciprocal responsiveness to signals. In *The integration of the child into a social world,* ed. M. P. M. Richards. Cambridge, Cambridge Univ. Press.

Ainsworth, M. D. S., M. C. Blehar, et al., eds. 1978. *Patterns of attachment: A psychological study of the strange situation.* Hillsdale, NJ, Erlbaum.

Ainsworth, M. S. 1979. "Infant-mother attachment." *American Psychologist* 34(10): 932–37.

Akhtar, N., M. Carpenter, et al. 1996. "The role of discourse novelty in early word learning." *Child Development* 67: 635–45.

Akhtar, N., and M. A. Gernsbacher. 2008. "On privileging the role of gaze in infant social cognition." *Child Development Perspectives* 2(2): 59–65.

Akiskal, H. S., and W. T. McKinney, Jr. 1973. "Depressive disorders: Toward a unified hypothesis." *Science* 182: 20–29.

Alberch, P., and M. J. Blanco. 1996. "Evolutionary patterns in ontogenetic transformation: From laws to regularities." *International Journal of Developmental Biology* 40(4): 845–58.

Alberts, J. R., and D. J. Gubernick. 1990. Functional organization of dyadic and triadic parent-offspring systems. In *Mammalian parenting,* ed. N. A. Krasnegor and R. S. Bridges. New York, Oxford Univ. Press: 416–40.

Alcock, J. 2001. *The triumph of sociobiology.* New York, Oxford Univ. Press.

Alcorta, C. 1982. "Paternal behavior and group competition." *Behavior Science Research* 17: 3–23.

Aldrich, C. A., and E. S. Hewitt. 1947. "A self-regulating feeding program for infants." *Journal of the American Medical Association* 135: 340–42.

Aldridge, M. A., E. S. Braga, et al. 1999. "The intermodal representation of speech in newborns." *Developmental Science* 2(1): 42–46.

Alemseged, Z., F. Spoor, et al. 2006. "A juvenile early hominin skeleton from Dikika, Ethiopia." *Nature* 443(7109): 296–301.

Alexander, G. E. 1994. "Basal ganglia-thalamocortical circuits: Their role in control of movements." *Journal of Clinical Neurophysiology* 11(4): 420–31.

Alexander, G. E., M. D. Crutcher, et al. 1990. "Basal ganglia-thalamocortical circuits: Parallel substrates for motor, oculomotor, 'prefrontal' and 'limbic' functions." *Progress in Brain Research* 85: 119–46.

Alexander, R. 1979. *Darwinism and human affairs.* Seattle, Univ. of Washington Press.

Alkon, A., S. Lippert, et al. 2006. "The ontogeny of autonomic measures in 6- and 12-month-old infants." *Developmental Psychobiology* 48(3): 197–208.

Allen, L. S., M. Hines, et al. 2007. Two sexually dimorphic cell groups in the human brain. In *Sex and the brain*, ed. G. Einstein. Cambridge, MA, MIT Press: 327–37.

Allen, L. S., M. F. Richey, et al. 1991. "Sex differences in the corpus callosum of the living human being." *Journal of Neuroscience* 11: 933–42.

Allman, J. 1999. *Evolving brains*. New York, W. H. Freeman.

Allman, J., A. Rosin, et al. 1998. "Parenting and survival in anthropoid primates: Caretakers live longer." *Proceedings of the National Academy of Sciences* 95(12): 6866–69.

Allman, J. M., T. McLaughlin, et al. 1993. "Brain structures and life-span in primate species." *Proceedings of the National Academy of Sciences* 90: 3559–63.

Almagor, U. 1989. Dual organization reconsidered. In *The attraction of opposites: Thought and society in the dualistic mode*, ed. D. Maybury-Lewis and U. Almagor. Ann Arbor, Univ. of Michigan Press: 19–32.

Almli, C. R., M. J. Rivkin, et al. 2007. "The NIH MRI study of normal brain development (Objective-2): Newborns, infants, toddlers, and preschoolers." *Neuroimage* 35(1): 308–25.

Als, H. 1977. "The newborn communicates." *Journal of Communication* 27: 66–73.

Altmann, J. 1979. *Baboon mothers and infants*. Cambridge, MA, Harvard Univ. Press.

Altmann, J., G. Hausfater, et al. 1988. Determinants of reproductive success in Savannah baboons, *Papio cynocephalus*. In *Reproductive success: Studies of individual variation in contrasting breeding systems*, ed. T. H. Clutton-Brock. Chicago, Univ. of Chicago Press: 403–18.

Altmann, M. 1958. "Social integration of the moose calf." *Animal Behaviour* 6: 155–59.

Altmann, S. 1998. *Foraging for survival: Yearling baboons in Africa*. Chicago, Univ. of Chicago Press.

Amaral, P. P., M. E. Dinger, et al. 2008. "The eukaryotic genome as an RNA machine." *Science* 319(5871): 1787–89.

Ambalavanar, R., B. J. McCabe, et al. 1999. "Learning-related fos-like immunoreactivity in the chick brain: Time-course and co-localization with GABA and parvalbumin." *Neuroscience* 93(4): 1515–24.

Ambrose, J. A. 1959. The development of the smiling response in early infancy. In *Determinants of infant behavior I*, ed. B. M. Foss. London, Methuen.

American Academy of Pediatrics. 1978. "Breast-feeding: A commentary in celebration of the International Year of the Child, 1979." *Pediatrics* 62: 591–601.

American Academy of Pediatrics Committee on Adolescence, American College of Obstetricians and Gynecologists Committee on Adolescent Health Care, et al. 2006. "Menstruation in girls and adolescents: Using the menstrual cycle as a vital sign." *Pediatrics* 118(5): 2245–50.

American Academy of Pediatrics Committee on Adolescence, M. J. Blythe, et al. 2007. "Contraception and adolescents." *Pediatrics* 120(5): 1135–48.

Amores, A., A. Force, et al. 1998. "Zebrafish *hox* clusters and vertebrate genome evolution." *Science* 282: 1711–14.

Anastasi, A. 1958. "Heredity, environment, and the question "How?" *Psychological Review* 65: 197–208.

Anderson, C. M. 1989. "Neandertal pelves and gestation length: Hypotheses and holism in paleoanthropology." *American Anthropologist* 91(2): 327–40.

Anderson, P. W. 1972. "More is different: Broken symmetry and the nature of the hierarchical structure of science." *Science* 177(4047): 393–96.

Andersson, K., G. Bohlin, et al. 1999. "Early temperament and stranger wariness as predictors of social inhibition in 2-year-olds." *British Journal of Developmental Psychology* 17(3): 421–34.

Andresen, P. A., and S. L. Telleen. 1992. "The relationship between social support and maternal behaviors and attitudes: A meta-analytic review." *American Journal of Community Psychology* 20(6): 753–74.

André-Thomas, Y. Chesni, et al., eds. 1960. *The neurological examination of the infant.* London, Heinemann.

Andrew, R. J. 1963. "Evolution of facial expressions." *Science* 142(3595): 1034–41.

Andrews, P., and J. Kelley. 2007. Middle Miocene dispersals of apes. *Folia Primatologica* 78(5–6): 328–43.

Anemone, R. L. 2002. Dental development and life history in hominid evolution. In *Human evolution through developmental change*, ed. N. Minugh-Purvis and K. J. McNamara. Baltimore, Johns Hopkins Univ. Press: 249–80.

Anemone, R. L., M. P. Mooney, et al. 1996. "Longitudinal study of dental development in chimpanzees of known chronological age: Implications for understanding the age at death of Plio-Pleistocene hominids." *American Journal of Physical Anthropology* 99(1): 119–33.

Angelini, D. R., and T. C. Kaufman. 2005. "Comparative developmental genetics and the evolution of arthropod body plans." *Annual Review of Genetics* 39: 95–119.

Angier, N. 1997. Sexual identity not pliable after all, report says. *New York Times,* Mar. 14, 1997, 1.

———. 2008. About death, just like us or pretty much unaware? *New York Times,* Sept. 2, 2008, D1.

Angold, A., E. J. Costello, et al. 1999. "Pubertal changes in hormone levels and depression in girls." *Psychological Medicine* 29(5): 1043–53.

Angold, A., and C. W. Worthman. 1993. "Puberty onset of gender differences in rates of depression: A developmental, epidemiologic and neuroendocrine perspective." *Journal of Affective Disorders* 29(2–3): 145–58.

Angulo y Gonzalez, A. W. 1929. "Is myelinogeny an absolute index of behavioral capability?" *Journal of Comparative Neurology* 48: 459–64.

Annick Ledebt, B. B. 2000. "Acquisition of upper body stability during walking in toddlers." *Developmental Psychobiology* 36(4): 311–24.

Anonymous. 1951. "Letter from Freud." *American Journal of Psychiatry* 107(10): 786–87: http://www.lettersofnote.com/2009/10/homosexuality-is-nothing-to-be-ashamed.html (as of Dec. 4, 2009).

Arborelius, L., M. J. Owens, et al. 1999. "The role of corticotropin-releasing factor in depression and anxiety disorders." *Journal of Endocrinology* 160(1): 1–12.

Arcadi, A. C., and R. W. Wrangham. 1999. "Infanticide in chimpanzees: Review of cases and a new within-group observation from the Kanyawara study group in Kibale National Park." *Primates* 40(2): 337–57.

Archer, J. 1999. *The nature of grief: The evolution and psychology of reactions to loss.* New York, Routledge.

———. 2000. "Sex differences in aggression between heterosexual partners: A meta-analytic review." *Psychological Bulletin* 126(5): 651–80.

———. 2004. "Sex differences in aggression in real-world settings: A meta-analytic review." *Review of General Psychology* 8(4): 291–322.

Ardrey, R. 1961. *African genesis.* New York, Atheneum.

Ariëns Kappers, C. U., G. C. Huber, et al. 1936. *The comparative anatomy of the nervous system of vertebrates, including man.* New York, Macmillan.

Ariès, P. 1962. *Centuries of childhood.* New York, Vintage.

Arling, G. L., and H. F. Harlow. 1967. "Effects of social deprivation on maternal behavior of Rhesus monkeys." *Journal of Comparative and Physiological Psychology* 64(3): 371–77.

Arlt, W., F. Callies, et al. 2000. "DHEA replacement in women with adrenal insufficiency—pharmacokinetics, bioconversion and clinical effects on well-being, sexuality and cognition." *Endocrine Research* 26(4): 505–11.

Armstrong, E. 1983. "Relative brain size and metabolism in mammals." *Science* 220(4603): 1302–4.

———. 1985. "Allometric considerations of the adult mammalian brain, with special emphasis on primates." In *Size and scaling in primate biology,* ed. W. L. Jungers. New York, Plenum: 115–46.

Aron, A., H. E. Fisher, et al. 2005. "Reward, motivation, and emotion systems associated with early-stage intense romantic love." *Journal of Neurophysiology* 160(1): 1–12.

Aronov, D., A. S. Andalman, et al. 2008. "A specialized forebrain circuit for vocal babbling in the juvenile songbird." *Science* 320(5876): 630–34.

Asbury, K., J. F. Dunn, et al. 2006. "Birthweight-discordance and differences in early parenting relate to monozygotic twin differences in behaviour problems and academic achievement at age 7." *Developmental Science* 9(2): F22–31.

Asch, S. E. 1955. "Opinions and social pressure." *Scientific American* 193(5): 31–35.

———. 1956. "Studies of independence and conformity: I. A minority of one against a unanimous majority." *Psychological Monographs* 70(9): 70.

Asendorpf, J. B., J. J. A. Denissen, et al. 2008. "Inhibited and aggressive preschool children at 23 years of age: Personality and social transitions into adulthood." *Developmental Psychology* 44(4): 997–1011.

Astington, J. W. 2004. "Bridging the gap between theory of mind and moral reasoning." *New Directions for Child and Adolescent Development* (103): 63–72.

Aubrey, J. S., and K. Harrison. 2004. "The gender-role content of children's favorite television programs and its links to their gender-related perceptions." *Media Psychology* 6(2): 111–46.

Avis, J., and P. L. Harris. 1991. "Belief-desire reasoning among Baka children: Evidence for a universal conception of mind." *Child Development* 62(3): 460–67.

Axelrod, R. 1984. *The evolution of cooperation.* New York, Basic Books.

Axelrod, R., and W. D. Hamilton. 1981. "The evolution of cooperation." *Science* 211: 1390–96.

Axia, V. D., and T. S. Weisner. 2002. "Infant stress reactivity and home cultural ecology of Italian infants and families." *Infant Behavior and Development* 25(3): 255–68.

Ayala, F. J. 1995. "The myth of Eve: Molecular biology and human origins." *Science* 270: 1930–36.

Ayala, F. J., and A. A. Escalante. 1996. "The evolution of human populations: A molecular perspective." *Molecular Phylogenetics and Evolution* 5(1): 188–201.

Ayatollahi, S. M. T., S. Pourahmad, et al. 2006. "Trend in physical growth among children in southern Iran, 1988–2003." *Annals of Human Biology* 33(4): 510–14.

Ayoub, D. M., W. T. Greenough, et al. 1983. "Sex differences in dendritic structure in the preoptic area of the juvenile macaque monkey brain." *Science* 219: 197–98.

Ayres, B. 1974. "Bride theft and raiding for wives in cross-cultural perspective." *Anthropological Quarterly* 47: 238–52.

Bacon, M. K., I. L. Child, et al. 1963. "A cross-cultural study of correlates of crime." *Journal of Abnormal and Social Psychology* 66: 241–300.

Badcock, C. 1990. *Oedipus in evolution: A new theory of sex.* Cambridge, Blackwell.

Badyaev, A. V. 2009. "Evolutionary significance of phenotypic accommodation in novel environments: An empirical test of the Baldwin effect." *Philosophical Transactions of the Royal Society of London B: Biological Sciences* 364(1520): 1125–41.

Bagby, R. M., P. T. Costa, Jr., et al. 1999. "Replicating the five factor model of personality in a psychiatric sample." *Personality and Individual Differences* 27(6): 1135–39.

Baglione, V., D. Canestrari, et al. 2003. "Kin selection in cooperative alliances of carrion crows [see comment]." *Science* 300(5627): 1947–49.

Bahuchet, S. 1999. Aka Pygmies. In *The Cambridge encyclopedia of hunters and gatherers,* ed. R. B. Lee and R. Daly. Cambridge, Cambridge Univ. Press: 190.

Bailey, D. B., M. R. Burchinal, et al. 1993. "Age of peers and early childhood development." *Child Development* 64(3): 848–62.

Bailey, D. B., R. A. McWilliam, et al. 1993. "Social interactions of toddlers and preschoolers in same-age and mixed-age play groups." *Journal of Applied Developmental Psychology* 14(2): 261–76.

Bailey, J. M. 2009. "What is sexual orientation and do women have one?" *Nebraska Symposium on Motivation* 54: 43–63.

Bailey, R. C. 1991. *The behavioral ecology of Efe Pygmy men in the Ituri Forest, Zaire.* Ann Arbor, Museum of Anthropology, Univ. of Michigan.

Baillargeon, R. H., M. Zoccolillo, et al. 2007. "Gender differences in physical aggression: A prospective population-based survey of children before and after 2 years of age." *Developmental Psychology* 43(1): 13–26.

Baird, A. A., and J. A. Fugelsang. 2004. "The emergence of consequential thought: Evidence from neuroscience." *Philosophical Transactions of the Royal Society of London B: Biological Sciences* 359(1451): 1797–1804.

Bakeman, R., L. Adamson, et al. 1990. "The social context of object exploration." *Child Development* 61: 794–809.

Baker, P. J., C. P. Robertson, et al. 1998. "Potential fitness benefits of group living in the red fox, Vulpes vulpes." *Animal Behaviour* 56(6): 1411–24.

Baldinotti, F., S. Majore, et al. 2008. "Molecular characterization of 6 unrelated Italian patients with 5alpha-reductase type 2 deficiency." *Journal of Andrology,* 29(1), 20–28.

Baldwin, J. M. 1896. "A new factor in evolution." *American Naturalist* 30: 441–553.

Baldwin, J. P., and J. I. Baldwin. 1974. "Exploration and social play in squirrel monkeys (Saimiri)." *American Zoologist* 14: 303.

Bales, K., J. Dietz, et al. 2000. "Effects of allocare-givers on fitness of infants and parents in callitrichid primates." *Folia Primatologica* 71(1–2): 27–38.

Bales, K. L., L. A. Pfeifer, et al. 2004. "Sex differences and developmental effects of manipulations of oxytocin on alloparenting and anxiety in prairie voles." *Developmental Psychobiology* 44(2): 123–31.

Balikci, A. 1970. *The Netsilik eskimo.* Garden City, Natural History Press.

Ball, G. F., and S. A. MacDougall-Shackleton. 2001. "Sex differences in songbirds 25 years later: What have we learned and where do we go?" *Microscopy Research and Technique* 54(6): 327–34.

Balter, M. 1999. "Restorers reveal 28,000-year-old artworks." *Science* 283: 1835.

Balzeau, A., D. Grimaud-Hervé, et al. 2005. "Internal cranial features of the Mojok-
erto child fossil (East Java, Indonesia)." *Journal of Human Evolution* 48(6): 535–53.

Bandura, A., D. Ross, et al. 1961. "Transmission of aggression through imitation of aggressive models." *Journal of Abnormal and Social Psychology* 63: 575–82.

Bandura, A., and R. H. Walters. 1963. *Social learning and personality development.* New
York, Holt, Rinehart and Winston.

Barash, D. P. 1975. "Ecology of parental behavior in the hoary marmot (*Marmota caligata*): An evolutionary interpretation." *Journal of Mammalogy* 56: 613–18.

Barber, N. 1991. "Play and energy regulation in mammals." *Quarterly Review of Biology*
66(2): 129–47.

Barbu-Roth, M., D. I. Anderson, et al. 2009. "Neonatal stepping in relation to terrestrial optic flow." *Child Development* 80(1): 8–14.

Bard, K. A. 2000. Crying in infant primates: Insights into the development of crying
in chimpanzees. In *Crying as a sign, a symptom, and a signal: Clinical emotional and
developmental aspects of infant and toddler crying. Clinics in Developmental Medicine
152*, ed. R. G. Barr and B. Hopkins. New York, Cambridge Univ. Press: 157–75.

———. 2002. "Should developmental psychologists imitate comparative psychologists?" *Developmental Science* 5(1): 14–15.

Barenbaum, J., V. Ruchkin, et al. 2004. "The psychosocial aspects of children exposed
to war: Practice and policy initiatives." *Journal of Child Psychology and Psychiatry
and Allied Disciplines* 45(1): 41–62.

Bar-Haim, Y., D. B. Sutton, et al. 2000. "Stability and change of attachment at 14, 24,
and 58 months of age: Behavior, representation, and life events." *Journal of Child
Psychology and Psychiatry and Allied Disciplines* 41(3): 381–88.

Barker, D. J. P. 2004. "The developmental origins of adult disease." *Journal of the American College of Nutrition,* 23(Suppl. 6): 588S–95S.

Barker, D. J. P., C. Osmond, et al. 1989. "Weight in infancy and death from ischaemic
heart disease." *Lancet* (Sept. 9): 577–80.

Barkovich, A. J. 2000. "Concepts of myelin and myelination in neuroradiology." *AJNR:
American Journal of Neuroradiology* 21(6): 1099–1109.

Barkow, J. H., L. Cosmides, et al., eds. 1992. *The adapted mind: Evolutionary psychology
and the generation of culture.* Oxford, Oxford Univ. Press.

Barlett, P. F. 1993. *American dreams, rural realities: Family farms in crisis.* Chapel Hill,
Univ. of North Carolina Press.

Barlow, K. 2001. "Working mothers and the work of culture in a Papua New Guinea
society." *Ethos* 29(1): 78–107.

Barnett, S. A. 1968. "The 'instinct to teach.'" *Nature* 220(5169): 747–49.

———. 1975. *The rat: A study in behavior.* Chicago, Univ. of Chicago Press.

Barr, C. S., T. K. Newman, et al. 2004. "Rearing condition and rh5-HTTLPR interact to
influence limbic-hypothalamic-pituitary-adrenal axis response to stress in infant
macaques." *Biological Psychiatry* 55(7): 733–38.

Barr, R. G. 1990a. "The early crying paradox: A modest proposal." *Human Nature* 1(4):
355–89.

———. 1990b. "The normal crying curve: What do we know?" *Developmental Medicine and Child Neurology* 32: 368–74.

Barr, R. G., S. Chen, et al. 1996. "Crying patterns in preterm infants." *Developmental
Medicine and Child Neurology* 38(4): 345–55.

Barr, R. G., M. Konner, et al. 1991. "Crying in !Kung San infants: A test of the cultural
specificity hypothesis." *Developmental Medicine and Child Neurology* 33: 601–10.

Barr, R. G., R. B. Trent, et al. 2006. "Age-related incidence curve of hospitalized Shaken Baby Syndrome cases: Convergent evidence for crying as a trigger to shaking." *Child Abuse and Neglect* 30(1): 7–16.

Barrett, R. 2008. *Aghor medicine: Pollution, death, and healing in Northern India.* Berkeley, Univ. of California Press.

Barry, H., M. K. Bacon, et al. 1957. "A cross-cultural survey of some sex differences in socialization." *Journal of Abnormal and Social Psychology* 55: 327–32.

Barry, H., III, M. K. Bacon, et al. 1967. Definitions, ratings and bibliographic sources for child training practices of 110 cultures. *Cross cultural approaches.* C. S. Ford. New Haven, HRAF Press.

Barry, H. I., L. Child, et al. 1959. "Relation of child training to subsistence economy." *American Anthropologist* 61: 51–63.

Barry, H. I., and L. M. Paxson. 1971. "Infancy and early childhood: Cross-cultural codes 2." *Ethnology* 10: 466–508.

Bartels, A., and S. Zeki. 2000. "The neural basis of romantic love." *Neuroreport* 11(17): http://journals.lww .com/neuroreport/Fulltext/2000/11270/The_neural_basis _of_romantic_love.46.aspx.

———. 2004. "The neural correlates of maternal and romantic love." *Neuroimage* 21(3): 1155–66.

Bartholomew, G. 1959. "Mother-young relations and the maturation of pup behavior in the Alaska fur seal." *Animal Behaviour* 7: 163–71.

Bartlett, E. E. 2004. "The effects of fatherhood on the health of men: A review of the literature." *Journal of Men's Health and Gender* 1(2–3): 159–69.

Barton, R. A. 1998. "Visual specialization and brain evolution in primates." *Proceedings of the Royal Society of London B: Biological Sciences* 26(1409): 1933–37.

Bateman, A. J. 1948. "Intrasexual selection in Drosophila." *Heredity* 2: 349–68.

Bates, E., and V. A. Marchman. 1987. What is and is not universal in language acquisition. In *Language, communication, and the brain,* ed. F. Plum. New York, Raven: 19–38.

Bates, E. A. 2004. "Explaining and interpreting deficits in language development across clinical groups: Where do we go from here?" *Brain and Language* 88(2): 248–53.

Bates, E. A., and J. L. Elman. 1993. Connectionism and the study of change. In *Brain development and cognition: A reader,* ed. M. H. Johnson. Oxford, Blackwell: 420–40.

Bates, G. 2003. "Huntingtin aggregation and toxicity in Huntington's disease." *Lancet* 361(9369): 1642–44.

Bateson, P. 1966. "The characteristics and context of imprinting." *Biological Review* 41: 177–220.

———. 1983. "Sensitive periods in behavioural development." *Archives of Disease in Childhood* 58: 85–86.

———. 1987. Imprinting as a process of competitive exclusion. In *Imprinting and cortical plasticity: Comparative aspects of sensitive periods,* ed. J. P. Rauschecker and P. Marler. New York, Wiley.

———. 1991. Are there principles of behavioural development? In *The development and integration of behaviour: Essays in honor of Robert Hinde,* ed. P. Bateson. Cambridge, Cambridge Univ. Press: 19–39.

———. 2003. "The promise of behavioural biology." *Animal Behaviour* 65(1): 11–17.

Bateson, P., and G. Horn. 1994. "Imprinting and recognition memory: A neural net model." *Animal Behaviour* 48(3): 695–715.

Battaglia, M., A. Ogliari, et al. 2005. "Influence of the serotonin transporter promoter gene and shyness on children's cerebral responses to facial expressions." *Archives of General Psychiatry* 62(1): 85–94.

Bauer, E. B., and B. B. Smuts. 2007. "Cooperation and competition during dyadic play in domestic dogs, *Canis familiaris.*" *Animal Behaviour* 73(Part 3): 489–99.

Baumrind, D. 1993. "The average expectable environment is not enough: A response to Scarr." *Child Development* 64: 1299–1317.

Baxter, L. R., Jr. 2003. "Basal ganglia systems in ritualistic social displays: Reptiles and humans; function and illness." *Physiology and Behavior* 79(3): 451–60.

Bayer, C. P., F. Klasen, et al. 2007. "Association of trauma and PTSD symptoms with openness to reconciliation and feelings of revenge among former Ugandan and Congolese child soldiers." *Journal of the American Medical Association* 298(5): 555–59.

Bayley, N. 1965. "Comparisons of mental and motor test scores for ages 1–15 months by sex, birth order, race, geographical location and education of parents." *Child Development* 36: 379–411.

———. 1969. *Manual for the Bayley Scales of infant development.* New York, Psychological Corporation.

Beach, F. A. 1950. "The snark was a boojum." *American Psychologist* 5: 115–24.

Becker, J. B., K. J. Berkley, et al., eds. 2008. *Sex differences in the brain: From genes to behavior.* New York, Oxford Univ. Press.

Beintema, D. J. 1968. *A neurological study of newborn infants.* Lavenham, Eng., Spastics International Medical Publications.

Bekoff, M. 1974. "Social play and play-soliciting by infant canids." *American Zoologist* 14: 323.

Bekoff, M., and J. A. Byers, eds. 1998. *Animal play: Evolutionary, comparative, and ecological perspectives.* Cambridge, Cambridge Univ. Press.

Belizan, J. M., F. Althabe, et al. 2007. "Health consequences of the increasing caesarean section rates." *Epidemiology* 18(4): 485–86.

Bel'kovich, V. M., E. E. Ivanova, et al. 1991. Dolphin play behavior in the open sea. In *Dolphin societies: Discoveries and puzzles,* ed. K. Pryor and K. S. Norris. Berkeley, Univ. of California Press: 67–77.

Bell, A. P., and M. S. Weinberg. 1978. *Homosexualities: A study of diversity among men and women.* New York, Simon and Schuster.

Bell, A. P., M. S. Weinberg, et al. 1981. *Sexual preference: Its development in men and women.* Bloomington, Indiana Univ. Press.

Bell, M. A., and N. A. Fox. 1994. Brain development over the first year of life: Relations between electroencephalographic frequency and coherence and cognitive and affective behaviors. In *Human behavior and the developing brain,* ed. G. Dawson and K. W. Fischer. New York, Guilford: 314–45.

Bell, R., and N. S. Costello. 1964. "Three tests for sex differences in tactile sensitivity in the newborn." *Biologia Neonatorum (Basel)* 7: 335–47.

Bell, R., G. M. Weller, et al. 1971. "Newborn and preschooler: Organization of behavior and relations between periods." *Monographs of the Society for Research in Child Development* 36: 1–145.

Bell, S. M., and M. D. S. Ainsworth. 1972. "Infant crying and maternal responsiveness." *Child Development* 43: 1171–90.

Bellugi, U., S. Marks, et al. 1993. Dissociation between language and cognitive functions in Williams syndrome. In *Language development in exceptional circumstances,* ed. D. Bishop and K. Mogford. Hove, Eng., Erlbaum: 177–89.

Belsky, J. 1999. Interactional and contextual determinants of attachment security. In

Handbook of attachment: Theory, research, and clinical applications, ed. J. S. P. R. Cassidy. New York, Guilford: 249–64.

———. 2001. "Emanuel Miller lecture developmental risks (still) associated with early child care." *Journal of Child Psychology and Psychiatry and Allied Disciplines* 42(7): 845–59.

———. 2002. "Quantity counts: Amount of child care and children's socioemotional development." *Journal of Developmental and Behavioral Pediatrics* 23(3): 167–70.

Belsky, J., M. J. Bakermans-Kranenburg, et al. 2007. "For better and for worse: Differential susceptibility to environmental influences." *Current Directions in Psychological Science* 16(6): 300–4.

Belsky, J., and R. M. P. Fearon. 2008. Precursors of attachment security. In *Handbook of attachment,* 2nd ed., ed. J. Cassidy and P. R. Shaver. New York, Guilford: 295–316.

Belsky, J., L. Steinberg, et al. 1991. "Childhood experience, interpersonal development, and reproductive strategy: An evolutionary theory of socialization." *Child Development* 62: 647–70.

Belsky, J., L. D. Steinberg, et al. 2007. "Family rearing antecedents of pubertal timing." *Child Development* 78(4): 1302–21.

Ben Shaul, D. M. 1962. "The composition of the milk of wild animals." *International Zoological Year Book* 4: 333–42.

Benedict, R. 1934. *Patterns of culture.* Boston, Houghton Mifflin.

———. 1946. *The chrysanthemum and the sword: Patterns of Japanese culture.* Boston, Houghton Mifflin.

Benes, F. M. 1998. "Brain development, VII. Human brain growth spans decades." *American Journal of Psychiatry* 155(11): 1489.

Benes, F. M., M. Turtle, et al. 1994. "Myelination of a key relay zone in the hippocampal formation occurs in the human brain during childhood, adolescence, and adulthood." *Archives of General Psychiatry* 51: 477–84.

Bengtsson, S. L., Z. Nagy, et al. 2005. "Extensive piano practicing has regionally specific effects on white matter development." *Nature Neuroscience* 8(9): 1148–50.

Benjamin, J., L. Li, et al. 1996. "Population and familial association between the D4 dopamine receptor gene and measures of novelty seeking." *Nature Genetics* 12: 81–84.

Bennett, A. J., K. P. Lesch, et al. 2002. "Early experience and serotonin transporter gene variation interact to influence primate CNS function." *Molecular Psychiatry* 7(1): 118–22.

Bennett, E. L., M. C. Diamond, et al. 1996. "Chemical and anatomical plasticity of brain." *Journal of Neuropsychiatry and Clinical Neurosciences* 8(4): 459–70.

Benus, R. F. 1999. "Differential effect of handling on adult aggression in male mice bidirectionally selected for attack latency." *Aggressive Behavior* 25(5): 365–68.

Benzer, S. 1973. "Genetic dissection of behavior." *Scientific American* 229(6): 24–37.

Berenbaum, S. A., and M. Hines. 1992. "Early androgens are related to childhood sex-typed toy preferences." *Psychological Science* 3(3): 203–6.

Berenbaum, S. A., C. L. Martin, et al. 2008. Sex differences in children's play. In *Sex differences in the brain: From genes to behavior,* ed. J. B. Becker et al. New York, Oxford Univ. Press: 275–90.

Berg, S. J., and K. E. Wynne-Edwards. 2001. "Changes in testosterone, cortisol, and estradiol levels in men becoming fathers." *Mayo Clinic Proceedings* 76(6): 582–92.

Berge, C. 1998. "Heterochronic processes in human evolution: An ontogenetic analysis of the hominid pelvis." *American Journal of Physical Anthropology* 105(4): 441–59.

———. 2002. Peramorphic processes in the evolution of the hominid pelvis and femur. In *Human evolution through developmental change,* ed. N. Minugh-Purvis and K. J. McNamara. Baltimore, Johns Hopkins Univ. Press.

Berger, J. 1986. *Wild horses of the Great Basin: Social competition and population size.* Chicago, Univ. of Chicago Press.

Berger, L. R. 2006. "Brief communication: Predatory bird damage to the Taung type-skull of Australopithecus afficanus Dart 1925." *American Journal of Physical Anthropology* 131(2): 166–68.

Berger, T. D., and E. Trinkhaus. 1995. "Patterns of trauma among the Neandertals." *Journal of Archeological Science* 22: 841–52.

Bergmuller, R., D. Heg, et al. 2005. "Helpers in a cooperatively breeding cichlid stay and pay or disperse and breed, depending on ecological constraints." *Proceedings of the Royal Society of London B: Biological Sciences* 272(1560): 325–31.

Berkowitz, L. 1993. *Aggression: Its causes, consequences and control.* Philadelphia, Temple Univ. Press.

Berkowitz, M., and J. Gibbs. 1983. "Measuring the developmental features of moral discussion." *Merrill-Palmer Quarterly* 29: 399–410.

Berlin, B. 2006. "The First Congress of Ethnozoological Nomenclature." *Journal of the Royal Anthropological Institute,* 12(S1): 23–44.

Berlin, B., and P. Kay. 1969. *Basic color terms: Their universality and evolution.* Berkeley, Univ. of California Press.

Berlin, L. J., J. Cassidy, et al. 2008. The influence of early attachments on other relationships. In *Handbook of attachment,* 2nd ed., ed. J. Cassidy and P. R. Shaver. New York, Guilford: 333–47.

Berman, C. M., K. L. R. Rasmussen, et al. 1997. "Group size, infant development and social networks in free-ranging rhesus." *Animal Behaviour* 53(2): 405–21.

Bermudez de Castro, J. M., A. Rosas, et al. 1999. "A modern human pattern of dental development in lower pleistocene hominids from Atapuerca-TD6 (Spain)." *Proceedings of the National Academy of Sciences* 96(7): 4210–13.

Bernal, J. 1973. "Night waking in infants during the first fourteen months." *Developmental Medicine and Child Neurology* 20: 760.

Bernheim, S. M., and E. Mayeri. 1995. "Complex behavior induced by egg-laying hormone in Aplysia." *Journal of Comparative Physiology A: Sensory Neural and Behavioral Physiology* 176(1): 131–36.

Berns, G. S., S. Moore, et al. 2009. "Adolescent engagement in dangerous behaviors is associated with increased white matter maturity of frontal cortex." *PLoS One* 4(8): e6773.

Berrebi, A. S., R. H. Fitch, et al. 1988. "Corpus callosum: Region-specific effects of sex, early experience and age." *Brain Research* 438: 216–24.

Berrigan, D., E. L. Charnov, et al. 1993. "Phylogenetic contrasts and the evolution of mammalian life histories." *Evolutionary Ecology* 7(3): 270–78.

Betzig, L. 1986. *Despotism and differential reproduction: A Darwinian view of history.* New York, Aldine.

———. 1992. "Roman polygyny." *Ethology and Sociobiology* 13: 309–49.

———, ed. 1997. *Human nature: A critical reader.* New York, Oxford Univ. Press.

Betzig, L., M. B. Mulder, et al., eds. 1988. *Human reproductive behaviour: A Darwinian perspective.* Cambridge, Cambridge Univ. Press.

Biben, M. 1992. "Allomaternal vocal behavior in squirrel monkeys." *Developmental Psychobiology* 25(2): 79–92.

Bickerton, D. 1990. *Language and species.* Chicago, Univ. of Chicago Press.

———. 1998. The creation and re-creation of language. In *Handbook of evolutionary psychology,* ed. C. Crawford and D. L. Krebs. Mahwah, NJ, Erlbaum: 613–34.

Biederman, J., and S. V. Faraone. 2005. "Attention-deficit hyperactivity disorder." *Lancet* 366(9481): 237–48.

Biervert, C., B. C. Schroeder, et al. 1998. "A potassium channel mutation in neonatal human epilepsy." *Science* 279: 403–6.

Biesele, M. 1993. *Women like meat: The folklore and foraging ideology of the Kalahari Ju/'hoan.* Bloomington, S.A., Witwatersrand Univ. Press.

———. 1997. An ideal of unassisted birth: Hunting, healing, and transformation among the Kalahari Ju/'hoansi. In *Childbirth and authoritative knowledge: Cross-cultural perspectives,* ed. R. E. Davis-Floyd and C. F. Sargent. Berkeley, Univ. of California Press: 474–93.

Billman, B. R., P. M. Lambert, et al. 2000. "Cannibalism, warfare, and drought in the Mesa Verde Region during the twelfth century AD." *American Antiquity* 65(1): 145–78.

Bird, D. W., and R. Bliege Bird. 2000. "The ethnoarchaeology of juvenile foragers: Shellfishing strategies among Meriam children." *Journal of Anthropological Archaeology* 19: 461–76.

———. 2002. "Children on the reef: Slow learning or strategic foraging?" *Human Nature* 13(2): 269–97.

———. 2005. Martu children's hunting strategies in the Western Desert, Australia. In *Hunter-gatherer childhoods: Evolutionary, developmental and cultural perspectives,* ed. B. S. Hewlett and M. E. Lamb. New Brunswick, NJ, AldineTransaction: 129–46.

Birdwhistell, R. L. 1970. *Kinesics and context: Essays on body motion communication.* Philadelphia, Univ. of Pennsylvania Press.

Bischof, H.-J., E. Geissler, et al. 2002. "Limitations of the sensitive period for sexual imprinting: Neuroanatomical and behavioral experiments in the zebra finch *(Taeniopygia guttata)*." *Behavioural Brain Research* 133(2): 317–22.

Bishop, D., and K. Mogford, eds. 1993. *Language development in exceptional circumstances.* Hove, Eng., Erlbaum.

Bishop, J. M., J. U. M. Jarvis, et al. 2004. "Molecular insight into patterns of colony composition and paternity in the common mole-rat *Cryptomys hottentotus hottentotus*." *Molecular Ecology* 13(5): 1217–29.

Bitterman, M. E. 1965. "Phyletic differences in learning." *American Psychologist* 20: 396–410.

———. 1975. "The comparative analysis of learning." *Science* 188: 699–709.

Bixler, R. H. 1982a. "Comment on the incidence and purpose of royal sibling incest." *American Ethnologist* 9: 580–82.

———. 1982b. "Sibling incest in the royal families of Egypt, Peru, and Hawaii." *Journal of Sex Research* 18: 264–81.

Bjorklund, D. F., and A. D. Pellegrini. 2002. *The origins of human nature: Evolutionary developmental psychology.* Washington, DC, American Psychological Association.

Bjorklund, D. F., and P. K. Smith. 2003. "Evolutionary developmental psychology: Introduction to the special issue." *Journal of Experimental Child Psychology* 85(3): 195–98.

Black, D. 2004. "Sex, lies and a quest for identity: The boy raised as a girl suffered for social experiment." *Toronto Star,* May 11, 2004, A3.

Blakemore, S.-J. 2008. "The social brain in adolescence." *Nature Reviews Neuroscience* 9(4): 267–77.

Blakemore, S.-J., and S. Choudhury. 2006. "Development of the adolescent brain: Implications for executive function and social cognition." *Journal of Child Psychology and Psychiatry and Allied Disciplines* 47(3–4): 296–312.

Blanchard, M., and M. Main. 1979. "Avoidance of the attachment figure and socioemotional adjustment in day-care infants." *Developmental Psychology* 4: 445–46.

Blasco, P. M., D. B. Bailey, et al. 1993. "Dimensions of mastery in same-age and mixed-age integrated classrooms." *Early Childhood Research Quarterly* 8(2): 193–206.

Blinkov, S. M., and I. I. Glezer, eds. 1968. *The human brain in figures and tables: A quantitative handbook.* New York, Basic/Plenum.

Block, J., and J. H. Block. 2006. "Venturing a 30-year longitudinal study." *American Psychologist* 61(4): 315–27.

Block, J. H., J. Block, et al. 1986. "The personality of children prior to divorce: A prospective study." *Child Development* 57: 827–40.

Bloom, S. L., D. D. McIntire, et al. 1998. "Lack of effect of walking on labor and delivery." *New England Journal of Medicine* 339(2): 76–79.

Blurton Jones, N. G. 1971. "Criteria for use in describing facial expressions." *Human Biology* 43: 365–413.

———. 1972a. Comparative aspects of mother-child contact. In *Ethological studies of child behaviour,* ed. N. G. B. Jones. Cambridge, Cambridge Univ. Press.

———, ed. 1972b. *Ethological studies of child behaviour.* Cambridge, Cambridge Univ. Press.

———. 1978. "Natural selection and birthweight." *Annals of Human Biology* 5(5): 487–89.

———. 1990. The costs of children and the adaptive scheduling of births: Towards a sociobiological perspective of demography. In *The sociobiology of sexual and reproductive strategies,* ed. A. E. Rasa, C. Vogel, and E. Voland. London, Chapman and Hall.

———. 1993. The lives of hunter-gatherer children: Effects of parental behavior and parental reproductive strategy. In *Juvenile primates: Life history, development, and behavior,* ed. M. E. Pereira and L. A. Fairbanks. New York, Oxford: 309–26.

Blurton Jones, N., K. Hawkes, et al. 1994. "Foraging patterns of !Kung adults and children: Why didn't !Kung children forage?" *Journal of Anthropological Research* 50: 217–48.

Blurton Jones, N. G., K. Hawkes, et al. 1989. Modeling and measuring costs of children in two foraging societies. In *Comparative socioecology: The behavioral ecology of humans and other mammals,* ed. V. Standen and R. A. Foley. Oxford, Blackwell Scientific.

———. 1997. Why do Hadza children forage? In *Uniting psychology and biology: Integrative perspectives on human development,* ed. N. L. Segal, G. E. Weisfeld, and C. C. Weisfeld. Washington, DC, American Psychological Association: 279–313.

Blurton Jones, N. G., and M. J. Konner. 1973. Sex differences in behavior of two-to-five-year-olds in London and amongst the Kalahari Desert Bushmen. In *Comparative ecology and behavior of primates,* ed. R. P. Michael and J. H. Crook. London, Academic.

———. 1976. !Kung knowledge of animal behavior (or: the proper study of mankind is animals). In *Kalahari hunter-gatherers: Studies of the !Kung San and their neigh-bors,* ed. R. B. Lee and I. DeVore. Cambridge, MA, Harvard Univ. Press: 325–48.

Boas, F. 1889. "Notes on the Snanaimuq." *American Anthropologist* 2(4): 321–28.

———. 1938. *The mind of primitive man.* New York, Macmillan.

———. 1940a. Remarks on the anthropological study of children 1912. In *Race, language and culture.* New York, Free Press: 94–102.

———. 1940b. The methods of ethnology 1920. In *Race, language and culture.* New York, Free Press: 281–89.

———, ed. 1940c. *Race, language and culture.* New York, Free Press.

Boccia, M. L., and C. A. Pedersen. 2001. "Brief vs. long maternal separations in infancy: Contrasting relationships with adult maternal behavior and lactation levels of aggression and anxiety." *Psychoneuroendocrinology* 26(7): 657–72.

Bock, J. 2005. What makes a competent adult forager? In *Hunter-gatherer childhoods: Evolutionary, developmental and cultural perspectives,* ed. B. S. Hewlett and M. E. Lamb. New Brunswick, NJ, AldineTransaction: 109–28.

Bock, J., and K. Braun. 1999a. "Blockade of N-methyl-D-aspartate receptor activation suppresses learning-induced synaptic elimination." *Proceedings of the National Academy of Sciences* 96(5): 2485–90.

———. 1999b. "Filial imprinting in domestic chicks is associated with spine pruning in the associative area, dorsocaudal neostriatum." *European Journal of Neuroscience* 11(7): 2566–70.

Boeckx, C., and M. Piattelli-Palmarini. 2005. "Language as a natural object—Linguistics as a natural science." *Linguistic Review* 22(2–4): 447–66.

Boesch, C. 1991. "Teaching among wild chimpanzees." *Animal Behaviour* 41: 530–32.

Boespflug-Tanguy, O., P. Labauge, et al. 2008. "Genes involved in leukodystrophies: A glance at glial functions." *Current Neurology and Neuroscience Reports* 8(3): 217–29.

Boggess, J. 1979. "Troop male membership changes and infant killing in Langurs (*Presbytis entellus*)." *Folia Primatologica* 32: 65–107.

———. 1984. Infant killing and male reproductive strategies in Langurs (*Presbytis entellus*). In *Infanticide: Comparative and evolutionary perspectives,* ed. G. Hausfater and S. B. Hrdy. New York, Aldine de Gruyter: 283–310.

Bogin, B. 1994. "Adolescence in evolutionary perspective." *Acta Paediatrica* 406(Suppl.): 29–35.

———. 1997. "Evolutionary hypotheses for human childhood." *Yearbook of Physical Anthropology* 40: 63–89.

———. 1999. *Patterns of human growth.* 2nd ed. New York, Cambridge Univ. Press.

———. 2006. Modern human life history: The evolution of human childhood and fertility. In *Evolution of Human Life History,* ed. K. Hawkes and R. R. Paine. Santa Fe, NM, School of American Research: 197–230.

Bohannon, J. N. I., B. MacWhinney, et al. 1990. "No negative evidence revisited: Beyond learnability or who has to prove what to whom." *Developmental Psychology* 26(2): 221–26.

Bohannon, J. N. I., and L. Stanowicz. 1988. "The issue of negative evidence: Adult responses to children's language errors." *Developmental Psychology* 24(5): 684–89.

Bohn, A., and D. Berntsen. 2008. "Life story development in childhood: The development of life story abilities and the acquisition of cultural life scripts from late middle childhood to adolescence." *Developmental Psychology* 44(4): 1135–47.

Bokhorst, C. L., M. J. Bakermans-Kranenburg, et al. 2003. "The importance of shared environment in mother-infant attachment security: A behavioral genetic study." *Child Development* 74(6): 1769–82.

Bolhuis, J. J. 1999. "Early learning and the development of filial preferences in the chick." *Behavioural Brain Research* 98(2): 245–52.

Bolk, L. 1926. *Das problem der menschwerdung.* Jena, Gustav Fischer.

Bolter, D., and A. Zihlman. 2007. Primate growth and development: A functional and evolutionary approach. In *Primates in perspective,* ed. C. J. Campbell et al. New York, Oxford Univ. Press: 408–22.

Bonner, J. T. 1965. *Size and cycle.* Princeton, Princeton Univ. Press.

———. 1980. *The evolution of culture in animals.* Princeton, Princeton Univ. Press.

———. 1993. *Life cycles: Reflections of an evolutionary biologist.* Princeton, Princeton Univ. Press.

Boone, J. L. 2002. "Subsistence strategies and early human population history: An evolutionary ecological perspective." *World Archaeology* 34(1): 6–25.

Booth, C. 1973. "Early social experience and the gregariousness in the rat." *Developmental Psychology* 8: 360–68.

Bordes, F. 1968. *The old stone age.* New York, McGraw-Hill.

Borgerhoff Mulder, M. 1988. Kipsigis bridewealth payments. In *Human reproductive behavior: A Darwinian perspective,* ed. L. Betzig, M. B. Mulder, and P. Turke. New York, Cambridge Univ. Press.

———. 1990. "Kipsigis women's preferences for wealthy men: Evidence for female choice in mammals?" *Behavioral Ecology and Sociobiology* 27: 255–64.

———. 1991. Human behavioural ecology. In *Behavioural ecology: An evolutionary approach,* ed. J. R. Krebs and N. B. Davies. Oxford, Blackwell: 69–98.

Borgerhoff-Mulder, M., S. Bowles, et al. 2009. "Intergenerational wealth transmission and the dynamics of inequality in small-scale societies." *Science* 326(5953): 682–88.

Bornstein, M. H., O. M. Haynes, et al. 1999. "Play in two societies: Pervasiveness of process, specificity of structure." *Child Development* 70(2): 317–31.

Bornstein, M. H., P. Venuti, et al. 2002. "Mother-child play in Italy: Regional variation, individual stability, and mutual dyadic influence." *Parenting: Science and Practice* 2(3): 273–301.

Boswell, J. 1988. *The kindness of strangers: The abandonment of children in Western Europe from late antiquity to the Renaissance.* New York, Pantheon.

Bouchard, T. J., Jr. 1994. "Genes, environment, and personality." *Science* 264: 1700–1.

Bouchard, T. J., Jr., and M. McGue. 2003. "Genetic and environmental influences on human psychological differences." *Journal of Neurobiology* 54(1): 4–45.

Bourgeois, J. P., and P. Rakic. 1996. "Synaptogenesis in the occipital cortex of macaque monkey devoid of retinal input from early embryonic stages." *European Journal of Neuroscience* 8(5): 942–50.

Bourliere, F. 1955. *Natural history of mammals.* London, Harrap.

Bowlby, J. 1969. *Attachment.* New York, Basic Books.

———. 1973. *Separation: Anxiety and anger.* New York, Basic Books.

———. 1980. *Loss.* New York, Basic Books.

Bowles, S. 2009. "Did warfare among ancestral hunter-gatherers affect the evolution of human social behaviors? [see comment]." *Science* 324(5932): 1293–98.

Boyce, W. T., and B. J. Ellis. 2005. "Biological sensitivity to context: I. An evolutionary-developmental theory of the origins and functions of stress reactivity." *Development and Psychopathology* 17(2): 271–301.

Boyd, R., and P. J. Richerson. 1985. *Culture and the evolutionary process.* Chicago, Univ. of Chicago Press.

———. 1988. An evolutionary model of social learning: The effects of spatial and temporal variation. In *Social learning: Psychological and biological perspectives,* ed. T. R. Zentall and B. G. Galef. Hillsdale, NJ, Erlbaum: 29–48.

Brachfeld, S., S. Goldberg, et al. 1980. "Parent-infant interaction in free play at 8 and 12 months: Effects of prematurity and immaturity." *Infant Behavior and Development* 3(4): 289–305.

Brackbill, Y. 1958. "Extinction of the smiling response in infants as a function of reinforcement schedule." *Child Development* 39: 114–24.

Bradley, P., and G. Horn. 1987. Neural consequences of imprinting. In *Imprinting and cortical plasticity: Comparative aspects of sensitive periods,* ed. J. P. Rauschecker and P. Marler. New York, Wiley.

Bradley, S. J., G. D. Oliver, et al. 1998. "Experiment of nurture: Ablatio penis at 2 months, sex reassignment at 7 months, and a psychosexual follow-up in young adulthood." *Pediatrics* 102(1): e9.

Bradshaw, G. A., and A. N. Schore. 2007. "How elephants are opening doors: Developmental neuroethology, attachment and social context." *Ethology* 113(5): 426–36.

Brain, L., and J. N. Walton. 1969. *Brain's diseases of the nervous system.* Oxford, Oxford Univ. Press.

Brain, R., and M. Wilkinson. 1959. "Observations on the extensor plantar reflex and its relationship to the functions of the pyramidal tract." *Brain* 82: 297–320.

Braun, K., J. Bock, et al. 1999. "The dorsocaudal neostriatum of the domestic chick: A structure serving higher associative functions." *Behavioural Brain Research* 98(2): 211–18.

Braungart-Rieker, J., S. Courtney, et al. 1999. "Mother- and father-infant attachment: Families in context." *Journal of Family Psychology* 13(4): 535–53.

Brazelton, T. B. 1973. *Neonatal behavioral assessment scale.* Philadelphia, Lippincott.

———. 1990. "Commentary: Parent-infant co-sleeping revisited." *Ab Initio* 2(1): 1–7.

Bredesen, D. E. 1995. "Neural apoptosis." *Annals of Neurology* 38: 839–51.

Breedlove, S. M. 1992. Sexual differentiation of the brain and behavior. In *Behavioral endocrinology,* ed. J. B. Becker, S. M. Breedlove, and D. Crews. Cambridge, MA, MIT Press.

———. 1994. "Sexual differentiation of the human nervous system." *Annual Review of Psychology* 45: 389–418.

Breedlove, S. M., and A. P. Arnold. 1980. "Hormone accumulation in a sexually dimorphic nucleus of the rat spinal cord." *Science* 210: 564–66.

Breese, G. R., R. D. Smith, et al. 1973. "Induction of adrenal catecholamine synthesizing enzymes following mother-infant separation." *Nature New Biology* 246(Nov. 21): 94–96.

Breidbach, O., and M. T. Ghiselin. 2007. "Evolution and development: Past, present, and future." *Theory in Biosciences* 125(2): 157–71.

Brent, L., T. Koban, et al. 2002. "Abnormal, abusive, and stress-related behaviors in baboon mothers." *Biological Psychiatry* 52(11): 1047–56.

Bretherton, I. 1992. "The origins of attachment theory: John Bowlby and Mary Ainsworth." *Developmental Psychology* 28(5): 759–75.

Bretherton, I., and M. D. S. Ainsworth. 1974. Responses of one-year-olds to a stranger in a strange situation. In *Origins of fear,* ed. M. Lewis and L. A. Rosenblum. New York, Wiley.

Breton, G. 1996. "Punctualism and gradualism acting together within one lineage: Thoughts about the complexity and the unpredictability of the evolutionary processes." *Geobios (Lyon)* 29(1): 125–30.

Breuggeman, L. A. 1973. "Parental care in a group of free-ranging rhesus monkeys (*Macaca mulatta*)." *Folia Primatologica* 20: 178–210.

Breuil, H. 1952. *Four hundred centuries of cave art.* Montignac, France, Centre d'etudes et de documentation prehistoriques.

Brewster, A. L., J. P. Nelson, et al. 1998. "Victim, perpetrator, family, and incident characteristics of 32 infant maltreatment deaths in the United States Air Force." *Child Abuse and Neglect* 22(2): 91–101.

Bridges, R. S. 1990. Endocrine regulation of parental behavior in rodents. In *Mammalian parenting,* ed. N. A. Krasnegor and R. S. Bridges. New York, Oxford Univ. Press: 93–117.

Briggs, A. W., J. M. Good, et al. 2009. "Targeted retrieval and analysis of five Neandertal mtDNA genomes." *Science* 325: 318–21.

Bright, P., and M. D. Kopelman. 2001. "Learning and memory: Recent findings." *Current Opinion in Neurology* 14(4): 449–55.

Brindis, C. D. 2006. "A public health success: Understanding policy changes related to teen sexual activity and pregnancy." *Annual Review of Public Health* 27(1): 277–95.

Broadhurst, C. L., Y. Wang, et al. 2002. "Brain-specific lipids from marine, lacustrine, or terrestrial food resources: Potential impact on early African Homo sapiens." *Comparative Biochemistry and Physiology Part B, Biochemistry and Molecular Biology* 131(4): 653–73.

Broca, P. 1861. "Remarques sur le siege de la faculté du langage articule, suive d'une observation d'aphemie." *Bulletin of the Society of Anatomists (Paris)* 36: 330–57.

———. 1878. "Le grand lobe limbique et la scissure limbique dans le série de mammifères." *Revue d'Anthropologie,* 2e série I: 385–498.

Brody, B. A., H. C. Kinney, et al. 1987. "Sequence of central nervous system myelination in human infancy. I. An autopsy study of myelination." *Journal of Neuropathology and Experimental Neurology* 46(3): 283–301.

Brody, J. 1993. "Personal health: Modern certified midwives are leading a revolution in high-quality obstetric care." *New York Times,* April 28, 1993, C11.

Bronson, G. W. 1972. "Infants' reactions to unfamiliar persons and novel objects." *Monographs of the Society for Research in Child Development* 37(2): 46.

———. 1974. "The postnatal growth of visual capacity." *Child Development* 45: 873–90.

———. 1982. Structure, status and characteristics of the nervous system at birth. In *Psychobiology of the human newborn,* ed. P. Stratton. New York, Wiley.

Brookhart, J., and E. Hock. 1976. "The effects of experimental context and experiential background on infants' behavior toward the mother and a stranger." *Child Development* 47: 333–40.

Broude, G. J. 1975. "Norms of premarital sexual behavior: A cross-cultural study." *Ethos* 3(3): 381–402.

———. 1988. "Rethinking the couvade: Cross-cultural evidence." *American Anthropologist* 90(4): 902–11.

———. 1990. "Protest masculinity: A further look at the causes and the concept." *Ethos* 18: 103–22.

Broude, G. J., and S. J. Greene. 1976. "Cross-cultural codes on twenty sexual attitudes and practices." *Ethnology* 15: 409–29.

Brown, C. H., and S. R. Witkowski. 1981. "Figurative language in a universalist perspective." *American Ethnologist* 8: 596–615.

Brown, D. E. 1991. *Human universals*. Philadelphia, Temple Univ. Press.

Brown, J. H., and R. M. Sibly. 2006. "Life-history evolution under a production constraint." *Proceedings of the National Academy of Sciences* 103(47): 17595–99.

Brown, J. R., H. Ye, et al. 1996. "A defect in nurturing in mice lacking the immediate early gene fosB." *Cell* 86(2): 297–309.

Brown, K. S., C. W. Marean, et al. 2009. "Fire as an engineering tool of early modern humans." *Science* 325: 859–62.

Brown, P., T. Sutikna, et al. 2004. "A new small-bodied hominin from the Late Pleistocene of Flores, Indonesia [see comment]." *Nature* 431(7012): 1055–61.

Brown, P. J., and M. Konner. 1987. An anthropological perspective on obesity. In *Human obesity*, ed. R. J. Wurtman and J. J. Wurtman. *Annals of the New York Academy of Sciences* 499: 29–46.

Brown, R. 1973. *A first language: The early stages*. Cambridge, MA, Harvard Univ. Press.

Brown, R., and G. J. Armelagos. 2001. "Apportionment of racial diversity: A review." *Evolutionary Anthropology* 10: 15–20.

Brownell, W., C. Bader, et al. 1985. "Evoked mechanical responses of isolated cochlear hair cells." *Science* 227: 194–96.

Bruce, L. L., and T. J. Neary. 1995. "The limbic system of tetrapods: A comparative analysis of cortical and amygdalar populations." *Brain, Behavior and Evolution* 46: 224–34.

Bruell, J. H. 1970. Heritability of emotional behavior. In *Physiological correlates of emotion*, ed. P. Black. New York, Academic.

Bruer, J. T. 1999. *The myth of the first three years: A new understanding of early brain development and lifelong learning*. New York, Free Press.

Brunelli, S. A., and M. A. Hofer. 1990. Parental behavior in juvenile rats: Environmental and biological determinants. In *Mammalian parenting*, ed. N. A. Krasnegor and R. S. Bridges. New York, Oxford Univ. Press: 372–99.

Brunelli, S. A., R. D. Shindledecker, et al. 1987. "Behavioral responses of juvenile rats (*Rattus norvegicus*) to neonates after infusion of maternal blood plasma." *Journal of Comparative Psychology* 101(1): 47–59.

Bruner, J. 1981. "The social context of language acquisition." *Language and Communication* 1(2/3): 155–78.

———. 1997. "Celebrating divergence: Piaget and Vygotsky." *Human Development* 40(2): 63–73.

———. 1999. Prologue. In *Lev Vygotsky: Critical assessments. Vol. IV: Future directions*, ed. P. Lloyd and C. Fernyhough. Oxford, Routledge: 421–41.

Bruner, J. S. 1972. "Nature and uses of immaturity." *American Psychologist* 27(8): 28–60.

———. 1983. *Child's talk: Learning to use language*. New York, Norton.

Bruner, J. S., A. Jolly, et al., eds. 1976. *Play—its role in development and evolution*. New York, Basic Books.

Buchan, J. C., S. C. Alberts, et al. 2003. "True paternal care in a multi-male primate society." *Nature (London)* 425(6954): 179–81.

Buchanan, C. M., J. S. Eccles, et al. 1992. "Are adolescents the victims of raging hormones: Evidence for activational effects of hormones on moods and behavior at adolescence." *Psychological Bulletin* 111(1): 62–107.

Buchwald, J. S. 1985. A comparative model of infant cry. In *Infant crying: Theoretical and research perspectives,* ed. B. M. Lester and C. F. Z. Boukydis. New York, Plenum: 279–305.

Bullmore, E. T., P. W. Woodruff, et al. 1998. "Does dysplasia cause anatomical dysconnectivity in schizophrenia?" *Schizophrenia Research* 30(2): 127–35.

Bullock, T. H. 1999. "Neuroethology has pregnant agendas." *Journal of Comparative Physiology A: Sensory Neural and Behavioral Physiology,* 185(4): 291–95.

Buntin, J. D. 1996. Neural and hormonal control of parental behavior in birds. In *Parental care: Evolution, mechanisms, and adaptive significance. Advances in the study of behavior,* vol. 25, ed. J. S. Rosenblatt and C. T. Snowdon. San Diego, Academic: 161–213.

Burchinal, M. R., A. Follmer, et al. 1996. "The relations of maternal social support and family structure with maternal responsiveness and child outcomes among African American families." *Developmental Psychology* 32(6): 1073–83.

Burda, H., R. L. Honeycutt, et al. 2000. "Are naked and common mole-rats eusocial and if so, why?" *Behavioral Ecology and Sociobiology* 47(5): 293–303.

Burghardt, G. M. 1988. Precocity, play, and the ectotherm-endotherm transition: Profound reorganization or superficial adaptation? In *Handbook of Behavioral Neurobiology,* vol. 9: *Developmental psychobiology and behavioral ecology,* ed. E. M. Blass. New York, Plenum: 107–48.

———. 2005. *The genesis of animal play.* Cambridge, MA, MIT Press/Bradford.

Burkhalter, J. E., and R. L. Balster. 1979. "Effects of phencyclidine on isolation-induced aggression in mice." *Psychological Reports* 45(2): 571–76.

Burn, J., J. A. Birkbeck, et al. 1975. "Early fetal brain growth." *Human Biology* 47(4): 511–22.

Burnett, S., and S.-J. Blakemore. 2009. "The development of adolescent social cognition." *Annals of the New York Academy of Sciences* 1167: 51–56.

Burnham, T. C., J. F. Chapman, et al. 2003. "Men in committed, romantic relationships have lower testosterone." *Hormones and Behavior* 44(2): 119–22.

Burt, A., and R. L. Trivers. 2006. *Genes in conflict: The biology of selfish genetic elements.* Cambridge, MA, Harvard Univ. Press/Belknap.

Burt, S. A. 2009. "Are there meaningful etiological differences within antisocial behavior? Results of a meta-analysis." *Clinical Psychology Review* 29(2): 163–78.

Burt, S. A., and J. M. Neiderhiser. 2009. "Aggressive versus nonaggressive antisocial behavior: Distinctive etiological moderation by age." *Developmental Psychology* 45(4): 1164–78.

Burton, F. D. 1972. The integration of biology and behavior in the socialization of *Macaca sylvana* of Gibraltar. In *Primate socialization,* ed. F. E. Poirier. New York, Random House.

Burton, R. V., and J. W. M. Whiting. 1961. "The absent father and cross-sex identity." *Merrill-Palmer Quarterly of Behavior and Development* 7: 85–95.

Bush, E. C., and J. M. Allman. 2004. "The scaling of frontal cortex in primates and carnivores." *Proceedings of the National Academy of Sciences* 101(11): 3962–66.

Bushnell, I. W., F. Sai, et al. 1989. "Neonatal recognition of the mother's face." *British Journal of Developmental Psychology* 7(1): 3–15.

Buss, D. 1989. "Sex differences in human mate preferences: Evolutionary hypotheses tested in 37 cultures." *Behavioral and Brain Sciences* 12: 1–49.

Buss, D., and D. P. Schmitt. 1993. "Sexual strategies theory: An evolutionary perspective on human mating." *Psychological Review* 100: 204–32.

Buss, D. M. 1998. "Sexual strategies theory: Historical origins and current status." *Journal of Sex Research* 35(1): 19–31.

Buss, K. A., J. R. M. Schumacher, et al. 2003. "Right frontal brain activity, cortisol, and withdrawal behavior in 6-month-old infants." *Behavioral Neuroscience* 117(1): 11–20.

Butcher, L. M., J. K. Kennedy, et al. 2006. "Generalist genes and cognitive neuroscience." *Current Opinion in Neurobiology* 16(2): 145–51.

Butler, A., A. Wei, et al. 1989. "A family of putative potassium channel genes in *Drosophila.*" *Science* 243: 943–47.

Butler, A. B. 1995. "The dorsal thalamus of jawed vertebrates: A comparative viewpoint." *Brain, Behavior and Evolution* 46: 209–23.

Butler, A. B., and W. Hodos. 2005. *Comparative vertebrate neuroanatomy: Evolution and adaptation.* 2nd ed. New York, Wiley-Liss.

Butterworth, B., B. Comrie, et al., eds. 1984. *Explanations for language universals.* Berlin, Mouton.

Buxhoeveden, D. P., and M. F. Casanova. 2002. "The minicolumn and evolution of the brain." *Brain Behavior and Evolution* 60(3): 125–51.

Buzsaki, G., and A. Draguhn. 2004. "Neuronal oscillations in cortical networks." *Science* 304(5679): 1926–29.

Byers, J. A. 1998. Biological effects of locomotor play: Getting into shape, or something more specific? In *Animal play: Evolutionary, comparative, and ecological perspectives,* ed. M. Bekoff and J. A. Byers. Cambridge, Cambridge Univ. Press: 205–20.

———. 1999a. "Play's the thing." *Natural History* 108(6): 40–45.

———. 1999b. "The distribution of play behaviour among Australian marsupials." *Journal of Zoology (London)* 247(3): 349–56.

Byers, J. A., and C. Walker. 1995. "Refining the motor training hypothesis for the evolution of play." *American Naturalist* 146(1): 25–40.

Byrd, K. R., and W. E. Briner. 1999. "Fighting, nonagonistic social behavior, and exploration in isolation-reared rats." *Aggressive Behavior* 25(3): 211–23.

Cabrera, N. J., C. S. Tamis-LeMonda, et al. 2000. "Fatherhood in the twenty-first century." *Child Development* 71(1): 127–36.

Caceres, M., J. Lachuer, et al. 2003. "Elevated gene expression levels distinguish human from non-human primate brains." *Proceedings of the National Academy of Sciences* 100(22): 13030–35.

Cai, L. Q., Y. S. Zhu, et al. 1996. "5 alpha-reductase-2 gene mutations in the Dominican Republic." *Journal of Clinical Endocrinology and Metabolism* 81(5): 1730–35.

Caillois, R. 1961. *Man, play, and games.* New York, Free Press.

Cairns, R. B. 1993. Belated but bedazzling: Timing and genetic influence in social development. In *Developmental time and timing,* ed. G. Turkewitz and D. A. Devenny. Hillsdale, NJ: Erlbaum: 61–84.

Caldji, C., B. Tannenbaum, et al. 1998. "Maternal care during infancy regulates the development of neural systems mediating the expression of fearfulness in the rat." *Proceedings of the National Academy of Sciences* 95(9): 5335–40.

Caldwell, B. 1964. The effects of infant care. In *Review of child development research,* vol. 1, ed. M. L. Hoffman and L. W. Hoffman. New York, Russell Sage: 9–87.

Caldwell, B. M., C. Wright, et al. 1970. "Infant day-care and attachment." *American Journal of Orthopsychiatry* 40: 397–412.

Caldwell, C. A., and A. Whiten. 2002. "Evolutionary perspectives on imitation: Is a comparative psychology of social learning possible?" *Animal Cognition* 5(4): 193–208.

Calegaro-Marques, C., and J. C. Bicca-Marques. 1993. "Allomaternal care in the black howler monkey *(Alouatta caraya)*." *Folia Primatologica* 61(2): 104–9.

Calhoun, F., and K. Warren. 2007. "Fetal alcohol syndrome: Historical perspectives." *Neuroscience and Biobehavioral Reviews* 31(2): 168–71.

Caliandro, P., E. Stalberg, et al. 2007. "Sensitivity of conventional motor nerve conduction examination in detecting patchy demyelination: A simulated model." *Clinical Neurophysiology* 118(7): 1577–85.

Call, J., M. Carpenter, et al. 2005. "Copying results and copying actions in the process of social learning: Chimpanzees *(Pan troglodytes)* and human children *(Homo sapiens)*." *Animal Cognition* 8(3): 151–63.

Callaghan, T., P. Rochat, et al. 2005. "Synchrony in the onset of mental-state reasoning: Evidence from five cultures." *Psychological Science* 16(5): 378–84.

Callier, V., J. A. Clack, et al. 2009. "Contrasting developmental trajectories in the earliest known tetrapod forelimbs." *Science* 324: 364–367.

Cameron, P., and K. Cameron. 2002. "Children of homosexual parents report childhood difficulties." *Psychological Reports* 90(1): 71–82.

Campbell, A. 2004. "Female competition: Causes, constraints, content, and contexts." *Journal of Sex Research* 41(1): 16–26.

Campbell, B. 2006. "Adrenarche and the evolution of human life history." *American Journal of Human Biology* 18(5): 569–89.

Campbell, B. A., and J. Jaynes. 1966. "Reinstatement." *Psychological Review* 73: 487–90.

Campbell, C. J. 2007. Primate sexuality and reproduction. In *Primates in perspective,* ed. C. J. Campbell et al. New York, Oxford Univ. Press: 571–91.

Campbell, C. J., A. Fuentes, et al., eds. 2007. *Primates in perspective.* New York, Oxford Univ. Press.

Campbell, P. S., M. X. Zarrow, et al. 1973. "The effect of infantile stimulation upon hypothalamic CRF levels following adrenalectomy in the adult rat." *Proceedings of the Society for Experimental Biology and Medicine* 142: 781–83.

Campos, J. J., D. Witherington, et al. 2008. "Rediscovering development in infancy." *Child Development* 79(6): 1625–32.

Can, S., Y. S. Zhu, et al. 1998. "The identification of 5 alpha-reductase-2 and 17 beta-hydroxysteroid dehydrogenase-3 gene defects in male pseudohermaphrodites from a Turkish kindred." *Journal of Clinical Endocrinology and Metabolism* 83(2): 560–69.

Canivet, C., I. Jakobsson, et al. 2000. "Infantile colic. Follow-up at four years of age: Still more 'emotional' [see comments]." *Acta Paediatrica* 89(1): 13–17.

Cannon, W. B. 1963 [1915]. *Bodily changes in pain, hunger, fear, and rage.* New York, Harper & Row.

———. 1927. "The James-Lange theory of emotions: A critical examination and an alternative theory." *American Journal of Psychology* 39: 106–24.

Canto, P., F. Vilchis, et al. 1997. "Mutations of the 5 alpha-reductase type 2 gene in eight Mexican patients from six different pedigrees with 5 alpha-reductase-2 deficiency." *Clinical Endocrinology (Oxf)* 46(2): 155–60.

Caputo, D., W. Essman, et al. 1968. "Housing modification as a variable in fasting-induced ulcerogenesis." *Journal of Psychosomatic Research* 12: 129–35.

Carey, S., R. Diamond, et al. 1980. "Development of face recognition: A maturational component?" *Developmental Psychology* 16(4): 257–69.

Carlier, J. G., and O. P. Steeno. 1985. "Oigarche: The age at first ejaculation." *Andrologia* 17(1): 104–6.

Carmody, D. P., S. M. Dunn, et al. 2004. "A quantitative measure of myelination development in infants, using MR images." *Neuroradiology* 46(9): 781–86.

Carnow, T. B., J. H. Carson, et al. 1984. "Myelin basic protein gene expression in quaking, jumpy, and myelin synthesis-deficient mice." *Developmental Biology* 106: 38–44.

Caro, T. M. 1981. "Predatory behaviour and social play in kittens." *Behaviour* 76: 1–24.

———. 1994. *Cheetahs of the Serengeti plains.* Chicago, Univ. of Chicago Press.

Caro, T. M., and M. D. Hauser. 1992. "Is there teaching in nonhuman animals?" *Quarterly Review of Biology* 67(2): 151–74.

Carpendale, J. I., and M. J. Chandler. 1996. "On the distinction between false belief understanding and subscribing to an interpretive theory of mind." *Child Development* 67(4): 1686–1706.

Carpenter, C. R. 1965. The howlers of Barro Colorado island. In *Primate behavior,* ed. I. DeVore. New York, Holt, Rinehart, and Winston.

Carpenter, K., K. F. Hirsch, et al., eds. 1994. *Dinosaur eggs and babies.* New York, Cambridge Univ. Press.

Carre, J. M., S. K. Putnam, et al. 2009. "Testosterone responses to competition predict future aggressive behaviour at a cost to reward in men." *Psychoneuroendocrinology* 34(4): 561–70.

Carrer, H. F., and M. J. Cambiasso. 2002. "Sexual differentiation of the brain: Genes, estrogen, and neurotrophic factors." *Cellular and Molecular Neurobiology* 22(5–6): 479–500.

Carroll, R. L. 1988. *Vertebrate paleontology and evolution.* New York, Freeman.

———. 1997. *Patterns and processes of vertebrate evolution.* New York, Cambridge Univ. Press.

Carroll, S. B. 2003. "Genetics and the making of Homo sapiens." *Nature* 422(6934): 849–57.

———. 2005. "Evolution at two levels: On genes and form." *PLoS Biology* 3(7): e245.

Carter, C. S. 1998. "Neuroendocrine perspectives on social attachment and love." *Psychoneuroendocrinology* 23(8): 779–818.

———. 2003. "Developmental consequences of oxytocin." *Physiology and Behavior* 79(3): 383–97.

Carter, C. S., J. R. Williams, et al. 1992. "Oxytocin and social bonding." *Annals of the New York Academy of Sciences* 652: 204–11.

Cartmill, M. 1993. *A view to death in the morning: Hunting and nature through history.* Cambridge, MA, Harvard Univ. Press.

Carver, K., K. Joyner, et al. 2003. National estimates of adolescent romantic relationships. In *Adolescent romantic relations and sexual behavior: Theory, research, and practical implications,* ed. P. Florsheim. Mahwah, NJ, Erlbaum: 23–56.

Cascio, C. J., G. Gerig, et al. 2007. "Diffusion tensor imaging: Application to the study of the developing brain." *Journal of the American Academy of Child and Adolescent Psychiatry* 46(2): 213–23.

Caselli, M. C., E. Bates, et al. 1995. "A cross-linguistic study of early lexical development." *Cognitive Development* 10: 159–99.

Cashdan, E. 1989. Hunters and gatherers: Economic behavior in bands. In *Economic anthropology*, ed. S. Plattner. Stanford, Stanford Univ. Press: 21–48.

Caspi, A., and Moffitt, T. E. 2006. "Gene-environment interactions in psychiatry: Joining forces with neuroscience." *Nature Reviews Neuroscience* 7(7): 583–90.

Caspi, A., B. W. Roberts, et al. 2005. "Personality development: Stability and change." *Annual Review of Psychology* 56(1): 453–84.

Caspi, A., K. Sugden, et al. 2003. "Influence of life stress on depression: Moderation by a polymorphism in the 5-HTT gene [see comment]." *Science* 301(5631): 386–89.

Cassem, E. H. 1999. The person confronting death. In *The Harvard guide to modern psychiatry*, ed. A. M. J. Nicholi. Cambridge, MA, Harvard Univ. Press: 699–731.

Cassidy, J. 2008. The nature of the child's tie. In *Handbook of attachment*, 2nd ed., ed. J. Cassidy and P. R. Shaver. New York, Guilford: 3–22.

Caudill, W., and D. Plath. 1966. "Who sleeps by whom? Parent-child involvement in urban Japanese families." *Psychiatry* 29: 344–66.

Cavalli-Sforza, L. L., and M. W. Feldman. 1981. *Cultural transmission and evolution: A quantitative approach*. Princeton, Princeton Univ. Press.

———. 2003. "The application of molecular genetic approaches to the study of human evolution." *Nature Genetics* 33(Suppl.): 266–75.

Cavalli-Sforza, L. L., P. Menozzi, et al. 1994. *The history and geography of human genes*. Princeton, Princeton Univ. Press.

Caviness, V. S., and R. Sidman. 1973. "Time of origin of corresponding cell classes in the cerebral cortex of normal and reeler mutant mice: An autoradiographic analysis." *Journal of Comparative Neurology* 148: 141–52.

Caviness, V. S., Jr., and P. Rakic. 1978. "Mechanisms of cortical development: A view from mutations in mice." *Annual Review of Neuroscience* 1: 296–326.

Cefalo, R. C., and W. A. J. Bowes. 1998. "Managing labor—never walk alone." *New England Journal of Medicine* 339(2): 117–18.

Chagnon, N. A. 1988. "Life histories, blood revenge, and warfare in a tribal population." *Science* 239: 985–92.

Chaloupkova, H., G. Illmann, et al. 2007. "The effect of pre-weaning housing on the play and agonistic behaviour of domestic pigs." *Applied Animal Behaviour Science* 103(1–2): 25–34.

Chambers, R. A., J. R. Taylor, et al. 2003. "Developmental neurocircuitry of motivation in adolescence: A critical period of addiction vulnerability." *American Journal of Psychiatry* 160(6): 1041–52.

Chamove, A. 1973. "Rearing infant rhesus together." *Behaviour* 67: 48–66.

Chamove, A., H. Harlow, et al. 1967. "Sex differences in the infant-directed behavior of preadolescent rhesus monkeys." *Child Development* 38: 329–35.

Chamove, A., L. Rosenblum, et al. 1973. "Monkeys (*Macaca mulatta*) raised only with peers: A pilot study." *Animal Behaviour* 21: 316–25.

Champagne, F. A., and J. P. Curley. 2009. "Epigenetic mechanisms mediating the long-term effects of maternal care on development." *Neuroscience and Biobehavioral Reviews* 33(4): 593–600.

Champagne, F. A., D. D. Francis, et al. 2003. "Variations in maternal care in the rat as a mediating influence for the effects of environment on development." *Physiology and Behavior* 79(3): 359–71.

Champoux, M., E. Byrne, et al. 1992. "Motherless mothers revisited: Rhesus maternal behavior and rearing history." *Primates* 33(2): 251–55.

Chanas-Sacre, G., B. Rogister, et al. 2000. "Radial glia phenotype: Origin, regulation, and transdifferentiation." *Journal of Neuroscience Research* 61(4): 357–63.

Changeux, J. P. 1985. *Neuronal man: The biology of mind.* New York, Oxford Univ. Press.

Changeux, J.-P., and A. Danchin. 1976. "Selective stabilisation of developing synapses as a mechanism for the specification of neuronal networks." *Nature* 264: 705–12.

Chapais, B. 2008. *Primeval kinship: How pair-bonding gave birth to human society.* Cambridge, MA, Harvard Univ. Press.

Chapais, B., C. Gathier, et al. 1997. "Relatedness threshold for nepotism in Japanese macaques." *Animal Behaviour* 53(5): 1089–1101.

Charnov, E. L. 1993. *Life history invariants: Some explorations of symmetry in evolutionary ecology.* New York, Oxford Univ. Press.

Charnov, E. L., and D. Berrigan. 1991. "Dimensionless numbers and the assembly rules for life histories." *Philosophical Transactions of the Royal Society of London B: Biological Sciences* 332(1262): 41–48.

———. 1993. "Why do female primates have such long lifespans and so few babies? Or life in the slow lane." *Evolutionary Anthropology* 1: 191–94.

Chauvet, J.-M., E. B. Deschamps, et al. 1996. *Dawn of art: The Chauvet cave, the oldest known paintings in the world.* New York, Abrams.

Chen, C.-Y., and K.-G. Wang. 2006. "Are routine interventions necessary in normal birth?" *Taiwanese Journal of Obstetrics and Gynecology* 45(4): 302–6.

Cheney, D., R. Seyfarth, et al. 1986. "Social relationships and social cognition in nonhuman primates." *Science* 234: 1361.

Cheney, D. L., and R. M. Seyfarth. 1990. *How monkeys see the world: Inside the mind of another species.* Chicago, Univ. of Chicago Press.

Cheng, M.-F. 1992. "For whom does the female dove coo? A case for the role of vocal self-stimulation." *Animal Behaviour* 43: 1035–44.

Cheng, M.-F., J. P. Peng, et al. 1998. "Hypothalamic neurons preferentially respond to female nest coo stimulation: Demonstration of direct acoustic stimulation of luteinizing hormone release." *Journal of Neuroscience* 18(14): 5477–89.

Chesler, P. 1969. "Maternal influence in learning by observation in kittens." *Science* 166: 901–3.

Chi, J. G., E. C. Dooling, et al. 1977. "Left-right asymmetries of the temporal speech areas of the human fetus." *Archives of Neurology* 34(6): 346–48.

Chick, G., and J. W. Loy. 2001. "Making men of them: Male socialization for warfare and combative sports." *World Cultures* 12(1): 2–17.

Childs, A. M., L. A. Ramenghi, et al. 2001. "Cerebral maturation in premature infants: Quantitative assessment using MR imaging." *American Journal of Neuroradiology* 22(8): 1577–82.

Chisholm, J. S. 1983. *Navajo infancy: An ethological study of child development.* New York, Aldine.

———. 1993. "Death, hope, and sex: Life-history theory and the development of reproductive strategies." *Current Anthropology* 34(1): 1–24.

———. 1999a. "Attachment and time preference: Relations between early stress and sexual behavior in a sample of American university women." *Human Nature* 10(1): 51–83.

———. 1999b. *Death, hope and sex: Steps to an evolutionary ecology of mind and morality.* New York, Cambridge Univ. Press.

Chisholm, J. S., V. K. Burbank, et al. 2005. Early stress: Perspectives from developmental evolutionary ecology. In *Origins of the social mind: Evolutionary psychology and child development,* ed. B. J. Ellis and D. F. Bjorklund. New York, Guilford: 76–107.

Chivers, D. P., and R. J. F. Smith. 1995. "Chemical recognition of risky habitats is culturally transmitted among fathead minnows, *Pimephales promelas* (Osteichthyes, Cyprinidae)." *Ethology* 99: 286–96.

Chivers, M. L., and J. M. Bailey. 2005. "A sex difference in features that elicit genital response." *Biological Psychology* 70(2): 115–20.

Chomsky, N. 1957. *Syntactic structures.* The Hague, Mouton.

———. 1959. "Review of B. F. Skinner's verbal behavior." *Language* 35: 26–58.

———. 1988. *Language and problems of knowledge: The Managua lectures.* Cambridge, MA, MIT Press.

———. 1993. *Language and thought.* Wakefield, RI, Moyer Bell.

Chorover, S. L. 1979. *From genesis to genocide: The meaning of human nature and the power of behavior control.* Cambridge, MA, MIT Press.

Christie, J. F., S. J. Stone, et al., eds. 2002. *Play in same-age and multiage grouping arrangements.* Westport, CT, Ablex.

Chrousos, G. P. 1998. "Stressors, stress, and neuroendocrine integration of the adaptive response. The 1997 Hans Selye Memorial Lecture." *Annals of the New York Academy of Sciences* 851: 311–35.

Chugani, H. T. 1994. Development of regional brain glucose metabolism in relation to behavior and plasticity. In *Human behavior and the developing brain,* ed. G. Dawson and K. W. Fischer. New York, Guilford: 153–75.

———. 1998. "Biological basis of emotions: Brain systems and brain development." *Pediatrics* 102(5 Suppl. E): 1225–29.

Chugani, H. T., M. E. Behen, et al. 2001. "Local brain functional activity following early deprivation: A study of postinstitutionalized Romanian orphans." *Neuroimage* 14(6): 1290–1301.

Chugani, D. C., and H. T. Chugani. 2000. "PET: Mapping of serotonin synthesis." *Advances in Neurology* 83: 165–71.

Chugani, D. C., O. Muzik, et al. 2001. "Postnatal maturation of human GABAA receptors measured with positron emission tomography." *Annals of Neurology* 49(5): 618–26.

Chugani, H. T., M. E. Phelps, et al. 1987. "Positron emission tomography study of human brain functional development." *Annals of Neurology* 22: 487–97.

Chwalisz, K., E. Diener, et al. 1988. "Autonomic arousal feedback and emotional experience: Evidence from the spinal cord injured." *Journal of Personality and Social Psychology* 54: 820–28.

Clark, C. B. 1977. A preliminary report on weaning among chimpanzees of the Gombe National Park, Tanzania. In *Primate bio-social development: Biological, social and ecological determinants,* ed. S. Chevalier-Skolnikoff and F. E. Poirier. New York, Garland.

Clark, M. M., and B. G. Galef. 1999. "A testosterone-mediated trade-off between parental and sexual effort in male Mongolian gerbils (*Meriones unguiculatus*)." *Journal of Comparative Psychology* 113(4): 388–95.

———. 2000. "Why some male Mongolian gerbils may help at the nest: Testosterone, asexuality and alloparenting." *Animal Behaviour* 59(4): 801–6.

———. 2001. "Age-related changes in paternal responses of gerbils parallel changes in their testosterone concentrations." *Developmental Psychobiology* 39(3): 179–87.

Clendinen, D., and A. Nagourney. 1999. *Out for good: The struggle to build a gay rights movement in America.* New York, Simon and Schuster.

Cline, H. T. 2001. "Dendritic arbor development and synaptogenesis." *Current Opinion in Neurobiology* 11(1): 118–26.

Clutton-Brock, T. H., ed. 1988. *Reproductive success: Studies of individual variation in contrasting breeding systems.* Chicago, Univ. of Chicago Press.

———. 1989. "Female transfer and inbreeding avoidance in social mammals." *Nature* 337: 70–72.

———. 1991. *The evolution of parental care.* Princeton, Princeton Univ. Press.

———. 2002. "Breeding together: Kin selection and mutualism in cooperative vertebrates." *Science* 296(5565): 69–72.

———. 2007. "Sexual selection in males and females." *Science* 318(5858): 1882–85.

———. 2009. "Cooperation between non-kin in animal societies." *Nature* 462: 51–57.

Clutton-Brock, T. H., and C. Godfray. 1991. Parental investment. In *Behavioural ecology: An evolutionary approach,* ed. J. R. Krebs and N. B. Davies. Oxford, Blackwell: 234–62.

Clutton-Brock, T. H., F. E. Guinness, et al. 1982. *Red deer: Behavior and ecology of two sexes.* Chicago, Univ. of Chicago Press.

Clutton-Brock, T. H., S. J. Hodge, et al. 2006. "Intrasexual competition and sexual selection in cooperative mammals." *Nature* 444(7122): 1065–68.

Cochran, M., and S. Niego. 2002. Parenting and social networks. In *Handbook of parenting,* vol. 4: *Social conditions and applied parenting,* 2nd ed., ed. Marc H. Bornstein. Mahwah, NJ, Erlbaum: 123–46.

Coco, M. L., and J. M. Weiss. 2005. "Neural substrates of coping behavior in the rat: Possible importance of mesocorticolimbic dopamine system." *Behavioral Neuroscience* 119(2): 429–45.

Codere, H. 1950. *Fighting with property: A study of Kwakiutl potlatching and warfare, 1792–1930.* Seattle, Univ. of Washington Press.

Codner, M. A., and R. D. Nadler. 1984. "Mother-infant separation and reunion in the great apes." *Primates* 25(2): 204–17.

Coe, C., S. Mendoza, et al. 1978. "Mother-infant attachment in the squirrel monkey: Adrenal response to separation." *Behavioral Biology* 22: 256–63.

Coe, C. L., and C. M. Erickson. 1997. "Stress decreases lymphocyte cytolytic activity in the young monkey even after blockade of steroid and opiate hormone receptors." *Developmental Psychobiology* 30(1): 1–10.

Coe, C. L., and N. R. Hall. 1996. "Psychological disturbance alters thymic and adrenal hormone secretion in a parallel but independent manner." *Psychoneuroendocrinology* 21(2): 237–47.

Coe, C. L., S. G. Wiener, et al. 1985. Endocrine and immune responses to separation and maternal loss in nonhuman primates. In *Psychobiology of attachment and separation,* ed. M. Reite and T. Field. Orlando, Academic: 163–99.

Cohen, A. A. 2004. "Female post-reproductive lifespan: A general mammalian trait." *Biological Reviews of the Cambridge Philosophical Society* 79(4): 733–50.

Cohen-Bendahan, C. C. C., C. van de Beek, et al. 2005. "Prenatal sex hormone effects on child and adult sex-typed behavior: Methods and findings." *Neuroscience and Biobehavioral Reviews* 29(2): 353–84.

Cohen Kadosh, K., and M. H. Johnson. 2007. "Developing a cortex specialized for face perception." *Trends in Cognitive Sciences* 11(9): 367–69.

Cohen-Kettenis, P. T. 2005. "Gender change in 46,XY persons with 5alpha-reductase-2 deficiency and 17beta-hydroxysteroid dehydrogenase-3 deficiency." *Archives of Sexual Behavior* 34(4): 399–410.

Colbert, E. H. 1958. Morphology and behavior. In *Behavior and evolution,* ed. A. Roe and G. G. Simpson. New Haven, Yale Univ. Press: 27–47.

Colby, A., L. Kohlberg, et al. 1983. "A longitudinal study of moral judgment." *Monographs of the Society for Research in Child Development* 48(1–2): 124.

Cole, M. 1997. Cultural mechanisms of cognitive development. In *Change and development: Issues of theory, method, and application,* ed. E. Amsel and K. A. Renninger. Mahwah, NJ, Erlbaum: 245–63.

Coles, C. D., K. A. Platzman, et al. 1997. "A comparison of children affected by prenatal alcohol exposure and attention deficit, hyperactivity disorder." *Alcoholism, Clinical and Experimental Research* 21(1): 150–61.

Coles, R. 1990. *The spiritual life of children.* Boston, Houghton Mifflin.

Collaer, M. L., and M. Hines. 1995. "Human behavioral sex differences: A role for gonadal hormones during early development?" *Psychological Bulletin* 118(1): 55–107.

Collias, N. 1956. "The analysis of socialization in sheep and goats." *Ecology* 37: 228–39.

Collias, N. E. 2000. "Filial imprinting and leadership among chicks in family integration of the domestic fowl." *Behaviour* 137(2): 197–211.

Collins, W. A., and L. Steinberg. 2006. Adolescent development in interpersonal context. In *Handbook of child psychology,* vol. 3: *Social, emotional, and personality development,* ed. N. Eisenberg, W. Damon, and R. M. Lerner. Hoboken, NJ, Wiley: 1003–67.

Comrie, B. 1989. *Language universals and linguistic typology: Syntax and morphology,* 2nd ed. Chicago, Univ. of Chicago Press.

Conard, N. J., M. Malina, et al. 2009. "New flutes document the earliest musical tradition in southwestern Germany." *Nature* 460(7256): 737–40.

Conel, J. 1939–63. *The postnatal development of the human cerebral cortex,* vols. 1–8. Cambridge, MA, Harvard Univ. Press.

Conroy, G. C., and K. Kuykendall. 1995. "Paleopediatrics: Or when did human infants really become human?" *American Journal of Physical Anthropology* 98: 121–31.

Conroy, G. C., and M. W. Vannier. 1987. "Dental development of the Taung skull from computerized tomography." *Nature* 329(6140): 625–27.

———. 1988. "The nature of Taung dental maturation continued." *Nature* 333(6176): 808.

Conway Morris, S. 1989. "Burgess shale faunas and the Cambrian explosion." *Science* 246: 339–46.

———. 1998. *The crucible of creation: The Burgess Shale and the rise of animals.* Oxford, Oxford Univ. Press.

Cook, R. J., B. M. Dickens, et al. 2002. "Female genital cutting (mutilation/circumcision): Ethical and legal dimensions." *International Journal of Gynaecology and Obstetrics* 79(3): 281–87.

Coplan, J. D., M. W. Andrews, et al. 1996. "Persistent elevations of cerebrospinal fluid concentrations of corticotropin-releasing factor in adult nonhuman primates exposed to early-life stressors: Implications for the pathophysiology of mood and anxiety disorders." *Proceedings of the National Academy of Sciences* 93(4): 1619–23.

Coplan, J. D., L. A. Rosenblum, et al. 1995. "Primate models of anxiety: Longitudinal perspectives." *Psychiatric Clinics of North America* 18(4): 727–43.

Coplan, J. D., E. L. P. Smith, et al. 2006. "Maternal-infant response to variable foraging demand in nonhuman primates: Effects of timing of stressor on cerebrospinal fluid corticotropin-releasing factor and circulating glucocorticoid concentrations." *Annals of the New York Academy of Sciences* 1071: 525–33.

Coplan, J. D., R. C. Trost, et al. 1998. "Cerebrospinal fluid concentrations of somatostatin and biogenic amines in grown primates reared by mothers exposed to manipulated foraging conditions." *Archives of General Psychiatry* 55(5): 473–77.

Coqueugniot, H., J. J. Hublin, et al. 2004. "Early brain growth in *Homo erectus* and implications for cognitive ability." *Nature* 431(7006): 299–302.

Corbett, K. S. 2000. "Explaining infant feeding style of low-income black women." *Journal of Pediatric Nursing* 15(2): 73–81.

Corbett, L., and A. Newsome. 1975. Dingo society and its maintenance: A preliminary analysis. In *The wild canids*, ed. M. W. Fox. New York, Van Nostrand Reinhold.

Coser, L. 1956. *The functions of social conflict.* New York, Free Press.

Cosgrove, K. P., C. M. Mazure, et al. 2007. "Evolving knowledge of sex differences in brain structure, function, and chemistry." *Biological Psychiatry* 62(8): 847–55.

Cosmides, L., and J. Tooby. 2000. The cognitive neuroscience of social reasoning. In *The new cognitive neurosciences*, 2nd ed., ed. M. Gazzaniga. Cambridge, MA, MIT Press: 1259–70.

Coss, R. G., and A. Globus. 1978. "Spine stems on tectal interneurons in jewel fish are shortened by social stimulation." *Science* 200: 787–90.

Costa, P. T., Jr., R. R. McCrae, et al. 2000. Personality development from adolescence through adulthood: Further cross-cultural comparisons of age differences. In *Temperament and personality development across the life span*, ed. V. J. Molfese. Mahwah, NJ, Erlbaum: 235–52.

Costabile, A., P. K. Smith, et al. 1991. "Cross-national comparison of how children distinguish serious and playful fighting." *Developmental Psychology* 27(5): 881–87.

Costello, E. J., A. Angold, et al. 1996. "The Great Smoky Mountains study of youth: Goals, design, methods, and the prevalence of DSM-III-R disorders." *Archives of General Psychiatry* 53(12): 1129–36.

Costello, E. J., H. Egger, et al. 2005. "10-year research update review: The epidemiology of child and adolescent psychiatric disorders: I. Methods and public health burden." *Journal of the American Academy of Child and Adolescent Psychiatry* 44(10): 972–86.

Costello, E. J., D. L. Foley, et al. 2006. "10-year research update review: The epidemiology of child and adolescent psychiatric disorders: II. Developmental epidemiology." *Journal of the American Academy of Child and Adolescent Psychiatry* 45(1): 8–25.

Couillard-Despres, S., J. Winkler, et al. 2001. "Molecular mechanisms of neuronal migration disorders, quo vadis?" *Current Molecular Medicine* 1(6): 677–88.

Counsell, S. J., E. F. Maalouf, et al. 2002. "MR imaging assessment of myelination in the very preterm brain." *AJNR: American Journal of Neuroradiology* 23(5): 872–81.

Count, E. W. 1947. "Brain and body weight in man: Their antecedents in growth and evolution." *Annals of the New York Academy of Sciences* 46: 993–1122.

———. 1973. *Being and becoming human: Essays on the biogram.* New York, Van Nostrand Reinhold.

Cournoyer, D. E., and B. P. Malcolm. 2004. "Evaluating claims for universals: A method analysis approach." *Cross-Cultural Research* 38(4): 319–42.

Crabbe, J. C. 2002. "Genetic contributions to addiction." *Annual Review of Psychology* 53(1): 435–62.

Craig, I. W., and K. E. Halton. 2009. "Genetics of human aggressive behaviour." *Human Genetics* 126(1): 101–13.

Crain, S. 1991. "Language acquisition in the absence of experience (article with commentary and response)." *Behavioral and Brain Sciences* 14: 597–650.

Crawford, S. S., and E. K. Balon. 1996. Cause and effect of parental care in fishes: An epigenetic perspective. In *Parental care: Evolution, mechanisms, and adaptive significance. Advances in the study of behavior,* vol. 25, ed. J. S. Rosenblatt and C. T. Snowdon. San Diego, Academic: 53–107.

Crispo, E. 2007. The Baldwin effect and genetic assimilation: Revisiting two mechanisms of evolutionary change mediated by phenotypic plasticity. *Evolution* 61(11): 2469–79.

Crognier, E., A. Baali, and M. K. Hilali. 2001. "Do 'helpers at the nest' increase their parents' reproductive success?" *American Journal of Human Biology* 13(3): 365–73.

Crognier, E., M. Villena, et al. 2002. "Helping patterns and reproductive success in Aymara communities." *American Journal of Human Biology* 14(3): 372–79.

Cronk, L. 1994. Reciprocity and the power of giving. In *The developing structure of temperament and personality from infancy to adulthood,* ed. C. F. J. Halverson, G. A. Kohnstamm, and R. P. Martin. Hillsdale, NJ, Erlbaum: 157–63.

Crouter, A. C., and A. Booth, eds. 2006. *Romance and sex in adolescence and emerging adulthood: Risks and opportunities.* Mahwah, NJ, Erlbaum.

Crump, M. L. 1996. Parental care among the amphibia. In *Parental care: Evolution, mechanisms, and adaptive significance. Advances in the study of behavior,* vol. 25, ed. J. S. Rosenblatt and C. T. Snowdon. San Diego, Academic: 109–44.

Cummins, S. F., A. E. Nichols, et al. 2005. "Aplysia seductin is a water-borne protein pheromone that acts in concert with attractin to stimulate mate attraction." *Peptides* 26(3): 351–59.

Cunnane, S. C., and M. A. Crawford. 2003. "Survival of the fattest: Fat babies were the key to evolution of the large human brain." *Comparative Biochemistry and Physiology Part A, Molecular and Integrative Physiology* 136(1): 17–26.

Curran, L. 1999. "An analysis of cycles in skirt lengths and widths in the UK and Germany, 1954–1990." *Clothing and Textiles Research Journal* 17(2): 65–72.

Curran, T., and G. D'Arcangelo. 1998. "Role of reelin in the control of brain development." *Brain Research—Brain Research Reviews* 26(2–3): 285–94.

Cutright, P. 1972. "The teenage sexual revolution and the myth of an abstinent past." *Family Planning Perspectives* 4: 24–31.

D'Andrade, R. 1966. Sex differences and cultural institutions. In *The development of sex differences,* ed. E. E. Maccoby. Stanford, Stanford Univ. Press: 174–204.

———. 1987. A folk model of the mind. In *Cultural models of language and thought,* ed. D. Holland and N. Quinn. New York, Cambridge Univ. Press: 112–48.

———. 1990. Some propositions about the relations between culture and human cognition. In *Cultural psychology: Essays on comparative human development,* ed. J. W. Stigler, R. A. Schweder, and G. Herdt. New York, Cambridge Univ. Press: 67–129.

———. 1992. Schemas and motivation. In *Human motives and cultural models,* ed. R. D'Andrade and C. Strauss. New York, Cambridge Univ. Press: 23–44.

———. 2000. "The sad story of anthropology, 1950–99." *Cross-Cultural Research: The Journal of Comparative Social Science* 34(3): 219–32.

———. 2001. "A cognitivist's view of the units debate in cultural anthropology." *Cross-Cultural Research: The Journal of Comparative Social Science* 35(2): 242–57.

Dabbs, J. M., and R. Morris. 1990. "Testosterone, social class, and antisocial behavior in a sample of 4,462 men." *Psychological Science* 1(3): 209–11.

Dabbs, J. M., T. S. Carr, et al. 1995. "Testosterone, crime, and misbehavior among 692 male prison inmates." *Personality and Individual Differences* 18(5): 627–33.

Daenen, E. W. P. M., G. Wolterink, et al. 2002. "The effects of neonatal lesions in the amygdala or ventral hippocampus on social behaviour later in life." *Behavioural Brain Research* 136(2): 571–82.

Dahlberg, F., ed. 1981. *Woman the gatherer.* New Haven, Yale Univ. Press.

Daly, M., and M. Wilson. 1983. *Sex, evolution and behavior,* 2nd ed. Boston: Willard Grant Press.

———. 1985. "Child abuse and other risks of not living with both parents." *Ethology and Sociobiology* 6(4): 197–210.

———. 1988a. "Evolutionary social psychology and family homicide." *Science* 242(4878): 519–24.

———. 1988b. *Homicide.* New York, Aldine de Gruyter.

———. 1996. "Violence against stepchildren." *Current Directions in Psychological Science* 5(3): 77–81.

———. 1998. *The truth about Cinderella: A Darwinian view of parental love.* New Haven, Yale Univ. Press.

Damasio, A. 1994. *Descartes' error: Emotion, reason, and the human brain.* New York, Putnam's.

Damasio, A. R. 1995. "On some functions of the human prefrontal cortex." *Annals of the New York Academy of Sciences* 769: 241–51.

———. 1998. "Emotion in the perspective of an integrated nervous system." *Brain Research Reviews* 26(2–3): 83–86.

Damasio, H., T. Grabowski, et al. 1994. "The return of Phineas Gage: Clues about the brain from the skull of a famous patient." *Science* 264: 1102–5.

Damiani, R., S. Modesto, et al. 2003. "Earliest evidence of cynodont burrowing." *Proceedings of the Royal Society of London B: Biological Sciences* 270(1525): 1747–51.

Damon, W. 1980. "Patterns of change in children's social reasoning: A two-year longitudinal study." *Child Development* 51: 1010–17.

Dart, R. 1925. "Australopithecus Africanus: The man ape of South Africa." *Nature* 115: 195–99.

Darwin, C. 1958 [1859]. *The origin of species by means of natural selection, or the preservation of favored races in the struggle for life.* New York, New American Library.

———. 1981 [1871]. *The descent of man, and selection in relation to sex.* Princeton, Princeton Univ. Press.

———. 1872. *Expression of the emotions in man and animals.* London, J. Murray.

———. 1956 [1877]. Biographical sketch of an infant. In *The Darwin reader,* ed. M. Bates and P. S. Humphrey. New York, Scribner's: 403–11.

———. 1998. *The expression of the emotions in man and animals.* 3rd ed. New York, Oxford Univ. Press.

Davatzikos, C., and S. M. Resnick. 1998. "Sex differences in anatomic measures of interhemispheric connectivity: Correlations with cognition in women but not men." *Cerebral Cortex* 8(7): 635–40.

Davenport, R. K. 1979. Some behavioral disturbances of great apes in captivity. In *The great apes,* ed. D. A. Hamburg and E. R. McCown. Reading, MA, Benjamin/Cummings: 356.

David, M., H. K. von Schwarzenfeld, et al. 1999. "Perinatal outcome in hospital and birth center obstetric care." *International Journal of Gynaecology and Obstetrics* 65(2): 149–56.

Davidson, J. M., C. A. Camargo, et al. 1979. "Effects of androgen on sexual behavior in hypogonadal men." *Journal of Clinical Endocrinology and Metabolism* 48(6): 955–58.

Davidson, R. J. 2000. The neuroscience of affective style. In *The new cognitive neurosciences*, ed. M. S. Gazzaniga et al. Cambridge, MA, MIT Press: 1149.

———. 2004. "Well-being and affective style: Neural substrates and biobehavioural correlates." *Philosophical Transactions of the Royal Society of London B: Biological Sciences* 359(1449): 1395–1411.

Davidson, R. J., D. A. Lewis, et al. 2002. "Neural and behavioral substrates of mood and mood regulation." *Biological Psychiatry* 52(6): 478–502.

Davies, W., and L. S. Wilkinson. 2006. "It is not all hormones: Alternative explanations for sexual differentiation of the brain." *Brain Research* 1126(1): 36–45.

Davis, J. N., and M. Daly. 1997. "Evolutionary theory and the human family." *Quarterly Review of Biology* 72(4): 407–35.

Davis, M., and P. J. Whalen. 2001. "The amygdala: Vigilance and emotion." *Molecular Psychiatry* 6(1): 13–34.

Davis, P. G., and P. Gandleman. 1972. "Pup-killing produced by the administration of testosterone propionate to adult female mice." *Hormones and Behavior* 3: 169–73.

Davis-Floyd, R. E. 1992. *Birth as an American rite of passage.* Berkeley, Univ. of California Press.

Dawkins, R. 1971. Selective neurone death as a possible memory mechanism. *Nature* 229: 118–19.

———. 1976. *The selfish gene.* Oxford: Oxford Univ. Press.

———. 1987. *The blind watchmaker.* New York, Norton.

De Benedictis, R. 1973. "The behavior of young primates during adult copulation." *American Anthropologist* 75: 1469–85.

De Bono, M. 2003. "Molecular approaches to aggregation behavior and social attachment." *Journal of Neurobiology* 54(1): 78–92.

De Jong, T. R., M. Chauke, et al. 2009. "From here to paternity: Neural correlates of the onset of paternal behavior in California mice *(Peromyscus californicus)*." *Hormones and Behavior* 56(2): 220–31.

De Jong, B. M., K. L. Leenders, et al. 2002. "Right parieto-premotor activation related to limb-independent antiphase movement." *Cerebral Cortex* 12(11): 1213–17.

De Lacoste-Utamsing, M. C., and R. L. Holloway. 1982. "Sexual dimorphism in human corpus callosum." *Science* 216: 1431–32.

De la Rosa, E. J., and F. de Pablo. 2000. "Cell death in early neural development: Beyond the neurotrophic theory." *Trends in Neurosciences* 23(10): 454–58.

De Oliveira, C. R., C. R. Ruiz-Miranda, et al. 2003. "Play behavior in juvenile golden lion tamarins (Callitrichidae: Primates): Organization in relation to costs." *Ethology* 109(7): 593–612.

De Waal, F. 1989a. *Chimpanzee politics: Power and sex among apes.* Baltimore, Johns Hopkins Univ. Press.

De Waal, F. B. M. 1989b. Behavioral contrasts between bonobo and chimpanzee. In *Understanding chimpanzees*, ed. P. G. Heltne and L. A. Marquardt. Cambridge, MA, Harvard Univ. Press/Chicago Academy of Sciences: 154–75.

De Waal, F. B. M., and P. L. Tyack, eds. 2003. *Animal social complexity: Intelligence, culture, and individualized societies.* Cambridge, MA: Harvard Univ. Press.

Deacon, T. W. 1990. "Rethinking mammalian brain evolution." *American Zoologist* 30: 629–705.

———. 1997a. *The symbolic species: The co-evolution of language and the brain.* New York, Norton.

———. 1997b. "What makes the human brain different?" *Annual Review of Anthropology* 26: 337–514.

Deag, J. M. 1980. "Interactions between males and unweaned barbary macaques: Testing the agonistic buffering hypothesis." *Behaviour* 75: 54–81.

Deag, J., and J. Crook. 1971. "Social behavior and agonistic buffering in the Wild Barbary Macaque *Macaca sylvana.*" *Folia Primatologica* 15: 183–200.

Dean, C. 2000. "Progress in understanding hominoid dental development." *Journal of Anatomy* 197(Pt. 1): 77–101.

Dean, C., M. G. Leakey, et al. 2001. "Growth processes in teeth distinguish modern humans from *Homo erectus* and earlier hominins." *Nature (London)* 414(6864): 628–31.

DeCasper, A. J., and W. P. Fifer. 1980. "Of human bonding: Newborns prefer their mothers' voices." *Science* 208(4448): 1174–76.

DeCasper, A. J., J.-P. Lecanuet, et al. 1994. "Fetal reactions to recurrent maternal speech." *Infant Behavior and Development* 17(2): 159–64.

DeCasper, A. J., and P. A. Prescott. 1984. "Human newborns' perception of male voices: Preference, discrimination, and reinforcing value." *Developmental Psychobiology* 17(5): 481–91.

DeCasper, A. J., and M. J. Spence. 1986. "Prenatal maternal speech influences newborns' perception of speech sounds." *Infant Behavior and Development* 9: 133–50.

Defleur, A., T. White, et al. 1999. "Neanderthal cannibalism at Moula-Guercy, Ardeche, France [see comment]." *Science* 286(5437): 128–31.

DeGusta, D. 1999. "Fijian cannibalism: Evidence from Navatu." *American Journal of Physical Anthropology* 110(October): 215–41.

Dehaene-Lambertz, G. 2004. "Bases cérébrale de l'acquisition du langage: Apport de la neuro-imagerie." *Neuropsychiatrie de l'Enfance et de l'Adolescence* 52(7): 45259.

Dehaene-Lambertz, G., L. Hertz-Pannier, et al. 2006. "Functional organization of perisylvian activation during presentation of sentences in preverbal infants." *Proceedings of the National Academy of Sciences* 103(38): 14240–45.

Dekaban, A. S., and D. Sadowsky. 1978. "Changes in brain weights during the span of human life: Relation of brain weight to body heights and body weights." *Annals of Neurology* 4: 345–56.

Delemarre-van de Waal, H. A. 2005. "Secular trend of timing of puberty." *Endocrine Development* 8: 1–14.

Demaree, H. A., D. E. Everhart, et al. 2005. "Brain lateralization of emotional processing: Historical roots and a future incorporating 'dominance.'" *Behavioral and Cognitive Neuroscience Reviews* 4(1): 3–20.

Demaret, A. 1999. "Some ethological data about violence." *Cahiers d'Ethologie* 19(2): 137–60.

DeMause, L., ed. 1974. *The history of childhood.* New York, Harper & Row.

Demerens, C., B. Stankoff, et al. 1996. "Induction of myelination in the central nervous system by electrical activity." *Proceedings of the National Academy of Sciences* 93(18): 9887–92.

Demos, John. 2009. Personal communication, Sept. 18, 2009, used by permission.

Denenberg, V., J. Woodcock, et al. 1968. "Long-term effects of preweaning and postweaning free environment experience on rats' problem-solving behavior." *Journal of Comparative and Physiological Psychology* 66: 533–35.

Denenberg, V. H. 1964. "Critical periods, stimulus input, and emotional reactivity: A theory of infantile stimulation." *Psychological Review* 71: 355–51.

———. 1999. "Commentary: Is maternal stimulation the mediator of the handling effect in infancy?" *Developmental Psychobiology* 34(1): 1–3.

Denenberg, V. H., R. Paschke, et al. 1968. "Killing of mice by rats prevented by early interaction between the two species." *Psychonomic Science* 11(1): 39.

Denenberg, V. H., K. M. Rosenberg, et al. 1969. "Mice reared with rat aunts: Effects on plasma corticosterone and open field activity." *Nature* 221: 73–74.

Denenberg, V. H., G. Sherman, et al. 1996. "Effects of embryo transfer and cortical ectopias upon the behavior of BXSB-Yaa and BXSB-Yaa + mice." *Brain Research: Developmental Brain Research* 93(1–2): 100–8.

Denes, A. S., G. Jekely, et al. 2007. "Molecular architecture of annelid nerve cord supports common origin of nervous system centralization in bilateria [see comment]." *Cell* 129(2): 277–88.

Dennis, W. 1940. "Does culture appreciably affect patterns of infant behavior?" *Journal of Social Psychology* 12: 305–17.

Dennis, W., and M. G. Dennis. 1940. "The effect of cradling practices upon the onset of walking in Hopi children." *Pedagogical Seminary and Journal of Genetic Psychology* 56: 77–86.

Derluyn, I., E. Broekaert, et al. 2004. "Post-traumatic stress in former Ugandan child soldiers [see comment]." *Lancet* 363(9412): 861–63.

Dessureau, B. K., C. O. Kurowski, et al. 1998. "A reassessment of the role of pitch and duration in adults' responses to infant crying." *Infant Behavior and Development* 21(2): 367–71.

DeThorne, L. S., S. A. Petrill, et al. 2008. "Genetic effects on children's conversational language use." *Journal of Speech Language and Hearing Research* 51(2): 423–35.

Dettling, A., C. R. Pryce, et al. 1998. "Physiological responses to parental separation and a strange situation are related to parental care received in juvenile Goeldi's monkeys (*Callimico goeldii*)." *Developmental Psychobiology* 33(1): 21–31.

Dettling, A. C., J. Feldon, et al. 2002. "Early deprivation and behavioral and physiological responses to social separation/novelty in the marmoset." *Pharmacology, Biochemistry and Behavior* 73(1): 259–69.

Devane, D., J. Lalor, et al. 2007. "The use of intrapartum electronic fetal heart rate monitoring: A national survey." *Irish Medical Journal* 100(2): 360–62.

DeVore, I. 1963. Mother-infant relations in free-ranging baboons. In *Maternal behavior in mammals,* ed. H. L. Rheingold. New York, Wiley.

———. 1965a. Male dominance and mating behavior in baboons. In *Sexual behavior,* ed. F. Beach. New York, Wiley.

———. 1965b. *Primate behavior.* New York, Holt, Rinehart, and Winston.

DeVore, I., and M. J. Konner. 1974. Infancy in hunter-gatherer life: An ethological perspective. In *Ethology and psychiatry,* ed. N. White. Toronto, Univ. of Toronto Press.

Diamond, A. 1991. Frontal lobe involvement in cognitive changes during the first year of life. In *Brain maturation and cognitive development: Comparative and cross-cultural perspectives,* ed. K. R. Gibson and A. C. Petersen. New York, Aldine de Gruyter: 127–80.

Diamond, A., J. F. Werker, et al. 1994. Toward understanding commonalities in the development of object search, detour navigation, categorization, and speech perception. In *Human behavior and the developing brain,* ed. G. Dawson and K. W. Fischer. New York, Guilford: 380–426.

Diamond, J., and A. Bond. 2003. "A comparative analysis of social play in birds." *Behaviour* 140: 1091–1115.

Diamond, L. M. 2008. "Female bisexuality from adolescence to adulthood: Results from a 10-year longitudinal study." *Developmental Psychology* 44(1): 5–14.

Diamond, M., and H. K. Sigmundson. 1997. "Sex reassignment at birth: Long-term review and clinical implications." *Archives of Pediatric and Adolescent Medicine* 151(3): 298–304.

Diamond, M. C. 1991. Environmental influences on the young brain. In *Brain maturation and cognitive development: Comparative and cross-cultural perspectives*, ed. K. R. Gibson and A. C. Petersen. New York, Aldine de Gruyter: 107–24.

Diamond, M. C., R. E. Johnson, et al. 1985. "Plasticity in the 904-day-old male rat cerebral cortex." *Experimental Neurology* 87: 309–17.

Diamond, M. C., D. Krech, et al. 1964. "The effects of an enriched environment on the histology of the rat cerebral cortex." *Journal of Comparative Neurology* 123: 111–20.

Diamond, R., S. Carey, et al. 1983. "Genetic influences on the development of spatial skills during early adolescence." *Cognition* 13(2): 167–85.

Dickeman, M. 1975. "Demographic consequences of infanticide in man." *Annual Review of Ecology and Systematics* 6: 107–37.

Dietrich, R. B., W. G. Bradley, et al. 1988. "MR evaluation of early myelination patterns in normal and developmentally delayed infants." *American Journal of Neuroradiology* 9: 69–76.

Dirix, C. E. H., J. G. Nijhuis, et al. 2009. "Aspects of fetal learning and memory." *Child Development* 80(4): 1251–58.

Divale, W., and M. Harris. 1976. "Population, warfare, and the male supremacist complex." *American Anthropologist* 78: 521–38.

Dixon, A. F. 1980. "Androgen and aggressive behavior in primates: A review." *Aggressive Behavior* 6: 37–68.

Dixon, L., K. Browne, et al. 2005. "Risk factors of parents abused as children: A mediational analysis of the intergenerational continuity of child maltreatment (Part I)." *Journal of Child Psychology and Psychiatry* 46(1): 47–57.

Dixon, L., C. Hamilton-Giachritsis, et al. 2005. "Attributions and behaviours of parents abused as children: A mediational analysis of the intergenerational continuity of child maltreatment (Part II)." *Journal of Child Psychology and Psychiatry* 46(1): 58–68.

Dobbing, J., and J. Sands. 1979. "Comparative aspects of the brain growth spurt." *Early Human Development* 3(1): 79–83.

Dobson, F. S. 2007. A lifestyle view of life-history evolution. *Proceedings of the National Academy of Sciences* 104(45): 17565–66.

Dodge, K. A., M. T. Greenberg, et al. 2008. "Testing an idealized dynamic cascade model of the development of serious violence in adolescence." *Child Development* 79(6): 1907–27.

Dolan, R. J. 2002. "Emotion, cognition, and behavior." *Science* 298(5596): 1191–94.

Dolhinow, P., J. J. McKenna, et al. 1979. "Rank and reproduction among female langur monkeys: Aging and improvement (they're not just getting older, they're getting better)." *Aggressive Behavior* 5(1): 19–30.

Dolhinow, P., and M. A. Taff. 1993. "Immature and adult langur monkey (*Presbytis entellus*) males: Infant-initiated adoption in a colony group." *International Journal of Primatology* 14(6): 919–26.

Dollard, J., L. W. Doob, et al. 1939. *Frustration and aggression.* New Haven, Yale Univ. Press.

Domjan, M. 1996. *The essentials of conditioning and learning.* Pacific Grove, CA, Brooks/ Cole.

———. 1998. *The principles of learning and behavior.* Pacific Grove, CA, Brooks/Cole.

———. 2005. "Pavlovian conditioning: A functional perspective." *Annual Review of Psychology* 56(1): 179–206.

Donald, M. 1991. *Origins of the modern mind: Three stages in the evolution of culture and cognition.* Cambridge, MA, Harvard Univ. Press.

Dondi, M., F. Simion, et al. 1999. "Can newborns discriminate between their own cry and the cry of another newborn infant?" *Developmental Psychology* 35(2): 418–26.

Donoghue, P. C. J., S. Bengtson, et al. 2006. "Synchrotron X-ray tomographic microscopy of fossil embryos." *Nature* 442(7103): 680–83.

Dorn, L. D., S. F. Hitt, et al. 1999. "Biopsychological and cognitive differences in children with premature vs. on-time adrenarche." *Archives of Pediatrics and Adolescent Medicine* 153(2): 137–46.

Dorus, S., E. J. Vallender, et al. 2004. "Accelerated evolution of nervous system genes in the origin of *Homo sapiens.*" *Cell* 119(7): 1027–40.

Dou, C. L., S. Li, et al. 1999. "Dual role of brain factor-1 in regulating growth and patterning of the cerebral hemispheres." *Cerebral Cortex* 9(6): 543–50.

Doudet, D., D. Hommer, et al. 1995. "Cerebral glucose metabolism, CSF 5-HIAA, and aggressive behavior in the rhesus monkey." *American Journal of Psychiatry* 152: 1782–87.

Douglas-Hamilton, I., and O. Douglas-Hamilton. 1978. *Among the elephants.* New York: Penguin.

Doussard-Roosevelt, J. A., L. A. Montgomery, et al. 2003. "Short-term stability of physiological measures in kindergarten children: Respiratory sinus arrhythmia, heart period, and cortisol." *Developmental Psychobiology* 43(3): 230–42.

Doyle, D. A., J. Morais Cabral, et al. 1998. "The structure of the potassium channel: Molecular basis of K+ conduction and selectivity." *Science* 280: 69–77.

Draper, P. 1973. "Crowding among hunter-gatherers: The !Kung bushmen." *Science* 182(4109): 301–3.

———. 1976. Social and economic constraints on child life among the !Kung. In *Kalahari hunter-gatherers,* ed. R. B. Lee and I. DeVore. Cambridge, MA, Harvard Univ. Press.

———. 1978. The learning environment for aggression and anti-social behavior among the !Kung (Kalahari Desert, Botswana, Africa). In *Learning non-aggression: The experience of non-literate societies,* ed. A. Montagu. New York, Oxford Univ. Press: 31–53.

Draper, P., and E. Cashdan. 1988. "Technological change and child behavior among the !Kung." *Ethnology* 27(4): 339–65.

Draper, P., and H. Harpending. 1982. "Father absence and reproductive strategy: An evolutionary perspective." *Journal of Anthropological Research* 38: 255–73.

———. 1987. Parent investment and the child's environment. In *Parenting across the life span: Biosocial dimensions,* ed. J. B. Lancaster et al. New York, Aldine de Gruyter: 207–36.

Draper, P., and N. Howell. 2005. The growth and kinship resources of Ju/'oansi children. In *Hunter-gatherer childhoods: Evolutionary, developmental and cultural per-*

spectives, ed. B. S. Hewlett and M. E. Lamb. New Brunswick, NJ, AldineTransaction: 262–81.

Drea, C. M. 2005. "Bateman revisited: The reproductive tactics of female primates." *Integrative and Comparative Biology* 45(5): 915–23.

Dru, A., ed. 1959. *The journals of Kierkegaard.* New York, Harper Torchbooks.

Drummond, K. D., S. J. Bradley, et al. 2008. "A follow-up study of girls with gender identity disorder." *Developmental Psychology* 44(1): 895–900.

Duberman, M., M. Vicinus, et al., eds. 1989. *Hidden history: Reclaiming the gay and lesbian past.* New York, Penguin/Meridian.

DuBois, C. 1944. *The people of Alor.* New York, Harper.

Dubois, J., G. Dehaene-Lambertz, et al. 2008a. Asynchrony of the early maturation of white matter bundles in healthy infants: Quantitative landmarks revealed noninvasively by diffusion tensor imaging. *Human Brain Mapping* 29(1): 14–27.

———. 2008b. Microstructural correlates of infant functional development: Example of the visual pathways. *Journal of Neuroscience* 28(8): 1943–48.

Dubois, J., L. Hertz-Pannier, et al. 2009. Structural asymmetries in the infant language and sensori-motor networks. *Cerebral Cortex* 19(2): 414–23.

Dudley, D. 1974a. "Contributions of paternal care to the growth and development of the young in *Peromyscus californicus.*" *Behavioral Biology* 11: 155–66.

———. 1974b. "Paternal behavior in the California mouse, *Peromyscus californicus.*" *Behavioral Biology* 11: 247–52.

Duffy, J. E., C. L. Morrison, et al. 2000. "Multiple origins of eusociality among sponge-dwelling shrimps (Synalpheus)." *Evolution* 54(2): 503–16.

Duffy, K. G., R. W. Wrangham,, et al. 2007. "Male chimpanzees exchange political support for mating opportunities." *Current Biology* 17(15): R586–87.

Dugatkin, L. 1992. "Sexual selection and imitation: Females copy the mate choice of others." *American Naturalist* 139: 1384–89.

———. 1996. Copying and mate choice. In *Social learning in animals: The roots of culture,* ed. B. G. Galef, Jr., and C. M. Heyes. San Diego, Academic: 85–105.

Dunbar, R. I. M. 1995. "Neocortex size and group size in primates: A test of the hypothesis." *Journal of Human Evolution* 28: 287–96.

———. 1998. "The social-brain hypothesis." *Evolutionary Anthropology* 6(5): 178–90.

———. 2003. Why are apes so smart? In *Primate life histories and socioecology,* ed. P. M. Kappeler and M. E. Pereira. Chicago, Univ. of Chicago Press: 285–98.

Dunger, D. B., M. L. Ahmed, et al. 2005. "Effects of obesity on growth and puberty." *Best Practice and Research Clinical Endocrinology and Metabolism* 19(3): 375–90.

Dunn, J. 1996. Sibling relationships and perceived self-competence: Patterns of stability between childhood and early adolescence. In *The five to seven year shift: The age of reason and responsibility,* ed. A. J. Sameroff and M. M. Haith. Chicago, Univ. of Chicago Press: 253–70.

Dunn, J., and C. Hughes. 2001. "'I got some swords and you're dead!': Violent fantasy, antisocial behavior, friendship, and moral sensibility in young children." *Child Development* 72(2): 491–505.

Dupanloup, I., L. Pereira, et al. 2003. "A recent shift from polygyny to monogamy in humans is suggested by the analysis of worldwide Y-chromosome diversity." *Journal of Molecular Evolution* 57(1): 85–97.

Durham, W. H. 1990. "Advances in evolutionary culture theory." *Annual Review of Anthropology* 19: 187–210.

———. 1991. *Coevolution: Genes, culture, and human diversity.* Stanford, Stanford Univ. Press.

———. 1992. "Applications of evolutionary culture theory." *Annual Review of Anthropology* 21: 331–55.

Eagly, A. H., and V. J. Steffen. 1986. "Gender and aggressive behavior: A meta-analytic review of the social psychological literature." *Psychological Bulletin* 100(3): 309–30.

Eakins, P. S. 1989. "Free-standing birth centers in California. Program and medical outcome." *Journal of Reproductive Medicine* 34(12): 960–70.

Early, J. D., and T. N. Headland. 1998. *Population dynamics of a Philippine rain forest people: The San Ildefonso Agta.* Gainesville, Univ. Press of Florida.

Eaton, S. B., S. B. Eaton III, et al. 1997. "Paleolithic nutrition revisited: A twelve-year retrospective on its nature and implications [see comment]." *European Journal of Clinical Nutrition* 51(4): 207–16.

Eaton, S. B., and S. B. I. Eaton. 1999. The evolutionary context of chronic degenerative diseases. In *Evolution in health and disease,* ed. S. C. Stearns. Oxford, Oxford Univ. Press: 251–59.

Eaton, S. B., and M. Konner. 1985. "Paleolithic nutrition. A consideration of its nature and current implications." *New England Journal of Medicine* 312(5): 283–89.

Eaton, S. B., M. Konner, et al. 1988. "Stone agers in the fast lane: Chronic degenerative disease in evolutionary perspective." *American Journal of Medicine* 84: 739–49.

Eaton, S. B., M. C. Pike, et al. 1994. "Women's reproductive cancers in evolutionary perspective." *Quarterly Review of Biology* 69(3): 353–67.

Eaves, L. J., H. J. Eysenck, et al. 1989. *Genes, culture and personality: An empirical approach.* New York, Academic.

Eaves, L. J., J. L. Silberg, et al. 1997. "Genetics and developmental psychopathology: 2. The main effects of genes and environment on behavioral problems in the Virginia Twin Study of Adolescent Behavioral Development." *Journal of Child Psychology and Psychiatry and Allied Disciplines* 38(8): 965–80.

Ebbesson, S. O. 1984. "Evolution and ontogeny of neural circuits." *Behavioral and Brain Sciences* 7: 321–66.

Edelman, G. M. 1987. *Neural Darwinism: The theory of neuronal group selection.* New York, Basic.

Edelman, G. M., and F. S. Jones. 1998. "Gene regulation of cell adhesion: A key step in neural morphogenesis." *Brain Research Reviews* 26(2–3): 337–52.

Edwards, C. P. 1975. "Societal complexity and moral development: A Kenyan study." *Ethos* 3(4): 505–27.

———. 1978. "Social experience and moral judgment in East African young adults." *Journal of Genetic Psychology* 133(1): 19–29.

———. 1981. The comparative study of the development of moral judgement and reasoning. In *Handbook of cross-cultural human development,* ed. R. H. Munroe, R. L. Munroe, and B. B. Whiting. New York, Garland STPM: 501–28.

———. 1993. Behavioral sex differences in children of diverse cultures: The case of nurturance to infants. In *Juvenile primates: Life history, development, and behavior,* ed. M. E. Pereira and L. A. Fairbanks. New York, Oxford Univ. Press: 327–38.

———. 1994. Cross-cultural research on Kohlberg's stages: The basis for consensus. In *New research in moral development,* ed. B. E. Puka. New York, Garland: 373–84.

Edwards, D. A. 1968. "Mice: Fighting by neonatally androgenized females." *Science* 161(3845): 1027–28.

———. 1969. "Early androgen stimulation and aggressive behavior in male and female mice." *Physiology and Behavior* 4(3): 333–38.

———. 1970. "Post-neonatal androgenization and adult aggressive behavior in female mice." *Physiology and Behavior* 5(4): 465–67.

Edwards, D. A., and K. G. Burge. 1971. "Early androgen treatment and male and female sexual behavior in mice." *Hormones and Behavior* 2(1): 49–58.

Edwards, D. A., F. R. Nahai, et al. 1993. "Pathways linking the olfactory bulbs with the medial preoptic anterior hypothalamus are important for intermale aggression in mice." *Physiology and Behavior* 53: 611–15.

Egeland, B., D. Jacobvitz, et al. 1987. Intergenerational continuity of abuse. In *Child abuse and neglect: Biosocial dimensions,* ed. R. J. Gelles and J. B. Lancaster. New York, Aldine de Gruyter: 255–76.

Eggan, D. 1943. "The general problem of Hopi adjustment." *American Anthropologist* 45: 357–73.

Ehrhardt, A. A., and H. F. L. Meyer-Bahlburg. 1981. "Effects of prenatal sex hormones on gender-related behavior." *Science* 211: 1312–18.

Ehrlich, A. 1974. "Infant development in two prosimian species: Greater galago and slow loris." *Developmental Psychobiology* 7(5): 439–54.

Eibl-Eibesfeldt, I. 1960. *Versuche über den Nestbau erfahrunsloser Ratten* (film). Institute für den Wissenschaftlichen.

———. 1971a. "Vorprogrammierung in menschlichen sozialverhalten." *Mitteilungen an die Max-Planck-Gesellschaft* 5: 307–38.

———. 1971b. "Zur Ethologie menschlichen Grussverhaltens: II. Das Gruss verhalten und einige andere Muster freundlicher Kontaktaufnahme der Waika-Indianer (Yanoama)." *Zeitschrift für Tierpsychologie* 29: 196–213.

———. 1973a. *Der vorprogrammierte mensch: Das Ererbte als bestimmender Faktor im menschlichen Verhalten.* Wien, Verlag Fritz Molden.

———. 1973b. The expressive behavior of the deaf-and-blind-born. In *Social communication and movement,* ed. M. V. Cranach and I. Vine. London, Academic: 163–94.

———. 1975. *Ethology.* New York, Holt, Rinehart and Winston.

———. 1979. "Human ethology: Concepts and implications for the sciences of man." *Behavioral and Brain Sciences* 2: 1–57.

———. 1983. Patterns of parent-child interaction in a cross-cultural perspective. In *The behavior of human infants,* ed. A. Oliverio and M. Zappella. New York, Plenum.

———. 1988. *Human ethology.* New York, Aldine de Gruyter.

Eimas, P. D. 1997. Infant speech perception: Processing characteristics, representational units, and the learning of words. In *Perceptual learning: The psychology of learning and motivation,* ed. R. Goldstone. San Diego, Academic: 127–69.

———. 2000. "Infant perception and cognition and the initial architecture of constructivist models." *Behavioral and Brain Sciences* 23(5): 791–92.

Eiseley, L. 1958. *Darwin's century.* New York, Doubleday.

Eisenberg, J. F. 1981. *The mammalian radiations: An analysis of trends in evolution, adaptation, and behavior.* Chicago, Univ. of Chicago Press.

———. 1988. Reproduction in polyprotodont marsupials and similar-sized eutherians with a speculation concerning the evolution of litter size in mammals. In *Evolution of life histories of mammals: Theory and pattern,* ed. M. S. Boyce. New Haven, Yale Univ. Press: 291–310.

Ekman, P. 1973. Cross-cultural studies of facial expression. In *Darwin and facial expression,* ed. P. Ekman. New York, Academic.

Ekman, P., W. V. Friesen, et al. 1987. "Universals and cultural differences in the judgements of facial expressions of emotion." *Journal of Personality and Social Psychology* 53: 712–17.

Ekman, P., and E. L. Rosenberg, eds. 1997. *What the face reveals: Basic and applied studies of spontaneous expression using the facial action coding system (FACS)*. New York, Oxford Univ. Press.

Elbedour, S., R. ten Bensel, et al. 1993. "Ecological integrated model of children of war: Individual and social psychology." *Child Abuse and Neglect* 17(6): 805–19.

Elder, G. H. 1974. *Children of the great depression: Social change in the life course*. Chicago, Univ. of Chicago Press.

Elder, G. H., V. King, et al. 1996. "Intergenerational continuity and change in rural lives: Historical and developmental insights." *International Journal of Behavioral Development* 19(2): 433–56.

Elder, G. H., T. V. Nguyen, et al. 1985. "Linking family hardship to children's lives." *Child Development* 56(2): 361–75.

Eldredge, N., and S. J. Gould. 1972. Punctuated equilibria: An alternative to phyletic gradualism. In *Models in paleobiology*, ed. T. J. M. Schopf. San Francisco, Freeman, Cooper: 82–115.

Elias, M. 1981. "Serum cortisol, testosterone, and testosterone-binding globulin responses to competitive fighting in human males." *Aggressive Behavior* 7(3): 215–24.

Elias, M., J. Teas, et al. 1986. "Nursing practices and lactation amenorrhoea." *Journal of Biosocial Science* 18: 1–10.

Elias, M. F. 1977. "Relative maturity of Cebus and squirrel monkeys at birth and during infancy." *Developmental Psychobiology* 10: 519–28.

Elias, M. F., N. A. Nicholson, et al. 1986. Two sub-cultures of maternal care in the United States. In *Current perspectives in primate social dynamics*, ed. D. M. Taub and F. A. King. New York, Van Nostrand Reinhold.

Elkin, A. P. 1964. *The Australian aborigines*. Garden City, NY, Doubleday/Anchor/American Museum of Natural History.

Elliman, D., and M. A. Lynch. 2000. "The physical punishment of children." *Archives of Disease in Childhood* 83(3): 196–98.

Ellis, B. J. 2004. "Timing of pubertal maturation in girls: An integrated life history approach." *Psychological Bulletin* 130(6): 920–58.

Ellis, B. J., and D. F. Bjorklund, eds. 2005. *Origins of the social mind: Evolutionary psychology and child development*. New York: Guilford.

Ellis, B. J., and W. Boyce. 2008. "Biological sensitivity to context." *Current Directions in Psychological Science* 17(3): 183–87.

Ellis, B. J., and M. J. Essex. 2007. "Family environments, adrenarche, and sexual maturation: A longitudinal test of a life history model." *Child Development* 78(6): 1799–1817.

Ellis, B. J., M. J. Essex, et al. 2005. "Biological sensitivity to context: II. Empirical explorations of an evolutionary-developmental theory." *Development and Psychopathology* 17(2): 303–28.

Ellis, B. J., J. J. Jackson, et al. 2006. "The stress response systems: Universality and adaptive individual differences." *Developmental Review* 26(2): 175–212.

Ellison, P. T., N. R. Peacock, et al. 1986. "Salivary progesterone and luteal function in two low-fertility populations of northeast Zaire." *Human Biology* 58: 473–83.

Elman, J. L. 2005. "Connectionist models of cognitive development: Where next?" *Trends in Cognitive Sciences* 9(3): 111–17.

Elman, J. L., E. A. Bates, et al. 1996. *Rethinking innateness: A connectionist perspective on development.* Cambridge, MA, MIT Press.

Eluvathingal, T. J., H. T. Chugani, et al. 2006. "Abnormal brain connectivity in children after early severe socioemotional deprivation: A diffusion tensor imaging study." *Pediatrics* 117(6): 2093–2100.

Ember, C. R. 1973. "Feminine task assignment and the social behavior of boys." *Ethos* 1: 424–39.

———. 1994. "Improvements in cross-cultural research methods." *Cross-Cultural Research* 28(4): 364–70.

Ember, C. R., and M. Ember. 1994. "War, socialization, and interpersonal violence: A cross-cultural study." *Journal of Conflict Resolution* 38(4): 620–46.

———. 2005. "Explaining corporal punishment of children: A cross-cultural study." *American Anthropologist* 107(4): 609–19.

Emde, R. N., T. J. Gaensbauer, et al. 1976. *Emotional expression in infancy: A biobehavioral study.* New York, International Universities Press.

Emde, R. N., and R. J. Harmon. 1972. "Endogenous and exogenous smiling systems in early infancy." *Journal of the American Academy of Child Psychiatry* 11: 177–200.

Emde, R. N., R. Plomin, et al. 1992. "Temperament, emotion, and cognition at fourteen months: The MacArthur Longitudinal Twin Study." *Child Development* 63: 1437–55.

Emlen, S. T. 1995. "An evolutionary view of the family." *Proceedings of the National Academy of Sciences* 92: 8092–99.

Emmons, S. W., and J. Lipton. 2003. "Genetic basis of male sexual behavior." *Journal of Neurobiology* 54(1): 93–110.

Enard, W., S. Gehre, et al. 2009. "A humanized version of *Foxp2* affects cortico-basal ganglia circuits in mice." *Cell* 137(5): 800–2.

Enard, W., M. Przeworski, et al. 2002. "Molecular evolution of FOXP2, a gene involved in speech and language." *Nature* 418: 869–72.

Endler, J. A. 1986. *Natural selection in the wild.* Princeton, Princeton Univ. Press.

Engh, A. L., J. C. Beehner, et al. 2006. "Behavioural and hormonal responses to predation in female chacma baboons (*Papio hamadryas ursinus*)." *Proceedings of the Royal Society of London B: Biological Sciences* 273(1587): 707–12.

Engle, P. L., and C. Breaux. 1998. "Fathers' involvement with children: Perspectives from developing countries." *Social Policy Report, Society for Research in Child Development* XII(1): 1–23.

Enns, G. M., D. R. Martinez, et al. 1999. "Molecular correlations in phenylketonuria: Mutation patterns and corresponding biochemical and clinical phenotypes in a heterogeneous California population." *Pediatric Research* 46(5): 594–602.

Enoch, M.-A., and D. Goldman. 2000. Genetics of alcoholism. In *Genetic influences on neural and behavioral functions,* ed. D. W. Pfaff et al. Boca Raton, CRC: 147–57.

Epple, G. 1975. "Parental behavior in *Saguinus fuscicollis ssp.* (Callithricidae)." *Folia Primatologica* 24: 221–38.

Epstein, H. T. 1973. "Possible metabolic constraints on human brain weight at birth." *American Journal of Physical Anthropology* 39: 135–36.

Erchak, G. M. 1980. "The acquisition of cultural rules by Kpelle children." *Ethos* 8(1): 40–48.

Erickson, C. J. 1986. Social induction of the ovarian response in the female ring dove. In *Reproduction: A behavioral and neuroendocrine perspective,* ed. B. R. Komisaruk et al. *Annals of the New York Academy of Sciences* 474: 13–19.

Erickson, K., W. Drevets, and J. Schulkin. 2003. "Glucocorticoid regulation of diverse cognitive functions in normal and pathological emotional states." *Neuroscience and Biobehavioral Reviews* 27(3): 233–46.

Erickson, M. T. 1993. "Rethinking Oedipus: An evolutionary perspective of incest avoidance." *American Journal of Psychiatry* 150(3, March): 411–16.

Erikson, E. H. 1959. "Identity and the life cycle." *Psychological Issues* 1(1): 1–171.

———. 1968. *Identity: Youth and crisis.* New York, Norton.

Ernst, C., and J. Angst. 1983. *Birth order: Its influence on personality.* Berlin, Springer-Verlag.

Essman, W., and J. Frisone. 1966. "Isolation-induced facilitation of gastric ulcerogenesis in mice." *Journal of Psychosomatic Research* 10: 183–88.

Essman, W. B. 1966. "The development of activity differences in isolated and aggregated mice." *Animal Behaviour* 14(4): 406–9.

———. 1968. "Differences in locomotor activity and brain-serotonin metabolism in differentially housed mice." *Journal of Comparative and Physiological Psychology* 66(1): 244–46.

Estes, R. D., and J. Goddard. 1967. "Prey selection and hunting behavior of the African wild dog." *Journal of Wildlife Management* 31: 52–70.

Estioko-Griffin, A. 1985. Women as hunters: The case of an eastern Cagayan Agta group. In *The Agta of Northeastern Luzon: Recent studies,* ed. P. B. Griffin and A. Estioko-Griffin. Cebu City, Philippines, San Carlos: 18–32.

Estioko-Griffin, A., and P. B. Griffin. 1981. Woman the hunter: The Agta. In *Woman the gatherer,* ed. F. Dahlberg. New Haven, Yale Univ. Press: 121–51.

Ethell, I. M., and E. B. Pasquale. 2005. "Molecular mechanisms of dendritic spine development and remodeling." *Progress in Neurobiology* 75(3): 161–205.

Evans, A. C., and Brain Development Cooperative Group. 2006. "The NIH MRI study of normal brain development." *Neuroimage* 30(1): 184–202.

Evans, P. D., J. R. Anderson, et al. 2004. "Reconstructing the evolutionary history of microcephalin, a gene controlling human brain size." *Human Molecular Genetics* 13(11): 1139–45.

Evans-Pritchard, E. E. 1940. *The Nuer.* New York, Oxford Univ. Press.

Evarts, E. B. 1975. "The third Stevenson lecture. Changing concepts of central control of movement." *Canadian Journal of Physiology* 53: 191–201.

Eveleth, P., and J. M. Tanner, eds. 1976. *Worldwide variation in human growth.* London, Cambridge Univ. Press.

Ewer, R. F. 1968. *Ethology of mammals.* London, Elek.

Eysenck, H. J., and M. W. Eysenck. 1985. *Personality and individual differences: A natural science approach.* New York, Plenum.

Fagen, R. 1981. *Animal play behavior.* New York, Oxford Univ. Press.

———. 1987. Play, games and innovation: Sociobiological findings and unanswered questions. In *Sociobiology and psychology: Ideas, issues and applications,* ed. C. Crawford, M. Smith, and D. Krebs. Hillsdale, NJ, Erlbaum: 253–68.

———. 1993. Primate juveniles and primate play. In *Juvenile primates: Life history, development, and behavior,* ed. M. E. Pereira and L. A. Fairbanks. New York, Oxford: 182–96.

Fagen, R., and J. Fagen. 2004. "Juvenile survival and benefits of play behaviour in brown bears, *Ursus arctos.*" *Evolutionary Ecology Research* 6(1): 89–102.

Fagen, R. M. 1977. "Selection for optimal age-dependent schedules of play behavior." *American Naturalist* 111: 395–414.

Fairbanks, L., and M. McGuire. 1986. "Age, reproductive value, and dominance-related behaviour in vervet monkey females: Cross-generational influences on social relationships and reproduction." *Animal Behaviour* 34(6): 1710–21.

Fairbanks, L. A. 1990. "Reciprocal benefits of allomothering for female vervet monkeys." *Animal Behaviour* 40(3): 553–62.

———. 2000. The developmental timing of primate play: A neural selection model. In *Biology, brains, and behavior: The evolution of human development,* ed. S. T. Parker, J. Langer, and M. L. McKinney. Santa Fe, School of American Research Press.

Fairbanks, L. A., and M. E. Pereira. 2002. Juvenile primates: Dimensions for future research. In *Juvenile primates: Life history, development, and behavior,* ed. M. E. Pereira and L. A. Fairbanks. Chicago, Univ. of Chicago Press: 359–66.

Fairgrieve, C. 1995. "Infanticide and infant eating in the blue monkey (*Cercopithecus mitis stuhlmanni*) in the Budongo Forest Reserve, Uganda." *Folia Primatology* 64: 69–72.

Falk, D. 1983. "The Taung endocast: A reply to Holloway." *American Journal of Physical Anthropology* 60(4): 479–89.

———. 1989. "Ape-like endocast of 'ape-man' Taung." *American Journal of Physical Anthropology* 80(3): 335–39.

———. 2004a. "Hominin brain evolution—new century, new directions." *Collegium Antropologicum* 28(Suppl. 2): 59–64.

———. 2004b. "Prelinguistic evolution in early hominins: Whence motherese?" *Behavioral and Brain Sciences* 27(4): 491–541.

Falk, D., and R. Clarke. 2007. "Brief communication: New reconstruction of the Taung endocast." *American Journal of Physical Anthropology* 134(4): 529–34.

Falk, D., J. C. Redmond, Jr., et al. 2000. "Early hominid brain evolution: A new look at old endocasts." *Journal of Human Evolution* 38(5): 695–717.

Falkowski, P. G., M. E. Katz, et al. 2005. "The rise of oxygen over the past 205 million years and the evolution of large placental mammals." *Science* 309(5744): 2202–4.

Fantz, R. L. 1958. "Pattern vision in young infants." *Psychological Record* 8: 43–47.

———. 1963. "Pattern vision in newborn infants." *Science* 140: 296–97.

Farris, M. R. 2000. "Smiling of male and female infants to mother vs. stranger at 2 and 3 months of age." *Psychological Reports* 87(3): 723–28.

Farver, J. A., and C. Howes. 1988. "Cross-cultural differences in social interaction: A comparison of American and Indonesian children." *Journal of Cross-Cultural Psychology* 19(2): 203–15.

Fauber, R., R. Forehand, et al. 1990. "A mediational model of the impact of marital conflict on adolescent adjustment in intact and divorced families: The role of disrupted parenting." *Child Development* 61: 1112–23.

Featherstone, D. E., and K. Broadie. 2000. "Surprises from Drosophila: Genetic mechanisms of synaptic development and plasticity." *Brain Research Bulletin* 53(5): 501–11.

Fecteau, S., L. Carmant, et al. 2004. "A motor resonance mechanism in children? Evidence from subdural electrodes in a 36-month-old child." *Neuroreport* 15(17): 2625–27.

Federman, D. D. 1967. *Abnormal sexual development: A genetic and endocrine approach to differential diagnosis.* Philadelphia, W. B. Saunders.

———. 1982. "The Jeremiah Metzger Lecture. The determinants of human sexuality." *Transactions of the American Clinical and Climatological Association* 94: 162–72.

———. 2004. "Three facets of sexual differentiation." *New England Journal of Medicine* 350(4): 323–24.

———. 2006. "The biology of human sex differences." *New England Journal of Medicine* 354(14): 1507–14.

Fedigan, L. M., and M. S. M. Pavelka. 2007. Reproductive cessation in female primates: Comparison of Japanese macaques and humans. In *Primates in perspective*, ed. C. J. Campbell et al. New York, Oxford Univ. Press: 437–47.

Feeney, J. A. 2008. Adult romantic attachment: Developments in the study of couple relationships. In *Handbook of attachment*, 2nd ed., ed. J. Cassidy and P. R. Shaver. New York, Guilford: 456–81.

Feher, O., H. Wang, et al. 2009. "*De novo* establishment of wild-type song culture in the zebra finch." *Nature* 459: 564–68.

Feldman, J. F., N. Brody, et al. 1980. "Sex differences in non-elicited neonatal behaviors." *Merrill-Palmer Quarterly* 26: 63–73.

Feldman, M., L. L. Cavalli-Sforza, et al. 1994. On the complexity of cultural transmission and evolution. In *Complexity: Metaphors, models, and reality*, ed. G. A. Cowan, D. Pines, and D. Meltzer. Reading, MA, Addison-Wesley: vol. 19, 47–64.

Feldman, R., and A. I. Eidelman. 2003. "Skin-to-skin contact (Kangaroo Care) accelerates autonomic and neurobehavioural maturation in preterm infants." *Developmental Medicine and Child Neurology* 45(4): 274–81.

Feldman, R., C. W. Greenbaum, et al. 1997. "Change in mother-infant interactive behavior: Relations to change in the mother, the infant, and the social context." *Infant Behavior and Development* 20(2): 151–64.

Felson, R. B. 2000. A social psychological approach to interpersonal aggression. In *Aggression and violence: An introductory text*, ed. V. B. Van Hasselt and H. Michel. Boston, Allyn and Bacon: 9–22.

Ferber, R. 2006. *Solve your child's sleep problems*. New York, Fireside.

Ferguson, C. A. 1964. "Baby talk in six languages." *American Anthropologist* 66(6): 103–14.

Ferguson, J. N., L. J. Young, et al. 2000. "Social amnesia in mice lacking the oxytocin gene." *Nature Genetics* 25: 284–88.

Fernald, A. 1985. "Four-month-old infants prefer to listen to motherese." *Infant Behavior and Development* 8: 181–95.

———. 1992. Human maternal vocalizations to infants as biologically relevant signals: An evolutionary perspective. In *The adapted mind: Evolutionary psychology and the generation of culture*, ed. J. H. Barkow, L. Cosmides, and J. Tooby. New York, Oxford Univ. Press: 391–428.

Fernald, A., T. Taeschner, et al. 1989. "A cross-language study of prosodic modifications in mothers' and fathers' speech to preverbal infants." *Journal of Child Language* 16: 477–501.

Fernandes, C., J. L. Paya-Cano, et al. 2004. "Hippocampal gene expression profiling across eight mouse inbred strains: Towards understanding the molecular basis for behaviour." *European Journal of Neuroscience* 19(9): 2576–82.

Fernandez, B., D. Cardebat, et al. 2004. "Functional MRI follow-up study of language processes in healthy subjects and during recovery in a case of aphasia." *Stroke* 35(9): 2171–76.

Fernandez-Jalvo, Y., J. Carlos Diez, et al. 1999. "Human cannibalism in the Early Pleistocene of Europe (Gran Dolina, Sierra de Atapuerca, Burgos, Spain)." *Journal of Human Evolution* 37(3–4): 591–622.

Fernyhough, C. 1997. Vygotsky's sociocultural approach: Theoretical issues and implications for current research. In *The development of social cognition*, ed. S. Hala. East Sussex, Eng., Psychology Press: 65–92.

Ferraro, T. N., and R. J. Buono. 2000. The genetics of epilepsy: Mouse and human studies. In *Genetic influences on neural and behavioral functions,* ed. D. W. Pfaff et al. Boca Raton, CRC: 117–46.

Field, T. M. 1982. "Same-sex preferences of preschool children: An artifact of same-age grouping?" *Child Study Journal* 12(3): 151–59.

Field, T. M., A. M. Sostek, et al. 1981. *Culture and early interactions.* Hillsdale, NJ, Erlbaum.

Fields, W. M., P. Segerdahl, et al. 2007. *The material practices of ape language research.* New York, Cambridge Univ. Press.

Finch, C. E., and M. R. Rose. 1995. "Hormones and the physiological architecture of life history evolution." *Quarterly Review of Biology* 70(1): 1–52.

Finlay, B. L., and P. Brodsky. 2007. Cortical evolution as the expression of a program for disproportionate growth and the proliferation of areas. In *Evolution of nervous systems,* ed. J. H. Kaas. New York, Elsevier/Oxford Univ. Press: 73–96.

Finlay, B. L., and R. B. Darlington. 1995. "Linked regularities in the development and evolution of mammalian brains." *Science* 268: 1578–84.

Finlay, B. L., R. B. Darlington, et al. 2001. "Developmental structure in brain evolution." *Behavioral and Brain Sciences* 24(2): 263–308.

Finlay, B. L., M. N. Hersman, et al. 1998. "Patterns of vertebrate neurogenesis and the paths of vertebrate evolution." *Brain, Behavior and Evolution* 52(4–5): 232–42.

Firth, R. 1957. *We, the Tikopia: A sociological study of kinship in primitive Polynesia.* Boston, Beacon.

Fishbein, H. D. 1976. *Evolution, development, and children's learning.* Pacific Palisades, CA, Goodyear.

Fisher, H. E. 1989. "Evolution of human serial pairbonding." *American Journal of Physical Anthropology* 78(3): 331–54.

———. 2006. Broken hearts: The nature and risks of romantic rejection. In *Romance and sex in adolescence and emerging adulthood: Risks and opportunities,* ed. A. C. Crouter and A. Booth. Mahwah, NJ, Erlbaum: 3–28.

Fisher, R. A. 1958. *The genetical theory of natural selection.* New York, Dover.

Fisher, S. E. and C. Scharff. 2009. "FOXP2 as a molecular window into speech and language." *Trends in Genetics* 25(4): 166–77.

Fite, J. E., J. A. French, et al. 2005. "Elevated urinary testosterone excretion and decreased maternal caregiving effort in marmosets when conception occurs during the period of infant dependence." *Hormones and Behavior* 47(1): 39–48.

Flack, J. C., L. A. Jeannotte, et al. 2004. "Play signaling and the perception of social rules by juvenile chimpanzees *(Pan troglodytes)*." *Journal of Comparative Psychology* 118(2): 149–59.

Flavell, J. H. 1963. *The developmental psychology of Jean Piaget.* Princeton, D. Van Nostrand.

———. 1986. "The development of children's knowledge about the appearance-reality distinction." *American Psychologist* 41(4): 418–25.

Flavell, J. H., E. R. Flavell, et al. 2001. "Development of children's understanding of connections between thinking and feeling." *Psychological Science* 12(5): 430–32.

Flavell, J. H., F. L. Green, et al. 1997. "The development of children's knowledge about inner speech." *Child Development* 68(1): 39–47.

———. 1999. "Development of children's knowledge about unconsciousness." *Child Development* 70(2): 396–412.

Fleagle, J. G. 1985. Size and adaptation in primates. In *Size and scaling in primate biology,* ed. W. L. Jungers. New York, Plenum: 1–19.

―――. 1999. *Primate adaptation and evolution,* 2nd ed. New York, Academic.

Flechsig, P. 1920. *Anatomie des Menschlichen Gehirns und Röckenmarks auf myeloge-netischer Grundlage.* Leipzig, Georg Thieme.

Fleming, A. S. 1990. Hormonal and experiential correlates of maternal responsiveness in human mothers. In *Mammalian parenting,* ed. N. A. Krasnegor and R. S. Bridges. New York, Oxford Univ. Press: 184–208.

Fleming, A. S., C. Corter, et al. 1993. "Postpartum factors related to mothers' attraction to newborn infant odors." *Developmental Psychology* 26: 115–32.

Fleming, A. S., H. D. Morgan, et al. 1996. Experiential factors in postpartum regulation of maternal care. In *Parental care: Evolution, mechanisms, and adaptive significance. Advances in the study of behavior,* vol. 25, ed. J. S. Rosenblatt and C. T. Snowdon. San Diego, Academic: 295–332.

Fleming, A. S., M. Steiner, et al. 1997. "Cortisol, hedonics, and maternal responsiveness in human mothers." *Hormones and Behavior* 32(2): 85–98.

Fletcher, C. F., and W. N. Frankel. 1999. "Ataxic mouse mutants and molecular mechanisms of absence epilepsy." *Human Molecular Genetics* 8(10): 1907–12.

Flinn, M. V. 1988. Parent-offspring interactions in a Caribbean village: Daughter guarding. In *Human reproductive behaviour: A Darwinian perspective,* ed. L. Betzig, M. B. Mulder, and P. Turke. Cambridge, Cambridge Univ. Press: 189–200.

―――. 2006. "Evolution and ontogeny of stress response to social challenges in the human child." *Developmental Review* 26(2): 138–74.

Flinn, M. V., and B. G. England. 1997. "Social economics of childhood glucocorticoid stress response and health." *American Journal of Physical Anthropology* 102(1): 33–53.

Flint, J. 2003. "Analysis of quantitative trait loci that influence animal behavior." *Journal of Neurobiology* 54(1): 46–77.

Floeter, M. K., and W. T. Greenough. 1979. "Cerebellar plasticity: Modification of purkinje cell structure by differential rearing in monkeys." *Science* 206: 227–29.

Flombaum, J. I., and L. R. Santos. 2005. "Rhesus monkeys attribute perceptions to others." *Current Biology* 15(5): 447–52.

Florsheim, P., ed. 2003. *Adolescent romantic relations and sexual behavior: Theory, research, and practical implications.* Mahwah, NJ, Erlbaum.

Fogel, A., G. C. Nelson-Goens, et al. 2000. "Do different infant smiles reflect different positive emotions?" *Social Development* 9(4) 2000: 497–520.

Fomon, S. J. 1974. *Infant nutrition,* 2nd ed. Philadelphia, W. B. Saunders.

Forbes, E. E., and R. E. Dahl. 2010. "Pubertal development and behavior: Hormonal activation of social and motivational tendencies." *Brain and Cognition,* 72(1, February): 66–72: http://dx.doi.org (doi:10.1016/j.bandc.2009.10.007).

Ford, C. S. 1945. *A comparative study of human reproduction.* Publications in Anthropology, No. 32. New Haven, Yale Univ. Press.

Forster, S., and M. Cords. 2005. "Socialization of infant blue monkeys (*Cercopithecus mitis stuhlmanni*): Allomaternal interactions and sex differences." *Behaviour* 142(7): 869–96.

Fossey, D. 1983. *Gorillas in the mist.* Boston, Houghton Mifflin.

―――. 1984. Infanticide in mountain gorillas (*Gorilla gorilla beringei*) with comparative notes on chimpanzees. In *Infanticide: Comparative and evolutionary perspectives,* ed. G. Hausfater and S. B. Hrdy. New York, Aldine de Gruyter: 217–36.

Fouts, H. N., B. S. Hewlett, et al. 2001. "Weaning and the nature of early childhood interactions among Bofi foragers in Central Africa." *Human Nature* 12(1): 27–46.

———. 2005. "Parent-offspring weaning conflicts among the Bofi farmers and for-agers of central Africa." *Current Anthropology* 46(1): 29–50.

Fouts, H. N., and M. E. Lamb. 2005. Weanling emotional patterns among the Bofi for-agers of Central Africa: The role of maternal availability and sensitivity. In *Hunter-gatherer childhoods: Evolutionary, developmental and cultural perspectives,* ed. B. S. Hewlett and M. E. Lamb. New Brunswick, NJ, AldineTransaction: 309–21.

Fouts, H. N., M. E. Lamb, et al. 2004. "Infant crying in hunter-gatherer cultures." *Behavioral and Brain Sciences* 27(4): 462–63.

Fouts, R. S. 1974. "Language: Origins, definitions, and chimpanzees." *Journal of Human Evolution* 3: 475–82.

Fowler, J. 1995. *Stages of faith: The psychology of human development and the quest for meaning.* San Francisco, HarperSanFrancisco.

Fowler, J. W. 1989. Strength for the journey: Early childhood development in selfhood and faith. In *Faith development in early childhood,* ed. E. Blazer. Lanham, MD, Sheed and Ward: 1–36.

———. 1991. "Stages in faith consciousness." *New Directions for Child Development* 52: 27–45.

Fowler, J. W., and M. L. Dell. 2004. "Stages of faith and identity: Birth to teens." *Child and Adolescent Psychiatric Clinics of North America* 13(1): 17–33.

———. 2006. Stages of faith from infancy through adolescence: Reflections on three decades of faith development theory. In *The handbook of spiritual development in childhood and adolescence,* ed. E. C. Roehlkepartain et al. Thousand Oaks, CA, Sage: 34–45.

Fox, J. R. 1962. "Sibling incest." *British Journal of Sociology* 13: 128–50.

Fox, M. W., ed. 1975. *The wild canids: Their systematics, behavioral ecology and evolution.* New York, Van Nostrand Reinhold.

Fox, N. A. 1998. "Temperament and regulation of emotion in the first years of life." *Pediatrics* 102(5): 1230–35.

Fox, N. A., and J. A. Card. 1999. Psychophysiological measures in the study of attach-ment. In *Handbook of attachment: Theory, research, and clinical applications,* ed. J. Cassidy and P. R. Shaver. New York, Guilford: 226–45.

Fox, N. A., and R. J. Davidson. 1991. Hemispheric specialization and attachment be-haviors: Developmental processes and individual differences in separation pro-test. In *Intersections with attachment,* ed. J. L. Gewirtz and W. M. Kurtines. Hills-dale, NJ, Erlbaum: 147–64.

Fox, N. A., and A. A. Hane. 2008. Studying the biology of human attachment. In *Hand-book of attachment,* 2nd ed., ed. J. Cassidy and P. R. Shaver. New York, Guilford: 217–40.

Fox, N. A., K. H. Rubin, et al. 1995. "Frontal activation asymmetry and social compe-tence at four years of age." *Child Development* 66: 1770–84.

Fox, R. 1967. *Kinship and marriage.* Harmondsworth, Eng., Penguin.

———. 1980. *The red lamp of incest.* New York, Dutton.

Fraiberg, S. 1977. *Insights from the blind: Developmental studies of blind children.* New York, Basic.

Francis, D. D., K. Szegda, et al. 2003. "Epigenetic sources of behavioral differences in mice." *Nature Neuroscience* 6(5): 445–46.

Franklin, M. S., G. W. Kraemer, et al. 2000. "Gender differences in brain volume and size of corpus callosum and amygdala of rhesus monkey measured from MRI im-ages." *Brain Research* 852(2): 263–67.

Franzen, J. L., P. D. Gingerich, et al. 2009. "Complete primate skeleton from the Middle Eocene of Messel in Germany: Morphology and paleobiology." *PLoS One* 4(5): e5723.

Frayser, S. G. 1994. "Defining normal childhood sexuality: An anthropological approach." *Annual Review of Sex Research* V: 173–217.

Fredrickson, B. L. 1998. "What good are positive emotions?" *Review of General Psychology* 2(3): 300–19.

Freedman, D. G. 1958. "Constitutional and environmental interactions in rearing 4 breeds of dogs." *Science* 127: 585–86.

——. 1964. "Smiling in blind infants and the issue of innate vs. acquired." *Journal of Psychology and Psychiatry* 5: 171–84.

——. 1974. *Human infancy: An evolutionary perspective.* New York, Wiley.

——. 1997. My three mentors. In *Uniting psychology and biology: Integrative perspectives on human development,* ed. N. L. Segal, G. E. Weisfeld, and C. C. Weisfeld. Washington, DC, American Psychological Association: 19–45.

Freedman, D. G., and N. A. Freedman. 1969. "Differences in behavior between Chinese-American and European-American newborns." *Nature* 224: 1227.

Freeman, D. 1982. *Margaret Mead and Samoa: The making and unmaking of an anthropological myth.* Cambridge, MA, Harvard Univ. Press.

Freeman, J. M. 2007. "Beware: The misuse of technology and the law of unintended consequences." *Neurotherapeutics* 4(3): 549–54.

Freeman, R. 1990. "Intrapartum fetal monitoring—a disappointing story [comment]." *New England Journal of Medicine* 322(9): 624–26.

Freud, S. 1905. Three essays on the theory of sexuality. In *The standard edition of the complete psychological works of Sigmund Freud,* vol. 7. London, Hogarth: 135–245.

——. 1913. *Totem and taboo.* New York, Norton.

——. 1920. *A general introduction to psychoanalysis.* New York, Washington Square.

——. 1924. The passing of the Oedipus-complex. In *Collected Papers,* vol. 2. London, Hogarth.

——. 1933. *New introductory lectures.* New York, Norton.

——. 1949. *An outline of psycho-analysis.* New York, Norton.

——. 1961a. *Civilization and its discontents.* New York, Norton.

——. 1961b. *The future of an illusion.* New York, Doubleday Anchor.

Frias-Armenta, M. 2002. "Long-term effects of child punishment on Mexican women: A structural model." *Child Abuse and Neglect* 26(4): 371–86.

Friederici, A. D. 2005. "Neurophysiological markers of early language acquisition: From syllables to sentences." *Trends in Cognitive Sciences* 9(10): 481–88.

Frisch, R., G. Wyshak, et al. 1980. "Delayed menarche and amenorrhea in ballet dancers." *New England Journal of Medicine* 303: 17–19.

Frisch, R. E. 1978. "Population, food intake, and fertility." *Science* 199: 22–30.

——. 2004. *Female fertility and the body fat connection.* Chicago, Univ. of Chicago Press.

Frisch, R. E., and R. Revelle. 1970. "Height and weight at menarche and a hypothesis of critical body weights and adolescent events." *Science* 169: 397–98.

——. 1971. "Height and weight at menarche and a hypothesis of menarche." *Archives of Disease in Childhood* 46(249): 695–701.

Fromberg, D., and D. Bergen, eds. 1998. *Play from birth to twelve and beyond: Contexts, perspectives, and meanings.* New York, Garland.

Fry, D. P. 1987. "Differences between playfighting and serious fighting among Zapotec children." *Ethology and Sociobiology* 8(4): 285–306.

———. 1990. "Play aggression among Zapotec children: Implications for the practice hypothesis." *Aggressive Behavior* 16: 321–40.

———. 1998. "Anthropological perspectives on aggression: Sex differences and cultural variation." *Aggressive Behavior* 24(2): 81–95.

———. 2005. Rough-and-tumble social play in humans. In *The nature of play: Great apes and humans*, ed. A. D. Pellegrini and P. K. Smith. New York, Guilford.

Fuentes, A. 1999. "Re-evaluating primate monogamy." *American Anthropologist* 100(4): 890–907.

Fuller, G. B., M. X. Zarrow, et al. 1970. "Testosterone propionate during gestation in the rabbit: Effect on subsequent maternal behavior." *Journal of Reproductive Fertility* 23: 285–90.

Fuller, J. L. 1955. "Hereditary differences in trainability of purebred dogs." *Journal of Genetic Psychology* 87: 229–38.

Fuller, J. L., and L. D. Clark. 1968. "Genotype and behavioral vulnerability to isolation in dogs." *Journal of Comparative and Physiological Psychology* 66(1): 151–56.

Galaburda, A. M. 1993. "Neuroanatomic basis of developmental dyslexia." *Neurologic Clinics* 11(1): 161–73.

Galaburda, A. M., M. LeMay, et al. 1978. "Right-left asymmetries in the brain." *Science* 199: 852–56.

Galaburda, A. M., G. F. Sherman, et al. 1985. "Developmental dyslexia: Four consecutive patients with cortical anomalies." *Annals of Neurology* 18(2): 222–33.

Galdikas, B. M. F. 1979. Orangutan adaptation at Tanjung Puting Reserve. In *The great apes*, ed. D. A. Hamburg and E. R. McCown. Menlo Park, CA, Benjamin/Cummings.

———. 1985a. "Adult male sociality and reproductive tactics among orangutans at Tanjung-Puting Borneo Indonesia." *Folia Primatologica* 45(1): 9–24.

———. 1985b. "Orangutan pongo-pygmaeus-pygmaeus sociality at Tanjung-Puting Indonesia." *American Journal of Primatology* 9(2): 101–20.

Galef, B. G., Jr., and C. M. Heyes. 2004. "Social learning and imitation: Introduction." *Learning and Behavior* 32(1): 1–3.

Galef, B. G., Jr., E. E. Whiskin, et al. 2005. "A new way to study teaching in animals: Despite demonstrable benefits, rat dams do not teach their young what to eat." *Animal Behaviour* 70(Part 1): 91–96.

Gallese, V., M. Rochat, et al. 2009. "Motor cognition and its role in the phylogeny and ontogeny of action understanding." *Developmental Psychology* 45(1): 103–13.

Gannon, P. J., R. L. Holloway, et al. 1998. "Asymmetry of chimpanzee planum temporale: Humanlike pattern of Wernicke's brain language area homolog." *Science* 279: 220–22.

Gans, C. 1996. An overview of parental care among the reptilia. In *Parental care: Evolution, mechanisms, and adaptive significance. Advances in the study of behavior*, vol. 25, ed. J. S. Rosenblatt and C. T. Snowdon. San Diego, Academic: 145–57.

Garber, P. A. 1994. "Phylogenetic approach to the study of tamarin and marmoset social systems." *American Journal of Primatology* 34: 199–219.

Garcia, J., and R. A. Koelling. 1966. "Relation of cue to consequence in avoidance learning." *Psychonomic Science* 4: 123–24.

Garcia, J., L. P. Brett, et al. 1989. Limits of Darwinian conditioning. In *Contemporary learning theories: Instrumental conditioning theory and the impact of biological con-*

straints on learning, ed. S. B. Klein and R. R. Mowrer. Hillsdale, NJ, Erlbaum: 181–203.

Garcia, J., D. F. Quick, et al. 1984. Conditioned disgust and fear from mollusk to monkey. In *Primary neural substrates of learning and behavioral change,* ed. D. L. Alkon and J. Farley. Cambridge, Cambridge Univ. Press.

Gardner, B. T., and R. A. Gardner. 1974. Comparing the early utterances of child and chimpanzee. In *Minnesota Symposia on Child Psychology,* ed. A. Pick. Minneapolis, Univ. of Minnesota Press: 8.

———. 1975. "Evidence for sentence constituents in the early utterances of child and chimpanzee." *Journal of Experimental Psychology* 104: 244–67.

Gardner, H. 2004. *Frames of mind: The theory of multiple intelligences.* New York, Basic.

Gardner, R., and H. Harris. 1971. *The Nuer.* Watertown, MA, Documentary Educational Resources: 73 min.

Gardner, R. A., B. T. Gardner, et al. 1989. *Teaching sign language to chimpanzees.* Albany, State Univ. of New York Press.

Garmezy, N., A. S. Masten, et al. 1984. "The study of stress and competence in children: A building block for developmental psychopathology." *Child Development* 55: 97–111.

Garmezy, N., and A. Tellegen. 1984. Studies of stress-resistant children: Methods, variables, and preliminary findings. In *Applied developmental psychology,* ed. F. J. Morrison, C. Lord, and D. P. Keating. Orlando, FL, Academic: vol. 1, 231–87.

Garn, S. M. 1987. "The secular trend in size and maturational timing and its implications for nutritional assessment." *Journal of Nutrition* 117(5): 817–23.

Gatti, C. 1989. Culla a tre piazze. *Europeo:* 112–16.

Gaulin, S. J. C. 1992. Evolution of sex differences in spatial ability. *Yearbook of Physical Anthropology* 35: 125–51.

Gazzaniga, M. S. 2008. *Human: The science behind what makes us unique.* New York, HarperCollins/Ecco.

Gazzaniga, M. S., J. E. Bogen, et al. 1992. Some functional effects of sectioning the cerebral commissures in man. In *Frontiers in cognitive neuroscience,* ed. S. M. Kosslyn and R. A. Andersen. Cambridge, MA, MIT Press: 609–11.

Geary, D. C. 2000. "Evolution and proximate expression of human paternal investment." *Psychological Bulletin* 126(1): 55–77.

Geber, M., and R. F. A. Dean. 1957. "The state of development of new born African children." *Lancet* 1: 1216–19.

Gebo, D. L., M. Dagosto, et al. 2000. "The oldest known anthropoid postcranial fossils and the early evolution of higher primates." *Nature* 404(6775): 276–78.

Geertz, C. 1971. Deep play: Notes on the Balinese cockfight. In *Myth, symbol, and culture,* ed. C. Geertz. New York, Norton: 1–37.

———. 1973. *The interpretation of cultures.* New York, Basic.

Gehring, W. 1999. *Master control genes in development and evolution: The homeobox story.* New Haven, Yale Univ. Press.

Gehring, W. J. 2005. "New perspectives on eye development and the evolution of eyes and photoreceptors." *Journal of Heredity* 96(3): 171–84.

Geist, V. 1971. *Mountain sheep: A study in behavior and evolution.* Chicago, Univ. of Chicago Press.

Gentry, R. 1974. "The development of social behavior through play in the Steller sea lion." *American Zoologist* 14: 391.

Gerall, H. D., I. L. Ward, et al. 1967. "Disruption of the male rat's sexual behaviour induced by social isolation." *Animal Behaviour* 15: 54–58.

Geschwind, N. 1972. "Language and the brain." *Scientific American* 226: 76–83.

———. 1992. The organization of language and the brain. In *Frontiers in cognitive neuroscience,* ed. S. M. Kosslyn and R. A. Andersen. Cambridge, MA, MIT Press: 634–40.

Geschwind, N., and A. M. Galaburda. 1985. "Cerebral lateralization: Biological mechanisms, associations, and pathology: A hypothesis and a program for research (Parts I, II, and III)." *Archives of Neurology* 42: 428–59; 521–52; 634–54.

———. 1987. *Cerebral lateralization.* Cambridge, MA, MIT Press.

Gesell, A., and C. S. Amatruda, eds. 1947. *Developmental diagnosis,* 2nd ed. New York, Hoeber.

———, eds. 1965. *Developmental diagnosis, normal and abnormal child development.* New York, Harper & Row.

Gewirtz, J. L. 1965. The course of infant smiling in four child-rearing environments in Israel. In *Determinants of infant behavior III,* ed. B. M. Foss. London, Methuen.

———. 1972. *Attachment and dependency.* Washington, DC, V. H. Winston.

Gewirtz, J. L., and M. Pelaez-Nogueras. 2000. *Infant emotions under the positive-reinforcer control of caregiver attention and touch.* Reno, NV, Context Press.

Ghez, C., and J. Krakauer. 2000. The organization of movement. In *Principles of neural science,* ed. E. R. Kandel, J. H. Schwartz, and T. M. Jessell. New York, McGraw-Hill: 653.

Ghiglieri, M. P. 1999. *The dark side of man: Tracing the origins of male violence.* Reading, MA, Perseus.

Gibbons, A. 1998a. "Genes put mammals in age of dinosaurs." *Science* 280: 675–76.

———. 1998b. "New study points to Eurasian ape as great ape ancestor." *Science* 281: 622–23.

Gibson, K. R. 1977. Brain structure and intelligence in macaques and human infants from a Piagetian perspective. In *Primate biosocial development,* ed. S. Chevalier-Skolnikoff and F. E. Poirier. New York, Garland.

———. 1981. Comparative neuro-ontogeny: Its implications for the development of human intelligence. In *Infancy and epistemology,* ed. G. Butterworth. Brighton, Eng., Harvester.

———. 1991. Myelination and behavioral development: A comparative perspective on questions of neoteny, altriciality, and intelligence. In *Brain maturation and cognitive development: Comparative and cross-cultural perspectives,* ed. K. R. Gibson and A. C. Petersen. New York, Aldine de Gruyter: 29–63.

———. 2002. "Customs and cultures in animals and humans: Neurobiological and evolutionary considerations." *Anthropological Theory* 2(3): 323–39.

Giedd, J. N., F. X. Castellanos, et al. 1997. "Sexual dimorphism of the developing human brain." *Progress in Neuro-Psychopharmacology and Biological Psychiatry* 21(8): 1185–1201.

Giedd, J. N., L. S. Clasen, et al. 2006. "Puberty-related influences on brain development." *Molecular and Cellular Endocrinology* 254–255: 154–62.

Gier, H. T. 1975. Ecology and social behavior of the coyote. In *The wild canids,* ed. M. W. Fox. New York, Van Nostrand Reinhold.

Gilbert, S. L., W. B. Dobyns, et al. 2005. "Genetic links between brain development and brain evolution." *Nature Reviews Genetics* 6(7): 581–90.

Gilles, F. H., A. Leviton, et al., eds. 1983. *The developing human brain: Growth and epidemiologic neuropathology.* Boston, John Wright/PSG.

Gilligan, C. 1982. *In a different voice: Psychological theory and women's development.* Cambridge, MA, Harvard Univ. Press.

Gilmer, W. S., and W. T. McKinney. 2003. "Early experience and depressive disorders: Human and non-human primate studies." *Journal of Affective Disorders* 75(2): 97–113.

Gilmore, D. D. 1990. *Manhood in the making: Cultural concepts of masculinity.* New Haven, Yale Univ. Press.

Gilmore, D. P. 2002. "Sexual dimorphism in the central nervous system of marsupials." *International Review of Cytology* 214: 193–224.

Gingerich, P. D. 1983. "Rates of evolution: Effects of time and temporal scaling." *Science* 222: 159–61.

Gingrich, B., Y. Liu, et al. 2000. "Dopamine D2 receptors in the nucleus accumbens are important for social attachment in female prairie voles (*Microtus ochrogaster*)." *Behavioral Neuroscience* 114(1): 173–83.

Giordano, P. C., W. D. Manning, et al. 2006. Adolescent romantic relationships: An emerging portrait of their nature and developmental significance. In *Romance and sex in adolescence and emerging adulthood: Risks and opportunities,* ed. A. C. Crouter and A. Booth. Mahwah, NJ, Erlbaum: 127–50.

Gittleman, J. L. 1981. "The phylogeny of parental care in fishes." *Animal Behaviour* 29: 936–41.

———. 1986. "Carnivore life history patterns: Allometric, phylogenetic, and ecological associations." *American Naturalist* 127(6): 744–71.

Gittleman, J. L., H.-K. Luh, et al. 2000. Evolutionary development, life histories, and brain size: Finding connections via a multivariate method. In *Biology, brains, and behavior: The evolution of human development,* ed. S. T. Parker, J. Langer, and M. L. McKinney. Santa Fe, School of American Research Press: 159–79.

Gladwin, T. 1970. *East is a big bird: Navigation and logic on Pulawat Atoll.* Cambridge, MA, Harvard Univ. Press.

Gleitman, L. R. 2006. *A human universal: The capacity to learn a language.* Mahwah, NJ, Erlbaum.

Gleitman, L. R., and E. L. Newport. 1995. The invention of language by children: Environmental and biological influences on the acquisition of language. In *Language: An invitation to cognitive science,* ed. L. R. Gleitman and M. Liberman. Cambridge, MA, MIT Press: vol. 1, 1–24.

Globus, A., M. R. Rosensweig, et al. 1973. "Effects of differential experience on dendritic spine counts." *Journal of Comparative and Physiological Psychology* 82: 175–81.

Gluckman, P. D., and M. A. Hanson. 2006. "Evolution, development and timing of puberty." *Trends in Endocrinology and Metabolism* 17(1): 7–12.

Godfrey, L. R., and M. R. Sutherland. 1996. "Paradox of peramorphic paedomorphosis: Heterochrony and human evolution." *American Journal of Physical Anthropology* 99(1): 17–42.

Godfrey, L. R., K. E. Samonds, et al. 2001. "Teeth, brains, and primate life histories." *American Journal of Physical Anthropology* 114(3): 192–214.

Goin, R. P. 1998. "A review of peer social development in early childhood." *Early Child Development and Care* 142: 1–8.

Goldberg, H. E. 2003. *Jewish passages: Cycles of Jewish life.* Berkeley, Univ. of California Press.

Goldberg, S., and M. Lewis. 1969. "Play behavior in the year-old infant: Early sex differences." *Child Development* 40: 21–31.

Golden, M. H. N. 1998. Catch-up growth in height. In *The Cambridge encyclopedia of human growth and development,* ed. S. J. Ulijaszek, F. E. Johnston, and M. A. Preece. Cambridge, Cambridge Univ. Press: 346–47.

Goldfrank, E. 1945. "Socialization, personality, and the structure of Pueblo society." *American Anthropologist* 47: 516–39.

Goldin-Meadow, S. 2005a. *The resilience of language: What gesture creation in deaf children can tell us about how all children learn language.* New York, Psychology Press.

———. 2005b. "Watching language grow [comment]." *Proceedings of the National Academy of Sciences* 102(7): 2271–72.

Goldin-Meadow, S., and H. Feldman. 1977. "The development of language-like communication without a language model." *Science* 197: 401–3.

Goldin-Meadow, S., and C. Mylander. 1983. "Gestural communication in deaf children: Noneffect of parental input on language development." *Science* 221: 372–74.

———. 1991. Levels of structure in a communication system developed without a language model. In *Brain maturation and cognitive development: Comparative and cross-cultural perspectives,* ed. K. R. Gibson and A. C. Petersen. New York, Aldine de Gruyter: 315–44.

Goldman, J. A. 1981. "Social participation of preschool children in same- versus mixed-age groups." *Child Development* 52(2): 644–50.

Goldstein, A., and E. B. Foa. 1979. *Handbook of behavioral interventions.* Somerset, NJ, Wiley.

Goleman, D. 1994. *Emotional intelligence: Why it can matter more than IQ.* New York, Bantam.

Golombok, S., and R. Fivush. 1994. *Gender development.* Cambridge, Cambridge Univ. Press.

Golombok, S., J. Rust, et al. 2008. "Developmental trajectories of sex-typed behavior in boys and girls: A longitudinal general population study of children aged 2.5–8 years." *Child Development* 79(5): 1583–93.

Golub, H. L., and M. J. Corwin. 1985. A physioacoustic model of the infant cry. In *Infant crying: Theoretical and research perspectives,* ed. B. M. Lester and C. F. Z. Boukydis. New York, Plenum: 59–82.

Gomber, J., and G. Mitchell. 1974. "Preliminary report on adult male isolation-reared rhesus monkeys caged with infants." *Developmental Psychology* 10: 298.

Gombrich, E. H. 1996. "The miracle at Chauvet." *New York Review of Books* (Nov. 14): 8–12.

Gomez, C., E. M. Özbudak, et al. 2008. "Control of segment number in vertebrate embryos." *Nature* 454: 335–39.

Gómez, J.-C., and B. Martin-Andrade. 2004. Fantasy play in apes. In *The nature of play: Great apes and humans,* ed. A. D. Pellegrini and P. K. Smith. New York, Guilford.

Gonsiorek, J. C., and J. D. Weinrich, eds. 1991. *Homosexuality: Research implications for public policy.* Newbury Park, CA, Sage Publications.

Good, M., and T. Willoughby. 2008. "Adolescence as a sensitive period for spiritual development." *Child Development Perspectives* 2(1): 32–37.

Goodall, J. 1967. Mother-offspring relationships in chimpanzees. In *Primate ethology,* ed. D. Morris. Chicago, Aldine.

———. 1968. "The behavior of free-living chimpanzees in the Gombe Stream Reserve." *Animal Behaviour Monographs* 1: 161–311.

———. 1971. *In the shadow of man.* New York, Dell.

———. 1977. "Infant killing and cannibalism in free-living chimpanzees." *Folia Primatologica* 28: 259–82.

———. 1986. *The chimpanzees of Gombe: Patterns of behavior.* Cambridge, MA, Harvard
Univ. Press.

Goodfellow, P. N., and R. L. Badge. 1993. "SRY and sex determination in mammals."
Annual Review of Genetics 27: 71–92.

Goodlin-Jones, B. L., M. M. Burnham, et al. 2001. "Night waking, sleep-wake organization, and self-soothing in the first year of life." *Journal of Developmental and Behavioral Pediatrics* 22(4): 226–33.

Goodman, M. J., A. Estioko-Griffin, et al. 1985. "Menarche, pregnancy, birth spacing
and menopause among the Agta women foragers of Cagayan province, Luzon, the
Philippines." *Annals of Human Biology* 12(2): 169–77.

Goodrich, F. W., and H. Thoms. 1948. "A clinical study of natural childbirth: A preliminary report from a teaching ward service." *American Journal of Obstetrics and
Gynecology* 56(5): 875–83.

Goodson, J. L., and A. H. Bass. 2001. "Social behavior functions and related anatomical
characteristics of vasotocin/vasopressin systems in vertebrates." *Brain Research
—Brain Research Reviews* 35(3): 246–65.

Goodson, J. L., S. E. Schrock, et al. 2009. "Mesotocin and nonapeptide receptors promote estrildid flocking behavior." *Science* 325: 862–66.

Goodwin, B. 1993. Development as a robust natural process. In *Thinking about biology: An invitation to current theoretical biology,* ed. W. D. Stein and F. J. Varela. Reading, MA, Addison-Wesley: Lecture Notes vol. 3, 123–46.

———. 1994a. Developmental complexity and evolutionary order. In *Complexity:
Metaphors, models, and reality,* ed. G. A. Cowan, D. Pines, and D. Meltzer. Reading,
MA, Addison-Wesley: vol. 19, 205–22.

———. 1994b. *How the leopard changed its spots.* New York, Scribner's.

Goody, E. N. 1973. *Contexts of kinship.* Cambridge, Cambridge Univ. Press.

———. 1982. *Parenthood and social reproduction.* Cambridge, Cambridge Univ. Press.

Goody, J. 1983. *The development of the family and marriage in Europe.* Cambridge, Cambridge Univ. Press.

Gopnik, M., and M. Crago. 1993. "Familial aggregation of a genetic language disorder."
Cognition 39: 1–50.

Gordon, N. S., S. Kollack-Walker, et al. 2002. "Expression of c-fos gene activation
during rough and tumble play in juvenile rats." *Brain Research Bulletin* 57(5): 651–
59.

Gordon, P. 1990. "Learnability and feedback." *Developmental Psychology* 26(2): 221–
26.

Goren-Inbar, N., N. Alperson, et al. 2004. "Evidence of hominin control of fire at
Gesher Benot Ya'aqov, Israel [see comment]." *Science* 304(5671): 725–27.

Gorer, G. 1962. *The people of Great Russia.* New York, Norton.

Gormally, S., R. G. Barr, et al. 2001. "Contact and nutrient caregiving effects on newborn infant pain responses." *Developmental Medicine and Child Neurology* 43(1):
28–38.

Gorski, R. A. 1996. Gonadal hormones and the organization of brain structure and
function. In *The lifespan development of individuals: Behavioral, neurobiological, and*

psychosocial perspectives, ed. D. Magnusson. New York, Cambridge Univ. Press: 315–40.

———. 2000. Sexual differentiation of the nervous system. In *Principles of neural science,* ed. E. R. Kandel, J. H. Schwartz, and T. M. Jessell. New York, McGraw-Hill: 1131.

———. 2002. "Hypothalamic imprinting by gonadal steroid hormones." *Advances in Experimental Medicine and Biology* 511: 57–70; discussion 70–73.

Gorski, R. A., J. H. Gordon, et al. 1978. "Evidence for a morphological sex difference within the medial preoptic area of the rat brain." *Brain Research* 148: 333–46.

Gosso, Y., E. Otta, et al. 2004. *Play in hunter-gatherer society.* New York, Guilford.

Gottesman, I. I., and D. R. Hanson. 2005. "Human development: Biological and genetic processes." *Annual Review of Psychology* 56(1): 263–86.

Gottlieb, G. 1980. "Development of species identification in ducklings: VII. Highly specific early experience fosters species-specific perception in wood ducklings." *Journal of Comparative and Physiological Psychology* 94: 1019–27.

———. 1982. Conceptions of prenatal development: Behavioral embryology. In *Learning, development, and culture: Essays in evolutionary epistemology,* ed. H. C. Plotkin. New York, Wiley: 363–90.

———. 1991a. "Experiential canalization of behavioral development: Theory." *Developmental Psychology* 27(1): 4–13.

———. 1991b. "Experiential canalization of behavioral development: Results." *Developmental Psychology* 27(1): 35–39.

———. 1992. *Individual development and evolution: The genesis of novel behavior.* New York, Oxford Univ. Press.

———. 1995. "Some conceptual deficiencies in 'developmental' behavior genetics." *Human Development* 38: 131–41.

———. 1996. Commentary: A systems view of psychobiological development. In *The lifespan development of individuals: Behavioral, neurobiological, and psychosocial perspectives,* ed. D. Magnusson. New York, Cambridge Univ. Press: 76–103.

Gough, K. 1975. The origin of the family. In *Toward an anthropology of women,* ed. R. R. Reiter. New York, Monthly Review: 51–76.

Gould, E., P. Tanapat, et al. 1998. "Proliferation of granule cell precursors in the dentate gyrus of adult monkeys is diminished by stress." *Proceedings of the National Academy of Sciences* 95(6): 3168–71.

Gould, K. G., M. Flint, et al. 1981. "Chimpanzee reproductive senescence: A possible model for evolution of the menopause." *Maturitas* 3(2): 157–66.

Gould, L. 1992. "Alloparental care in free-ranging Lemur catta at Berenty Reserve, Madagascar." *Folia Primatologica* 58(2): 72–83.

Gould, S. J. 1977. *Ontogeny and phylogeny.* Cambridge, MA, Harvard Univ. Press/ Belknap.

———. 1981. *The mismeasure of man.* New York, Norton.

———. 1988. The uses of heterochrony. In *Heterochrony in evolution: A multidisciplinary approach,* ed. M. L. McKinney. New York, Plenum.

———. 1989. *Wonderful life: The Burgess Shale and the nature of history.* New York, Norton.

———. 2000. "Of coiled oysters and big brains: How to rescue the terminology of heterochrony, now gone astray." *Evolution and Development* 2(5): 241–48.

———. 2002. *The structure of evolutionary theory.* Cambridge, MA, Harvard Univ. Press.

Gould, S. J., and E. S. Vrba. 1982. "Exaptation—a missing term in the science of form." *Paleobiology* 8(1): 4–15.

Gouldner, A. W. 1960. "The norm of reciprocity: A preliminary statement." *American Sociological Review* 25: 161–78.

Gouzoules, H. 1974a. "Harassment of sexual behavior in the stumptail macaque, *Macaca arctoides*." *Folia Primatologica* 22(2–3): 208–17.

———. 1974b. "Group responses to parturition in *Macaca arctoides*." *Primates* 15(2–3): 287–92.

———. 1975. "Maternal rank and early social interactions of infant stumptail macaques, *Macaca arctoides*." *Primates* 16(4): 405–18.

Gouzoules, H., and S. Gouzoules. 1995. "Recruitment screams of pigtail monkeys (*Macaca nemestrina*): Ontogenetic perspectives." *Behaviour* 132(5–6): 431–50.

———. 2000. "Agonistic screams differ among four species of macaques: The significance of motivation-structural rules." *Animal Behaviour* 59(3): 501–12.

———. 2002. "Primate communication: By nature honest, or by experience wise?" *International Journal of Primatology* 23(4): 821–48.

Gouzoules, H., S. Gouzoules, et al. 1998. "Agonistic screams and the classification of dominance relationships: Are monkeys fuzzy logicians?" *Animal Behaviour* 55: 51–60.

Gouzoules, S., and H. Gouzoules. 1987. Kinship. In *Primate societies,* ed. B. B. Smuts et al. Chicago, Univ. of Chicago Press: 299–305.

Gowaty, P. A. 1996. Field studies of parental care in birds: New data focus questions on variation among females. In *Parental care: Evolution, mechanisms, and adaptive significance. Advances in the study of behavior,* vol. 25, ed. J. S. Rosenblatt and C. T. Snowdon. San Diego, Academic: 477–531.

Goy, R., and D. Goldfoot. 1974. Experiential and hormonal factors influencing the development of sexual behavior in the male rhesus monkey. In *The neurosciences: Third study program,* ed. F. O. Schmitt and F. G. Worden. Cambridge, MA, MIT Press.

Goy, R. W. 1970. "Experimental control of psychosexuality." *Philosophical Transactions of the Royal Society of London B: Biological Sciences* 259: 149–62.

———. 1996. Patterns of juvenile behavior following early hormonal interventions. In *The lifespan development of individuals: Behavioral, neurobiological, and psychosocial perspectives,* ed. D. Magnusson. New York, Cambridge Univ. Press: 296–314.

Goy, R. W., F. B. Bercovitch, et al. 1988. "Behavioral masculinization is independent of genital masculinization in prenatally androgenized female Rhesus Macaques." *Hormones and Behavior* 22: 552–71.

Grabe, H. J., M. Lange, et al. 2005. "Mental and physical distress is modulated by a polymorphism in the 5-HT transporter gene interacting with social stressors and chronic disease burden." *Molecular Psychiatry* 10(2): 220–24.

Grafen, A. 1991. Modelling in behavioural ecology. In *Behavioural ecology: An evolutionary approach,* ed. J. R. Krebs and N. B. Davies. Oxford, Blackwell: 5–31.

Granqvist, P., and L. A. Kirkpatrick. 2008. Attachment and religious representations and behavior. In *Handbook of attachment,* 2nd ed., ed. J. Cassidy and P. R. Shaver. New York, Guilford: 906–33.

Grant, B. R., and P. R. Grant. 1989. *Evolutionary dynamics of a natural population: The large cactus finch of the Galapagos.* Chicago, Univ. of Chicago Press.

Grant, P. R., and B. Grant. 2008. *How species multiply: The radiation of Darwin's finches.* Princeton, Princeton Univ. Press.

Gray, L., L. Watt, et al. 2000. "Skin-to-skin contact is analgesic in healthy newborns." *Pediatrics* 105(1): e14.

Gray, P., and J. Feldman. 2004. "Playing in the zone of proximal development: Qualities of self-directed age mixing between adolescents and young children at a democratic school." *American Journal of Education* 110(2): 108–45.

Gray, P. A., H. Fu, et al. 2004. "Mouse brain organization revealed through direct genome-scale TF expression analysis." *Science* 306(5705): 2255–57.

Graybiel, A. M. 2005. "The basal ganglia: Learning new tricks and loving it." *Current Opinion in Neurobiology* 15(6): 638–44.

Graziano, W., D. French, et al. 1976. "Peer interaction in same- and mixed-age triads in relation to chronological age and incentive condition." *Child Development* 47(3): 707–14.

Green, R. 1987. *The "sissy boy syndrome" and the development of homosexuality.* New Haven, Yale Univ. Press.

Greenberg, J. H., ed. 1963. *Universals of language.* Cambridge, MA, MIT Press.

———. 1975. "Research on language universals." *Annual Review of Anthropology* 4: 75–94.

Greenberg, J. H., C. A. Ferguson, et al., eds. 1978. *Universals of human language.* Stanford, Stanford Univ. Press.

Greenberg, L. 1979. "Genetic component of bee odor in kin recognition." *Science* 206: 1095–97.

Greenfield, P. M. 1999. "Historical change and cognitive change: A two-decade follow-up study in Zinacantan, a Maya community in Chiapas, Mexico." *Mind, Culture, and Activity* 6: 92–108.

———. 2000a. "What psychology can do for anthropology, or why anthropology took postmodernism on the chin." *American Anthropologist* 102(3): 564–76.

———. 2000b. Culture and universals: Integrating social and cognitive development. In *Culture, thought, and development,* ed. L. P. Nucci, G. B. Saxe, and E. Turiel. Mahwah, NJ, Erlbaum: 231–77.

———. 2005. *Weaving generations together: Evolving creativity in the Maya of Chiapas.* Santa Fe, School of American Research.

Greenfield, P. M., H. Keller, et al. 2003. "Cultural pathways through universal development." *Annual Review of Psychology* 54: 461–90.

Greenfield, P. M., and E. S. Savage-Rumbaugh. 1991. Imitation, grammatical development, and the invention of protogrammar by an ape. In *Biological and behavioral determinants of language development,* ed. N. A. Krasnegor et al. Hillsdale, NJ, Erlbaum: 235–58.

———. 1994. Grammatical combination in *Pan paniscus:* Processes of learning and invention in the evolution and development of language. In *"Language" and intelligence in monkeys and apes: Comparative developmental perspectives,* ed. S. T. Parker and K. R. Gibson. New York, Cambridge Univ. Press: 540–78.

Greenfield, P. M., and L. K. Suzuki. 1997. Culture and human development: Implications for parenting, education, pediatrics, and mental health. In *Handbook of child psychology,* vol. 4, ed. I. Sigel. New York, Wiley: 1059–1109.

Greenough, W. F. 1977. "Sex differences in dendritic patterns in hamster preoptic area." *Brain Research* 126: 63–72.

Greenspan, R. J., and J. F. Ferveur. 2000. "Courtship in Drosophila." *Annual Review of Genetics* 34: 205–32.

Gregg, C. L., M. E. Haffner, et al. 1976. "The relative efficacy of vestibular-proprioceptive stimulation and the upright position in enhancing visual pursuit in neonates." *Child Development* 47: 309–14.

Greven, P. 1990. *Spare the child: The religious roots of punishment and the psychological impact of physical abuse.* New York, Random House.

Griffin, A. S. 2004. "Social learning about predators: A review and prospectus." *Learning and Behavior* 32(1): 131–40.

Griffin, A. S., and S. A. West. 2003. "Kin discrimination and the benefit of helping in cooperatively breeding vertebrates." *Science* 302(5645): 634–36.

Griffin, P. B., and A. Estioko-Griffin, eds. 1985. *The Agta of Northeastern Luzon: Recent studies.* Humanities Series No 16. Cebu City, Philippines, San Carlos Publications.

Griffin, P. B., M. Goodman, et al. 1992. Agta women hunters: Subsistence, child care and reproduction. In *Man and his culture: A resurgence,* ed. P. Bellwood. New Delhi, Vedams: 173–79.

Griffin, P. B., and M. B. Griffin. 1992. Fathers and childcare among the Cagayan Agta. In *Father-child relations: Cultural and biosocial contexts,* ed. B. S. Hewlett. New York, Aldine de Gruyter: 297–320.

———. 1999. The Agta of eastern Luzon, Philippines. In *The Cambridge encyclopedia of hunters and gatherers,* ed. R. B. Lee and R. Daly. Cambridge, Cambridge Univ. Press: 289.

Griffiths, I. R. 1996. "Myelin mutants: Model systems for the study of normal and abnormal myelination." *Bioessays* 18(10): 789–97.

Griffiths, P. E. 1997. *What emotions really are: The problem of psychological categories.* Chicago, Univ. of Chicago Press.

Griffiths, R. 1954. *The abilities of babies.* London, Univ. of London Press.

Grigorenko, E. L. 2001. "Developmental dyslexia: An update on genes, brains, and environments." *Journal of Child Psychology and Psychiatry and Allied Disciplines* 42(1): 91–125.

Grosbras, M.-H., M. Jansen, et al. 2007. "Neural mechanisms of resistance to peer influence in early adolescence." *Journal of Neuroscience* 27(30): 8040–45.

Gross, C. G. 2005. "Processing the facial image: A brief history." *American Psychologist* 60(8): 755–63.

Grosshans, H., and W. Filipowicz. 2008. "Molecular biology: The expanding world of small RNAs." *Nature* 451(7177): 414–16.

Grossmann, K., K. E. Grossmann, et al. 2008. A wider view of attachment and exploration: The influence of mothers and fathers on the development of psychological security from infancy to young adulthood. In *Handbook of attachment,* 2nd ed., ed. J. Cassidy and P. R. Shaver. New York, Guilford: 857–79.

Grossmann, T., and M. H. Johnson. 2007. "The development of the social brain in human infancy." *European Journal of Neuroscience* 25(4): 909–19.

Grossmann, T., M. H. Johnson, et al. 2008. "Early cortical specialization for face-to-face communication in human infants." *Proceedings of the Royal Society of London B: Biological Sciences* 275(1653): 2803–11.

Gruss, M., J. Bock, et al. 2003. "Haloperidol impairs auditory filial imprinting and modulates monoaminergic neurotransmission in an imprinting-relevant forebrain area of the domestic chick." *Journal of Neurochemistry* 87(3): 686–96.

Gubernick, D. J. 1981. Parent and infant attachment in mammals. In *Parental care in mammals,* ed. D. J. Gubernick. New York, Plenum: 243–305.

Guemple, L. 1979. *Inuit adoption.* Canadian Ethnology Service No. 47. Ottawa, National Museums of Canada.

Gunnar, M. R. 1998. "Quality of early care and buffering of neuroendocrine stress reactions: Potential effects on the developing human brain." *Preventive Medicine* 27(2): 208–11.

———. 2003. "Integrating neuroscience and psychological approaches in the study of early experiences." *Annals of the New York Academy of Sciences* 1008: 238–47.

Gunnar, M. R., L. Brodersen, et al. 1996. "Stress reactivity and attachment security." *Developmental Psychobiology* 29(3): 191–204.

Gunnar, M. R., and C. L. Cheatham. 2003. "Brain and behavior interfaces: Stress and the developing brain." *Infant Mental Health Journal* 24(3): 195–211.

Gunnar, M. R., and B. Donzella. 2002. "Social regulation of the cortisol levels in early human development." *Psychoneuroendocrinology* 27(1–2): 199–220.

Gunnar, M. R., S. J. Morison, et al. 2001. "Salivary cortisol levels in children adopted from Romanian orphanages." *Development and Psychopathology* 13(3): 611–28.

Gur, R. C. 2005. "Brain maturation and its relevance to understanding criminal culpability of juveniles." *Current Psychiatry Reports* 7(4): 292–96.

Gursky, S. 2000. "Allocare in a nocturnal primate: Data on the spectral tarsier, *Tarsius spectrum.*" *Folia Primatologica* 71(1–2): 39–54.

Gurven, M., H. Kaplan, et al. 2006. "How long does it take to become a proficient hunter? Implications for the evolution of extended development and long life span." *Journal of Human Evolution* 51(5): 454–70.

Gusella, J. F., R. E. Tanzi, et al. 1984. "DNA markers for nervous system diseases." *Science* 225: 1320–26.

Guttman, N., and D. R. Zimmerman. 2000. "Low-income mothers' views on breast-feeding." *Social Science and Medicine* 50(10): 1457–73.

Gyllensten, L., and T. Malmfors. 1963. "Myelination of the optic nerve and dependence on visual function." *Journal of Embryology and Experimental Morphology* 11: 255–66.

Habermas, T., and S. Bluck. 2000. "Getting a life: The emergence of the life story in adolescence." *Psychological Bulletin* 126(5): 748–69.

Habermas, T., and C. de Silveira. 2008. "The development of global coherence in life narratives across adolescence: Temporal, causal, and thematic aspects." *Developmental Psychology* 44(3): 707–21.

Hadfield, M. G., and C. Milio. 1988. "Isolation-induced fighting in mice and regional brain monoamine utilization." *Behavioural Brain Research* 31(1): 93–96.

Haeckel, E. 1905. *Evolution of man.* London, Watts.

Hahn, M. E., J. K. Hewitt, et al., eds. 1990. *Developmental behavior genetics: Neural, biometrical, and evolutionary approaches.* New York: Oxford Univ. Press.

Haig, D. 2007. "Weissman rules! OK? Epigenetics and the Lamarckian temptation." *Biology and Philosophy* 22: 415–28.

Haines-Stiles, G., and P. Montagnon. 1992a. Louder than words (episode 2). In *Childhood,* produced by Thirteen/WNET-NY. New York, Ambrose Video.

———. 1992b. Life's lessons (episode 5). In *Childhood,* produced by Thirteen/WNET-NY. New York, Ambrose Video.

———. 1992c. Among equals. In *Childhood,* produced by Thirteen/WNET-NY. New York, Ambrose Video.

Haith, M. M. 1973. Visual scanning in infants. In *The competent infant: Research and commentary,* ed. L. J. Stone, H. T. Smith, and L. B. Murphy. New York, Basic: 320–23.

Haith, M. M., T. Bergman, et al. 1977. "Eye contact and face scanning in early infancy." *Science* 198: 853–55.

Haldane, J. B. S. 1990 [1953]. *On being the right size and other essays.* New York, Oxford Univ. Press.

Hall, E. T. 1959. *The silent language.* New York, Doubleday.

———. 1963. "A system for the notation of proxemic behavior." *American Anthropologist* 65(5): 1003–26.

Hall, J. C., S. J. Kulkarni, et al. 1990. Genetic and molecular analysis of neural development and behavior in *Drosophila.* In *Developmental behavior genetics: Neural, biometrical, and evolutionary approaches,* ed. M. E. Hahn et al. New York, Oxford Univ. Press: 100–12.

Hall, K., and I. DeVore. 1965. Baboon social behavior. In *Primate behavior,* ed. I. DeVore. New York, Holt, Rinehart, and Winston.

Haller, J., G. B. Makara, et al. 1996. "The effect of alpha 2 adrenoceptor blockers on aggressive behavior in mice: Implications for the actions of adrenoceptor agents." *Psychopharmacology* 126(4): 345–50.

Halpern, C. T. 2003. Biological influences on adolescent romantic and sexual behavior. In *Adolescent romantic relations and sexual behavior: Theory, research, and practical implications,* ed. P. Florsheim. Mahwah, NJ, Erlbaum: 57–84.

Halpern, C. T., J. R. Udry, et al. 1997. "Testosterone predicts initiation of coitus in adolescent females." *Psychosomatic Medicine* 59(2): 161–71.

Halpern, C. T., J. R. Udry, et al. 1998. "Monthly measures of salivary testosterone predict sexual activity in adolescent males." *Archives of Sexual Behavior* 27(5): 445–65.

Hames, R., and P. Draper. 2004. "Women's work, child care, and helpers-at-the-nest in a hunter-gatherer society." *Human Nature* 15(4): 319–41.

Hames, R. B. 1988. The allocation of parental care among the Ye'kwana. In *Human reproductive behaviour: A Darwinian perspective,* ed. L. Betzig, M. B. Mulder, and P. Turke. Cambridge, Cambridge Univ. Press: 237–51.

Hamilton, A. F. d. C., R. M. Brindley, et al. 2007. "Imitation and action understanding in autistic spectrum disorders: How valid is the hypothesis of a deficit in the mirror neuron system?" *Neuropsychologia* 45(8): 1859–68.

Hamilton, W. D. 1964. "The genetical evolution of social behavior, I and II." *Journal of Theoretical Biology* 7: 1–52.

Hammock, E. A. D., and L. J. Young. 2004. "Functional microsatellite polymorphism associated with divergent social structure in vole species." *Molecular Biology and Evolution* 21(6): 1057–63.

Hampson, E. 2008. Endocrine contributions to sex differences in visuospatial perception and cognition. In *Sex differences in the brain: From genes to behavior,* ed. J. B. Becker et al. New York, Oxford Univ. Press: 15–33.

Hampson, S. E., and L. R. Goldberg. 2006. "A first large cohort study of personality trait stability over the 40 years between elementary school and midlife." *Journal of Personality and Social Psychology,* 91(4): 763–79.

Hane, A. A., S. Feldstein, et al. 2003. "The relation between coordinated interpersonal timing and maternal sensitivity in four-month-old infants." *Journal of Psycholinguistic Research* 32(5): 525–39.

Hanks, C., and F. Rebelsky. 1977. "Mommy and the midnight visitor: A study of occasional co-sleeping." *Psychiatry* 40: 277–80.

Harcourt-Smith, W. E. H., and L. C. Aiello. 2004. "Fossils, feet and the evolution of human bipedal locomotion." *Journal of Anatomy* 204(5): 403–16.

Hardin, M. G., and M. Ernst. 2009. "Functional brain imaging of development-related risk and vulnerability for substance use in adolescents." *Journal of Addiction Medicine* 3: 47–54.

Harding, C. F., and B. K. Follett. 1979. "Hormone changes triggered by aggression in a natural population of blackbirds." *Science* 203: 918–20.

Harding, R. S. O., and S. C. Strum. 1998. The predatory baboons of Kekopey. In *The primate anthology: Essays on primate behavior, ecology, and conservation from natural history,* ed. R. L. Ciochon and R. A. Nisbett. Upper Saddle River, NJ, Prentice Hall: 99–105.

Hardy-Brown, K., and R. Plomin. 1985. "Infant communicative development: Evidence from adoptive and biological families for genetic and environmental influences on rate differences." *Developmental Psychology* 21(2): 378–85.

Hardy-Brown, K., R. Plomin, et al. 1981. "Genetic and environmental influences on the rate of communicative development in the first year of life." *Developmental Psychology* 17(6): 704–17.

Hare, B., J. Call, et al. 2000. "Chimpanzees know what conspecifics do and do not see." *Animal Behaviour* 59(4): 771–85.

Harkness, S., C. P. Edwards, et al. 1981. "Social roles and moral reasoning: A case study in a rural African community." *Developmental Psychology* 17(5): 593–603.

Harkness, S., U. Moscardino, et al. 2006. "Mixed methods in international collaborative research: The experiences of the international study of parents, children, and schools." *Cross-Cultural Research: The Journal of Comparative Social Science* 40(1): 65–82.

Harlan, W. R., G. P. Grillo, et al. 1979. "Secondary sex characteristics of boys 12 to 17 years of age: The U.S. Health Examination Survey." *Journal of Pediatrics* 95(2): 293–97.

Harlan, W. R., E. A. Harlan, et al. 1980. "Secondary sex characteristics of girls 12 to 17 years of age: The U.S. Health Examination Survey." *Journal of Pediatrics* 96(6): 1074–78.

Harley, J. K. 1963. *Adolescent youths in peer groups: A cross-cultural study.* Department of Anthropology, Harvard Univ.

Harlow, H., and M. Harlow. 1965. The affectional systems. In *Behavior of nonhuman primates,* vol. 2., ed. A. M. Schrier, H. F. Harlow, and F. Stollnitz. New York, Academic.

Harlow, H., and S. Suomi. 1971. "Social recovery by isolation-reared monkeys." *Proceedings of the National Academy of Sciences* 68: 1534–38.

Harlow, H. F. 1958a. The evolution of learning. In *Behavior and evolution,* ed. A. Roe and G. G. Simpson. New Haven, Yale Univ. Press: 269–90.

———. 1958b. "The nature of love." *American Psychologist* 13: 673–85.

———. 1959. "Love in infant monkeys." *Scientific American* (June): 1–8.

———. 1963. The maternal affectional system. In *Determinants of infant behavior,* vol. 2, ed. B. M. Foss. London, Methuen.

———. 1971. *Learning to love.* San Francisco, Albion.

———. 1975. "Lust, latency and love: Simian secrets of successful sex." *Journal of Sex Research* 11(2): 79–90.

Harlow, H. F., R. O. Dodsworth, et al. 1965. "Total social isolation in monkeys." *Proceedings of the National Academy of Sciences* 54: 90–97.

Harlow, H. F., and C. Mears. 1978. The nature of complex, unlearned responses. In *The development of affect,* vol. 1, ed. M. Lewis and L. A. Rosenblum. New York, Plenum Press: 257–74.

Harris, H. 1995. Rethinking heterosexual relationships in Polynesia: A case study of Mangaia, Cook Island. In *Romantic passion: A universal experience?,* ed. W. Jankowiak. New York, Columbia Univ. Press: 95–127.

Harris, J. R. 1995. "Where is the child's environment? A group socialization theory of development." *Psychological Review* 102: 458–89.

———. 1998. *The nurture assumption: Why children turn out the way they do: Parents matter less than you think and peers matter more.* New York, Free Press.

Harris, M. 1968. *The rise of anthropological theory: A history of theories of culture.* New York, Crowell.

Harter, S. 1996. Developmental changes in self-understanding across the 5 to 7 shift. In *The five to seven year shift: The age of reason and responsibility,* ed. A. J. Sameroff and M. M. Haith. Chicago, Univ. of Chicago Press: 207–36.

Hartung, J. 1982. "Polygyny and inheritance of wealth." *Current Anthropology* 23: 1–12.

Harvey, P., and M. Pagel. 1991. *The comparative method in evolutionary biology.* New York, Oxford Univ. Press.

Harvey, P. H., M. D. Pagel, et al. 1991. "Mammalian metabolism and life histories." *American Naturalist* 137(4): 556–66.

Hasegawa, M., S. Houdou, et al. 1992. "Development of myelination in the human fetal and infant cerebrum: A myelin basic protein immunohistochemical study." *Brain and Development* 14(1): 1–6.

Hasler, A. D. 1960. "Guideposts of migrating fishes." *Science* 131: 785–92.

Hatchwell, B. J., A. F. Russell, et al. 2000. "Divorce in cooperatively breeding long-tailed tits: A consequence of inbreeding avoidance?" *Proceedings of the Royal Society of London B: Biological Sciences* 267(1445): 813–19.

Hatfield, E., R. L. Rapson, et al. 2007. Passionate love and sexual desire. In *Handbook of cultural psychology,* ed. S. Kitayama and D. Cohen. New York, Guilford: 760–79.

Hauser, M. D. 2006. *Moral minds: How nature designed our universal sense of right and wrong.* New York, HarperCollins/Ecco.

Hausfater, G. 1984. Infanticide in langurs: Strategies, counterstrategies, and parameter values. In *Infanticide: Comparative and evolutionary perspectives,* ed. G. Hausfater and S. B. Hrdy. New York, Aldine de Gruyter: 257–82.

Hausfater, G., and S. B. Hrdy, eds. 1984. *Infanticide: Comparative and evolutionary perspectives.* New York, Aldine de Gruyter.

Havelock, J. C., R. J. Auchus, et al. 2004. "The rise in adrenal androgen biosynthesis: Adrenarche [see comment]." *Seminars in Reproductive Medicine* 22(4): 337–47.

Hawes, J. M., and N. R. Hiner, eds. 1985. *American childhood: A research guide and historical handbook.* Westport, CT, Greenwood.

———. 1991. *Children in historical and comparative perspective: An international handbook and research guide.* New York, Greenwood.

Hawkes, K. 1991a. "Hunting income patterns among the Hadza: Big game, common goods, foraging goals, and the evolution of the human diet." *Philosophical Transactions of the Royal Society of London B: Biological Sciences* 334: 243–51.

———. 1991b. "Showing off: Tests of an hypothesis about men's foraging goals." *Ethology and Sociobiology* 12(1): 29–54.

———. 2003. "Grandmothers and the evolution of human longevity." *American Journal of Human Biology* 15(3): 380–400.

———. 2004. "Human longevity: The grandmother effect [comment]." *Nature* 428(6979): 128–29.

———. 2006. Slow life histories and human evolution. In *Evolution of human life history,* ed. K. Hawkes and R. R. Paine. Santa Fe, School of American Research Press: 95–126.

Hawkes, K., J. F. O'Connell, et al. 1998. "Grandmothering, menopause, and the evolution of human life histories." *Proceedings of the National Academy of Sciences* 95(3): 1336–39.

Hay, D. F., A. Payne, et al. 2004. "Peer relations in childhood." *Journal of Child Psychology and Psychiatry* 45(1): 84–108.

Haynes, C. A., J. N. V. Miles, et al. 2000. "A confirmatory factor analysis of two models of sensation seeking." *Personality and Individual Differences* 29(5): 823–39.

Haynes, R. L., N. S. Borenstein, et al. 2005. "Axonal development in the cerebral white matter of the human fetus and infant." *Journal of Comparative Neurology* 484(2): 156–67.

He, Z., K. C. Wang, et al. 2002. "Knowing how to navigate: Mechanisms of semaphorin signaling in the nervous system." *Science's Stke [Electronic Resource]: Signal Transduction Knowledge Environment* 2002(119): RE1.

Headland, T. N. 1989. "Population decline in a Philippine Negrito hunter-gatherer society." *American Journal of Human Biology* 1: 59–72.

Heath, K. M. 2003. "The effects of kin propinquity on infant mortality." *Social Biology* 50(3–4): 270–80.

Hebb, D. O. 1946. "On the nature of fear." *Psychological Review* 53: 259–76.

———. 1949. *The organization of behavior: A neuropsychological theory.* New York, Wiley.

Heider, K. 1991. *Landscapes of emotion: Mapping three cultures of emotion in Indonesia.* New York, Cambridge Univ. Press.

Heim, C., and C. B. Nemeroff. 2001. "The role of childhood trauma in the neurobiology of mood and anxiety disorders: Preclinical and clinical studies." *Biological Psychiatry* 49(12): 1023–39.

Heim, C., P. M. Plotsky, et al. 2004. "Importance of studying the contributions of early adverse experience to neurobiological findings in depression." *Neuropsychopharmacology* 29(4): 641–48.

Heimer, L., and G. W. Van Hoesen. 2006. "The limbic lobe and its output channels: Implications for emotional functions and adaptive behavior." *Neuroscience and Biobehavioral Reviews* 30(2): 126–47.

Heinz, A., J. D. Higley, et al. 1998. "In vivo association between alcohol intoxication, aggression, and serotonin transporter availability in nonhuman primates." *American Journal of Psychiatry* 155(8): 1023–28.

Hendrix, L. 1985. "Economy and child training reexamined." *Ethos* 13(3): 246–61.

Henneberger, C., T. Papouin, et al. 2010. "Long-term potentiation depends on release of d-serine from astrocytes." *Nature* 463(8673): 232–36.

Hennenlotter, A., U. Schroeder, et al. 2005. "A common neural basis for receptive and expressive communication of pleasant facial affect." *Neuroimage* 26(2): 581–91.

Henrich, J. 2001. "Cultural transmission and the diffusion of innovations: Adoption dynamics indicate that biased cultural transmission is the predominate force in behavioral change." *American Anthropologist* 103(4): 992–1013.

———. 2004. "Demography and cultural evolution: Why adaptive cultural processes produced maladaptive losses in Tasmania." *American Antiquity* 69(2): 197–214.

Henry, J., and S. Herrero. 1974. "Social play in the American black bear." *American Zoologist* 14: 371.

Herculano-Houzel, S., C. E. Collins, et al. 2007. "Cellular scaling rules for primate brains." *Proceedings of the National Academy of Sciences* 104(9): 3562–67.

Herdt, G. 1987. *Guardians of the flutes: Idioms of masculinity.* New York, Columbia Univ. Press.

———. 1994a. Mistaken sex: Culture, biology and the third sex in New Guinea. In *Third sex, third gender: Beyond sexual dimorphism in culture and history,* ed. G. Herdt. New York, Zone: 419–45.

———, ed. 1994b. *Third sex, third gender: Beyond sexual dimorphism in culture and history.* New York, Zone.

Hereford, J., T. F. Hansen, et al. 2004. "Comparing strengths of directional selection: How strong is strong?" *Evolution* 58(10): 2133–43.

Hergenhahn, B. R., and M. H. Olson. 1996. *An introduction to theories of learning.* Upper Saddle River, NJ, Prentice Hall.

Herman-Giddens, M. E. 2006. "Recent data on pubertal milestones in United States children: The secular trend toward earlier development." *International Journal of Andrology* 29(1): 241–6; discussion 286–90.

Herrera, A. A., and Y. Zeng. 2003. "Activity-dependent switch from synapse formation to synapse elimination during development of neuromuscular junctions." *Journal of Neurocytology* 32(5–8): 817–33.

Herrmann, E., J. Call, et al. 2007. "Humans have evolved specialized skills of social cognition: The cultural intelligence hypothesis [see comment]." *Science* 317(5843): 1360–66.

Herrnstein, R. J., and C. Murray. 1994. *The bell curve: Intelligence and class structure in American life.* New York, Free Press.

Herschkowitz, N., J. Kagan, et al. 1997. "Neurobiological bases of behavioral development in the first year." *Neuropediatrics* 28(6): 296–306.

Herschkowitz, N., J. Kagan, et al. 1999. "Neurobiological bases of behavioral development in the second year." *Neuropediatrics* 30(5): 221–30.

Hertz, R. 1960. The pre-eminence of the right hand: A study in religious polarity. In *Death and the right hand.* Glencoe, IL, Free Press.

Hertz-Pannier, L., C. Chiron, et al. 2002. "Late plasticity for language in a child's non-dominant hemisphere: A pre- and post-surgery fMRI study." *Brain* 125(Pt. 2): 361–72.

Hess, E. 1974. *Imprinting.* Chicago, Univ. of Chicago Press.

Hetherington, E. M. 1993. "An overview of the Virginia Longitudinal Study of Divorce and Remarriage with a focus on early adolescence." *Journal of Family Psychology* 7(1): 39–56.

Hetherington, E. M., and R. Cox. 1982. Effects of divorce on parents and children. In *Nontraditional families: Parenting and child development,* ed. M. E. Lamb. Hillsdale, NJ, Erlbaum: 233–88.

Hetherington, E. M., and A. M. Elmore. 2003. Risk and resilience in children coping with their parents' divorce and remarriage. In *Resilience and vulnerability: Adaptation in the context of childhood adversities,* ed. Suniya S. Luthar. Cambridge, Cambridge Univ. Press.

Hetherington, E. M., and J. Kelly. 2002. *For better or for worse: Divorce reconsidered.* New York, Norton.

Hetherington, E. M., and M. Stanley-Hagan. 1999. "The adjustment of children with divorced parents: A risk and resiliency perspective." *Journal of Child Psychology and Psychiatry and Allied Disciplines* 40(1): 129–40.

Hewlett, B. 2005. Vulnerable lives: The experience of death and loss among the Aka and Ngandu adolescents of the Central African Republic. In *Hunter-gatherer childhoods: Evolutionary, developmental and cultural perspectives,* ed. B. S. Hewlett and M. E. Lamb. New Brunswick, NJ, AldineTransaction: 322–42.

Hewlett, B. S. 1988. Sexual selection and paternal investment among Aka pygmies. In

Human reproductive behaviour: A Darwinian perspective, ed. L. Betzig, M. B. Mulder, and P. Turke. Cambridge, Cambridge Univ. Press: 263–76.

———. 1991a. "Demography and childcare in preindustrial societies." *Journal of Anthropological Research* 47(1): 1–37.

———. 1991b. *Intimate fathers: The nature and context of Aka Pygmy paternal infant care.* Ann Arbor, Univ. of Michigan Press.

———. 2004. Personal communication.

Hewlett, B. S., and M. E. Lamb, eds. 2005. *Hunter-gatherer childhoods: Evolutionary, developmental, and cultural perspectives.* New Brunswick, NJ, AldineTransaction.

Hewlett, B. S., M. E. Lamb, et al. 1998. "Culture and early infancy among central African foragers and farmers." *Developmental Psychology* 34(4): 653–61.

———. 2000. "Internal working models, trust, and sharing among foragers." *Current Anthropology* 41(2): 287–97.

Heywood, C. 2001. *A history of childhood: Children and childhood in the West from medieval to modern times.* Hoboken, NJ, Polity/ Wiley.

Higgens, K. 1985. "Ritual and symbol in Baka life history." *Anthropology and Humanism Quarterly* 10(4): 100–6.

Higley, J. D., S. J. Suomi, et al. 1996. "A nonhuman primate model of type II alcoholism? Part 2. Diminished social competence and excessive aggression correlates with low cerebrospinal fluid 5-hydroxyindoleacetic acid concentrations." *Alcoholism: Clinical and Experimental Research* 20(4): 643–50.

Hilger, S. M. I. 1957. *Araucanian child life and its cultural background.* Smithsonian Miscellaneous Collections, vol. 133. Washington, Smithsonian Institution.

Hill, K., and A. M. Hurtado. 1996. *Ache life history: The ecology and demography of a foraging people.* New York, Aldine de Gruyter.

———. 1999. The Aché of Paraguay. In *The Cambridge encyclopedia of hunters and gatherers,* ed. R. B. Lee and R. Daly. Cambridge, Cambridge Univ. Press: 92–96.

Hill, K., A. M. Hurtado, et al. 2007. "High adult mortality among Hiwi hunter-gatherers: Implications for human evolution." *Journal of Human Evolution* 52(4): 443–54.

Hill, V., and B. H. Pillow. 2006. "Children's understanding of reputations." *Journal of Genetic Psychology* 167(2): 137–57.

Hille, B. 1992. *Ionic channels of excitable membranes,* 2nd ed. Sunderland, MA, Sinauer.

Hinde, R. A., and L. Davies. 1972a. "Removing infant rhesus from mother for 13 days compared with removing mother from infant." *Journal of Child Psychology and Psychiatry* 13: 227–37.

———. 1972b. "Changes in mother-infant relationship after separation in rhesus monkeys." *Nature* 239: 41–42.

Hinde, R. A., and N. Tinbergen. 1958. The comparative study of species-specific behavior. In *Behavior and evolution,* ed. A. Roe and G. G. Simpson. New Haven, Yale Univ. Press: 251–68.

Hinde, R. A., and Y. Spencer-Booth. 1967. "The behaviour of socially living rhesus monkeys in their first two and a half years." *Animal Behaviour* 15: 169–96.

———. 1971. "Effects of brief separation from mother on rhesus monkeys." *Science* 173: 111–18.

Hindley, C. B., A. M. Filliozat, et al. 1966. "Differences in the age of walking in five European longitudinal samples." *Human Biology* 38: 364–79.

Hiner, N. R., and J. M. Hawes, eds. 1985. *Growing up in America: Children in historical perspective.* Urbana, Univ. of Illinois Press.

Hines, M. 2003. "Sex steroids and human behavior: Prenatal androgen exposure and sex-typical play behavior in children." *Annals of the New York Academy of Sciences* 1007: 272–82.

———. 2004. "Psychosexual development in individuals who have female pseudo-hermaphroditism." *Child and Adolescent Psychiatric Clinics of North America* 13(3): 641–56.

———. 2005. *Brain gender.* New York, Oxford Univ. Press.

———. 2006. "Prenatal testosterone and gender-related behaviour." *European Journal of Endocrinology* 155(Suppl. 1): S115–21.

Hines, M., L. S. Allen, et al. 1992. "Sex differences in the subregions of the medial nucleus of the amygdala and the bed nucleus of the stria terminalis of the rat." *Brain Research* 579: 321–26.

Hines, M., and F. R. Kaufman. 1994. "Androgen and the development of human sex-typical behavior: Rough-and-tumble play and sex of preferred playmates in children with congenital adrenal hyperplasia (CAH)." *Child Development* 65: 1042–53.

Hinshaw, S. P. 2003. "Impulsivity, emotion regulation, and developmental psychopathology: Specificity versus generality of linkages." *Annals of the New York Academy of Sciences* 1008: 149–59.

Hirasawa, A. 2005. Infant care among the sedentarized Baka hunter-gatherers in Southeastern Cameroon. In *Hunter-gatherer childhoods: Evolutionary, developmental and cultural perspectives,* ed. B. S. Hewlett and M. E. Lamb. New Brunswick, NJ, AldineTransaction: 365–84.

Hirsch, H., and D. Spinnelli. 1970. "Visual experience modified distribution of horizontally and vertically oriented receptive fields in cats." *Science* 168: 869–71.

Hirschi, T., and M. Gottfredson. 1983. "Age and the explanation of crime." *American Journal of Sociology* 89(3): 552–85.

Hiscock, M., N. Perachio, et al. 2001. "Is there a sex difference in human laterality? IV. An exhaustive survey of dual-task interference studies from six neuropsychology journals." *Journal of Clinical and Experimental Neuropsychology* 23(2): 137–48.

Hittelman, J. H., and R. Dickes. 1979. "Sex differences in neonatal eye contact time." *Merrill-Palmer Quarterly* 25: 171–84.

Hobert, O. 2003. "Behavioral plasticity in *C. elegans:* Paradigms, circuits, genes." *Journal of Neurobiology* 54(1): 203–23.

Hobson, J. A., and E. F. Pace-Schott. 2002. The cognitive neuroscience of sleep: Consciousness, neuronal systems, and learning. *Nature Reviews Neuroscience* 3(9): 679–93.

Hochberg, Z. 2008. "Juvenility in the context of life history theory." *Archives of Disease in Childhood* 93(6): 534–39.

Hockett, C. 1960. "The origin of speech." *Scientific American* 203(3): 88–111.

Hockett, C. F. 1973. *Man's place in nature.* New York, McGraw-Hill.

Hofer, M. A. 1975. "Studies on how early maternal separation produces behavioral change in young rats." *Psychosomatic Medicine* 37(3): 245–64.

———. 2002. "Unexplained crying: An evolutionary perspective." *Acta Paediatrica* 91(5): 491–96.

Hofstede, G., and R. R. McCrae. 2004. "Personality and culture revisited: Linking traits and dimensions of culture." *Cross-Cultural Research* 38(1): 52–88.

Hoksbergen, R. A. C., J. ter Laak, et al. 2003. "Posttraumatic stress disorder in adopted children from Romania." *American Journal of Orthopsychiatry* 73(3): 255–65.

———. 2005. "Post-institutional autistic syndrome in Romanian adoptees." *Journal of Autism and Developmental Disorders* 35(5): 615–23.

Holloway, R. L. 1966. "Cranial capacity and neuron number: A critique and proposal." *American Journal of Physical Anthropology* 25: 305–14.

———. 1974. "The casts of fossil hominid brains." *Scientific American* 231(1): 106–15.

———. 1984. "The Taung endocast and the lunate sulcus: A rejection of the hypothesis of its anterior position." *American Journal of Physical Anthropology* 64(3): 285–87.

———. 1991. "On Falk's 1989 accusations regarding Holloway's study of the Taung endocast: A reply." *American Journal of Physical Anthropology* 84(1): 87–91.

Holloway, R. L., D. C. Broadfield, et al. 2003a. "The lunate sulcus and early hominid brain evolution: Toward the end of a controversy." *American Journal of Physical Anthropology* 36(Suppl.): 117.

Holloway, R. L., D. C. Broadfield, et al. 2003b. "Morphology and histology of chimpanzee primary visual striate cortex indicate that brain reorganization predated brain expansion in early hominid evolution." *Anatomical Record* 273A(1): 594–602.

Holmberg, A. R. 1969. *Nomads of the long bow: The Siriono of eastern Bolivia.* Garden City, NY, Natural History Press.

Holmes, W. G., and P. W. Sherman. 1982. "The ontogeny of kin recognition in two species of ground squirrels." *American Zoologist* 22: 491–517.

Holt, A. B., D. B. Cheek, et al. 1975. Brain size and the relation of the primate to the nonprimate. In *Fetal and postnatal cellular growth: Hormones and nutrition,* ed. D. B. Cheek. New York, Wiley.

Holt, N. L., and J. G. H. Dunn. 2004. "Toward a grounded theory of the psychosocial competencies and environmental conditions associated with soccer success." *Journal of Applied Sport Psychology* 16(3): 199–219.

Hoogland, J. L. 1982. "Prairie dogs avoid extreme inbreeding." *Science* 215: 1639–41.

Hopkins, K. 1980. "Brother-sister marriage in Roman Egypt." *Comparative Studies in Society and History* 22: 303–54.

Hopkins, W. D., and L. Marino. 2000. "Asymmetries in cerebral width in nonhuman primate brains as revealed by magnetic resonance imaging (MRI)." *Neuropsychologia* 38: 493–99.

Horn, G. 1991. Cerebral function and behaviour investigated through a study of filial imprinting. In *The development and integration of behaviour: Essays in honor of Robert Hinde,* ed. P. Bateson. Cambridge, Cambridge Univ. Press: 121–48.

Horton, H. L., and P. Levitt. 1988. "A unique membrane protein is expressed on early developing limbic system axons and cortical targets." *Journal of Neuroscience* 8(12): 4653–61.

Horwich, R. H. 1974. "Regressive periods in primate behavioral development with reference to other mammals." *Primates* 15: 141–49.

Hotta, Y., and S. Benzer. 1972. "Mapping of behavior in *Drosophila* mosaics." *Nature* 240: 527–35.

Howell, F. C. 1968. *Early man.* New York, Time-Life.

Howell, N. 1979. *Demography of the Dobe !Kung.* New York, Academic.

———. 2010. *Life histories of the Dobe !Kung: Food, fatness, and well-being over the life span.* Berkeley, Univ. of California Press.

Howes, C., and J. Farver. 1987. "Social pretend play in 2-year-olds: Effects of age of partner." *Early Childhood Research Quarterly* 2(4): 305–14.

Hoyle, C. H. 1999. "Neuropeptide families and their receptors: Evolutionary perspectives." *Brain Research* 848(1–2): 1–25.

Hrdy, S. B. 1974. "Male-male competition and infanticide among the langurs (*Presbytis entellus*) of Abu, Rajasthan, India." *Folia Primatologica* 22(1): 19–58.

———. 1976. The care and exploitation of non-human primate infants by conspecifics other than the mother. In *Advances in the study of behavior,* vol. 6. New York, Academic: 101–58.

———. 1977. *The langurs of Abu: Female and male strategies of reproduction.* Cambridge, MA, Harvard Univ. Press.

———. 1979. "Infanticide among animals: A review, classification, and examination of the implications for the reproductive strategies of females." *Ethology and Sociobiology* 1: 13–40.

———. 1981. *The woman that never evolved.* Cambridge, MA, Harvard Univ. Press.

———. 1992. "Fitness tradeoffs in the history and evolution of delegated mothering with special reference to wet-nursing, abandonment, and infanticide." *Ethology and Sociobiology* 13(5–6): 409–42.

———. 1999. *Mother nature: A history of mothers, infants, and natural selection.* New York, Pantheon.

———. 2000. "The optimal number of fathers. Evolution, demography, and history in the shaping of female mate preferences." *Annals of the New York Academy of Sciences* 907: 75–96.

———. 2001. "The past, present, and future of the human family." Tanner Lectures on Human Values: http://www.tannerlectures.utah.edu/lectures/Hrdy_02.pdf.

———. 2003. "Comment on 'Maternal Sentiments: How Strong Are They?' by Arthur P. Wolf." *Current Anthropology* 44(Suppl.): S41–42.

———. 2005. Comes the child before man: How cooperative breeding and prolonged postweaning dependence shaped human potential. In *Hunter-gatherer childhoods: Evolutionary, developmental and cultural perspectives,* ed. B. S. Hewlett and M. E. Lamb. New Brunswick, NJ, Transaction: 65–91.

———. 2009. *Mothers and others: The evolutionary origins of mutual understanding.* Cambridge, MA, Harvard Univ. Press.

Hrdy, S. B., and G. Hausfater. 1984. Comparative and evolutionary perspectives on infanticide: Introduction and overview. In *Infanticide: Comparative and evolutionary perspectives,* ed. G. Hausfater and S. B. Hrdy. New York, Aldine: xiii–xxxix.

Hsu, M.-T. 2003. "Comment on 'Maternal sentiments: How strong are they?' by Arthur P. Wolf." *Current Anthropology* 44(Suppl.): S42–43.

Hubel, D., T. Wiesel, et al. 1977. "Plasticity of ocular dominance columns in monkey striate cortex." *Philosophical Transactions of the Royal Society of London B: Biological Sciences* (278): 377–409.

Hubel, D. H. 1987. *Eye, brain, and vision.* New York, Scientific American Library.

Huizinga, J. 1949. *Homo ludens: A study of the play-element in culture.* London, Routledge.

Hunt, P. J., E. M. Gurnell, et al. 2000. "Improvement in mood and fatigue after dehydroepiandrosterone replacement in Addison's disease in a randomized, double blind trial." *Journal of Clinical Endocrinology and Metabolism* 85(12): 4650–56.

Hunziker, U. A., and R. G. Barr. 1986. "Increased carrying reduces infant crying: A randomized controlled trial." *Pediatrics* 77: 641–48.

Huot, R. L., P. M. Plotsky, et al. 2002. "Neonatal maternal separation reduces hippocampal mossy fiber density in adult Long Evans rats." *Brain Research* 950(1–2): 52–63.

Hurtado, A., K. Hill, et al. 1992. "Trade-offs between female food acquisition and child care among Hiwi and Ache foragers." *Human Nature* 3(3): 185–216.

Hurtado, A. M., and K. R. Hill, eds. 1992. *Paternal effect on offspring survivorship among*

Ache and Hiwi hunter-gatherers: Implications for modeling pair-bond stability. Hawthorne, NY, Aldine de Gruyter.

Hutchinson, R., R. Ulrich, et al. 1965. "Effects of age and related factors on the pain-aggression reaction." *Journal of Comparative and Physiological Psychology* 59: 365–69.

Huttenlocher, P. R. 1970. "Myelination and the development of function in immature pyramidal tract." *Experimental Neurology* 29: 405–15.

———. 1979. "Synaptic density in human frontal cortex—developmental changes and effects of aging." *Brain Research* 163: 195–205.

———. 1994. Synaptogenesis in human cerebral cortex. In *Human behavior and the developing brain,* ed. G. Dawson and K. W. Fischer. New York, Guilford: 137–52.

———. 2003. "Basic neuroscience research has important implications for child development [comment]." *Nature Neuroscience* 6(6): 541.

Huttenlocher, P. R., and A. S. Dabholkar. 1997. "Regional differences in synaptogenesis in human cerebral cortex." *Journal of Comparative Neurology* 387(2): 167–78.

Huttenlocher, P. R., and C. de Courten. 1987. "The development of synapses in striate cortex of man." *Human Neurobiology* 6: 1–9.

Huttenlocher, P. R., C. de Courten, et al. 1982. "Synaptic development in human cerebral cortex." *International Journal of Neurology* 16–17: 144–54.

Huxley, J. 1942. *Evolution, the modern synthesis.* London, Allen and Unwin.

Huxley, J. S. 1993 [1932]. *Problems of relative growth.* Baltimore, Johns Hopkins Univ. Press.

Hyde, J. S. 2005. "The gender similarities hypothesis." *American Psychologist* 60(6): 581–92.

Iacoboni, M. 2005. "Neural mechanisms of imitation." *Current Opinion in Neurobiology* 15(6): 632–37.

Ibanez, L., and F. de Zegher. 2006. "Puberty after prenatal growth restraint." *Hormone Research* 65(Suppl. 3): 112–15.

Iijima, M., O. Arisaka, et al. 2001. "Sex differences in children's free drawings: A study on girls with congenital adrenal hyperplasia." *Hormones and Behavior* 40(2): 99–104.

Imai, T., T. Yamazaki, et al. 2009. "Pre-target axon sorting establishes the neural map topography." *Science* 325(5940): 585–90.

Immelman, K. 1966. "Zur irreversibilität der prägung." *Naturwissenschaft* 53: 209.

Imperato-McGinley, J. 1997. "5 alpha-reductase-2 deficiency." *Current Therapy in Endocrinology and Metabolism* 6: 384–87.

———. 2002. "5alpha-reductase-2 deficiency and complete androgen insensitivity: Lessons from nature." *Advances in Experimental Medicine and Biology* 511: 121–31; discussion 131–34.

Imperato-McGinley, J., R. E. Peterson, et al. 1979. "Androgens and the evolution of male-gender identity among male pseudohermaphrodites with 5a-reductase deficiency." *New England Journal of Medicine* 300: 1233–70.

———. 1980. Androgens and the evolution of male-gender identity among male pseudohermaphrodites with 5a-reductase deficiency. In *Annual progress in child psychology and child development.* New York, Brunner/Mazel.

Imperato-McGinley, J., and Y. S. Zhu. 2002. "Androgens and male physiology: The syndrome of 5alpha-reductase-2 deficiency." *Molecular and Cellular Endocrinology* 198(1–2): 51–59.

Inhelder, B., and J. Piaget. 1958. *The growth of logical thinking: From childhood to adolescence.* New York, Basic.

Insel, T. R. 1997. "A neurobiological basis of social attachment." *American Journal of Psychiatry* 154(6): 726–35.

———. 2000. "Toward a neurobiology of attachment." *Review of General Psychology* 4(2): 176–85.

———. 2003. "Is social attachment an addictive disorder?" *Physiology and Behavior* 79(3): 351–57.

Insel, T. R., and J. T. Winslow. 1991. "Central administration of oxytocin modulates the infant rat's response to social isolation." *European Journal of Pharmacology* 203: 149–52.

Insel, T. R., L. Young, et al. 1997. "Molecular aspects of monogamy." *Annals of the New York Academy of Sciences* 807: 302–16.

International Schizophrenia Consortium. 2009. "Common polygenic variation contributes to risk of schizophrenia and bipolar disorder." *Nature* 460: 748–52.

Irons, W. 1997. Cultural and biological success. In *Human nature: A critical reader,* ed. L. Betzig. New York, Oxford Univ. Press: 36–45.

Isaac, F. L., and E. R. McCown. 1976. *Human origins: Louis Leakey and the East African evidence.* Menlo Park, CA, Benjamin.

Isaacson, R. L. 2002. "Unsolved mysteries: The hippocampus." *Behavioral and Cognitive Neuroscience Reviews* 1(2): 87–107.

Isen, A. M. 1993. Positive affect and decision making. In *Handbook of emotions,* ed. M. Lewis and J. M. Haviland. New York, Guilford: 261–77.

Ishikawa, F., and D. F. Hay. 2006. "Triadic interaction among newly acquainted 2-year-olds." *Social Development* 15(1): 145–68.

Isler, K., E. Kirk Christopher, et al. 2008. "Endocranial volumes of primate species: Scaling analyses using a comprehensive and reliable data set." *Journal of Human Evolution* 55(6): 967–78.

Itani, J. 1958. "On the acquisition and propagation of a new food habit in the troop of Japanese monkeys at Takasakiyama." *Primates* 1: 131–48.

———. 1959. "Paternal care in the wild Japanese monkey, *Macaca fuscata fuscata.*" *Primates* 2: 61–93.

———. 1963. Paternal care in the wild Japanese monkey. In *Primate social behavior,* ed. C. H. Southwick. Princeton, NJ, Van Nostrand.

Ivanenko, Y. P., N. Dominici, et al. 2007. "Development of independent walking in toddlers." *Exercise and Sport Sciences Reviews* 35(2): 67–73.

Ivey, P. K. 2000. "Cooperative reproduction in Ituri Forest hunter-gatherers: Who cares for Efe infants?" *Current Anthropology* 41(5): 856–66.

Iwaniuk, A. N., J. E. Nelson, et al. 2001. "Do big-brained animals play more? Comparative analyses of play and relative brain size in mammals." *Journal of Comparative Psychology* 115(1): 29–41.

Izard, C. E. 1977. *Human emotions.* New York, Plenum.

Jablonska, E., and M. J. Lamb. 2005. *Evolution in four dimensions: Genetic, epigenetic, behavioral, and symbolic variation in the history of life.* Cambridge, MA, MIT Press.

Jackson, J. F. 1993. "Human behavioral genetics, Scarr's theory, and her views on interventions: A critical review and commentary on their implications for African American children." *Child Development* 64: 1318–32.

Jaffee, S. R., A. Caspi, et al. 2004. "The limits of child effects: Evidence for genetically mediated child effects on corporal punishment but not on physical maltreatment." *Developmental Psychology* 40(6): 1047–58.

———. 2005. "Nature x nurture: Genetic vulnerabilities interact with physical maltreatment to promote conduct problems." *Development and Psychopathology* 17(1): 67–84.

Jakobson, R., and M. Halle. 1971 [1956]. *Fundamentals of language,* 2nd ed. The Hague, Mouton.

James, W. 1902. *Varieties of religious experience.* London, Longman, Green.

Jamison, C. S., L. L. Cornell, et al. 2002. "Are all grandmothers equal? A review and a preliminary test of the 'grandmother hypothesis' in Tokugawa, Japan." *American Journal of Physical Anthropology* 119(1): 67–76.

Janis, C., and P. J. Jarman. 1984. The hoofed mammals. In *Encyclopedia of mammals,* ed. D. I. Macdonald. New York, Facts on File: 468–79.

Jankowiak, W. 1995a. Romantic passion in the People's Republic of China. In *Romantic passion: A universal experience?,* ed. W. Jankowiak. New York, Columbia Univ. Press: 166–84.

————, ed. 1995b. *Romantic passion: A universal experience?* New York, Columbia Univ. Press.

Jankowiak, W., and E. Fischer. 1992. "A cross-cultural perspective on romantic love." *Ethnology* 31(2): 149–55.

Janowsky, J. S. 1996. Is there a neural basis for cognitive transitions in school-age children? In *The five to seven year shift: The age of reason and responsibility,* ed. A. J. Sameroff and M. M. Haith. Chicago, Univ. of Chicago Press: 33–60.

Janson, C. H., and C. P. van Schaik. 1993. Ecological risk aversion in juvenile primates: Slow and steady wins the race. In *Juvenile primates: Life history, development, and behavior,* ed. M. E. Pereira and L. A. Fairbanks. New York, Oxford Univ. Press: 57–74.

Janszky, J., A. Fogarasi, et al. 2005. "Hyperorality in epileptic seizures: Periictal incomplete Klüver-Bucy syndrome." *Epilepsia* 46(8): 1235–40.

Jarvis, J. U. M. 1981. "Eusociality in a mammal: Cooperative breeding in naked mole-rat colonies." *Science* 212: 571–73.

Jenkins, W. M., M. M. Merzenich, et al. 1990. "Functional reorganization of primary somatosensory cortex in adult owl monkeys after behaviorally controlled tactile stimulation." *Journal of Neurophysiology* 63(1): 82–104.

Jensen, A. R. 1969. "How much can we boost IQ and scholastic achievement?" *Harvard Educational Review* 39: 1–123.

————. 2000. "The g factor: Psychometrics and biology." *Novartis Foundation Symposium* 233: 37–47; discussion 47–57.

Jerison, H. J. 1973. *Evolution of the brain and intelligence.* New York, Academic.

————. 1976. "Paleoneurology and the evolution of mind." *Scientific American* 234(1): 90–101.

————. 2001. Epilogue: The study of primate brain evolution: Where do we go from here? In *Evolutionary anatomy of the primate cerebral cortex,* ed. D. Falk and K. R. Gibson. New York, Cambridge Univ. Press: 305–37.

Jha, S., and R. Patel. 2004. "Klüver-Bucy syndrome—an experience with six cases." *Neurology India* 52(3): 369–71.

Ji, C. Y. 2001. "Age at spermarche and comparison of growth and performance of pre- and post-spermarcheal Chinese boys." *American Journal of Human Biology* 13(1): 35–43.

Ji, Q., Z.-X. Luo, et al. 2002. "The earliest known eutherian mammal [comment]." *Nature* 416(6883): 816–22.

Ji, Q., Z.-X. Luo, et al. 2009. "Evolutionary development of the middle ear in Mesozoic therian mammals [see comment]." *Science* 326(5950): 278–81.

Johanson, D., B. Edgar, et al. 2006. *From Lucy to language: Revised, updated, and expanded.* New York, Simon and Schuster.

Johanson, D. C., and T. D. White. 1979. "A systematic assessment of early African hominids." *Science* 203: 321–30.

Johnson, A. W., and T. Earle. 1987. *The evolution of human societies: From foraging group to agrarian state.* Stanford, Stanford Univ. Press.

Johnson, C. 1966. "Environmental modification of habitat selection in adult damselflies." *Ecology* 47: 674–76.

Johnson, L. R., C. Farb, et al. 2005. "Localization of glucocorticoid receptors at postsynaptic membranes in the lateral amygdala." *Neuroscience* 136(1): 289–99.

Johnson, M. H., ed. 1993. *Brain development and cognition.* Cambridge, Blackwell.

———. 2007. "Developing a social brain." *Acta Paediatrica* 96(1): 3–5.

Johnson, M. H., T. Grossmann, et al. 2009. "Mapping functional brain development: Building a social brain through interactive specialization." *Developmental Psychology* 45(1): 151–59.

Johnson, M. H., and A. Karmiloff-Smith. 1992. "Can neural selectionism be applied to cognitive development and its disorders?" *New Ideas in Psychology* 10(1): 35–46.

Johnson, S. C., C. S. Dweck, et al. 2007. "Evidence for infants' internal working models of attachment." *Psychological Science* 18(6): 501–2.

Johnson, W., M. McGue, et al. 2005. "Personality stability in late adulthood: A behavioral genetic analysis." *Journal of Personality* 73(2): 523–52.

Jolly, A. 1966. *Lemur behavior: A Madagascar field study.* Chicago, Univ. of Chicago Press.

———. 1999. *Lucy's legacy: Sex and intelligence in human evolution.* Cambridge, MA, Harvard Univ. Press.

Jones, G. H., C. A. Marsden, et al. 1990. "Increased sensitivity to amphetamine and reward-related stimuli following social isolation in rats: Possible disruption of dopamine-dependent mechanisms of the nucleus accumbens." *Psychopharmacology* 102: 364–72.

Jones, J. S. 1981. "An uncensored page of fossil history." *Nature* 293: 427–28.

Jones, S. S., and H.-W. Hong. 2001. "Onset of voluntary communication: Smiling looks to mother." *Infancy* 2(3): 353–70.

Jordan, A. 2004. "The role of media in children's development: An ecological perspective." *Journal of Developmental and Behavioral Pediatrics* 25(3): 196–206.

Jossin, Y. 2004. "Neuronal migration and the role of reelin during early development of the cerebral cortex." *Molecular Neurobiology* 30(3): 225–51.

Juarez, J., I. del Rio-Portilla, et al. 1998. "Effects of prenatal testosterone on sex and age differences in behavior elicited by stimulus pups in the rat." *Developmental Psychobiology* 32(2): 121–29.

Juchau, M. R. 1993. "Chemical teratogenesis." *Progress in Drug Research* 41: 9–50.

Junod, H. A. 1927. *The life of a South African tribe,* 2nd ed., 2 vols. London, Macmillan.

Kaare, B., and J. Woodburn. 1999. The Hadza of Tanzania. In *The Cambridge encyclopedia of hunters and gatherers,* ed. R. B. Lee and R. Daly. Cambridge, Cambridge Univ. Press: 200.

Kaas, J. H. 1995. "The evolution of isocortex." *Brain, Behavior and Evolution* 46: 187–96.

———. 2005. "From mice to men: The evolution of the large, complex human brain." *Journal of Biosciences* 30(2): 155–65.

———. 2006. "Evolution of the neocortex." *Current Biology* 16(21): R910–14.

Kagan, J. 1970. "Attention and psychological change in the young child." *Science* 170: 826–32.

———. 1971. *Change and continuity in infancy.* New York, Wiley.

———. 1974. Discrepancy, temperament, and infant distress. In *The origins of fear,* ed. M. Lewis and L. A. Rosenblum. New York, Wiley: 229–48.

———. 1976. "Emergent themes in human development." *American Scientist* 64: 186–196.

———. 1977. The uses of cross-cultural research in early development. In *Culture and infancy,* ed. T. H. Leiderman, S. R. Tulkin, and A. Rosenfeld. New York, Academic.

———. 1981a. *The second year: The emergence of self-awareness.* Cambridge, MA, Harvard Univ. Press.

———. 1981b. Universals in human development. In *Handbook of cross-cultural human development,* ed. R. H. Munroe, R. L. Munroe, and B. B. Whiting. New York, Garland STPM: 53–62.

———. 1984. *The nature of the child.* New York, Basic.

———. 1994. *Galen's prophecy: Temperament in human nature.* New York, Basic.

———. 1996. Temperamental contributions to the development of social behavior. In *The lifespan development of individuals: Behavioral, neurobiological, and psychosocial perspectives,* ed. D. Magnusson. New York, Cambridge Univ. Press: 376–93.

———. 2003. "Biology, context, and developmental inquiry." *Annual Review of Psychology* 54: 1–23.

———. 2008. "In defense of qualitative changes in development." *Child Development* 79(6): 1606–24.

Kagan, J., R. B. Kearsley, et al., eds. 1978. *Infancy: Its place in human development.* Cambridge, MA, Harvard Univ. Press.

Kagan, J., and R. E. Klein. 1973. "Cross-cultural perspectives on early development." *American Psychologist* 28: 947–61.

Kagan, J., R. E. Klein, et al. 1979. *A cross-cultural study of child development.* Chicago, Univ. of Chicago Press.

Kagan, J., J. S. Reznick, et al. 1988. "Biological bases of childhood shyness." *Science* 240: 167–71.

———. 1999. "Infant temperament and anxious symptoms in school age children." *Development and Psychopathology* 11(2): 209–24.

———. 2007. "The preservation of two infant temperaments into adolescence." *Monographs of the Society for Research in Child Development* 72(2): 1–75.

Kagan, S. L., and P. R. Neville. 1996. Combining endogenous and exogenous factors in the shift years: The transition to school. In *The five to seven year shift: The age of reason and responsibility,* ed. A. J. Sameroff and M. M. Haith. Chicago, Univ. of Chicago Press: 387–405.

Kalin, N. H. 1993. "The neurobiology of fear." *Scientific American* 268(5): 94–101.

———. 2003. "Nonhuman primate studies of fear, anxiety, and temperament and the role of benzodiazepine receptors and GABA systems." *Journal of Clinical Psychiatry* 64(Suppl. 3): 41–44.

Kalin, N. H., C. Larson, et al. 1998. "Asymmetric frontal brain activity, cortisol, and behavior associated with fearful temperament in rhesus monkeys." *Behavioral Neuroscience* 112(2): 286–92.

Kalin, N. H., and S. E. Shelton. 1989. "Defensive behaviors in infant rhesus monkeys: Environmental cues and neurochemical regulation." *Science* 243: 1718–21.

————. 2003. "Nonhuman primate models to study anxiety, emotion regulation, and psychopathology." *Annals of the New York Academy of Sciences* 1008: 189–200.

Kalin, N. H., S. E. Shelton, et al. 1998. "Individual differences in freezing and cortisol in infant and mother rhesus monkeys." *Behavioral Neuroscience* 112(1): 251–54.

Kamholz, J. A. 1996. "Regulation of myelin development." *Multiple Sclerosis* 2(5): 236–40.

Kaminen, N., K. Hannula-Jouppi, et al. 2003. "A genome scan for developmental dyslexia confirms linkage to chromosome 2p11 and suggests a new locus on 7q32." *Journal of Medical Genetics* 40(5): 340–45.

Kamyshev, N. G., K. G. Iliadi, et al. 1999. "Drosophila conditioned courtship: Two ways of testing memory." *Learning and Memory* 6(1): 1–20.

Kandel, E. R. 2001. "The molecular biology of memory storage: A dialogue between genes and synapses." *Science* 294(5544): 1030–38.

Kano, T. 1992. *The last ape: Pygmy chimpanzee behavior and ecology.* Stanford, Stanford Univ. Press.

————. 1996. Male rank order and copulation rate in a unit-group of bonobos at Wamba, Zaire. In *Great ape societies,* ed. W. C. McGrew, L. F. Marchant, and T. Nishida. Cambridge, Cambridge Univ. Press: 135–45.

Kaplan, H., and H. Dove. 1987. "Infant development among the Ache of eastern Paraguay." *Developmental Psychology* 23(2): 190–98.

Kaplan, H., K. Hill, et al. 2000. "A theory of human life history evolution: Diet, intelligence, and longevity." *Evolutionary Anthropology* 9(4): 156–85.

Kaplan, J. 1972. "Differences in the mother-infant relations of squirrel monkeys housed in social and restricted environments." *Developmental Psychology* 5: 43–52.

Kaplan, J., and R. J. Schusterman. 1972. "Social preferences of mother and infant squirrel monkeys following different rearing experiences." *Developmental Psychobiology* 5(1): 53–59.

Kaplowitz, P. 2006. "Pubertal development in girls: Secular trends." *Current Opinion in Obstetrics and Gynecology* 18(5): 487–91.

Kappeler, P. M., M. E. Pereira, et al. 2003. Primate life histories and socioecology. In *Primate life histories and socioecology,* ed. P. M. Kappeler and M. E. Pereira. Chicago, Univ. of Chicago Press: 1–20.

Kardiner, A., and R. Linton, eds. 1939. *The individual and his society.* New York, Columbia Univ. Press.

————, eds. 1945. *The psychological frontiers of society.* New York, Columbia Univ. Press.

Karlberg, J. 2002. "Secular trends in pubertal development." *Hormone Research* 57(Suppl. 2): 19–30.

Karmiloff-Smith, A. 1995. *Beyond modularity: A developmental perspective on cognitive science.* Cambridge, MIT Press.

————. 2009. "Nativism versus neuroconstructivism: Rethinking the study of developmental disorders." *Developmental Psychology* 45(1): 56–63.

Karmiloff-Smith, A., E. Klima, et al. 1995. "Is there a social module? Language, face processing and theory of mind in individuals with Williams syndrome." *Journal of Cognitive Neuroscience* 7: 196–208.

Karp, H. 2002. *The happiest baby on the block.* New York, Bantam.

Karraker, K. H., D. A. Vogel, et al. 1995. "Parents' gender-stereotyped perceptions of newborns: The eye of the beholder revisited." *Sex Roles* 33(9/10): 687–701.

Karsten, K. B., L. N. Andriamandimbiarisoa, et al. 2008. "A unique life history among tetrapods: An annual chameleon living mostly as an egg." *Proceedings of the National Academy of Sciences* 105(26): 8980–84.

Kaskan, P. M., and B. L. Finlay. 2001. Encephalization and its developmental structure: How many ways can a brain get big? In *Evolutionary anatomy of the primate cerebral cortex,* ed. D. Falk and K. R. Gibson. Cambridge, Cambridge Univ. Press: 14–29.

Katsikitis, M., and I. Pilowsky. 1988. "A study of facial expression in Parkinson's disease using a novel microcomputer-based method." *Journal of Neurology, Neurosurgery and Psychiatry* 51(3): 362–66.

Katz, M. M. 1981. Gaining sense at age two in the outer Fiji Islands: A crosscultural study of cognitive development. EdD diss., Harvard Univ. Ann Arbor, MI, University Microfilms International.

Katz, M. M., and M. J. Konner. 1981. The role of the father: An anthropological perspective. In *The role of the father in child development,* ed. M. E. Lamb. New York, Wiley: 155–86.

Katz, P. A., and K. R. Ksansnak. 1994. "Developmental aspects of gender role flexibility and traditionality in middle childhood and adolescence." *Developmental Psychology* 30(2): 272–82.

Katz, R. 1982. *Boiling energy: Community healing among the Kalahari !Kung.* Cambridge, MA, Harvard Univ. Press.

Kauffman, S. A. 1987. "Developmental logic and its evolution." *BioEssays* 6(2): 82–87.
———. 1993. *The origins of order: Self-organization and selection in evolution.* New York, Oxford Univ. Press.

Kaufman, J., and E. F. Zigler. 1988. "Do abused children become abusive parents?" *Annual Progress in Child Psychiatry and Child Development:* 591–600.

Kay, R. F., and E. C. Kirk. 2000. "Osteological evidence for the evolution of activity pattern and visual acuity in primates." *American Journal of Physical Anthropology* 113(2): 235–62.

Keeley, L. H. 1996. *War before civilization: The myth of the peaceful savage.* New York, Oxford Univ. Press.

Keene, M. F. L., and E. E. Hewer. 1931. "Some observations on myelination in the human central nervous system." *Journal of Anatomy* 66(pt. 1): 1–13.

Keller, M., M. Meurisse, et al. 2004. "Mapping the neural substrates involved in maternal responsiveness and lamb olfactory memory in parturient ewes using Fos imaging." *Behavioral Neuroscience* 118(6): 1274–84.

Kelley, J., and T. M. Smith. 2003. "Age at first molar emergence in early Miocene *Afropithecus turkanensis* and life-history evolution in the Hominoidea." *Journal of Human Evolution* 44(3): 307–29.

Kempe, C. H., F. N. Silverman, et al. 1984 [1962]. "Landmark article July 7, 1962: The battered-child syndrome." *Journal of the American Medical Association* 251(24): 3288–94.

Kendler, K. S., J. W. Kuhn, et al. 2005. "The interaction of stressful life events and a serotonin transporter polymorphism in the prediction of episodes of major depression: A replication." *Archives of General Psychiatry* 62(5): 529–35.

Kendrick, K. M., and R. F. Drewitt. 1979. "Testosterone reduces refractory period of stria terminalis neurons in the rat brain." *Science* 204: 877–79.

Kennedy, G. E. 2005. "From the ape's dilemma to the weanling's dilemma: Early weaning and its evolutionary context." *Journal of Human Evolution* 48(2): 123–45.

Kennell, J., M. Klaus, et al. 1991. "Continuous emotional support during labor in a US hospital: A randomized controlled trial." *Journal of the American Medical Association* 265(17): 2197–2201.

Kenyon, C. 1994. "If birds can fly, why can't we? Homeotic genes and evolution." *Cell* 78: 175–80.

Ketterson, E. D., and V. Nolan, Jr. 1994. "Male parental behavior in birds." *Annual Review of Ecological Systems* 25: 601–28.

———. 1992. "Hormones and life histories: An integrative approach." *American Naturalist* 140: S33–62.

Keverne, E. B., and J. P. Curley. 2008. "Epigenetics, brain evolution, and behaviour." *Frontiers in Neuroendocrinology* 29(3): 398–412.

Kevles, D. J. 1985. *In the name of eugenics: Genetics and the uses of human heredity*. New York, Knopf.

Kidd, K. K., A. J. Pakstis, et al. 2004. "Understanding human DNA sequence variation." *Journal of Heredity* 95(5): 406–20.

Kikusui, T., and Y. Mori. 2009. "Behavioural and neurochemical consequences of early weaning in rodents." *Journal of Neuroendocrinology* 21(4): 427–31.

Kilbride, J. E., and P. L. Kilbride. 1975. "Sitting and smiling behavior of Baganda infants: The influence of culturally constituted experience." *Journal of Cross-Cultural Psychology* 6: 88–107.

Kilbride, P. L. 1980. "Sensorimotor behavior of Baganda and Samia infants: A controlled comparison." *Journal of Cross-Cultural Psychology* 11: 131–52.

Killeen, P. R. 1994. "Frustration: Theory and practice." *Psychonomic Bulletin and Review* 1(3): 323–26.

Killgore, W. D. S., and D. A. Yurgelun-Todd. 2007. "Neural correlates of emotional intelligence in adolescent children." *Cognitive, Affective and Behavioral Neuroscience* 7(2): 140–51.

Kim, J., J. Q. Kerr, et al. 2000. "Molecular heterochrony in the early development of Drosophila." *Proceedings of the National Academy of Sciences* 97(1): 212–16.

Kim-Cohen, J., A. Caspi, et al. 2006. "MAOA, maltreatment, and gene-environment interaction predicting children's mental health: New evidence and a meta-analysis." *Molecular Psychiatry* 11(10): 903–13.

King, J. A., J. Tenney, et al. 2003. "Neural substrates underlying impulsivity." *Annals of the New York Academy of Sciences* 1008: 160–69.

Kingsley, J. R., G. H. Collins, et al. 1970. "Effect of sciatic neurectomy on myelinogenesis in the rat spinal cord." *Experimental Neurology* 26: 498–508.

Kinney, H. C., B. A. Brody, et al. 1988. "Sequence of central nervous system myelination in human infancy. II. Patterns of myelination in autopsied infants." *Journal of Neuropathology and Experimental Neurology* 47(3): 217–34.

Kinney, H. C., J. Karthigasan, et al. 1994. "Myelination in the developing human brain: Biochemical correlates." *Neurochemical Research* 19(8): 983–96.

Kisilevsky, B. S., S. M. Hains, et al. 1998. "The still-face effect in Chinese and Canadian 3- to 6-month-old infants." *Developmental Psychology* 34(4): 629–39.

Kitayama, S., and D. Cohen. 2007. *Handbook of cultural psychology*. New York, Guilford.

Kitayama, S., H. R. Markus, et al. 2000. "Culture, emotion, and well-being: Good feelings in Japan and the United States." *Cognition and Emotion* 14(1): 93–124.

Kitihara, M. 1984. "Female physiology and female puberty." *Ethos* 12(2): 132–50.

Klaus, M. H., and J. H. Kennell, eds. 1976. *Maternal-infant bonding*. St. Louis, Mosby.

Kleiman, D. G., and J. R. Malcolm. 1981. The evolution of male parental investment in mammals. In *Parental care in mammals,* ed. D. J. Gubernick. New York, Plenum: 347–87.

Klein, R. 2009. *The human career,* 3rd ed. Chicago, Univ. of Chicago Press.

Klein, R. A. 1999. Treating fear of flying with virtual reality exposure therapy. In *Innovations in clinical practice: A source book,* vol. 17, ed. L. VandeCreek and T. L. Jackson. Sarasota, FL, Professional Resource Press/Professional Resource Exchange: 449–65.

Klein, R. E., R. E. Lasky, et al. 1977. Relationship of infant/caretaker interaction, social class, and nutritional status to developmental test performance among Guatemalan infants. In *Culture and infancy,* ed. P. H. Leiderman, S. R. Tulkin, and A. Rosenfeld. New York, Academic.

Klein, R. G., and B. Edgar. 2002. *The dawn of human culture.* New York, Wiley.

Klima, E., and U. Bellugi. 1979. *The signs of language.* Cambridge, MA, Harvard Univ. Press.

Klinger, L. J., J. A. Hamilton, et al. 2001. "Children's perceptions of aggressive and gender-specific content in toy commercials." *Social Behavior and Personality* 29(1): 11–20.

Klopfer, D. H., and M. Klopfer. 1968. "Maternal 'imprinting' in goats: Fostering of alien young." *Zeitschrift für Tierpsychologie* 25: 862–66.

Klüver, H., and P. C. Bucy. 1939. "Preliminary analysis of the temporal lobes in monkeys." *Archives of Neurological Psychiatry* 42: 1979–2000.

Kmita, M., and D. Duboule. 2003. "Organizing axes in time and space; 25 years of colinear tinkering." *Science* 301(5631): 331–33.

Knafo, A., and R. Plomin. 2006. "Parental discipline and affection and children's prosocial behavior: Genetic and environmental links." *Journal of Personality and Social Psychology* 90(1): 147–64.

Knauft, B. M. 1993. *South Coast New Guinea cultures: History, comparison, dialectic.* New York, Cambridge Univ. Press.

Knickmeyer, R. C., and S. Baron-Cohen. 2006. "Fetal testosterone and sex differences." *Early Human Development* 82(12): 755–60.

Knoll, J. G., C. A. Wolfe, et al. 2007. "Estrogen modulates neuronal movements within the developing preoptic area-anterior hypothalamus." *European Journal of Neuroscience* 26(5): 1091–99.

Knott, C. D., and S. M. Kahlenberg. 2007. Orangutans in perspective: Forced copulations and female mating resistance. In *Primates in perspective,* ed. C. J. Campbell et al. New York, Oxford Univ. Press: 290–305.

Koch, S. V., A. M. T. Kejs, et al. 2004. "Educational attainment among survivors of childhood cancer: A population-based cohort study in Denmark." *British Journal of Cancer* 91(5): 923–28.

Koenig, W. D., F. A. Pitelka, et al. 1992. "The evolution of delayed dispersal in cooperative breeders." *Quarterly Review of Biology* 67(2): 111–50.

Koff, E., J. Rierdan, et al. 1978. "Changes in representation of body image as a function of menarcheal status." *Developmental Psychology* 14(6): 635–42.

———. 1981. "The personal and interpersonal significance of menarche." *Journal of the American Academy of Child Psychiatry* 20: 148–58.

Kohlberg, L. 1963. "The development of children's orientations toward a moral order: Sequence in the development of moral thought." *Vita Humana* 6: 11–33.

Kohrt, B. A., M. J. D. Jordans, et al. 2008. "Comparison of mental health between for-

mer child soldiers and children never conscripted by armed groups in Nepal."
Journal of the American Medical Association 300(6): 691–702.

Kokko, H., R. A. Johnstone, et al. 2001. "The evolution of cooperative breeding through group augmentation." *Proceedings of the Royal Society of London B: Biological Sciences* 268(1463): 187–96.

Komlos, J., and A. Breitfelder. 2008. "Differences in the physical growth of US-born black and white children and adolescents ages 2–19, born 1942–2002." *Annals of Human Biology* 35(1): 11–21.

Konner, M. J. 1972. Aspects of the developmental ethology of a foraging people. In *Ethological studies of child behavior,* ed. N. G. B. Jones. Cambridge, Cambridge Univ. Press: 285–304.

———. 1973. "Newborn walking: Additional data." *Science* 179: 307.

———. 1975. Relations among infants and juveniles in comparative perspective. In *Friendship and peer relations,* ed. M. Lewis and L. A. Rosenblum. New York, Wiley: 99–129.

———. 1976. Maternal care, infant behavior and development among the !Kung. In *Kalahari hunter-gatherers,* ed. R. B. Lee and I. DeVore. Cambridge, MA, Harvard Univ. Press: 218–45.

———. 1977a. "Adolescent pregnancy." *New York Times.*

———. 1977b. Infancy among the Kalahari Desert San. In *Culture and infancy,* ed. P. H. Leiderman, S. R. Tulkin, and A. Rosenfeld. New York, Academic: 287–328.

———. 1981. Evolution of human behavior development. In *Handbook of cross-cultural human development,* ed. R. H. Munroe, R. L. Munroe, and B. B. Whiting. New York, Garland STPM: 3–51.

———. 1982. Biological aspects of mother-infant bond. In *Development of attachment and affiliative processes,* ed. R. Emde and R. Harmon. New York, Plenum.

———. 1985. "Transcendental medication." *The Sciences* 25(3): 2–4.

———. 1991. Universals of behavioral development in relation to brain myelination. In *Brain maturation and cognitive development: Comparative and cross-cultural perspectives,* ed. K. R. Gibson and A. C. Petersen. New York, Aldine de Gruyter: 181–223.

———. 1995. Anthropology and psychiatry. In *Comprehensive textbook of psychiatry,* 6th ed., ed. H. Kaplan and B. Sadock. Baltimore, Williams and Wilkins: 283–99.

———. 2002. *The tangled wing: Biological constraints on the human spirit,* rev. ed. New York, Holt/Times.

———. 2005. Hunter-gatherer infancy and childhood: The !Kung and others. In *Hunter-gatherer childhoods: Evolutionary, developmental and cultural perspectives,* ed. B. S. Hewlett and M. E. Lamb. New Brunswick, NJ, AldineTransaction: 19–64.

———. 2006. Human nature, ethnic violence, and war. In *The psychology of resolving global conflicts: From war to peace,* vol. 1, ed. M. Fitzduff and C. E. Stout. Westport, CT, Praeger Security International.

———. 2007. Trauma, adaptation, and resilience: A cross-cultural and evolutionary perspective. In *Understanding trauma: Cultural, psychological, and biological perspectives on terror and its aftermath,* ed. L. J. Kirmayer, R. Lemelson, and M. Barad. New York, Cambridge Univ. Press.

Konner, M. J., and M. J. Shostak. 1986. Adolescent pregnancy and childbearing: An anthropological perspective. In *School-age pregnancy and childbearing: Biosocial dimensions,* ed. J. B. Lancaster and B. A. Hamburg. New York, Aldine.

———. 1987. "Timing and management of birth among the !Kung: Biocultural inter-action in reproductive adaptation." *Cultural Anthropology* 2: 11–28.

Konner, M. J., and C. M. Super. 1987. Sudden infant death syndrome: An anthropological hypothesis. In *The role of culture in developmental disorder,* ed. C. M. Super. New York, Academic: 95–108.

Konner, M. J., and C. Worthman. 1980. "Nursing frequency, gonadal function, and birth spacing among !Kung hunter-gatherers." *Science* 207: 788–91.

Korja, R., J. Maunu, et al. 2008. "Mother-infant interaction is influenced by the amount of holding in preterm infants." *Early Human Development* 84(4): 257–67.

Kornack, D., and P. Rakic. 1998. "Changes in cell-cycle kinetics during the development and evolution of primate neocortex." *Proceedings of the National Academy of Sciences* 95: 1242–46.

Korner, A. 1969. "Neonatal startles, smiles, erections, and reflex sucks as related to state, sex and individuality." *Child Development* 40: 1039–53.

———. 1972. "State as variable, as obstacle and as mediator of stimulation in infant research." *Merrill-Palmer Quarterly* 18: 77–94.

———. 1973. "Sex differences in newborns with special reference to differences in the organization of oral behavior." *Journal of Child Psychology and Psychiatry* 14: 19–29.

Korner, A. F. 1979. Maternal rhythms and waterbeds: A form of intervention with premature infants. In *Origins of the infant's social responsiveness,* ed. E. B. Thoman. Hillsdale, NJ, Erlbaum.

Korner, A. F., and R. Grobstein. 1966. "Visual alertness as related to soothing in neonates: Implications for maternal stimulation and early deprivation." *Child Development* 37(4): 867–76.

Korner, A. F., H. C. Kraemer, et al. 1975. "Effects of waterbed flotation on premature infants: A pilot study." *Pediatrics* 56: 361.

Korner, A. F., and E. B. Thoman. 1970. "Visual alertness in neonates as evoked by maternal care." *Journal of Experimental Child Psychology* 10: 67–78.

Kortlandt, A. 1967. Experimentation with chimpanzees in the wild. In *Progress in primatology,* ed. D. Starck, R. Schneider, and H. J. Kuhn. Stuttgart, Fischeri.

Kovács, Á. M., and J. Mehler. 2009. "Flexible learning of multiple speech structures in bilingual infants." *Science* 325: 611–12.

Kovas, Y., M. E. Hayiou-Thomas, et al. 2005. "Genetic influences in different aspects of language development: The etiology of language skills in 4.5-year-old twins." *Child Development* 76(3): 632–51.

Kowalski, H. S., S. R. Wyver, et al. 2004. "Toddlers' emerging symbolic play: A first-born advantage?" *Early Child Development and Care* 174(4): 389–400.

Kraemer, G. W. 1992. "A psychobiological theory of attachment." *Behavioral and Brain Sciences* 15(3): 493–541.

Kraemer, G. W., and A. S. Clarke. 1990. "The behavioral neurobiology of self-injurious behavior in rhesus monkeys." *Progress in Neuro-Psychopharmacology and Biological Psychiatry* 14(Suppl.): S141–68.

Kraemer, G. W., M. H. Ebert, et al. 1984. "Hypersensitivity to d-amphetamine several years after early social deprivation in rhesus monkeys." *Psychopharmacology* 82(3): 266–71.

———. 1989. "A longitudinal study of the effect of different social rearing conditions on cerebrospinal fluid norepinephrine and biogenic amine metabolites in rhesus monkeys." *Neuropsychopharmacology* 2(3): 175–89.

Kraemer, G. W., and W. T. McKinney. 1985. "Social separation increases alcohol consumption in rhesus monkeys." *Psychopharmacology* 86(1–2): 182–89.

Kraemer, G. W., D. E. Schmidt, et al. 1997. The behavioral neurobiology of self-injurious behavior in rhesus monkeys: Current concepts and relations to impulsive behavior in humans. In *The neurobiology of suicide: From the bench to the clinic,* ed. D. M. Stoff and J. J. Mann. *Annals of the New York Academy of Sciences* 836: 12–38.

Kramer, K. L. 2005. "Children's help and the pace of reproduction: Cooperative breeding in humans." *Evolutionary Anthropology: Issues, News, and Reviews* 14(6): 224–37.

Kramer, M. S. 2003. "The epidemiology of adverse pregnancy outcomes: An overview." *Journal of Nutrition* 133(5 Suppl. 2): 1592S–96.

Krarup, C. 2002. "Nerve conduction studies in selected peripheral nerve disorders." *Current Opinion in Neurology* 15(5): 579–93.

Krasnegor, N. A., and R. S. Bridges, eds. 1990. *Mammalian parenting.* New York, Oxford Univ. Press.

Krause, J. 2008. "SPECT and PET of the dopamine transporter in attention-deficit/hyperactivity disorder." *Expert Review of Neurotherapeutics* 8(4): 611–25.

Krause, J., C. Lalueza-Fox, et al. 2007. "The derived FOXP2 variant of modern humans was shared with Neandertals [see comment]." *Current Biology* 17(21): 1908–12.

Krebs, J. R., and N. B. Davies, eds. 1991. *Behavioral ecology: An evolutionary approach.* Oxford, Blackwell.

Krech, D., M. Rosensweig, et al. 1966. "Environmental impoverishment, social isolation and changes in brain chemistry and anatomy." *Physiology and Behavior* 1: 99–109.

Kreppner, J. M., T. G. O'Connor, et al. 2001. "Can inattention/overactivity be an institutional deprivation syndrome?" *Journal of Abnormal Child Psychology* 29(6): 513–28.

Kreuz, L., and R. Rose. 1972. "Assessment of aggressive behavior and plasma testosterone in a young criminal population." *Psychosomatic Medicine* 34: 321–32.

Kreuz, L. E., R. M. Rose, et al. 1972. "Suppression of plasma testosterone levels and psychological stress." *Archives of General Psychiatry* 26: 479–82.

Kringelbach, M. L. 2005. "The human orbitofrontal cortex: Linking reward to hedonic experience." *Nature Reviews Neuroscience* 6(9): 691–702.

Kringelbach, M. L., A. Lehtonen, et al. 2008. "A specific and rapid neural signature for parental instinct." *PLoS ONE [Electronic Resource]* 3(2): e1664.

Kroeber, A. L. 1901. "Decorative symbolism of the Arapaho." *American Anthropologist* 3: 308–36.

———. 1909. "Classificatory systems of relationship." *Journal of the Royal Anthropological Institute* 39: 77–84.

———. 1928. "Sub-human culture beginnings." *Quarterly Review of Biology* 3: 325–342.

———. 1948. *Anthropology.* New York, Harcourt, Brace.

Kroodsma, D. 2004. The diversity and plasticity of birdsong. In *Nature's music: The science of birdsong,* ed. P. Marler and H. Slabbekoorn. San Diego, Elsevier Scientific: 108–31.

Krubitzer, L. 1998. "What can monotremes tell us about brain evolution?" *Philosophical Transactions of the Royal Society of London B: Biological Sciences* 3531372: 1127–46.

Krueger, C., D. Holditch-Davis, et al. 2004. "Recurring auditory experience in the 28- to 34-week-old fetus." *Infant Behavior and Development* 27(4): 537–43.

Kruger, A., and M. Konner. Forthcoming. "Who responds to infant crying? An analysis of !Kung multiple caregiving." *Human Nature.*

Kruger, A. C. 1992. "The effect of peer and adult-child transactive discussions on moral reasoning." *Merrill-Palmer Quarterly* 38(2): 191–211.

———. 1993. "Peer collaboration: Conflict, cooperation, or both." *Social Development* 2(3): 165–82.

———. 2009. Personal communication, July 12, 2009.

———. 2010. Communion and culture. In *Mimesis and science,* ed. S. Garrels. East Lansing, Michigan State Univ. Press.

Kruger, A. C., and M. Tomasello. 1986. "Transactive discussions with peers and adults." *Developmental Psychology* 22: 681–85.

———. 1996. Cultural learning and learning culture. In *Handbook of education and human development: New models of learning, teaching, and schooling,* ed. D. Olson and N. Torrance. Oxford, Blackwell: 369–87.

Kubovy, M. 1999. On the pleasures of the mind. In *Well-being: The foundations of hedonic psychology,* ed. D. Kahneman, D. Ed, and N. Schwarz. New York, Russell Sage: 134–54.

Kuhl, P. K. 2007. "Is speech learning 'gated' by the social brain?" *Developmental Science* 10(1): 110–20.

Kuhl, P. K., and A. N. Meltzoff. 1996. "Infant vocalizations in response to speech: Vocal imitation and developmental change." *Journal of the Acoustical Society of America* 100(4 Pt. 1): 2425–38.

Kuida, K., T. F. Hayda, et al. 1998. "Reduced apoptosis and sytochrome-C-mediated caspase activation in mice lacking caspase-9." *Cell* 94: 325–37.

Kumar, A., G. S. Solanki, et al. 2005. "Observations on parturition and allomothering in wild capped langur *(Trachypithecus pileatus)*." *Primates* 46(3): 215–17.

Kumar, S., and S. B. Hedges. 1998. "A molecular timescale for vertebrate evolution." *Nature* 392(6679): 917–20.

Kummer, H. 1968. *Social organization of Hamadryas baboons.* Chicago, Univ. of Chicago Press.

Kunzle, H., and S. Radtke-Schuller. 2001. "Hippocampal fields in the hedgehog tenrec. Their architecture and major intrinsic connections." *Neuroscience Research* 41(3): 267–91.

Kupferman, I. 1991. Hypothalamus and limbic system: Peptidergic neurons, homeostasis, and emotional behavior. In *Principles of neural science,* ed. E. R. Kandel, J. H. Schwartz, and T. M. Jessell. New York, Elsevier: 296–308.

Kurjak, A., M. Stanojevic, et al. 2004. "Behavioral pattern continuity from prenatal to postnatal life—a study by four-dimensional (4D) ultrasonography." *Journal of Perinatal Medicine* 32(4): 346–53.

Kurland, J. 1977. *Kin selection in the Japanese monkey.* Basel, Karger.

Kurz, E. M., D. R. Sengelaub, et al. 1986. "Androgens regulate the dendritic length of mammalian motoneurons in adulthood." *Science* 232: 395–97.

Kuwert, P., C. Spitzer, et al. 2008. "Trauma and post-traumatic stress symptoms in former German child soldiers of World War II." *International Psychogeriatrics* 20(5): 1014–18.

Kyriacou, C. P., and J. C. Hall. 1986. "Interspecific genetic control of courtship song production and reception in *Drosophila*." *Science* 232: 494–97.

Lacbawan, F. L., and M. Muenke. 2002. "Central nervous system embryogenesis and its failures." *Pediatric and Developmental Pathology* 5(5): 425–47.

Lack, D. 1943. *The life of the robin.* London, Witherby.

Lacruz, R. S., F. Ramirez Rozzi, et al. 2005. "Dental enamel hypoplasia, age at death, and weaning in the Taung child." *South African Journal of Science* 101: 567–69.

LaFrance, M., M. A. Hecht, et al. 2003. "The contingent smile: A meta-analysis of sex differences in smiling." *Psychological Bulletin* 129(2): 305–34.

Lahdenpera, M., V. Lummaa, et al. 2004. "Fitness benefits of prolonged post-reproductive lifespan in women [see comment]." *Nature* 428(6979): 178–81.

Lahdenpera, M., A. F. Russell, et al. 2007. "Selection for long lifespan in men: Benefits of grandfathering?" *Proceedings of the Royal Society of London B: Biological Sciences* 274(1624): 2437–44.

Lahey, B. B., M. Schwab-Stone, et al. 2000. "Age and gender differences in oppositional behavior and conduct problems: A cross-sectional household study of middle childhood and adolescence." *Journal of Abnormal Psychology* 109(3): 488–503.

Lai, C. S., S. E. Fisher, et al. 2001. "A forkhead-domain gene is mutated in a severe speech and language disorder [see comment]." *Nature* 413(6855): 519–23.

Lai, C. S. L., D. Gerrelli, et al. 2003. "FOXP2 expression during brain development coincides with adult sites of pathology in a severe speech and language disorder." *Brain* 126(Pt. 11): 2455–62.

Lailvaux, S. P., and D. J. Irschick. 2006. "A functional perspective on sexual selection: Insights and future prospects." *Animal Behaviour* 72(Pt. 2): 263–73.

Lakoff, G. 1987. *Women, fire, and dangerous things: What categories reveal about the mind.* Chicago, Univ. of Chicago Press.

Lakoff, G., and M. Johnson. 1999. *Philosophy in the flesh: The embodied mind and its challenge to Western thought.* New York, Basic.

Laland, K. N. 2004. "Social learning strategies." *Learning and Behavior: A Psychonomic Society Publication* 32(1): 4–14.

Lamb, M. E. 1975. "Fathers: Forgotten contributors to child development." *Human Development* 18(4): 245–66.

———. 1981. *The role of the father in child development,* 2nd ed. New York, Wiley.

Lamb, M. E., and L. Ahnert, eds. 2006. *Nonparental child care: Context, concepts, correlates, and consequences.* Hoboken, NJ, Wiley.

Lamb, M. E., and C. Lewis. 2004. The development and significance of father-child relationships in two-parent families. In *The role of the father in child development,* 4th ed., ed. M. E. Lamb. Hoboken, NJ, Wiley: 272–306.

Lamprecht, J. 1979. "Field observations on the behaviour and social system of the bat-eared fox (Otocyon megalotis Desmarest)." *Zeitschrift für Tierpsychologie* 49: 262–84.

Lamprecht, R., C. R. Farb, et al. 2006. "Fear conditioning drives profilin into amygdala dendritic spines." *Nature Neuroscience* 9(4): 481–83.

Lancaster, J. B. 1968. "On the evolution of tool-using behavior." *American Anthropologist* 70: 56–66.

———. 1971. "Play-mothering: The relations between juvenile females and young infants among free-ranging vervet monkeys." *Folia Primatology* 15: 161–82.

———. 1991. "A feminist and evolutionary biologist looks at women." *Yearbook of Physical Anthropology* 34: 1–11.

Lancaster, J., and R. Lee. 1965. The annual reproductive cycle in monkeys and apes. In *Primate behavior,* ed. I. DeVore. New York, Holt, Rinehart, and Winston.

Lancaster, J. B., and C. S. Lancaster. 1983. Parental investment: The hominid adaptation. In *How humans adapt,* ed. D. Ortner. Washington, DC, Smithsonian Institution Press: 35–56.

———. 1987. The watershed: Change in parental-investment and family-formation strategies in the course of human evolution. In *Parenting across the life span: Biosocial dimensions,* ed. J. B. Lancaster et al. New York, Aldine de Gruyter: 187–205.

Lancy, D. F. 2007. "Accounting for variability in mother-child play." *American Anthropologist* 109(2): 273–84.

Landauer, T. K., and J. W. M. Whiting. 1964. "Infantile stimulation and adult stature of human males." *American Anthropologist* 66: 1007–28.

———. 1981. Correlates and consequences of stress in infancy. In *Handbook of cross-cultural development,* ed. R. H. Munroe, R. L. Munroe, and B. B. Whiting. New York, Garland.

Lande, R. 1980. "Microevolution in relation to macroevolution." *Paleobiology* 6: 235–38.

———. 1982. "A quantitative genetic theory of life history evolution." *Ecology* 63(3): 607–15.

Lane, H. 1976. *The wild boy of Aveyron.* Cambridge, MA, Harvard Univ. Press.

Langworthy, O. 1933. "Development of behavior patterns and myelination of the nervous system in the human fetus and infant." *Contributions to Embryology* 139: 1–57.

Lanuza, E., M. Belekhova, et al. 1998. "Identification of the reptilian basolateral amygdala: An anatomical investigation of the afferents to the posterior dorsal ventricular ridge of the lizard *Podarcis hispanica*." *European Journal of Neuroscience* 10(11): 3517–34.

Laron, Z., J. Arad, et al. 1980. "Age at first conscious ejaculation: A milestone in male puberty." *Helvetica Pediatrica Acta* 35: 13–20.

Larson, M. S. 2003. "Gender, race, and aggression in television commercials that feature children." *Sex Roles* 48(1–2): 67–75.

Laska-Mierzejewska, T., and E. Olszewska. 2006. "Changes in the biological status of Polish girls from a rural region associated with economic and political processes in the period 1967–2001." *Journal of Biosocial Science* 38(2): 187–202.

Laslett, P. 1965. *The world we have lost.* New York, Scribner's.

———. 1977. *Family life and illicit love in earlier generations.* Cambridge, Cambridge Univ. Press.

Lave, J., and E. Wenger. 1991. *Situated learning: Legitimate peripheral participation.* New York, Cambridge Univ. Press.

Layzer, D. 1974. "Heritability analyses of I.Q. scores: Science or numerology?" *Science* 183: 1259–66.

Leakey, L., P. V. Tobias, et al. 1964. "A new species of the genus Homo from Olduvai Gorge." *Nature* 202: 7–9.

Leakey, M. D., and R. L. Hay. 1979. "Pliocene footprints in the Laetolil beds at Laetolil, Northern Tanzania." *Nature* 278: 308–12.

Leakey, M. G., C. S. Feibel, et al. 1998. "New specimens and confirmation of an early age for *Australopithecus anamensis*." *Nature* 393(May 7): 62–66.

LeBlanc, S., and K. E. Register. 2003. *Constant battles: The myth of the peaceful, noble savage.* New York, St. Martin's.

Le Boeuf, B., and R. Peterson 1969. "Social status and mating activity in elephant seals." *Science* 163: 91–93.

Le Boeuf, B. J. 1974. "Male-male competition and reproductive success in elephant seals." *American Zoologist* 14: 163–76.

Lecanuet, J. P., C. Granier-Deferre, et al. 1987. "Perception et discrimination foetales de stimuli langagiers; mise en évidence à partir de la réactivité cardiaque; résultats préliminaires." *Comptes Rendus de L'Académie des Sciences III: Sciences de la Vie* 305(5): 161–64.

Lecours, A. R. 1975. Myelogenetic correlates of the development of speech and language. In *Foundations of language development,* vol. 1, ed. E. Lenneberg and E. Lenneberg. New York, Academic: 121–35.

Lecours, A. R., F. Lhermitte, et al. 1983. *Aphasiology.* London, Balliere Tindall.

LeDoux, J. 2003. "The emotional brain, fear, and the amygdala." *Cellular and Molecular Neurobiology* 23(4–5): 727–38.

Lee, C., R. G. Barr, et al. 2007. "Age-related incidence of publicly reported shaken baby syndrome cases: Is crying a trigger for shaking?" *Journal of Developmental and Behavioral Pediatrics* 28(4): 288–93.

Lee, C.-Y., C.-W. Lam, et al. 2003. "Steroid 5alpha-reductase 2 deficiency in two generations of a non-consanguineous Chinese family." *Journal of Pediatric Endocrinology* 16(8): 1197–1201.

Lee, P. 1987. "Allomothering among African elephants." *Animal Behaviour* 35(1): 278–91.

Lee, P. C., and M. D. Hauser. 1998. "Long-term consequences of changes in territory quality on feeding and reproductive strategies of vervet monkeys." *Journal of Animal Ecology* 67(3): 347–58.

Lee, R. 1968. The sociology of !Kung bushman trance performances. In *Trance and possession states,* ed. R. Prince. Montreal, Bucke Memorial Society.

Lee, R. B. 1979. *The !Kung San: Men, women and work in a foraging society.* Cambridge, Cambridge Univ. Press.

———. 1982. The sociology of !Kung bushman trance performances. In *Trance and possession states,* ed. R. Prince. Montreal, Bucke Memorial Society.

Lee, R. B., and I. DeVore. 1968. *Man the hunter.* Chicago, Aldine.

Lee, S.-H., and M. H. Wolpoff. 2003. "The pattern of evolution in Pleistocene human brain size." *Paleobiology* 29(2): 186–96.

Lee, Y.-C., B.-W. Soong, et al. 2005. "Median nerve motor conduction velocity is concordant with myelin protein zero gene mutation." *Journal of Neurology* 252(2): 151–55.

Leeb, R. T., and F. G. Rejskind. 1998. *Paternal sex-typing of newborns: Standardization of a rating scale.* International Conference on Infant Studies, Atlanta, Georgia.

Legg, A. T., and T. P. O'Connor. 2003. "Gradients and growth cone guidance of grasshopper neurons." *Journal of Histochemistry and Cytochemistry* 51(4): 445–54.

LeGros Clark, W. E. 1959. *The antecedents of man: An introduction to the evolution of the primates.* Chicago, Quadrangle.

Lehrman, D. S. 1953. "A critique of Konrad Lorenz's theory of instinctive behavior." *Quarterly Review of Biology* 28: 337–63.

———. 1955. "The physiological basis of parental feeding behavior in the ring dove." *Behavior* 7: 241–86.

———. 1965. Interaction between internal and external environments in the regulation of the reproductive cycle of the ring dove. In *Sex and behavior,* ed. F. A. Beach. New York, Wiley.

Lehrman, D. S. 1971. Experiential background for the induction of reproductive behavior patterns by hormones. In *The biopsychology of development,* ed. E. Tobach, L. R. Aronson, and E. Shaw. New York, Academic: 297–302.

Lehrman, D. S., P. N. Brody, et al. 1961. "The presence of the mate and of nesting material as stimuli for the development of incubation behavior and for gonadotropin secretion in the ring dove *(Streptopelia risoria)." Endocrinology* 68: 507–16.

Leiderman, P. H., and G. F. Leiderman. 1974. Affective and cognitive consequences of polymatric infant care in the East African highlands. In *Minnesota Symposia on Child Psychology,* ed. A. Pick. Minneapolis, Univ. of Minnesota Press: 8.

———. 1977. Economic change and infant care in an East African agricultural community. In *Culture and infancy,* ed. P. H. Leiderman, S. R. Tulkin, and A. Rosenfeld. New York, Academic.

Leiderman, P. H., and M. J. Seashore. 1975. "Mother-infant neonatal separation: Some delayed consequences." *Ciba Foundation Symposium* (33): 213–39.

Leiderman, P. H., S. R. Tulkin, et al., eds. 1977. *Culture and infancy.* New York, Academic.

Leigh, H., and M. A. Hofer. 1975. "Long-term effects of preweaning isolation from littermates in rats." *Behavioral Biology* 15(2): 173–81.

Leigh, S. R. 1994. "Ontogenetic correlates of diet in anthropoid primates." *American Journal of Physical Anthropology* 94(4): 499–522.

———. 1996. "Evolution of human growth spurts." *American Journal of Physical Anthropology* 101(4): 455–74.

———. 2004. "Brain growth, life history, and cognition in primate and human evolution." *American Journal of Primatology* 62(3): 139–64.

———. 2006. "Brain ontogeny and life history in *Homo erectus." Journal of Human Evolution* 50(1): 104–8.

Leigh, S. R., and G. E. Blomquist. 2007. Life history. In *Primates in perspective,* ed. C. J. Campbell et al. New York, Oxford Univ. Press: 396–407.

Leigh, S. R., and B. T. Shea. 1996. "Ontogeny of body size variation in African apes." *American Journal of Physical Anthropology* 99(1): 43–65.

LeMay, M. 1976. "Morphological cerebral asymmetries of modern man, fossil man, and nonhuman primate." *Annals of the New York Academy of Sciences* 280: 349–66.

LeMay, M., and N. Geschwind. 1975. "Hemispheric differences in the brains of great apes." *Brain, Behavior and Evolution* 11: 48–52.

Lenneberg, E. 1967. *Biological foundations of language.* New York, Wiley.

Lenneberg, E. H. 1964. A biological perspective of language. In *New directions in the study of language,* ed. E. H. Lenneberg. Cambridge, MA, MIT Press.

———. 1967. *Biological foundations of language.* New York, Wiley.

Lenroot, R. K., and J. N. Giedd. 2006. "Brain development in children and adolescents: Insights from anatomical magnetic resonance imaging." *Neuroscience and Biobehavioral Reviews* 30(6): 718–29.

———. 2008. "The changing impact of genes and environment on brain development during childhood and adolescence: Initial findings from a neuroimaging study of pediatric twins." *Development and Psychopathology* 20(4): 1161–75.

Lenti Boero, D., C. Volpe, et al. 1998. "Newborns crying in different contexts: Discrete or graded signals?" *Perceptual and Motor Skills* 86(3 Pt. 2): 1123–40.

Lenzi, D., C. Trentini, et al. 2009. "Neural basis of maternal communication and emotional expression processing during infant preverbal stage." *Cerebral Cortex* 19(5): 1124–33.

Leonard, W. R., M. L. Robertson, et al. 2003. "Metabolic correlates of hominid brain evolution." *Comparative Biochemistry and Physiology Part A, Molecular and Integrative Physiology* 136(1): 5–15.

Lepage, J.-F., and H. Theoret. 2007. "The mirror neuron system: grasping others' actions from birth? [see comment]." *Developmental Science* 10(5): 513–23.

Lessells, C. M. 1991. The evolution of life histories. In *Behavioral ecology: An evolutionary approach,* ed. J. R. Krebs and N. B. Davies. Oxford, Blackwell: 32–68.

Lester, B. M., and C. F. Z. Boukydis. 1985. *Infant crying: Theoretical and research perspectives.* New York, Plenum.

Leung, A. K., J. E. Fagan, et al. 1994. "Children and television." *American Family Physician* 50(5): 909–12.

LeVay, S. 1991. "A difference in hypothalamic structure between heterosexual and homosexual men." *Science* 253: 1034–37.

———. 1997. *Queer science: The use and abuse of research into homosexuality.* Cambridge, MA, MIT Press.

LeVay, S., T. N. Wiesel, et al. 1980. "The development of ocular dominance columns in normal and visually deprived monkeys." *Journal of Comparative Neurology* 191: 1–51.

Leveno, K. J., F. G. Cunningham, et al. 1986. "A prospective comparison of selective and universal electronic fetal monitoring in 34,995 pregnancies." *New England Journal of Medicine* 315(10): 615–19.

Levine, C., L. Kohlberg, et al. 1985. "The current formulation of Kohlberg's theory and a response to critics." *Human Development* 28(2): 94–100.

Levine, J. B., R. M. Youngs, et al. 2007. "Isolation rearing and hyperlocomotion are associated with reduced immediate early gene expression levels in the medial prefrontal cortex." *Neuroscience* 145(1): 42–55.

LeVine, R. A. 1982. *Culture, behavior, and personality,* 2nd ed. Chicago, Aldine.

———. 1997. Mother-infant interaction in cross-cultural perspective. In *Uniting psychology and biology: Integrative perspectives on human development,* ed. N. L. Segal, G. E. Weisfeld, and C. C. Weisfeld. Washington, DC, American Psychological Association: 339–54.

———. 2007. Anthropological foundations of cultural psychology. In *Handbook of cultural psychology,* ed. S. Kitayama and D. Cohen. New York, Guilford: 40–58.

LeVine, R. A., S. Dixon, et al. 1994. *Child care and culture: Lessons from Africa.* New York, Cambridge Univ. Press.

LeVine, R. A., P. M. Miller, et al. 1988. "Parental behavior in diverse societies." *New Directions in Child Development* 40: 65–74.

Levine, S. 2001. "Primary social relationships influence the development of the hypothalamic—pituitary—adrenal axis in the rat." *Physiology and Behavior* 73(3): 255–60.

Levine, S., G. C. Haltmeyer, et al. 1967. "Physiological and behavioral effects of infantile stimulation." *Physiology of Behavior* 2: 55–59.

Levine, S., and T. Mody. 2003. "The long-term psychobiological consequences of intermittent postnatal separation in the squirrel monkey." *Neuroscience and Biobehavioral Reviews* 27(1–2): 83–89.

Levine, S., and J. R. F. Mullins. 1964. "Hormonal influences in brain organization in infant rats." *Science* 152: 1585–92.

Levins, R. 1964. "Theory of fitness in a heterogeneous environment. 3. The response to selection." *Journal of Theoretical Biology* 110: 224–40.

Lévi-Strauss, C. 1963. *Totemism.* Boston, Beacon.

———. 1968. *The savage mind.* Chicago, Univ. of Chicago Press.

———. 1969. *The elementary structures of kinship (Les structures élémentaires de la parenté).* London, Eyre and Spottiswoode.

Levy, D. A., and P. R. Nail. 1993. "Contagion: A theoretical and empirical review and reconceptualization." *Genetic, Social, and General Psychology Monographs* 119(2): 233–84.

Lévy, F., M. Keller, et al. 2004. "Olfactory regulation of maternal behavior in mammals." *Hormones and Behavior* 46(3): 284–302.

Levy, R. 1973. *Tahitians: Mind and experience in the Society Islands.* Chicago, Univ. of Chicago Press.

Lewis, B. A., and L. A. Thompson. 1992. "A study of developmental speech and language disorders in twins." *Journal of Speech and Hearing Research* 35: 1086–94.

Lewis, K. P. 2000. "A comparative study of primate play behaviour: Implications for the study of cognition." *Folia Primatologica* 71(6): 417–21.

———. 2005. Social play in the great apes. In *The nature of play: Great apes and humans,* ed. A. D. Pellegrini and P. K. Smith. New York, Guilford: 27–53.

Lewis, K. P., and R. A. Barton. 2004. "Playing for keeps: Evolutionary relationships between social play and the cerebellum in nonhuman primates." *Human Nature* 15(1): 5–21.

Lewis, M. 2003. "The emergence of consciousness and its role in human development." *Annals of the New York Academy of Sciences* 1001: 104–33.

Lewis, M., and P. Ban. 1977. Biological variations and cultural diversity: An exploratory study. In *Culture and infancy: Variations in the human experience,* ed. P. H. Leiderman, S. R. Tulkin, and A. Rosenfeld. New York, Academic.

Lewis, M., and D. P. Carmody. 2008. "Self-representation and brain development." *Developmental Psychology* 44(5): 1329–34.

Lewis, M., and L. Rosenblum. 1974. *The origins of fear.* New York, Wiley.

Lewis, M., and L. A. Rosenblum, eds. 1975. *Friendship and peer relations.* New York, Wiley.

Lewis, S. E., and A. E. Pusey. 1997. Factors influencing the occurrence of communal care in plural breeding mammals. In *Cooperative breeding in mammals,* ed. N. G. Solomon and J. A. French. New York, Cambridge Univ. Press.

Lewontin, R. 1972. "The apportionment of human diversity." *Evolutionary Biology* 6: 381–98.

Lewontin, R. C. 1974. *The genetic basis of evolutionary change.* New York, Columbia Univ. Press.

Leyhausen, P. 1979. *Cat behavior: The predatory and social behavior of domestic and wild cats.* New York, Garland STPM.

Li, L., R. W. Garden, et al. 1999. "Egg-laying hormone peptides in the aplysiidae family." *Journal of Experimental Biology* 202(Pt. 21): 2961–73.

Li, S. 2003. "Comment on 'Maternal Sentiments: How Strong Are They?' by Arthur P. Wolf." *Current Anthropology* 44(Suppl.): S43–44.

Lichtneckert, R., and H. Reichert. 2005. "Insights into the urbilaterian brain: Conserved genetic patterning mechanisms in insect and vertebrate brain development." *Heredity* 94(5): 465–77.

Liddell, B. J., K. J. Brown, et al. 2005. "A direct brainstem-amygdala-cortical 'alarm' system for subliminal signals of fear." *Neuroimage* 24(1): 235–43.

Lidz, J., H. Gleitman, et al. 2003. "Understanding how input matters: Verb learning and the footprint of universal grammar." *Cognition* 87(3): 151–78.

Lidz, J., and L. R. Gleitman. 2004. "Yes, we still need Universal Grammar: Reply." *Cognition* 94(1): 85–93.

Lieberman, D. 2009. "Rethinking the Taiwanese minor marriage data: Evidence the mind uses multiple kinship cues to regulate inbreeding avoidance." *Evolution and Human Behavior* 30(3): 153–60.

Lieberman, D., and E. Hatfield. 2006. Passionate love: Cross-cultural and evolutionary perspectives. In *The new psychology of love,* ed. R. J. Sternberg and K. Weis. New Haven, CT, Yale Univ. Press: 274–97.

Liegeois, F., T. Baldeweg, et al. 2003. "Language fMRI abnormalities associated with FOXP2 gene mutation." *Nature Neuroscience* 6(11): 1230–37.

Lieshoff, C., J. Grosse-Ophoff, et al. 2004. "Sexual imprinting leads to lateralized and non-lateralized expression of the immediate early gene zenk in the zebra finch brain." *Behavioural Brain Research* 148(1–2): 145–55.

Lillard, A. 2001. "Pretend play as twin earth: A social-cognitive analysis." *Developmental Review* 21(4): 495–531.

Lim, M. M., E. A. D. Hammock, et al. 2004. "The role of vasopressin in the genetic and neural regulation of monogamy." *Journal of Neuroendocrinology* 16(4): 325–32.

Lim, M. M., Z. Wang, et al. 2004. "Enhanced partner preference in a promiscuous species by manipulating the expression of a single gene [see comment]." *Nature* 429(6993): 754–57.

Lim, M. M., and L. J. Young. 2004. "Vasopressin-dependent neural circuits underlying pair bond formation in the monogamous prairie vole." *Neuroscience* 125(1): 35–45.

Lindburg, D. G., and L. D. Hazell. 1972. "Licking of the neonate and duration of labor in great apes and man." *American Anthropologist* 74(3): 318–25.

Linton, R. 1939. Marquesan culture. In *The individual and his society: The psychodynamics of primitive social organization,* ed. A. Kardiner. New York, Columbia Univ. Press: 197–250.

Lipsitt, L. P. 1990. "Learning and memory in infants." *Merrill-Palmer Quarterly* 36(1): 53–65.

Lipton, E. L., A. Steinschneider, et al. 1965. "Swaddling, a child care practice: Historical, cultural and experimental observations." *Pediatrics* 35(Suppl.): 519–67.

Liu, D., J. Diorio, et al. 1997. "Maternal care, hippocampal glucocorticoid receptors, and hypothalamic-pituitary-adrenal responses to stress [see comments]." *Science* 277(5332): 1659–62.

Locke, J. L. 1993. *The child's path to spoken language.* Cambridge, MA, Harvard Univ. Press.

Locke, J. L., and P. L. Mather. 1989. "Genetic factors in the ontogeny of spoken language: Evidence from monozygotic and dizygotic twins." *Journal of Child Language* 16: 553–59.

Lockhard, R. 1971. "Reflections on the fall of comparative psychology: Is there a message for us all?" *American Psychologist* 26: 168–79.

Loeber, R., and M. Stouthamer-Loeber. 1998. "Development of juvenile aggression and violence: Some common misconceptions and controversies." *American Psychologist* 53(2): 242–59.

Loehlin, J. C., R. R. McCrae, et al. 1998. "Heritabilities of common and measure-specific components of the Big Five personality factors." *Journal of Research in Personality* 32(4): 431–53.

Loftus, E. F. 2003. "Make-believe memories." *American Psychologist* 58(11): 867–73.

Loizos, C. 1966. Play in mammals. In *Play, territoriality and exploration in mammals*, ed. P. Jewell and C. Loizos. New York, Academic.

Lomax, A. 1968. *Folk song style and culture*. Washington, DC, American Association for the Advancement of Science.

Long, J. A. 1990. "Heterochrony and the origin of tetrapods." *Lethaia* 23(2): 157–66.

Lonsdorf, E. V. 2006. "What is the role of mothers in the acquisition of termite-fishing behaviors in wild chimpanzees *(Pan troglodytes schweinfurthii)*?" *Animal Cognition* 9(1): 36–46.

Lonsdorf, E. V., L. E. Eberly, et al. 2004. "Sex differences in learning in chimpanzees [erratum appears in *Nature*, 429(6988, May 13, 2004):154]." *Nature* 428(6984): 715–16.

Lonstein, J. S., and G. J. De Vries. 2000. "Sex differences in the parental behavior of rodents." *Neuroscience and Biobehavioral Reviews* 24(6): 669–86.

Lonstein, J. S., B. D. Rood, et al. 2002. "Parental responsiveness is feminized after neonatal castration in virgin male prairie voles, but is not masculinized by perinatal testosterone in virgin females." *Hormones and Behavior* 41(1): 80–87.

Lorenz, K. 1970 [1935]. Companions as factors in the bird's environment. In *Studies in animal and human behavior*, ed. K. Lorenz. Cambridge, MA, Harvard Univ. Press: vol. 1, 101–258.

———. 1971 [1941]. Comparative studies of the motor patterns of Anatinae. In *Studies in animal and human behavior*, ed. K. Lorenz. Cambridge, MA, Harvard Univ. Press: vol. 2, 14–114.

———. 1965. *Evolution and modification of behavior*. Chicago, Univ. of Chicago Press.

———. 1970. *Studies in animal and human behavior*. Cambridge, MA, Harvard Univ. Press.

———. 1981. *The foundations of ethology*. New York, Springer-Verlag.

Lott, D., D. S. Scholz, et al. 1967. "Exteroceptive stimulation of the reproductive system of the female ring dove *(Streptopelia rosoria)* by the mate and the colony milieu." *Animal Behaviour* 15: 433–37.

Lougee, M. D., R. Grueneich, et al. 1977. "Social interaction in same- and mixed-age dyads of preschool children." *Child Development* 48(4): 1353–61.

Lovejoy, C. O. 1981. "The origin of man." *Science* 211: 341–50.

———. 2009. "Reexamining human origins in light of *Ardipithecus ramidus*." *Science* 326(5949): 74e1–74e8.

Low, B. S. 2003. Ecological and social complexities in human monogamy. In *Monogamy: Mating strategies and partnerships in birds, humans, and other mammals*, ed. U. H. Reichard and C. Boesch. Cambridge, Cambridge Univ. Press: 161–76.

Lowie, R. H. 1954. *Indians of the Plains*. New York, American Museum of Natural History.

Lozoff, B., and G. Brittenham. 1978. Infant care: Cache or carry. In *Meeting of the Society for Pediatric Research*. New York, Society for Pediatric Research.

Lummaa, V., T. Vuorisalo, et al. 1998. "Why cry? Adaptive significance of intensive crying in human infants." *Evolution and Human Behavior* 19(3): 193–202.

Lumsden, C. J. 1989. The gene-culture connection: Interactions across levels of analysis. In *Interaction in human development*, ed. M. H. Bornstein and J. S. Bruner. Hillsdale, NJ, Erlbaum: 261–77.

Lumsden, C. J., and E. O. Wilson. 1981. *Genes, mind, and culture: The coevolutionary process*. Cambridge, MA, Harvard Univ. Press.

Lundqvist, C., and K. G. Sabel. 2000. "Brief report: The Brazelton Neonatal Behavioral

Assessment Scale detects differences among newborn infants of optimal health." *Journal of Pediatric Psychology* 25(8): 577–82.

Lundy, B. L., N. A. Jones, et al. 1999. "Prenatal depression effects on neonates." *Infant Behavior and Development* 22(1): 119–29.

Luo, Z.-X., A. W. Crompton, et al. 2001. "A new mammaliaform from the early Jurassic and evolution of mammalian characteristics [comment]." *Science* 292(5521): 1535–40.

Luo, Z.-X. 2007. "Transformation and diversification in early mammals." *Nature* 450(13): 1011–19.

Luskin, M. B. 1993. "Mammalian forebrain development: How cells find a permanent home and establish their identity." *Journal of NIH Research* 5: 60–64.

———. 1994. "Neuronal cell lineage in the vertebrate central nervous system." *Journal of the Federation of American Societies for Experimental Biology* 8: 722–30.

Lutz, C. A. 1988. *Unnatural emotions: Everyday sentiments on a Micronesian atoll and their challenge to western theory.* Chicago, Univ. of Chicago Press.

Lutz, W., and R. Qiang. 2002. "Determinants of human population growth." *Philosophical Transactions of the Royal Society of London B: Biological Sciences* 357(1425): 1197–1210.

Luyendijk, W., and P. D. Treffers. 1992. "The smile in anencephalic infants." *Clinical Neurology and Neurosurgery* 94(Suppl.): S113–17.

Lyon, G., A. Fattal-Valevski, and E. H. Kolodny. 2006. "Leukodystrophies: Clinical and genetic aspects." *Topics in Magnetic Resonance Imaging* 17(4): 219–42.

Lyons, D. E., L. R. Santos, et al. 2006. "Reflections of other minds: How primate social cognition can inform the function of mirror neurons." *Current Opinion in Neurobiology* 16(2): 230–34.

Lyons, M. J., W. R. True, et al. 1995. "Differential heritability of adult and juvenile antisocial traits." *Archives of General Psychiatry* 52(11): 906–15.

Lyytinen, P., M. L. Laakso, et al. 1999. "The development and predictive relations of play and language across the second year." *Scandinavian Journal of Psychology* 40(3): 177–86.

MacArthur, R. H., and E. O. Wilson. 1967. *Theory of island biogeography.* Princeton, Princeton Univ. Press.

Maccoby, E. E., ed. 1966. *The development of sex differences.* Stanford, Stanford Univ. Press.

———. 1998. *The two sexes: Growing up apart, coming together.* Cambridge, MA, Harvard Univ. Press.

Maccoby, E. E., and C. C. Lewis. 2003. "Less day care or different day care? [comment]." *Child Development* 74(4): 1069–75.

Maccoby, E. E., and C. N. Jacklin. 1974. *The psychology of sex differences.* Stanford, Stanford Univ. Press.

———. 1980. "Sex differences in aggression: A rejoinder and reprise." *Child Development* 51(4): 964–80.

MacDonald, G., P. R. Nail, et al. 2004. "Expanding the scope of the social response context model." *Basic and Applied Social Psychology* 26(1): 77–92.

Mace, R. 2009. "Anthropology. On becoming modern [comment]." *Science* 324(5932): 1280–81.

Mace, R., and M. Pagel. 1994. "The comparative method in anthropology." *Current Anthropology* 35(5): 549–64.

———. 1997. Tips, branches, and nodes: Seeking adaptation through comparative studies. In *Human nature: A critical reader,* ed. L. Betzig. New York, Oxford Univ. Press: 297–310.

MacFarlane, A. 1974. Olfaction in the development of social preferences in the human neonate. In *CIBA Foundation Symposium on the Parent-Infant Relationship.* London, CIBA Foundation.

———. 1977. *The psychology of childbirth.* Cambridge, MA, Harvard Univ. Press.

MacKinnon, K. C. 2007. Social beginnings: The tapestry of infant and adult interactions. In *Primates in perspective,* ed. C. J. Campbell et al. New York, Oxford Univ. Press: 571–91.

MacLarnon, A. 1993. The vertebral canal. In *The Nariokotome Homo erectus skeleton,* ed. A. Walker and R. Leakey. Cambridge, MA, Harvard Univ. Press: 359–90.

MacLarnon, A. M., and G. P. Hewitt. 1999. "The evolution of human speech: The role of enhanced breathing control." *American Journal of Physical Anthropology* 109(3): 341–63.

MacLean, K. 2003. "The impact of institutionalization on child development." *Development and Psychopathology* 15(4): 853–84.

MacLean, P. D. 1952. "Some psychiatric implications of physiological studies on frontotemporal portion of limbic system (visceral brain)." *Electroencephalography and Clinical Neurophysiology* 4: 407–18.

———. 1973. A triune concept of brain and behavior. In *The Hincks Memorial Lectures.* Toronto, Univ. of Toronto Press.

———. 1985. "Brain evolution relating to family, play, and the separation call." *Archives of General Psychiatry* 42: 405–17.

———. 1990. *The triune brain in evolution: Role in paleocerebral functions.* New York, Plenum.

———. 1993. Cerebral evolution of emotion. In *Handbook of emotions,* ed. M. Lewis and J. M. Haviland. New York, Guilford: 67–83.

MacLean, P. D., and D. Ploog. 1978. "Effects of lesions of globus pallidus on species-typical display behavior of squirrel monkeys." *Brain Research* 149: 175–96.

MacLeod, C. E., K. Zilles, et al. 2003. "Expansion of the neocerebellum in Hominoidea." *Journal of Human Evolution* 44(4): 401–29.

MacNeilage, P. F., L. J. Rogers, et al. 2009. "Origins of the left and right brain." *Scientific American* 301(1): 60–67.

Madden, J. R., H.-J. P. Kunc, et al. 2009. "Why do meerkat pups stop begging?" *Animal Behaviour* 75(1): 85–89.

Maestripieri, D. 1995. "First steps in the macaque world: Do rhesus mothers encourage their infants' independent locomotion?" *Animal Behaviour* 49: 1–9.

———. 1998. "Parenting styles of abusive mothers in group-living rhesus macaques." *Animal Behaviour* 55(1): 1–11.

———. 2001. "Is there mother-infant bonding in primates?" *Developmental Review* 21(1): 93–120.

Maestripieri, D., C. L. Hoffman, et al. 2009. "Mother-infant interactions in free-ranging rhesus macaques: Relationships between physiological and behavioral variables." *Physiology and Behavior* 96(4–5): 613–19.

Maestripieri, D., T. Jovanovic, et al. 2000. "Crying and infant abuse in rhesus monkeys." *Child Development* 71(2): 301–9.

Maestripieri, D., K. Wallen, et al. 1997a. "Genealogical and demographic influences on infant abuse and neglect in group-living sooty mangabeys (*Cercocebus atys*)." *Developmental Psychobiology* 31(3): 175–80.

————. 1997b. "Infant abuse runs in families of group-living pigtail macaques." *Child Abuse and Neglect* 21(5): 465–71.

Maestu, J., J. Allik, et al. 2008. "Associations between an alpha 2A adrenergic receptor gene polymorphism and adolescent personality." *American Journal of Medical Genetics, Part B, Neuropsychiatric Genetics* 147B(4): 418–23.

Maital, S. L., and M. H. Bornstein. 2003. "The ecology of collaborative child rearing: A systems approach to child care on the kibbutz." *Ethos* 31(2): 274–306.

Malinowski, B. 1961 [1927]. *Sex and repression in savage society.* Cleveland, World.

————. 1935. *Coral gardens and their magic: A study of tilling the soil and of agricultural rites in the Trobriand Islands.* London, Allen and Unwin.

Mallamaci, A., and A. Stoykova. 2006. "Gene networks controlling early cerebral cortex arealization." *European Journal of Neuroscience* 23(4): 847–56.

Manlove, J., K. Franzetta, et al. 2006. Adolescent sexual relationships: Contraceptive consistency and pregnancy prevention approaches. In *Romance and sex in adolescence and emerging adulthood: Risks and opportunities,* ed. A. C. Crouter and A. Booth. Mahwah, NJ, Erlbaum: 181–212.

Manlove, J., E. Ikramullah, et al. 2009. "Trends in sexual experience, contraceptive use, and teenage childbearing: 1992–2002." *Journal of Adolescent Health* 44(5): 413–24.

Mann, A. 1988. "The nature of Taung dental maturation." *Nature* 333(6169): 123.

Mann, A., M. Lampl, et al. 1990. "Patterns of ontogeny in human evolution: Evidence from dental development." *Yearbook of Physical Anthropology* 33: 111–50.

Mann, J., and B. B. Smuts. 1998. "Natal attraction: Allomaternal care and mother-infant separations in wild bottlenose dolphins." *Animal Behaviour* 55(5): 1097–1113.

Manning, C. J., E. K. Wakeland, et al. 1992. "Communal nesting patterns in mice implicate MHC genes in kin recognition." *Nature* 360: 581–83.

Manning, W. D., P. C. Giordano, et al. 2006. "Hooking up: The relationship contexts of 'nonrelationship' sex." *Journal of Adolescent Research* 21(5): 459–83.

Manoli, D. S., M. Foss, et al. 2005. "Male-specific fruitless specifies the neural substrates of Drosophila courtship behaviour [see comment]." *Nature* 436(7049): 395–400.

Manuck, S. B., M. E. Bleil, et al. 2005. "The socio-economic status of communities predicts variation in brain serotonergic responsivity." *Psychological Medicine* 35(4): 519–28.

Marazziti, D., H. S. Akiskal, et al. 1999. "Alteration of the platelet serotonin transporter in romantic love [comment]." *Psychological Medicine* 29(3): 741–45.

Marcus, G. F. 1993. "Negative evidence in language acquisition." *Cognition* 46: 53–85.

Marcus, G. F., S. Pinker, et al. 1992. "Overregularization in language development." *Monographs of the Society for Research in Child Development* 57(4, Serial No. 228).

Marean, C. W., M. Bar-Matthews, et al. 2007. "Early human use of marine resources and pigment in South Africa during the Middle Pleistocene [see comment]." *Nature* 449(7164): 905–8.

Maringer, J. 1960. *The gods of prehistoric man.* New York, Knopf.

Marler, P. 1991. "Song-learning behavior: The interface with neuroethology." *Trends in Neurosciences* 14(5): 199–206.

————. 1993. The instinct to learn. In *Brain development and cognition,* ed. M. H. Johnson. Cambridge, Blackwell: 454–80.

Marler, P., and H. Slabbekoorn, eds. 2004. *Nature's music: The science of birdsong.* San Diego, Elsevier Scientific.

Marlowe, F. W. 1999. "Male care and mating effort among Hadza foragers." *Behavioral Biology and Sociobiology* 46: 57–64.

———. 2000. "Paternal investment and the human mating system." *Behavioural Processes* 51: 45–61.

———. 2001. "Male contribution to diet and female reproductive success among foragers." *Current Anthropology* 42(5): 755–60.

———. 2002. Why the Hadza are still hunter-gatherers. In *Ethnicity, hunter-gatherers, and the "Other,"* ed. S. Kent. Washington, DC, Smithsonian Institution Press: 247–75.

———. 2003. "A critical period for provisioning by Hadza men: Implications for pair bonding." *Evolution and Human Behavior* 24(3): 217–29.

———. 2005a. Who tends Hadza children? In *Hunter-gatherer childhoods: Evolutionary, developmental and cultural perspectives,* ed. B. S. Hewlett and M. E. Lamb. New Brunswick, NJ, AldineTransaction: 177–90.

———. 2005b. "Hunter-gatherers and human evolution." *Evolutionary Anthropology* 14: 54–67.

Marsh, R., A. J. Gerber, and B. S. Peterson. 2008. "Neuroimaging studies of normal brain development and their relevance for understanding childhood neuropsychiatric disorders." *Journal of the American Academy of Child and Adolescent Psychiatry* 47(11): 1233–51.

Marshall, L. 1960. "!Kung Bushman bands." *Africa* 30(4): 325–55.

Marshall, L. J. 1959. "Marriage among !Kung bushmen." *Africa* 29: 335–65.

———. 1976. *The !Kung of Nyae Nyae.* Cambridge, MA, Harvard Univ. Press.

———. 1981. "The medicine dance of the !Kung bushmen." *Africa* 39: 347–81.

———. 1999. *Nyae Nyae !Kung: Beliefs and rites.* Cambridge, MA, Harvard Univ. Press.

Marshall, W. A., and J. M. Tanner. 1969. "Variations in pattern of pubertal changes in girls." *Archives of Disease in Childhood* 44(235): 291–303.

———. 1970. "Variations in the pattern of pubertal changes in boys." *Archives of Disease in Childhood* 45(239): 13–23.

Marsiglio, W., P. Amato, et al. 2000. "Scholarship on fatherhood in the 1990s and beyond." *Journal of Marriage and the Family* 62(4): 1173–91.

Martin, J. B. 1999. "Molecular basis of the neurodegenerative disorders." *New England Journal of Medicine* 340(June 24): 1970–80.

Martin, J. E. 2005. "The influence of rearing on personality ratings of captive chimpanzees (*Pan troglodytes*)." *Applied Animal Behaviour Science* 90(2): 167–81.

Martin, M. V., J. D. Churchill, et al. 2009. "Genetic influences on hippocampal structure and function in recombinant inbred mice." *Behavioural Brain Research* 196(1): 78–83.

Martin, P., and T. M. Caro. 1985. "On the functions of play and its role in behavioral development." *Advances in the Study of Behavior* 15: 59–103.

Martin, R. 1990. *Primate origins and evolution: A phylogenetic reconstruction.* Princeton, Princeton Univ. Press.

Martin, R. D. 1981. "Relative brain size and basal metabolic rate in terrestrial vertebrates." *Nature* 293: 57–60.

———. 1983. *Human brain evolution in an ecological context.* 52nd James Arthur Lecture on the Evolution of the Human Brain. New York, American Museum of Natural History.

———. 1995. Phylogenetic aspects of primate reproduction: The context of advanced maternal care. In *Motherhood in human and nonhuman primates. Biosocial determinants,* ed. C. R. Pryce and R. D. Martin. Basel, Karger: 16–26.

———. 2007. "The evolution of human reproduction: A primatological perspective." *American Journal of Physical Anthropology* (Suppl. 45): 59–84.

Martin, R. D., M. Genoud, et al. 2005. "Problems of allometric scaling analysis: Examples from mammalian reproductive biology." *Journal of Experimental Biology* 208(Pt9): 1731–47.

Martin, R. D., and P. H. Harvey. 1985. Brain size allometry: Ontogeny and phylogeny. In *Size and scaling in primate biology,* ed. W. L. Jungers. New York, Plenum: 147–73.

Martin, R. D., C. Soligo, et al. 2007. "Primate origins: Implications of a cretaceous ancestry." *Folia Primatologica* 78(5–6): 277–96.

Martin-Krumm, C. P., P. G. Sarrazin, et al. 2003. "Explanatory style and resilience after sports failure." *Personality and Individual Differences* 35(7): 1685–95.

Martineau, J., S. Cochin, et al. 2008. "Impaired cortical activation in autistic children: Is the mirror neuron system involved?" *International Journal of Psychophysiology* 68(1): 35–40.

Martinez, A. J., and R. L. Friede. 1970. "Changes in nerve cell bodies during the myelination of their axons." *Journal of Comparative Neurology* 138: 329–38.

Masataka, N., and M. Kohda. 1988. "Primate play vocalizations and their functional significance." *Folia Primatologica* 50(1–2): 152–56.

Mashour, G. A., E. E. Walker, et al. 2005. "Psychosurgery: Past, present, and future." *Brain Research—Brain Research Reviews* 48(3): 409–19.

Mason, A. J., J. S. Hayflick, et al. 1987. "A deletion truncating the gonadotropin-releasing hormone gene is responsible for hypogonadism in the hpg mouse." *Science* 234: 1366–71.

Mason, A. J., S. L. Pitts, et al. 1986. "The hypogonadal mouse: Reproductive functions restored by gene therapy." *Science* 234: 1372–78.

Mason, W. A. 1966. "Social organization of the South American monkey, *Callicebus moloch*: A preliminary report." *Tulane Studies in Zoology* 13: 23–28.

Mason, W. A., and M. D. Kenney. 1974. "Redirection of filial attachments in rhesus monkeys: Dogs as mother surrogates." *Science* 183: 1209–11.

Matheny, A. P., Jr. 1990. Developmental behavior genetics: Contributions from the Louisville Twin Study. In *Developmental behavior genetics: Neural, biometrical, and evolutionary approaches,* ed. M. E. Hahn et al. New York, Oxford Univ. Press: 25–39.

Mather, P. L., and K. N. Black. 1984. "Hereditary and environmental influences on preschool twins' language skills." *Developmental Psychology* 20(2): 303–8.

Mathews, J. J., and K. Zadak. 1991. "The alternative birth movement in the United States: History and current status." *Women and Health* 17(1): 39–56.

Matisoo-Smith, E., S. Watt, et al. 1997. "Genetic relatedness and alloparental behaviour in a captive group of spider monkeys (*Ateles geoffroyi*)." *Folia Primatologica* 68(1): 26–30.

Matthews, M. A. 1968. "An electron microscopic study of the relationship between axon diameter and the initiation of myelin production in the peripheral nervous system." *Anatomical Record* 161: 337–52.

Mattick, J. S. 2005. "The functional genomics of noncoding RNA [comment]." *Science* 309(5740): 1527–28.

Mauss, M., and W. D. Halls. 1990. *The gift: The form and reason for exchange in archaic societies.* New York, Norton.

Maxfield, M. G., and C. S. Widom. 1996. "The cycle of violence. Revisited 6 years later." *Archives of Pediatrics and Adolescent Medicine* 150(4): 390–95.

Maxson, S. C. 2000. Genetic influences on aggressive behavior. In *Genetic influences on neural and behavioral functions,* ed. D. W. Pfaff et al. New York, CRC: 405–16.

Maynard, A. E. 2002. "Cultural teaching: The development of teaching skills in Maya sibling interactions." *Child Development* 73(3): 969–82.

———. 2004. "Cultures of teaching in childhood: Formal schooling and Maya sibling teaching at home." *Cognitive Development* 19(4): 517–35.

Maynard, A. E., and P. M. Greenfield. 2003. "Implicit cognitive development in cultural tools and children: Lessons from Maya Mexico." *Cognitive Development* 18(4): 489–510.

Maynard, A. E., P. M. Greenfield, et al. 1999. "Culture, history, biology, and body: Native and non-native acquisition of technological skill." *Ethos* 27(3): 379–402.

Maynard Smith, J. 1982. *Evolution and the theory of games.* Cambridge, Cambridge Univ. Press.

———. 1993. *The theory of evolution.* Cambridge, Cambridge Univ. Press.

Maynard Smith, J., R. Burian, et al. 1985. "Developmental constraints and evolution." *Quarterly Review of Biology* 60: 265–87.

Mayr, E. 1958. Behavior and systematics. In *Behavior and evolution,* ed. A. Roe and G. G. Simpson. New Haven, Yale Univ. Press: 341–62.

———. 1963. *Animal species and evolution.* Cambridge, MA, Harvard Univ. Press.

———. 1982. *The growth of biological thought: Diversity, evolution, and inheritance.* Cambridge, MA, Harvard Univ. Press/Belknap.

———. 1988. *Toward a new philosophy of biology.* Cambridge, MA, Harvard Univ. Press.

———. 1997. *This is biology.* Cambridge, MA, Harvard Univ. Press/Belknap.

———. 2001. *What evolution is.* New York, Basic.

Mayson, T. A., S. R. Harris, et al. 2007. "Gross motor development of Asian and European children on four motor assessments: A literature review." *Pediatric Physical Therapy* 19(2): 148–53.

Mazur, A., and A. Booth. 1998. "Testosterone and dominance in men." *Behavioral and Brain Sciences* 21(3): 353–63; discussion 363–97.

Mazur, A., and T. A. Lamb. 1980. "Testosterone, status, and mood in human males." *Hormones and Behavior* 14: 236–46.

McBrearty, S. 2007. Down with the revolution. In *Rethinking the human revolution,* ed. P. Mellars, K. Boyle, O. Bar-Yosef, and C. Stringer. Cambridge, UK, McDonald Institute for Archeological Research: 131–51.

McBrearty, S., and A. S. Brooks. 2000. "The revolution that wasn't: A new interpretation of the origin of modern human behavior." *Journal of Human Evolution* 39(5): 453–563.

McBride, B. A., and M. M. Lutz. 2004. Intervention: Changing the nature and extent of father involvement. In *The role of the father in child development,* 4th ed., ed. M. E. Lamb. Hoboken, NJ, Wiley: 446–75.

McCabe, J. 1983. "FBD marriage: Further support for the Westermarck hypothesis of the incest taboo?" *American Anthropologist* 85: 50–69.

McCarley, R. W., and J. Hobson. 1975. "Discharge patterns of cat brain stem neurons during desynchronized sleep." *Journal of Neurophysiology* 38: 751–66.

McCarthy, M. M., and A. P. Arnold. 2008. Sex differences in the brain: What's old and what's new? In *Sex differences in the brain: From genes to behavior,* ed. J. B. Becker et al. New York, Oxford Univ. Press: 15–33.

McClintock, M. K. 1998. "Whither menstrual synchrony?" *Annual Review of Sex Research* 9: 77–95.

McClintock, M. K., and G. Herdt. 1996. "Rethinking puberty: The development of sexual attraction." *Current Directions in Psychological Science* 5(6): 178–83.

McCrae, R. R., and P. T. Costa, Jr. 1997. "Personality trait structure as a human universal." *American Psychologist* 52(5): 509–16.

———. 1999. A five-factor theory of personality. In *Handbook of personality: Theory and research,* 2nd ed., ed. L. A. Pervin and O. P. John. New York, Guilford: 139–53.

McCrae, R. R., P. T. Costa, Jr., et al. 2000. "Nature over nurture: Temperament, personality, and life span development." *Journal of Personality and Social Psychology* 78: 173–86.

McCrae, R. R., A. Terracciano, et al. 2005. "Universal features of personality traits from the observer's perspective: Data from 50 cultures." *Journal of Personality and Social Psychology* 88(3): 547–61.

McDonnell, S. M., and A. Poulin. 2002. "Equid play ethogram." *Applied Animal Behaviour Science* 78(2–4): 263–90.

McElreath, R., R. Boyd, et al. 2003. "Shared norms and the evolution of ethnic markers." *Current Anthropology* 44(1): 122–29.

McEwen, B. 1978. "Sexual maturation and differentiation: The role of the gonadal steroids." *Progress in Brain Research* 48: 291–308.

———. 2000. Stress, sex, and the structural and functional plasticity of the hippocampus. In *The new cognitive neurosciences,* 2nd ed., ed. M. S. Gazzaniga. Cambridge, MA, MIT Press: 171–97.

McFarlane, S. 2000. "Attraction vs. repulsion: The growth cone decides." *Biochemistry and Cell Biology* 78(5): 563–68.

McGill, T. E., and G. R. Tucker. 1964. "Genotype and sex drive in intact and castrated mice." *Science* 145: 514–15.

McGinnis, M. Y. 2004. "Anabolic androgenic steroids and aggression: Studies using animal models." *Annals of the New York Academy of Sciences* 1036: 399–415.

McGivern, R. F., J. Andersen, et al. 2002. "Cognitive efficiency on a match to sample task decreases at the onset of puberty in children." *Brain and Cognition* 50(1): 73–89.

McGraw, M. 1935. *Growth: A study of Johnny and Jimmy.* New York, Appleton-Century-Crofts.

———. 1943. *The neuromuscular maturation of the human infant.* New York, Columbia Univ. Press.

McGrew, W. C. 1972. *An ethological study of children's behavior.* New York: Academic.

———. 1988. "Parental division of infant caretaking varies with family composition in cotton-top tamarins." *Animal Behaviour* 36: 285–310.

———. 1992. *Chimpanzee material culture: Implications for human evolution.* Cambridge, Cambridge Univ. Press.

———. 1994. Tools compared: The material of culture. In *Chimpanzee cultures,* ed. R. W. Wrangham et al. Cambridge, MA, Harvard Univ. Press: 25–39.

———. 2002. The nature of culture: Prospects and pitfalls of cultural primatology. In *Tree of origin: What primate behavior can tell us about human social evolution,* ed. F. B. M. de Waal. Cambridge, MA, Harvard Univ. Press: 229–54.

McGrew, W. C., L. F. Marchant, et al., eds. 1996. *Great ape societies.* Cambridge, Cambridge Univ. Press.

McHenry, H. M. 2004. Origin of human bipedality. *Evolutionary Anthropology: Issues, News, and Reviews* 13(3): 116–19.

McIntosh, J., H. Anisman, et al. 1999. "Short- and long-periods of neonatal maternal separation differentially affect anxiety and feeding in adult rats: Gender-dependent effects." *Brain Research: Developmental Brain Research* 113(1–2): 97–106.

McKenna, J., S. Mosko, et al. 1999. Breastfeeding and mother-infant cosleeping in relation to SIDS prevention. In *Evolutionary medicine,* ed. W. R. Trevathan, E. O. Smith, and J. J. McKenna. New York, Oxford Univ. Press: 53.

McKenna, J. J. 1981. Primate infant caregiving behavior: Origins, consequences, and variability with emphasis on the common Indian Langur monkey. In *Parental care in mammals,* ed. D. J. Gubernick. New York, Plenum: 389–416.

———. 1987. Parental supplements and surrogates among primates: Cross-species and cross-cultural comparisons. In *Parenting across the life span: Biosocial dimensions,* ed. J. B. Lancaster et al. New York, Aldine de Gruyter: 143–84.

McKenna, J. J., E. B. Thoman, et al. 1993. "Infant-parent co-sleeping in an evolutionary perspective: Implication for understanding infant sleep development in the Sudden Infant Death Syndrome." *Sleep* 16(3): 23–282.

McKeown, T. 1995. *The origins of human disease.* Oxford, Blackwell.

McKim, M. K., K. M. Cramer, et al. 1999. "Infant care decisions and attachment security: The Canadian Transition to Child Care Study." *Canadian Journal of Behavioural Science* 31(2): 92–106.

McKinney, M. L., and K. J. McNamara. 1991. *Heterochrony: The evolution of ontogeny.* New York, Plenum.

McKinney, W. T., Jr. 1974. "Animal models in psychiatry." *Perspectives in Biology and Medicine* 17(4): 529–42.

McKinney, W. T., Jr., R. Gardner, Jr., et al. 1994. "Conceptual basis of animal models in psychiatry: A conference summary." *Ethology and Sociobiology* 15: 369–82.

McKinney, W. T., L. D. Young, et al. 1973. "Chlorpromazine treatment of disturbed monkeys." *Archives of General Psychiatry* 29: 490–94.

McLean, K. C., and A. V. Breen. 2009. "Processes and content of narrative identity development in adolescence: Gender and well-being." *Developmental Psychology* 45(3): 702–10.

McMahon, T. A., and J. T. Bonner. 1983. *On size and life.* New York, Scientific American.

McNamara, K. J. 1997. *Shapes of time: The evolution of growth and development.* Baltimore, Johns Hopkins Univ. Press.

McNamara, K. J., and M. L. McKinney. 2005. "Heterochrony, disparity, and macroevolution." *Paleobiology* 31(2): 17–26.

McNutt, J. W., and J. B. Silk. 2008. "Pup production, sex ratios, and survivorship in African wild dogs, *Lycaon pictus.*" *Behavioral Ecology and Sociobiology* 62(7): 1061–67.

Mead, G. H. 1934. *Mind, self, and society.* Chicago, University of Chicago Press.

Mead, M. 1961 [1928]. *Coming of age in Samoa.* New York, Morrow.

———. 1930. *Growing up in New Guinea.* New York, New American Library.

———. 1932. "An investigation of the thought of primitive children, with special reference to animism." *Journal of the Royal Anthropological Institute* 62: 173–90.

———. 1935. *Sex and temperament in three primitive societies.* New York, Morrow.

———. 1949. *Male and female.* New York, Morrow.

————. 1962. A cultural anthropologist's approach to maternal deprivation. In *Deprivation of maternal care: A reassessment of its effects.* Public Health Papers 14. Geneva, World Health Organization.

————. 1964. *Continuities in cultural evolution.* New Haven, Yale Univ. Press.

————. 1965. *And keep your powder dry: An anthropologist looks at America.* New York, Morrow.

Mead, M., and M. Wolfenstein, eds. 1955. *Childhood in contemporary cultures.* Chicago, Univ. of Chicago Press.

Meaney, M. J. 2001. "Maternal care, gene expression, and the transmission of individual differences in stress reactivity across generations." *Annual Review of Neuroscience* 24: 1161–92.

Meaney, M. J., D. H. Aitken, et al. 1988. "Effect of neonatal handling on age-related impairments associated with the hippocampus." *Science* 239: 766–68.

Meaney, M. J., J. Diorio, et al. 1996. "Early environmental regulation of forebrain glucocorticoid receptor gene expression: Implications for adrenocortical responses to stress." *Developmental Neuroscience* 18(1–2): 49–72.

Meaney, M. J., E. Lozos, et al. 1990. "Infant carrying by nulliparous female vervet monkeys (*Cercopithecus aethiops*)." *Journal of Comparative Psychology* 104(4): 377–81.

Mega, M. S., J. L. Cummings, et al. 1997. The limbic system: An anatomic, phylogenetic, and clinical perspective. In *The neuropsychiatry of limbic and subcortical disorders,* ed. S. Salloway, P. Malloy, and J. L. Cummings. Washington, DC, American Psychiatric Press: 3–18.

Meggitt, M. J. 1965. *Desert people: A study of the Walbiri Aborigines of Central Australia.* Chicago, Univ. of Chicago Press.

Meiselman, K. C. 1978. *Incest: A psychological study of causes and effects with treatment recommendations.* San Francisco, Jossey-Bass.

Melli, G., and A. Hoke. 2007. "Canadian Association of Neurosciences review: Regulation of myelination by trophic factors and neuron-glial signaling." *Canadian Journal of Neurological Sciences* 34(3): 288–95.

Mellon, S. H. 2007. "Neurosteroid regulation of central nervous system development." *Pharmacology and Therapeutics* 116(1): 107–24.

Melnick, D. J., and M. C. Pearl. 1987. Cercopithecines in multimale groups: Genetic diversity and population structure. In *Primate societies,* ed. B. B. Smuts et al. Chicago, Univ. of Chicago Press: 121–34.

Melo, L., A. R. Mendes Pontes, et al. 2003. "Infanticide and cannibalism in wild common marmosets." *Folia Primatologica* 74(1): 48–50.

Meltzoff, A. N. 1990. "Towards a developmental cognitive science. The implications of cross-modal matching and imitation for the development of representation and memory in infancy." *Annals of the New York Academy of Sciences* 608: 1–31; discussion 31–37.

————. 2007. "'Like me': A foundation for social cognition." *Developmental Science* 10(1): 126–34.

Meltzoff, A., and R. W. Borton. 1979. "Intermodal matching by human neonates." *Nature* 282: 403–4.

Meltzoff, A., and M. K. Moore. 1977. "Imitation of facial and manual gestures by human neonates." *Science* 198: 75–78.

Meltzoff, A. N., and M. Moore. 1994. "Imitation, memory, and the representation of persons." *Infant Behavior and Development* 17(1): 83–99.

————. 1999. Resolving the debate about early imitation. In *The Blackwell reader in development psychology,* ed. A. M. D. Slater. Malden, MA, Blackwell: 151–55.

Meredith, H. V. 1978. *Human body growth in the first ten years of life.* Columbia, SC, State Printing.

Merzenich, M. 1998. "Long-term change of mind." *Science* 282: 1062–63.

Merzenich, M. M., J. H. Kaas, et al. 1983. "Topographic reorganization of somatosensory cortical areas 3b and 1 in adult monkeys following restricted deafferentation." *Neuroscience* 8(1): 33–55.

Messinger, D., M. Dondi, et al. 2002. "How sleeping neonates smile." *Developmental Science* 5(1): 48–54.

Messinger, D. S., A. Fogel, et al. 1999. "What's in a smile?" *Developmental Psychology* 35(3): 701–8.

Metzger, M., S. Jiang, et al. 2002. "A quantitative immuno-electron microscopic study of dopamine terminals in forebrain regions of the domestic chick involved in filial imprinting." *Neuroscience* 111(3): 611–23.

Meyer-Bahlburg, H. F., R. S. Gruen, et al. 1996. "Gender change from female to male in classical congenital adrenal hyperplasia." *Hormones and Behavior* 30(4): 319–32.

Meyer-Bahlburg, H. F. L. 2005. "Gender identity outcome in female-raised 46,XY persons with penile agenesis, cloacal exstrophy of the bladder, or penile ablation." *Archives of Sexual Behavior* 34(4): 423–38.

Meyer Zu Hörste, G., and K.-A. Nave. 2006. "Animal models of inherited neuropathies." *Current Opinion in Neurology* 19(5): 464–73.

Miachon, S., T. Rochet, et al. 1993. "Long-term isolation of Wistar rats alters brain monoamine turnover, blood corticosterone, and ACTH." *Brain Research Bulletin* 32: 611–14.

Michael, R. P., and E. Keverne. 1968. "Pheromones in the communication of sexual status in primates." *Nature* 218: 746–49.

————. 1969. "A male sex-attractant pheromone in rhesus monkey vaginal secretions." *Journal of Endocrinology* 46: xx–xxi.

Mickelson, R. A., ed. 2000. *Children on the streets of the Americas: Globalization, homelessness and education in the United States, Brazil, and Cuba.* London, Routledge.

Mikkelsen, T. S., M. J. Wakefield, et al. 2007. "Genome of the marsupial Monodelphis domestica reveals innovation in non-coding sequences." *Nature* 447(7141): 167–77.

Miller, A. M., and R. L. Harwood. 2002. "The cultural organization of parenting: Change and stability of behavior patterns during feeding and social play across the first year of life." *Parenting: Science and Practice* 2(3): 241–72.

Miller, J. G. 2007. Cultural psychology of moral development. In *Handbook of cultural psychology,* ed. S. Kitayama and D. Cohen. New York, Guilford: 77–99.

Miller, M. N., and J. A. Byers. 1998. Sparring as play in young pronghorn males. In *Animal play: Evolutionary, comparative, and ecological perspectives,* ed. M. Bekoff and J. A. Byers. Cambridge, Cambridge Univ. Press: 141–60.

Mills, D. L., and H. J. Neville. 1997. "Electrophysiological studies of language and language impairment." *Seminars in Pediatric Neurology* 4(2): 125–34.

Mineka, S., and M. Cook. 1988. Social learning and the acquisition of snake fear in monkeys. In *Social learning: Psychological and biological perspectives,* ed. T. R. Zentall and B. G. Galef, Jr. Hillsdale, NJ, Erlbaum: 51–73.

Mineka, S., J. L. Mystkowski, et al. 1999. "The effects of changing contexts on return of fear following exposure therapy for spider fear." *Journal of Consulting and Clinical Psychology* 67(4): 599–604.

References | 855

Minkowski, M. 1955. "Quelques réflexions sur la neurophysiologie du nouveau-né et du nourisson et ses relations avec celle du foetus." *Revue Neurologique* 93(1): 247–56.

Minoura, Y. 1992. "A sensitive period for the incorporation of a cultural meaning system: A study of Japanese children growing up in the United States." *Ethos* 20(3): 304–39.

Mintz, S. 2004. *Huck's raft: A history of American childhood.* Cambridge, MA, Harvard Univ. Press/Belknap.

Minugh-Purvis, N., and K. J. McNamara. 2002. *Human evolution through developmental change.* Baltimore, Johns Hopkins Univ. Press.

Mishra, R. C. 1997. Cognition and cognitive development. In *Handbook of cross-cultural psychology,* vol. 2: *Basic Processes and Human Development,* 2nd ed., ed. J. Berry et al. Boston, Allyn and Bacon: 143–76.

Mitani, J. C. 1994. Ethological studies of chimpanzee vocal behavior. In *Chimpanzee cultures,* ed. R. W. Wrangham et al. Cambridge, MA, Harvard Univ. Press: 195–210.

——. 2006. "Demographic influences on the behavior of chimpanzees." *Primates* 47(1): 6–13.

Mitani, J. C., and D. Watts. 1997. "The evolution of non-maternal caretaking among anthropoid primates: Do helpers help?" *Behavioral Ecology and Sociobiology* 40(4): 213–20.

Mitani, J. C., and D. P. Watts. 1999. "Demographic influences on the hunting behavior of chimpanzees." *American Journal of Physical Anthropology* 109(4): 439–54.

Mitani, J. C., D. P. Watts, and M. N. Muller. 2002. "Recent developments in the study of wild chimpanzee behavior." *Evolutionary Anthropology: Issues, News, and Reviews* 11(1): 9–25.

Mitchell, G. 1979. *Behavioral sex differences in nonhuman primates.* New York, Van Nostrand Reinhold.

Mitchell, G., and E. Brandt. 1972. Paternal behavior in primates. In *Primate socialization,* ed. F. E. Poirier. New York, Random House.

Mitchell, T. N., S. L. Free, et al. 2003. "Reliable callosal measurement: Population normative data confirm sex-related differences." *AJNR: American Journal of Neuroradiology* 24(3): 410–18.

Mitford, J. 1992. *The American way of birth.* London, Gollancz.

Mock, D. W. 2004. *More than kin and less than kind: The evolution of family conflict.* Cambridge, MA, Harvard Univ. Press.

Moffitt, T. E. 1993. "Adolescence-limited and life-course-persistent antisocial behavior: A developmental taxonomy." *Psychological Review* 100(4): 674–701.

Mohnot, S. M. 1980. "Intergroup infant kidnapping in Hanuman Langur." *Folia Primatologica* 34: 259–77.

Moles, A., B. L. Kieffer, et al. 2004. "Deficit in attachment behavior in mice lacking the mu-opioid receptor gene [see comment]." *Science* 304(5679): 1983–86.

Moller, A. P. 2003. The evolution of monogamy: Mating relationships, parental care, and sexual selection. In *Monogamy: Mating strategies and partnerships in birds, humans, and other mammals,* ed. U. H. Reichard and C. Boesch. Cambridge, Cambridge Univ. Press: 29–41.

Moltz, H., M. Lubin, et al. 1971. "Hormonal induction of maternal behavior in the ovariectomized nulliparous rat." *Physiology and Behavior* 5: 1373–77.

Moltz, H., and D. Robbins. 1965. "Maternal behavior of primiparous and multiparous rats." *Journal of Comparative and Physiological Psychology* 60: 417–21.

Money, J., J. Cawte, et al. 1970. "Sex training and traditions in Arnhem land." *British Journal of Medical Psychology* 47: 383–99.

Monnier, M. 1956. The behaviour of newborn anencephalics with varying degrees of anencephaly. In *Discussions on child development,* vol. 1, ed. J. M. Tanner and B. Inhelder. London, World Health Organization/Tavistock.

Monrad-Krohn, G. H. 1924. "On the dissociation of voluntary and emotional innervation in facial paralysis of central origin." *Brain* 47: 22–35.

———. 1927. "A few remarks on the question of dissociation between voluntary and emotional innervation in peripheral facial paralysis." *Acta Psychiatrica et Neurologica* 3: 35–39.

Montagnini, A., and A. Treves. 2003. "The evolution of mammalian cortex, from lamination to arealization." *Brain Research Bulletin* 60(4): 387–93.

Montagu, M. F. A. 1957. *The reproductive development of the female,* 2nd ed. New York, Julian.

Moon, C. M., and W. P. Fifer. 2000. "Evidence of transnatal auditory learning." *Journal of Perinatology* 20(8 Pt. 2): S37–44.

Moore, C. L. 1992. "The role of maternal stimulation in the development of sexual behavior and its neural basis." *Annals of the New York Academy of Sciences* 662: 160–77.

Moore, C. L., H. Dou, et al. 1992. "Maternal stimulation affects the number of motor neurons in a sexually dimorphic nucleus of the lumbar spinal cord." *Brain Research* 572: 52–56.

Moore, J. 1983. "Female transfer in primates." *International Journal of Primatology* 5(6): 537–89.

———. 1993. Inbreeding and outbreeding in primates: What's wrong with 'the dispersing sex'? In *The natural history of inbreeding and outbreeding: Theoretical and empirical perspectives,* ed. N. W. Thornhill. Chicago, Univ. of Chicago Press: 392–426.

———. 1996. Savanna chimpanzees, referential models and the last common ancestor. In *Great ape societies,* ed. W. C. McGrew, L. F. Marchant, and T. Nishida. Cambridge, Cambridge Univ. Press: 275–92.

Morelli, G. A., and E. Z. Tronick. 1992. "Efe fathers: One among many? A comparison of forager children's involvement with fathers and other males." *Social Development* 1(1): 36–54.

Morgan, G. A., and H. N. Ricciuti. 1969. Infants' responses to strangers during the first year. In *Determinants of infant behavior,* vol. 4, ed. B. M. Foss. London, Methuen.

Morgan, L. H. 1877. *Ancient society: Researches in the lines of human progress from savagery through barbarism to civilization.* London, MacMillan & Co.

Morgane, P. J., and D. J. Mokler. 2006. "The limbic brain: Continuing resolution." *Neuroscience and Biobehavioral Reviews* 30(2): 119–25.

Morland, H. S. 1990. "Parental behavior and infant development in ruffed lemurs (*Varecia variegata*) in a northeast Madagascar rain forest." *American Journal of Primatology* 20(4): 253–65.

Morley, D. 1973. *Pediatric priorities in the developing world.* London, Butterworth.

Moss, H. 1967. "Sex, age, and state as determinants of mother-infant interaction." *Merrill-Palmer Quarterly* 13: 19–36.

Mota, M. T. d. S., C. R. Franci, et al. 2006. "Hormonal changes related to paternal and alloparental care in common marmosets (*Callithrix jacchus*)." *Hormones and Behavior* 49(3): 293–302.

Moulson, M. C., N. A. Fox, et al. 2009. "Early adverse experiences and the neurobiology of facial emotion processing." *Developmental Psychology* 45(1): 17–30.

Moura, A. C. d. A., and P. C. Lee. 2004. "Capuchin stone tool use in Caatinga dry forest." *Science* 306(5703): 1909.

Moynihan, D. P., T. M. Smeeding, et al., eds. 2004. *The future of the family.* New York, Russell Sage.

Mukherjee, B., E. McCauley, et al. 2007. "Psychopathology, psychosocial, gender and cognitive outcomes in patients with cloacal exstrophy." *Journal of Urology* 178(2): 630–35; discussion 634–35.

Mulkens, S., S. M. Boegels, et al. 1999. "Attentional focus and fear of blushing: A case study." *Behavioural and Cognitive Psychotherapy* 27(2): 153–64.

Muller, R. A., M. E. Behen, et al. 1999. "Brain organization for language in children, adolescents, and adults with left hemisphere lesion: A PET study." *Progress in Neuro-Psychopharmacology and Biological Psychiatry* 23(4): 657–68.

Mummery, W., G. Schofield, et al. 2004. "Bouncing back: The role of coping style, social support and self-concept in resilience of sport performance." *Athletic Insight: Online Journal of Sport Psychology* 6(3): np.

Mundy, N. I., N. S. Badcock, et al. 2004. "Conserved genetic basis of a quantitative plumage trait involved in mate choice." *Science* 303(5665): 1870–73.

Munno, D. W., and N. I. Syed. 2003. "Synaptogenesis in the CNS: An odyssey from wiring together to firing together." *Journal of Physiology* 552(Pt. 1): 1–11.

Munroe, R. H., and R. L. Munroe. 1971. "Household density and infant care in an East African society." *Journal of Social Psychology* 83: 295–315.

———. 1984. "Infant experience and childhood cognition: A longitudinal study among the Logoli of Kenya." *Ethos* 12(4): 291–306.

Munroe, R. H., R. L. Munroe, et al. 1973. "The couvade: A psychological analysis." *Ethos* 1: 30–74.

———. 1997. "Infant experience and late-childhood dispositions: An eleven-year follow-up among the Logoli of Kenya." *Ethos* 25(3): 359–72.

———, eds. 1981. *Handbook of cross-cultural human development.* New York, Garland STPM.

Munroe, R. L. 1980. "Male transvestism and the couvade: A psycho-cultural analysis." *Ethos* 8(1): 49–59.

———. 2004. "Social structure and sex-role choices among children in four cultures." *Cross-Cultural Research* 38(4): 387–406.

Munroe, R. L., R. Hulefeld, et al. 2000. "Aggression among children in four cultures." *Cross-Cultural Research* 34(1): 3–25.

Murakami, Y., K. Uchida, et al. 2005. "Evolution of the brain developmental plan: Insights from agnathans." *Developmental Biology* 280(2): 249–59.

Murdock, G. P. 1945. The common denominator of cultures. In *The science of man in the world crisis,* ed. R. Linton. New York, Columbia Univ. Press: 123–42.

———. 1949. *Social structure.* London, Macmillan.

———. 1967. "Ethnographic atlas: A summary." *Ethnology* (6,2): 109–236.

Murdock, G. P., and D. O. Morrow. 1970. "Subsistence economy and supportive practices: Cross-cultural codes. 1." *Ethnology* 9: 302–30.

Murdock, G. P, and D. White. 1969. "Standard cross-cultural sample." *Ethnology* 8: 329–69.

Murdock, G. P., and S. F. Wilson. 1972. "Settlement patterns and community organization." *Ethnology* 11: 254–69.

Muris, P., H. Merckelbach, and R. Collaris. 1997. "Common childhood fears and their origins." *Behavioral Research and Therapy* 35(10): 929–37.

Murphy, M. R., P. D. MacLean, et al. 1981. "Species-typical behavior of hamsters deprived from birth of the neocortex." *Science* 213: 459–61.

Murray, S. O., and W. Roscoe, eds. 1998. *Boy-wives and female husbands: Studies in African homosexualities.* New York, Palgrave/St. Martin's.

Musil, A. 1928. *The manners and customs of the Rwala bedouin.* New York, American Geographical Society.

Myrskyla, M., H.-P. Kohler, et al. 2009. "Advances in development reverse fertility decline." *Nature* 460(8230): 741–43.

Nadler, R. D. 1983. Experiential influences on infant abuse of gorillas and some other nonhuman primates. In *Child abuse: The nonhuman primate data.* New York, Liss: 139–49.

Naftolin, F. 1994. "Brain aromatization of androgens." *Journal of Reproductive Medicine* 39(4): 257–61.

Naftolin, F., and E. Butz. 1981. "Sexual dimorphism." *Science* 211: 1263–24.

Nagy, E. 2008. "Innate intersubjectivity: Newborns' sensitivity to communication disturbance." *Developmental Psychology* 44(6): 1779–84.

Nagy, Z., H. Westerberg, et al. 2004. "Maturation of white matter is associated with the development of cognitive functions during childhood." *Journal of Cognitive Neuroscience* 16(7): 1227–33.

Nahm, F. K. D., D. Tranel, et al. 1993. "Cross-modal associations and the human amygdala." *Neuropsychologia* 31(8): 727–44.

Nail, P. R. 1986. "Toward an integration of some models and theories of social response." *Psychological Bulletin* 100(2): 190–206.

Nail, P. R., and W. B. Helton. 1999. "On the distinction between behavioral contagion, conversion conformity and compliance conformity." *North American Journal of Psychology* 1(1): 87–94.

Nanda, S. 1999. *Neither man nor woman: The hijras of India.* Belmont, CA, Wadsworth.

Napier, J. H., and P. H. Napier. 1985. *The natural history of the primates.* Cambridge, MA, MIT Press.

Nath, D. C., D. L. Leonetti, et al. 2000. "Analysis of birth intervals in a non-contracepting Indian population: An evolutionary ecological approach." *Journal of Biosocial Science* 32(3): 343–54.

National Institute of Child Health and Human Development (NICHD), Early Childhood Research Network. 1997. "The effects of infant child care on infant-mother attachment security: Results of the NICHD study of early child care." *Child Development* 68(5): 860–79.

Nations, M. K., L. A. Camino, et al. 1988. "'Nerves': Folk idiom for anxiety and depression?" *Social Science and Medicine* 26(12): 1245–59.

Nations, M. K., and L. A. Rebhun. 1988. "Angels with wet wings won't fly: Maternal sentiment in Brazil and the image of neglect." *Culture, Medicine and Psychiatry* 12(2): 141–200.

Natsuaki, M. N., X. Ge, et al. 2009. "Aggressive behavior between siblings and the development of externalizing problems: Evidence from a genetically sensitive study." *Developmental Psychology* 45: 1009–18.

Nauta, W. J. 1971. "The problem of the frontal lobe: A reinterpretation." *Journal of Psychiatric Research* 8: 167–87.

Nauta, W. J. H. 1958. "Hippocampal projections and related neural pathways to the midbrain in the cat." *Brain* 81: 319–40.

Nauta, W. J. H., and V. B. Domesic. 1978. Crossroads of limbic and striatal circuitry: Hypothalamonigral connections. In *Limbic mechanisms,* ed. K. E. Livingston and O. Hornykiewicz. New York, Plenum.

———. 1980. Neural associations of the limbic system. In *Neural substrates of behavior,* ed. A. Beckman. New York, Spectrum.

Nauta, W. J. H., and M. Feirtag 1986. *Fundamental neuroanatomy.* New York, Freeman.

Nauta, W. J. H., and W. R. Mehler 1966. "Projections of the lentiform nucleus in the monkey." *Brain Research* 1: 3–42.

Nave, K.-A., and J. L. Salzer. 2006. "Axonal regulation of myelination by neuregulin 1." *Current Opinion in Neurobiology* 16(5): 492–500.

Nee, S., A. Mooers, et al. 1992. "Tempo and mode of evolution revealed from molecular phylogenies." *Proceedings of the National Academy of Sciences* 89: 8322–26.

Nelson, E., and J. Panksepp. 1996. "Oxytocin mediates acquisition of maternally associated odor preferences in preweanling rat pups." *Behavioral Neuroscience* 110(3): 583–92.

Nelson, E. E., E. Leibenluft, et al. 2005. "The social re-orientation of adolescence: A neuroscience perspective on the process and its relation to psychopathology." *Psychological Medicine* 35(2): 163–74.

Nelson, K. 1996a. *Language in cognitive development: The emergence of the mediated mind.* Cambridge, MA, Harvard Univ. Press.

———. 1996b. Memory development from 4 to 7 years. In *The five to seven year shift: The age of reason and responsibility,* ed. A. J. Sameroff and M. M. Haith. Chicago, Univ. of Chicago Press: 141–60.

Neria, Y., Z. Solomon, et al. 2000. "Sensation seeking, wartime performance, and long-term adjustment among Israeli war veterans." *Personality and Individual Differences* 29(5): 921–32.

Nerlove, S. B. 1974. "Women's workload and infant feeding practices: A relationship with demographic implications." *Ethnology* 13: 207–14.

Nesse, R. M. 1999. Testing evolutionary hypotheses about mental disorders. In *Evolution in health and disease,* ed. S. C. Stearns. Oxford, Oxford Univ. Press: 260–66.

Nesse, R. M., and A. T. Lloyd. 1992. The evolution of psychodynamic mechanisms. In *The adapted mind: Evolutionary psychology and the generation of culture,* ed. J. H. Barkow, L. Cosmides, and J. Tooby. New York, Oxford Univ. Press: 601–24.

Neville, H. J. 1991. Neurobiology of cognitive and language processing: Effects of early experience. In *Brain maturation and cognitive development: Comparative and cross-cultural perspectives,* ed. K. R. Gibson and A. C. Petersen. New York, Aldine de Gruyter: 355–80.

———. 1995. Developmental specificity in neurocognitive development in humans. In *The cognitive neurosciences,* ed. M. S. Gazzaniga. Cambridge, MA, MIT Press: 219–31.

New, J., M. M. Krasnow, et al. 2007. "Spatial adaptations for plant foraging: Women excel and calories count." *Proceedings of the Royal Society of London B: Biological Sciences* 274(1626): 2679–84.

Newman, C. G. 1986. "The thalidomide syndrome: Risks of exposure and spectrum of malformations." *Clinics in Perinatology* 13(3): 555–73.

Newman, J. D. 1985. The infant cry of primates: An evolutionary perspective. In *Infant crying: Theoretical and research perspectives,* ed. B. M. Lester and C. F. Z. Boukydis. New York, Plenum: 307–23.

———. 2003. "Vocal communication and the triune brain." *Physiology and Behavior* 79(3): 495–502.

Newton, N., and M. Newton. 1972. Childbirth in crosscultural perspective. In *Modern perspectives in psycho-obstetrics,* ed. J. B. Howells. Edinburgh, Oliver and Boyd.

Ni, X., Y. Wang, et al. 2004. "A euprimate skull from the early Eocene of China [see comment]." *Nature* 427(6969): 65–68.

Niedźwiedzki, G., P. Szrek, et al. 2010. "Tetrapod trackways from the early Middle Devonian period of Poland." *Nature* 463(8623): 43–48.

Niewenhuys, O. 1996. "The paradox of child labor and anthropology." *Annual Review of Anthropology* 25: 237–51.

Nilsson, D.-E. 2004. "Eye evolution: A question of genetic promiscuity." *Current Opinion in Neurobiology* 14(4): 407–14.

Nilsson, D.-E., and S. Pelger. 1994. "A pessimistic estimate of the time required for an eye to evolve." *Proceedings of the Royal Society of London B: Biological Sciences* 256(1345): 53–58.

Nimchinsky, E. A., E. Gilissen, et al. 1999. "A neuronal morphologic type unique to humans and great apes." *Proceedings of the National Academy of Sciences* 96(9): 5268–73.

Nishida, T. 1983. "Alloparental behavior in wild chimpanzees of the Mahale Mountains, Tanzania." *Folia Primatologica* 41(1–2): 1–33.

———. 1987. Local traditions and cultural transmission. In *Primate societies,* ed. B. B. Smuts, D. L. Cheney, R. M. Seyfarth, R. W. Wrangham, and T. T. Struhsaker. Chicago, University of Chicago Press: 462–74.

Nishikawa, K. C. 2002. "Evolutionary convergence in nervous systems: Insights from comparative phylogenetic studies." *Brain, Behavior and Evolution* 59(5–6): 240–49.

Noirot, E. 1972. "The onset of maternal behavior in rats, hamsters, and mice: A selective review." *Advances in the Study of Behavior* 4: 107–45.

Noordam, C., V. Dhir, et al. 2009. "Inactivating *PAPSS2* mutations in a patient with premature pubarche." *New England Journal of Medicine* 360(22): 2310–18.

Nordeen, E. J., K. W. Nordeen, et al. 1985. "Androgens prevent normally occurring cell death in a sexually dimorphic spinal nucleus." *Science* 229: 671–73.

Norris, D. O. 1997. *Vertebrate endocrinology.* San Diego, Academic.

Norton-Griffiths, M. 1969. "The organization, control and development of parental feeding in the oystercatcher (*Haematopus ostralegus*)." *Behavior* 34: 55–114.

Noss, A. J., and B. S. Hewlett. 2001. "The contexts of female hunting in central Africa." *American Anthropologist* n.s. 103(4): 1024–40.

———. 2006b. *Evolutionary dynamics: Exploring the equations of life.* Cambridge, MA, Harvard Univ. Press/Belknap.

Novak, M. A., and H. F. Harlow. 1975. "Social recovery of monkeys isolated for the first year of life: 1. Rehabilitation and therapy." *Developmental Psychology* 11(4): 453–65.

Novak, M. F. S. X., and G. P. Sackett. 1997. "Pair-rearing infant monkeys (*Macaca nemestrina*) using a 'rotating-peer' strategy." *American Journal of Primatology* 41(2): 141–49.

Nowak, M. A. 2006a. "Five rules for the evolution of cooperation." *Science* 314(5805): 1560–63.

Nowak, R. M. 1991. *Walker's mammals of the world.* Baltimore, Johns Hopkins Univ. Press.

Nudo, R. J., and R. B. Masterton. 1990. "Descending pathways to the spinal cord, III: Sites of origin of the corticospinal tract." *Journal of Comparative Neurology* 296(4): 559–83.

Numan, M. 1994. "A neural circuitry analysis of maternal behavior in the rat." *Acta Paediatrica Supplement* 397: 19–28.

———. 1996. "A lesion and neuroanatomical tract-tracing analysis of the role of the bed nucleus of the stria terminalis in retrieval behavior and other aspects of maternal responsiveness in rats." *Developmental Psychobiology* 29(1): 23–51.

Numan, M., J. S. Rosenblatt, et al. 1977. "Medial preoptic area and onset of maternal behavior in the rat." *Journal of Comparative and Physiological Psychology* 91: 146–64.

Numan, M., and T. P. Sheehan. 1997. "Neuroanatomical circuitry for mammalian maternal behavior." *Annals of the New York Academy of Sciences* 807: 101–25.

Nunes, S., E.-M. Muecke, et al. 1999. "Endocrine and energetic mediation of play behavior in free-living Belding's ground squirrels." *Hormones and Behavior* 36: 153–65.

———. 2004. "Play behavior and motor development in juvenile Belding's ground squirrels (*Spermophilus beldingi*)." *Behavioral Ecology and Sociobiology* 56(2): 97–105.

Nüsslein-Volhard, C. 2008. *Coming to life: How genes drive development.* Carlsbad, CA, Kales.

Obler, L. K., and K. Gjerlow. 1999. *Language and the brain.* Cambridge, Cambridge Univ. Press.

O'Brien, T. G., and J. G. Robinson. 1991. "Allomaternal care by female wedge-capped capuchin monkeys: Effects of age, rank and relatedness." *Behaviour* 119(1–2): 30–50.

Ochs, E. 1986. Introduction. In *Language socialization across cultures,* ed. B. B. Schieffelin and E. Ochs. Cambridge, Cambridge Univ. Press: 1–13.

———. 1988. *Culture and language development: Language acquisition and language socialization in a Samoan village.* Cambridge, Cambridge Univ. Press.

O'Connell, J. F., K. Hawkes, et al. 1999. "Grandmothering and the evolution of homo erectus." *Journal of Human Evolution* 36(5): 461–85.

O'Connor, D., and R. Shine. 2003. "Lizards in 'nuclear families': A novel reptilian social system in *Egernia saxatilis* (Scincidae)." *Molecular Ecology* 12(3): 743–52.

O'Connor, T. G., D. Bredenkamp, et al. 1999. "Attachment disturbances and disorders in children exposed to early severe deprivation." *Infant Mental Health Journal* 20(1): 10–29.

O'Connor, T. G., K. Deater-Deckard, et al. 1998. "Genotype-environment correlations in late childhood and early adolescence: Antisocial behavioral problems and coercive parenting." *Developmental Psychology* 34(5): 970–81.

O'Connor, T. G., E. M. Hetherington, et al. 1995. "A twin-sibling study of observed parent-adolescent interactions." *Child Development* 66(3): 812–29.

O'Connor, T. G., R. S. Marvin, et al. 2003. "Child-parent attachment following early institutional deprivation." *Development and Psychopathology* 15(1): 19–38.

O'Connor, T. G., J. M. Neiderhiser, et al. 1998. "Genetic contributions to continuity, change, and co-occurrence of antisocial and depressive symptoms in adolescence." *Journal of Child Psychology and Psychiatry* 39(3): 323–36.

O'Connor, T. G., M. Rutter, et al. 2000. "The effects of global severe privation on cognitive competence: Extension and longitudinal follow-up. English and Romanian Adoptees Study Team." *Child Development* 71(2): 376–90.

O'Dell, K. M., and K. Kaiser. 1997. "Sexual behaviour: Secrets and flies." *Current Biology* 7(6): R345–7.

O'Driscoll, K., and M. Foley. 1983. "Correlation of decrease in perinatal mortality and increase in Cesarean section rates." *Obstetrics and Gynecology* 61(1): 1–5.

O'Driscoll, K., M. Foley, et al. 1984. "Active management of labor as an alternative to Cesarean section for dystocia." *Obstetrics and Gynecology* 63(4): 485–90.

Offidani, C., F. Pomini, et al. 1995. "Cocaine during pregnancy: A critical review of the literature." *Minerva Ginecologica* 47(9): 381–90.

Oftedal, O. T. 2002. "The mammary gland and its origin during synapsid evolution." *Journal of Mammary Gland Biology and Neoplasia* 7(3): 225–52.

Ogawa, S., C. J. Krebs, et al. 2000. Genes participating in reproductive behaviors. In *Genetic influences on neural and behavioral functions,* ed. D. W. Pfaff et al. New York, CRC: 417–30.

Ohman, A. 2005. "The role of the amygdala in human fear: Automatic detection of threat." *Psychoneuroendocrinology* 30(10): 953–58.

Okami, P., R. Olmstead, et al. 1998. "Early childhood exposure to parental nudity and scenes of parental sexuality ('primal scenes'): An 18-year longitudinal study of outcome." *Archives of Sexual Behavior* 27(4): 361–84.

Okami, P., and T. K. Shackelford. 2001. "Human sex differences in sexual psychology and behavior." *Annual Review of Sex Research* 12: 186–241.

Okami, P., T. Weisner, et al. 2002. "Outcome correlates of parent-child bedsharing: An eighteen-year longitudinal study [see comment]." *Journal of Developmental and Behavioral Pediatrics* 23(4): 244–53.

Okun, M. S., D. Bowers, et al. 2004. "What's in a 'smile'? Intra-operative observations of contralateral smiles induced by deep brain stimulation." *Neurocase* 10(4): 271–79.

Olazabal, D., and L. Young. 2006. "Species and individual differences in juvenile female alloparental care are associated with oxytocin receptor density in the striatum and the lateral septum." *Hormones and Behavior* 49(5): 681–87.

Olesen, P. J., Z. Nagy, et al. 2003. "Combined analysis of DTI and fMRI data reveals a joint maturation of white and grey matter in a fronto-parietal network." *Cognitive Brain Research* 18(1): 48–57.

Oliver, J. 1993. "Intergenerational transmission of child abuse: Rates, research, and clinical implications." *American Journal of Psychiatry* 150(9): 1315–24.

Olweus, D., A. Mattsson, et al. 1980. "Testosterone, aggression, physical, and personality dimensions in normal adolescent males." *Psychosomatic Medicine* 42: 253–69.

Olweus, D., A. Mattsson, et al. 1988. "Circulating testosterone levels and aggression in adolescent males: A causal analysis." *Psychosomatic Medicine* 50: 261–72.

Onuma, Y., S. Takahashi, et al. 2002. "Conservation of Pax 6 function and upstream activation by Notch signaling in eye development of frogs and flies." *Proceedings of the National Academy of Sciences* 99(4): 2020–25.

Opie, I., and P. Opie 1987. *The lore and language of schoolchildren.* New York, Oxford Univ. Press.

Opitz, B., and A. D. Friederici. 2004. "Brain correlates of language learning: The neuronal dissociation of rule-based versus similarity-based learning." *Journal of Neuroscience* 24(39): 8436–40.

Oppong, C. 1973. *Growing up in Dagbon.* Accra, Ghana Publishing.

Orgeur, P. 1994. "Sexual play behavior in lambs androgenized in utero." *Physiology and Behavior* 57(1): 185–87.

Orme, N. 2001. *Medieval children.* New Haven, Yale Univ. Press.

Ors, M., E. Ryding, et al. 2005. "SPECT findings in children with specific language impairment." *Cortex* 41(3): 316–26.

Osborne, K. A., A. Robichon, et al. 1997. "Natural behavior polymorphism due to a cGMP-Dependent protein kinase of drosophila." *Science* 277: 834–36.

Osgood, C. 1951. *The Koreans and their culture.* New York, Ronald Press.

Osofsky, J. D., and E. J. O'Connell. 1977. "Patterning of newborn behavior in an urban population." *Child Development* 48: 532–36.

Oster, H. 1978. Facial expression and affect development. In *The development of affect,* ed. M. Lewis and L. A. Rosenblum. New York, Plenum: 43–75.

Otterbein, K. F. 1968. "Internal war: A cross-cultural study." *American Anthropologist* 70: 277–89.

Otterbein, K. F., and C. S. Otterbein. 1965. "An eye for an eye, a tooth for a tooth: A cross-cultural study of feuding." *American Anthropologist* 67: 1470–82.

Ottoni, E. B., B. D. de Resende, and P. Izar. 2005. "Watching the best nutcrackers: What capuchin monkeys (*Cebus apella*) know about others' tool-using skills." *Animal Cognition* 8(4): 215–19.

Packer, C., L. Herbst, et al. 1988. Reproductive success of lions. In *Reproductive success: Studies of individual variation in contrasting breeding systems,* ed. T. H. Clutton-Brock. Chicago, Univ. of Chicago Press: 363–83.

Packer, C., A. E. Pusey, et al. 2001. "Egalitarianism in female African lions." *Science* 293(5530): 690–93.

Padua, L., P. Caliandro, et al. 2007. "A novel approach to the measurement of motor conduction velocity using a single fibre EMG electrode." *Clinical Neurophysiology* 118(9): 1985–90.

Pagel, M. D., and P. H. Harvey. 1988. "The taxon-level problem in the evolution of mammalian brain size: Facts and artifacts." *American Naturalist* 132(3): 344–59.

———. 1989. "Taxonomic differences in the scaling of brain on body weight among mammals." *Science* 244: 1589–93.

———. 1990. "Diversity in the brain sizes of newborn mammals: Allometry, energetics, or life history tactics?" *BioScience* 40(2): 116–22.

Paige, K. E., and J. M. Paige. 1973. "The politics of birth practices: A strategic analysis." *American Sociological Review* 38: 663–77.

———. 1981. *The politics of reproductive ritual.* Berkeley, Univ. of California Press.

Pajer, K., R. Tabbah, et al. 2006. "Adrenal androgen and gonadal hormone levels in adolescent girls with conduct disorder." *Psychoneuroendocrinology* 31(10): 1245–56.

Pak, T. R., and R. J. Handa. 2008. Steroid hormone receptors and sex differences in behavior. In *Sex differences in the brain: From genes to behavior,* ed. J. B. Becker et al. New York, Oxford Univ. Press: 109–38.

Pal, S. K. 2005. "Parental care in free-ranging dogs, *Canis familiaris.*" *Applied Animal Behaviour Science* 90(1): 31–47.

———. 2008. "Maturation and development of social behaviour during early ontogeny in free-ranging dog puppies in West Bengal, India." *Applied Animal Behaviour Science* 111(1–2): 95–107.

Palmert, M. R., D. L. Hayden, et al. 2001. "The longitudinal study of adrenal maturation during gonadal suppression: Evidence that adrenarche is a gradual process [comment]." *Journal of Clinical Endocrinology and Metabolism* 86(9): 4536–42.

Palthe, T., and J. Van Hooff. 1975. "A case of the adoption of an infant chimpanzee by a suckling foster chimpanzee." *Primates* 16(2): 231–34.

Pan, B. A., R. Y. Perlmann, et al. 2000. Food for thought: Dinner table as a context for observing parent-child discourse. In *Methods for studying language production,* ed. L. Menn and N. Bernstein Ratner. Mahwah, NJ, Erlbaum: 205–24.

Panger, M. 2007. Tool use and cognition in primates. In *Primates in perspective,* ed. C. J. Campbell et al. New York, Oxford Univ. Press: 665–77.

Panksepp, J. 1998. *Affective neuroscience: The foundations of human and animal emotions.* New York, Oxford Univ. Press.

———. 2003. "Neuroscience. Feeling the pain of social loss [comment]." *Science* 302(5643): 237–39.

———. 2005. "Why does separation distress hurt? Comment on MacDonald and Leary 2005 [comment]." *Psychological Bulletin* 131(2): 224–30; author reply 237–40.

Panksepp, J., L. Normansell, et al. 1994. "Effects of neonatal decortication on the social play of juvenile rats." *Physiology and Behavior* 56(3): 429–43.

Panter-Brick, C. 2002. "Street children, human rights, and public health: A critique and future directions." *Annual Review of Anthropology* 31: 147–71.

Panter-Brick, C., and M. T. Smith, eds. 2000. *Abandoned children.* New York, Cambridge Univ. Press.

Panter-Brick, C., A. Todd, et al. 1996. "Growth status of homeless Nepali boys: Do they differ from rural and urban controls?" *Social Science and Medicine* 43(4): 441–51.

Paoli, T., E. Palagi, et al. 2006. "Perineal swelling, intermenstrual cycle, and female sexual behavior in bonobos (*Pan paniscus*)." *American Journal of Primatology* 68(4): 333–47.

Papez, J. W. 1937. "A proposed mechanism of emotion." *Archives of Neurological Psychiatry* 38: 725–43.

Papousek, M., and N. von Hofacker. 1998. "Persistent crying in early infancy: A nontrivial condition of risk for the developing mother-infant relationship." *Child: Care, Health and Development* 24(5): 395–424.

Parazzini, C., C. Baldoli, et al. 2002. "Terminal zones of myelination: MR evaluation of children aged 20–40 months." *AJNR: American Journal of Neuroradiology* 23(10): 1669–73.

Parent, A.-S., G. Teilmann, et al. 2003. "The timing of normal puberty and the age limits of sexual precocity: Variations around the world, secular trends, and changes after migration." *Endocrine Reviews* 24(5): 668–93.

Pares, J. M., A. Perez-Gonzalez, et al. 2000. "On the age of the hominid fossils at the Sima de los Huesos, Sierra de Atapuerca, Spain: Paleomagnetic evidence." *American Journal of Physical Anthropology* 111(4): 451–61.

Parker, A. J., B. G. Cumming, et al. 2000. Binocular neurons and the perception of depth. In *The new cognitive neurosciences,* ed. M. S. Gazzaniga et al. Cambridge, MA, MIT Press: 263.

Parker, G. A., and P. L. Schwagmeyer. 2005. "Male 'mixed' reproductive strategies in biparental species: Trivers was probably right, but why?" *American Naturalist* 165(1): 95–106.

Parker, S. T. 2000. *Homo erectus* infancy and childhood: The turning point in the evolution of behavioral development in hominids. In *Biology, brains, and behavior: The evolution of human development,* ed. S. T. Parker, J. Langer, and M. L. McKinney. Santa Fe, School of American Research Press: 279–318.

———. 2002. Evolutionary relationships between molar eruption and cognitive development in anthropoid primates. In *Human evolution through developmental change,* ed. N. Minugh-Purvis and K. J. McNamara. Baltimore, MD, Johns Hopkins University Press: 305–16.

Parmelee, A. H., Jr. 1964. "A critical evaluation of the Moro reflex." *Pediatrics* 33: 773–88.

Parrott, W. G., and H. Gleitman. 1989. "Infants' expectations in play: The joy of peek-a-boo." *Cognition and Emotion* 3(4): 291–311.

Parsons, C. H., and L. J. Rogers. 2000. "NMDA receptor antagonists extend the sensitive period for imprinting." *Physiology and Behavior* 68(5): 749–53.

Pastner, C. M. 1986. "The Westermarck hypothesis and first cousin marriage: The cultural modification of negative sexual imprinting." *Journal of Anthropological Research* 42: 573–86.

Pasupathi, M., and T. Hoyt. 2009. "The development of narrative identity in late adolescence and emergent adulthood: The continued importance of listeners." *Developmental Psychology* 45(2): 558–74.

Patterson, C. J. 2008. "Sexual orientation across the life span: Introduction to the special section." *Developmental Psychology* 44(1): 1–4.

———. 2009. "Lesbian and gay parents and their children: A social science perspective." *Nebraska Symposium on Motivation* 54: 141–82.

Patterson, C. J., S. Hurt, et al. 1998. "Families of the lesbian baby boom: Children's contact with grandparents and other adults." *American Journal of Orthopsychiatry* 68(3): 390–99.

Paul, A., and J. Kuester. 1996. "Infant handling by female Barbary macaques (*Macaca sylvanus*) at Affenberg Salem: Testing functional and evolutionary hypotheses." *Behavioral Ecology and Sociobiology* 39(2): 133–45.

Paul, R. A. 1989. "Psychoanalytic anthropology." *Annual Review of Anthropology* 18: 177–202.

Paus, T. 2005. "Mapping brain maturation and cognitive development during adolescence." *Trends in Cognitive Sciences* 9(2): 60–68.

Paus, T., D. L. Collins, et al. 2001. "Maturation of white matter in the human brain: A review of magnetic resonance studies." *Brain Research Bulletin* 54(3): 255–66.

Pavelka, M. S. M., L. M. Fedigan, et al. 2002. "Availability and adaptive value of reproductive and postreproductive Japanese macaque mothers and grandmothers." *Animal Behaviour* 64(3): 407–14.

Pavlov, I. P. 1927. *Conditioned reflexes.* New York, Oxford Univ. Press.

Paylor, R., Y. Zhao, et al. 2001. "Learning impairments and motor dysfunctions in adult Lhx5-deficient mice displaying hippocampal disorganization." *Physiology and Behavior* 73(5): 781–92.

Payne, R. B., W. L. Thompson, et al. 1981. "Local song traditions in indigo buntings: Cultural transmission of behavior patterns across generations." *Behaviour* 77: 199–221.

Peacock, N. R. 1985. *Time allocation, work and fertility among Efe Pygmy women of northeast Zaire.* PhD diss., Dept. of Anthropology, Harvard Univ., Cambridge, MA.

———. 1991. Rethinking the sexual division of labor: Reproduction and women's work among the Efe. In *Gender at the crossroads of knowledge: Feminist anthropology in the postmodern era,* ed. M. di Leonardo. Berkeley, Univ. of California Press: 339–60.

Peccei, J. S. 2001. "A critique of the grandmother hypotheses: Old and new." *American Journal of Human Biology* 13(4): 434–52.

Peck, M. D., and D. Priolo-Kapel. 2002. "Child abuse by burning: A review of the literature and an algorithm for medical investigations." *Journal of Trauma-Injury Infection and Critical Care* 53(5): 1013–22.

Pedersen, C., J. D. Caldwell, et al. 1994. "Oxytocin activates the postpartum onset of rat maternal behavior in the ventral tegmental area and medial preoptic areas." *Behavioral Neuroscience* 108: 1163–71.

Pedersen, C. A., J. A. Ascher, et al. 1982. "Oxytocin induces maternal behavior in virgin female rats." *Science* 216: 648–49.

Pedersen, J. M., S. E. Glickman, et al. 1990. "Sex differences in the play behavior of immature spotted hyenas *Crocuta-Crocuta.*" *Hormones and Behavior* 24(3): 403–20.

Peiper, A. 1963. *Cerebral function in infancy and childhood,* 3rd ed. New York, Consultants Bureau.

Pelissier, C. 1991. "The anthropology of teaching and learning." *Annual Review of Anthropology* 20(1): 75–95.

Pellegrini, A. D., D. Dupuis, and P. K. Smith. 2007. Play in evolution and development. In *Developmental Review* 27(2): 261–76.

Pellegrini, A. D., and K. Gustafson. 2004. *Boys' and girls' uses of objects for exploration, play, and tools in early childhood.* New York, Guilford.

Pellegrini, A. D., and P. K. Smith. 2004. *The nature of play: Great apes and humans.* New York, Guilford.

Pellis, S. M., and A. N. Iwaniuk. 1999. "The roles of phylogeny and sociality in the evolution of social play in muroid rodents." *Animal Behaviour* 58(2): 361–73.

———. 2000. "Comparative analyses of the role of postnatal development on the expression of play fighting." *Developmental Psychobiology* 36(2): 136–47.

———. 2002. "Brain system size and adult-adult play in primates: A comparative analysis of the roles of the non-visual neocortex and the amygdala." *Behavioural Brain Research* 134(1–2): 31–39.

Pellis, S. M., E. Hastings, et al. 2006. "The effects of orbital frontal cortex damage on the modulation of defensive responses by rats in playful and nonplayful social contexts." *Behavioral Neuroscience* 120(1): 72–84.

Pellis, S. M., and V. C. Pellis. 1998. "Play fighting of rats in comparative perspective: A schema for neurobehavioral analyses." *Neuroscience and Biobehavioral Reviews* 23(1): 87–101.

Pelucchi, B., J. F. Hay, et al. 2009. "Statistical learning in a natural language by 8-month-old infants." *Child Development* 80(3): 674–85.

Pennington, B. F., P. A. Filipek, et al. 2000. "A twin MRI study of size variations in human brain." *Journal of Cognitive Neuroscience* 12(1): 223–32.

Peper, J. S., R. M. Brouwer, et al. 2009. "Sex steroids and brain structure in pubertal boys and girls." *Psychoneuroendocrinology* 34(3): 332–42.

Pepler, D., D. Jiang, et al. 2008. "Developmental trajectories of bullying and associated factors." *Child Development* 79(2): 325–38.

Peredery, O., M. A. Persinger, et al. 1992. "Absence of maternal behavior in rats with lithium/pilocarpine seizure-induced brain damage: Support of MacLean's triune brain theory." *Physiology and Behavior* 52(4): 665–71.

Pereira, M. E., and L. A. Fairbanks, eds. 1993. *Juvenile primates: Life history, development, and behavior.* New York, Oxford Univ. Press.

Pereira, M. E., and S. R. Leigh. 2003. Modes of primate development. In *Primate life histories and socioecology,* ed. P. M. Kappeler and M. E. Pereira. Chicago, Univ. of Chicago Press: 149–76.

Petersen, A. C. 1987. The nature of biological-psychosocial interactions: The sample case of early adolescence. In *Biological-psychosocial interactions in early adolescence,* ed. R. M. Lerner and T. T. Foch. Hillsdale, NJ, Erlbaum: 35–61.

———. 1988. "Adolescent development." *Annual Review of Psychology* 39: 583–607.

Petersen, A. C., and L. Crockett. 1986. Pubertal development and its relation to cognitive and psychosocial development in adolescent girls: Implications for parenting. In *School-age pregnancy and childbearing: Biosocial dimensions,* ed. J. B. Lancaster and B. A. Hamburg. New York, Aldine: 147–75.

Peterson, J. T. 1978. *The ecology of social boundaries: Agta foragers of the Philippines.* Urbana, Univ. of Illinois Press.

Petitto, L. A. 1993. On the ontogenetic requirements for early language acquisition. In *Developmental neurocognition: Speech and face processing in the first year of life,* ed. B. de Boysson-Bardies et al. NATO ASI series D: Behavioural and social sciences. Dordrecht, Neth., Kluwer: vol. 69, 365–83.

———. 2000. On the biological foundations of human language. In *The signs of language revisited: An anthology to honor Ursula Bellugi and Edward Klima,* ed. K. Emmorey and H. Lane. Mahwah, NJ, Erlbaum: 449–73.

Petitto, L. A., and P. F. Marentette. 1991. "Babbling in the manual mode: Evidence for the ontogeny of language." *Science* 251(5000): 1493–96.

Petrosini, L., A. Graziano, et al. 2003. "Watch how to do it! New advances in learning by observation." *Brain Research—Brain Research Reviews* 42(3): 252–64.

Pezawas, L., A. Meyer-Lindenberg, et al. 2005. "5-HTTLPR polymorphism impacts human cingulate-amygdala interactions: A genetic susceptibility mechanism for depression [see comment]." *Nature Neuroscience* 8(6): 828–34.

Pfaff, D. W., W. H. Berrettini, et al., eds. 2000. *Genetic influences on neural and behavioral functions.* New York, CRC.

Pfeifer, J. H., M. Iacoboni, et al. 2008. "Mirroring others' emotions relates to empathy and interpersonal competence in children." *Neuroimage* 39(4): 2076–85.

Pfeiffer, J. E. 1978. *The emergence of man.* New York, Harper & Row.

———. 1982. *The emergence of culture.* New York, Harper & Row.

Pfennig, D. W., and P. W. Sherman. 1995. "Kin recognition." *Scientific American* 272: 98–103.

Pflueger, H. J., and R. Menzel. 1999. "Neuroethology, its roots and future." *Journal of Comparative Physiology: A Sensory Neural and Behavioral Physiology* 185(4): 389–92.

Phelps, E. A., M. R. Delgado, et al. 2004. "Extinction learning in humans: Role of the amygdala and vmPFC." *Neuron* 43(6): 897–905.

Phelps, E. A., and J. E. LeDoux. 2005. "Contributions of the amygdala to emotion processing: From animal models to human behavior." *Neuron* 48(2): 175–87.

Piaget, J. 1951. *Play, dreams and imitation in childhood.* London, Routledge and Kegan Paul.

———. 1952. *The origins of intelligence in children.* New York, Norton.

———. 1965. *The moral judgment of the child.* New York, Free Press.

———. 1970. Piaget's theory. In *Carmichael's manual of child psychology*, ed. P. Mussen. New York, Wiley.

Pianka, E. R. 1970. "On 'r' and 'K' selection." *American Naturalist* 104: 453–64.

———. 1988. *Evolutionary ecology*. New York, Harper & Row.

Piattelli-Palmarini, M. 1989. "Evolution, selection and cognition: From 'learning' to parameter setting in biology and in the study of language." *Cognition* 31: 1–44.

———. 2002. "The barest essentials." *Nature* 416(6877): 129.

Pichaud, F., and C. Desplan. 2002. "Pax genes and eye organogenesis." *Current Opinion in Genetics and Development* 12(4): 430–34.

Picton, T. W., and M. J. Taylor. 2007. "Electrophysiological evaluation of human brain development." *Developmental Neuropsychology* 31(3): 249–78.

Pillard, R. C., and J. M. Bailey. 1998. "Human sexual orientation has a heritable component." *Human Biology* 70(2): 347–65.

Pillemer, D. B., E. Koff, et al. 1987. "Flashbulb memories of menarche and adult menstrual distress." *Journal of Adolescence* 10: 187–99.

Pillow, B. H. 1999. "Children's understanding of inferential knowledge." *Journal of Genetic Psychology* 160(4): 419–28.

———. 2008. "Development of children's understanding of cognitive activities." *Journal of Genetic Psychology* 169(4): 297–321.

Pillow, B. H., and A. J. Henrichon. 1996. "There's more to the picture than meets the eye: Young children's difficulty understanding biased interpretation." *Child Development* 67(3): 803–19.

Pimenta, A. F., and P. Levitt. 2004. "Characterization of the genomic structure of the mouse limbic system-associated membrane protein (Lsamp) gene." *Genomics* 83(5): 790–801.

Pimenta, A. F., B. S. Reinoso, et al. 1996. "Expression of the mRNAs encoding the limbic system-associated membrane protein (LAMP): II. Fetal rat brain." *Journal of Comparative Neurology* 375(2): 289–302.

Pin, T., B. Eldridge, et al. 2007. "A review of the effects of sleep position, play position, and equipment use on motor development in infants [see comment]." *Developmental Medicine and Child Neurology* 49(11): 858–67.

Pine, D. S., J. Costello, and A. Masten. 2005. "Trauma, proximity, and developmental psychopathology: The effects of war and terrorism on children." *Neuropsychopharmacology* 30(10): 1781–92.

Pinheiro, P. S. 2006. *World report on violence against children*. New York, United Nations.

Pinker, S. 1994. *The language instinct: How the mind creates language*. New York, Morrow.

———. 1995. Why the child holded the baby rabbits: A case study in language acquisition. In *An invitation to cognitive science: Language*, vol. 1, ed. L. R. Gleitman and M. Liberman. Cambridge, MA, MIT Press: 107–33.

———. 1997. "Language as a psychological adaptation." *Ciba Foundation Symposium* 208: 162–72.

Pinker, S., and P. Bloom. 1990. "Natural language and natural selection." *Behavioral and Brain Sciences* 13(4): 707–84.

Pizzari, T., T. R. Birkhead, et al. 2006. "Debating sexual selection and mating strategies [comment]." *Science* 312(5774): 689–97; author reply 689–97.

Plattner, S., ed. 1989. *Economic anthropology*. Stanford, Stanford Univ. Press.

Plavcan, J. M., and C. P. van Schaik. 1997. "Interpreting hominid behavior on the basis of sexual dimorphism." *Journal of Human Evolution* 32(4): 345–74.

Pleck, J. H., and B. P. Masciadrelli. 2004. Paternal involvement by U.S. residential fathers: Levels, sources, and consequences. In *The role of the father in child development,* 4th ed., ed. M. E. Lamb. Hoboken, NJ, Wiley: 222–71.

Plomin, R. 2005. "Finding genes in child psychology and psychiatry: When are we going to be there?" *Journal of Child Psychology and Psychiatry and Allied Disciplines* 46(10): 1030–38.

Plomin, R., and D. Daniels. 1987. "Why are children in the same family so different from one another?" *Behavioral and Brain Sciences* 10: 1–60.

Plomin, R., R. N. Emde, et al. 1993. "Genetic change and continuity from fourteen to twenty months: The MacArthur Longitudinal Twin Study." *Child Development* 64: 1354–76.

Plomin, R., and D. W. Fulker. 1987. Behavioral genetics and development in early adolescence. In *Biological-psychosocial interactions in early adolescence,* ed. R. M. Lerner and T. T. Foch. Hillsdale, NJ, Erlbaum.

Plomin, R., and P. McGuffin. 2003. "Psychopathology in the postgenomic era." *Annual Review of Psychology* 54: 205–28.

Plomin, R., M. J. Owen, et al. 1994. "The genetic basis of complex human behaviors." *Science* 264: 1733–39.

Plomin, R., S. A. Petrill, et al. 1996. "What genetic research on intelligence tells us about the environment." *Journal of Biosocial Science* 28(4): 587–606.

Plomin, R., and D. C. Rowe. 1978. Genes, environment and development of temperament in young human twins. In *The development of behavior: Comparative and evolutionary aspects,* ed. G. M. Burhardt and M. Bekoff. New York, Garland.

Plomin, R., and M. Rutter. 1998. "Child development, molecular genetics, and what to do with genes once they are found." *Child Development* 69(4): 1223–42.

Ploog, D. 1988. "An outline of human neuroethology." *Human Neurobiology* 6: 227–38.

Plotsky, P. M., and M. J. Meaney. 1993. "Early, postnatal experience alters hypothalamic corticotropin-releasing factor (CRF) mRNA, median eminence CRF content and stress-induced release in adult rats." *Molecular Brain Research* 18: 195–200.

Plotsky, P. M., M. J. Owens, et al. 1998. "Psychoneuroendocrinology of depression. Hypothalamic-pituitary-adrenal axis." *Psychiatric Clinics of North America* 21(2): 293–307.

Pobiner, B. L., J. DeSilva, et al. 2007. "Taphonomic analysis of skeletal remains from chimpanzee hunts at Ngogo, Kibale National Park, Uganda." *Journal of Human Evolution* 52(6): 614–36.

Podos, J., S. K. Huber, et al. 2004. "Bird song: The interface of evolution and mechanism." *Annual Review of Ecology, Evolution, and Systematics* 35(1): 55–87.

Polanczyk, G., M. S. de Lima, et al. 2007. "The worldwide prevalence of ADHD: A systematic review and metaregression analysis [see comment]." *American Journal of Psychiatry* 164(6): 942–48.

Pollard, K. S., S. R. Salama, et al. 2006. "An RNA gene expressed during cortical development evolved rapidly in humans." *Nature* 443(7108): 167–72.

Pollock, G., and L. A. Dugatkin. 1992. "Reciprocity and the emergence of reputation." *Journal of Theoretical Biology* 159: 25–37.

Pollock, L. 1984. *Forgotten children: Parent-child relations from 1500 to 1900.* New York, Cambridge Univ. Press.

———. 1987. *A lasting relationship: Parents and children over three centuries.* New York, Cambridge Univ. Press.

Pope, H. G., Jr., E. M. Kouri, et al. 2000. "Effects of supraphysiologic doses of testosterone on mood and aggression in normal men: A randomized controlled trial." *Archives of General Psychiatry* 57(2): 133–40.

Popma, A., R. Vermeiren, et al. 2007. "Cortisol moderates the relationship between testosterone and aggression in delinquent male adolescents." *Biological Psychiatry* 61(3): 405–11.

Popp, J. L., and I. DeVore. 1979. Aggressive competition and social dominance theory: Synopsis. In *The great apes,* ed. D. A. Hamburg and E. McCown. Menlo Park, CA, Benjamin/Cummings.

Porges, S. W. 2003. "The polyvagal theory: Phylogenetic contributions to social behavior." *Physiology and Behavior* 79(3): 503–13.

———. 2007. "The polyvagal perspective." *Biological Psychology* 74(2): 116–43.

Porges, S. W., T. C. Riniolo, et al. 2003. "Heart rate and respiration in reptiles: Contrasts between a sit-and-wait predator and an intensive forager." *Brain and Cognition* 52(1): 88–96.

Portmann, A. 1945. "Die Ontogenese des Menschen als Problem der Evolutionsforschung." *Verhandlungen der Schweizerischen Naturforschenden Gesellschaft* 125: 44–53.

Potts, R. 1998. "Environmental hypotheses of hominin evolution." *American Journal of Physical Anthropology* (Suppl. 27): 93–136.

———. 2004. "Paleoenvironmental basis of cognitive evolution in great apes." *American Journal of Primatology* 62(3): 209–28.

Powdermaker, H. 1933. *Life in Lesu.* New York, Norton.

Powell, A., S. Shennan, et al. 2009. "Late Pleistocene demography and the appearance of modern human behavior [see comment]." *Science* 324(5932): 1298–1301.

Powell, R. A. 1993. *The fisher: Life history, ecology, and behavior,* 2nd ed. Minneapolis, Univ. of Minnesota Press.

Power, C., and I. Watts. 1997. "The woman with the zebra's penis: Gender, mutability and performance." *Journal of the Royal Anthropological Institute* 3(3): 537–60.

Prather, J. F., S. Peters, et al. 2008. "Precise auditory-vocal mirroring in neurons for learned vocal communication [see comment]." *Nature* 451(7176): 305–10.

Pratt, D. M., and V. H. Anderson. 1979. "Giraffe cow-calf relationships and social development of the calf in the Serengeti." *Zeitschrift für Tierpsychologie* 51: 233–51.

Praveen, E. P., A. K. Desai, et al. 2008. "Gender identity of children and young adults with 5alpha-reductase deficiency." *Journal of Pediatric Endocrinology* 21(2): 173–79.

Prechtl, H., and D. Beintema. 1964. *The neurological examination of the full-term newborn infant.* Little Club Clinics in Developmental Medicine No. 12. London, Heinemann.

Prechtl, H. F. R. 1965. "Problems of behavioral studies in the newborn infant." *Advances in the Study of Behavior* 1: 75–98.

———, ed. 1984. *Continuity of neural functions from prenatal to postnatal life.* London, Spastics International Medical Publications.

Premack, D. 1971. "Language in chimpanzee?" *Science* 172: 808–22.

Premack, D., and A. J. Premack. 1996. Why animals lack pedagogy and some cultures have more of it than others. In *The handbook of education and human development: New models of learning, teaching and schooling,* ed. D. R. Olson and N. Torrance. Malden, MA, Blackwell: xii, 804.

Prescott, J. 1985. "Play and the development of individual and social behavior in the American bison *Bison bison.*" *Biology of Behaviour* 10(3): 261–76.

Preston, S. H., and M. R. Haines. 1991. *Fatal years: Child mortality in late nineteenth-century America.* Princeton, Princeton Univ. Press.

Preuschoft, S. 1995. *"Laughter" and "smiling" in macaques: An evolutionary perspective.* Utrecht, Universiteit Utrecht, Faculteit Biologie.

Preuschoft, S., and J. A. R. A. M. van Hooff. 1995. "Homologizing primate facial displays: A critical review of methods." *Folia Primatologica* 65: 121–37.

———. 1997. The social function of "smile" and "laughter": Variations across primate species and societies. In *Nonverbal communication: Where nature meets culture,* ed. U. Segerstràle and P. Molnàr. Mahwah, NJ, Erlbaum: 171–89.

Preuss, T. M. 2001. The discovery of cerebral diversity: An unwelcome scientific revolution. In *Evolutionary anatomy of the primate cerebral cortex,* ed. D. Falk and K. R. Gibson. New York, Cambridge Univ. Press: 138–64.

———. 2009. "The human brain: Rewired and running hot." Lecture presented at the conference "The Evolution of Brain, Mind, and Culture," Emory University, Nov. 13, 2009.

Preuss, T. M., M. Caceres, et al. 2004. "Human brain evolution: Insights from microarrays." *Nature Reviews Genetics* 5(11): 850–60.

Propper, C., G. A. Moore, et al. 2008. "Gene-environment contributions to the development of infant vagal reactivity: The interaction of dopamine and maternal sensitivity." *Child Development* 79(5): 1377–94.

Prud'homme, B., N. Gompel, et al. 2007. "Emerging principles of regulatory evolution." *Proceedings of the National Academy of Sciences* 104(Suppl. 1): 8605–12.

Pruett, K. D. 1998. "Role of the father." *Pediatrics* 102(5 Suppl. E): 1253–61.

Pryce, C. R. 1996. Socialization, hormones, and the regulation of maternal behavior in nonhuman simian primates. In *Parental care: Evolution, mechanisms, and adaptive significance. Advances in the study of behavior,* vol. 25, ed. J. S. Rosenblatt and C. T. Snowdon. San Diego, Academic: 423–73.

Pujol, J., C. Soriano-Mas, et al. 2006. "Myelination of language-related areas in the developing brain [see comment]." *Neurology* 66(3): 339–43.

Purpura, D. P. 1979. Pathobiology of cortical neurons in metabolic and unclassified amentias. In *Congenital and acquired cognitive disorders,* ed. R. Katzman. New York, Raven: 55.

Purpura, D. P., and K. Susuki. 1976. "Distortion of neuronal geometry and formation of aberrant synapses in neuronal storage disease." *Brain Research* 116: 1–21.

Purves, D. 1988. *Body and brain: A trophic theory of neural connections.* Cambridge, MA, Harvard Univ. Press.

Purves, D., and J. W. Lichtman. 1980. "Elimination of synapses in the developing nervous system." *Science* 210: 153–57.

———. 1985. *Principles of neural development.* Sunderland, MA, Sinauer.

Putnam, N. H., T. Butts, et al. 2008. "The amphioxus genome and the evolution of the chordate karyotype." *Nature* 453: 1064–71.

Pysh, J. J., and G. M. Weiss. 1979. "Exercise during development induces an increase in purkinje cell dendritic tree size." *Science* 206: 230–32.

Quadagno, D. M., and J. Rockwell. 1972. "The effect of gonadal hormones in infancy on maternal behavior in the adult rat." *Hormones and Behavior* 3: 55–62.

Quinn, N. 2006a. "The self." *Anthropological Theory* 6(3): 362–84.

———. 2006b. "Universals of child rearing." *Anthropological Theory* 5(4): 477–516.

Quinn, N., and D. Holland. 1987. Culture and cognition. In *Cultural models of language and thought,* ed. D. Holland and N. Quinn. New York, Cambridge Univ. Press: 3–42.

Rabain-Jamin, J., A. E. Maynard, et al. 2003. "Implications of sibling caregiving for sibling relations and teaching interactions in two cultures." *Ethos* 31(2): 204–31.

Rabinovich, R., S. Gaudzinski-Windheuser, et al. 2008. "Systematic butchering of fallow deer (Dama) at the early middle Pleistocene Acheulian site of Gesher Benot Ya'aqov (Israel)." *Journal of Human Evolution* 54(1): 134–49.

Rabinowicz, T. 1979. The differentiate maturation of the human cerebral cortex. In *Human growth,* vol. 3, ed. F. Falkner and J. M. Tanner. New York, Plenum.

Radinsky, L. 1974. "The fossil evidence of anthropoid brain evolution." *American Journal of Physical Anthropology* 41(1): 15–28.

———. 1975. "Primate brain evolution." *American Scientist* 63: 656–63.

———. 1977. "Early primate brains: Fact and fiction." *Journal of Human Evolution* 6: 79–86.

Radinsky, L. B. 1970. The fossil evidence of prosimian brain evolution. In *The primate brain,* ed. C. R. Noback and W. Montagna. New York, Appleton-Century-Crofts.

———. 1987. *The evolution of vertebrate design.* Chicago, Univ. of Chicago Press.

Raff, R. A. 1996. *The shape of life: Genes, development, and the evolution of animal form.* Chicago, Univ. of Chicago Press.

———. 2007. "Written in stone: Fossils, genes and evo-devo." *Nature Reviews Genetics* 8(12): 911–20.

Raff, R. A., and E. C. Raff. 1987. *Development as an evolutionary process.* New York, Liss.

Rahman, Q. 2005. "The neurodevelopment of human sexual orientation." *Neuroscience and Biobehavioral Reviews* 29(7): 1057–66.

Rahman, Q., A. Collins, et al. 2008. "Maternal inheritance and familial fecundity factors in male homosexuality." *Archives of Sexual Behavior* 37(6): 962–69.

Rainey, W. E., B. R. Carr, et al. 2002. "Dissecting human adrenal androgen production." *Trends in Endocrinology and Metabolism* 13(6): 234–39.

Raisman, G., and P. M. Field. 1971. "Sexual dimorphism in the pre-optic area of the rat." *Science* 173: 731–33.

———. 1973. "Sexual dimorphism in the neuropil of the preoptic area of the rat and its dependence on neonatal androgen." *Brain Research* 54: 1–29.

Rajpurohit, L. S., and V. Sommer. 1993. Juvenile male emigration from natal one-male troops in hanuman langurs. In *Juvenile Primates: Life history, development, and behavior,* ed. M. E. Pereira and L. A. Fairbanks. New York, Oxford Univ. Press: 86–103.

Rakic, P. 1972. "Mode of cell migration to the superficial layers of fetal monkey neocortex." *Journal of Comparative Neurology* 145(1): 61–83.

———. 1985. "Limits of neurogenesis in primates." *Science* 227: 1054–56.

———. 1988. "Specification of cerebral cortical areas." *Science* 241: 170–76.

———. 2000. Section I: Development (with contributions by 13 authors). In *The new cognitive neurosciences,* ed. M. S. Gazzaniga. Cambridge, MA, MIT Press: 5–115.

Rakic, P., and V. S. Caviness, Jr. 1995. "Cortical development: View from neurological mutants two decades later." *Neuron* 14(6): 1101–4.

Rakic, P., and D. R. Kornack. 2001. Neocortical expansion and elaboration during primate evolution: A view from neuroembryology. In *Evolutionary anatomy of the primate cerebral cortex,* ed. D. Falk and K. R. Gibson. Cambridge, Cambridge Univ. Press: 30–56.

Ralls, K. 1976. "Mammals in which females are larger than males." *Quarterly Review of Biology* 51: 245–76.

Ralls, K., B. Cypher, et al. 2007. "Social monogamy in kit foxes: Formation, association, duration, and dissolution of mated pairs." *Journal of Mammalogy* 88(6): 1439–46.

Ramsay, M., and P. M. Fitzhardinge. 1977. "A comparative study of two developmental scales: The Bayley and the Griffiths." *Early Human Development* 1(2): 151–57.

Rankin, C. H. 2002. "From gene to identified neuron to behaviour in *Caenorhabditis elegans*." *Nature Reviews Genetics* 3(8): 622–30.

Rannala, B., and Z. Yang. 2007. "Inferring speciation times under an episodic molecular clock." *Systematic Biology* 56(3): 453–66.

Ransel, D. L. 1988. *Mothers of misery: Child abandonment in Russia*. Princeton, Princeton Univ. Press.

Ransom, T. W., and B. S. Ransom. 1971. "Adult male-infant relations among baboons *(Papio anubis)*." *Folia Primatologica* 16: 179–95.

Rasoloharijaona, S., B. Rakotosamimanana, et al. 2003. "Pair-specific usage of sleeping sites and their implications for social organization in a nocturnal Malagasy primate, the Milne Edwards' sportive lemur *(Lepilemur edwardsi)*." *American Journal of Physical Anthropology* 122(3): 251–58.

Rauschecker, J. P., and P. Marler. 1987. *Imprinting and cortical plasticity*. New York, Wiley.

Ray, O. S., and R. J. Barrett. 1975. "Behavioral, pharmacological and biochemical analysis of genetic differences in rats." *Behavioral Biology* 15: 391–417.

Raymond, J. L., S. G. Lisberger, et al. 1996. "The cerebellum: A neuronal learning machine?" *Science* 272: 1126–31.

Reader, S. M., and K. N. Laland. 2002. "Social intelligence, innovation, and enhanced brain size in primates." *Proceedings of the National Academy of Sciences* 99(7): 4436–41.

Rebelsky, F., and S. Hanks. 1971. "Fathers' verbal interaction with infants in the first three months of life." *Child Development* 42: 63–68.

Reburn, C. J., and K. E. Wynne-Edwards. 1999. "Hormonal changes in males of a naturally biparental and a uniparental mammal." *Hormones and Behavior* 35(2): 163–76.

Reddy, G. 2005. *With respect to sex: Negotiating hijra identity in South India*. Chicago, Univ. of Chicago Press.

Reddy, V. 2000. "Coyness in early infancy." *Developmental Science* 3(2): 186–92.

Redican, W. K., and G. Mitchell. 1973. "A longitudinal study of parental behavior in adult male rhesus monkeys: 1. Observations on the first dyad." *Developmental Psychology* 8: 135–36.

Reep, R. L., B. L. Finlay, et al. 2007. "The limbic system in mammalian brain evolution." *Brain, Behavior, and Evolution* 70(1): 57–70.

Reeve, H. K. 1998. Acting for the good of others: Kinship and reciprocity with some new twists. In *Handbook of evolutionary psychology*, ed. C. Crawford and D. L. Krebs. Mahwah, NJ, Erlbaum: 43–83.

Reichard, U. H. 2003. Monogamy: Past and present. In *Monogamy: Mating strategies and partnerships in birds, humans, and other mammals*, ed. U. H. Reichard and C. Boesch. Cambridge, Cambridge Univ. Press: 3–25.

Reichard, U. H., and C. Boesch. 2003. *Monogamy: Mating strategies and partnerships in birds, humans, and other mammals*. Cambridge, Cambridge Univ. Press.

Reichert, H. 2002. "Conserved genetic mechanisms for embryonic brain patterning." *International Journal of Developmental Biology* 46(1): 81–87.

———. 2009. "Evolutionary conservation of mechanisms for neural regionalization, proliferation and interconnection in brain development." *Biology Letters* 5(1): 112–16.

Reichert, H., and A. Simeone. 1999. "Conserved usage of gap and homeotic genes in patterning the CNS." *Current Opinion in Neurobiology* 9(5): 589–95.

Reiner, W. G. 2005. Gender identity and sex-of-rearing in children with disorders of sexual differentiation. *Journal of Pediatric Endocrinology* 18(6): 549–53.

Reiner, W. G., and J. P. Gearhart. 2004. "Discordant sexual identity in some genetic males with cloacal exstrophy assigned to female sex at birth." *New England Journal of Medicine* 350(4): 333–41.

Reinisch, J. M. 1974. "Fetal hormones, the brain, and sex differences: A heuristic integrative review of the literature." *Archives of Sexual Behavior* 3: 51–90.

———. 1981. "Prenatal exposure to synthetic progestins increases potential for aggression in humans." *Science* 211: 1171–73.

Reinisch, J. M., and S. A. Sanders. 1992. Prenatal hormonal contributions to sex differences in human cognitive and personality development. In *Handbook of behavioral neurobiology*, vol. 11: *Sexual differentiation*, ed. A. A. Gerall, H. Moltz, and I. L. Ward. New York, Plenum: 221–43.

Reiss, A. L., S. Eliez, et al. 2000. "Brain imaging in neurogenetic conditions: Realizing the potential of behavioral neurogenetics research." *Mental Retardation and Developmental Disabilities Research Reviews* 6(3): 186–97.

Reiss, D., E. M. Hetherington, et al. 1995. "Genetic questions for environmental studies. Differential parenting and psychopathology in adolescence." *Archives of General Psychiatry* 52(11): 925–36.

Reiss, D., J. M. Neiderhiser, et al. 2000. *The relationship code: Deciphering genetic and social influences on adolescent development*. Cambridge, MA, Harvard Univ. Press.

Reite, M. 1987. "Infant abuse and neglect: Lessons from the primate laboratory." *Child Abuse and Neglect* 11(3): 347–55.

Reite, M., and J. P. Capitanio. 1985. On the nature of social separation and social attachment. In *The psychobiology of attachment and separation*, ed. M. Reite and T. Field. New York, Academic: 223–55.

Reite, M., I. C. Kaufman, et al. 1974. "Depression in infant monkeys: Physiological correlates." *Psychosomatic Medicine* 36(4): 363–67.

Reite, M., R. Short, et al. 1978. "Physiological correlates of maternal separation in surrogate-reared infants: A study in altered attachment bonds." *Developmental Psychobiology* 11: 427–35.

———. 1981. "Attachment, loss, and depression." *Journal of Child Psychology and Psychiatry* 22: 141–69.

Reiter, E. O. 1987. Normal and abnormal sexual development and puberty. In *Clinical endocrinology: An illustrated text*, ed. G. M. Besser and A. G. Cudworth. Philadelphia, Lippincott: 12.1–12.28.

Remer, T., and F. Manz. 2001. "The midgrowth spurt in healthy children is not caused by adrenarche." *Journal of Clinical Endocrinology and Metabolism* 86(9): 4183–86.

Rendell, L., and H. Whitehead. 2001. "Culture in whales and dolphins." *Behavioral and Brain Sciences* 24(2): 309–82.

Reno, P. L., R. S. Meindl, et al. 2003. "Sexual dimorphism in *Australopithecus afarensis* was similar to that of modern humans [see comment]." *Proceedings of the National Academy of Sciences* 100(16): 9404–9.

Reznick, D., M. J. Bryant, et al. 2002. "r- and K-selection revisited: The role of population regulation in life-history evolution." *Ecology* 83(6): 1509–20.

Ricciuti, H. N. 1974. Fear and the development of social attachments in the first year of life. In *The origins of fear*, ed. M. Lewis and L. A. Rosenblum. New York, Wiley.

Rice, D., and S. Barone, Jr. 2000. "Critical periods of vulnerability for the developing nervous system: Evidence from humans and animal models." *Environmental Health Perspectives* 108(Suppl. 3): 511–33.

Rich, A. 1978. *The dream of a common language.* New York, Norton.

Richard, A. F. 1978. *Behavioral variation: Case study of a Malagasy lemur.* Lewisburg, PA, Bucknell Univ. Press.

Richards, M. 1965. Some effects of experience on maternal behavior in rodents. In *Determinants of infant behavior,* vol. 4, ed. B. Foss. London, Methuen.

Richards, M., and A. C. Petersen. 1987. Biological theoretical models of adolescent development. In *Handbook of adolescent psychology,* ed. V. B. Van Hasselt and M. Hersen. New York, Pergamon: 34–52.

Richards, P. G., G. E. Bertocci, et al. 2006. "Shaken baby syndrome." *Archives of Disease in Childhood* 91(3): 205–6.

Richardson, D. S., T. Burke, et al. 2002. "Direct benefits and the evolution of female-biased cooperative breeding in Seychelles warblers." *Evolution* 56(11): 2313–21.

Richardson, J., and A. L. Kroeber. 1940. *Three centuries of women's dress fashions: A quantitative analysis.* Berkeley, Univ. of California Press.

Richmond, B. G., and D. S. Strait. 2000. "Evidence that humans evolved from a knuckle-walking ancestor [see comment]." *Nature* 404(6776): 382–85.

Ricklefs, R. E. 2004. "The cognitive face of avian life histories—The 2003 Margaret Morse Nice Lecture." *Wilson Bulletin* 116(2): 119–33.

Ridley, M., ed. 1997. *Evolution.* Oxford, Oxford Univ. Press.

Rieger, G., J. A. W. Linsenmeier, et al. 2008. "Sexual orientation and childhood gender nonconformity: Evidence from home videos." *Developmental Psychology* 44(1): 46–58.

———. 2009. "Childhood gender nonconformity remains a robust and neutral correlate of sexual orientation: Reply to Hegarty 2009." *Developmental Psychology* 45(4): 901–3.

Rierdan, J., E. Koff, et al. 1989a. "A longitudinal analysis of body image as a predictor of the onset and persistence of adolescent girls' depression." *Journal of Early Adolescence* 9(4): 454–66.

———. 1989b. "Timing of menarche, preparation, and initial menstrual experience: Replication and further analyses in a prospective study." *Journal of Youth and Adolescence* 18(5): 413–26.

Rightmire, G. P. 2004. "Brain size and encephalization in early to Mid-Pleistocene Homo." *American Journal of Physical Anthropology* 124(2): 109–23.

Rilling, J. K. 2006. "Human and nonhuman primate brains: Are they allometrically scaled versions of the same design?" *Evolutionary Anthropology: Issues, News, and Reviews* 15(2): 65–77.

———. 2008. "Neuroscientific approaches and applications within anthropology." *American Journal of Physical Anthropology* (Suppl. 47): 2–32.

Rilling, J. K., J. E. Dagenais, et al. 2008. "Social cognitive neural networks during in-group and out-group interactions." *Neuroimage* 41(4): 1447–61.

Rilling, J. K., M. F. Glasser, et al. 2008. "The evolution of the arcuate fasciculus revealed with comparative DTI [see comment]." *Nature Neuroscience* 11(4): 426–428.

Rilling, J. K., and T. R. Insel. 1998. "Evolution of the cerebellum in primates: Differences in relative volume among monkeys, apes and humans." *Brain, Behavior, and Evolution* 52(6): 308–14.

———. 1999. "The primate neocortex in comparative perspective using magnetic resonance imaging." *Journal of Human Evolution* 37: 191–223.

Rilling, J. K., and R. A. Seligman. 2002. "A quantitative morphometric comparative analysis of the primate temporal lobe." *Journal of Human Evolution* 42(5): 505–33.

Rilling, J. K., J. T. Winslow, et al. 2001. "Neural correlates of maternal separation in rhesus monkeys." *Biological Psychiatry* 49(2): 146–57.

Rinn, W. E. 1984. "The neuropsychology of facial expression: A review of the neurological and psychological mechanisms for producing facial expressions." *Psychological Bulletin* 95: 52–77.

Risch, N., R. Herrell, et al. 2009. "Interaction between the serotonin transporter gene (5-HTTLPR), stressful life events, and risk of depression: A meta-analysis." *Journal of the American Medical Association* 301(23): 2462–71.

Riska, B., and W. B. Atchley. 1985. "Genetics of growth predict patterns of brain-size evolution." *Science* 229: 668–71.

Ritchie, J. M. 1984. Pathophysiology of conduction in demyelinated nerve fibers. In *Myelin,* 2nd ed., ed. P. Morell. New York, Plenum.

Rizzolatti, G., and L. Craighero. 2004. "The mirror-neuron system." *Annual Review of Neuroscience* 27(1): 169–92.

Rizzuto, A.-M. 1981. *The birth of the living god: A psychoanalytic study.* Chicago, Univ. of Chicago Press.

Roberts, R., K. T. Jenkins, et al. 2001. "Prolactin levels are elevated after infant carrying in parentally inexperienced common marmosets." *Physiology and Behavior* 72(5): 713–20.

Robinson, P., and H. Tajfel, eds. 1997. *Social groups and identities: Developing the legacy of Henri Tajfel.* International Series in Social Psychology. London, Butterworth-Heinemann.

Robson, K. S. 1967. "The role of eye-to-eye contact in maternal-infant attachment." *Journal of Pediatrics* 77: 976–85.

Robson, K. S., and H. A. Moss. 1970. "Patterns and determinants of maternal attachment." *Journal of Pediatrics* 77(6): 976–85.

Robson, S. L., C. P. van Schaik, et al. 2006. The derived features of human life history. In *Evolution of human life history,* ed. K. Hawkes and R. R. Paine. Santa Fe, School of American Research Press: 45–93.

Rochat, P. 2007. "Intentional action arises from early reciprocal exchanges." *Acta Psychologica* 124(1): 8–25.

———, ed. 1999. *Early social cognition: Understanding others in the first months of life.* Mahwah, NJ, Erlbaum.

Rochat, P., M. D. Dias, et al. 2009. "Fairness in distributive justice by 3- and 5-year-olds across seven cultures." *Journal of Cross-Cultural Psychology* 40(3): 416–42.

Roch-Lecours, A. R. 1975. Myelogenetic correlates of the development of speech and language. In *Foundations of language development: A multidisciplinary approach,* ed. E. H. Lenneberg and E. Lenneberg. New York, Academic: 121–35.

Rodgers, J. L. 2001. "What causes birth order-intelligence patterns? The admixture hypothesis, revived." *American Psychologist* 56(6–7): 505–10.

Rodman, P. S., and H. M. McHenry. 1980. "Bioenergetics of hominid bipedalism." *American Journal of Physical Anthropology* 52: 103–6.

Rodman, P. S., and J. C. Mitani. 1987. Orangutans: Sexual dimorphism in a solitary species. In *Primate societies,* ed. B. B. Smuts et al. Chicago, Univ. of Chicago Press: 146–54.

Rodriguez, F., J. C. Lopez, et al. 2002. "Spatial memory and hippocampal pallium through vertebrate evolution: Insights from reptiles and teleost fish." *Brain Research Bulletin* 57(3–4): 499–503.

Roe, A., and G. G. Simpson. 1958. *Behavior and evolution.* New Haven, Yale Univ. Press.

Roff, D. A. 2007. "Contributions of genomics to life-history theory." *Nature Reviews Genetics* 8(2): 116–25.

Rogers, D. A., G. E. McClearn, et al. 1963. "Alcohol preference as a function of its caloric utility in mice." *Journal of Comparative and Physiological Psychology* 56: 666–72.

Rogoff, B. 1990. *Apprenticeship in thinking: Cognitive development in social context.* New York, Oxford Univ. Press.

———. 1996. Developmental transitions in children's participation in sociocultural activities. In *The five to seven year shift: The age of reason and responsibility,* ed. A. J. Sameroff and M. M. Haith. Chicago, Univ. of Chicago Press: 273–94.

———. 1997. Evaluating development in the process of participation: Theory, methods, and practice building on each other. In *Change and development: Issues of theory, method, and application,* ed. E. Amsel and K. A. Renninger. Mahwah, NJ, Erlbaum: 265–85.

Rogoff, B., M. J. Sellers, et al. 1975. "Age of assignment of roles and responsibilities to children: A cross-cultural survey." *Human Development* 18: 353–69.

Rohner, R. P. 1975. *They love me, they love me not: A worldwide study of the effects of parental acceptance and rejection.* New Haven, HRAF.

Rohner, R. P., A. Khaleque, et al. 2005. "Parental acceptance-rejection: Theory, methods, cross-cultural evidence, and implications." *Ethos* 33(3): 299–334.

Romeo, R. D. 2003. "Puberty: A period of both organizational and activational effects of steroid hormones on neurobehavioural development." *Journal of Neuroendocrinology* 15(12): 1185–92.

Romer, A. 1958. Phylogeny and behavior with special reference to vertebrate evolution. In *Behavior and evolution,* ed. A. Roe and G. G. Simpson. New Haven, Yale Univ. Press: 48–75.

Rooks, J. P., N. L. Weatherby, et al. 1989. "Outcomes of care in birth centers. The National Birth Center Study [see comment]." *New England Journal of Medicine* 321(26): 1804–11.

Roopnarine, J. L., H. N. Fouts, et al. 2005. "Mothers' and fathers' behaviors toward their 3- to 4-month-old infants in lower, middle, and upper socioeconomic African American families." *Developmental Psychology* 41(5): 723–32.

Rorke, L. B., and H. E. Riggs. 1969. *Myelination of the brain in the newborn.* Philadelphia, Lippincott.

Rosaldo, R. 1993. *Culture and truth: The remaking of social analysis.* Boston, Beacon.

Roscoe, W. 1991. *The Zuni man-woman.* Albuquerque, Univ. of New Mexico Press.

———. 1994. How to become a Berdache: Toward a unified analysis of gender diversity. In *Third sex, third gender: Beyond sexual dimorphism in culture and history,* ed. G. Herdt. New York, Zone: 329–72.

Rose, A. J., and K. D. Rudolph. 2006. "A review of sex differences in peer relationship processes: Potential trade-offs for the emotional and behavioral development of girls and boys." *Psychological Bulletin* 132(1): 98–131.

Rose, D. H. 1976. Dentate gyrus granule cells and cognitive development: Explorations in the substrates of behavioral change. Harvard Univ.

Rose, M. B. 1991. *Evolutionary biology of aging.* New York, Oxford Univ. Press.

Rose, R. J., M. Koskenvuo, et al. 1988. "Shared genes, shared experiences, and similarity of personality: Data from 14,288 adult Finnish co-twins." *Journal of Personality and Social Psychology* 54(1): 161–71.

Rose, R. M., P. G. Bourne, et al. 1969. "Androgen responses to stress: II. Excretion of testosterone, epitestosterone, androsterone, and etiocholanolone during basic combat training and under threat of attack." *Psychosomatic Medicine* 31: 418–36.

Rosen, G. D., G. F. Sherman, et al. 1996. "Birthdates of neurons in induced microgyria." *Brain Research* 727(1–2): 71–78.

Rosen, J., and F. Hart. 1963. "Effects of early social isolation upon adult timidity and dominance in Peromyscus." *Psychological Reports* 13: 47–50.

Rosenberg, K., and W. Trevathan. 2002. "Birth, obstetrics and human evolution." *BJOG: An International Journal of Obstetrics and Gynaecology* 109(11): 1199–1206.

Rosenberg, K. R. 1992. "The evolution of modern human childbirth." *Yearbook of Physical Anthropology* 35: 89–124.

Rosenberg, K. R., and W. R. Trevathan. 2001. "The evolution of human birth." *Scientific American* 285(5): 72–77.

Rosenblatt, J. S. 1967. "Nonhormonal basis of maternal behavior in the rat." *Science* 156: 1512–14.

———. 2003. "Outline of the evolution of behavioral and nonbehavioral patterns of parental care among the vertebrates: Critical characteristics of mammalian and avian parental behavior." *Scandinavian Journal of Psychology* 44(3): 265–71.

Rosenblatt, J. S., and C. T. Snowdon, eds. 1996. *Parental care: Evolution, mechanisms, and adaptive significance. Advances in the study of behavior.* San Diego, Academic.

Rosenblum, L. A. 1971. Infant attachment in monkeys. In *The origins of human social relations,* ed. R. Schaffer. New York, Academic.

———. 1978. Affective maturation and the mother-infant relationship. In *The development of affect,* ed. M. Lewis and L. A. Rosenblum. New York, Plenum: 275–92.

———. 1998. Effective mothering in a familial context: A nonhuman primate perspective. In *Families, risk, and competence,* ed. M. Lewis and F. Candice. Mahwah, NJ, Erlbaum: 71–87.

Rosenblum, L. A., and S. Alpert. 1974. Fear of strangers and specificity of attachment in monkeys. In *The origins of fear,* ed. M. Lewis and L. A. Rosenblum. New York, Wiley: 165–93.

Rosenblum, L. A., and M. Andrews. 1995. Primate developmental models of stress. In *Chronic diseases: Perspectives in behavioral medicine,* ed. M. Stein and A. Baum. Mahwah, NJ, Erlbaum: 97–113.

Rosenblum, L. A., and H. F. Harlow. 1963. "Generalization of affectional responses in rhesus monkeys." *Perceptual and motor skills* 16: 561–64.

Rosenblum, L. A., and I. C. Kaufman. 1968. "Variations in infant development and response to maternal loss in monkeys." *American Journal of Orthopsychiatry* 38(3): 418–26.

———. 1971. Laboratory observations of early mother-infant relations in pigtail and bonnet macaques. In *Social communication among primates,* ed. S. A. Altmann. Chicago, Univ. of Chicago Press: 33–41.

Rosenblum, L. A., and G. S. Paully. 1987. "Primate models of separation-induced depression." *Psychiatric Clinics of North America* 10(3): 437–47.

Rosenfeld, A. A., A. O. R. Wenegrat, et al. 1982. "Sleeping patterns in upper-middle-class families when the child awakens ill or frightened." *Archives of General Psychiatry* 39(August): 943–47.

Rosenzweig, M. 1966. "Environmental complexity, cerebral change, and behavior." *American Psychologist* 21: 321–32.

Rosenzweig, M. R., E. L. Bennett, et al. 1972. "Brain changes in response to experience." *Scientific American* 226(2): 22–29.

Ross, C., and A. MacLarnon. 2000. "The evolution of non-maternal care in anthropoid primates: A test of the hypotheses." *Folia Primatologica* 71(1–2): 93–113.

Ross, C., and G. Regan. 2000. "Allocare, predation risk, social structure and natal coat colour in anthropoid primates." *Folia Primatologica* 71(1–2): 67–76.

Ross, E. D. 1993. "Nonverbal aspects of language." *Behavioral Neurology* 11(1): 9–23.

Ross, E. D., and M.-M. Mesulam. 1979. "Dominant language functions of the right hemisphere? Prosody and emotional gesturing." *Archives of Neurology* 36: 144–48.

Ross, M. D., M. J. Owren, et al. 2009. "Reconstructing the evolution of laughter in great apes and humans." *Current Biology* 19(13): 1106–11.

Rothbart, M. K., and E. E. Maccoby. 1966. "Parents' differential reactions to sons and daughters." *Journal of Personality and Social Psychology* 4: 237–43.

Rovee, C. K., and D. T. Rovee. 1969. "Conjugate reinforcement of infant exploratory behavior." *Journal of Experimental Child Psychology* 8: 33–39.

Rovee-Collier, C., L. Earley, et al. 1989. "Ontogeny of early event memory: III. Attentional determinants of retrieval at 2 and 3 months." *Infant Behavior and Development* 12: 147–61.

Rovee-Collier, C. K., and J. B. Capatides. 1979. "Positive behavioral contrast in 3-month-old infants on multiple conjugate reinforcement schedules." *Journal of the Experimental Analysis of Behavior* 32(1): 15–27.

Rovee-Collier, C. K., and L. P. Lipsitt. 1982. Learning, adaptation, and memory in the newborn. In *Psychobiology of the human newborn,* ed. P. Stratton. New York, Wiley: 147–90.

Rowe, D. C. 1994. *The limits of family influence: Genes, experience, and behavior.* New York, Guilford.

Rowe, R., B. Maughan, et al. 2004. "Testosterone, antisocial behavior, and social dominance in boys: Pubertal development and biosocial interaction." *Biological Psychiatry* 55(5): 546–52.

Rubenstein, D. R., and I. J. Lovette. 2009. "Reproductive skew and selection on female ornamentation in social species." *Nature* 462: 786–89.

Rubin, J. Z., F. J. Provenzano, et al. 1974. "The eye of the beholder: Parents' views on the sex of newborns." *American Journal of Orthopsychiatry* 44: 512–19.

Ruddle, F. H., J. L. Bartels, et al. 1994. "Evolution of *Hox* genes." *Annual Review of Genetics* 28: 423–42.

Ruelle, D. 1991. *Chance and chaos.* Princeton, Princeton Univ. Press.

———. 2001. "Here be no dragons." *Nature* 411(6833): 27.

Rumbaugh, D., E. von Glaserfeld, et al. 1974. "Lana (chimpanzee) learning language: A progress report." *Brain and Language* 1: 205–12.

Rumbaugh, D. M. 1997. Competence, cortex, and primate models. In *Development of the prefrontal cortex: Evolution, neurobiology, and behavior,* ed. N. A. Krasnegor, G. R. Lyon, and P. Goldman-Rakic. Baltimore, Brookes: 117–39.

Rumbaugh, D. M., and T. V. Gill. 1974. *Language, apes, and the apple which-is orange, please.* Symposium of the Fifth Congress of the International Primate Society.

Rumbaugh, D. M., T. V. Gill, et al. 1975. "Conversations with a chimpanzee in a computer-controlled environment." *Biological Psychiatry* 10: 627–41.

Rumbaugh, D. M., and J. L. Pate. 1984. The evolution of cognition in primates: A comparative perspective. In *Animal cognition,* ed. H. L. Roitblat, T. G. Bever, and H. S. Terrace. Hillsdale, NJ, Erlbaum.

Rumbaugh, D. M., and D. A. Washburn. 2003. *Intelligence of apes and other rational beings.* New Haven, Yale Univ. Press.

Rumbaugh, D. M., D. A. Washburn, et al. 1996. Respondents, operants, and emergents: Toward an integrated perspective on behavior. In *Learning as a self-organizing process*, ed. K. Pribram and J. King. Hillsdale, NJ, Erlbaum: 57–73.

———. 1998. Discrimination learning set and transfer. In *Comparative psychology: A handbook*, ed. G. Greenberg and M. M. Haraway. New York, Garland: 562–65.

Russell, C. M. 1999. "A meta-analysis of published research on the effects of nonmaternal care on child development." *Dissertation Abstracts International Section A: Humanities and Social Sciences*, vol. 59(9-A), Mar. 1999, pp. 3362. Calgary, Canada, University of Calgary.

Russell, G., and C. P. Hwang. 2004. The impact of workplace practices on father involvement. In *The role of the father in child development*, 4th ed., ed. M. E. Lamb. Hoboken, NJ, Wiley: 476–503.

Russon, A. E., and B. M. F. Galdikas. 1993. "Imitation in free-ranging rehabilitant orangutans (*Pongo pygmaeus*)." *Journal of Comparative Psychology* 107(2): 147–61.

———. 1995. Imitation and tool use in rehabilitant orangutans. In *The neglected ape*, ed. R. D. Nadler et al. New York, Plenum: 191–97.

Rutberg, A. T. 1983. "The evolution of monogamy in primates." *Journal of Theoretical Biology* 104: 93–112.

Rutter, M. 1990. Changing patterns of psychiatric disorders during adolescence. In *Adolescence and puberty*, ed. J. Bancroft and J. M. Reinisch. New York, Oxford Univ. Press: 124–45.

———. 1998. "Developmental catch-up, and deficit, following adoption after severe global early privation. English and Romanian Adoptees (ERA) Study Team." *Journal of Child Psychology and Psychiatry* 39(4): 465–76.

———. 2006. *Genes and behavior: Nature-nurture interplay explained*, Malden, MA, Blackwell.

———. 2007. "Gene-environment interdependence." *Developmental Science* 10(1): 12–18.

Rutter, M., T. G. O'Connor, et al. 2004. "Are there biological programming effects for psychological development? Findings from a study of Romanian adoptees." *Developmental Psychology* 40(1): 81–94.

Rutter, M., and R. Plomin. 1997. "Opportunities for psychiatry from genetic findings [see comments]." *British Journal of Psychiatry* 171: 209–19.

Rutter, M. L., J. M. Kreppner, et al. 2001. "Specificity and heterogeneity in children's responses to profound institutional privation." *British Journal of Psychiatry* 179: 97–103.

Ryang, S. 2003. "Comment on 'Maternal Sentiments: How Strong Are They?' by Arthur P. Wolf." *Current Anthropology* 44(Suppl.): S44–45.

Sacher, G. A. 1982. The role of brain maturation in the evolution of the primates. In *Primate brain evolution*, ed. E. Armstrong and D. Falk. New York, Plenum: 97–112.

Sacher, G. A., and E. F. Staffeldt. 1974. "Relation of gestation time to brain weight for placental mammals: Implication for the theory of vertebrate growth." *American Naturalist* 108: 593–615.

Saenger, P., and J. DiMartino-Nardi. 2001. "Premature adrenarche." *Journal of Endocrinological Investigation* 24(9): 724–33.

Saffran, J. R., R. N. Aslin, et al. 1996. "Statistical learning by 8-month-old infants." *Science* 274: 1926–28.

Safron, A., B. Barch, et al. 2007. "Neural correlates of sexual arousal in homosexual and heterosexual men." *Behavioral Neuroscience* 121(2): 237–48.

Sagerstråle, U. 2000. *Defenders of the truth: The battle for science in the sociobiology debate and beyond.* Oxford, Oxford Univ. Press.

Sagi, A., M. H. van Ijzendoorn, et al. 1994. "Sleeping out of home in a kibbutz communal arrangement: It makes a difference for infant-mother attachment." *Child Development* 65: 992–1004.

———. 1995. "Attachments in a multiple-caregiver and multiple-infant environment: The case of the Israeli kibbutzim." Monographs of the Society for Research in Child Development, vol. 60(2–3), no. 244: 71–91.

Sagi-Schwartz, A., M. H. van Ijzendoorn, et al. 2003. "Attachment and traumatic stress in female holocaust child survivors and their daughters." *American Journal of Psychiatry* 160(6): 1086–92.

Sahakian, B. 1974. "Social isolation rearing and drug-induced stereotypy." *Meeting of the International Society for the Study of Brain and Behavior.* Paris, International Society for the Study of Brain and Behavior.

Sahin, N. T., S. Pinker, et al. 2009. "Sequential processing of lexical, grammatical, and phonological information within Broca's area [see comment]." *Science* 326(5951): 445–49.

Sakai, K. L. 2005. "Language acquisition and brain development." *Science* 310: 815–19.

Salahpour, A., I. O. Medvedev, et al. 2007. "Local knockdown of genes in the brain using small interfering RNA: A phenotypic comparison with knockout animals." *Biological Psychiatry* 61(1): 65–69.

Salloway, S., P. Malloy, et al., eds. 1997. *The neuropsychiatry of limbic and subcortical disorders.* Washington, DC, American Psychiatric Press.

Salvini-Plawen, L., and E. Mayr. 1977. "On the evolution of photoreceptors and eyes." *Evolutionary Biology* 10: 207–63.

Sameroff, A. J., ed. 1978. *Organization and stability of newborn behavior: A commentary on the Brazelton neonatal behavior assessment scale.* Monographs of the Society for Research in Child Development. Chicago, Univ. of Chicago Press.

Sameroff, A. J., and M. M. Haith, eds. 1996. *The five to seven year shift: The age of reason and responsibility.* Chicago, Univ. of Chicago Press.

Sánchez, M. M., F. Aguado, et al. 1998. "Neuroendocrine and immunocytochemical demonstrations of decreased hypothalamo-pituitary-adrenal axis responsiveness to restraint stress after long-term social isolation." *Endocrinology* 139(2): 579–87.

Sánchez, M. M., E. F. Hearn, et al. 1998. "Differential rearing affects corpus callosum size and cognitive function of rhesus monkeys." *Brain Research* 812(1–2): 38–49.

Sánchez, M. M., C. O. Ladd, et al. 2001. "Early adverse experience as a developmental risk factor for later psychopathology: Evidence from rodent and primate models." *Development and Psychopathology* 13(3): 419–49.

Sánchez, M. M., P. M. Noble, et al. 2005. "Alterations in diurnal cortisol rhythm and acoustic startle response in nonhuman primates with adverse rearing." *Biological Psychiatry* 57(4): 373–81.

Sánchez-Toscano, F., M. d. M. Sánchez, et al. 1991. "Changes in the number of dendritic spines in the medial preoptic area during a premature long-term social isolation in rats." *Neuroscience Letters* 122: 1–3.

Sanday, P. R. 1986. *Divine hunger: Cannibalism as a cultural system.* Cambridge, Cambridge Univ. Press.

Sanders, W. B. 1970. *Juvenile offenders for a thousand years.* Chapel Hill, Univ. of North Carolina Press.

Sandler, W., I. Meir, et al. 2005. "The emergence of grammar: Systematic structure in a new language [see comment]." *Proceedings of the National Academy of Sciences* 102(7): 2661–65.

Sandmire, H. F. 1990. "Whither electronic fetal monitoring?" *Obstetrics and Gynecology* 76(6): 1130–34.

Sanes, D. H., T. A. Reh, et al. 2006. *Development of the nervous system.* Burlington, MA, Elsevier Academic.

Sapir, E. 1994. *The psychology of culture: A course of lectures.* Berlin, Mouton de Gruyter.

Sapolsky, R. M. 1992. Neuroendocrinology of the stress-response. In *Behavioral endocrinology,* ed. J. B. Becker, S. M. Breedlove, and D. Crews. Cambridge, MA, MIT Press: 287–324.

———. 1995. "Social subordinance as a marker of hypercortisolism. Some unexpected subtleties." *Annals of the New York Academy of Sciences* 71: 626–39.

———. 1997. "The importance of a well-groomed child [comment]." *Science* 277(5332): 1620–21.

———. 2001. *A primate's memoir: A neuroscientist's unconventional life among the baboons.* New York, Scribner.

Sapolsky, R. M., L. M. Romero, et al. 2000. "How do glucocorticoids influence stress responses? Integrating permissive, suppressive, stimulatory, and preparative actions." *Endocrine Reviews* 21(1): 55–89.

Sarich, V. 1970. The origin of the hominids: An immunological approach. In *Perspectives on human evolution,* ed. S. L. Washburn and P. Jay. New York, Holt, Rinehart and Winston.

———. 1992. Immunological evidence on primates. In *The Cambridge encyclopedia of human evolution,* ed. S. Jones, R. Martin, and D. Pilbeam. New York, Cambridge Univ. Press: 303–6.

Satta, Y., and N. Takahata. 2002. "Out of Africa with regional interbreeding? Modern human origins." *Bioessays* 24(10): 871–75.

Savage-Rumbaugh, E. S. 1991. Language learning in the bonobo: How and why they learn. In *Biological and behavioral determinants of language development,* ed. N. A. Krasnegor et al. Hillsdale, NJ, Erlbaum: 209–33.

———. 1994. Hominid evolution: Looking to modern apes for clues. In *Hominid culture in primate perspective,* ed. D. Quiatt and J. Itani. Niwot, Univ. Press of Colorado: 7–49.

Savage-Rumbaugh, E. S., J. Murphy, et al. 1993. "Language comprehension in ape and child." Monographs of the Society for Research in Child Development 58(3–4, serial no. 333).

Savage-Rumbaugh, E. S., and D. M. Rumbaugh. 1998. Perspectives on consciousness, language, and other emergent processes in apes and humans. In *Toward a science of consciousness II: The second Tucson discussions and debates,* ed. S. R. Hameroff, A. W. Kaszniak, and A. C. Scott. Cambridge, MA, MIT Press: 533–49.

Savage-Rumbaugh, E. S., D. M. Rumbaugh, et al. 1978. "Symbolic communication between two chimpanzees." *Science* 201: 641–44.

Savage-Rumbaugh, S., and R. Lewin. 1994. *Kanzi: The ape at the brink of the human mind.* New York, Wiley.

Savage-Rumbaugh, S., S. G. Shanker, et al. 1998. *Apes, language, and the human mind.* New York, Oxford Univ. Press.

Saver, J. L., and A. R. Damasio. 1991. "Preserved access and processing of social knowl-

edge in a patient with acquired sociopathy due to ventromedial frontal damage."
Neuropsychologia 29(12): 1241–49.

Sbrzesny, H. 1976. *Die spiele der !Ko-Bushleute: Unter besonderer berücksichtigung ihrer sozialisierenden und gruppenbindenden funktionen.* Munich, Piper.

Scaife, M., and J. S. Bruner. 1975. "The capacity for joint visual attention in the infant." *Nature* 253: 265–66.

Scarr, S. 1992. "Developmental theories for the 1990s: Development and individual differences." *Child Development* 63: 1–19.

———. 1993. "Biological and cultural diversity: The legacy of Darwin for development." *Child Development* 64: 1333–53.

Scarr, S., and K. McCartney. 1983. "How people make their own environments: A theory of genotype environment effects." *Child Development* 54: 424–35.

Scelza, B. A. 2009. "The grandmaternal niche: Critical caretaking among Martu Aborigines." *American Journal of Human Biology* 21: 448–54.

Schacter, F. F., M. L. Fuchs, et al. 1989. "Co-sleeping and sleep problems in Hispanic-American urban young children." *Pediatrics* 84: 522–30.

Schafe, G. E., V. Doyere, et al. 2005. "Tracking the fear engram: The lateral amygdala is an essential locus of fear memory storage." *Journal of Neuroscience* 25(43): 10010–14.

Schaffer, H. R. 1966. "The onset of fear of strangers and the incongruity hypothesis." *Journal of Child Psychology and Psychiatry and Allied Disciplines* 7(2): 95–106.

Schaller, G. 1963. *The mountain gorilla: Ecology and behavior.* Chicago, Univ. of Chicago Press.

———. 1967. *The deer and the tiger.* Chicago, Univ. of Chicago Press.

———. 1972. *The Serengeti lion: A study of predator-prey relations.* Chicago, Univ. of Chicago Press.

Schanberg, S. M., and T. M. Field. 1987. "Sensory deprivation stress and supplemental stimulation in the rat pup and preterm human neonate." *Child Development* 58(6): 1431–47.

Scheibel, A. B. 1991. Some structural and developmental correlates of human speech. In *Brain maturation and cognitive development: Comparative and cross-cultural perspectives,* ed. K. R. Gibson and A. C. Petersen. New York, Aldine de Gruyter: 345–53.

Scheller, R. H., R. Kaldany, et al. 1984. "Neuropeptides: Mediators of behavior in aplysia." *Science* 225: 1300–8.

Schenkel, R. 1966. Play, exploration and territoriality in the wild lion. In *Play, exploration and territory in mammals,* ed. P. Jewell and C. Loizos. Symposium of the Zoological Society of London, No. 18. London, Zoological Society of London.

Scheper-Hughes, N. 1985. "Culture, scarcity, and maternal thinking: Maternal detachment and infant survival in a Brazilian shantytown." *Ethos* 13(4): 291–317.

———. 1992. *Death without weeping: The violence of everyday life in Brazil.* Berkeley, Univ. of California Press.

Scherer, S. S., and L. Wrabetz. 2008. "Molecular mechanisms of inherited demyelinating neuropathies." *GLIA* 56(14): 1578–89.

Schick, K. D., and N. Toth. 1993. *Making silent stones speak.* New York, Simon and Schuster.

Schiefenhövel, W. 1988. *Geburtsverhalten und reproduktive Strategien der Eipo: Ergebnisse humanethologische un ethnomedizinischer Untersuchungen im zentralen Bergland von Irian Jaya (West-Neuguinea), Indonesien.* Berlin, Dietrich Reimer Verlag.

Schieffelin, B. B. 1990. *The give and take of everyday life: Language socialization of Kaluli children.* Cambridge, Cambridge Univ. Press.

Schieffelin, B. B., and E. Ochs, eds. 1986. *Language socialization across cultures.* Cambridge, Cambridge Univ. Press.

Schiller, F. 1979. *Paul Broca: Founder of French anthropology, explorer of the brain.* Berkeley, Univ. of California Press.

Schinke, R. J., and W. C. Jerome. 2002. "Understanding and refining the resilience of elite athletes: An intervention strategy." *Athletic Insight: Online Journal of Sport Psychology* 4(3): np.

Schino, G., L. Speranza, et al. 2003. "Infant handling and maternal response in Japanese macaques." *International Journal of Primatology* 24(3): 627–38.

Schlegel, A., and H. Barry. 1979. "Adolescent initiation ceremonies: A cross-cultural code." *Ethnology* 18: 199–210.

———, eds. 1991. *Adolescence: An anthropological inquiry.* New York, Free Press.

Schleidt, W. M. 1961. "Reaktionen von Truthühnern auf fliegende Raubvögel und Versuche zur Analyse ihrer AAMs." *Zeitschrift für Tierpsychologie* 18: 534–60.

Schlesinger, K., and B. J. Griek. 1970. The genetics and biochemistry of audiogenic seizures. In *Contributions to behavior-genetic analysis: The mouse as a prototype,* ed. G. Linzey and D. D. Thiessen. New York, Appleton-Century-Crofts: 219–57.

Schleussner, E., U. Schneider, et al. 2004. "Prenatal evidence of left-right asymmetries in auditory evoked responses using fetal magnetoencephalography." *Early Human Development* 78(2): 133–36.

Schlottman, R. S., and B. Seay. 1972. "Mother-infant separation in the Java monkey (*Macaca irus*)." *Journal of Comparative and Physiological Psychology* 79(2): 334–40.

Schmidt, L. A., N. A. Fox, et al. 1997. "Behavioral and neuroendocrine responses in shy children." *Developmental Psychobiology* 30(2): 127–40.

Schmidt, L. A., V. Miskovic, et al. 2008. "Shyness and timidity in young adults who were born at extremely low birth weight." *Pediatrics* 122(1): e181–87.

Schmithorst, V. J., M. Wilke, et al. 2002. "Correlation of white matter diffusivity and anisotropy with age during childhood and adolescence: A cross-sectional diffusion-tensor MR imaging study." *Radiology* 222(1): 212–18.

Schneider, M., and M. Koch. 2005. "Deficient social and play behavior in juvenile and adult rats after neonatal cortical lesion: Effects of chronic pubertal cannabinoid treatment." *Neuropsychopharmacology* 30(5): 944–57.

Schneider, T. S. 2007. *Someone must survive to tell the world.* Montreal: Polish-Jewish Heritage Foundation of Canada.

Scholz, H., M. Franz, et al. 2005. "The *hangover* gene defines a stress pathway required for ethanol tolerance development." *Nature* 436: 845–47.

Schradin, C., and N. Pillay. 2004. "The striped mouse (*Rhabdomys pumilio*) from the succulent karoo, South Africa: A territorial group-living solitary forager with communal breeding and helpers at the nest." *Journal of Comparative Psychology* 118(1): 37–47.

Schuiling, G. A. 2003. "The benefit and the doubt: Why monogamy?" *Journal of Psychosomatic Obstetrics and Gynecology* 24(1): 55–61.

Schultz, A. H. 1963. Age changes, sex differences, and variability as factors in the classification of primates. In *Classification and human evolution,* ed. S. L. Washburn. Chicago, Aldine.

Schwartz, C. E., C. I. Wright, et al. 2003. "Inhibited and uninhibited infants 'grown up': Adult amygdalar response to novelty." *Science* 300(5627): 1952–53.

Scott, E. S., and L. Steinberg. 2008. *Rethinking juvenile justice.* Cambridge, MA, Harvard Univ. Press.

Scott, J. P. 1963. "'Critical periods' in the development of behavior." *Science* 139: 1110–16.

Scott, J. P., and J. L. Fuller. 1965. *Genetics and the social behavior of the dog.* Chicago, Univ. of Chicago Press.

Scott, R. S., P. S. Ungar, et al. 2005. "Dental microwear texture analysis shows within-species diet variability in fossil hominins." *Nature* 436: 693–95.

Scrimshaw, S. C. M. 1984. Infanticide in human populations: Societal and individual concerns. In *Infanticide: Comparative and evolutionary perspectives,* ed. G. Hausfater and S. B. Hrdy. New York, Aldine de Gruyter: 439–62.

Scriver, C. R., and C. L. Clow. 1980. "Phenylketonuria: Epitome of human biochemical genetics." *New England Journal of Medicine* 303: 1336–42, 1394–1400.

Sear, R., R. Mace, et al. 2000. "Maternal grandmothers improve nutritional status and survival of children in rural Gambia." *Proceedings of the Royal Society of London B: Biological Sciences* 267(1453): 1641–47.

Sear, R., F. Steele, et al. 2002. "The effects of kin on child mortality in rural Gambia." *Demography* 39(1): 43–63.

Sears, W., and M. White. 1999. *Nighttime parenting: A La Leche League International book.* New York, Penguin Putnam/Plume.

Seashore, M. J., A. D. Leifer, et al. 1973. "The effects of denial of early mother-infant interaction on maternal self-confidence." *Journal of Personality and Social Psychology* 26(3): 369–78.

Segall, M. H., D. T. Campbell, and M. J. Herskovits. 1966. *The influence of culture on visual perception.* Indianapolis, Bobbs-Merrill.

Seger, J. 1991. Cooperation and conflict in social insects. In *Behavioural ecology: An evolutionary approach,* ed. J. R. Krebs and N. B. Davies. Oxford, Blackwell: 338–73.

Segerstråle, U. 2000. *Defenders of the truth: The battle for science in the sociobiology debate and beyond.* Oxford, Oxford Univ. Press.

Seidman, L. J., E. M. Valera, et al. 2005. "Structural brain imaging of attention-deficit/hyperactivity disorder." *Biological Psychiatry* 57(11): 1263–72.

Sekiguchi, M., R. S. Nowakowski, et al. 1995. "Morphological abnormalities in the hippocampus of the weaver mutant mouse." *Brain Research* 696(1–2): 262–67.

Selby, J. M., and B. S. Bradley. 2003. "Infants in groups: A paradigm for the study of early social experience." *Human Development* 46(4): 197–221.

Seligman, M. E. P., and J. L. Hager, eds. 1972. *Biological boundaries of learning.* New York, Meredith.

Selkoe, D. J. 2004. "Alzheimer disease: Mechanistic understanding predicts novel therapies." *Annals of Internal Medicine* 140(8): 627–38.

Sellen, D. W. 1998. "Infant and young child feeding practices among African pastoralists: The Datoga of Tanzania." *Journal of Biosocial Science* 30(4): 481–99.

———. 2001. "Comparison of infant feeding patterns reported for nonindustrial populations with current recommendations." *Journal of Nutrition* 131(10): 2707–15.

———. 2007. "Evolution of infant and young child feeding: Implications for contemporary public health." *Annual Review of Nutrition* 27: 123–48.

Sellen, D. W., and R. Mace. 1997. "Fertility and mode of subsistence: A phylogenetic analysis." *Current Anthropology* 38(5): 878–89.

Selye, H. 1936. "A syndrome produced by diverse nocuous agents." *Nature* 138: 32.

———. 1946. "The general adaptation syndrome and the diseases of adaptation." *Journal of Clinical Endocrinology and Metabolism* 6: 117–230.

———. 1975. "Stress and distress." *Comprehensive Therapy* 1(8): 9–13.

———. 1976. "Forty years of stress research: Principal remaining problems and misconceptions." *CMAJ: Canadian Medical Association Journal* 115(1): 53–56.

———. 1998. "A syndrome produced by diverse nocuous agents. 1936." *Journal of Neuropsychiatry and Clinical Neurosciences* 10(2): 230–31.

Semaw, S., M. J. Rogers, et al. 2003. "2.6-million-year-old stone tools and associated bones from OGS-6 and OGS-7, Gona, Afar, Ethiopia." *Journal of Human Evolution* 45(2): 169–77.

Semaw, S., S. W. Simpson, et al. 2005. "Early Pliocene hominids from Gona, Ethiopia." *Nature* 433(7023): 301–5.

Semendeferi, K., and H. Damasio. 2000. "The brain and its main anatomical subdivisions in living hominoids using magnetic resonance imaging." *Journal of Human Evolution* 38(2): 317–32.

Sendtner, M., G. Pei, et al. 2000. "Developmental motoneuron cell death and neurotrophic factors." *Cell and Tissue Research* 301(1): 71–84.

Senghas, A. 2005. "Language emergence: Clues from a new bedouin sign." *Current Biology* 15(12): R463–65.

Senghas, A., S. Kita, et al. 2004. "Children creating core properties of language: Evidence from an emerging sign language in Nicaragua [see comment]." *Science* 305(5691): 1779–82.

Seraphin, S. B. 2004. "The neuroendocrinology, neuroanatomy, and behavior-pharmacology of dopamine in juvenile nursery-reared and mother-reared rhesus monkeys (*Macaca mulatta*)." PhD diss., Emory Univ., Atlanta, GA.

Serbin, L. A., and J. Karp. 2004. "The intergenerational transfer of psychosocial risk: Mediators of vulnerability and resilience." *Annual Review of Psychology* 55(1): 333–63.

Serbin, L. A., K. K. Powlishta, et al. 1993. "The development of sex typing in middle childhood." *Monographs of the Society for Research in Child Development* 58(2): 1–74.

Seymour, S. 1981. Cooperation and competition: Some issues and problems in cross-cultural analysis. In *Handbook of cross-cultural human development,* ed. R. H. Munroe, R. L. Munroe, and B. B. Whiting. New York, Garland STPM: 717–38.

———. 2004. "Multiple caretaking of infants and young children: An area in critical need of a feminist psychological anthropology." *Ethos* 32(4): 538–56.

Seymour, S. C. 2001. "Child care in India: An examination of the 'household size/infant indulgence' hypothesis." *Cross-Cultural Research* 35(1): 3–22.

Shanley, D. P., and T. B. Kirkwood. 2001. "Evolution of the human menopause." *Bioessays* 23(3): 282–87.

Sharpe, L. L., T. H. Clutton-Brock, et al. 2002. "Experimental provisioning increases play in free-ranging meerkats." *Animal Behaviour* 64(1): 113–21.

Shea, B. T. 1983a. "Allometry and heterochrony in the African apes." *American Journal of Physical Anthropology* 62(3): 275–89.

———. 1983b. "Paedomorphosis and neoteny in the pygmy chimpanzee." *Science* 222: 521–22.

———. 1988. Heterochrony in primates. In *Heterochrony in evolution: A multidisciplinary approach,* ed. M. L. McKinney. New York, Plenum.

———. 1992. Neoteny. In *The Cambridge encyclopedia of human evolution,* ed. S. Jones, R. Martin, and D. Pilbeam. New York, Cambridge Univ. Press: 104.

———. 2000. Current issues in the investigation of evolution by heterochrony, with emphasis on the debate over human neoteny. In *Biology, brains, and behavior: The evolution of human development,* ed. S. T. Parker, J. Langer, and M. L. McKinney. Santa Fe, School of American Research Press: 161–213.

Sheldon, B. C. 2002. "Relating paternity to paternal care." *Philosophical Transactions of the Royal Society of London B: Biological Sciences* 357(1419): 341–50.

Sheldon, P. R. 1997. The plus ça change model: Explaining stasis and evolution in response to abiotic stress over geological timescales. In *Experientia Supplementum (Basel); Environmental stress, adaptation and evolution,* ed. R. Bijlsma and V. Loeschcke, 83: 307–19. Basel, Switz., Birkhaeuser Boston.

Shepher, J. 1971. "Mate selection among second generation kibbutz adolescents and adults: Incest avoidance and negative imprinting." *Archives of Sexual Behavior* 1: 293–307.

———. 1983. *Incest: A biosocial view.* New York, Academic.

Shergill, S. S., R. M. Murray, et al. 1998. "Auditory hallucinations: A review of psychological treatments." *Schizophrenia Research* 32(3): 137–50.

Sherif, M., O. J. Harvey, et al. 1961. *Intergroup conflict and cooperation: The Robbers Cave experiment.* Norman, OK, Institute of Group Relations.

Sherman, P. W. 1977. "Nepotism and the evolution of alarm calls." *Science* 197: 1246–53.

Sherman, P. W., J. U. M. Jarvis, et al. 1991. *The biology of the naked mole rat.* Princeton, Princeton Univ. Press.

Sherwood, C. C. 2005. "Comparative anatomy of the facial motor nucleus in mammals, with an analysis of neuron numbers in primates." *Anatomical Record, Part A, Discoveries in Molecular, Cellular, and Evolutionary Biology* 287(1): 1067–79.

Sherwood, C. C., R. L. Holloway, et al. 2003. "Neuroanatomical basis of facial expression in monkeys, apes, and humans." *Annals of the New York Academy of Sciences* 1000: 99–103.

Sherwood, C. C., R. L. Holloway, et al. 2004. "Cortical orofacial motor representation in Old World monkeys, great apes, and humans. II. Stereologic analysis of chemoarchitecture." *Brain, Behavior and Evolution* 63(2): 82–106.

Sherwood, C. C., C. D. Stimpson, et al. 2006. "Evolution of increased glia-neuron ratios in the human frontal cortex." *Proceedings of the National Academy of Sciences* 103(37): 13606–11.

Sherwood, C. C., F. Subiaul, et al. 2008. "A natural history of the human mind: Tracing evolutionary changes in brain and cognition." *Journal of Anatomy* 212(4): 426–54.

Shi, J., D. F. Levinson, et al. 2009. "Common variants on chromosome 6p22.1 are associated with schizophrenia." *Nature* 460: 753–57.

Shimada, S., and K. Hiraki. 2006. "Infant's brain responses to live and televised action." *Neuroimage* 32(2): 930–39.

Shiner, R. L., A. S. Masten, et al. 2003. "Childhood personality foreshadows adult personality and life outcomes two decades later." *Journal of Personality* 71(6): 1145–70.

Shipman, P. 1986. "Scavenging or hunting in early hominids: Theoretical framework and tests." *American Anthropologist* 88: 27–43.

Shore, B. 1996. *Culture in mind: Cognition, culture, and the problem of meaning.* New York, Oxford Univ. Press.

Shostak, M. 1981. *Nisa: The life and words of a !Kung woman.* Cambridge, MA, Harvard Univ. Press.

Shu, D. G., S. C. Morris, et al. 2003. "Head and backbone of the Early Cambrian verte-brate Haikouichthys." *Nature* 421(6922): 526–29.

Shu, S. Y., X. M. Bao, et al. 2003. "New component of the limbic system: Marginal division of the neostriatum that links the limbic system to the basal nucleus of Meynert." *Journal of Neuroscience Research* 71(5): 751–57.

Shubin, N., C. Tabin, et al. 2009. "Deep homology and the origins of evolutionary novelty." *Nature* 457(7231): 818–23.

Shweder, R. A. 1993. The cultural psychology of the emotions. In *Handbook of emotions,* ed. M. Lewis and J. M. Haviland. New York, Guilford: 417–31.

Shy, K. K., D. A. Luthy, et al. 1990. "Effects of electronic fetal-heart-rate monitoring, as compared with periodic auscultation, on the neurologic development of premature infants [see comment]." *New England Journal of Medicine* 322(9): 588–93.

Sibly, R. M., and J. H. Brown. 2007. "Effects of body size and lifestyle on evolution of mammal life histories." *Proceedings of the National Academy of Sciences* 104(45): 17707–12.

———. 2009. "Mammal reproductive strategies driven by offspring mortality-size relationships." *American Naturalist* 173(6): E185–99.

Sichel, D. 2003. Neurohormonal aspects of postpartum depression and psychosis. In *Infanticide: Psychosocial and legal perspectives on mothers who kill,* ed. M. Spinelli. Washington, DC, American Psychiatric Publishing: 61–79.

Sidman, R. 1970. Autoradiographic methods and principles for study of the nervous system with thymidine-H^3. In *Contemporary research techniques of neuroanatomy,* ed. S. O. E. Ebbesson and W. Nauta. New York, Springer.

Siegler, R. S. 1995. "How does change occur: A microgenetic study of number conservation." *Cognitive Psychology* 28: 1–49.

———. 1996. *Emerging minds: The process of change in children's thinking.* New York, Oxford Univ. Press.

Siegler, R. S., and K. Crowley. 1991. "The microgenetic method: A direct means for studying cognitive development." *American Psychologist* 46(6): 606–20.

Sieveking, A. 1979. *The cave artists.* London, Thames and Hudson.

Sigle-Rushton, W., and S. McLanahan. 2004. Father absence and child well-being: A critical review. In *The future of the family,* ed. D. P. Moynihan, T. M. Smeeding, and L. Rainwater. New York, Russell Sage: 116–55.

Silberg, J. L., N. G. Martin, et al. 1987. "Genetic and environmental factors in primary dysmenorrhea and its relationship to anxiety, depression, and neuroticism." *Behavior Genetics* 17(4): 363–83.

Silk, J. B. 1980. "Adoption and kinship in Oceania." *American Anthropologist* 82: 799–820.

———. 1987a. "Adoption among the Inuit." *Ethos* 15(3): 320–30.

———. 1987b. "Adoption and fosterage in human societies: Adaptations or enigmas?" *Cultural Anthropology* 2: 39–49.

———. 1999. "Why are infants so attractive to others? The form and function of infant handling in bonnet macaques." *Animal Behaviour* 57(5): 1021–32.

Silk, J. B., S. C. Alberts, et al. 2003. "Social bonds of female baboons enhance infant survival [see comment]." *Science* 302(5648): 1231–34.

Silk, J. B., S. F. Brosnan, et al. 2005. "Chimpanzees are indifferent to the welfare of unrelated group members." *Nature* 437(7063): 1357–59.

Silverman, I., and M. Eals. 1992. Sex differences in spatial abilities: Evolutionary theory and data. From *The adapted mind: Evolutionary psychology and the generation of culture,* ed. J. H. Barkow, L. Cosmides, and J. Tooby. New York, Oxford Univ. Press: 601–24.

Simmel, G. 1903. "The sociology of conflict: I." *American Journal of Sociology* 9: 490–525.

Simmons, L. W. 2005. "The evolution of polyandry: Sperm competition, sperm selection, and offspring viability." *Annual Review of Ecology Evolution and Systematics* 36: 125–46.

Simon, V. A., J. W. Aikins, et al. 2008. "Romantic partner selection and socialization during early adolescence." *Child Development* 79(6): 1676–92.

Simons, E. 1995. "Egyptian oligocene primates: A review." *Yearbook of Physical Anthropology* 38: 199–238.

Simons, E. L., E. R. Seiffert, et al. 2007. "A remarkable female cranium of the early Oligocene anthropoid *Aegyptopithecus zeuxis* (Catarrhini, Propliopithecidae)." *Proceedings of the National Academy of Sciences* 104(21): 8731–36.

Simons, M., and K. Trajkovic. 2006. "Neuron-glia communication in the control of oligodendrocyte function and myelin biogenesis." *Journal of Cell Science* 119: 4381–89.

Simons, M., and J. Trotter. 2007. "Wrapping it up: The cell biology of myelination." *Current Opinion in Neurobiology* 17(5): 533–40.

Simpson, G. G. 1944. *Tempo and mode in evolution.* New York, Columbia Univ. Press.

———. 1953. "The Baldwin effect." *Evolution* 7: 110–17.

Simpson, J. A., and J. Belsky. 2008. Attachment theory within a modern evolutionary framework. In *Handbook of attachment,* 2nd ed., ed. J. Cassidy and P. R. Shaver. New York, Guilford: 131–57.

Simpson, J. A., and D. T. Kenrick, eds. 1997. *Evolutionary social psychology.* Mahwah, NJ, Erlbaum.

Sinclair, D. 1973. *Human growth after birth.* London, Oxford Univ. Press.

Sisk, C. L., and D. L. Foster. 2004. "The neural basis of puberty and adolescence." *Nature Neuroscience* 7(10): 1040–47.

Sisk, C. L., and J. L. Zehr. 2005. "Pubertal hormones organize the adolescent brain and behavior." *Frontiers in Neuroendocrinology* 26(3–4): 163–74.

Siskind, J. 1973. *To hunt in the morning.* New York, Oxford Univ. Press.

Siviy, S. M., and D. M. Atrens. 1992. "The energetic costs of rough-and-tumble play in the juvenile rat." *Developmental Psychobiology* 25(2): 137–48.

Siviy, S. M., K. A. Harrison, et al. 2006. "Fear, risk assessment, and playfulness in the juvenile rat." *Behavioral Neuroscience* 120(1): 49–59.

Siviy, S. M., and J. Panksepp. 1985. "Dorsomedial diencephalic involvement in the juvenile play of rats." *Behavioral Neuroscience* 99(6): 1103–13.

———. 1987. "Juvenile play in the rat: Thalamic and brain stem involvement." *Physiology and Behavior* 41(2): 103–14.

Skinner, B. F. 1938. *The behavior of organisms.* New York, Appleton-Century-Crofts.

———. 1948. *Walden two.* New York, Macmillan.

———. 1957. *Verbal behavior.* New York, Appleton-Century-Crofts.

———. 1966. "The phylogeny and ontogeny of behavior." *Science* 153: 1205–13.

Skinner, M. M., and B. Wood. 2006. The evolution of modern human life history: A paleontological perspective. In *Evolution of human life history,* ed. K. Hawkes and R. R. Paine. Santa Fe, School of American Research Press: 331–64.

Sklar, C. A., S. L. Kaplan, et al. 1980. "Evidence for dissociation between adrenarche and gonadarche: Studies in patients with idiopathic precocious puberty, gonadal dysgenesis, isolated gonadotropin deficiency, and constitutionally delayed growth and adolescence." *Journal of Clinical Endocrinology and Metabolism* 51(3): 548–56.

Skuse, D. H. 1984. "Extreme deprivation in early childhood: I. Diverse outcomes for three siblings from an extraordinary family." *Journal of Child Psychology and Psychiatry and Allied Disciplines* 25(4): 523–41.

———. 1985. "Extreme deprivation in early childhood: A reply." *Journal of Child Psychology and Psychiatry and Allied Disciplines* 26(5): 827–28.

———. 1993. Extreme deprivation in early childhood. In *Language development in exceptional circumstances,* ed. D. Bishop and K. Mogford. Hove, Eng., Erlbaum: 29–46.

Slack, F., and G. Ruvkun. 1997. "Temporal pattern formation by heterochronic genes." *Annual Review of Genetics* 31: 611–34.

Slater, A. 2000. Visual perception in the young infant: Early organization and rapid learning. In *Infant development: The essential readings,* ed. D. Muir and A. Slater. Malden, MA, Blackwell: 95–116.

Sleigh, M. J., and R. Lickliter. 1998. "Timing of presentation of prenatal auditory stimulation alters auditory and visual responsiveness in bobwhite quail chicks (*Colinus virginianus*)." *Journal of Comparative Psychology* 112(2): 153–60.

Slobin, D. I., ed. 1985–92. *The cross-linguistic study of language acquisition,* vols. 1–3. Hillsdale, NJ, Erlbaum.

Small, M. F. 1990. "Alloparental behaviour in Barbary macaques, *Macaca sylvanus.*" *Animal Behaviour* 39(2): 297–306.

Smith, B. A., T. J. Fillion, et al. 1990. "Orally mediated sources of calming in 1- to 3-day-old human infants." *Developmental Psychology* 26(5): 731–37.

Smith, C., and B. Lloyd. 1978. "Maternal behavior and perceived sex of infant: Revisited." *Child Development* 49: 1263–65.

Smith, C., and M. L. Denton. 2006. *Soul searching: The religious and spiritual lives of American teenagers.* Oxford, Oxford Univ. Press.

Smith, E. A., J. R. Udry, et al. 1985. "Pubertal development and friends: A biosocial explanation of adolescent sexual behavior." *Journal of Health and Social Behavior* 26: 183–92.

Smith, E. L., J. D. Coplan, et al. 1997. "Neurobiological alterations in adult nonhuman primates exposed to unpredictable early rearing. Relevance to posttraumatic stress disorder." *Annals of the New York Academy of Sciences* 821: 545–48.

Smith, H., T. Crumett, et al. 1994. "Ages of eruption of primate teeth." *Yearbook of Physical Anthropology* 37: 177–231.

Smith, P. K. 2005. Social and pretend play in children. In *The nature of play: Great apes and humans,* ed. A. D. Pellegrini and P. K. Smith. London, Guilford: 173–209.

Smith, P. K., R. Smees, et al. 2004. "Play fighting and real fighting: Using video playback methodology with young children." *Aggressive Behavior* 30(2): 164–73.

Smith Fernandez, A., C. Pieau, et al. 1998. "Expression of the *Emx-1* and *Dlx-1* homeobox genes define three molecularly distinct domains in the telencephalon of mouse, chick, turtle, and frog embryos: Implications for the evolution of telencephalic subdivisions in amniotes." *Development* 125: 2099–2111.

Smuts, B. 1985. *Sex and friendship in baboons.* New York, Aldine.

———. 1992. "Male aggression against women: An evolutionary perspective." *Human Nature* 3(1): 1–44.

Smuts, B. B., and R. W. Smuts. 1993. "Male aggression and sexual coercion of females in nonhuman primates and other mammals: Evidence and theoretical implications." *Advances in the Study of Behavior* 22: 1–63.

Snarey, J., and K. Keljo. 1991. In a *Gemeinschaft* voice: The cross-cultural expansion of moral development theory. In *Handbook of moral behavior and development*, vol. 1, ed. W. M. Kurtines and J. L. Gewirtz. Mahwah, NJ, Erlbaum: 1–41.

Snarey, J., and P. Samuelson. 2008. Moral education in the cognitive developmental tradition: Lawrence Kohlberg's revolutionary ideas. In *Handbook on moral character and education*, ed. L. Nucci. Mahwah, NJ, Erlbaum: 53–79.

Snarey, J. R. 1985. "Cross-cultural universality of social-moral development: A critical review of Kohlbergian research." *Psychological Bulletin* 97(2): 202–32.

Snarey, J. R., J. Reimer, et al. 1985. "The kibbutz as a model for moral education: A longitudinal cross-cultural study." *Journal of Applied Developmental Psychology* 6(2–3): 151–72.

Snyder, P. A. 1972. Behavior of *Leontopithecus rosalia* (the golden lion marmoset) and related species: A review. In *Saving the lion marmoset*, ed. D. D. Bridgewater. Wheeling, WV, Wild Animal Propagation Trust.

Soderquist, T. R., and M. Serena. 2000. "Juvenile behaviour and dispersal of chuditch (*Dasyurus geoffroii*) (Marsupialia: Dasyuridae)." *Australian Journal of Zoology* 48(5): 551–60.

Sokol, B. W., M. J. Chandler, et al. 2004. "From mechanical to autonomous agency: The relationship between children's moral judgments and their developing theories of mind." *New Directions for Child and Adolescent Development*(103): 19–36.

Sokolowski, M. B. 2001. "Drosophila: Genetics meets behaviour." *Nature Reviews Genetics* 2(11): 879–90.

Solecki, R. S. 1971. *Shanidar, the first flower people.* New York, Knopf.

Solomon, J., and C. George. 2008. The measurement of attachment security and related constructs in infancy and early childhood. In *Handbook of attachment*, 2nd ed., ed. J. Cassidy and P. R. Shaver. New York, Guilford: 383–416.

Solomon, N. G., and J. A. French. 1997. *Cooperative breeding in mammals.* New York, Cambridge Univ. Press.

Soltis, J. 2004. "The signal functions of early infant crying." *Behavioral and Brain Sciences* 27(4): 443–90.

Soltis, J., R. Boyd, et al. 1995. "Can group-functional behaviors evolve by cultural group selection? An empirical test." *Current Anthropology* 36(3): 473–94.

Soltis, J., R. Thomsen, et al. 2000. "Infanticide by resident males and female counterstrategies in wild Japanese macaques (*Macaca fuscata*)." *Behavioral Ecology and Sociobiology* 48(3): 195–202.

Soriano-Mas, C., J. Pujol, et al. 2009. "Age-related brain structural alterations in children with specific language impairment." *Human Brain Mapping* 30(5): 1626–36.

Southwick, C., M. Beg, et al. 1965. Rhesus monkeys in North India. In *Primate behavior,* ed. I. DeVore. New York, Holt, Rinehart, and Winston.

Southwick, C. H., and L. H. Clark. 1968. "Interstrain differences in aggressive behavior and exploratory activity of inbred mice." *Communications in Behavioral Biology* 1: 49–59.

Sowell, E. R., B. S. Peterson, et al. 2003. "Mapping cortical change across the human life span." *Nature Neuroscience* 6(3): 309–15.

Spadoni, A. D., C. L. McGee, et al. 2007. "Neuroimaging and fetal alcohol spectrum disorders." *Neuroscience and Biobehavioral Reviews* 31(2): 239–45.

Spain, D. H. 1987. "The Westermarck-Freud incest-theory debate." *Current Anthropology* 28: 623–45.

Spangler, G., and K. Grossman. 1999. Individual and physiological correlates of attachment disorganization in infancy. In *Attachment disorganization,* ed. J. G. C. Solomon. New York, Guilford: 95–124.

Spear, L. P. 2000. "Neurobehavioral changes in adolescence." *Current Directions in Psychological Science* 9(4): 111–14.

Spelke, E. S., and K. D. Kinzler. 2009. "Innateness, learning, and rationality." *Child Development Perspectives* 3(2): 96–98.

Spencer, J. P., M. S. Blumberg, et al. 2009. "Short arms and talking eggs: Why we should no longer abide the nativist–empiricist debate." *Child Development Perspectives* 3: 79–87.

Spevak, A. M., D. M. Quadagno, et al. 1973. "The effects of isolation on sexual and social behavior in the rat." *Behavioral Biology* 8(1): 63–73.

Spindler, G. D. 1979. *The making of psychological anthropology.* Berkeley, Univ. of California Press.

Spinka, M., R. C. Newberry, et al. 2001. "Mammalian play: Training for the unexpected." *Quarterly Review of Biology* 76(2): 141–68.

Spiro, M. E. 1979. *Gender and culture: Kibbutz women revisited.* Durham, NC, Duke Univ. Press.

———. 1982. *Oedipus in the Trobriands.* Chicago, Univ. of Chicago Press.

Spiteri, E., G. Konopka, et al. 2007. "Identification of the transcriptional targets of FOXP2, a gene linked to speech and language, in developing human brain." *American Journal of Human Genetics* 81(6): 1144–57.

Spitz, R. A., and K. M. Wolf. 1946. "The smiling response: A contribution to the ontogenesis of social relations." *Genetic Psychology Monographs* 34: 57–125.

Spock, B. 1968. *Baby and child care.* New York, Simon and Schuster/Pocket.

Spock, B., and S. J. Parker. 1998. *Dr. Spock's baby and child care,* 7th ed. New York, Simon and Schuster/Pocket.

Sroufe, L. A., E. A. Carlson, et al. 1999. "Implications of attachment theory for developmental psychopathology." *Development and Psychopathology* 11(1): 1–13.

Sroufe, L. A., and E. Waters. 1976. "The ontogenesis of smiling and laughter: A perspective on the organization of development in infancy." *Psychological Review* 83: 173–89.

———. 1977. "Attachment as an organizational construct." *Child Development* 48: 1184–99.

St. James-Roberts, I., S. Conroy, et al. 1998. "Stability and outcome of persistent infant crying." *Infant Behavior and Development* 21(3): 411–35.

St. James-Roberts, I., and P. Menon-Johansson. 1999. "Predicting infant crying from fetal movement data: An exploratory study." *Early Human Development* 54(1): 55–62.

Stack, C. B. 1974. *All our kin: Strategies for survival in a black community.* New York, Harper & Row.

Stambrook, M., and K. C. H. Parker. 1987. "The development of the concept of death in childhood: A review of the literature." *Merrill-Palmer Quarterly* 33(2): 133–57.

Stanford, C. 1996. "The hunting ecology of wild chimpanzees: Implications for the evolutionary ecology of Pliocene hominids." *American Anthropologist* 98: 96–113.

Stanford, C. B. 1992. "Costs and benefits of allomothering in wild capped langurs (*Presbytis pileata*)." *Behavioral Ecology and Sociobiology* 30(1): 29–34.

———. 1999. *The hunting apes: Meat eating and the origins of human behavior.* Princeton, Princeton Univ. Press.

———. 2002. The ape's gift: Meat-eating, meat-sharing, and human evolution. In *Tree of origin: What primate behavior can tell us about human social evolution,* ed. F. B. M. de Waal. Cambridge, MA, Harvard Univ. Press: 95–117.

Stanford, C. B., and R. Wrangham. 1998. *Chimpanzee and red colobus: Ecology of predator and prey.* Cambridge, MA, Harvard Univ. Press.

Stanton, W. 1960. *The leopard's spots: Scientific attitudes toward race in America 1815–1859.* Chicago, Univ. of Chicago Press.

Staudt, M., I. Krageloh-Mann, et al. 2000. "[Normal myelination in childhood brains using MRI—a meta analysis]." *Rofo: Fortschritte auf dem Gebiete der Rontgenstrahlen und der Nuklearmedizin* 172(10): 802–11.

Stearns, S. C. 1992. *The evolution of life histories.* Oxford, Oxford Univ. Press.

Stearns, S. C., M. E. Pereira, et al. 2003. Primate life histories and future research. In *Primate life histories and socioecology,* ed. P. M. Kappeler and M. E. Pereira. Chicago, Univ. of Chicago Press: 301–12.

Steelman, L. C., B. Powell, et al. 2002. "Reconsidering the effects of sibling configurations: Recent advances and challenges." *Annual Review of Sociology* 28(1): 243–69.

Stefansson, H., R. A. Ophoff, et al. 2009. "Common variants conferring risk of schizophrenia." *Nature* 460: 744–47.

Steinberg, L. 2005. "Cognitive and affective development in adolescence." *Trends in Cognitive Sciences* 9(2): 69–74.

———. 2007. "Risk taking in adolescence: New perspectives from brain and behavioral science." *Current Directions in Psychological Science* 16(2): 55–59.

———. 2008. "A social neuroscience perspective on adolescent risk-taking." *Developmental Review* 28(1): 78–106.

———. 2009. "A behavioral scientist looks at adolescent brain development." *Brain and Cognition:* http://dx.doi.org (doi:10.1016/j.bandc.2009.11.003) [introduction to a special issue on the adolescent brain].

Steinberg, L., D. Albert, et al. 2008. "Age differences in sensation seeking and impulsivity as indexed by behavior and self-report: Evidence for a dual systems model." *Developmental Psychology* 44(6): 1764–78.

Steinberg, L., S. Graham, et al. 2009. "Age differences in future orientation and delay discounting." *Child Development* 80(1): 28–44.

Steinberg, L., and E. S. Scott. 2003. "Less guilty by reason of adolescence: Developmental immaturity, diminished responsibility, and the juvenile death penalty." *American Psychologist* 58(12): 1009–18.

Stern, D. 1974. Mother and infant at play: The dyadic interaction involving facial, vocal, and gaze behaviors. In *The effect of the infant on its caregiver,* ed. M. Lewis and L. A. Rosenblum. New York, Wiley.

———. 1977. *The first relationship: Mother and infant.* Cambridge, MA, Harvard Univ. Press.

Stern, D., and J. Gibbon. 1977. Temporal expectancies of social behaviors in mother-infant play. In *The origins of the infant's responsiveness,* ed. E. Thomas. New York, Erlbaum.

Stern, D. L. 1990. Joy and satisfaction in infancy. In *Pleasure beyond the pleasure principle,* ed. R. A. Glick and S. Bone. New Haven, Yale Univ. Press: 13–25.

Stern, D. N. 1985. *The interpersonal world of the infant: A view from psychoanalysis and developmental psychology.* New York, Basic.

Stern, J. M. 1986. "Licking, touching, and suckling: Contact stimulation and maternal psychobiology in rats and women." *Annals of the New York Academy of Sciences* 474: 95–107.

———. 1996. Somatosensation and maternal care in Norway rats. In *Parental care: Evolution, mechanisms, and adaptive significance. Advances in the study of behavior,* vol. 25, ed. J. S. Rosenblatt and C. T. Snowdon. San Diego, Academic: 243–94.

Stern, J. M., M. Konner, et al. 1986. "Nursing behavior, prolactin and postpartum amenorrhoea during prolonged lactation in American and !Kung mothers." *Clinical Endocrinology* 25: 247–58.

Stern, K. and M. K. McClintock. 1998. "Regulation of ovulation by human pheromones." *Nature* 392(March 12): 177–79.

Stern, M., and K. H. Karraker. 1989. "Sex-stereotyping of infants: A review of gender-labeling studies." *Sex Roles* 20(9/10): 501–22.

Stevens, C. F. 2001. "An evolutionary scaling law for the primate visual system and its basis in cortical function." *Nature (London)* 411(6834): 193–95.

Stiver, K. A., P. Dierkes, et al. 2005. "Relatedness and helping in fish: Examining the theoretical predictions." *Proceedings of the Royal Society of London B: Biological Sciences* 272(1572): 1593–99.

Stocking, G. W. 1968. *Race, culture, and evolution: Essays in the history of anthropology.* New York, Free Press.

Stocking, G. W., Jr., ed. 1974. *The shaping of American anthropology 1883–1911: A Franz Boas reader.* New York, Basic.

Stoel-Gammon, C., and K. Otomo. 1986. "Babbling development of hearing-impaired and normally hearing subjects." *Journal of Speech and Hearing Disorders* 51: 33–41.

Stolk, J., R. Conner, et al. 1974. "Social environment and brain biogenic amine metabolism in rats." *Journal of Comparative and Physiological Psychology* 87: 203–7.

Stoller, R. J., and G. H. Herdt. 1985. "Theories of origins of male homosexuality." *Archives of General Psychiatry* 42: 399.

Stone, V. E., and P. Gerrans. 2006. "What's domain-specific about theory of mind?" *Social Neuroscience* 1(3–4): 309–19.

Stout, D. 2002. "Skill and cognition in stone tool production: An ethnographic case study from Irian Jaya." *Current Anthropology* 45(3): 693–722.

Stout, D., and T. Chaminade. 2007. "The evolutionary neuroscience of tool making." *Neuropsychologia* 45(5): 1091–1100.

Stout, D., N. Toth, et al. 2008. "Neural correlates of Early Stone Age toolmaking: Technology, language and cognition in human evolution." *Philosophical Transactions of the Royal Society of London B: Biological Sciences* 363(1499): 1939–49.

Strathearn, L., J. Li, et al. 2008. "What's in a smile? Maternal brain responses to infant facial cues [erratum appears in *Pediatrics* 122(3, Sept. 2008):689]." *Pediatrics* 122(1): 40–51.

Strauss, C. 1992. Models and motives. In *Human motives and cultural models,* ed. R. D'Andrade and C. Strauss. New York, Cambridge Univ. Press: 1–20.

Strauss, C., and N. Quinn. 1997. *A cognitive theory of cultural meaning.* New York, Cambridge Univ. Press.

Striano, T., P. A. Brennan, et al. 2002. "Maternal depressive symptoms and 6-month-old infants' sensitivity to facial expressions." *Infancy* 3(1): 115–26.

Striedter, G. 1997. "The telencephalon of tetrapods in evolution." *Brain, Behavior and Evolution* 49: 179–213.

Striedter, G. F. 2005. *Principles of brain evolution.* Sunderland, MA, Sinauer.

Strier, K. B. 1993. Growing up in a patrifocal society: Sex differences in the spatial relations of immature muriquis. In *Juvenile primates: Life history, development, and behavior,* ed. M. E. Pereira and L. A. Fairbanks. New York, Oxford: 138–47.

———. 1994. "Myth of the typical primate." *Yearbook of Physical Anthropology* 37: 233–71.

———. 1996. "Male reproductive strategies in New World primates." *Human Nature* 7(2): 105–23.

———. 2006. *Primate behavioral ecology,* 3rd ed. Boston, Allyn and Bacon.

Stringer, C., and C. Gamble. 1993. *In search of the Neanderthals: Solving the puzzle of human origins.* New York, Thames and Hudson.

Stromswold, K. 2000. The cognitive neuroscience of language acquisition. In *The new cognitive neurosciences,* ed. M. S. Gazzaniga. Cambridge, MA, MIT Press.

Stroud, L. R., R. L. Paster, et al. 2009. "Maternal smoking during pregnancy and neonatal behavior: A large-scale community study." *Pediatrics* 123(5): e842–48.

Struhsaker, T. 1967. "Behavior of vervet monkeys (*Cercopithecus aethiops*)." Univ. of California Publication in Zoology 82.

Struhsaker, T. T., and L. Leland. 1987. Colobines: Infanticide by adult males. In *Primate societies,* ed. B. B. Smuts et al. Chicago, Univ. of Chicago Press: 83–97.

Stumpf, R. 2007. Chimpanzees and bonobos: Diversity within and between species. In *Primates in perspective,* ed. C. J. Campbell et al. New York, Oxford Univ. Press: 321–44.

Styne, D. M. 1994. "Physiology of puberty." *Hormone Research* 41(Suppl. 2): 3–6.

Sugiyama, L. S., and R. Chacon. 2005. Juvenile responses to household ecology among the Yora of Peruvian Amazonia. In *Hunter-gatherer childhoods: Evolutionary, developmental and cultural perspectives,* ed. B. S. Hewlett and M. E. Lamb. New Brunswick, NJ, AldineTransaction: 237–61.

Sugiyama, Y., H. Kurita, et al. 2009. "Carrying of dead infants by Japanese macaque (*Macaca fuscata*) mothers." *Anthropological Science* 117(2): 113–19.

Sulak, P. J. 2003. "Sexually transmitted diseases." *Seminars in Reproductive Medicine* 21(4): 399–413.

Sulloway, F. 1997. *Born to rebel: Birth order, family dynamics, and creative lives.* New York: Vintage.

Sun, T., C. Patoine, et al. 2005. "Early asymmetry of gene transcription in embryonic human left and right cerebral cortex." *Science* 308(5729): 1794–98.

Sun, X.-Z., S. Takahashi, et al. 2002. "Normal and abnormal neuronal migration in the developing cerebral cortex." *Journal of Medical Investigation* 49(3–4): 97–110.

Suomi, S., H. Harlow, et al. 1972. "Monkey psychiatrists." *American Journal of Psychiatry* 128: 927–32.

Suomi, S. J. 1997. "Early determinants of behaviour: Evidence from primate studies." *British Medical Bulletin* 53(1): 170–84.

———. 2003. "Gene-environment interactions and the neurobiology of social conflict." *Annals of the New York Academy of Sciences* 1008: 132–39.

———. 2008. Attachment in rhesus monkeys. In *Handbook of attachment: Theory, research, and clinical applications,* 2nd ed., ed. J. S. Cassidy and P. R. Shaver. New York, Guilford: 173–91.

Suomi, S. J., and H. F. Harlow. 1972. "Social rehabilitation of isolate-reared monkeys." *Developmental Psychology* 6: 487–96.

————. 1975. The role and reason of peer relationships in rhesus monkeys. In *Friendship and peer relations,* ed. M. Lewis and L. A. Rosenblum. New York, Wiley: 153–85.

Super, C. M. 1976. "Environmental effects on motor development: The case of 'African infant precocity.'" *Developmental Medicine and Child Neurology* 18: 561–67.

————. 1981. Behavioral development in infancy. In *Handbook of cross-cultural human development,* ed. R. H. Munroe, R. L. Munroe, and B. B. Whiting. New York, Garland STPM: 181–270.

————, ed. 1987. *The role of culture in developmental disorder.* New York, Academic.

————. 1991. Developmental transitions of cognitive functioning in rural Kenya and metropolitan America. In *Brain maturation and cognitive development: Comparative and cross-cultural perspectives,* ed. K. R. Gibson and A. C. Petersen. New York, Aldine de Gruyter: 225–51.

Super, C. M, J. Kagan, et al. 1972. "Discrepancy and attention in the five-month infant." *Genetic Psychology Monographs* 85: 305–31.

Sutton-Smith, B. 1997. *The ambiguity of play.* Cambridge, MA, Harvard Univ. Press.

Suzuki, M., L. S. Wright, et al. 2004. "Mitotic and neurogenic effects of dehydroepiandrosterone (DHEA) on human neural stem cell cultures derived from the fetal cortex." *Proceedings of the National Academy of Sciences* 101(9): 3202–7.

Svejda, M. J., B. J. Pannabecker, et al. 1982. Parent-to-infant attachment: A critique of the early "bonding" model. In *The development of attachment and affiliative systems,* ed. R. N. Emde and R. J. Harmon. New York, Plenum: 83–93.

Swaab, D. F. 2004. "Sexual differentiation of the human brain: Relevance for gender identity, transsexualism and sexual orientation." *Gynecological Endocrinology* 19(6): 301–12.

Swaab, D. F., W. C. J. Chung, et al. 2001. "Structural and functional sex differences in the human hypothalamus." *Hormones and Behavior* 40(2): 93–98.

Swaab, D. F., and E. Fliers 1985. "A sexually dimorphic nucleus in the human brain." *Science* 228: 1112–15.

Swadesh, M. 2006. *The origin and diversification of languages,* ed. Joel F. Sherzer. Somerset, NJ, Transaction.

Swanson, G. E. 1964. *The birth of the gods: The origin of primitive beliefs.* Ann Arbor, Univ. of Michigan Press.

Swartz, K. B., D. Sarauw, et al. 1999. Comparative aspects of mirror self-recognition in great apes. In *The mentalities of gorillas and orangutans: Comparative perspectives,* ed. S. T. Parker, R. W. Mitchell, et al. New York, Cambridge Univ. Press: 283–94.

Szamalek, J. M., V. Goidts, et al. 2006. "The chimpanzee-specific pericentric inversions that distinguish humans and chimpanzees have identical breakpoints in *Pan troglodytes* and *Pan paniscus.*" *Genomics* 87(1): 39–45.

Tager-Flusberg, H. 1994. *Constraints on language acquisition: Studies of atypical children.* Hillsdale, NJ, Erlbaum.

Tajfel, H. 1982. *Social identity and intergroup relations.* New York, Cambridge Univ. Press.

Takada, A. 2005. Mother-infant interactions among the !Xun: Analysis of gymnastic and breastfeeding behaviors. In *Hunter-gatherer childhoods: Evolutionary, developmental and cultural perspectives,* ed. B. S. Hewlett and M. E. Lamb. New Brunswick, NJ, AldineTransaction: 289–308.

Takahashi, K., F.-C. Liu, et al. 2003. "Expression of Foxp2, a gene involved in speech and language, in the developing and adult striatum." *Journal of Neuroscience Research* 73(1): 61–72.

Takahata, Y., H. Ihobe, et al. 1996. Comparing copulations of chimpanzees and bonobos: Do females exhibit proceptivity or receptivity? In *Great ape societies*, ed. W. C. McGrew, L. F. Marchant, and T. Nishida. Cambridge, Cambridge Univ. Press: 146–55.

Tallal, P., R. Ross, et al. 1989. "Familial aggregation in specific language impairment." *Journal of Speech and Hearing Disorders* 54: 167–73.

Talmon, Y. 1964. "Mate selection in collective settlements." *American Sociological Review* 29: 491–508.

Tang, Y.-P., E. Shimizu, et al. 1999. "Genetic enhancement of learning and memory in mice." *Nature* 401(6748): 63–69.

Tanner, J. 1968. "Earlier maturation in man." *Scientific American* 218(1): 21–27.

———. 1975. "Trend toward earlier menarche in London, Oslo, Copenhagen, the Netherlands and Hungary." *Nature* 243: 95–96.

Tanner, J. M. 1962. *Growth at adolescence.* Oxford, Blackwell Scientific.

———. 1981. *A History of the study of human growth.* Cambridge, Cambridge Univ. Press.

Tardif, S. D., R. L. Carson, et al. 1990. "Infant-care behavior of mothers and fathers in a communal-care primate the cotton-top tamarin (*Saguinus-oedipus*)." *American Journal of Primatology* 22(2): 73–86.

Tasker, F. 2005. "Lesbian mothers, gay fathers, and their children: A review." *Journal of Developmental and Behavioral Pediatrics* 26(3): 224–40.

Tattersall, I. 1998. *Becoming human: Evolution and human uniqueness.* New York, Harcourt Brace.

Taub, D., and P. Mehlman. 1991. Primate paternalistic investment: A cross-species view. In *Understanding behavior: What primate studies tell us about human behavior,* ed. J. D. Loy and C. B. Peters. New York, Oxford Univ. Press: 51–89.

Tavare, S., C. R. Marshall, et al. 2002. "Using the fossil record to estimate the age of the last common ancestor of extant primates." *Nature* 416(6882): 726–29.

Taylor, M., B. M. Esbensen, et al. 1994. "Children's understanding of knowledge acquisition: The tendency for children to report that they have always known what they have just learned." *Child Development* 65(6): 1581–1604.

Taylor, W. D., D. C. Steffens, et al. 2005. "Influence of serotonin transporter promoter region polymorphisms on hippocampal volumes in late-life depression." *Archives of General Psychiatry* 62(5): 537–44.

Teleki, G. 1973. *The predatory behavior of wild chimpanzees.* Lewisburg, PA, Bucknell Univ. Press.

Tellegen, A., D. T. Lykken, et al. 1988. "Personality similarity in twins reared apart and together." *Journal of Personality and Social Psychology* 54: 1031–39.

Tennes, K. H., and E. E. Lampl. 1964. "Stranger and separation anxiety in infancy." *Journal of Nervous and Mental Disorders* 139: 247–54.

Terkel, J., and J. S. Rosenblatt. 1972. "Humoral factors underlying maternal behavior at parturition: Cross transfusion between freely moving rats." *Journal of Comparative and Physiological Psychology* 80: 365–71.

Terracciano, A., A. M. Abdel-Khalek, et al. 2005. "National character does not reflect mean personality trait levels in 49 cultures." *Science* 310(5745): 96–100.

Terracciano, A., P. T. Costa, Jr., and R. R. McCrae. 2006. "Personality plasticity after age 30." *Personality and Social Psychology Bulletin* 32(8): 999–1009.

Terrace, H. S., L. A. Petitto, et al. 1979. "Can an ape create a sentence?" *Science* 206(4421): 891–902.

Terranova, M. L., G. Laviola, et al. 1998. "A description of the ontogeny of mouse agonistic behavior." *Journal of Comparative Psychology* 112(1): 3–12.

Textor, R. B. 1967. *A cross-cultural summary.* New Haven, HRAF Press.

Thelen, E. 1993. Self-organization in developmental processes: Can systems approaches work? In *Brain development and cognition,* ed. M. H. Johnson. Cambridge, Blackwell: 555–91.

———. 1995. "Motor development. A new synthesis." *American Psychologist* 50(2): 79–95.

Thelen, E., and L. B. Smith. 1994. *A dynamic systems approach to the development of cognition and action.* Cambridge, MA, MIT Press.

Thierry, B., and J. R. Anderson. 1986. "Adoption in anthropoid primates." *International Journal of Primatology* 7(2): 191–216.

Thiery, J. P. 2003. "Cell adhesion in development: A complex signaling network." *Current Opinion in Genetics and Development* 13(4): 365–71.

Thiessen, D. D. 1964. "Amphetamine toxicity, population density, and behavior." *Psychological Bulletin* 62(6): 401–10.

Thoman, E., S. Levine, et al. 1967. "Effects of maternal deprivation and incubator rearing on adrenocortical activity in the adult rat." *Developmental Psychobiology* 1(1): 21–23.

Thompson, D. W. 1992 [1942]. *On growth and form: The complete revised edition.* New York, Dover.

Thompson, J. 1941. "Development of facial expression of emotion in blind and seeing children." *Archives of Psychology* 264: 1–47.

Thompson, K. V. 1998. Self-assessment in juvenile play. In *Animal play: Evolutionary, comparative, and ecological perspectives,* ed. M. Bekoff and J. A. Byers. Cambridge, Cambridge Univ. Press: 183–204.

Thompson, R. A. 2005. "Multiple relationships multiply considered." *Human Development* 48(1–2): 102–7.

Thornhill, N. W. 1991. "An evolutionary analysis of rules regulating human inbreeding and marriage." *Behavioral and Brain Sciences* 14: 247–93.

Thorpe, W. H. 1939. "Further studies on pre-imaginal olfactory conditioning in insects." *Proceedings of the Royal Society of London B: Biological Sciences* 127: 424–33.

Tiger, L. 1984. *Men in groups.* New York, Marion Boyars.

Tiger, L., and R. Fox. 1971. *The imperial animal.* New York, Holt, Rinehart and Winston.

Tiger, L. and J. Shepher. 1975. *Women in the kibbutz.* New York, Harcourt Brace.

Tilney, F., and L. Casamajor. 1924. "Myelinogenesis as applied to the study of behavior." *Archives of Neurology and Psychiatry* 12: 1–66.

Tinbergen, N. 1960. *The herring gull's world: A study of the social behavior of birds.* New York, Basic.

———. 1963. "On aims and methods of ethology." *Zeitschrift für Tierpsychologie* 20: 410–33.

———. 1964. The evolution of signalling devices. In *Social behavior and organization among vertebrates,* ed. W. Etkin. Chicago, Univ. of Chicago Press.

———. 1975. *The animal in its world: Explorations of an ethologist 1932–1972.* Cambridge, MA, Harvard Univ. Press.

Tinbergen, E., and N. Tinbergen. 1972. Early childhood autism: An ethological approach. In *Advances in ethology* (suppl. to *Zeitschrift für Tierpsychologie*). Berlin, Parey: 1–53.

Tither, J. M., and B. J. Ellis. 2008. "Impact of fathers on daughters' age at menarche: A genetically and environmentally controlled sibling study." *Developmental Psychology* 44(5): 1409–20.

Tizabi, Y., V. J. Massari, et al. 1980. "Isolation induced aggression and catecholamine variations in discrete brain areas of the mouse." *Brain Research Bulletin* 5(1): 81–86.

Tobias, P. V., and D. Falk. 1988. "Evidence for a dual pattern of cranial venous sinuses on the endocranial cast of Taung (*Australopithecus africanus*)." *American Journal of Physical Anthropology* 76(3): 309–12.

Tomasello, M. 1999a. "The human adaptation for culture." *Annual Review of Anthropology* 28(1): 509–29.

———. 1999b. *The cultural origins of human cognition.* Cambridge, MA, Harvard Univ. Press.

Tomasello, M., J. Call, et al. 2003. "Chimpanzees understand psychological states—the question is which ones and to what extent." *Trends in Cognitive Sciences* 7(4): 153–56.

Tomasello, M., and M. Carpenter. 2005. "The emergence of social cognition in three young chimpanzees." *Monographs of the Society for Research in Child Development* 70(1).

Tomasello, M., and M. J. Farrar. 1986. "Joint attention and early language." *Child Development* 57: 1454–63.

Tomasello, M., and A. C. Kruger. 1992. "Joint attention on actions: Acquiring verbs in ostensive and nonostensive contexts." *Journal of Child Language* 19: 311–34.

Tomasello, M., A. C. Kruger, et al. 1993. "Cultural learning." *Behavioral and Brain Sciences* 16(3): 495–552.

Tomita, K., T. Kubo, et al. 2007. "The neurotrophin receptor p75NTR in Schwann cells is implicated in remyelination and motor recovery after peripheral nerve injury." *GLIA* 55(11): 1199–1208.

Tönnies, F. 1957. *Community and society.* New York, Harper & Row.

Tooby, J., and I. DeVore. 1987. The reconstruction of hominid behavioral evolution through strategic modeling. In *The evolution of human behavior: Primate models,* ed. W. G. Kinzey. New York, State Univ. of New York Press: 183–237.

Tooby, J., and L. Cosmides. 2000. Toward mapping the evolved functional organization of mind and brain. In *The new cognitive neurosciences,* 2nd ed., ed. M. S. Gazzaniga. Cambridge, MA, MIT Press: 1167–78.

Toran-Allerand, C. D. 1976. "Sex steroids and the development of the newborn mouse hypothalamus and preoptic area in vitro: Implications for sexual differentiation." *Brain Research* 106: 407–12.

———. 1996. "Mechanisms of estrogen action during neural development: Mediation by interactions with the neurotrophins and their receptors?" *Journal of Steroid Biochemistry and Molecular Biology* 56(1–6 Spec. No.): 169–78.

Toth, N., K. D. Schick, et al. 1993. "Pan the tool-maker: Investigations into the stone tool-making and tool-using capabilities of a bonobo (*Pan paniscus*)." *Journal of Archaeological Science* 20: 81–91.

Tramo, M. J., W. C. Loftus, et al. 1998. "Brain size, head size, and intelligence quotient in monozygotic twins." *Neurology* 50(5): 1246–52.

Tremblay, R. E., W. W. Hartup, et al. 2005. *Developmental origins of aggression.* New York, Guilford.

Tremlin, T. 2005. *Minds and gods: The cognitive foundations of religion.* Oxford, Oxford Univ. Press.

Trevarthen, C. 1979. Communication and cooperation in early infancy: A description of primary intersubjectivity. In *Before speech: The beginning of human communication,* ed. M. Bullowa. London, Cambridge Univ. Press: 321–47.

Trevarthen, C., and K. J. Aitken. 2001. "Infant intersubjectivity: Research, theory, and clinical applications." *Journal of Child Psychology and Psychiatry and Allied Disciplines* 42(1): 3–48.

Trevathan, W. R. 1987. *Human birth: An evolutionary perspective.* New York, Aldine de Gruyter.

Trinkhaus, E. 1978. "Hard times among the Neanderthals." *Natural History* 87: 58–63.

———. 1995. *The Shanidar Neandertals.* New York, Academic.

Trinkhaus, E., and W. W. Howells. 1979. "The Neanderthals." *Scientific American* 241(6): 118–33.

Trinkhaus, E., and P. Shipman. 1993. *The Neandertals: Changing the image of mankind.* New York, Knopf.

Trivers, R. L. 1971. "The evolution of reciprocal altruism." *Quarterly Review of Biology* 46: 35–57.

———. 1972. Parental investment and sexual selection. In *Sexual selection and the descent of man, 1871–1971,* ed. B. Campbell. Chicago, Aldine: 136–79.

———. 1974. "Parent-offspring conflict." *American Zoologist* 14: 249–64.

———. 1985. *Social evolution.* Menlo Park, CA, Benjamin Cummings.

———. 1991. Deceit and self-deception: The relationship between communication and consciousness. In *Man and beast revisited,* ed. M. H. Robinson and L. Tiger. Washington, DC, Smithsonian Institution Press: 175–91.

Trivers, R. L., and D. E. Willard. 1973. "Natural selection of parental ability to vary the sex ratio of offspring." *Science* 179: 90–92.

Tronick, E. Z., G. A. Morelli, et al. 1987. "Multiple caretaking of Efe (Pygmy) infants." *American Anthropologist* 89: 96–106.

———. 1992. "The Efe forager infant and toddler's pattern of social relationships: Multiple and simultaneous." *Developmental Psychology* 28(4): 568–77.

Tronick, E. Z., and S. A. Winn. 1992. "The neurobehavioral organization of Efe (pygmy) infants." *Journal of Developmental and Behavioral Pediatrics* 13(6): 421–24.

Trull, T. J., and C. A. Durrett. 2005. "Categorical and dimensional models of personality disorder." *Annual Review of Clinical Psychology* 1(1): 355–80.

Trumbo, S. T. 1996. Parental care in invertebrates. In *Parental care: Evolution, mechanisms, and adaptive significance. Advances in the study of behavior,* vol. 25, ed. J. S. Rosenblatt and C. T. Snowdon. San Diego, Academic: 3–51.

Tucker, B., and A. G. Young. 2005. Growing up Mikea: Children's time allocation and tuber foraging in southwestern Madagascar. In *Hunter-gatherer childhoods: Evolutionary, developmental and cultural perspectives,* ed. B. S. Hewlett and M. E. Lamb. New Brunswick, NJ, AldineTransaction: 147–71.

Tulkin, S., and J. Kagan. 1972. "Mother-child interaction in the first year of life." *Child Development* 43: 31–41.

Tulkin, S. R. 1977. Social class differences in maternal and infant behavior. In *Culture and infancy,* ed. P. H. Leiderman, S. R. Tulkin, and A. Rosenfeld. New York, Academic.

Tullberg, B. S., M. Ah-King, et al. 2002. "Phylogenetic reconstruction of parental-care systems in the ancestors of birds." *Philosophical Transactions of the Royal Society of London B: Biological Sciences* 357(1419): 251–57.

Turke, P. W. 1988. Helpers at the nest: Childcare networks on Ifaluk. In *Human reproductive behaviour: A Darwinian perspective,* ed. L. Betzig, M. B. Mulder, and P. Turke. Cambridge, Cambridge Univ. Press: 173–88.

Turnbull, C. M. 1962. *The forest people: A study of the Pygmies of the Congo.* New York, Simon and Schuster.

———. 1965. *Wayward servants: The two worlds of the African Pygmies.* Garden City, NY, Natural History Press.

———. 1981. Mbuti womanhood. In *Woman the gatherer,* ed. F. Dahlberg. New Haven, Yale Univ. Press: 205–19.

Tutin, L., and W. McGrew. 1973. "Sexual behavior of group-living adolescent chimpanzees." *American Journal of Physical Anthropology* 38: 195–99.

Tylor, E. B. 1871. *Primitive culture: Researches into the development of mythology, philosophy, religion, language, art and custom.* London, Murray.

Tymicki, K. 2004. "Kin influence on female reproductive behavior: The evidence from reconstitution of the Bejsce parish registers, 18th to 20th centuries, Poland." *American Journal of Human Biology* 16(5): 508–22.

Udry, J. R. 1990. Hormonal and social determinants of adolescent sexual initiation. In *Adolescence and puberty,* ed. J. Bancroft and J. M. Reinisch. New York, Oxford Univ. Press: 70–87.

Udry, J. R., and L. M. Talbert. 1988. "Sex hormone effects on personality at puberty." *Journal of Personality and Social Psychology* 54(2): 291–95.

Uehara, S., and R. Nyuno. 1983. "One observed case of temporary adoption of an infant by unrelated nulliparous females among wild chimpanzees in the Mahale Mountains, Tanzania." *Primates* 24(4): 456–66.

Ullman, M. T., R. A. Miranda, et al. 2008. Sex differences in the neurocognition of language. In *Sex differences in the brain: From genes to behavior,* ed. J. B. Becker et al. New York, Oxford Univ. Press: 291–309.

Ursin, H., and B. R. Kaada. 1960. "Functional localization with the amygdaloid complex in the cat." *Electroencephalography and Clinical Neurology* 12: 1–20.

Utsunomiya, H., K. Takano, et al. 1999. "Development of the temporal lobe in infants and children: Analysis by MR-based volumetry." *AJNR: American Journal of Neuroradiology* 20(4): 717–23.

Uyemura, K., H. Asou, et al. 1996. "Cell-adhesion proteins of the immunoglobulin superfamily in the nervous system." *Essays in Biochemistry* 31: 37–48.

Vaillancourt, T., D. deCatanzaro, et al. 2009. "Androgen dynamics in the context of children's peer relations: An examination of the links between testosterone and peer victimization." *Aggressive Behavior* 35(1): 103–13.

Vaillant, G. E. 1977. *Adaptation to life.* Boston, Little, Brown.

———. 1993. *The wisdom of the ego.* Cambridge, MA, Harvard Univ. Press.

Vaina, L. M., J. W. Belliveau, et al. 1998. "Neural systems underlying learning and representation of global motion." *Proceedings of the National Academy of Sciences* 95(21): 12657–62.

Valdés-Pérez, R. E., and V. Pericliev. 1999. "Computer enumeration of significant implicational universals of kinship terminology." *Cross-Cultural Research* 33(2): 162–74.

Valenstein, E., W. Riss, et al. 1955. "Experimental and genetic factors in the organization of sexual behavior in male guinea pigs." *Journal of Comparative and Physiological Psychology* 48: 397–403.

Vallortigara, G., and L. J. Rogers. 2005. "Survival with an asymmetrical brain: Advantages and disadvantages of cerebral lateralization." *Behavioral and Brain Sciences* 28(4): 575–89; discussion 589–633.

Valsiner, J. 2000. *Culture and human development.* London, Sage.

Valverde, F. 1967. "Apical dendritic spines of the visual cortex and light deprivation in the mouse." *Experimental Brain Research* 3: 337–52.

Valzelli, L. 1973. "The 'isolation syndrome' in mice." *Psychopharmacologia* 31(4): 305–20.

Van Bokhoven, I., S. H. M. van Goozen, et al. 2006. "Salivary testosterone and aggression, delinquency, and social dominance in a population-based longitudinal study of adolescent males." *Hormones and Behavior* 50(1): 118–25.

Van den Berg, S. M., A. Setiawan, et al. 2006. "Individual differences in puberty onset in girls: Bayesian estimation of heritabilities and genetic correlations." *Behavior Genetics* 36(2): 261–70.

Van den Bergh, B. R. H., and A. Marcoen. 2004. "High antenatal maternal anxiety is related to ADHD symptoms, externalizing problems, and anxiety in 8- and 9-year-olds." *Child Development* 75(4): 1085–97.

Van den Berghe, P. L. 1987. Incest taboos and avoidance: Some African applications. In *Sociobiology and psychology: Ideas, issues and applications,* ed. C. Crawford, M. Smith, and D. Krebs. Hillsdale, NJ, Erlbaum: 353–71.

Van den Boom, D. C. 1995. "Do first-year intervention effects endure? Followup during toddlerhood of a sample of Dutch irritable infants." *Child Development* 66: 1798–1816.

Van der Meij, L., A. P. Buunk, et al. 2008. "The presence of a woman increases testosterone in aggressive dominant men." *Hormones and Behavior* 54(5): 640–44.

Van Heyningen, V., and K. A. Williamson. 2002. "PAX6 in sensory development." *Human Molecular Genetics* 11(10): 1161–67.

Van Hooff, J. A. R. A. M. 1972. A comparative approach to the phylogeny of laughter and smiling. In *Non-verbal communications,* ed. R. A. Hinde. Cambridge, Cambridge Univ. Press.

Van Ijzendoorn, M., and A. Sagi-Schwartz. 2008. Cross-cultural patterns of attachment. In *Handbook of attachment,* 2nd ed., ed. J. Cassidy and P. R. Shaver. New York, Guilford: 880–905.

Van Ijzendoorn, M. H., and M. J. Bakermans-Kranenburg. 2006. "DRD4 7-repeat polymorphism moderates the association between maternal unresolved loss or trauma and infant disorganization." *Attachment and Human Development* 8(4): 291–307.

Van Ijzendoorn, M. H., M. J. Bakermans-Kranenburg, et al. 2003. "Are children of Holocaust survivors less well-adapted? A meta-analytic investigation of secondary traumatization." *Journal of Traumatic Stress* 16(5): 459–69.

Van Lawick, H., and J. van Lawick-Goodall, eds. 1971. *Innocent killers.* Boston, Houghton Mifflin.

Van Noordwijk, M. A., and C. P. van Schaik. 2004. Sexual selection and the careers of primate males: Paternity concentration, dominance-acquisition tactics and transfer decisions. In *Sexual selection in primates,* ed. P. M. Kappeler and C. P. van Schaik. New York, Cambridge Univ. Press: 208–29.

Van Roo, B. L., E. D. Ketterson, et al. 2003. "Testosterone and prolactin in two songbirds that differ in paternal care: The blue-headed vireo and the red-eyed vireo." *Hormones and Behavior* 44(5): 435–41.

Van Schaik, C. P., N. Barrickman, et al. 2006. Primate life histories and the role of brains. In *Evolution of human life history,* ed. K. Hawkes and R. R. Paine. Santa Fe, School of American Research Press: 127–54.

Van Schaik, C. P., and C. H. Janson, eds. 2000. *Infanticide by males and its implications.* Cambridge: Cambridge Univ. Press.

Van Schaik, C. P., and P. M. Kappeler. 2003. The evolution of social monogamy in primates. In *Monogamy: Mating strategies and partnerships in birds, humans, and other mammals,* ed. U. H. Reichard and C. Boesch. Cambridge, Cambridge Univ. Press: 59–80.

Van Schaik, C. P., G. R. Pradhan, et al. 2004. Mating conflict in primates: Infanticide, sexual harassment and female sexuality. In *Sexual selection in primates,* ed. P. M. Kappeler and C. P. van Schaik. New York, Cambridge Univ. Press: 131–50.

Van Schaik, C. P., and J. A. R. A. M. Van Hooff. 1996. Toward an understanding of the orangutan's social system. In *Great ape societies,* ed. W. C. McGrew, L. F. Marchant, and T. Nishida. Cambridge, Cambridge Univ. Press: 3–15.

Van Sleuwen, B. E., A. C. Engelberts, et al. 2007. "Swaddling: A systematic review." *Pediatrics* 120(4): e1097–e1106.

Van Velsen, H. H. E. T., and W. V. Weterling. 1960. "Residence, power groups and intra-society aggression." *International Archives of Ethnology* 49: 169–200.

Van Weissenbruch, M. M., and H. A. Delemarre-van de Waal. 2006. "Early influences on the tempo of puberty." *Hormone Research* 65(Suppl. 3): 105–11.

van Wingen, G., C. Mattern, et al. 2010. "Testosterone reduces amygdala-orbitofrontal cortex coupling." *Psychoneuroendocrinology* 35: 105–13.

Vanderschuren, L. J. M. J., R. J. M. Niesink, et al. 1997. "The neurobiology of social play behavior in rats." *Neuroscience and Biobehavioral Reviews* 21(3): 309–26.

Vanderschuren, L. J. M. J., E. A. Stein, et al. 1995. "Social play alters regional brain opioid receptor binding in juvenile rats." *Brain Research* 680(1–2): 148–56.

Vannatta, K., C. A. Gerhardt, et al. 2007. "Intensity of CNS treatment for pediatric cancer: Prediction of social outcomes in survivors." *Pediatric Blood and Cancer* 49(5): 716–22.

Vargha-Khadem, F., D. G. Gadian, et al. 2005. "FOXP2 and the neuroanatomy of speech and language." *Nature Reviews Neuroscience* 6(2): 131–38.

Varki, A., and D. L. Nelson. 2007. "Genomic comparisons of humans and chimpanzees." *Annual Review of Anthropology* 36: 191–209.

Venditti, C., A. Meade, et al. 2010. "Phylogenies reveal new interpretation of speciation and the Red Queen." *Nature* 463(8623): 349–52.

Verweij, K. J. H., B. P. Zietsch, et al. 2009. "Shared aetiology of risky sexual behaviour and adolescent misconduct: Genetic and environmental influences." *Genes, Brain, and Behavior* 8(1): 107–13.

Vinden, P. G. 1996. "Junin Quechua children's understanding of mind." *Child Development* 67(4): 1707–16.

———. 1999. "Children's understanding of mind and emotion: A multi-culture study." *Cognition and Emotion* 13(1): 19–48.

———. 2002. "Understanding minds and evidence for belief: A study of Mofu children in Cameroon." *International Journal of Behavioral Development* 26(5): 445–52.

Vining, D. R. 1986. "Social versus reproductive success: The central theoretical problem of human sociobiology." *Behavioral and Brain Sciences* 9(1): 167–216.

Vinovskis, M. A. 1988. *An "epidemic" of adolescent pregnancy? Some historical and policy considerations.* New York, Oxford Univ. Press.

Vogel, C., and H. Loch. 1984. Reproductive parameters, adult-male replacements, and infanticide among free-ranging Langurs (*Presbytis entellus*) at Jodhpur (Rajasthan), India. In *Infanticide: Comparative and Evolutionary Perspectives,* ed. G. Hausfater and S. B. Hrdy. New York, Aldine de Gruyter: 237–56.

Von Frisch, K. 1967. *The dance language and orientation of bees.* Cambridge, MA, Harvard Univ. Press.

Vrba, E. S. 1998. "Multiphasic growth models and the evolution of prolonged growth exemplified by human brain evolution." *Journal of Theoretical Biology* 190(3): 227–39.

Vygotsky, L. 1962. *Thought and language.* Cambridge, MA, MIT Press.

Waber, D. P., C. De Moor, et al. 2007. "The NIH MRI study of normal brain development: Performance of a population based sample of healthy children aged 6 to 18 years on a neuropsychological battery." *Journal of the International Neuropsychological Society* 13(5): 729–46.

Wada, J. A., R. Clarke, et al. 1975. "Cerebral hemispheric asymmetry in humans. Cortical speech zones in 100 adults and 100 infant brains." *Archives of Neurology* 32(4): 239–46.

Wada, J. A., and A. E. Davis. 1977. "Fundamental nature of human infant's brain asymmetry." *Canadian Journal of Neurological Sciences* 4(3): 203–7.

Waddell, C., D. R. Offord, et al. 2002. "Child psychiatric epidemiology and Canadian public policy-making: The state of the science and the art of the possible." *Canadian Journal of Psychiatry/Revue Canadienne de Psychiatrie* 47(9): 825–32.

Waddington, C. H. 1942. "Canalisation of development and the inheritance of acquired characters." *Nature* 150: 563–65.

——. 1953. "Genetic assimilation of an acquired character." *Evolution* 7: 118.

——. 1957. *The strategy of the genes.* London, Allen & Unwin.

——. 1971. Concepts of development. In *The biopsychology of development,* ed. E. Tobach, L. R. Aronson, and E. Shaw. New York, Academic: 17–23.

Wade, M. J., and S. M. Shuster. 2002. "The evolution of parental care in the context of sexual selection: A critical reassessment of parental investment theory." *American Naturalist* 160(3): 285–92.

Wade, N. 1998. "Of men and mice: Here they come to save the day." *New York Times,* May 10, 1998, 45.

Wainright, J. L., and C. J. Patterson. 2009. "Peer relations among adolescents with female same-sex parents." *Developmental Psychology* 44(1): 117–26.

Wainright, J. L., S. T. Russell, et al. 2004. "Psychosocial adjustment, school outcomes, and romantic relationships of adolescents with same-sex parents." *Child Development* 75(6): 1886–98.

Walker, A. 1997. *Proconsul:* Function and phylogeny. In *Function, phylogeny, and fossils: Miocene hominoid evolution and adaptation,* ed. D. R. Begun, C. V. Ward, and M. D. Rose. New York, Plenum: 209–24.

Walker, A., and C. B. Ruff. 1993. The reconstruction of the pelvis. In *The Nariokotome Homo erectus skeleton,* ed. A. Walker and R. Leakey. Cambridge, MA, Harvard Univ. Press: 221–33.

Wallace, A. F. C. 1952. "The modal personality structure of the Tuscarora Indians as revealed by the Rorschach test." *Bulletin of the Bureau of American Indian Ethnology* 150.

————. 1961. *Culture and personality,* 2nd ed. New York, Random House.

Wallerstein, J. S. 1987. "Children of divorce: Report of a ten-year follow-up of early latency-age children." *American Journal of Orthopsychiatry* 57(2): 199–211.

Wallerstein, J. S., and J. M. Lewis. 2004. "The unexpected legacy of divorce: Report of a 25-year study." *Psychoanalytic Psychology* 21(3): 353–70.

Wallis, J. 1992. Socioenvironmental effects on timing of first postpartum cycles in chimpanzees. In *Topics in primatology,* vol. 1, *Human origins,* ed. T. Nishida et al. Tokyo, Univ. of Tokyo Press: 119–30.

Walsh, D., and S. M. Downe. 2004. "Outcomes of free-standing, midwife-led birth centers: A structured review." *Birth* 31(3): 222–29.

Walsh, R. N., R. A. Cummins, et al. 1973. "Environmentally induced changes in the dimensions of the rat cerebrum: A replication and extension." *Developmental Psychobiology* 6(1): 3–7.

Wang, K., H. Zhang, et al. 2009. "Common genetic variants on 5p14.1 associate with autism spectrum disorders." *Nature* 459: 528–33.

Wang, W., Y.-X. Wu, et al. 2000. "Test of sensation seeking in a Chinese sample." *Personality and Individual Differences* 28(1): 169–79.

Wang, Y., Z. Wang, et al. 2003. "Evolution of endothermy in animals: A review." *Zoological Research* 24(6): 480–87.

Wang, Z., T. J. Hulihan, et al. 1997. "Sexual and social experience is associated with different patterns of behavior and neural activation in male prairie voles." *Brain Research* 767: 321–32.

Wang, Z., K. Moody, et al. 1997. "Vasopressin and oxytocin immunoreactive neurons and fibers in the forebrain of male and female common marmosets (*Callithrix jacchus*)." *Synapse* 27(1): 14–25.

Wang, Z., D. Toloczko, et al. 1997. "Vasopressin in the forebrain of common marmosets (*Callithrix jacchus*): Studies with in situ hybridization, immunocytochemistry and receptor autoradiography." *Brain Research* 768(1–2): 147–56.

Wang, Z., L. Zhou, et al. 1996. "Immunoreactivity of central vasopressin and oxytocin pathways in microtine rodents: A quantitative comparative study." *Journal of Comparative Neurology* 366(4): 726–37.

Want, S. C., and P. L. Harris. 2002. "How do children ape? Applying concepts from the study of non-human primates to the developmental study of 'imitation' in children." *Developmental Science* 5(1): 1–13.

Wappner, R., S. Cho, et al. 1999. "Management of phenylketonuria for optimal outcome: A review of guidelines for phenylketonuria management and a report of surveys of parents, patients, and clinic directors." *Pediatrics* 104(6): e68.

Ward, S., B. Brown, et al. 1999. "*Equatorius*: A new hominoid genus from the middle Miocene of Kenya." *Science* 285(5432): 1382–86.

Warren, M. P., and J. Brooks-Gunn. 1989. "Mood and behavior at adolescence: Evidence for hormonal factors." *Journal of Clinical Endocrinology and Metabolism* 69(1): 77–83.

Warren, N. 1972. "African infant precocity." *Psychological Bulletin* 78: 353–67.

Warren, W., and A. Ivinskis. 1973. "Isolation-rearing effects of emotionality of hooded rats." *Psychological Reports* 32: 1011–14.

Warren, W. C., L. W. Hillier, et al. 2008. "Genome analysis of the platypus reveals unique signatures of evolution." *Nature* 453(7192): 175–83.

Washburn, S. L. 1960. "Tools and human evolution." *Scientific American* 203(3): 3–15.

Watamura, S. E., B. Donzella, et al. 2003. "Morning-to-afternoon increases in cortisol concentrations for infants and toddlers at child care: Age differences and behavioral correlates [see comment]." *Child Development* 74(4): 1006–20.

Waters, E., D. Corcoran, et al. 2005. "Attachment, other relationships, and the theory that all good things go together." *Human Development* 48(1–2): 80–84.

Watson, D. M. 1998. Kangaroos at play: Play behaviour in the Macropodoidea. In *Animal play: Evolutionary, comparative, and ecological perspectives,* ed. M. Bekoff and J. A. Byers. Cambridge, Cambridge Univ. Press: 61–95.

Watson, J. S. 1972. "Smiling, cooing, and 'The game.'" *Merrill-Palmer Quarterly* 18: 323–39.

Watson, J. S., and C. T. o. P. Ramey. 1972. "Reactions to response-contingent stimulation in early infancy." *Merrill-Palmer Quarterly* 18: 219–28.

Watson, M. W., and T. Amgott-Kwan. 1983. "Transitions in children's understanding of parental roles." *Developmental Psychology* 19(5): 659–66.

Watson, M. W., K. W. Fischer, et al. 2004. "Pathways to aggression in children and adolescents." *Harvard Educational Review* 74(4): 404–30.

Watt, S. L. 1994. "Alloparental behavior in a captive group of spider monkeys (*Ateles geoffroyi*) at the Auckland Zoo." *International Journal of Primatology* 15(1): 135–51.

Watts, D. P., M. Muller, et al. 2006. "Lethal intergroup aggression by chimpanzees in Kibale National Park, Uganda." *American Journal of Primatology* 68(2): 161–80.

Watts, D. P., and A. E. Pusey. 1993. Behavior of juvenile and adolescent great apes. In *Juvenile primates: Life history, development, and behavior,* ed. M. E. Pereira and L. A. Fairbanks. New York, Oxford Univ. Press: 148–67.

Waxman, S. G., and L. Bangalore. 2004. Electrophysiologic consequences of myelination. In *Myelin biology and disorders,* vol. 1, ed. R. Lazzarini. San Diego, Elsevier: 117–42.

Wayne, N. L., W. Lee, et al. 2004. "Activity-dependent regulation of neurohormone synthesis and its impact on reproductive behavior in aplysia." *Biology of Reproduction* 70(2): 277–81.

Weaver, I. C. G., N. Cervoni, et al. 2004. "Epigenetic programming by maternal behavior [see comment]." *Nature Neuroscience* 7(8): 847–54.

Weddell, R. A., J. D. Miller, et al. 1990. "Voluntary emotional facial expressions in patients with focal cerebral lesions." *Neuropsychologia* 28(1): 49–60.

Weglage, J., M. Pietsch, et al. 1999. "Regression of neuropsychological deficits in early-treated phenylketonurics during adolescence." *Journal of Inherited Metabolic Disease* 22(6): 693–705.

Wegner, M. 2000. "Transcriptional control in myelinating glia: The basic recipe." *GLIA* 29(2): 118–23.

Weinberg, M. K., E. Z. Tronick, et al. 1999. "Gender differences in emotional expressivity and self-regulation during early infancy." *Developmental Psychology* 35(1): 175–88.

Weintraub, S., and M.-M. Mesulam. 1983. "Developmental learning disabilities of the right hemisphere: Emotional, interpersonal, and cognitive components." *Archives of Neurology* 40: 463–68.

Weintraub, S., M.-M. Mesulam, et al. 1981. "Disturbances in prosody: A right-hemisphere contribution to language." *Archives of Neurology* 38: 742–44.

Wiesel, T., and D. Hubel. 1965. "Extent of recovery from the effects of visual deprivation in kittens." *Journal of Neurophysiology* 28: 1060–72.

Weisfeld, G. E. 1999. *Evolutionary principles of human adolescence.* New York, Basic.

Weisfeld, G. E., and L. Woodward. 2004. "Current evolutionary perspectives on adolescent romantic relations and sexuality [see comment]." *Journal of the American Academy of Child and Adolescent Psychiatry* 43(1): 11–19; discussion 20–23.

Weisner, T. S. 1987. Socialization for parenthood in sibling caretaking societies. In *Parenting across the life span: Biosocial dimensions,* ed. J. B. Lancaster et al. New York, Aldine de Gruyter: 237–70.

———. 1996. The 5 to 7 transition as an ecocultural project. In *The five to seven year shift: The age of reason and responsibility,* ed. A. J. Sameroff and M. M. Haith. Chicago, Univ. of Chicago Press: 295–326.

———. 1997. "The ecocultural project of human development: Why ethnography and its findings matter." *Ethos* 25(2): 177–90.

———. 2002. "Ecocultural understanding of children's developmental pathways." *Human Development* 45(4): 275–81.

Weisner, T. S., R. Gallimore, et al. 1977. "My brother's keeper: Child and sibling caretaking [and comments and reply]." *Current Anthropology* 18(2): 169–90.

Weiss, J. 1972. "Psychological factors in stress and disease." *Scientific American* 226: 104–13.

Weiss, J. M., and C. D. Kilts. 1995. Animal models of depression and schizophrenia. In *The American Psychiatric Press textbook of psychopharmacology,* ed. A. F. Shatzberg and C. B. Nemeroff. Washington, DC, American Psychiatric Press: 81–123.

Welch, A. S., and B. L. Welch. 1971. Isolation, reactivity and aggression: Evidence for an involvement of brain catecholamines and serotonin. In *The physiology of aggression and defeat,* ed. G. Eleftheriou and J. Scott. New York, Plenum: 91–137.

Welch, B., and A. Welch. 1965. "Effect of grouping on the level of brain norepinephrine in white swiss mice." *Life Sciences* 4: 1011.

Welch, B. L., and A. S. Welch. 1966. "Graded effect of social stimulation upon d-amphetamine toxicity, aggressiveness and heart and adrenal weight." *Journal of Pharmacology and Experimental Therapeutics* 151(3): 331–38.

Wells, D. A. 2005. "Tool culture, the Baldwin effect and the evolution of the human hand." *Journal and Proceedings of the Royal Society of New South Wales* 138(Pt. 3–4): 85–92.

Werner, E. E. 1972. "Infants around the world: Cross-cultural studies of psychomotor development from birth to two years." *Journal of Cross-Cultural Psychology* 3: 111–34.

———. 1989. "High-risk children in young adulthood: A longitudinal study from birth to 32 years." *American Journal of Orthopsychiatry* 59(1): 72–81.

———. 1996. "Vulnerable but invincible: High risk children from birth to adulthood." *European Child and Adolescent Psychiatry* 5(Suppl. 1): 47–51.

Werner, E. E., and R. S. Smith. 2001. *Journeys from childhood to midlife: Risk, resilience, and recovery.* Ithaca, NY, Cornell Univ. Press.

Werner, H. 1948. *Comparative psychology of mental development.* Chicago, Follet.

Wernicke, C. 1874. *Der aphasische symptom-complex. Eine psychologische studie auf anatomischer basis.* Breslau, Cohn and Weigert.

Wertsch, J. V., and R. Sohmer. 1995. "Vygotsky on learning and development." *Human Development* 38: 332–37.

Wertz, R., and D. Wertz. 1989. *Lying in: A history of childbirth in America,* expanded edition. New Haven, Yale Univ. Press.

West, M. 1974. "Social play in the domestic cat." *American Zoologist* 14: 427.

West, G., J. Brown, et al. 1997. "A general model for the origin of allometric scaling laws in biology." *Science* 276: 122–26.

West, M. J., and A. P. King. 1987. "Settling nature and nurture into an ontogenetic niche." *Developmental Psychobiology* 20(5): 549–62.

West-Eberhard, M. J. 1989. "Phenotype plasticity and the origins of diversity." *Annual Review of Ecology and Systematics* 20: 249–78.

———. 2003. *Developmental plasticity and evolution.* New York, Oxford Univ. Press.

———. 2005. "Phenotypic accommodation: Adaptive innovation due to developmental plasticity." *Journal of Experimental Zoology Part B: Molecular and Developmental Evolution* 304B(6): 610–18.

Westendorp, R. G., and T. B. Kirkwood. 1998. "Human longevity at the cost of reproductive success [see comment]." *Nature* 396(6713): 743–46.

Westermarck, E. 1922. *The history of human marriage.* New York, Allerton.

Whalen, P. J., J. Kagan, et al. 2004. "Human amygdala responsivity to masked fearful eye whites." *Science* 306(5704): 2061.

Whitcome, K. K., L. J. Shapiro, et al. 2007. "Fetal load and the evolution of lumbar lordosis in bipedal hominins." *Nature* 450(7172): 1075–78.

White, C. R., T. M. Blackburn, et al. 2009. "Phylogenetically informed analysis of the allometry of mammalian basal metabolic rate supports neither geometric nor quarter-power scaling." *Evolution* 63(10): 2658–67.

White, F. J., and K. D. Wood. 2007. "Female feeding priority in bonobos, *Pan paniscus,* and the question of female dominance." *American Journal of Primatology* 69(8): 837–50.

White, S. 1970. "Some general outlines of the matrix of developmental changes between five and seven years." *Bulletin of the Orton Society* 20: 41–57.

White, S. H. 1965. Evidence for a hierarchical arrangement of learning processes. *Advances in child development and behavior,* vol. 2, ed. L. P. Lipsitt and C. C. Spiker. New York, Academic.

———. 1968. *Changes in learning processes in the late preschool years.* Chicago, American Educational Research Association.

———. 1996. The child's entry into the "age of reason." In *The five to seven year shift: The age of reason and responsibility,* ed. A. J. Sameroff and M. M. Haith. Chicago, Univ. of Chicago Press: 17–30.

White, T. D. 1980. "Evolutionary implications of pliocene hominid footprints." *Science* 208: 175–76.

———, T. D. 1992. *Prehistoric cannibalism at Mancos 5Mtumr-2346.* Princeton, Princeton Univ. Press.

White, T. D., B. Asfaw, et al. 2009. "*Ardipithecus ramidus* and the paleobiology of early hominids." *Science* 326(5949): 75–86.

White, T. D., G. Suwa, et al. 1994. "*Australopithecus ramidus,* a new species of early hominid from Aramis, Ethiopia." *Nature* 371: 306–12.

Whitehead, H. 1996. "Babysitting, dive synchrony, and indications of alloparental care in sperm whales." *Behavioral Ecology and Sociobiology* 38(4): 237–44.

Whiteman, E. A., and I. M. Cote. 2004. "Monogamy in marine fishes." *Biological Reviews of the Cambridge Philosophical Society* 79(2): 351–75.

Whiten, A. 1988. *Machiavellian intelligence: Social expertise and the evolution of intellect in monkeys, apes, and humans.* Oxford, Clarendon.

———. 1999. Parental encouragement in gorilla in comparative perspective: Implications for social cognition and the evolution of teaching. In *The mentalities of gorillas and orangutans: Comparative perspectives,* ed. S. T. Parker, R. W. Mitchell, and H. L. Miles. New York, Cambridge Univ. Press.

———. 2002. "Primatology and developmental science: Who's aping whom?" *Developmental Science* 5(1): 36–38.

———. 2007. "Pan African culture: Memes and genes in wild chimpanzees." *Proceedings of the National Academy of Sciences* 104(45): 17559–60.

Whiten, A., and C. Boesch. 2001. "The cultures of chimpanzees." *Scientific American* 284(1): 60–67.

Whiten, A., and R. W. Byrne, eds. 1997. *Machiavellian intelligence II: Extensions and evaluations.* Cambridge, Cambridge Univ. Press.

Whiten, A., J. Goodall, et al. 1999. "Cultures in chimpanzees." *Nature* 399(6737): 682–85.

———. 2001. "Charting cultural variation in chimpanzees." *Behaviour* 138(11–12): 1481–1516.

Whiten, A., and C. P. van Schaik. 2007. "The evolution of animal 'cultures' and social intelligence." *Philosophical Transactions of the Royal Society of London B: Biological Sciences* 362(1480): 603–20.

Whiting, B. B. 1963. *Six cultures: Studies of child rearing.* New York, Wiley.

———. 1965. "Sex identity conflict and physical violence: A comparative study." *American Anthropologist* 67(Pt. 2): 123–40.

———. 1972. Work and the family: Cross-cultural perspectives. In *Women: Resource for a changing world.* Cambridge, MA, Radcliffe Institute.

Whiting, B. B., and C. P. Edwards. 1973. "A cross-cultural analysis of sex differences in the behavior of children aged three through eleven." *Journal of Social Psychology* 91: 171–88.

———. 1988. *Children of different worlds: The formation of social behavior.* Cambridge, MA, Harvard Univ. Press.

Whiting, B. B., and J. W. M. Whiting. 1975a. *Children of six cultures: A psychocultural analysis.* Cambridge, MA, Harvard Univ. Press.

Whiting, J. W. M., and B. B. Whiting. 1975b. "Aloofness and intimacy between husbands and wives." *Ethos* 3: 183–207.

Whiting, J. W. 1941. *Becoming a Kwoma.* New Haven, Yale Univ. Press.

Whiting, J. W. M. 1961. Socialization process and personality. In *Psychological anthropology: Approaches to culture and personality,* ed. F. L. K. Hsu. Homewood, IL, Dorsey: 355–80.

———. 1965. Menarcheal age and infant stress in humans. In *Sex and behavior,* ed. F. A. Beach. New York, John Wiley.

———. 1971. Causes and consequences of the amount of body contact between mother and infant. In *70th Annual Meeting of the American Anthropological Association.* New York, American Anthropological Association.

———. 1981. Environmental constraints on infant care practices. In *Handbook of cross-cultural human development,* ed. R. H. Munroe, R. L. Munroe, and B. B. Whiting. New York, Garland STPM: 155–79.

———. 1990. Adolescent rituals and identity conflicts. In *Cultural psychology: Essays on comparative human development,* ed. J. W. Stigler, R. A. Shweder, and G. Herdt. Cambridge, Cambridge Univ. Press: 357–65.

Whiting, J. W. M., and I. L. Child. 1953. *Child training and personality: A cross-cultural study.* New Haven, Yale Univ. Press.

Whiting, J. W. M., I. L. Child, et al. 1966. *Field guide for a study of socialization.* New York, Wiley.

Whiting, J. W. M., R. Kluckholn, et al. 1958. The function of male initiation ceremonies at puberty. In *Readings in social psychology,* ed. E. E. Maccoby, T. Newcomb, and E. Hartley. New York, Holt.

Whiting, J. W. M., T. K. Landauer, et al. 1968. "Infantile vaccination and adult stature." *Child Development* 39: 59–67.

Whitten, P. L. 1987. Infants and adult males. In *Primate societies,* ed. B. B. Smuts et al. Chicago, Univ. of Chicago Press: 343–57.

Whittle, S., M. B. Yap, et al. 2008. "Prefrontal and amygdala volumes are related to adolescents' affective behaviors during parent—adolescent interactions." *PNAS Proceedings of the National Academy of Sciences* 105(9): 3652–57.

Whittle, S., M. B. H. Yap, et al. 2009. "Maternal responses to adolescent positive affect are associated with adolescents' reward neuroanatomy." *Social Cognitive and Affective Neuroscience* 4(3): 247–56.

Whittle, S., M. Yücel, et al. 2008. "Neuroanatomical correlates of temperament in early adolescents." *Journal of the American Academy of Child and Adolescent Psychiatry* 47(6): 682–93.

Whorf, B. L. 1988. The relation of habitual thought and behavior to language. In *High points in anthropology,* ed. B. Paul and M. Glazer. 2nd ed. New York, McGraw-Hill: 152–71.

Wibom, R., F. M. Lasorsa, et al. 2009. "Mutation in AGC1 deficiency associated with global cerebral hypomyelination." *New England Journal of Medicine* 361(5): 489–95.

Wich, S. A., S. S. Utami-Atmoko, et al. 2004. "Life history of wild Sumatran orangutans (*Pongo abelii*)." *Journal of Human Evolution* 47(6): 385–98.

Wicher, D., C. Walther, et al. 2001. "Non-synaptic ion channels in insects—basic properties of currents and their modulation in neurons and skeletal muscles." *Progress in Neurobiology* 64(5): 431–525.

Wicht, H., and R. G. Northcutt. 1992. "The forebrain of the Pacific Hagfish: A cladistic reconstruction of the ancestral craniate forebrain." *Brain, Behavior and Evolution* 40: 25–64.

Wickler, W., and U. Seibt. 1977. *Das Prinzip Eigennutz: Ursachen und Konsequenzen sozialen Verhaltens.* Hamburg, Hoffmann und Campe.

Widom, C. S. 1989. "The cycle of violence." *Science* 244(4901): 160–66.

Wierzbicka, A. 2005. "Empirical universals of language as a basis for the study of other human universals and as a tool for exploring cross-cultural differences." *Ethos* 33(2): 256–91.

Wiessner, P. 1982. Risk, reciprocity and social influences on !Kung San economics. In *Politics and history in band societies,* ed. E. Leacock and R. B. Lee. Cambridge, Cambridge Univ. Press: 61–84.

———. 2002. "Hunting, healing, and hxaro exchange: A long-term perspective on !Kung (Ju/'hoansi) large-game hunting." *Evolution and Human Behavior* 23(6): 407–36.

Wiessner, P., and A. Tumu. 1998. *Historical vines: Enga networks of exchange, ritual, and warfare in Papua New Guinea.* Washington, DC, Smithsonian Institution Press.

Wikan, U. 1990. *Managing turbulent hearts: A Balinese formula for living.* Chicago, Univ. of Chicago Press.

Wild, B., M. Erb, et al. 2003. "Why are smiles contagious? An fMRI study of the interaction between perception of facial affect and facial movements." *Psychiatry Research: Neuroimaging* 123(1): 17–36.

Wilhelm, D., and P. Koopman. 2006. "The makings of maleness: Towards an integrated view of male sexual development." *Nature Reviews Genetics* 7(8): 620–31.

Will, J. A., P. A. Self, et al. 1976. "Maternal behavior and perceived sex of infant." *American Journal of Orthopsychiatry* 46: 135–39.

Williams, G. C. 1966. *Adaptation and natural selection.* Princeton, Princeton Univ. Press.

———. 1992. *Natural selection: Domains, levels, and challenges.* New York, Oxford Univ. Press.

Williams, M. A., F. McGlone, et al. 2005. "Differential amygdala responses to happy and fearful facial expressions depend on selective attention." *Neuroimage* 24(2): 417–25.

Williams, R. 2006. "The psychosocial consequences for children and young people who are exposed to terrorism, war, conflict and natural disasters." *Current Opinion in Psychiatry* 19(4): 337–49.

Williamson, D. E., K. Coleman, et al. 2003. "Heritability of fearful-anxious endophenotypes in infant rhesus macaques: A preliminary report." *Biological Psychiatry* 53(4): 284–91.

Wilmot, C. A., T. A. Fico, et al. 1989. "Dopamine autoreceptor agonists attenuate spontaneous motor activity but not spontaneous fighting in individually-housed mice." *Pharmacology, Biochemistry and Behavior* 33(2): 387–91.

Wilson, A. B., I. Ahnesjo, et al. 2003. "The dynamics of male brooding, mating patterns, and sex roles in pipefishes and seahorses (family Syngnathidae)." *Evolution* 57(6): 1374–86.

Wilson, E. O. 1971. *The insect societies.* Cambridge, MA, Harvard Univ. Press.

———. 1975. *Sociobiology: The new synthesis.* Cambridge, MA, Harvard Univ. Press.

———. 1978. *On human nature.* Cambridge, MA, Harvard Univ. Press.

———. 1998. *Consilience: The unity of knowledge.* New York, Knopf.

Wilson, R. S. 1978. "Synchronies in mental development: An epigenetic perspective." *Science* 202: 939–48.

Wimer, C. 1990. Genetic studies of brain development. In *Developmental behavior genetics: Neural, biometrical, and evolutionary approaches,* ed. M. E. Hahn et al. New York, Oxford Univ. Press: 85–99.

Windle, W. F., M. W. Fish, et al. 1934. "Myelogeny of the cat as related to development of fiber tracts and prenatal behavior patterns." *Journal of Comparative Neurology* 59: 139–65.

Winslow, J. T., N. Hastings, et al. 1993. "A role for central vasopressin in pair bonding in monogamous prairie voles." *Nature* 365(6446): 545–48.

Winslow, J. T., E. F. Hearn, et al. 2000. "Infant vocalization, adult aggression, and fear behavior of an oxytocin null mutant mouse." *Hormones and Behavior* 37(2): 145–55.

Wisner, K. L., B. L. Gracious, et al. 2003. Postpartum disorders: Phenomenology, treatment approaches, and relationship to infanticide. In *Infanticide: Psychosocial and legal perspectives on mothers who kill,* ed. M. Spinelli. Washington, DC, American Psychiatric Publishing: 35–60.

Wisniewski, A. B. 1998. "Sexually-dimorphic patterns of cortical asymmetry, and the role for sex steroid hormones in determining cortical patterns of lateralization." *Psychoneuroendocrinology* 23(5): 519–47.

Witelson, S. F., and W. Pallie. 1973. "Left hemispheric specialization for language in the newborn: Neuroanatomical evidence of asymmetry." *Brain* 96: 641–46.

Wolf, A. 1970. "Childhood association and sexual attraction: A further test of the Westermarck hypothesis." *American Anthropologist* 72: 503.

————. 1993. "Westermarck redidivus." *Annual Review of Anthropology* 22: 157–75.

Wolf, A. P. 1995. *Sexual attraction and childhood association: A Chinese brief for Westermarck.* Stanford, Stanford Univ. Press.

————. 2003. "Maternal sentiments: How strong are they?" *Current Anthropology* 44(Suppl.): S31–49.

Wolff, P. H. 1963. Observations on the early development of smiling. In *Determinants of infant behavior II,* ed. B. M. Foss. London, Methuen.

————. 1968. "Sucking patterns of infant mammals." *Brain, Behavior and Evolution* 1: 354–67.

————. 1969. The natural history of crying and other vocalizations in early infancy. In *Determinants of infant behavior IV,* ed. B. M. Foss. London, Methuen: 81–109.

————. 1987. *The development of behavioral states and the expression of emotions in early infancy.* Chicago, Univ. of Chicago Press.

Wolfheim, J. H., G. D. Jensen, et al. 1970. "Effects of group environment on the mother-infant relationship in pigtailed monkeys *(Macaca nemestrina).*" *Primates* 11: 119–24.

Wolkowitz, O. M., L. Brizendine, et al. 2000. "The role of dehydroepiandrosterone (DHEA) in psychiatry." *Psychiatric Annals* 30(2): 123–28.

Wolpoff, M. H., J. M. Monge, et al. 1988. "Was Taung human or an ape?" *Nature* 335(6190): 501.

Wood, B. 2005. *Human evolution: A very short introduction.* Oxford, Oxford Univ. Press.

Wood, B., and B. G. Richmond. 2000. "Human evolution: Taxonomy and paleobiology." *Journal of Anatomy* 197(Pt. 1): 19–60.

Wood, W., and A. H. Eagly. 2002. "A cross-cultural analysis of the behavior of women and men: Implications for the origins of sex differences." *Psychological Bulletin* 128(5): 699–727.

Woodburn, J. 1968. An introduction to Hadza ecology. In *Man the hunter,* ed. R. B. Lee and I. DeVore. Chicago, Aldine: 49–55.

Woodruff, D. s. 1999. "Chimp cultural diversity (letter)." *Science* 285: 836.

Woolley, S. C., J. T. Sakata, et al. 2004. "Evolutionary insights into the regulation of courtship behavior in male amphibians and reptiles." *Physiology and Behavior* 83(2): 347–60.

Worlein, J. M., and G. P. Sackett. 1997. "Social development in nursery-reared pig-tailed macaques *(Macaca nemestrina).*" *American Journal of Primatology* 41(1): 23–35.

Worobey, J. 1990. Behavioral assessment of the neonate. In *Individual differences in infancy: Reliability, stability, prediction,* ed. J. Colombo and J. Fagen. Hillsdale, NJ, Erlbaum: 137–54.

Worthman, C. M. 1986. Developmental dyssynchrony as normative experience: Kikiyu adolescents. In *School-age pregnancy and childbearing: Biosocial dimensions,* ed. J. B. Lancaster and B. A. Hamburg. New York, Aldine: 95–112.

————. 1993. Biocultural interactions in human development. In *Juvenile primates: Life history, development, and behavior,* ed. M. E. Pereira and L. A. Fairbanks. New York, Oxford: 339–58.

————. 1999. Evolutionary perspectives on the onset of puberty. In *Evolutionary medicine,* ed. W. R. Trevathan, J. J. McKenna, and E. O. Smith. New York, Oxford Univ. Press: 135–64.

————. 2010a. Survival and health: A comparative global perspective. In *Handbook of cross-cultural developmental science,* ed. M. Bornstein. New York, Psychology Press: 39–60.

———. 2010b. "The ecology of human development: Evolving models for cultural psychology." *Journal of Cross-Cultural Psychology* 41.

Worthman, C. M., and M. J. Konner. 1987. "Testosterone levels change with subsistence hunting in !Kung San men." *Psychoneuroendocrinology* 12: 449–58.

Worthman, C. M., and J. Kuzara. 2005. "Life history and the early origins of health differentials." *American Journal of Human Biology* 17(1): 95–112.

Worthman, C. M., and C. Panter-Brick. 2008. "Homeless street children in Nepal: Use of allostatic load to assess the burden of childhood adversity." *Development and Psychopathology* 20(1): 233–55.

Worthman, C. M., and J. W. Whiting. 1987. "Social change in adolescent sexual behavior, mate selection, and premarital pregnancy rates in a Kikuyu community." *Ethos* 15(2): 145–65.

Wrangham, R. 1993. "The evolution of sexuality in chimpanzees and bonobos." *Human Nature* 4: 47–79.

———. 2009. *Catching fire: How cooking made us human.* New York, Basic.

Wrangham, R., and N. Conklin-Brittain. 2003. "Cooking as a biological trait." *Comparative Biochemistry and Physiology Part A, Molecular and Integrative Physiology* 136(1): 35–46.

Wrangham, R. W. 1999. "Evolution of coalitionary killing." *American Journal of Physical Anthropology* 110(S29): 1–30.

Wrangham, R. W., W. C. McGrew, et al., eds. 1994. *Chimpanzee cultures.* Cambridge, MA, Harvard Univ. Press/Chicago Academy of Sciences.

Wrangham, R. W., and D. Peterson. 1996. *Demonic males: Apes and the origins of human violence.* Boston, Houghton Mifflin.

Wright, L. 1994. *Remembering satan.* New York, Knopf.

Wright, P. C. 1990. "Patterns of paternal care in primates." *International Journal of Primatology* 11: 89–102.

Wright, S. 1922. "Coefficients of inbreeding and relationship." *American Naturalist* 56: 330–38.

———. 1932. "The roles of mutation, inbreeding, crossbreeding and selection in evolution." *Proceedings of the VI International Congress of Genetics* 1.

Wyckoff, G. J., W. Wang, et al. 2000. "Rapid evolution of male reproductive genes in the descent of man [see comments]." *Nature* 403(6767): 304–9.

Wynn, T. 2002. "Archaeology and cognitive evolution." *Behavioral and Brain Sciences* 25(3): 389–402; discussion 403–38.

Wyrwicka, W. 1976. The problem of motivation in feeding behavior. In *Hunger: Basic mechanisms and clinical implications,* ed. D. Novin, W. Wyrwicka, and G. Bray. New York, Raven.

Xie, J., and D. P. McCobb. 1998. "Control of alternative splicing of potassium channels by stress hormones." *Science* 280: 443–46.

Xuan, S., C. A. Baptista, et al. 1995. "Winged helix transcription factor BF-1 is essential for the development of the cerebral hemispheres." *Neuron* 14(6): 1141–52.

Yakovlev, P. I. 1962. "Morphological criteria of growth and maturation of the nervous system in man." *Mental Retardation* 39: 3–46.

———. 1970. Whole brain serial histological sections. In *Neuropathology: Methods and diagnosis,* ed. C. G. Tedeschi. Boston, Little, Brown.

Yakovlev, P. I., and A. R. Lecours. 1967. The myelogenetic cycles of regional maturation of the brain. In *Regional development of the brain in early life*, ed. A. Minkowski. Oxford, Blackwell Scientific: 3–70.

Yang, G., F. Pan, et al. 2009. "Stably maintained dendritic spines are associated with lifelong memories." *Nature* 462: 920–24.

Yang, J., R. R. McCrae, et al. 1999. "Cross-cultural personality assessment in psychiatric populations: The NEO-PI—R in the People's Republic of China." *Psychological Assessment* 11(3): 359–68.

Yarczower, M., and L. Hazlett. 1977. "Evolutionary scales and anagenesis." *Psychological Bulletin* 84(6): 1088–97.

Yogman, M. W. 1990. Male parental behavior in humans and nonhuman primates. In *Mammalian parenting*, ed. N. A. Krasnegor and R. S. Bridges. New York, Oxford Univ. Press: 461–84.

Young, A. J., A. A. Carlson, et al. 2005. "Trade-offs between extraterritorial prospecting and helping in a cooperative mammal." *Animal Behaviour* 70(4): 829–37.

Young, J. Z. 1981. *The life of vertebrates*, 3rd ed. Oxford, Oxford Univ. Press/Clarendon.

Young, L. J., and T. R. Insel. 2001. Hormones and parental behavior. In *Behavioral neuroendocrinology*, ed. J. Becker, D. Crews, and M. Breedlove. Cambridge, MA, MIT Press: 331–72.

Young, L. J., M. M. Lim, et al. 2001. "Cellular mechanisms of social attachment." *Hormones and Behavior* 40(2): 133–38.

Young, L. J., R. Nilsen, et al. 1999. "Increased affiliative response to vasopressin in mice expressing the V1a receptor from a monogamous vole." *Nature* 400: 766–68.

Young, L. J., J. T. Winslow, et al. 1997a. "Gene targeting approaches to neuroendocrinology: Oxytocin, maternal behavior, and affiliation." *Hormones and Behavior* 31(3): 221–31.

———. 1997b. "Species differences in V1a receptor gene expression in monogamous and nonmonogamous voles: Behavioral consequences." *Behavioral Neuroscience* 111(3): 599–605.

Young, L. J., Z. Wang, et al. 1998. "Neuroendocrine bases of monogamy." *Trends in Neuroscience* 21(2): 71–75.

Yu, Y.-Z., and J.-X. Shi. 2009. "Relationship between levels of testosterone and cortisol in saliva and aggressive behaviors of adolescents." *Biomedical and Environmental Sciences* 22(1): 44–49.

Zacharias, L., W. Rand, et al. 1976. "A prospective study of sexual development and growth in American girls." *Obstetrical and Gynecological Survey* 31: 325–37.

Zacharias, L., and R. Wurtman. 1968. "Age at menarche: Genetic and environmental influences." *New England Journal of Medicine* 280: 868–75.

Zahn-Waxler, C. 1993. "Warriors and worriers: Gender and psychopathology." *Development and Psychopathology* 5(1–2): 79–89.

Zahn-Waxler, C., E. A. Shirtcliff, and K. Marceau. 2008. Disorders of childhood and adolescence: Gender and psychopathology. In *Annual Review of Clinical Psychology*, vol. 4: 275–303.

Zald, D. H. 2003. "The human amygdala and the emotional evaluation of sensory stimuli." *Brain Research—Brain Research Reviews* 41(1): 88–123.

Zarrow, M. X., P. S. Campbell, et al. 1972. "Handling in infancy: Increased levels of the hypothalamic corticotropin releasing factor (CRF) following exposure to a novel situation." *Proceedings of the Society for Experimental Biology and Medicine* 141: 356–58.

Zarrow, M. X., V. H. Denenberg, et al. 1972. Hormones and maternal behavior in mammals. In *Hormones and behavior,* ed. S. Levine. New York, Academic.

Zeifman, D. M. 2001. "An ethological analysis of human infant crying: Answering Tinbergen's four questions." *Developmental Psychobiology* 39(4): 265–85.

Zeifman, D., and C. Hazan. 2008. Pair bonds as attachments: Reevaluating the evidence. In *Handbook of attachment,* 2nd ed., ed. J. Cassidy and P. R. Shaver. New York, Guilford: 436–55.

Zelazo, P. R. 1971. "Smiling to social stimuli: Eliciting and conditioning effects." *Developmental Psychology* 4: 32–42.

———. 1972. "Smiling and vocalizing: A cognitive emphasis." *Merrill-Palmer Quarterly of Behavior and Development* 18: 349–65.

Zelazo, P. R., and M. J. Komer. 1971. "Infant smiling to nonsocial stimuli and the recognition hypothesis." *Child Development* 42: 1327–39.

Zelazo, P. R., N. Zelazo, et al. 1972. "'Walking' in the newborn." *Science* 176: 314–15.

Zentall, T. R. 2003. "Imitation by animals: How do they do it?" *Current Directions in Psychological Science* 12(3): 91–95.

Zentall, T. R., and B. G. Galef, Jr., eds. 1988. *Social learning: Psychological and biological perspectives.* Hillsdale, NJ, Erlbaum.

Zera, A. J., and L. G. Harshman. 2001. "The physiology of life history trade-offs in animals." *Annual Review of Ecology and Systematics* 32(1): 95–126.

Zhang, J. 2006. "Parallel adaptive origins of digestive RNases in Asian and African leaf monkeys [see comment]." *Nature Genetics* 38(7): 819–23.

Zhao, Q.-K. 1994. "Mating competition and intergroup transfer of males in Tibetan macaques *(Macaca thibetana)* at Mt. Emei, China." *Primates* 35(1): 57–68.

Zhen-Wang, B., and J. Cheng-Ye. 2005. "Secular growth changes in body height and weight in children and adolescents in Shandong, China between 1939 and 2000." *Annals of Human Biology* 32(5): 650–65.

Zhong, Y., and C.-F. Wu. 1991. "Altered synaptic plasticity in Drosophila memory mutants with a defective cyclic AMP cascade." *Science* 251: 198–201.

Zhou, Y. H., J. B. Zheng, et al. 2002. "Novel PAX6 binding sites in the human genome and the role of repetitive elements in the evolution of gene regulation." *Genome Research* 12(11): 1716–22.

Zihlman, A. 1996. Reconstructions reconsidered: Chimpanzee models and human evolution. In *Great ape societies,* ed. W. C. McGrew, L. F. Marchant, and T. Nishida. Cambridge, Cambridge Univ. Press: 293–304.

Zihlman, A. L. 1981. Women as shapers of the human adaptation. In *Woman the gatherer,* ed. F. Dahlberg. New Haven, Yale Univ. Press: 75–120.

Zihlman, A. L., J. E. Cronin, et al. 1978. "Pygmy chimpanzee as a possible prototype for the common ancestor of humans, chimpanzees, and gorillas." *Nature* 275: 744–46.

Zucker, E. L., and J. Kaplan. 1981. "Allomaternal behavior in a group of free-ranging patas monkeys." *American Journal of Primatology* 1(1): 57–64.

Zuckerman, M. 1984. "Sensation seeking: A comparative approach to a human trait." *Behavioral and Brain Sciences* 7: 413–71.

———. 1990. "The psychophysiology of sensation seeking." *Journal of Personality* 58: 313–45.

Zuckerman, M., M. S. Buchsbaum, et al. 1980. "Sensation seeking and its biological correlates." *Psychological Bulletin* 88: 187–214.

Zuckerman, M., and D. M. Kuhlman. 2000. "Personality and risk-taking: Common biosocial factors." *Journal of Personality* 68(6): 999–1029.

Zwingman, T. A., P. E. Neumann, et al. 2001. "Rocker is a new variant of the voltage-dependent calcium channel gene Cacna1a." *Journal of Neuroscience* 21(4): 1169–78.

Acknowledgments

A book imagined, tinkered with, and transformed over thirty years owes much to many people, and I risk errors on both sides—omission of some who clearly deserve my thanks and inclusion of some to whom I owe a legitimate debt but who would rather not be mentioned in connection with this effort. For either error, I ask forgiveness.

Eric Wanner was the first person to believe in this book, when he signed me to my first contract with Harvard University Press in 1979. This was just as I became distracted by becoming a father, going to medical school, and in due course writing eight other books, beginning with *The Tangled Wing*. But *The Evolution of Childhood* was always to be my most ambitious, and remains so; thus the difficulty of completing it.

Others at Harvard University Press who have been kind and encouraging over the years have included Angela von der Lippe, Michael Fisher, and most importantly Elizabeth Knoll, the supremely literate, skilled, and gracious senior editor who has carried this effort through the last of its three decades.

Elizabeth has patiently cajoled, encouraged, nudged, praised, and defended me and promoted the book within and outside the press, even when it was still inchoate and growing apparently without end. I can say unequivocally that without her, this book would not exist. Meanwhile, our e-mail and conversations about Jane Austen, George Eliot, and Henry James have helped to keep me sane. Susan Boehmer made a helpful edit of the manuscript, and Kristin Sperber designed the book. Liz Duvall supervised the book's production, and Cindy Buck did the close editing in the final stages.

The skein of influence contains many strands and goes far back. My parents, Hannah and Irving Konner, although unable to hear and not high school graduates, spoke volumes to me about the resilience of parents and children and about how much is determined by character and love. Their values, refracted through an Orthodox Jewish prism, focused my attention on the family I would one day have as well as the one I came from. *L'dor vador,* says Jewish tradition—from generation to generation—a phrase that echoes within itself and infinitely forward; my parents took it seriously, and so did I. This as much as anything accounts for my fascination with origins, with explanation, and with culture.

An intense and dedicated high school history teacher, Dora Venit, gave me a passion for understanding human behavior and introduced me to thinkers in anthropology and psychiatry who viewed childhood as the key to deciphering culture and history. Paul Pavel helped me find my bearings as a young man. At Brooklyn College,

Gerald Henderson, who died far too young, taught me the basics of behavioral evolution and psychological anthropology, the foundation for all that is in these pages. Herbert Perluck, still a dear friend forty-six years later, taught me much of what I know about reading and writing and has been an unfailing ally and wise adviser.

My mentor at Harvard, Irven DeVore, gave me my anthropological career. I was his graduate student as well as a junior faculty member under his wing. He taught me behavioral primatology and evolutionary theory, showed me how to think rigorously about the evolution of behavior, and helped me negotiate countless obstacles in academic and real life. He and his wife Nancy opened their home to me and to my late wife Marjorie Shostak so many times and with such generosity as to make any attempt at thanking them seem small.

Together with Richard Lee, Nancy Howell, Patricia Draper, Henry Harpending, and John Yellen, the DeVores midwifed Marjorie and me into our first fieldwork experience, among the !Kung of Botswana, a formative episode of both our lives. The people we worked with in that setting, especially the mothers who allowed me to study their children, shaped my consciousness about human childhood and its likely original context. I refer to them throughout this book as the !Kung, which can be read without the click, because this is by far the most common name for them in the literature. Some anthropologists refer to them by their usual name for themselves, Ju/'hoansi, which roughly means "the real people."

In preparation for that first fieldwork, I was deeply influenced in my thinking about the anthropology of childhood by John and Beatrice Whiting, who held up a model of scientific method in this then-nascent field. Jerome Kagan and Steven Tulkin taught me methods of infant psychology, and Nicholas Blurton Jones welcomed me into his laboratory and (with his gracious wife Jill) his home in London to teach me ideas and methods of developmental ethology. Heinz Prechtl and T. Berry Brazelton taught me their methods for assessing newborn babies, and Neil Warren allowed me to watch him examine newborns in Kampala, Uganda.

Over the decades I have had the privilege of consulting with many people whose insight clarified domains of science relevant to this book. My postdoctoral years, supported by the Foundations Fund for Research in Psychiatry, included work in Richard Wurtman's laboratory at MIT, focused on neurotransmitters and neuroendocrine regulation. I also studied with the neuroanatomist Walle Nauta, who provoked my long love affair with the brain. Another great and courtly neuroanatomist, Paul Yakovlev, welcomed me into his research collection of human brain sections at the Armed Forces Institute of Pathology several times and guided me in exploring brain development through this uniquely valuable resource.

In medical school I was privileged to learn about clinical conditions from Walter Abelmann, Daniel Federman, Norman Geschwind, Hans Bode, Joseph Lipinski, and of course many patients—I delivered thirty-five babies and saw hundreds of children with a variety of illnesses and growth disorders. Ignorantly and feebly, I tried to help them, and my teachers saw to it that I did as little damage as possible. Nothing can substitute for such experience, and I wish I had had more of it. A very early version of Part II of this book emerged as an MD thesis in the Harvard-MIT Joint Program in Health Sciences and Technology, in which Walter Abelmann was my adviser.

My research collaborators have included Nicholas G. Blurton Jones, Carol Worthman, Mary Maxwell West (formerly Katz), Marjorie Elias, S. Boyd Eaton, Peter Brown, Ronald Barr, Lauren Adamson, Roger Bakeman, Ann Cale Kruger, and Sarah Davis.

James Chisholm welcomed Marjorie Shostak and me for a month at Navajo Mountain during his superb fieldwork on Navajo infants, and this affected the way I think about infancy. Co-teaching "Foundations of Behavior" for many years with Sarah Gouzoules has taught me much, and she has influenced my thinking in many ways. For most of the last decade my department at Emory has been headed by George Armelagos, a distinguished archaeologist and beloved teacher; his friendship, unflagging support, and esteem for me have meant a great deal.

Among the editors who have helped me learn to write over the course of three decades were Marian Wood, Elisabeth Sifton, Geoffrey Cowley, Robert Wright, Holcomb Noble, Bill Phillips, Dan Frank, John Michel, Jonathan Rosen, and the late novelist John Gardner. It's of course not their fault that I didn't learn their lessons better. Friends I have leaned on more than I care to say include Joe Beck, Steve Berman, Herb and Hazel Karp, David and Shlomit Ritz Finkelstein, Boyd and Daphne Eaton, John Felstiner, Jim Flannery, David Silbersweig, Misha Pless, Jeff Gavenman, and my brother, Lawrence Konner.

But Marjorie Shostak (the author of *Nisa: The Life and Words of a !Kung Woman* and *Return to Nisa*) was my first collaborator, both in field research and in life. We were together thirty years, and I cannot, even many years after her death, imagine how my life would have worked without her. Friend, lover, ally, critic, and more than occasional psychiatric nurse, she was a pillar of my young adult life and the loving and dedicated mother of my children.

Kathy Mote came to work for and with our family in 1987 and has stayed ever since. She was a mainstay of the children's lives while their mother was ill and dying, and after her death she became their second mom. As they grew away from us, I increasingly relied on her to assist me in all aspects of my professional life. I couldn't have done it without her.

Stefan Stein and Julian Gomez were two physicians who helped me survive the most difficult times in my life, and along the way taught me much about the lasting effects of childhood experience on emotions, as well as the limits on those effects.

I have been treated kindly by many thinkers and scientists who should by rights have been too busy to help me. They expressed support, gave advice, and answered hard questions in generous and often lengthy conversations.

They include, among others, Ernst Mayr, E. O. Wilson, Jon Seger, Robert Trivers, Sheldon White, Robert LeVine, Thomas Considine, Peter Rodman, Gerald Erchak, James Tanner, Jane Lancaster, Alice Rossi, David Hamburg, Bettye Hamburg, Carolyn Edwards, Victor Denenberg, Leonard Rosenblum, Michael Lewis, Stephen Suomi, William Mason, Walter Essman, Daniel Stern, Willard Hartup, Susan Goldberg, Charles Super, Sarah Harkness, Myrtle McGraw, Lionel Tiger, Robin Fox, James McKenna, Polly Wiessner, Irenäus Eibl-Eibesfeldt, Wulf Schiefenhövel, William McGrew, Sarah Blaffer Hrdy, Carol Worthman, Robert Sapolsky, Paul Ekman, Robert Paul, Bradd Shore, Duane Rumbaugh, Sue Savage-Rumbaugh, Robbie Case, William Durham, Ulric Neisser, Randolph Nesse, Michael Zeiler, James Rilling, Susan Seymour, Thomas Insel, Patricia Greenfield, Philippe Rochat, Roy D'Andrade, Naomi Quinn, Joan Silk, Charles Nemeroff, Vanna Axia, Barry Hewlett, Bonnie Hewlett, Michael Lamb, George Vaillant, Vittorio Gallese, and Harold Gouzoules. Geoffrey Haines-Stiles, producer of the *Childhood* series, included me in consultations involving Sandra Scarr, Robert Hinde, Alvin Poussaint, Urie Bronfenbrenner, and many others, and I learned a great deal. Sarah Blaffer Hrdy, André Nahmias, and Joel Stevenson made helpful comments on the manuscript. John Demos generously took time to locate and write me about unpublished material from his sensitive and revealing research on colonial witch trials. Barbara Finlay, Laurence Steinberg, and Dario Maestripieri took time to answer last-minute questions about their work.

I hope and trust that my errors of fact and opinion will not in any way reflect on these colleagues, from whom I learned much, but imperfectly.

My best friend and partner for the last part of my life is Ann Cale Kruger, who, to my great good fortune, agreed a few years ago to marry me. If rebirth is possible, I was reborn in her care and love, and it sustains me every day. As if that weren't enough, she happens to be a developmental psychologist on the cutting edge of research, and when she wasn't giving me sorely needed life lessons she was tutoring me in the new science of child development. She (emphatically) does not agree with all I say in this book, yet what little I understand about that science in its current form I owe largely to her. And her demurrals did not stop her from making one sacrifice after another during the years when the book took far too much of my time and mind. I don't know how I would live without her.

But to really learn about childhood, you have to have relationships with children. My "extra daughter" Becky Perry adopted our family at age eleven, and twenty-five years later (she is now a teacher) is still educating me, as is my "extra son" Ronald Kirby. My stepdaughter, Logan Kruger, admitted me into her life at a time when many a young teenager would simply have said *No way*. She was kind, smart, and funny, and she tolerated my designs on her mother with an open heart and mind. Closely watching her extraordinary life and growth has been a privilege.

My biological children, Susanna, Adam, and Sarah, did not have that choice; they were stuck with me, especially when their mother, after the bravest struggles, was taken away. They helped me to survive as much as I helped them. On my behalf I will say that I didn't just love them—that, I did to desperation—I also respected them, and that is why I learned so much. Today, all grown, they are not my accomplishments but my treasures and my friends. They taught me the most important things I know about childhood and parenthood; the rest is commentary.

Index

general theory of evolution and development, 720t, 722; middle childhood enculturation, 657, 658t, 659, 685
instrumental conditioning, 720t, 721
intermediate-level societies: carrying devices and practices, 405; Central American "soccer wars," 705, 707; childbirth, 401; cooperative breeding in extended family, 449–50; family organization, 471–78; HGC model, 567–68, 576–77; juvenile social relations, 495–96; male parental care, 466–75; NMC in, 495; nursing patterns, 405–6; observational learning, 647; play, and cultural transmission, 647–49; responsibilities in childhood, 517; sex play, 480; teaching, and cultural transmission, 517, 647; weaning age, 407–8, 408t; work, and cultural transmission, 647–49. *See also* Six Cultures Study
Inuit group. *See* Eskimo groups
in utero environmental effects, 28–29, 254, 624–28
invasion mechanism, and brain evolution, 81–82
Islam, 672, 724
Israeli kibbutzim: attachment growth and social fears, 230; communal living practice, 421, 534, 664, 695; incest avoidance taboo, 534, 695; moral development, 664; mother-infant bond, 411; smiling and mutual gaze in infancy, 220

James-Lange theory of emotions, 148, 151
Japanese cultures, 409, 418, 444, 626, 685, 724. *See also* Six Cultures Study
jewelfish, 1, 354, 367
Judaism and Jews: cultural coherence, 724; culture and personality construct during Holocaust, 599; infanticide during Holocaust, 216; middle childhood enculturation, 667; rites of passage, 683–84; sexuality beliefs, 331, 332; stress for children at war during Holocaust, 549; Tay-Sachs disease, 169–70
Junin Quechua group, 287
juvenile social relations: adaptations, 496–97; developmental mechanisms, 497–99; EEAs, 489; hunter-gatherer groups, 489–95; Indonesian culture,

498; intermediate-level societies, 495–96; kin selection theory, 484–85; mammals, 485–89; NMC in, 495; peer groups, 485, 493; primates, 488–89, 496–97; rites of passage, 493; sex play, 491, 496; summary, 484, 499; Western cultures, 498. *See also* middle childhood cognition shift; middle childhood enculturation; middle childhood transition
juvenility (paedomorphosis) condition, 61–63

Kaluli group, 247
Kanzi (bonobo), 108, 118, 342, 584, 588
Kenyan ethnic groups, 449–50, 478, 664. *See also* Six Cultures Study; *specific groups*
Kenyapithecus, 102
Kikuyu group, 302–3, 405–6, 449–50, 609–10
Kinderspiele (painting), 360, 508–9
Kipsigis group, 285–86, 609
Klüver-Bucy syndrome, 146, 236
!Ko group, 265, 507–8, 509, 679
Kohlberg's six-stage scheme, 664, 665
Kpelle group, 648
!Kung San group: attachment growth and social fears studies, 229–30; carrying devices and practices, 404; cooperative breeding in extended family, 387–88, 436–37, 442; crying levels in early infancy, 217, 388; cultural coherence, 724; designed learning, 659–60; development of large motor skills, 203; expected learning, 658, 658t; God concept, 673; HGC model, 566, 569–70, 571t, 572–73, 577–78; hunting beliefs and rituals, 121; infant crying responses, 387–88; infanticide, 424–25; in utero environmental effects, 627; LAD, 246, 247; male parental care, 462–63; maternal care, 385–88; menarche, 520, 521; mental capacity studies, 613; middle childhood cognition shift, 288; neonate neurobehavioral assessment, 210; NMC in, 435–37; nonphysical interactions, 387; nursing patterns, 385–86; "Oedipal" conflicts, 481–82; peer groups, 489, 490; physical contact with mother and caregivers, 387, 436–37; play, and